Multiscale Modeling and Simulation of Composite Materials and Structures

Young W. Kwon • David H. Allen
• Ramesh Talreja
Editors

Multiscale Modeling and Simulation of Composite Materials and Structures

Springer

Edited by:

Young W. Kwon
Naval Postgraduate School
Dept. of Mechanical and
Astronautical Engineering
Graduate School of Engineering
and Applied Sciences
Monterey, CA 93943
USA

David H. Allen
University of Nebraska-Lincoln
College of Engineering
114 OTHM
Lincoln, NE 68588
USA

Ramesh Talreja
Texas A&M University
Dept. Aerospace Engineering
Mail Stop 3141
College Station, TX 77843
USA

Library of Congress Control Number: 2007930778

ISBN 978-0-387-36318-9 e-ISBN 978-0-387-68556-4

Printed on acid-free paper.

9 8 7 6 5 4 3 2 1

springer.com

Preface

Ever since Democritus hypothesized the existence of the atom in the third century BC, evidence of the existence of physical phenomena on extremely small length scales has been accumulating. With the publication of Copernicus' book on the sun-centered solar system in 1543, evidence began to pile up that physical phenomena also occur on extremely large length scales. Today, it is not possible for us to put reliable bounds on either extreme. Indeed, just as the span of time is potentially infinite, the span of length may be unbounded.

Over the twentieth century, the most fundamental of sciences expanded the span of length by many orders of magnitude. Now, in the twenty-first century, one of the primary goals of science and technology seems to be the quest to develop reliable methods for linking physical phenomena that occur over multiple length scales. We now know that many vastly different fields of science such as biology, cosmology, paleontology, atmospheric physics, materials science, and even social sciences are faced with issues involving multiple length scales.

Until recently, our computational tools did not seem to be sufficiently powerful to engage in research activities focused on multiple length scales. However, with the rise of the high-speed digital computer, we now have the power to attempt to resolve scientific issues that were heretofore beyond our reach. Consider a single example – the growth of a crack in a ductile multigrain crystalline metal. Is this a problem in continuum mechanics, or is it molecular in scale? Or is it even smaller, perhaps even at the quantum scale? Evidence today seems to suggest that it is all of these, and that the physical phenomena observed at any of these length scales are indeed causatively linked to those on the neighboring length scales. Therefore, it would seem that the scientific profession is faced with a formidable challenge if predictive methodologies for such complicated phenomena are to be forthcoming. In the words of one of our colleagues, "A scientist loves a good challenge!"

In this text, we present recent work of 13 leaders in the field of multiscale mechanics, along with their coauthors, aimed at composite materials and structures. It is our hope that you the reader will in turn be challenged by these chapters, and that you will find your own paths to future developments in what amounts to one of the great problems of our time.

The editors thank Peter Beaumont because the inception of this book came out at the meeting "Advances in Multiscale Modeling of Composite Material Systems and Components" held in Monterey, California organized by him; Caitlin Wormesley, Greg Franklin, Carol Day, and Alex Greene of Springer for their support; and Patricia Worster at University of Nebrska for her dedicated service in editing all manuscripts.

Y.W. Kwon
D.H. Allen
R. Talreja

Contents

Chapter 1: Account for Random Microstructure in Multiscale Models

Vadim V. Silberschmidt

Wolfson School of Mechanical and Manufacturing Engineering,
Loughborough University, Ashby Road, Loughborough, Leics.,
LE11 3TU, UK

1.1 Introduction

The accumulated in last decades knowledge of fibre-reinforced composite materials, their effective properties as well as deformation and damage processes in them confirms a random (probabilistic) character of their failure (see, e.g. [1–4] and references therein). Such a character is determined by the specificity of microstructure of composites – a result of a manufacturing process of embedding of a huge number of reinforcing elements into a matrix. The resulting microscopic heterogeneity linked to randomness in positions of fibres, their bonding with the matrix, presence of microdefects, etc. causes a spatially and temporally non-uniform response to external loading even under macroscopically uniform loading conditions. The resulting pattern of deformation localisation and stress concentrations is neither uniform nor periodic; it defines macroscopic non-uniformity in evolution of various damage mechanisms.

At the current level of computational facilities, direct introduction of these stochastic microscopic features into computational models is prohibitive and counterproductive. A significantly better strategy is to employ multiscale models [5] that separate the levels of descriptions into (at least) local and global ones. The local level is used to incorporate details of a (real) microstructure of composites within a relatively small area (window) and to study the effect of its variability while the global one accounts for geometry of composite components/structures and loading/environmental conditions to study problems of their macroscopic behaviour, structural integrity and/or durability. But such separation of scales presupposes a necessity to bridge them within a framework of a single computational approach. Generally, various schemes to account for material's randomness can be employed both for various scales of modelling and bridging procedures.

The diversity of composites (in terms of constituents, their morphology and a type of reinforcement) makes a general analysis of their behaviour, including damage accumulation, practically infeasible. Hence this chapter is limited to analysis of the effect of randomness in distributions of filaments in matrix on damage evolution in two-phase fibrous composites under external load. A vast literature on composites that assumes a periodic character of reinforcement is not considered here (though some of its results are employed as an obvious comparison basis). Since only plies of unidirectional (continuous) fibre-reinforced composites are considered, the orientational randomness of inclusions is also not treated here. Though 3D studies and simulations are becoming a routine approach, and the respective experimental techniques, e.g. micro-X-ray computer tomography, can provide necessary volumetric data, for the sake of more 'transparency' a local modelling level in this chapter is limited to (predominantly) 2D analysis of unidirectional layers in the plane perpendicular to its fibres. This is due to the emphasis on transverse (matrix) cracking in cross-ply laminates, which is one of their main damage mechanisms under static and fatigue loading conditions [2, 4]. So, effectively, the (virginal) state of transverse cross-section of plies in such composites can be considered as a 2D distribution of circular inclusions in a matrix (Fig. 1.1).

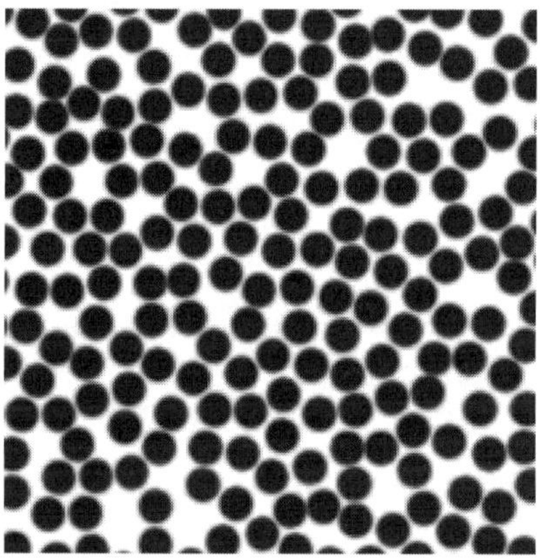

Fig. 1.1. Distribution of continuous graphite fibres in epoxy matrix in a transverse cross-section of a unidirectionally reinforced ply (digitalisation of a micrograph)

This chapter treats various aspects of randomness at various levels of modelling of fibre-reinforced cross-ply laminates – from the character of local distributions of fibres to non-uniformity of damage processes and cracking evolution and their influence on the composite's response to external loading.

1.2 Microstructures and Effective Properties

Though microstructural randomness of composites was obvious to researchers from the very beginning of the studies of such materials, the main emphasis of research was on the determination of their overall properties that could allow the use of deterministic continuous descriptions. In other words, an inhomogeneous material (discrete medium) is substituted by an equivalent homogenous one (continuous medium). This can be implemented by means of homogenisation procedures, 'smearing' microscopic features at the macroscopic level of modelling. In many cases, an assumption of a *coherent mixture* or *statistical homogeneity* is employed: The spatial distribution of the phases is assumed to be macroscopically homogeneous [6–11]. But even in this case, a full description of a composite with arbitrary geometry of phases and their volume fractions is cumbersome, so the emphasis is shifted to estimates of the effects of structural and microscopic features (volume fractions, shape of filaments, the extent of randomness in their distributions, variations in dimensions, etc.). The implementation of all of the mentioned factors within the framework of a single model is a rather complicated task, so historically effects of a single feature (or of a few ones) were studied separately. The research started for cases with so-called 'dilute dispersions' [6], i.e. low-volume fractions of reinforcement in a matrix, to exclude the effects due to their interactions, but later on it was extended to arbitrary volume fractions.

The main line of analysis was a use of periodic arrays of reinforcement in a matrix. Though micrographs of real microstructures vividly demonstrated deviations from regular patterns in distributions of inclusions (Fig. 1.1), (relative) simplicity of the approach made it very attractive. The notion of *representative volume*, used to estimate the effective properties, is also introduced early in the study of composites. According to Hill [6], it means a sample with two main properties:

1. Its structure is 'entirely typical' for the composite.
2. It contains a 'sufficient number' of microstructural elements so that boundary conditions at the surface of the composite do not affect its effective properties.

The main schemes used to determine the effective properties of composites are either the direct approaches, using, e.g. Voigt and Reuss estimates based on assumptions of uniform distributions of the stress and strain, respectively, or variational ones, employing, for instance, an elastic polarisation tensor [12]. The latter scheme allows one to obtain much closer bounds for the effective moduli than the Voigt and Reuss estimates. The well-known *Hashin–Shtrikman bounds* are determined on the basis of the original variational approach; the classical extremum principles of mechanics are used in [13] to obtain bounds for the overall elastic properties of an inhomogeneous system composed of various solid phases at arbitrary concentrations with ideal bonding.

The obtained results and bounds for elastic moduli explicitly depend on the volume fraction of constituents, or, for a two-phase composite, on the volume fraction of reinforcement due to an apparent relation

$$V_{\mathrm{f}} + V_{\mathrm{m}} = 1, \qquad (1.1)$$

where V_{f} and V_{m} are volume fractions of reinforcement (fibres) and matrix, respectively.

For a case of fibre-reinforced composites with continuous fibres, one of the first results for bounds of the effective elastic moduli for a case of a transversely isotropic composite with fibres of the same diameter, arranged in a hexagonal array, was obtained in [9]. More general results for a case of arbitrary geometry, restricted to the statistically transversal isotropy, are obtained in [14]. At the same time, Hashin [14] noted that it was 'not known how to use statistical details of phase geometry in prediction of macroscopic elastic behaviour'. The solution was based on the analysis performed for a cylindrical sub-region, extending from base to base of the fibre-reinforced specimen (Hashin introduced there the well-known now abbreviation RVE for *representative volume element*) with its transverse cross-section being, on the one hand, considerably smaller than that of the entire specimen but, on the other hand, considerably larger than that of the filament.

The Hashin's approach deals with a 'cylinder assemblage' by contrast with the 'concentric composite circular cylinders' of Hill [7]. Both approaches provide the same bounds for the transverse plain-strain bulk modulus for a two-phase fibre-reinforced composite. Still, these approaches predicted a relatively broad interval of effective properties important for various application magnitudes of the volume fraction of fibres $V_{\mathrm{f}} \approx 0.55$.

To improve the obtained bounds, approaches based on multi-point correlation functions were introduced. An example of such a function is the *n*-point probability function [15, 16]

$$S_n(\mathbf{x}_1, \mathbf{x}_2, \ldots, \mathbf{x}_n) = \left\langle \prod_{i=1}^{n} I(\mathbf{x}_i) \right\rangle, \tag{1.2}$$

where $I(\mathbf{x})$ is the *characteristic function* (known also as *indicator function* [17]) of the phase 1 (e.g. inclusions)

$$I(\mathbf{x}) = \begin{cases} 1, & \text{if } \mathbf{x} \text{ belongs to phase 1}, \\ 0, & \text{otherwise;} \end{cases} \tag{1.3}$$

angular brackets denote an ensemble average.

The volume fraction V_f is a one-point probability function. The two-point probability function $S_2^{(i)}(\mathbf{x}_1, \mathbf{x}_2)$ for a phase i of a composite can be interpreted as a probability that two points at positions $\mathbf{x}_1$ and $\mathbf{x}_2$ belong to this phase [18]. For statistically isotropic media, the two-point probability function depends only on the distance $r = |\mathbf{x}_1 - \mathbf{x}_2|$ between the points, and the simplified notation $S_2^{(i)}(r)$ can be used. For a statistically isotropic fibrous composite, two estimates hold

$$S_2^{\mathrm{f}}(0) = V_{\mathrm{f}} \tag{1.4}$$

and

$$\lim_{r \to \infty} S_2^{\mathrm{f}}(r) = V_{\mathrm{f}}^2. \tag{1.5}$$

Such correlation functions are normally referred to as *microstructural descriptors*, a thorough review of various types of which is given in [17].

To introduce the extent of connectedness of microstructural elements into consideration (that the two-point probability function lacks), another statistical measure – *lineal-path function* – is introduced in [19]. This parameter denoted $L_2^{(i)}(\mathbf{x}_1, \mathbf{x}_2)$ is linked to the probability that a line segment spanning from $\mathbf{x}_1$ and $\mathbf{x}_2$ is situated entirely in the phase i.

Three-point correlation functions are employed in [20] to obtain the bounds for elastic properties of composites. One disadvantage of the approach is the use of different correlation functions to define the upper and lower bounds of properties. So, Milton [21, 22] introduced 'simplified bounds' for two-component composites that depend on the volume fraction of two 'fundamental geometric parameters' $\xi_1 = 1 - \xi_2$ and $\eta_1 = 1 - \eta_2$ (ξ_1, $\eta_1 \in [0,1]$). These bounds are more restrictive than the Hashin–Shtrikman bounds (up to five times narrower according to Milton [21]); the latter correspond to cases $\xi_1 = \eta_1 = 0$ and $\xi_1 = \eta_1 = 1$. The self-consistent approximations of [11, 23] correspond – to the same order of approximation – to $\xi_1 = \eta_1 = V_f$. The fourth-order correlation functions for composites are suggested in [24].

An alternative approach to the self-consistent scheme is introduced in [25, 26] and coined *differential effective medium theory* in [26]. According to Norris [27], the suggested approach is rooted in the idea of Roscoe [28] that extended the famous Einstein's results on suspensions [29, 30]. One of the advantages of the differential scheme – as compared to the self-consistent one – is that it distinguishes between the two phases. One phase is taken as a matrix while the second – filament – is incrementally added to it from zero concentration to the final value [25, 27]. At each stage of the process, the added inclusions are considered to be embedded in a homogeneous material, corresponding to the composite formed by the matrix and all the previously added inclusions. This process is described by the tensorial differential equation of the following structure:

$$\frac{\mathrm{d}\mathbf{L}}{\mathrm{d}V_\mathrm{f}} = \frac{1}{1-V_\mathrm{f}}(\mathbf{L}_1 - \mathbf{L})\mathbf{E}_1, \tag{1.6}$$

with an obvious condition

$$\mathbf{L}(V_\mathrm{f} = 0) = \mathbf{L}_2. \tag{1.7}$$

Here $\mathbf{L}$ is the (fourth-order) tensor of effective moduli of the two-phase composite; $\mathbf{L}_1$ and $\mathbf{L}_2$ are moduli of inclusions and matrix, respectively; $\mathbf{E}_1 = [\mathbf{I} + \mathbf{P}(\mathbf{L}_1 - \mathbf{L})]$ is a strain concentration tensor; $\mathbf{I}$ is a unit tensor and tensor $\mathbf{P}$ was introduced by Hill [23]. A more generalised scheme is suggested in [27], where 'particles' of both matrix and inclusions can be added simultaneously to the initial material.

1.3 Microstructures and Their Descriptors

Since transversal arrangements of fibres in unidirectional layers of real composites are vividly random (Fig. 1.1), researchers trying to adequately describe them are confronted with several problems:

1. Characterisation of random microstructures
2. Comparison of random and periodic microstructures
3. Introduction of real microstructures into models

The first problem is traditionally solved with the help of the automatic image analysis (AIA) and various tessellation schemes. An attempt to quantify the random distribution of filaments (second phase) in a matrix by means of AIA and Dirichlet cell tessellation procedures was undertaken in [31, 32]. Voronoi tessellation, based on discretisation of a domain into multi-sided convex polygons (known as *Voronoi*) each containing no more than a single filament, is also used to estimate the character of distribution of distances between filaments [33, 34]. The distribution of cells is supposed to be of the Poisson type with the cumulative probability distribution function accounting for non-overlapping assemblage of filaments (known as *Gibbs hard-core process*)

$$P(\hat{V}_{\mathrm{f}} > V_{\mathrm{f}}) = 1 - \exp\left[-\frac{\overline{V}_{\mathrm{f}}}{1-\overline{V}_{\mathrm{f}}}\left(\frac{1}{V_{\mathrm{f}}} - 1\right)\right]. \tag{1.8}$$

It describes the cumulative probability that the local volume fraction of fibres $\hat{V}_{\mathrm{f}}$ exceeds a value V_{f}; $\overline{V}_{\mathrm{f}}$ denotes a mean volume fraction. In the case of the unidirectional 2D composite with random fibre spacing $V_{\mathrm{f}} = h/c$, where h and c are a fibre radius and a half-spacing between (centres of) neighbouring fibres, respectively (Fig. 1.2). The corresponding probability density function has the following form [34]:

$$p(V_{\mathrm{f}}) = \frac{\overline{V}_{\mathrm{f}}}{1-\overline{V}_{\mathrm{f}}}\frac{1}{V_{\mathrm{f}}^2}\exp\left[-\frac{\overline{V}_{\mathrm{f}}}{1-\overline{V}_{\mathrm{f}}}\left(\frac{1}{V_{\mathrm{f}}} - 1\right)\right]. \tag{1.9}$$

The *exact* relation for the probability density function for inter-fibre spacing x in the case of random impenetrable fibres of unit diameter is obtained in [35, 36]

$$p(x) = \frac{\overline{V}_{\mathrm{f}}}{1-\overline{V}_{\mathrm{f}}}\exp\left[-\frac{\overline{V}_{\mathrm{f}}}{1-\overline{V}_{\mathrm{f}}}(x-1)\right]. \tag{1.10}$$

Fig. 1.2. Longitudinal cross-section of unidirectional fibre-reinforced composite

A study of micrographs of a carbon fibre-reinforced PEEK prepreg, containing about 2,000 fibres with the volume fraction close to 50%, has shown that the distribution of Voronoi distances – distances in an arbitrary direction from the centroid of a fibre to the Voronoi cell boundary – can be assumed as a random one [37]. The Voronoi distance is also used as a random variable of the statistical description suggested in [38].

1.3.1 Parameters of Microstructure

Various parameters are introduced to quantify the extent of non-uniformity in distributions of filaments in composites. Several such parameters are suggested in [39]. The first one – *homogeneity distribution parameter* ξ – characterises the closeness of N particles (e.g. fibres in a transversal cross-section) within the window with area A

$$\xi = \frac{d_{\mathrm{p}}}{\sqrt{A/N}}. \tag{1.11}$$

This parameter is a ratio of two magnitudes of an inter-particle distance, one, d_{p}, corresponding to the peak of probability density diagram for this parameter and another being an effective average of it. Obviously, for a square lattice $\xi = 1$; its value diminishes with the increase in clusterisation. Another parameter – an *anisotropy parameter of the first kind* η – can also be applied to a distribution of cylindrical fibres in a transversal cross-section. It is introduced as [39]

$$\eta = \frac{1}{N}\sum_{i=1}^{N}\cos 2\theta_i, \tag{1.12}$$

where θ_i is an orientation angle for the direction from the centre of the window to the centroid of particle i. For a statistically isotropic distribution, this parameter should vanish.

Several parameters are suggested to characterise the extent of clustering and the properties of clusters (see, e.g. [40]). Still, in traditional carbon fibre-reinforced composites with $V_{\mathrm{f}} \geq 0.5$, the clusters are less obvious (if at all) than in metal matrix composites (MMCs).

As it is shown in [41], real distributions of fibres in unidirectional composites are neither periodic nor fully random, thus presupposing employment of measures that provide additional quantitative characteristics of the exact type of microstructures. So, based on the works of Ripley [42, 43], a *second-order intensity function* $K(r)$ was introduced to describe distributions of points in the following form [41]:

$$K(r) = \frac{A}{N^2} \sum_{k=1}^{N} \frac{I_k(r)}{w_k}. \tag{1.13}$$

This function characterises the expected number of further points (e.g. centres of fibres) within the distance r from an arbitrary point, normalised by their intensity (i.e. the number of points per unit area). Here, A is an area of the sampling window, containing N points, and $I_k(r)$ is the number of points situated within the distance r from the point k. The weighting factor w_k is introduced to account for the edge effects; it is equal to the ratio of the circumference of the circle situated within the window. If the entire circle with radius r is situated within the window, $w_k = 1$ and it is smaller than unity otherwise. The second-order function was applied to specimens of unidirectional fibre-reinforced composites exposed to different levels of external pressure during curing; also statistics for orientations and distances between fibres were used in terms of cumulative distribution functions. It was shown that these parameters, obtained with the use of image analysis from micrographs of real specimens, significantly differ from those of artificial microstructures with the same number of fibres, obtained by the Poisson process [41]. Unfortunately, second-order functions are not able to determine sub-patterns in distributions, so either parameters of a higher order or combinations of second-order functions with some other parameters should be used [44].

The second-order intensity function $K(r)$ can also be used to derive another quantitative parameter, characterising randomness in distribution of fibres (their centroids). It can be introduced in the following way [41, 44, 45]. The average number of fibre centroids located within a circular ring of radius r and thickness dr with a centre at a given fibre centroid is

$$dK(r) = K(r + dr) - K(r). \tag{1.14}$$

Dividing (1.14) by the area of the ring $2\pi r dr$, one can obtain the local spatial density of fibres. The ratio of the latter and the average spatial density N/A forms the *radial distribution function* [41, 45]

$$g(r) = \frac{A}{2\pi r N} \frac{dK(r)}{dr}. \tag{1.15}$$

Obviously, for a random Poisson process $g(r) = 1$. The value r_0, for which $g(r_0) = 1$, is a characteristic scale of the local disorder in an ensemble.

In parallel with statistical characterisation of distributions of microscopic features (e.g. filaments in a matrix) in composites, various topological characteristics are introduced. An obvious development in this direction is application of fractals [39, 46, 47].

A multifractal formalism can provide useful information on the type of the random distribution of fibres in the matrix [48]. It characterises the spatial scaling of non-uniform distributions: A local probability (number of fibres) P_i in the ith box (element) from a set of boxes, compactly covering the area of interest, scales with the box size l as

$$P_i(l) \propto l^{\alpha_i}, \tag{1.16}$$

where the scaling exponent α_i is known as *singularity strength*. According to the multifractal theory [49, 50], the number of elements with probability characterised by the same singularity strength is linked to the box size by the fractal (Hausdorff) dimension $f(\alpha)$

$$N(\alpha) \propto l^{-f(\alpha)}. \tag{1.17}$$

The function $f(\alpha)$, known as *multifractal spectrum*, describes the continuous (but finite) spectrum of scaling exponents for a random distribution. As it was shown in [48], the distribution of carbon fibres in epoxy matrix is multifractal; the respective multifractal spectrum was calculated.

1.3.2 Local Volume Fraction

Analysing the effects of microstructural randomness, an obvious idea is to consider the volume fraction of reinforcement not only in terms of a global description, i.e. as a parameter characterising the entire composite, but also as a field function, introducing the idea of a *local volume fraction*. A direct comparison of various parts of the composite (Fig. 1.3) vividly demonstrates that the volume fraction of fibres depends on a location in a composite. In Torquato [17], it is introduced as an average over a volume element (observation window) V_0 of the composite with the centroid at $\mathbf{x}$

$$V_{\mathrm{f}}(\mathbf{x}) = \frac{1}{V_0} \int_{V_0} I(\mathbf{x}) \theta(\mathbf{x} - \mathbf{z}) \mathrm{d}\mathbf{z}, \tag{1.18}$$

where $I(\mathbf{x})$ is the characteristic function (see (1.3)), $\mathbf{z}$ characterises any point in V_0 and $\theta(\mathbf{x} - \mathbf{z})$ is the indicator function

$$\theta(\mathbf{x} - \mathbf{z}) = \begin{cases} 1, & \mathbf{z} - \mathbf{x} \in V_0, \\ 0, & \text{otherwise}. \end{cases} \tag{1.19}$$

Fig. 1.3. Variations in local volume fraction of fibres (see Fig. 1.1)

Obviously, the size of V_0 will affect the level of the local volume fraction; but even for the same V_0 it depends on **x**. All the moments of this variable are studied for various systems in [51].

Two schemes are used in [52] to estimate variability of the volume fraction in carbon–epoxy T300/914 unidirectional composite. For macroscopic specimens with a cross-section 10 × 1 mm, a direct measurement of the Young's modulus of the composite is employed to calculate V_f using a linear rule of mixtures. An image analysis of fields 0.1 × 0.1 mm provides the data for direct estimation of the volume fraction of fibres (and its variation). In both cases, the distributions peak at (or close to) the nominal volume fraction $V_f = 0.6$. The increase in the window dimension results in the decrease in the scatter; still, even for a macroscopic specimens of the first method the measured interval of V_f was from 0.5 to 0.68.

Another analysis is performed for a micrograph of the ply's cross-sectional area of a carbon/epoxy composite, containing 603 fibres with diameter $d^f = 10$ μm [48]; the size of the window is 345 × 250 μm (its part is shown in Fig. 1.1). With the increase in the window size, the distribution of local magnitudes of V_f changes its shape and bounds – maximal (V_f^{max}) and minimal (V_f^{min}). The respective evolution of these bounds is shown in Fig. 1.4. For sufficiently small window size, these two bonds demonstrate mono-phase asymptotes: $V_f^{max} \to 1$ and $V_f^{min} \to 0$. With the increase in the window size, both bounds should converge to the average value

$$V_f^{max} \to \bar{V}_f \quad \text{and} \quad V_f^{min} \to \bar{V}_f. \tag{1.20}$$

Though this trend is distinct in Fig. 1.4, the full convergence of the bounds is not reached even at the length scale of 115 μm.

The spatial variation in the volume fraction of fibres causes considerable variations in the local values of stiffness: For the window size 30 μm, the axial and shear moduli demonstrate the scatter of more than 100% and the transverse module more than 40%.

An important parameter of local variations of the volume fraction of fibres V_f, treated as a random variable, can be linked to the standard deviation of its local magnitude. Such a parameter, named *coarseness C*, is introduced in [15, 36] as a the standard deviation of the volume fraction of filaments normalised by its mean value $\overline{V}_f$

$$C = \frac{1}{\overline{V}_f} \sqrt{\langle V_f^2 \rangle - \overline{V}_f^2}. \tag{1.21}$$

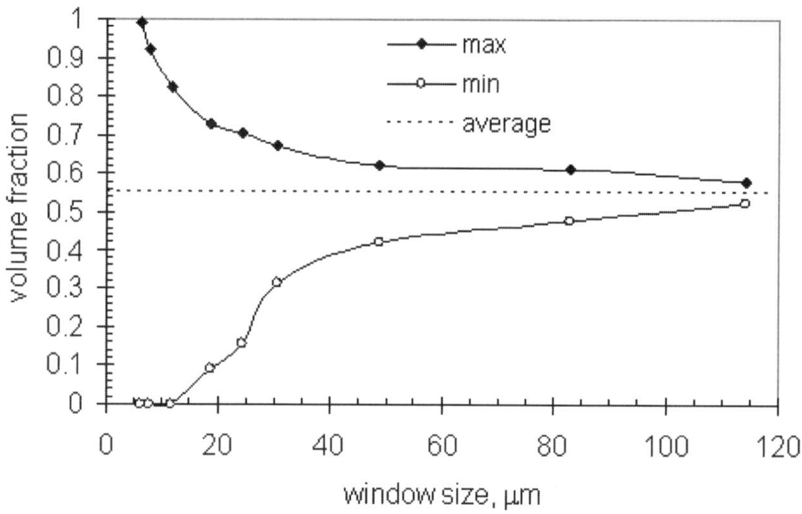

Fig. 1.4. Evolution of bounds for local volume fraction with window size

The change of coarseness C with the window size, calculated for the arrangement of 603 fibres that was treated before (see Fig. 1.4), is presented in Fig. 1.5. Obviously, $C = 0$ for an infinite area.

1.4 RVE Size

The problem of transferability of results obtained for some part of the composite's cross-section cannot be solved without the knowledge of representativeness of these results. This, in its turn, necessitates determination of the RVE size (or sizes). Obviously, a problem of the minimum size of the RVE for a random media – one of the central questions in the study of such materials – is linked to the analysed property/process and is not a universal parameter, depending purely on morphology of reinforcement. It is defined by the type of the property, the property's contrast in a composite and the volume fraction; the chosen precision of approximation plays a very important part in this as well as the type of the boundary conditions [53].

A diversity of ways to characterise and quantify microstructures and their randomness leads to various schemes for definition of the minimum RVE size. For instance, the last parameter from Sect. 1.3.2 – coarseness C – can be used for this purpose, linking the minimum RVE size to implementation of the condition of closeness of C to zero. Some notions of RVE (see one by Hill above) either explicitly or implicitly introduce ways to determine the respective minimum size. Still, in many cases this process leads to very large dimensions of RVEs, so that they could contain a sufficient number of microstructural elements to ensure the statistical representativeness.

The radial distribution function $g(r)$ (see Fig. 1.15) was also suggested as a basis to define the minimal RVE size. The latter is considered to be the value of radius r_0, for which $g(r_0) = 1$.

Another notion, employed in [54], is introduced with regard to the overall modulus $\widehat{\mathbf{L}}$ that will provide a sufficiently correct link between average (macroscopic) stresses and strains

$$\langle \boldsymbol{\sigma} \rangle = \widehat{\mathbf{L}} \langle \boldsymbol{\varepsilon} \rangle, \tag{1.22}$$

where angular brackets denote averaging. Normalising contribution of the non-local terms for simple cases of materials (non-overlapping spheres in a matrix) and deformation (varying normal strain and shear strain), it is possible to obtain explicit estimates. In the cases with extreme contrasts – rigid inclusions or voids – the RVE size practically does not exceed two reinforcement diameters for 5% error of $\widehat{\mathbf{L}}$ [54]. Even for a case of higher precision (error 1%), this size for the most demanding cases of matrix–reinforcement combinations is less than 5 diameters. This is confirmed by numerical studies of overall properties of non-linear composites with random distributions of fibres [55].

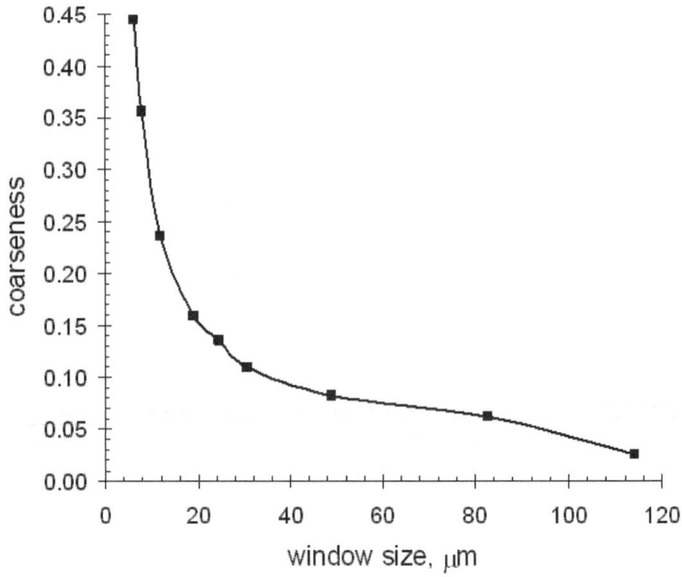

Fig. 1.5. Evolution of coarseness with window size

1.5 Periodic vs. Random

Determination of the RVE size allows considering two inter-linked problems: comparison of periodic arrangements of reinforcements with random ones and introduction of random microstructures into modelling schemes. The latter can be implemented by a direct incorporation of microstructural morphology from obtained micrographs into various computational schemes (e.g. by means of discretisation of digitalised images into finite elements) with a subsequent numerical solution of the problem. But since transferability of this ad hoc solution to other areas even of the same composite is, at least, questionable, special procedures of the microstructure's *reconstruction* were developed; they will be treated in Sect. 1.6.

Generally, a vivid deviation of real microstructures from periodic ones initiated their comparison at the early stages of the history of micromechanics of composites [56]. Obvious advantages of the use of periodic arrangements of fibres in studies and simulations – from existence of analytical estimates to possibility of high-refined meshes for very small windows – impelled researchers to compare those arrangements with real (random) microstructures to study reducibility of the latter to the former.

Since the manufacturing of periodic arrangements for large number of microscopic reinforcing elements is cumbersome, the main tool for such comparisons is computational analysis. Hence, in studies of composite materials, a considerable emphasis has been on a comparison of random and non-random models and their closeness to experimental results for composites.

The data obtained by combination of the image analysis and Dirichlet cell tessellation procedures [31, 32] are used to numerically generate distributions of reinforcements with the same average magnitudes and standard deviations. Numerical simulations, performed for various periodic and random distributions of fibres with different shapes of their cross-sections for boron fibre-reinforced aluminium, have vividly demonstrated a significant influence of microstructural features on the composite's effective elastic moduli as well as on the plastic flow/localisation. The case with a random distribution of circular fibres (30 fibres were used in a statistical realisation) in a transverse cross-section provided the closest match to experimentally measured values [57]. It is worth mentioning that the effect of fibres' distributions on the material's response is relatively stronger than that of their shape.

For a unidirectional fibre-reinforced boron–aluminium composite ($V_f = 0.48$), models for square and hexagonal arrays and a random distribution were compared with experimental data for components of stiffness and compliance [58]. Two experimental methods were used to estimate those components: the ultrasonic-velocity method and the resonance method. It is demonstrated that the random distribution model provides the best approximations to the measured parameters.

An extensive study of the periodic and random distributions of fibres in matrix and their effect on transverse properties is implemented in [55] with the employment of Fast Fourier Transforms as an alternative to finite element analysis. A window of 1,024 × 1,024 pixels is used to resolve an area of non-linear composite (with elasto-plastic matrix) containing 64 fibres ($V_f = 0.475$) with a resolution 128 × 128 pixels per fibre. The main result for effective properties is that the scatter in the transversal Young's moduli is small (standard deviation is less than 1%) while the flow stress and hardening modulus demonstrate higher fluctuations. Another approach is used in [44] based on the perturbed periodic square arrays of fibres with clustered or staggered patterns that are compared with a square one. These less ordered patterns are generated by a shift of the centre of a circular cross-section of a fibre in a unit cell (each containing a single fibre) from its fully symmetrical position in a square array. Combined with a special type of boundary conditions, this scheme allows simulations of unidirectional continuously reinforced MMCs with the volume fraction of fibres, similar to the previous example. The obtained results for a boron–aluminium MMC demonstrate that the axial overall response is practically independent of the arrangement of fibres, while the transverse parameters (both the elastic moduli and yield limits) are highly sensitive to it.

Another study [59], based on the boundary element method, compared square and hexagonal arrangement of fibres with random ones, using the *embedded cell approach.* The latter is implemented by embedding the core, containing a discrete arrangement of fibres (60 fibres in random cores), into a homogeneous media with properties obtained, e.g. by the self-consistent scheme, to which the far-field boundary conditions are applied. It was shown that the fibre packing arrangement significantly affects the overall stress distribution as well as the extent of stress localisation, with considerably higher stress concentrations observed in random sets. For instance, for carbon fibre/epoxy composite ($V_f = 0.56$) 90% higher local radial stresses are reported in [60] for a random arrangement as compared to a hexagonal one. This can be naturally explained by the fact that some morphological parameters of random microstructures differ from those of periodic ones. An obvious example is the nearest-neighbour distances for random and clustered distributions are smaller than those for square or hexagonal patterns [61].

A detailed analysis of the effect of arrangement of fibres on stress concentration was undertaken in [62]. The studied microstructures included 600 fibres in a matrix with the contrast in the Young's modulus 6.7 and close Poisson's ratios. The volume concentration of reinforcement was 0.1 and the minimum distance between the centroids of reinforcements was three times their radius. The main conclusion is that the proximity of reinforcements along the direction of loading results in the highest stress concentration. Alignments of fibres, close to each other, along directions at large angles to the loading one cause the stress reduction.

To account for the effect of the relative positions of, and orientations between, neighbouring fibres, a special *stress interaction parameter* was introduced in [63]. This is an additional descriptor used to quantify a short-range configuration of fibres

$$c = \min_i \min_j \left[d_{ij} \left(1 + \frac{\theta_{ij}}{\alpha} \right) \right]. \tag{1.23}$$

Here d_{ij} is a distance between centroids of two fibres i and j; θ_{ij} is the angle between the line connecting them and the loading direction; $\alpha = \pi/3$ is a normalising factor. The lower c, the higher the radial stress concentration.

An attempt is also made to determine parameters of a periodic microstructure, similar to a random two-phase microstructure [64]. It employs discretisation of the image of the actual microstructure and calculation of its power spectral density. Then parameters of the unit cell are chosen: number of reinforcements, their geometry, initial dimensions of the cell and positions of reinforcements. Since the discrete power spectral densities of the original and equivalent systems are obtained under various conditions (e.g. frequencies), the so-called 'rebinning' [64] process is used to match the frequencies. The final positions of particles in the unit cell are found with the use of the minimisation of a specially constructed function.

1.6 Direct Introduction of Microstructure

Understanding of (sometimes) considerable deviations of microstructures of real composites from periodic ones resulted in an obvious idea of a direct introduction of such a microstructure into numerical schemes. Two principal ways to implement this are possible (1) direct use of a scanned microstructural images and (2) generation of artificial microstructures. While the former procedure is relatively straightforward, the latter one raises the question of reproducibility of main features of the material. A somewhat simplified way for a 2D case of a transversal cross-section is to introduce a 'fully random' system using, e.g. a planar Poisson process for hard-core (i.e. impenetrable) disks [65]. An obvious deficiency of this approach is that it is not linked to the exact type of stochasticity of the microstructure that can deviate from the ideal randomness. Hence, it is necessary to reflect the features (at least principal ones) of an original microstructure in the artificially generated microstructure of the composite.

Naturally, a direct account for random features of the microstructure began in micromechanical models. One of the typical examples is the introduction of the random fibre spacing [66, 67]. A random number generator is used to produce a set of random numbers that, after re-ordering in an ascending order, present the set of transversal co-ordinates z_m for parallel fibres in a 2D set. The fibre spacing d_m is then defined in an obvious way

$$d_m = z_{m+1} - z_m \tag{1.24}$$

with the average spacing being

$$d_0 = \frac{1}{N-1}\sum_{1}^{N-1} d_m = \frac{1}{N-1}(z_{m+1} - z_m), \tag{1.25}$$

where N is the total number of fibres.

Several statistical realisations for sets of 12 fibres are used in calculations of the stress concentration factors (SCFs); both the average and the variance of the distribution of spacing are equal to unity. To exclude an overlap of neighbouring fibres, a lower cut-off at some arbitrary chosen level is introduced. A transition from the deterministic case with a uniform distribution of fibres in a transversal direction results in a statistical character of magnitudes of SCFs in the composite.

A more consistent approach to reflect the original microstructure and its features in a model was named *reconstruction* (e.g. [18]) and is understood as generation of the microstructure employing the correlation functions, characterising its morphology.

The suggested variant of this procedure for a two-phase statistically isotropic material is based on two-point correlation functions, introduced for two digitized representations of the microstructure – original, with the reference correlation function $f_0(r)$, and 'reconstructed' (i.e. generated), with $f_s(r)$. Manipulation of the generated image, using interchanges of the states for pairs of arbitrarily selected pixels of two phases to preserve the volume fraction, is controlled by minimisation of a variable E that plays the role of energy in this approach [18]

$$E = \sum_i [f_s(r_i) - f_0(r_i)], \tag{1.26}$$

where r is the distance between two points. The probability of acceptance of such phase interchange, estimated after each step, is determined by the energy change for two subsequent steps ΔE

$$p(\Delta E) = \begin{cases} 1, & \Delta E < 0, \\ \exp(-\Delta E / T), & \Delta E > 0. \end{cases} \tag{1.27}$$

Here T plays the role of temperature.

The suggested scheme converges to the reference correlation function $f_0(r)$. It is worth mentioning that the general framework can be used with a variety of correlation functions. The authors used two functions – the two-point probability function and lineal-path function – since they characterise various facets of real microstructures, e.g. the former is good in catching the short-range information while the latter contains information on connectedness. One of the suggested ways to expand this description is to account for various correlation functions in the expression for energy, modifying it to accommodate various functions with some weights [18]. Reconstructions of random microstructures, based on two mentioned functions, though demonstrated a rather good match for the reference and artificial microstructure, still resulted in some deviations from the original, as measurements with other correlation functions have shown. Another way for reconstruction of the microstructure is suggested in [68], based on the maximum entropy principle.

An alternative variant to generate of a periodic unit cell, based on the data on statistic features of the composite's microstructure, is suggested in [69]. For a fibre-reinforced composite material characterised by a given second-order intensity function $\bar{K}(r)$ (see (1.13)), which is evaluated in points r_i, $i = \overline{1, N_m}$, positions of centroids of N fibres in a 2D unit cell with dimensions $H_1 \times H_2$ are defined by means of the following relation

$$\mathbf{x}^N_{H_1,H_2} = \arg\min_{\mathbf{x}\in\mathbf{S}} F\left(\mathbf{x}^N_{H_1,H_2}\right), \tag{1.28}$$

where

$$F\left(\mathbf{x}^N_{H_1,H_2}\right) = \sum_{i=1}^{N_m} \left(\frac{\bar{K}(r_i) - K(r_i)}{\pi r_i^2} \right)^2 .$$

Here, vector $\mathbf{x} = \{x_i, y_i\}$, $i = \overline{1, N}$, defines positions of the centroids in the unit cell and $\mathbf{S}$ denotes a set of admissible $\mathbf{x}$. A numerical realisation can be implemented with the use of genetic algorithms [69].

A so-called *mean-window technique* [70], based on the exact averaging over the volume fraction, can also be used as a basis to generate model microstructures with statistical functions, similar to observed/measured ones. It allows estimating effective properties of a composite by averaging randomly selected windows from a real structure, as obtained from a microtomographic analysis. The suggested method exploits the hypothesis of ergodicity, according to which an ensemble average of a property obtained on smaller volumes is equal to the average over an infinite one [71]. Authors also distinguish between the physical and geometrical RVEs. The latter has a standard sense as a material volume of specific size (e.g. equal to the correlation length of the two-point probability function [71]). In contrast, the *physical RVE* (PRVE) is defined by the level of variation of the physical property, which should become insignificant for a PRVE. One of the possible formulations for the PRVE is that of a minimum material volume for which the standard deviation for different statistical realisations is smaller than the measurement error for a chosen parameter. Applicability of the mean-window technique is restricted to fulfilment of two conditions [71] (1) symmetry of the probability distribution function and (2) linearity of the variation of the chosen property within the respective interval of the local volume fraction, for a given window size. Another feature affecting the size of this PRVE is the contrast of properties (i.e. the ratio of elastic moduli of the matrix and reinforcement): An increase in the contrast will request a larger window.

1.7 From Micro to Macro: A Way to Multiple Scales

The choice of the adequate dimensions for the RVE and introduction of the (principal) microstructural features, though complying with requirements of representativeness, still do not always allow a direct transfer of the results, obtained for some area (window) to a real component/structure. There are several reasons for this:

1. The results, obtained for an RVE, will depend on the type of the employed boundary conditions: homogeneous strain or stress, periodic, etc. [53, 72].
2. A transfer of loads/environmental conditions, externally applied to a macroscopic component, to the RVE can be implemented in various ways.
3. Gradients in stress and/or strain fields due to non-uniform loading or stress concentration as well as edge effects affect the results.

Still, RVE with periodic boundary conditions is broadly used to estimate the effective properties of heterogeneous materials with a given type of microstructure. More advanced schemes have been introduced to overcome the discussed limitations and to develop more adequate modelling tools. One possibility is to use a *meso-scale window* as an alternative to the RVE [73, 74]. The window is placed within a two-phase domain and its size L (and the respective non-dimensional window scale $\delta = L/d$, where d is the size of inclusions) is being changed to determine a convergence condition for two responses of an elastic heterogeneous media, controlled by either stresses or strains. Calculations, based on two respective types of boundary conditions, preclude the necessity to use periodic boundary conditions within the RVE formalism, still providing bounds for stiffness for any δ. Random simulations (using Monte Carlo sampling) employ approximations of the planar continuum by a very fine spring network, with respective stiffness magnitudes depending on coordinates. A detailed analysis of probability density distributions shows that the beta distribution can properly describe variations in the local stiffness without truncations; if truncations are acceptable then Chi, Gumbel max, Rayleigh and Gauss distributions can be used for this purpose [74]. Generally, RVE is considered as a deterministic limit of a statistical volume element (SVE) in [75].

An alternative to the RVE formalism is an approach base on lattice models that was initially developed for problems of statistical physics [76]. In application to materials with spatial randomness in their properties, linked to their microstructure, it employs not a single RVE but a set of them – each with its own properties – compactly covering the area of interest [48, 77–79].

Another approach is the scheme named *Voronoi cell finite element method* [80, 81] that is an extension of the Dirichlet tessellation. Each of the Voronoi polygons of such tessellation, containing a single filament, is treated as a finite element. This representation of microstructure, which can be linked to image analysis data, is used as a basis for a two-level computational model for heterogeneous materials [81]. A two-way linkage between the levels is introduced in the following way:

1. Data from the microscopic level are used to estimate the global effective parameters of the composite at each point of the macroscopic level.
2. Macroscopically calculated data are used to determine the distribution of stresses and strains in the microstructure.

This scheme goes back to the first attempts to introduce multiscale modelling schemes (though the term was suggested considerably later) for composites. Their necessity has always been obvious: numerical simulations based on the finite elements method (or any comparable schemes), accounting for the exact position of all the fibres in the whole specimen is not practical since dimensions of the elements should be considerably smaller than the diameter of a fibre. With the latter being, e.g. 5–10 μm for carbon fibres, the simulation of a standard composite specimen becomes prohibitive at the current level of computational power. Hence, to overcome this obstacle, the problem was separated into solvable sub-tasks. One of the typical approaches is to limit a microscopic analysis with a sufficient resolution of microstructural elements to a relatively small area of interest of the component, with the component itself being analysed within the framework of a macroscopic mechanics, i.e. employing of the effective properties. In Akbarzadeh and Adams [82], the following two-step procedure is used:

1. A macroscopic analysis of the whole region (homogeneous and transversely isotropic in the case of [82])
2. A micromechanical analysis of the region of interest containing several fibres with the boundary conditions obtained from the previous stage

The results for a notch area of the Charpy specimen for a random distribution of fibres (in total, four fibres are situated in the studied microscopic area) have demonstrated that the microscopic features can have a profound effect on the global behaviour: It can be more important for stress localisation than the curvature of the notch [82].

Still, the random character of materials and of deformational and failure processes in them should not be always limited to the microscopic level of their description.

1.8 Randomness at Macroscale

Microstructural (local) randomness of heterogeneous materials does affect their response at the global (macroscopic) level. Here two principal trends are vivid. On the one hand, the increase in the considered area (volume) 'smoothens' local fluctuations of properties linked with the microstructure – similar to an averaging procedure for larger sets – thus resulting in the decreasing variability of the global effective properties for macrovolumes (or *twin specimens*). On the other hand, mechanisms causing spatial localisation of deformation and/or fracture processes, for instance, plastic flow and crack nucleation and/or propagation, inherit some of the randomness of the underlying microscopic structure.

Even in the latter case, there is no direct mapping of the microstructural stochasticity onto the macroscopic patterns of behaviour. This effect is more pronounced at the stage of the onset of localisation, but weakens with its development due to a strong interaction with additional mechanisms. Fluctuations in the spatial distribution of constituents result in a non-uniform distribution of microscopic stresses. This factor, together with – also non-uniform – distributions of defects, results in considerable variations in plastic flow and/or damage accumulation and transition to formation of macroscopic defects. This scenario is additionally complicated by multiplicity of damage mechanisms and their interactions.

The material's behaviour at the initial *pre-critical stage* of deformation is characterised by a bulk response of the entire specimen, demonstrating relatively low spatial fluctuations of stresses and/or strains that are roughly proportional to fluctuations in material's properties caused by its microstructure. With the onset of the *critical* and *post-critical stages*, characterised by the localisation of deformation and failure processes, the extent of macroscopic non-uniformity can significantly increase. This localisation can also, in its turn, affect the global material's behaviour. For instance, in composites with elastic–ideal plastic matrix in many cases, only a small part of the matrix participates in the plastic flow, the spatial pattern of which directly affects the overall flow stress of the composite, as shown in [55].

Thus, the macroscopic randomness can be considered as a given feature of a composite only at some stages of its life; in many cases it changes as due to its interaction with loading/environmental conditions. One of the striking examples is evolution of matrix cracking in cross-ply laminates under axial tensile loading. This process starts relatively early in the loading history of composites: at low external stretching under quasi-static loading or during initial cycles of tensile fatigue. Distributions of transverse cracks along the longitudinal axis of cross-ply composites are random: Twin specimens demonstrate various numbers and positions of matrix cracks [83–86]. A typical character of evolution of matrix cracking in a single specimen during its loading history is given in Fig. 1.6. Even in the same specimens, with matrix cracks not crossing their entire width, the crack distribution along two longitudinal edges is different – and random [87].

1.8.1 Evolution of Matrix Cracking

The evolution of matrix cracking, reflected in a pattern of transverse cracks, is a result of interplay of processes of cracks' nucleation and their interaction. At the initial stage of loading history, the macroscopically applied external load accelerates damage evolution in places of stress concentrations and/or largest or preferably oriented microdefects (microcracks, voids, fibre debonding, etc.). This development causes a practically uncorkrelated generation of matrix cracks with inter-crack spacing demonstrating a high extent of randomness (Fig. 1.6a,b).

Generation of matrix cracks, crossing the entire thickness of a weak 90° layer, causes a significant change in the pattern of stress distributions, especially in this layer. Instead of the longitudinally uniform – if temporarily to neglect variations due to the microstructural randomness for the simplicity of our analysis – field of the axial stress, a new stress pattern arises: Unloaded zones appear in the direct vicinity of the transverse cracks. They are due to the so-called *shielding effect* caused by the traction-free surfaces of the matrix cracks.

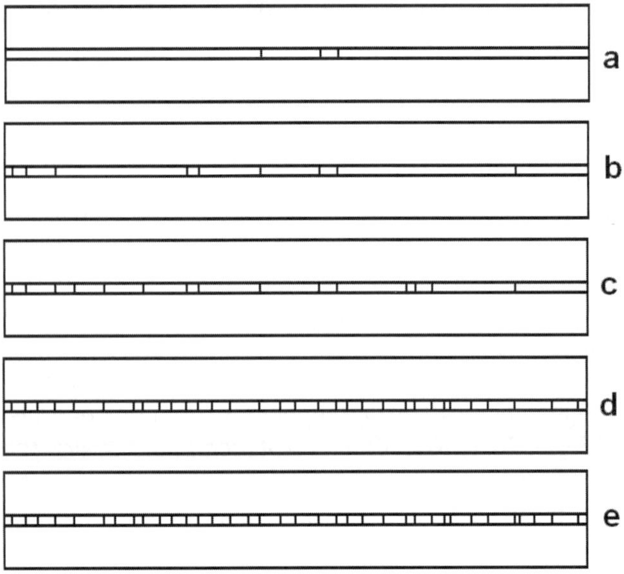

Fig. 1.6. Matrix cracking in a 20 mm long part of $[0_7/90]_s$ carbon–epoxy T300/914 specimen at various moments of loading history: (**a**) 10 cycles; (**b**) 100 cycles; (**c**) 1,000 cycles; (**d**) 10,000 cycles; (**e**) 100,000 cycles (*different horizontal and vertical scales*)

These unloaded zones have a considerable length in the axial direction – up to several mm. At the initial stages of matrix cracking, characterised by a low crack density, zones from neighbouring cracks practically do not overlap. Still, the areas of the 90° layer, unaffected by matrix cracking (i.e. not partially unloaded), decrease with the increase in the external load or number of cycles. It means that initially spatially uniform conditions with regard to crack nucleation change: Areas situated somewhere in the middle between two neighbouring cracks with a larger spacing become more preferable for matrix cracking. With the further increase in the number of cycles (or the external load/stretching for quasi-static conditions), the number of cracks increases, and the neighbouring unloading zones begin overlapping (interacting). A resulting distribution of the axial stress component, normalised by its far-field magnitude, in a weak layer that was calculated for a part of a real random distribution of matrix cracks with the use of detailed finite element simulations (see [88]) is given in Fig. 1.7.

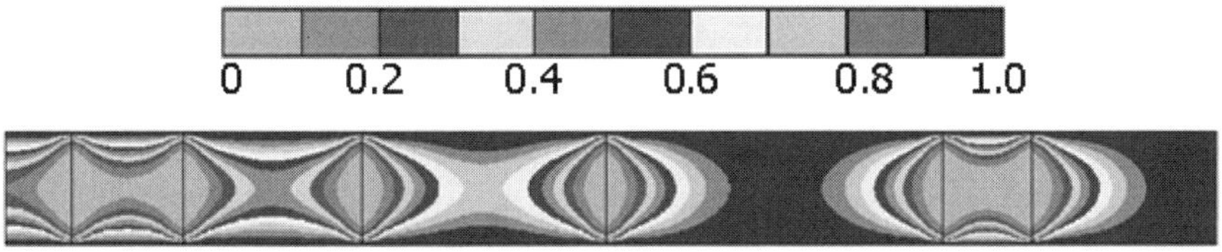

Fig. 1.7. Distribution of the normalised axial stress in a 10 mm long part of 90° layer of $[0_1/90_2]_s$ carbon–epoxy T300/914 composite

Superposition of stress reduction in the areas of overlap causes the additional decline in the probability to initiate a matrix crack in these areas (in some cases the axial stress state can even change from tension to compression). Hence, the stage of random matrix cracking is gradually changed by a more ordered character of transverse crack nucleation, with mid-spacing areas becoming preferable places for this process. Retrieving now the notion of microstructure-induced material's randomness, it is obvious that two major processes determine the pattern of transverse cracks in the 90° layer: The randomness in a spatial distribution of flaws is responsible for spatially non-uniform nucleation of cracks while the re-distribution of macroscopic axial stress orders this process, limiting it to a diminishing part of the composite. Obviously, for the material with a considerable scatter discussed in the previous sections, an event of matrix crack nucleation close to the existing one is still possible if the extent of local stress concentration near a strong flaw is not fully compensated by unloading due to the shielding effect (compare successive stages of the loading history in Fig. 1.6).

The above considerations are based on a one-dimensional interpretation of the process (similar to many modelling schemes that deal with axial stresses, averaged over the thickness of 90° layers). Obviously, the cracking processes in relatively thick layers demonstrate their own specificity as well as in wide specimens; in the latter, under tensile fatigue, matrix cracks do not instantly occupy the entire width of the specimen but grow with random rates [89, 90] (the difference between two schemes also depends on laminate thickness [4]).

Various parameters are used to characterise the randomness of distributions of matrix cracks: the Weibull's distribution [85], number of cracks in bands [2], etc. The multifractal analysis has been found to be a very convenient tool for this purpose [89–91]. It was found that twin specimens, exposed to the same loading history, demonstrate a considerable difference in patterns of transverse cracking (in numbers of cracks and their positions) but their multifractal spectra are very close (see Fig. 1.8).

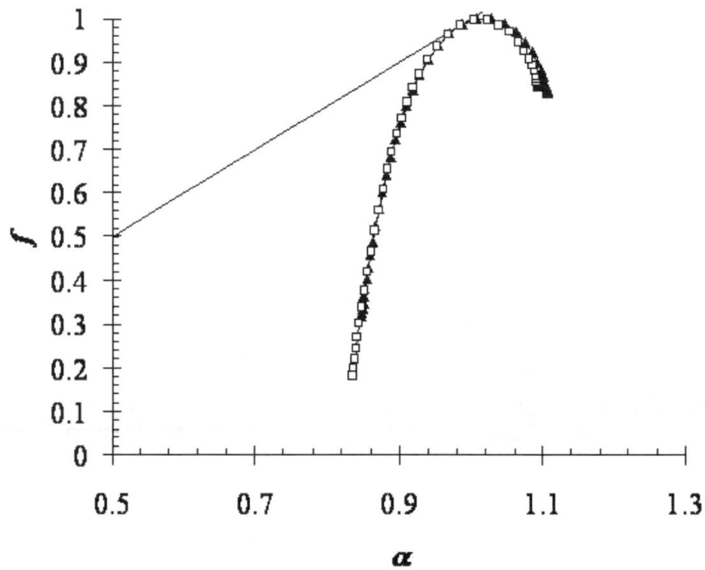

Fig. 1.8. Multifractal spectra of matrix crack sets in two twin specimens of carbon–epoxy T300/914 $[0_2/90_3/0_2]$ laminate loaded with 100,000 cycles (1,050 N mm^{-})

Besides, multifractal spectra depend on both loading conditions and structure of cross-ply laminates (stacking order) (see [92] for a detailed analysis). A typical transformation of $f(\alpha)$ functions with the loading history is given in Fig. 1.9. The width of a multifractal spectrum is linked to an extent of the distribution's randomness: For a fully uniform distribution it is reduced to a single point. Hence, it is possible to interpret the effects of various factors on randomness in matrix cracking.

The initial stage of the tensile fatigue, characterised by a nearly non-restricted (i.e. random) nucleation of matrix cracks, has a rather wide multifractal spectrum (Fig. 1.8). With the increase in the number of cycles, the above described ordering mechanism results in less random patterns of cracks. This is reflected in considerably narrower multifractal spectra. Our results [91, 93] have shown that, for long loading histories when the process of matrix cracking attains the so-called *characteristic damage state* with nearly non-changing distributions of transverse cracks, the respective $f(\alpha)$ functions are very close to each other.

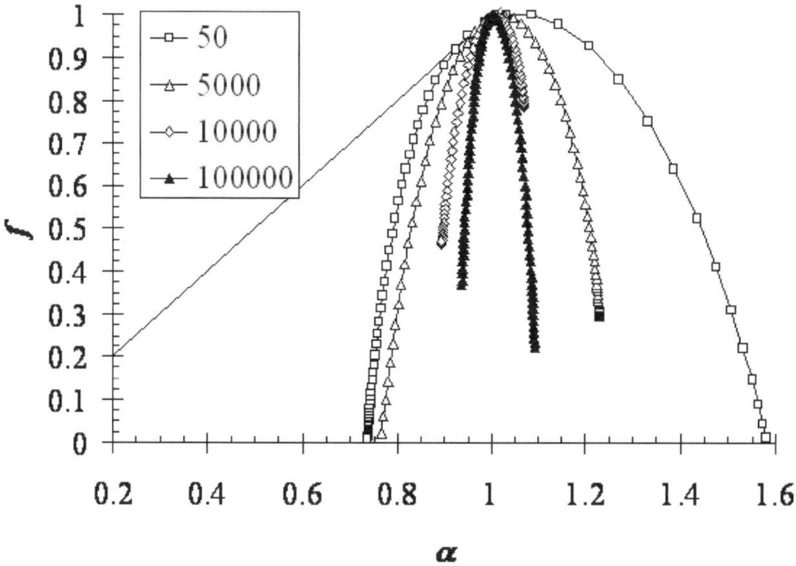

Fig. 1.9. Multifractal spectra of axial distributions of transverse cracks for different number of load cycles

1.8.2 Multi-Mechanism Damage

A detailed analysis of the single damage mechanism in cross-ply laminates allows a better understanding of the macroscopic manifestation of micro-structural randomness when one neglects a rather non-trivial interaction between various mechanisms [1]. Here let us consider some of the effects of this mechanism on the macroscopic behaviour of these composites. Firstly, contrary to ideas of the standard approaches, dealing with an area between two neighbouring transverse cracks as a macroscopic RVE, the load (stress) transfer depends also on the exact type of the axial distribution of cracks. It is shown [93] that, at the advanced stages of loading (i.e. in cases of high crack density) for the same spacing, the level of axial stresses both in weak (90°) and stiff (0°) layers differs for different types of crack distributions.

The pattern of matrix cracking affects the distribution of axial cracks and delamination zones in composites. The former are nucleated near 0°/90° interfaces close to tips of matrix cracks, while the latter are centred on intersections of transverse and axial cracks [1]. Delamination zones effectively reduce the spacing between neighbouring transverse cracks [92], thus additionally diminishing zones of preferable nucleation for new cracks (i.e. acting as another ordering mechanism for an ensemble of matrix cracks). Delamination also affects the macroscopic mechanic properties of laminates (e.g. the flexural modulus) to a considerably higher degree than transverse cracking. This effect also depends on the type of the distribution of matrix cracking, showing a considerable scatter in the extent of the modulus' deterioration for the same total length of delamination zones.

Tips of matrix cracks at the 0°/90° interfaces cause significant stress concentrations in stiff layers, adjacent to them. Such local overloading can result in fibre breakage and final rapture of the laminate. Here, the randomness in the pattern of stress concentration due to the underlying character of distribution of transverse cracks interacts with other random microstructural features, e.g. non-uniform longitudinal distributions of fibre's strength and fibre debonding areas. Such interaction additionally complicates the scenario of macroscopic damage evolution in composites.

Thus the result of interacting processes linked to microstructural randomness and ordering due to the load re-distributions 'percolates' to other damage mechanisms, affecting its macroscopic response to external loading. This complicated scenario of multi-mechanism damage is hard to adequately reflect in modelling schemes, which in many cases are reduced to either a single-mechanism studies or an explicit analysis of the interaction between mechanisms at the local level.

1.9 Conclusions

The above analysis of only a few features and damage mechanisms in cross-ply laminates – non-uniformity in the distribution of fibres in plies as well as matrix cracking and delamination – vividly demonstrates a challenge facing researchers who are developing modelling schemes for these materials. Though estimation of the effective macroscopic properties of these composites in a virginal state (i.e. without macroscopic defects) is a relatively simple task which can be solved analytically, an adequate description of damage evolution, especially at the stage of nucleation of macroscopic defects, presupposes a totally different strategy. Some elements of analytical schemes (e.g. load transfer rules, etc.) can be effectively used also in this strategy, forming one part of the computational analysis [94, 95].

In general, modelling of damage in composites can be implemented with the use of various multiscale strategies [5]. Not all the suggested multiscale schemes take random features of the microstructure into consideration. Generally, they combine partial solutions (e.g. for specific local areas, single damage mechanisms or local interactions of a few mechanisms), based on refined descriptions of the microstructure and/or geometry of the induced damage, with global ones that lack the detailedness at the microscopic level but reproduce the exact structure of a composite and macroscopic loading conditions.

A lattice-model approach [48, 78] employs ideas of continuum damage mechanics to link micro and macro levels. This is achieved by introducing a damage parameter as an additional variable at the macroscopic (continuous) level. This parameter characterises evolution of ensemble of microdefects and its macroscopic manifestation. The effect of spatially random microstructure (e.g. due to a stochastic spatial distribution of fibres) is accounted for, in terms of the varying local stiffness, reflecting the experimentally observed scatter. In this case, application of an externally uniform tensile load results in non-uniform distributions of the axial stress and, subsequently, different rates of damage accumulation for different parts of a 90° layer. The attainment of the critical damage concentration (i.e. implementation of the local failure criterion) in any point of this layer results in initiation of a matrix crack. Matrix cracking causes stress re-distribution, including formation of unloaded zones due to the shielding effect. The latter (as some other factors, for instance, a through-thickness stress variation linked to the effects of the stacking order and resin-rich zones) is incorporated by mapping of a (dynamic) matrix of stress coefficients onto the current stress levels in elements.

This approach allows a natural reflection of the interaction of microstructural randomness and additional ordering, imposed by matrix cracking at advanced stages of the loading history. Though this approach has been used for cross-ply laminates under tensile fatigue, it can be expanded to more complicated cases of both structure of laminates and loading conditions. This can be achieved by introducing of additional load transfer mechanisms, as it was implemented in [94].

Inclusion of additional damage mechanisms, necessary to describe multi-mechanism failure in fibre-reinforced cross-ply composites exposed to conditions of high-cycle fatigue, can be achieved by combining physical and continuum modelling tools within another multiscale formalism [96]. It is based on the use of various damage parameters for respective mechanisms – transverse cracking, delamination and fibre breaks – each with its own damage accumulation law, reflecting experimental observations and measurements [96, 97]. In some cases, an additional scale between microscopic and macroscopic ones – a meso-scale – could be introduced to incorporate several damage entities and capture their interaction for a more precise description of respective local perturbations of the stress field [1].

References

1. Talreja R (2006) Damage analysis for structural integrity and durability of composite materials. Fatigue Fract Eng Mater Struct 29: 481–506
2. Berthelot JM (2003) Transverse cracking and delamination in cross-ply glass-fiber and carbon-fiber reinforced plastic laminates: static and fatigue loading. Appl Mech Rev 56: 1–37
3. Reifsnider KL, Case SW (2002) Damage tolerance & durability in material systems. Wiley-Interscience, New York
4. Ladevèze P, Lubineau G, Violeau D (2006) A computational damage micromodel of laminated composites. Int J Fracture 137: 139–150
5. Soutis C, Beaumont PWR (eds) (2005) Multiscale modelling of composite material systems: the art of predictive damage modelling. Woodhead, Cambridge
6. Hill R (1963) Elastic properties of reinforced solids: some theoretical principles. J Mech Phys Solids 11: 357–372
7. Hill R (1964) Theory of mechanical properties of fibre-strengthened materials. I. Elastic behaviour. J Mech Phys Solids 12: 199–212
8. Hashin Z (1964) Theory of mechanical behaviour of heterogeneous media. Appl Mech Rev 17: 1–9
9. Hashin Z, Rosen BW (1964) The elastic moduli of fiber-reinforced materials. J Appl Mech 31: 223–232
10. Hashin Z (1965) Elasticity of random media. Trans Soc Rheol 9: 381–406
11. Budiansky B (1965) On the elastic moduli of some heterogeneous materials. J Mech Phys Solids 13: 223–227
12. Hashin Z, Shtrikman S (1963) A variational approach to the theory of the elastic behaviour of multiphase materials. J Mech Phys Solids 11: 127–140
13. Walpole LJ (1966) On bounds for the overall elastic moduli of inhomogeneous systems I. J Mech Phys Solids 14: 151–162
14. Hashin Z (1965) On elastic behaviour of fibre reinforced materials of arbitrary transverse phase geometry. J Mech Phys Solids 13: 119–134
15. Lu B, Torquato S (1990) Local volume fraction fluctuations in heterogeneous media. J Chem Phys 93: 3452–3459
16. Gibiansky LV, Torquato S (1995) Geometrical-parameter bounds on the effective moduli of composites. J Mech Phys Solids 43: 1587–1613
17. Torquato S (2001) Random heterogeneous materials. Microstructure and macroscopic properties. Springer, Berlin Heidelberg New York
18. Yeong CLY, Torquato S (1998) Reconstructing random media. Phys Rev E 57: 495–506
19. Lu B, Torquato S (1992) Lineal-path function for random heterogeneous materials. Phys Rev A 45: 922–929
20. Beran M, Molyneux J (1966) Use of classical variational principles to determine bounds for the effective bulk modulus in heterogeneous media. Q Appl Math 24: 107–118
21. Milton GW (1981) Bounds on the electromagnetic, elastic, and other properties of two-component composites. Phys Rev Lett 46: 542–545

22. Milton GW (1982) Bounds on the elastic and transport properties of two-component composites. J Mech Phys Solids 30: 177–191
23. Hill R (1965) A self-consistent mechanics of composite materials. J Mech Phys Solids 13: 213–222
24. Milton GW, Phan-Thien N (1982) New bounds on the effective moduli of two-component materials. Proc R Soc A 380: 305–331
25. Boucher S (1974) On the effective moduli of isotropic two-phase elastic composites. J Compos Mater 8: 82–89
26. Mclaughlin R (1977) A study of the differential scheme for composite materials. Int J Eng Sci 15: 237–244
27. Norris AN (1985) A differential scheme for the effective moduli of composites. Mech Mater 4: 1–16
28. Roscoe R (1952) The viscosity of suspensions of rigid spheres. Br J Appl Phys 3: 267–269
29. Einstein A (1906) Eine neue Bestimmung der Molek ldimensionen. Ann Phys 19: 289–306
30. Einstein A (1911) Berichtigung zu meiner Arbeit "Eine neue Bestimmung der Moleküldimensionen". Ann Phys 34: 591–592
31. Spitzig WA, Kelly JF, Richmond O (1985) Quantitative characterization of second-phase populations. Metallography 18: 235–261
32. Brockenbrough JR, Hunt WH Jr, Richmond O (1992) Reinforced material model using actual microstructural geometry. Scripta Metall Mater 27: 385–390
33. Davy PJ, Guild FJ (1988) Distribution of interparticle distance and its application in finite-element modeling of composite materials. Proc R Soc A 418: 95–112
34. Chung I, Weitsman Y (1994) A mechanics model for the compressive response of fiber reinforced composites. Int J Solids Struct 31: 2519–2536
35. Torquato S (1998) Morphology and effective properties of disordered heterogeneous media. Int J Solids Struct 35: 2385–2406
36. Torquato S, Lu B (1993) Chord-length distribution function for two-phase random media. Phys Rev E 47: 2950–2953
37. Green D, Guild FJ (1991) Quantitative microstructural analysis of a continuous fibre composite. Composites 22: 239–242
38. Davy PJ, Guild FJ (1988) The distribution of interparticle distance and its application in finite-element modelling of composite materials. Proc R Soc A 418: 95–112
39. Taya M, Muramatsu K, Lloyd DJ, Watanabe R (1991) Determination of distribution patterns of fillers in composites by micromorphological parameters. JSME Int J: Ser I 34: 198–206
40. Yotte S, Breysse D, Riss J, Ghosh S (2001) Cluster characterisation in a metal matrix composite. Mater Charact 46: 211–219
41. Pyrz R (1994) Quantitative description of the microstructure of composites. Part I. Morphology of unidirectional composite systems. Compos Sci Technol 50: 197–208
42. Ripley BD (1976) The second-order analysis of stationary point processes. J Appl Probab 13: 255–266

43. Ripley BD (1977) Modelling spatial patterns (with discussion). J R Stat Soc B 39: 172–212
44. Pyrz R (2004) Microstructural description of composites, Statistical methods. In: B hm HJ (ed) Mechanics of microstructured materials. Springer, Berlin Heidelberg New York, pp 173–233
45. Bulsara VN, Talreja R, Qu J (1999) Damage initiation under transverse loading of unidirectional composites with arbitrarily distributed fibers. Compos Sci Technol 59: 673–682
46. Cross SS (1994) The application of fractal geometric analysis to microscopic images. Micron 25: 101–113
47. Summerscales J, Guild FJ, Pearce NRL, Russell PM (2001) Voronoi cells, fractal dimensions and fibre composites. J Microsc 201: 153–162
48. Silberschmidt VV (2005) Multiscale modelling of cracking in cross-ply laminates. In: Soutis C, Beaumont PWR (eds) Multiscale modelling of composite material systems: the art of predictive damage modelling. Woodhead Publishing, Cambridge, pp 196–216
49. Chhabra AB, Meneveau C, Jensen RV, Sreenivasan KR (1989) Direct determination of the $f(\alpha)$ singularity spectrum and its application to fully developed turbulence. Phys Rev A 40: 5284–5293
50. Harte D (2001) Multifractals: theory and applications. Chapman & Hall/CRC, Boca Raton
51. Quintanilla J, Torquato S (1997) Local volume fraction fluctuations in random media. J Chem Phys 106: 2741–2751
52. Baxevanakis C, Jeulin D, Renard J (1995) Fracture statistics of a unidirectional composite. Int J Fracture 73: 149–181
53. Kanit T, Forest S, Galliet I, Mounoury V, Jeulin D (2003) Determination of the size of the representative volume element for random composites: statistical and numerical approach. Int J Solids Struct 40: 3647–3679
54. Drugan WJ, Willis JR (1996) A micromechanics-based nonlocal constitutive equation and estimates of representative volume element size for elastic composites. J Mech Phys Solids 44: 497–524
55. Moulinec H, Suquet P (1998) A numerical method for computing the overall response of nonlinear composites with complex microstructure. Comput Meth Appl Mech Eng 157: 69–94
56. Adams DF, Tsai SW (1969) The influence of random filament packing on the transverse stiffness of unidirectional composites. J Compos Mater 3: 368–381
57. Brockenbrough JR, Suresh S, Wienecke HA (1991) Deformation of metal–matrix composites with continuous fibers: geometrical effects of fiber distribution and shape. Acta Metall Mater 39: 735–752
58. Datta SK, Ledbetter HM (1983) Elastic constants of fiber-reinforced boron–aluminum: observation and theory. Int J Solids Struct 19: 885–894
59. Knight MG, Wrobel LC, Henshall JL (2003) Micromechanical response of fibre-reinforced materials using the boundary element technique. Compos Struct 62: 341–352

60. Matsuda T, Ohno N, Tanaka H, Shimizu T (2003) Effects of fiber distribution on elastic–viscoplastic behavior of long fiber-reinforced laminates. Int J Mech Sci 45: 1583–1598
61. Everett RK, Chu JH (1993) Modeling of non-uniform composite microstructures. J Compos Mater 27: 1128–1144
62. Ganguly P, Poole WJ (2004) Influence of reinforcement arrangement on the local reinforcement stresses in composite materials. J Mech Phys Solids 52: 1355–1377
63. Pyrz R, Bochenek B (1998) Topological disorder of microstructure and its relation to the stress field. Int J Solids Struct 35: 2413–2427
64. Povirk GL (1995) Incorporation of microstructural information into models of two-phase materials. Acta Metall Mater 43: 3199–3206
65. Alzebdeh K, Ostoja-Starzewski M (1996) Micromechanically based stochastic finite elements: length scales and anisotropy. Probab Eng Mech 11: 205–214
66. Smith RL (1983) The random variation of stress concentration factors in fibrous composites. J Mater Sci Lett 2: 385–387
67. Fukuda H (1985) Stress concentration factors in unidirectional composites with random fiber spacing. Compos Sci Technol 22: 153–163
68. Sankaran S, Zabaras N (2006) A maximum entropy approach for property prediction of random microstructures. Acta Mater 54: 2265–2276
69. Zeman J, Šejnoha M (2001) Numerical evaluation of effective elastic properties of graphite fiber tow impregnated by polymer matrix. J Mech Phys Solids 49: 69–90
70. Xi Y (1996) Analysis of internal structures of composite materials by second-order property of mosaic patterns. Mater Charact 36: 11–25
71. Borbély A, Kenesei P, Biermann H (2006) Estimation of the effective properties of particle-reinforced metal–matrix composites from microtomographic reconstructions. Acta Mater 54: 2735–2744
72. Ostoja-Starzewski M, Wang X (1999) Stochastic finite elements as a bridge between random material microstructure and global response. Comput Meth Appl Mech Eng 168: 35–49
73. Ostoja-Starzewski M (1993) Micromechanics as a basis of random elastic continuum approximations. Probab Eng Mech 8: 107–114
74. Ostoja-Starzewski M (1998) Random field models of heterogeneous materials. Int J Solids Struct 35: 2429–2455
75. Ostoja-Starzewski M (2002) Microstructural randomness versus representative volume element in thermomechanics. J App Mech 69: 25–35
76. Hermann HJ, Roux S (eds) (1990) Statistical models for the fracture of disordered media. North-Holland/Elsevier Science, Amsterdam/New York
77. Silberschmidt VV, Chaboche J-L (1994) The effect of material stochasticity on crack–damage interaction and crack propagation. Eng Fract Mech 48: 379–387
78. Silberschmidt VV (1997) Model of matrix cracking in carbon fiber-reinforced cross-ply laminates. Mech Compos Mater Struct 4: 23–37

79. Silberschmidt VV (2003) Crack propagation in random materials: computational analysis. Comput Mater Sci 26: 159–166
80. Ghosh S, Mukhopadhyay SN (1993) A material based finite element analysis of heterogeneous media involving Dirichlet tessellations. Comput Meth Appl Mech Eng 104: 211–247
81. Ghosh S, Lee K, Moorthy S (1995) Multiple scale analysis of heterogeneous elastic structures using homogenization theory and Voronoi cell finite element method. Int J Solids Struct 32: 27–62
82. Akbarzadeh A, Adams DF (1976) A hybrid finite element micromechanical analysis of composite materials. Fibre Sci Technol 9: 277–295
83. Manders PW, Chou T-W, Jones FR, Rock JW (1983) Statistical analysis of multiple fracture in 0°/90°/0° glass fibre/epoxy resin laminates. J Mater Sci 18: 2876–2889
84. Boniface L, Smith PA, Ogin SL, Bader MG (1987) Observations on transverse ply crack growth in $(0/90_2)_s$ CFRP laminate under monotonic and cyclic loading. In: Matthews FL, Buskell NCR (eds) Proceedings of 6th International Conference on Composite Materials (ICCM-6) and 2nd European Conference on Composite Materials (ECCM-2), Vol. 3. Elsevier Applied Science, Oxford, pp 156–165
85. Bergmann HW, Block J (1992) Fracture/damage mechanics of composites – static and fatigue properties. Institut f r Stukturmechanik DLR, Braunschweig (DLR-Mitteilung 92-03)
86. Berthelot J-M, Le Corre J-F (2000) Statistical analysis of the progression of transverse cracking and delamination in cross-ply laminates. Compos Sci Technol 60: 2659–2669
87. Lafarie-Frenot MC, Hénaff-Gardin C (1991) Formation and growth of 90 ply fatigue cracks in carbon/epoxy laminates. Compos Sci Technol 40: 307–324
88. Silberschmidt VV (2005) Matrix cracking in cross-ply laminates: effect of randomness. Compos A: Appl Sci Manuf 36: 129–135
89. Silberschmidt VV, Hénaff-Gardin C (1996) Multifractality of transverse cracking and cracks' length distribution in a cross-ply laminate under fatigue. In: Petit J (ed) ECF 11. Mechanisms and mechanics of damage and failure, Vol. 3. EMAS, London, pp 1609–1614
90. Silberschmidt VV (1998) Multifractal characteristics of matrix cracking in laminates under T-fatigue. Comput Mater Sci 13: 154–159
91. Silberschmidt VV (1995) Scaling and multifractal character of matrix cracking in carbon fibre-reinforced cross-ply laminates. Mech Compos Mater Struct 2: 243–255
92. Akshantala NV, Talreja R (1998) A mechanistic model for fatigue damage evolution in composite laminates. Mech Mater 29: 123–140
93. Silberschmidt VV (1993) Models of stochastic fracture and analysis of matrix cracking in crossply laminates under cyclic loading. Deutsche Forschungsanstalt f r Luft- und Raumfahrt e.V., Braunschweig (Report DLR-IB-131-93/32)

94. McCartney LN, Schoeppner GA (2002) Predicting the effect of non-uniform ply cracking on the thermoelastic properties of cross-ply laminates. Compos Sci Technol 62: 1841–1856
95. McCartney LN (2005) Multiscale predictive modelling of cracking in laminate composites. In: Soutis C, Beaumont PWR (eds) Multiscale modelling of composite material systems: the art of predictive damage modelling. Woodhead Publishing, Cambridge, pp 65–98
96. Beaumont PWR (2003) Physical modelling of damage development in structural composite materials under stress. In: Harris B (ed) Fatigue in composites. Woodhead Publishing, Cambridge, pp 365–412
97. Poursartip A, Ashby MF, Beaumont PWR (1986) The fatigue damage mechanics of a carbon fibre composite laminate. I. Development of the model. Compos Sci Technol 25: 193–218

Chapter 2: Multiscale Modeling of Tensile Failure in Fiber-Reinforced Composites

Zhenhai Xia[1] and W.A. Curtin[2]

[1]Department of Mechanical Engineering, The University of Akron, Akron, OH 44325, USA
[2]Division of Engineering, Brown University, Providence, RI 02912, USA

2.1 Multiscale Damage and Failure of Fiber-Reinforced Composites

Fiber-reinforced composites can be engineered to exhibit high strength, high stiffness, and high toughness, and are, thus, attractive alternatives to monolithic polymer, metals, and ceramics in structural applications. To engineer the material for high performance, the relationship between material microstructure and its properties must be established to accurately predict the deformation and failure. Such a relationship between underlying constituent material properties and composite performance can also aid selection and/or optimization of new composite systems. Successful models can yield predictive insight into the origins of damage tolerance, size scaling, and reliability of existing composite systems and can be extended to investigate damage and failure under more complex loading and environmental conditions, such as fatigue and stress rupture.

Damage relevant to macroscopic failure of fiber-reinforced composite occurs at many length scales and by a variety of physical mechanisms. At the smallest scale, preexisting defects in the fibers propagate and form fiber cracks that impinge on the matrix and the interface. Debonding, sliding, and/or matrix yielding at the crack perimeter inhibit crack propagation into the matrix; but the ensuing deformations are complex. The load carried by the broken fiber is then redistributed among the remaining unbroken fibers and matrix as determined by the detailed conditions at the debonded fiber/matrix interface and in the matrix. Subsequent damage occurs in and around other fibers according to the statistical distribution of flaws in the fibers and the stresses acting on those flaws due to the applied stress and the stress redistribution. Eventually, macrocracks will form and grow, leading to failure of the composites. Figure 2.1 illustrates the damage evolution of fiber-reinforced composites at each length scale under different loading conditions.

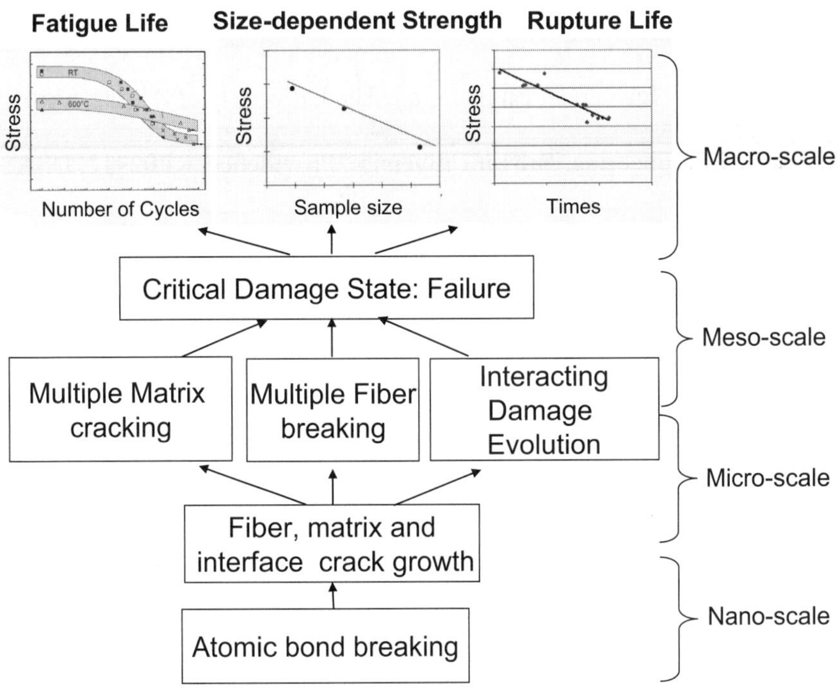

Fig. 2.1. Multiscale damage and failure in fiber-reinforced composites

Although the modeling path is conceptually clear, direct simulation of composite materials is still not a viable option despite advances in computational techniques and computing power. Finite element models that can capture micromechanical effects of cracks at the fiber/matrix/interface scale generally must employ mesh sizes of the order of the size of the microstructure and can result in an algebraic system with many millions of unknowns. It is insufficient, however, to focus only on one scale, i.e., a fiber break and the myriad details associated with it. On the other hand, homogenization and averaging techniques for analyzing heterogeneous materials, while possibly leading to manageable problem sizes, do not provide information about the microscopic fields needed, for example, to predict failure. Thus, there is a need for accurate and computationally efficient techniques that take into account the most important scales involved in the goal of the simulation while permitting the analyst to choose the level of accuracy and detail of description desired. Therefore, a *multiscale modeling* strategy is needed to accurately handle the evolution of damage at the larger scales while retaining important small-scale details and, thus, to accurately predict mechanical properties and performance of fiber-reinforced composites.

There are two main multiscale modeling techniques for materials: seamless coupling of methods in a single computational framework and hierarchical information transfer. Direct coupling methods are not viable for fiber composite problems because the damage spans a range of scales, and it is not possible to focus on one microscopic region in detail surrounded by a less-detailed description. Thus, the hierarchical multiscale modeling approach, in which the information of simulations at small length scales is processed and fed into larger-scale models, is preferable. The need for multiscale analyses has been well recognized; but until recently there has not been a direct connection made between the detailed structures at the fiber/matrix/interface scale, the multifiber damage problem, and large-scale component performance. Most work has assumed some approximate representation of the behavior at the smallest scale and pursued the larger-scale damage evolution. Such approaches are certainly warranted for understanding broad trends, identifying characteristic length scales associated with the damage, and for guiding the development of analytic models [5, 21, 31]. Other work has investigated the detailed stress states around damaged fibers, matrix, and/or interfaces but then employed only very simple models of overall composite behavior to indicate the important role of the microscale damage [12]. Specific system design and optimization requires attention to the detailed micromechanics of damage and load transfer around individual fiber breaks and the inclusion of such information directly into accurate larger-scale models.

In this chapter, one multiscale modeling approach for predicting tensile strengths of unidirectional fiber composites, including metal, polymer, and ceramic matrix composites will be reviewed. The quantitative success of this approach in predicting the tensile strength and its size dependence in a carbon fiber-reinforced plastic (CFRP), silicon carbide fiber/titanium matrix composites (TMCs), and alumina fiber/aluminum matrix composite (AMC) will be demonstrated. Finally, the approach will be extended to the prediction of strength and low-cycle fatigue life of TMCs.

This review emphasizes the published work of the present authors on multiscale modeling and simulation cast into a single overall framework. Progress in the field at one or several coupled scales has been made by many workers, with important insights and advances. In addition, analytic models for many problems in composite failure have been devised, but those works are not discussed here. Hence, the work presented here is not a comprehensive review of the literature. Interested readers can refer to several previous significant review articles [7, 22, 27] as well as other papers [19].

2.2 Multiscale Modeling via Information Transfer

2.2.1 Model Description and General Strategy

The fiber-reinforced composites considered here consist of continuous cylindrical fibers embedded in a matrix material in a unidirectional (aligned) arrangement. Such a composite can also be considered as a ply, a basic unit of a laminated composite structure. To develop a relationship between macroscopic properties and microstructure of the composite, a hierarchical set of models addressing physical phenomena at successive larger lengths scale, with coupling through information transfer, is introduced, as illustrated in Fig. 2.2. Figure 2.2 shows the full possible range of studies relevant to the problem. At the smallest scales, an atomistic or quantum analysis can assess features such as interface fracture energy and crack deflection at the bimaterial fiber/matrix interface. Key information on interfacial debonding and sliding is then passed into a continuum interface model, e.g., a cohesive zone, used in a micromechanical unit cell model consisting of matrix and a number of fibers to compute the stress redistribution around a fiber break for a particular material system. The stress redistribution is condensed into stress concentration factors on unbroken fibers, and, perhaps, stress intensity factors on matrix cracks, and this information is transferred to a larger-scale Monte Carlo model that tracks the evolution of fiber and/or matrix damage with increasing applied load. Details of the deformation around each fiber break are not retained at this scale, only their effects on stress concentrations. The Monte Carlo model is used to simulate damage up to the point of tensile failure, leading to a predicted average strength and statistical distribution for a composite sample that is small on the scale of practical samples but large compared to the critical damage size that drives failure. The tensile strength distribution calculated from the Monte Carlo model is then employed in analytic weak-link size-scaling models to predict ply strength and its statistical distribution as a function of physical size. Finally, the ply strength is used in standard laminated composite models to predict the strength and reliability of the composite component. In the last stage, other damage phenomena such as interply delamination can occur and change the local stresses in the plies themselves. In such cases, the ply strength vs. size can be used at smaller scales to assess the onset of local ply damage due to these other damage modes.

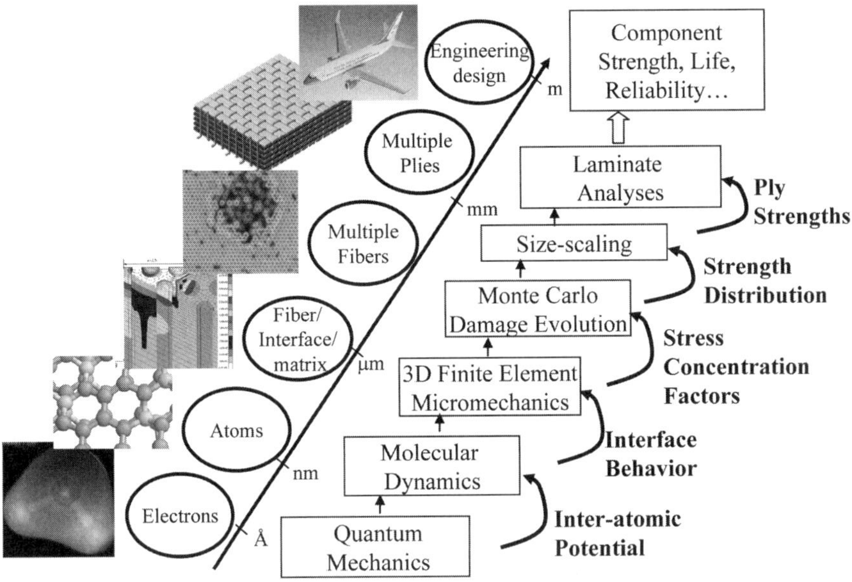

Fig. 2.2. Approach to multiscale modeling: scale coupling via information transfer

It is not necessary to always start from the quantum mechanical scale and progress upward. In fact, the goal of composite design is to shift the critical scale of damage from the nanoscale, e.g., the crack tip, to the much larger, observable, and detectable scale of collective fiber damage. Since a single fiber break does not initiate macroscopic failure, the details of the behavior at the smaller scales, while important, are not sufficient to predict failure. Therefore, one strategy is to envision possible modes of interface debonding and fiber/matrix constitutive behavior, as motivated by experiments or other theoretical models, and then use the multiscale modeling approach starting at the micromechanical scale. Parametric studies of the effect of interface and matrix behavior on the macroscopic fracture can then point to issues at smaller scales that would merit more detailed treatment. The work presented in this chapter focuses on the multiscale modeling of unidirectional fiber-reinforced composites starting from the micromechanical scale taking the interface behavior as a parametric input with quantities such as the interfacial coefficient of friction and interfacial strength obtained from experiments when applications to a particular material system are made.

2.2.2 Micromechanics at the Fiber/Matrix/Interface Scale

The goal of modeling at the micromechanics scale is to compute the detailed stress redistribution around broken fibers with various interfacial deformation models and extract from such studies the average stress concentrations induced in the surrounding unbroken fibers and the stress recovery along the broken fiber due to interface shear resistance. Since introduction of a fiber break or a matrix crack causes large stress changes only in the vicinity of the crack, a small-scale model with high spatial refinement is used. The model used consists of a hexagonal array of unidirectional fibers with a fiber volume fraction of V_f. Making use of symmetry, the model can be restricted to a 30° wedge, as shown in Fig. 2.3a. Each fiber in this wedge section represents a distinct set of neighbors relative to the central fiber. A 3D finite element representation of this model is then constructed to calculate the stress distributions around broken fibers (Fig. 2.3b,c). The axial length of the model depends on the interface and matrix behavior and is generally chosen such that the stress distribution at the end of the model is not affected by the stress redistributions caused by the introduction of fiber or matrix damage at the midplane. The size of the model in the radial direction (perpendicular to the fibers) is chosen so that the deformation of fibers at the outer perimeter is not affected by the imposed fiber damage. For example, with a single central broken fiber, we use the nearest eight sets of neighbors (43 fibers total). With seven broken fibers (fibers 1 and 2 broken in the 30° wedge section), a larger model extending out to tenth neighbors and containing a total of 91 fibers is used. The mesh sizes are selected to obtain converged results, for which there is no a priori guidance except that there should be at least several elements in the matrix region between the fibers and within the fibers themselves. The model is subjected to tensile loading along the axis of the fibers, and the appropriate boundary conditions are shown in Fig. 2.3b. The nodes of uncracked material at the crack plane ($z = 0$) have fixed displacements in the z-direction while the outer surface of the model is traction free.

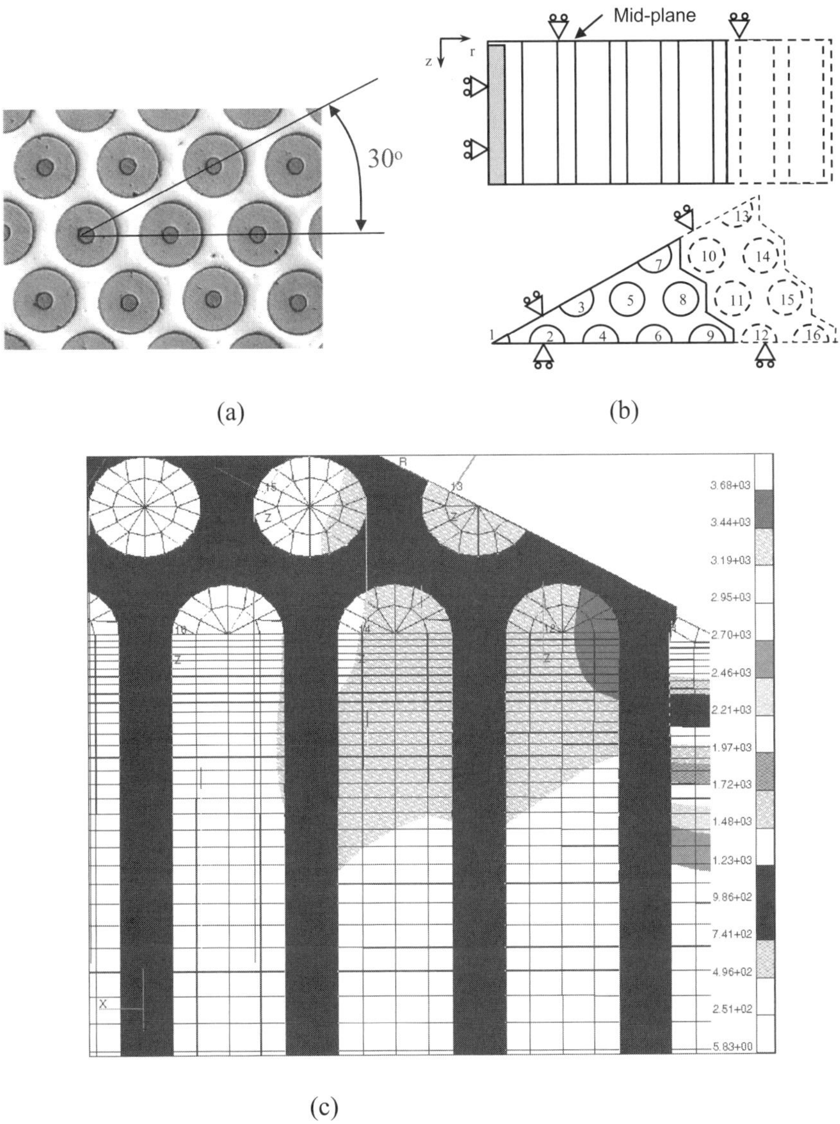

Fig. 2.3. (**a**) Optical image of Ti/SiC composite microstructure, (**b**) wedge section of model hexagonal distribution of fiber composite with boundary conditions for finite element analysis, and (**c**) a 30° wedge of finite element model showing the axial stress distribution in the fibers and matrix around a central broken fiber (reprinted with permission from [38])

Modeling of loading transfer through a fiber/matrix interface is a key step to properly simulate the stress distributions in the fibers. The interfaces can be classified into weak and strong bond interfaces according to interfacial bonding strength. If the fiber/matrix interface is strong, no interfacial debonding occurs. Modeling of such an interface is simple. Since there is no sliding between the fiber and matrix, the matrix and fiber elements are compatible and shear the same nodes at the interface in the finite element model. However, if the interfacial bonding is weak, the interface will debond, leading to sliding during loading. In this case, contact elements can be used to simulate stress transfer across the fiber/matrix interface. If the residual thermal stresses (axial tension in the matrix, axial compression in the fibers, and radial compression σ_r at the interface) are high, the fiber/matrix bond strength is usually assumed to be zero for simplification. Interfacial stress transfer is then realized by Coulomb friction at the interface so that the friction shear stress τ along the interface in the slip zone is simply $\tau = -\mu\sigma_r$, where μ is the coefficient of friction.

The introduction of a fiber break in the central fiber at the midplane of the model induces significant changes in the local stresses around the break (e.g., Fig. 2.3c). The stress distribution around a broken fiber is very complex. Multiscale modeling progresses by assuming that all of these details are not relevant to the desired macroscopic behavior. For the propagation of damage among fibers, the tensile stresses in the unbroken fibers drive the growth of preexisting flaws in those fibers if the tensile stress is large enough. It is assumed that it is sufficient to consider the average tensile stress through the cross-section of any fiber, rather than maintain the full spatial variation. While it is certainly true that any particular fiber can have a flaw that experiences a stress higher or lower than the average [20, 33], the influence of such an effect has not been considered. Condensing the detailed information from studies such as that shown in Fig. 2.3c, consider the stress in the broken fiber and the stresses in the surrounding fibers. The stress in the broken fiber is zero at the break point and recovers along the broken fiber, as shown in Fig. 2.4a. Shear deformation along the interface, by either shear yielding of a well-bonded plastically deforming matrix or frictional sliding along a debonded interface, leads to a nearly linear recovery of axial stress in the fiber. Figure 2.4b shows the average axial stress concentration factor (SCF = actual stress normalized by far-field applied fiber stress) in the plane of the fiber break on the successive sets of neighbors around the broken fiber. The stresses in the neighboring fibers are increased to compensate for the loss of load-carrying capacity in the broken fiber, with the SCF decreasing with increasing distance from

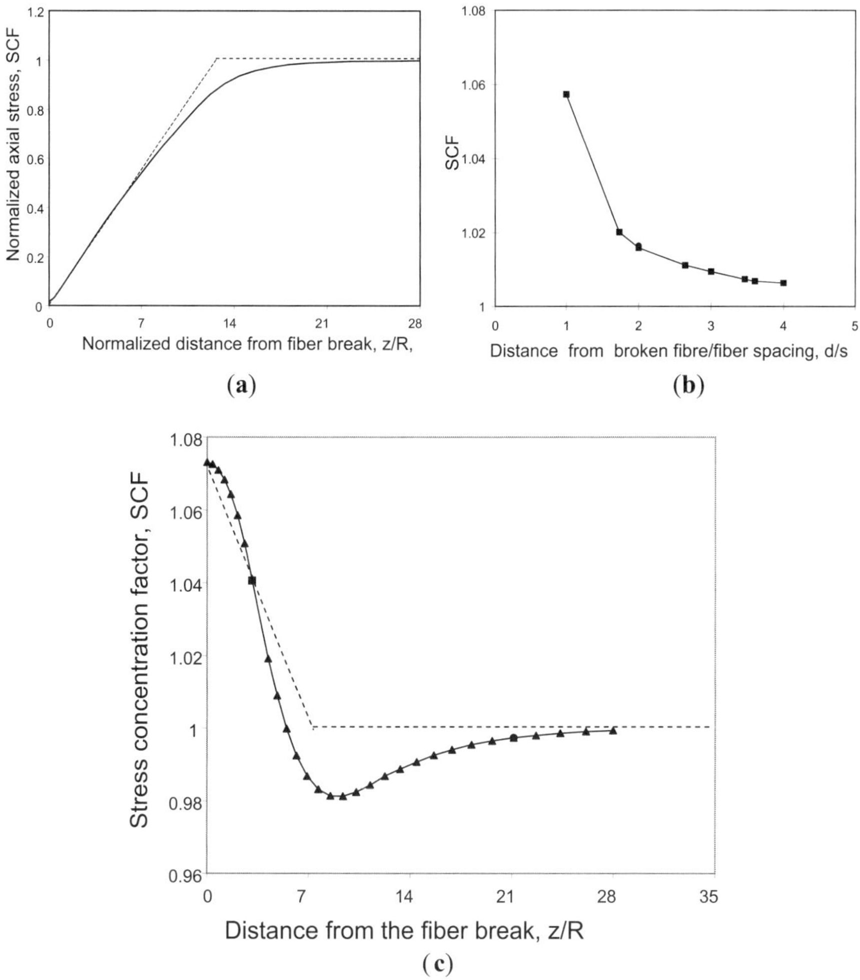

Fig. 2.4. (**a**) Axial stress distribution on the central broken fiber along the fiber direction z/R (R = fiber radius), normalized by the far-field fiber stress, (**b**) axial stress concentration factor (*SCF*) on the fibers as a function of the distance away from the broken fiber, normalized by fiber spacing s, and (**c**) average axial stress concentrations on the near-neighbor fibers along the fiber direction z. *Dashed lines* in (**a**) and (**c**) show the approximated stress concentrations using a constant interfacial shear stress τ model that is employed in one of the larger-scale models (Green's function model)

the broken fiber. The average stress concentration on the near-neighbor fibers vs. the distance z away from the crack plane is shown in Fig. 2.4c. Near the plane of the break, the neighboring fiber stresses are larger than in the far-field. Within increasing distance z, the broken fiber recovers its load-carrying capacity and the SCFs of the surrounding fibers thus decrease over a similar length scale. The SCF on the neighboring fibers can actually fall below unity before recovering to unity at larger distances, which is due to bending that arises from the need to satisfy compatibility. The details, such as those shown in Fig. 2.4, depend on the input constitutive properties: the fiber elastic modulus, the matrix elastic modulus and plastic flow behavior, if any, and the interface constitutive model. However, the results are generically those shown in Fig. 2.4, and the SCFs and length scales of stress recovery are the information derived from the detailed micromechanical model that is passed to a larger-scale damage accumulation model.

2.2.3 Mesoscale Modeling of Fiber Damage Evolution

The finite element (FE) models provide the detailed stress state around a single broken fiber. Larger clusters of broken fibers can be investigated, but such a direct numerical approach is limited to symmetric clusters of breaks due to the symmetry of the unit cell. Decreasing the symmetry of the unit cell is possible but computationally difficult. Furthermore, to understand the size scaling of the composite strength and, thus, predict strengths of very large samples, requires hundreds of simulations of failure in composites having several hundred fibers. Here, two alternative approaches to obtaining reasonably accurate but computationally more feasible results: the 3D shear-lag and Green's function methods are discussed. The goal of these methods is to reliably calculate the stress states in any surviving fibers given an arbitrary spatial distribution of fiber breaks, while capturing the proper SCFs and length scales computed from the detailed finite element method (FEM) models.

Shear-lag method

The shear-lag model (SLM) for fiber SCFs has a long history, dating back to the work of Hedgepeth and Hedgepeth and Van Dyke [4, 14, 15, 30]. In this model, the fibers are treated as one-dimensional extensional elements of modulus E_f while the matrix is treated as a material with modulus G_m that transfers tensile loads among fibers via shear deformation only and carries no tensile loads. Here we discuss a 3D SLM developed by Okabe and Takeda [25] that incorporates interface sliding due to friction and/or

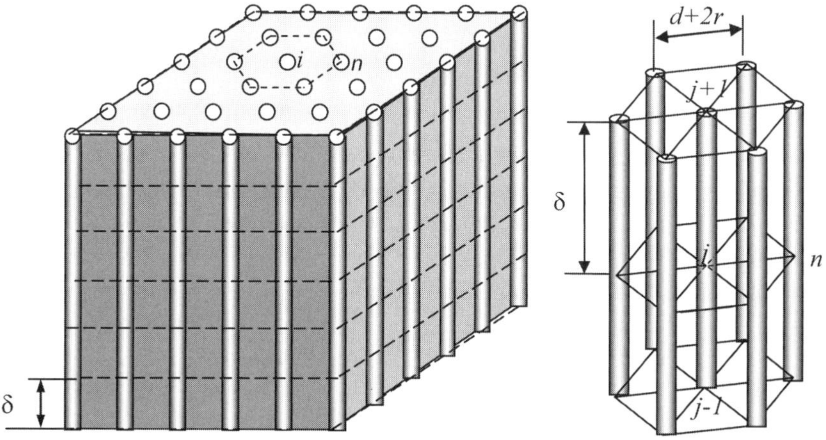

(a) (b)

Fig. 2.5. (**a**) 3D model with 1D fibers and load transfer calculation in shear-lag and Green's function models and (**b**) the nodes around *i*th fiber

matrix yielding as well as evolving fiber damage in a single, compact framework. A schematic view of the composite with a hexagonal fiber array and relevant notation is shown in Fig. 2.5. The SLM assumes that the local matrix shear stress is governed by the smaller of (1) the elastic shear stress associated with the neighboring fiber displacements

$$\tau_{\mathrm{n}}(z) = G_{\mathrm{m}}[u_n(z) - u_i(z)]/d, \tag{2.1}$$

where u_n is the displacement of the nth near-neighboring fiber to fiber i (see Fig. 2.5) and d is the fiber spacing or (2) $|\tau_{\mathrm{n}}| = \tau_{\mathrm{y}}$, where τ_{y} is the yield strength for an elastic/perfectly plastic matrix or the debonded interfacial shear stress for a sliding interface. Within this framework, force equilibrium on the ith fiber in a hexagonal array with an elastic/plastic matrix is, when discretized by a uniform mesh size $\Delta z = \delta$, given by

$$\begin{aligned} &3AE_{\mathrm{f}}\left[\frac{a_{i,j}(u_i(z_{j+1}) - u_i(z_j)) - a_{i,j-1}(u_i(z_j) - u_i(z_{j-1}))}{(2 + a_{i,j} + a_{i,j-1})\delta^2}\right] \\ &+ h\sum_{n=1}^{6}\left[\frac{G_{\mathrm{m}}}{d}(u_n(z_j) - u_i(z_j))b_n + \tau_{\mathrm{y}}(1 - b_n)\right] = 0, \end{aligned} \tag{2.2}$$

where $u_i(z_j)$ is the displacement of the jth node of the ith fiber located at longitudinal position z_j, with $h = \pi r/3$ and $A = \pi r^2$. In (2.2), $a_{i,j}$ are damage parameters: $a_{i,j-1} = 0$ if the element of fiber i between z_{j-1} and z_j is broken, and $a_{i,j-1} = 1$ if unbroken; similarly, $a_{i,j} = 0$ if the element between z_j and z_{j+1} is broken, and $a_{i,j} = 1$ if unbroken. The b_n ($n = 1$–6) in (2.2) are yield indicator parameters, with $b_n = 1$ if $|\tau_{\mathrm{n}}|$ is less than τ_{y} and $b_n = 0$ otherwise. Periodic boundary conditions are used on the lateral edges of the composite, so that all fibers have six neighbors. The boundary conditions for uniaxial loading are zero displacement at $z = 0$, $u_i(0) = 0$, and a constant applied displacement U at $z = L$, $u_i(L) = U$. The stresses in the unbroken fiber elements follow from Hooke's law as:

$$\sigma_i(z_j) = E_{\mathrm{f}}[u_i(z_j) - u_i(z_{j-1})]/\delta. \tag{2.3}$$

The SLM predicts the stress concentration and recovery around an arbitrary collection of broken fibers, but the assumptions in the model are not always appropriate. Comparison of the SLM predictions against full finite element modeling for the exact same problem shows that the standard SLM can accurately predict the stress recovery length along a broken fiber for a wide range of fiber/matrix stiffness ratios [37]. However, the SCFs are only accurate for systems with a high fiber/matrix stiffness ratio and high fiber volume fraction; practically, this corresponds to polymer matrix composites with high fiber fraction. For other material systems, in particular metal matrix composites, factors such as the neglect of shear across the finite fiber dimensions in the SLM, the matrix load-carrying capability, and/or the loading history, makes the SLM less accurate for the stress transfer [37]. The stress transfer is more diffuse in the SLM than in the full FEM studies, making the SLM less conservative in predictions of local damage evolution. Thus, care must be taken in using the SLM to model composite deformation and failure, although applications to polymer composites with stiff fibers and high fiber volume fractions should be accurate and realistic.

Green's function model

The Green's functional model (GFM) [36] uses the in-plane SCFs around a single fiber break as obtained from any detailed numerical model as a Green's function, makes a simple approximation for the SCFs along the length of the unbroken fibers, and then computes the 3D damage evolution due to multiple, interacting fiber breaks.

Specifically, the data for in-plane SCFs, as shown in Fig. 2.4b, define a Green's function G_{ij} for stress transfer from broken fiber j to surrounding fiber i. The stress distribution around a broken fiber is then modeled by the simple constant τ SLM. In other words, for fiber m broken at position z_m^{b}, the stress along the broken fiber is approximated as

$$\sigma_m^{\mathrm{b}}(z) = 2\tau\left|z - z_m^{\mathrm{b}}\right|/r, \quad \sigma_m^{\mathrm{b}}(z) \le \sigma_m(z) \tag{2.4}$$

as shown in Fig. 2.4a, where $\sigma_m(z)$ is the axial stress in the fiber existing before the break, so that (2.4) is operative only within the slip length around the fiber break. The stress lost by the broken fiber at position z,

$$p_m(z) = \sigma_{app,m}(z_m^b) - 2\tau\left|z - z_m^b\right|/r \qquad p_m(z) \ge \sigma_m(z) \tag{2.5}$$

is transferred to the surrounding fibers using the Green's function computed in the plane of the break. With these two features, the total stress $\sigma_i(z)$ on unbroken fiber i in plane z due to broken fibers $\{m\}$ is approximated as

$$\sigma_i(z) = \sigma_{\mathrm{app},i}(z) + [G_{ik}(1-G)_{kl}^{-1}G_{lm}]p_m(z), \tag{2.6}$$

where $\sigma_{\mathrm{app},i}(z)$ is the applied stress on fiber i at position z and there is an implied sum over the repeated indices k, l, m. Equations (2.4)–(2.6) predict that the stress transferred to surrounding fibers decreases linearly with distance from the fiber break until the slip region ends. This approximation is shown in Fig. 2.4c, from which it is evident that the model captures the basic features of the deformation but misses the subtle details associated with bending and compatibility that arise in the full FEM and also in the SLM. By construction, however, the GFM always satisfies equilibrium of the axial load, i.e., the sum of the forces over any cross-section of the fiber system is equal to the total force applied across the section. Equations (2.4)–(2.6) are solved at a discrete set of points z_j along each fiber and, thus, provide the analog of the stresses emerging from the solution of (2.1)–(2.3) in the SLM. Since the GFM takes the input directly from a more detailed calculation, it has a wider range of applicability than the SLM. However, for cases where the SLM is a good approximation, such as polymer matrix composites, the GFM contains some additional assumptions that could modify the predictions. A comparison of the GFM vs. the SLM, when the SCFs from the SLM are used as the input to the GFM, shows that the GFM predicts damage evolution and tensile strength in good agreement with the SLM for the systems considered [36], thus suggesting that the approximations made in the GFM model are reasonable.

2.2.4 Predictions of Tensile Strength in Small Samples

The ultimate tensile strength of the composite is determined by two contributions. The first contribution is the fiber bundle strength σ_{f}^{*}, which is determined via simulation of the evolution of fiber damage and stress transfer from broken to unbroken fibers using the shear-lag or Green's function method in a stochastic simulation model to be described below. The second contribution is the load-carrying capacity of the matrix. Since the fiber damage that drives ultimate failure is fairly localized in space, in both the longitudinal and transverse directions, most of the matrix is deforming as if in an undamaged composite. Thus, to a very good approximation, the average stress carried by the matrix is the axial stress in an undamaged composite at a stress equal to the composite strength. The ultimate strength can thus be expressed as

$$\sigma_{\mathrm{uts}} = V_{\mathrm{f}}\sigma_{\mathrm{f}}^{*} + (1 - V_{\mathrm{f}})\sigma_{\mathrm{m}}(\sigma_{\mathrm{uts}}), \tag{2.7}$$

where σ_{m} is the axial matrix stress and is a function of the applied stress. The main goal is to compute the fiber bundle strength σ_{f}^{*}.

For any fixed state of damage, i.e., spatial distribution of broken fibers, the SLM and the GFM compute the associated tensile stresses in all fibers in the system. Damage evolution then occurs by further failure of fibers due to the increasing stress concentrations. The progressive fiber damage occurs because the fibers have a statistical distribution of flaws within them, leading to a corresponding statistical distribution of strength on any set of fiber elements. Modeling of the damage evolution thus requires the appropriate fiber strength distribution as input. The cumulative probability of fiber failure $P_{\mathrm{f}}(\sigma, L)$ in a gauge length L at stress σ is usually modeled as a Weibull distribution that accounts for the flaw-sensitive, weak-link nature of the brittle fiber failure. In a two-parameter Weibull model, $P_{\mathrm{f}}(\sigma, L)$ is given by

$$P_{\mathrm{f}}(\sigma, L) = 1 - \exp\left[-\frac{L}{L_0}\left(\frac{\sigma}{\sigma_0}\right)^{m} \right], \tag{2.8}$$

where σ_0 is a characteristic fiber strength for fibers of length L_0 and m is the Weibull modulus describing the statistical spread in strengths. For most commercial fibers, the fiber strength properties are well characterized by the two-parameter Weibull strength model. The Weibull parameters σ_0 and m are usually obtained from experiments in which a large number of fibers of length L_0 are tested in tension prior to incorporation into the composite.

However, composite processing can damage the fibers, modifying the in situ strength distribution compared to the initial ex situ distribution. To address this problem, fibers can sometimes be extracted from as-processed composites and then tested to obtain the appropriate strength parameters [11]. Another approach is to examine the fracture mirrors on fibers protruding from the fracture surface of a tested composite, from which the fiber strength statistics can be derived [7]. In any case, simulations of composite tensile strength require accurate knowledge of the in situ fiber strength distribution.

Within the constant τ shear model for interface sliding, an analytic model that ignores local stress concentrations, the so-called *Global Load Sharing model*, permits for the identification of a characteristic stress that embodies most of the major dependencies of composite behavior on fiber and interfacial characteristics [6, 7]. This characteristic stress σ_c is the characteristic fiber strength at a characteristic length δ_c, $P_f(\sigma_c, \delta_c) = 0.632$, and these interrelated quantities are given by [6]

$$\sigma_c = \left(\frac{\sigma_0^m \tau L_0}{r} \right)^{1/(m+1)}, \quad \delta_c = \frac{r\sigma_c}{\tau}. \tag{2.9}$$

In a simulation model, it is often convenient to normalize all lengths by δ_c and stresses by σ_c, using an appropriate value of τ to define the length δ_c. Even if τ is approximate, (2.9) condenses some of the major physical dependencies of the composite failure into two key parameters.

With the above preliminaries, the computation of the fiber bundle strength σ_f^* is straightforward. The simulation algorithm proceeds as illustrated in Fig. 2.6. A simulation model contains a computationally tractable number of fibers (typically ~1,000) each of length $L \geq 2\delta_c$. Each fiber is discretized into a series of small elements of length $\bar{\delta} \ll \delta_c$, as illustrated in Fig. 2.5. Each fiber element is then assigned a tensile strength at random from the Weibull distribution, i.e., a random number R in the interval [0, 1] is selected; and the strength of the element is assigned to be $\sigma_c(\delta_c / \bar{\delta})^{1/m}(-\ln(1-R))^{1/m}$. An initial tensile load is applied to the fiber bundle, and fiber breaks are introduced into those fiber elements for which the applied stress exceeds the assigned element strength. After these fibers break, the stress redistribution is calculated with the shear-lag or Green's function model. Under the redistributed stress, some fiber elements may then exceed their assigned strengths and are broken; and the stress redistribution is computed again. This fiber break and stress redistribution

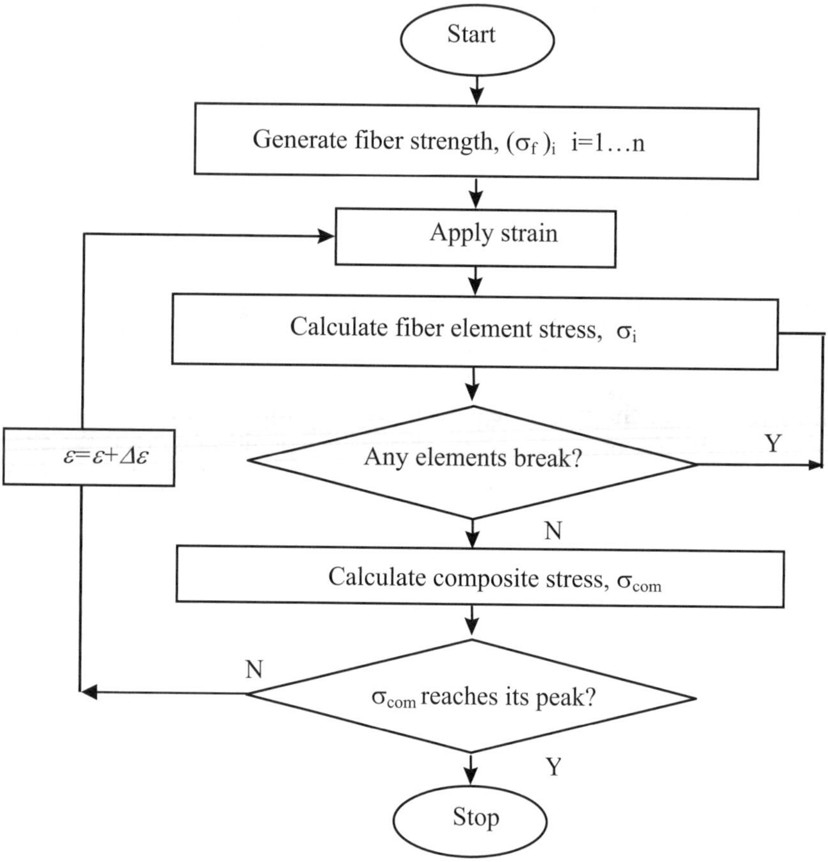

Fig. 2.6. Flow chart of simulation procedure for fiber damage evolution in fiber composites

process is repeated until no further fiber breaks are found; the damaged composite is then in a stable equilibrium state. The applied displacement or load is then increased by a small increment, and the above process is repeated. In the SLM, which is typically displacement controlled, the tensile strength is identified as the maximum stress. In the GFM, which is load controlled, the system undergoes catastrophic failure (all fibers break in some narrow range of the sample cross-section) at the tensile strength.

Figure 2.7 shows an example of the simulated stress–strain curve for an Al_2O_3/Al composite. The fiber damage evolution in the ultimate failure plane is shown via examination of the fiber SCFs in Fig. 2.7 at two stages: just at failure and just beyond failure. In Fig. 2.7, SCF values less than one indicate that the fiber is broken somewhere within a slip length of the failure plane and is carrying a reduced stress in the failure plane due to slip (2.4) while SCF values exceeding one indicate enhanced stresses on unbroken fibers in the plane of view. At low stress levels, isolated breaks occur at weak fiber elements throughout the material, and the stress concentrations are not sufficient to drive further failure. With increasing load, clusters of fiber breaks form due to both statistics and to enhanced local stresses. The stress concentrations around these clusters grow with the cluster size, driving further damage. When the load just reaches the tensile strength (Fig. 2.7a), a "critical" cluster of fiber breaks forms, consisting of a dispersed group of fiber breaks leading to local stress enhancements on the unbroken fibers in and around these breaks. With no further increase in applied load, fiber damage continues unabated spreading outward from the critical damage cluster. Figure 2.7b shows the damage configuration after some extent of unstable fiber damage. After some sporadic growth, the damage cluster becomes roughly penny shaped with very high-stress concentrations on its perimeter that drives the continued growth, similar to crack growth in a monolithic material.

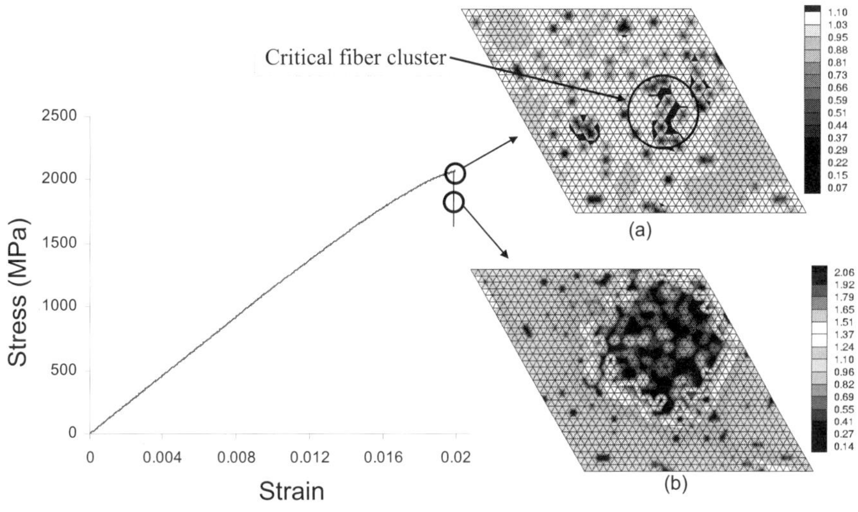

Fig. 2.7. Predicted stress–strain curve for an alumina fiber/aluminum composite with a matrix yield strength of 100 MPa, with schematics of fiber damage and stress concentrations in the plane of final fracture: (**a**) just at the failure strength, where a critical damage cluster can be identified and (**b**) after some unstable damage propagation at the failure strength, where the damage has formed a near penny shape crack. Each node corresponds to a single fiber (reprinted with permission from [35])

If either the fiber Weibull modulus or SCF is low, the composite can fail in a mode different from that described above. Figure 2.8a,b shows the failure process for an Al matrix composite with a lower matrix yield stress ($\sigma_y = 50$ MPa). In this case, the SCFs for individual breaks are lower leading to damage that is more uniformly distributed in the cross-section, as compared to Fig. 2.7a, such that a critical cluster cannot be identified in Fig. 2.8a. Even after unstable damage propagation (Fig. 2.8b), the damage still spreads quite uniformly through the cross-section. Thus, the SCF determined at the microscale by interface and matrix deformation plays a key role in determining the evolution of the damage, the formation of a critical damage cluster, the mode of damage, and, ultimately, the statistics and size scaling of the tensile strength.

The composite failure strength has a statistical distribution. By performing many simulations, with each simulation giving a different strength due to the different random fiber strengths and evolution of the damage clusters, the distribution can be determined numerically. The strength distribution depends mainly on the fiber Weibull modulus m and the SCFs. High m and/or high SCFs result in more localized damage (Fig. 2.7) and broader distributions while low Weibull modulus and/or low SCFs lead to more dispersed damage, as show in Fig. 2.8, and narrower distributions of strength. Thus, the strength is a combination of the spread in fiber strengths (m) and the SCFs [35]. The size scaling of the composite strength for material systems discussed below depends on the combination of fiber Weibull modulus and SCF.

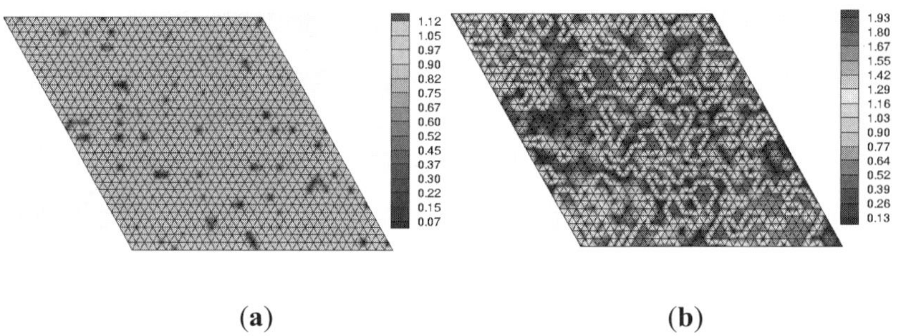

Fig. 2.8. Fiber damage and stress concentrations in the plane of final fracture in alumina fiber/aluminum composites. Each node point corresponds to a single fiber. (**a**) For $\sigma_y = 50$ MPa, just at the failure strength, where there is no clear localization of damage and (**b**) after some unstable damage propagation at the failure strength, with the damage still distributed across the entire sample cross-section (reprinted with permission from [35])

2.2.5 Size-Scaling Model at Large Scale

The composite tensile strength decreases with increasing sample size. This is due to the underlying dependence of the fiber strengths on length. However, since a number of fibers must fail locally in the composite to create a critical cluster capable of driving macroscopic failure, the statistical distribution of composite strengths at fixed size is much narrower than that of the single fiber. Similarly, the size scaling of the characteristic composite tensile strength is much weaker than that for the individual fibers [13]. Size scaling is an important issue because it bridges the scales between numerical simulation sizes and test specimen and/or component sizes. Size scaling is also intimately linked with reliability, i.e., the probability distribution of failure at any fixed size.

Since the composite strength is controlled by a weak-link failure, i.e., failure is driven by the formation of a localized cluster of damage somewhere in the material and much smaller than the sample size for large samples (see Fig. 2.7), information on the cumulative probability distribution of the fiber bundle strength at a fixed size $P_{n_s}(\sigma)$ can be used to obtain the characteristic $\tilde{\sigma}$ vs. the size as follows. First, the "size" involves the number of fibers n_f in the cross-section and the length of the sample L. Failure occurs within a longitudinal section length of $\sim\delta_c$, and so the sample length L can be viewed as consisting of a set of $L/(0.4\delta_c)$ independent "bundles," where the factor of 0.4 has been derived from detailed statistical analysis of simulations and analytic estimates [26]. The size of the composite is then $n = n_f L/(0.4\delta_c)$. Now, the characteristic strength $\tilde{\sigma}$ at any size n satisfies $P_n(\tilde{\sigma}) = 1 - e^{-1}$. Furthermore, weak-link scaling dictates that the probability distributions for samples of sizes n and n' are related via

$$P_n(\sigma) = 1 - \left(1 - P_{n'}(\sigma)\right)^{n/n'} . \tag{2.10}$$

Using the simulation data $P_{n_s}(\sigma)$ at size $n' = n_s$ on the right-hand side of (2.10) and setting the left-hand size equal to $1 - e^{-1}$, we find that the size n having characteristic fiber bundle strength $\tilde{\sigma}$ must satisfy

$$n = -n_s / \ln\left(1 - P_{n_s}(\tilde{\sigma})\right), \tag{2.11}$$

which then implicitly generates the strength vs. size, $\tilde{\sigma}(n)$.

To investigate the size scaling of composite strength numerically within the present model, we need to perform a large number of simulations on composites containing n_s fibers using the model described in Sect. 2.1. From these simulations, we directly obtain the cumulative probability distribution $P_{n_s}(\sigma)$ for failure of the fiber bundle at the simulated size n_s. The composite strength then follows directly from (2.11).

2.3 Case Studies: Prediction of Strength by the Multiscale Coupling Approach

An approach to the hierarchical modeling of composite tensile failure has been presented. The proposed multiscale modeling involves the passing of key information from smaller to larger scales. In this section, the general multiscale modeling approach will be implemented to predict the properties and performance of several different composite systems under tensile loadings. Although the microstructures in fiber-reinforced composites are similar, the fiber/matrix interfaces are quite different, and load transfer strongly depends on the interfacial bonding strength. In the case of a weak interface, debonding will occur when the fiber breaks with the subsequent deformation controlled by a frictional interfacial shear stress. Matrix yielding also plays an important role in the failure of the composites. High yielding matrices may bear significant loading, for instance, in metal matrix composites. In the absence of debonding, the matrix shear yield stress can determine the "sliding" behavior after a fiber breaks. Here, metal and polymer matrix composites will be used as examples to demonstrate how to predict the tensile strength of macroscopic composite samples from the detailed micromechanics.

2.3.1 Polymer Matrix Composites (PMCs)

Polymer composites reinforced by carbon or glass fibers have high strength and are widely used as high-performance materials in aerospace, electronics, and infrastructures. Here the multiscale modeling approach is used to predict the strength of fiber-reinforced polymer composites by linking composite microstructure and mesoscale fiber damage evolution to the mechanical properties at very large scale.

The stress concentration predicted by the shear-lag model and finite element model has been compared. The results show, for both the elastic and elastic–plastic cases, that the SLM agrees with the finite element predictions very well except for a region very near the fiber break, nearly independently of the fiber break load for polymer matrix composites with high fiber volume fraction [37]. In this case, the SLM is reliable for predicting the stress concentration in the fibers. Hence, the SLM is used with the mesoscale Monte Carlo damage evolution model following the standard procedure described in Sect. 2.2.

The composite material studied here is a plastic matrix reinforced by carbon fibers. The thermomechanical properties of the carbon fibers and polymer matrix are presented in Table 2.1. The fiber strength is described using a modified two-parameter Weibull model: $P_{\mathrm{f}}(\sigma, L) = 1-\exp[(L/L_0)^{\alpha}(\sigma/\sigma_0)^{m}]$, where $\alpha = 0.7$ is the fitting parameter from the experimental results for the carbon fiber [24]. The simulated composite is composed of 1,024 fibers of length $4\delta_{\mathrm{c}}$ in a hexagonal array with periodic boundary conditions. Each fiber is divided into 100 longitudinal elements to minimize discretization errors.In the absence of fiber damage, uniaxial loading determines the overall stress–strain response of the undamaged composite. Due to the very low matrix yield stress and high fiber strength and stiffness, the stress–strain behavior is nearly linear over the entire range of loading. If a fiber breaks during loading, the shear stress in the matrix near the fiber break may exceed the shear yield strength, leading to matrix yielding. The possibility of interface debonding, which can follow after matrix yielding, is neglected; debonding can be included and is neglected only for simplicity.

Table 2.1. Thermoelastic parameters of fiber and matrix [24]

Property	Fiber	Matrix
Fiber radius, r (μm)	2.5	–
Elastic modulus, E (GPa)	294	3.4
Fiber volume fraction, V_{f}	0.6	
Poisson's ratio, ν	0.22	0.345
Weibull modulus, m	3.8	–
Weibull strength, σ_0 (MPa) at L_0 = 50 mm	3,570	–
Yield shear strength, σ_{y} (MPa)	–	52.4

With the model and parameters noted above, 1,000 simulation studies of composite failure have been performed. From these simulations, the probability distribution $P_{n_{\mathrm{s}}}(\sigma)$ for failure of the fiber bundle at the simulated size n_{s} is directly obtained, as shown in Fig. 2.9. Applying the size-scaling theory (2.11), the strength of large composites comparable to the sizes tested experimentally, which contain 10^4–10^6 fibers with a gauge length of 10 mm is obtained. Figure 2.10 shows the experimental and predicted fiber

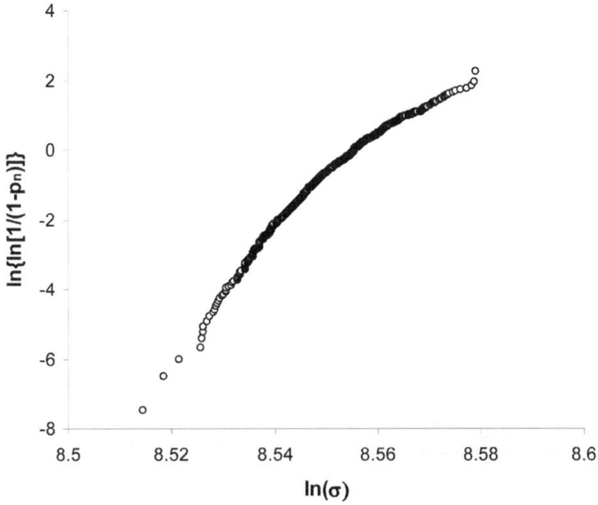

Fig. 2.9. Distribution of fiber bundle strengths (MPa) in a unidirectional CFRP composite containing 1,024 fibers of length $4\delta_c$, plotted in Weibull form

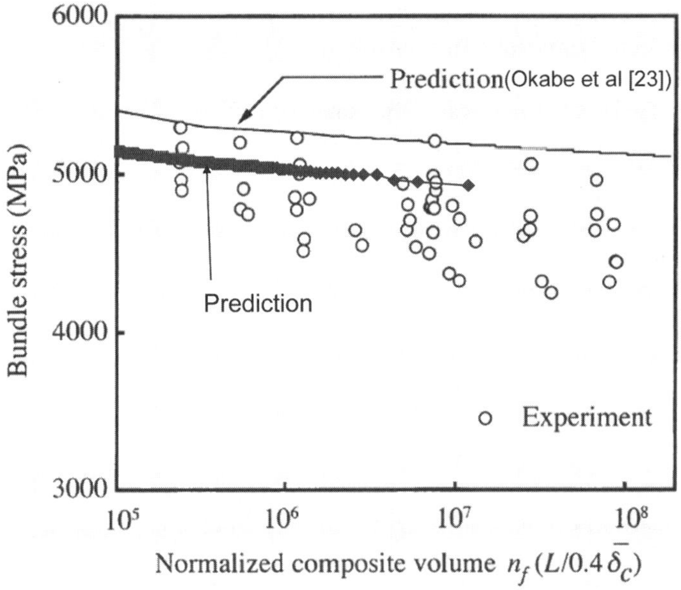

Fig. 2.10. Fiber bundle strength $\sigma_f^* = (\sigma_{uts} - \sigma_m)/V_f$ vs. linear composite size (number of fibers n_f times fiber length L), as predicted by simulations and as obtained experimentally (reprinted with permission from [24])

bundle strength vs. composite volume (n). The predicted strengths fit the experiments very well. Okabe et al. also predicted the strength using a similar method but with a modified Weibull distribution model (Weibull of Weibull (WOW) statistics) for fiber strength [24] and obtained strengths very close to those predicted.

2.3.2 Metal Matrix Composites (MMCs)

Al_2O_3/Al composite

We first consider an aluminum alloy reinforced by Al_2O_3 fibers. The thermomechanical properties of the fibers and aluminum alloy are presented in Table 2.2. Due to chemical bonding, the Al/Al_2O_3 interface is strong and debonding does not occur before matrix yielding. Unlike polymer matrix composites, however, the Al matrix has a much larger elastic modulus and, hence, can exhibit wide-spread yielding. The matrix can also carry significant loads around a broken fiber. Because of low fiber/matrix stiffness ratio, the discrepancy between the shear-lag model and the finite element model stress concentrations in the Al_2O_3/Al composite is significant [37]. Therefore, the Green's function method is used to accurately represent the stress concentrations derived from direct, small-scale finite element analyses results.

A range of yield strengths for the Al alloy (50, 100, and 200 MPa) is considered to examine possible effects of alloying, in situ aging, etc. that may prevail in the as-processed composite. Similar to the polymer matrix composites, uniaxial loading determines the overall stress–strain response of the undamaged composite in the absence of fiber damage. Above an applied composite stress of 500 MPa, the matrix is fully plastic.

Table 2.2. Thermoelastic parameters of fiber and matrix for Al MMC [35]

Property	Fiber	Matrix
Fiber radius, r (μm)	6	–
Elastic modulus, E (GPa)	390	70
Poisson's ratio, ν	0.22	0.345
Fiber volume fraction, V_f	0.65	
Thermal expansion coefficient, α (10^{-6} per °C)	6.5	24
Weibull modulus, m	9	–
Weibull strength, σ_0 (MPa) at $L_0 = 1$ m	2,060	–
Yield strength, σ_y (MPa)	–	50, 100, 200

The 3D FEM to obtain information on deformation around fiber breaks. Introducing a fiber break into the central fiber at the midplane of the FEM model induces significant changes in the local stresses. Figure 2.11a shows the shear stress distribution along the broken fiber for different matrix yield strengths at a load of 1,000 MPa. At a low yield strength ($\sigma_y = 50$ MPa), the shear stress in the plastic zone is essentially $\sigma_y / \sqrt{3}$. For higher σ_y, the shear stress shows more spatial variation but is still about $\sigma_y / \sqrt{3}$ on average. Figure 2.11b shows the average axial stress within the broken fiber along the fiber length at an applied stress of 2,000 MPa. As expected by equilibrium requirements, the stress recovers nearly linearly when the shear stress is nearly constant and then increases more slowly as the shear stress decreases to zero. The "slip" or "stress recovery" length around the broken fiber can be estimated using the simple shear-lag model and a constant interfacial shear stress τ, as indicated in Fig. 2.11b; the corresponding τ values are shown in Table 2.3. To capture the major effects of the in-plane stress redistribution, the average SCF (averaged over the fiber cross-section) is considered, as shown in Fig. 2.11c. The spatial extent of load redistribution varies with yield strength: the SCF of the near-neighbor fibers increases with increasing σ_y and the spatial range decreases. Since the matrix carries load, the SCFs are smaller than those in carbon fiber-reinforced polymers. Simulations are performed for a composite with 1,024 fibers of length $L = 10$ mm to obtain a statistical distribution of tensile strengths. The tensile strength of small size samples can be predicted directly via (2.7) using the mean strength obtained from many simulations of the fiber bundle strength. Such strengths, for the different matrix yield strengths, are shown in Table 2.3. The predictions are relatively insensitive to the value of the yield strengths due to a combination of factors although the mode of failure is quite different, as shown in Figs. 2.7 and 2.8. An increased σ_y, and hence increased τ, increases the characteristic strength and the fiber bundle strength as $\tau^{1/(m+1)}$, and increases the matrix contribution to the strength. However, increased τ also leads to increasing SCFs that are more localized on the nearest fibers, which drives the formation of larger damage clusters at lower loads and decreases the composite strength. In the present case, these competing factors cancel one another to a large degree, leading to a slow increase in composite strength with increasing τ. Since the SCF depends on other constitutive properties and the fiber damage evolution depends on the strength distribution, the cancellation is also a function of features such as the fiber/matrix elastic mismatch, the fiber diameter, and fiber Weibull modulus.

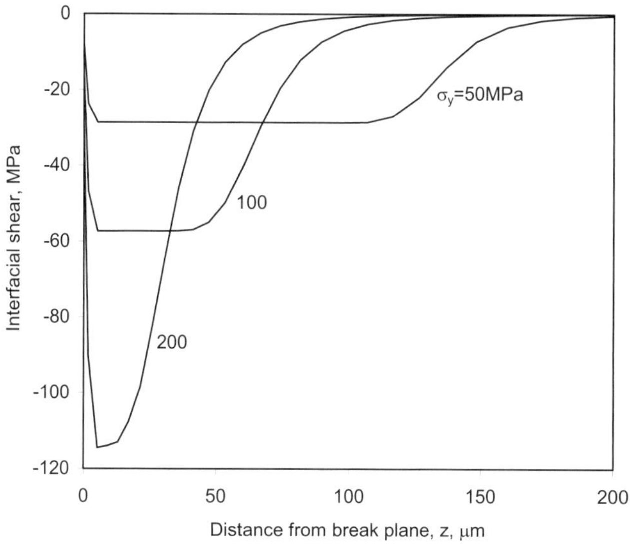

(a)

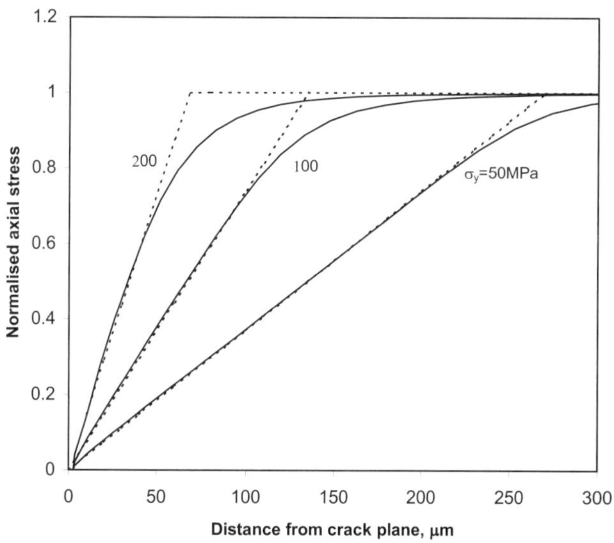

(b)

Fig. 2.11 (continue)

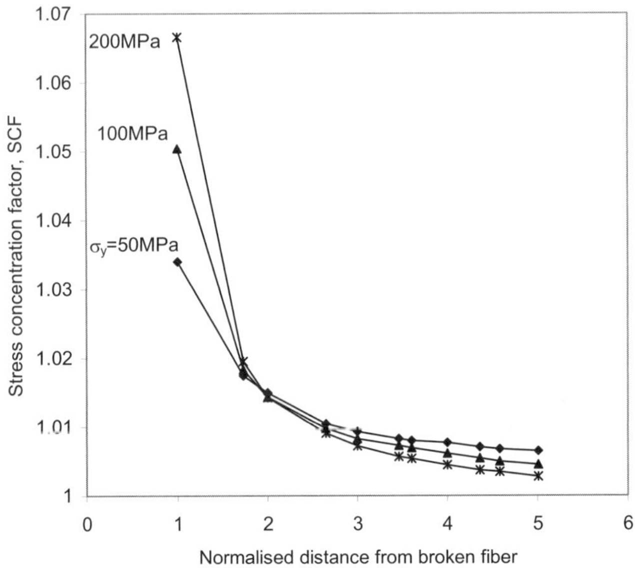

(c)

Fig. 2.11. (**a**) Interfacial shear stress along a broken fiber at an applied stress of 1,000 MPa for different matrix yield strengths, (**b**) axial stress distribution on a broken fiber along the fiber direction z, normalized by the far-field fiber stress, at an applied stress of 2,000 MPa for different matrix yield strengths, and (**c**) axial stress concentration factor (*SCF*) on nearby fibers vs. distance from the broken fiber, normalized by the fiber spacing s, at an applied stress of 2,000 MPa for different matrix yield strengths. *Dotted lines* show constant τ "shear-lag" fit to the data (reprinted with permission from [35])

Table 2.3. Parameters and tensile strength, as measured and as predicted [35]

Parameters	Prediction[a]			Experiment
Matrix yield strength, σ_y	50	100	200	100
Interfacial shear stress, τ (MPa)	32.5	65	125	–
Typical maximum stress, σ_c (MPa)	4,527	4,852	5,200	–
Average fragment length, δ_c (μm)	824	447	248.6	–
Tensile strength, σ_{uts} (MPa)	2,178	2,347	2,496	2,051 ± 141

[a]Average value of 20 results on 1,024 fibers of gauge length $L = 10$ mm.

Experiments on the current material system have been performed by Ramamurty et al. [28] using three-point, four-point, and tension loadings. Only the size of the tension test can be determined directly; the effective volumes tested in bending depend on the Weibull modulus of the composite strength distribution, which is not known a priori. Ramamurty et al. considered the measured scaling of the mean strengths to derive a composite Weibull modulus of about 55, which was then used to assign effective composite volumes to the three- and four-point bend test strengths. The four-point tests were deduced to have an effective size of about 12,000 mm of total fiber length (number of fibers times length of fibers). This matches the volume of 10,240 mm (1,024 fibers of length 10 mm) in the simulations performed rather closely. Hence, the quoted experimental strength in Table 2.3 is that for the four-point bend test. The agreement is quite reasonable, with a difference of ~10% for the yield stress of 100 MPa, which is close to that pertaining to the experiments. Some of the difference could be due to processing-induced fiber damage, such that the ex situ values are not directly applicable to the in situ fibers. Some of the difference may also be due to the influence of bending strain gradients, which is neglected in assuming that failure is driven by locally uniform tensile loading.

To investigate the size scaling of composite strength numerically, extensive simulation studies were performed on larger composite sizes. Specifically, 1,000 simulations were performed on composites containing 1,024 fibers of length 5 mm. From these simulations and (2.7), we directly obtain the probability distribution $P_{n_s}(\sigma)$ for failure of the composite at the simulated size n_s. Figure 2.12 shows the composite strength σ_{uts} vs. composite size $n_f L$ (n_f fibers each of length L), as obtained from the simulation data and (2.11), demonstrating the decreasing strength with increasing composite size. Also shown in Fig. 2.12 are the results of Ramamurty et al. at the estimated test sizes. The predicted strength decreases but more slowly than found experimentally, leading to a difference of ~27% at the largest size.

There are several possible reasons for this discrepancy, all of which lie at the micromechanical fiber/matrix/interface scale. First, the matrix deformation around the fibers is very large. Thus, the details of matrix hardening may be important in determining the stress redistribution. Moderate strain hardening typical of many Al alloys leads to an increase in the local SCF, which drives more localized damage, smaller critical clusters at failure, lower strengths, and an increasing size-scaling effect. This points to the necessity of understanding the matrix constitutive behavior in even more detail than done here to properly capture the SCFs. Okabe et al. recently used the spring element model (a microscale model similar to SLM) coupled with the size-scaling model to predict the composite strength. In their calculation, they used an elastic–plastic hardening matrix

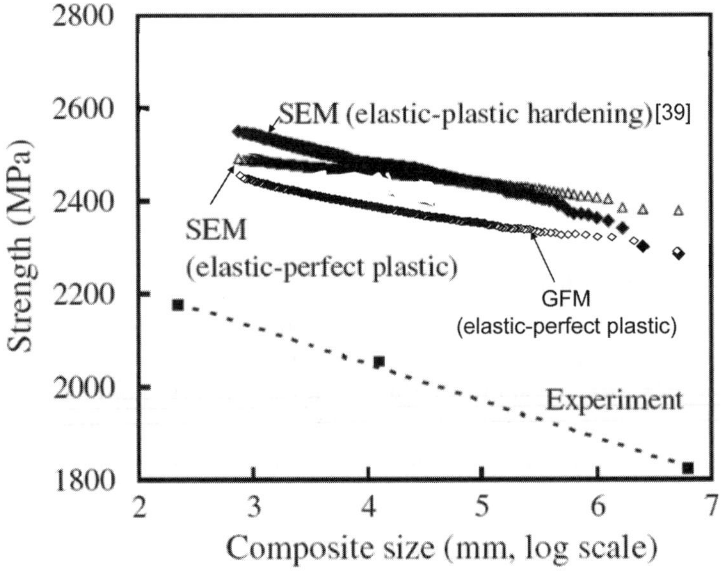

Fig. 2.12. Composite strength vs. linear composite size (number of fibers n_f times fiber length L) for an Al/Al_2O_3 metal matrix composite, as predicted by simulations (GFM and SEM (Spring element model) from [25]) and as obtained experimentally [28]

instead of a perfect plastic matrix [23]. The results in Fig. 2.12 show that, compared with the perfect plastic case, the strength with a strain-hardening matrix decreases quickly with size and is consistent with the experimental trend. Second, the Al matrix may undergo ductile failure locally, which would significantly affect the load transfer. In particular, if matrix "cracks" extend up to the neighboring fibers, then there will be additional local stress concentrations on regions of the neighboring fibers. Similar stress concentrations for sliding interfaces have been calculated in other materials and the influence on enhanced local fiber failure has been assessed [33]. For high interfacial sliding or shear yield stresses and high fiber Weibull moduli, the local stress concentrations can lead to a weaker composite and a more planar fracture surface (typical of Al MMCs). Methods to account for this effect within the mesoscale simulations described here have been addressed in [38]. Third, for MMCs, it is possible that matrix fracture could occur once the damage cluster gets close to critical and this might then trigger fracture. As a result, the strength would be different than predicted here. Such matrix fracture effects can be included in the Green's function model, as will be discussed in Chap. 4.

SiC/titanium composite

The SiC/Ti composites considered here were fabricated via magnetron sputtering of the matrix onto the fibers followed by isostatic pressing of the matrix-coated fibers into a composite [18]. This technique yields a nearly ideal hexagonal fiber distribution (Fig. 2.3a). The fiber volume fraction is 0.4. The interface is weak and debonds readily upon fiber fracture, and the interfacial sliding resistance in the as-fabricated composite is 55–75 MPa, as measured by pushout testing. Other thermomechanical properties are shown in Table 2.4. The SCS-6 fiber is represented by a homogeneous anisotropic material with elastic constants. The homogeneous Young's modulus E_f of the fiber is determined by fitting the rule of mixtures to a full 2D model. The elastic properties of the matrix are also shown in Table 2.4. The stress–strain behavior of the IMI834 Ti alloy as determined experimentally is very accurately represented by a Ramberg–Osgood relationship of the form

$$\varepsilon_{\mathrm{m}} = \sigma_{\mathrm{m}} / E_{\mathrm{m}} + (\sigma_{\mathrm{m}} / B)^{1/b}, \tag{2.12}$$

with $b = 0.0384$ the hardening exponent, $B = 1{,}229$ MPa the hardening coefficient, and E_{m} the matrix elastic modulus. The yield strength of the alloy is 950 MPa.

The multiscale modeling of composite strength starts from a micron-scale finite element model. A 3D finite element representation of the composite was constructed using symmetry to calculate the stress distributions around broken fibers. The model consists of 13,922 elements, including 3,255 gap elements and 13,888 nodes. The fiber/matrix bond strength is assumed to be zero. Coulomb friction at the interface $\tau = -\mu\sigma_{\mathrm{r}}$ is assumed. For the cooling range shown in Table 2.4, $\sigma_{\mathrm{r}} = 208$ MPa, leading to interfacial shear stresses ranging from 52 to 187 MPa for the range $0.25 \le \mu \le 0.9$. The finite element results show that the load transfer in metal matrix is affected by the matrix. Similar to the polymer matrix composites, the near-neighbor fibers bear most of the load and bear an increasing portion of the load as the friction coefficient increases. However, the surviving fibers do not take on all of the loads from the broken fiber. The fibers carry only 83% of the load at low coefficient of friction ($\mu = 0.25$) and only 64% of the load at high coefficient of friction ($\mu = 0.9$). The matrix carries the remainder of the load. Finite element results show that there is a clear axial stress concentration in the matrix near the broken fiber, which increases with increasing friction coefficient. The increased axial matrix stress occurs due to both hardening and constraint effects that govern the yielding. Similar calculations for an elastic/perfectly plastic matrix (no hardening) show similar increases in axial matrix load attributable purely to constraint effects. For a high coefficient of friction, $\mu = 0.9$, the stress in the matrix near the fiber break exceeds the tensile strength of the matrix so that a fiber break may actually cause matrix fracture. Matrix fracture can then induce a different and much more dangerous mode of composite fracture in which fiber and matrix fracture progress unstably from around a single break, since the load carried by the cracked matrix will be transferred predominantly onto the nearby fibers. This failure mode is in contrast to the distributed damage and failure that occurs when the matrix does not fail. Most existing models for stress transfer neglect the stress carried by the matrix and the possibility of matrix fracture.

Table 2.4. Thermoelastic parameters of fiber, matrix, and composite [38]

Material	E_{zz} (GPa)	$E_{rr} = E_{\theta\theta}$ (GPa)	$G_{r\theta}$ (GPa)	$G_{\theta z} = G_{zr}$ (GPa)	$\nu_{r\theta}$	$\nu_{\theta z} = \nu_{zr}$	α_{zz} (10^{-6} per °C)	$\alpha_{rr} = \alpha_{\theta\theta}$ (10^{-6} per °C)	ΔT[a] (°C)
Fiber	400	240	70	118	0.15	0.25	6.48	6.48	750
Matrix[b]	120				0.3		11.24		750
Composite	232	155	64	91	0.186	0.296	8.05	9.4	750

[a]Temperature difference for cooling from processing.
[b]Isotropic material.

Table 2.5. Thermoelastic parameters of SiC (SCS-6) fiber

Fiber radius, r (μm)	71
Elastic modulus, E_{zz} (GPa)	400
Weibull modulus, m	17
Weibull strength, σ_0 (MPa) at $L_0 = 25$ mm	4,580

Neglecting the possibility of matrix failure, the damage evolution during loading is simulated using the Green's function model with the SCFs around a broken fiber for different coefficient of friction calculated using the finite element model. The fiber parameters are listed in Table 2.5. The simulation method requires an appropriate value of τ to define the length δ_c and characteristic strength σ_c. The characteristic stress σ_c and characteristic length δ_c for different coefficient of friction are listed in Table 2.6.

Table 2.6. Parameters and tensile strength, as measured and as predicted [38]

Parameters	Prediction[a]			Experiment
Coefficient of friction, μ	0.25	0.5	0.9	–
Interfacial shear stress, τ (MPa)	51.7	103.5	186.3	55–75
Characteristic stress, σ_c (MPa)	4,946.4	5,141.6	5,313.0	–
Average fragment length, δ_c (μm)	6,792.9	3,527.1	2,024.8	–
Tensile strength, σ_{uts} (MPa)	2,268.0	2,317.5	2,344.5	2,200–2,300

[a]Average value of 20 results (Gauge length $L = 20$ mm, the number of fibers = 210).

The tensile strengths were predicted using the mean of many simulations of the fiber bundle strength σ_f^* and are shown in Table 2.6 for several different friction coefficients. The predictions contain no adjustable parameters and are in excellent agreement with the experimental data for the range of friction parameters consistent with the experimentally measured interfacial shear sliding stress [18]. The relative insensitivity of the predictions to the value of the friction coefficient is due to the same combination of factors as discussed previously for the Al MMCs.

Similar to Al matrix composites, the size scaling of composite strength was investigated numerically within the present model. One thousand simulations were performed on composites containing 1,000 fibers of length $4\delta_c$ (size $n_s = 4{,}000\delta_c$). The probability distribution $P_{n_s}(\sigma)$ for failure of the fiber bundle at the simulated size was directly obtained from these simulations. The results show that the tensile strength at the sizes of typical Ti MMC components ($\approx 10^6$ mm^3) is reduced by about 100 MPa below the value obtained on the small laboratory test coupons ($\approx 10^2$ mm^3). Currently, no experimental data are available to test the accuracy of the predicted scaling of strength with composite size.

2.4 Extension to Low-Cycle Fatigue of Titanium Matrix Composite

In highly stressed rotating components where Ti MMCs might be employed, the transient loadings associated with start-up/shutdown and maneuvering give rise to a situation where low-cycle fatigue dominates. Studies of low-cycle fatigue behavior of unnotched specimens indicate that matrix cracks in TMCs initiate at the matrix/fiber interface, fiber breaks, and surface flaws of the specimens and grow perpendicular to the fibers. The range of fatigue crack growth can be divided into a short crack range, a steady-state range, and composite failure [1, 2]. After an initial growth in the short crack range, the crack growth rate reaches a constant value (steady-state regime) due to fiber bridging that substantially shields the crack tip from the applied stresses. However, the high stresses in the bridging fibers can cause them to fail and ultimately drive the composite to fail catastrophically [3, 32].

Here, we extend the multiscale coupling approach to predict the low-cycle fatigue of Ti/SiC composites. We combine detailed finite element models of the stress states in and around small matrix fatigue cracks with the Green's function model to capture the stochastic fiber damage under the calculated stress states. Figure 2.13 illustrates the multiscale modeling procedure for low-cycle fatigue life predictions. The micromechanical finite element model provides both the crack tip stress intensity (ΔK_{eff}) governing fatigue crack growth and the stress distributions on the fibers in and around the fatigue crack. The Green's function model evaluates the failure of the fibers under the given stress state, distributes the stress from broken

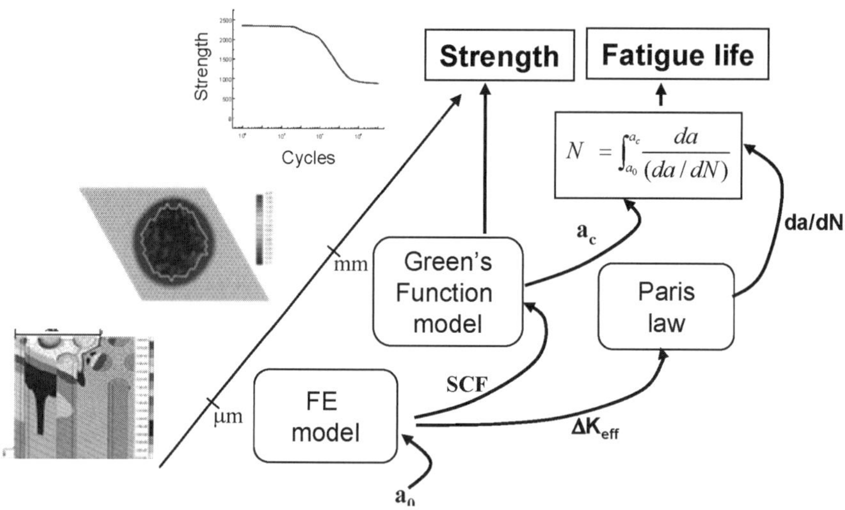

Fig. 2.13. Multiscale modeling of low-cycle fatigue of Ti/SiC composites

fibers onto the remaining fibers, and permits the evolution of fiber damage up to composite failure (critical crack length a_c). A Paris law links the crack growth rate (da/dN) with the ΔK_{eff}. The fatigue life is then predicted from da/dN and the computed critical size a_c.

2.4.1 Fatigue Failure Predictions

Model geometry and constitutive behavior

Under cyclic loading, the fatigue cracks in composites usually start at flaws such as interfacial reaction and broken fibers. Therefore, the simulation starts from the fiber/matrix interface at which there is an initial annular matrix crack of outer radius a_0. The initial annular crack width a_0 is taken to be the thickness of the brittle reaction layer formed during processing at the C/Ti interface of the SCS-6/Ti system, which is about 1 μm. A second initial state is considered in which the fiber inside the reaction layer crack has also failed, for which the initial crack is a penny crack of radius $R + a_0$. Probabilistic assessment of fiber fracture upon loading is used to determine which of these two initial states is relevant as a function of applied stress. We assume that the fiber/matrix (or more precisely the fiber/coating) interface is rather weak and debonds when the matrix crack impinges on the interfaces or when a fiber breaks. The interfacial shear stress τ along the debonded interface is controlled by Coulomb friction, as described in Sect. 3.2. The elastic constitutive properties of the matrix and fibers are shown in Tables 2.4 and 2.5. The elastic–plastic regime of the matrix is described by a Ramberg–Osgood relationship (2.12).

The finite element models for fatigue crack simulation are similar to those used in tensile simulation (Fig. 2.3) but an average material with composite properties is added such that the number of elements is reduced while the cracked area is kept below 1% of the model cross-section. Three FE models, each with a different matrix crack radius, were developed to predict the stress concentrations in the bridging fibers and the stress intensity factor at the crack tip, as shown in Fig. 2.14. The element sizes were selected to adequately determine the strain energy release rate along the matrix crack front, as described in "Failure simulations at fixed fatigue crack size."

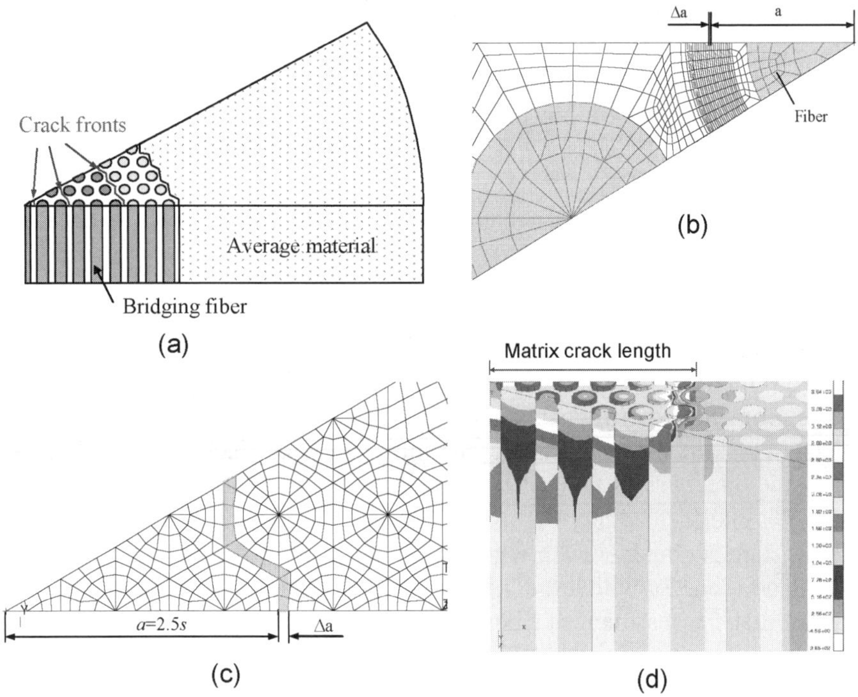

Fig. 2.14. The finite element models used here: (**a**) 30° wedge showing fibers, matrix, and average material with different crack lengths, (**b**) mesh distribution and crack tip propagation region for small matrix cracks of $a_m = 10$–40 μm, (**c**) mesh distribution and crack propagation region for an intermediate crack size of $a = 2.5s$, and (**d**) FE-predicted stress distribution in crack propagation region for a crack size of $a = 5s$ (reprinted with permission from [34])

Stress concentrations in bridging fibers

The presence of a matrix fatigue crack causes several important stress concentrations on the bridging fibers. First, there is a transfer of the matrix load onto the fibers. Second, there is an increased transfer of stress from broken fibers to unbroken fibers since the cracked matrix is not available to participate in the load sharing. Third, the matrix crack causes a stress concentration at the fiber surface that can drive "premature" fiber fracture. The calculation of the third stress concentration involves theoretical analysis and the details can be referred to in references [33, 34]. The first two factors calculated by the FE models are addressed as follows.

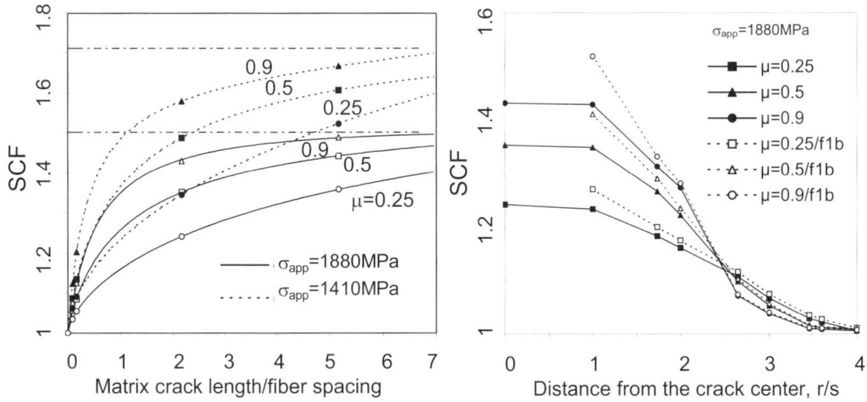

(a) **(b)**

Fig. 2.15. (**a**) Average stress concentration factor (*SCF*) on the central bridging fiber as a function of matrix crack length for applied stresses of 1,410 and 1,880 MPa. *Dashed line* shows the asymptotic SCF, (**b**) average axial SCF on the bridging fibers in and around a matrix crack of length $a = 2.5s$ (s is the fiber spacing) for an applied stress of 1,880 MPa with no broken fibers (*solid line*, *solid symbols*) and with a broken central fiber (*dashed lines*, *open symbols*) (reprinted with permission from [34])

Upon fatigue crack initiation and propagation, the fibers in the wake of the crack experience a greatly increased stress since the cracked matrix no longer carries any stress. In the limit of a large matrix crack, steady-state conditions deep within the crack dictate that the stresses on the surviving fibers in the crack plane attain the value $\sigma_f = \sigma_{comp}/V_f$. Figure 2.15a shows the average stress concentration on the central bridging fiber in the crack plane as a function of matrix crack length. For a high coefficient of friction, the maximum fiber stress becomes independent of the crack length after a relatively small amount of crack growth, i.e., steady-state conditions are reached for fairly small cracks. For a low coefficient of friction, considerable crack growth must occur before the central fiber stress becomes independent of length, i.e., steady-state conditions require rather longer crack lengths. The limiting stress concentration factor is SCF_{max} independent of the coefficient of friction.

When the central bridging fiber has failed, the stress concentrations on the surrounding surviving fibers are further elevated, as shown in Fig. 2.15b. Since the cracked matrix cannot carry any of the loads from the broken fiber, the stresses transferred to the nearby fibers are larger than those in the absence of matrix damage. Thus, the presence of the matrix fatigue crack enhances the stress on the fibers and the stress transferred from broken to unbroken fibers; both factors drive preferential damage of the fibers within the matrix crack region.

Failure simulations at fixed fatigue crack size

Fiber damage in a composite evolves stochastically due to the underlying statistical strength distribution of the brittle reinforcing fibers. Here, fiber damage in the composite is calculated using a numerical simulation technique based on the Green's function method described in Sect. 2.3. We obtain the Green's function G_{ij} in (2.5) from the full 3D finite element model, and specifically the resulting in-plane SCFs. In application to the present problem, the results of our detailed FE model with a specified matrix fatigue crack are also used to obtain the local applied field $\sigma_{\mathrm{app},i}(z)$ on every fiber in the cross-section (see Fig. 2.15, for example), which is the stress state prior to any fiber damage. Now, the effective applied stress $\sigma_{\mathrm{app},i}(z)$ in (2.6) is the stress due to applied fields plus matrix damage but not including any fiber damage.

The failure strength of composite sizes $n = 210$ and 1,024 at a gauge length of $L = 6$ mm, for various fatigue crack lengths was calculated. All calculations begin with a single initial matrix crack in the center of the composite. The calculated composite strength vs. fatigue crack size is shown in Fig. 2.16. For a fatigue crack size of zero, the simulations predict the ultimate tensile strength of the composite of about 2,300 MPa [18]. As the fatigue crack size increases, the composite strength decreases in a manner that depends strongly on the interfacial friction coefficient. For a high coefficient of friction, the higher fiber stress concentrations and surface stress concentrations drive fiber failure at smaller fatigue crack sizes. For lower coefficients of friction, the decrease in strength with increasing fatigue crack size is more gradual. The composite strength for systems with an initial fiber crack at the center of the model is also shown in Fig. 2.16. The strength of the composite is nearly independent of the existence of the initial fiber break for the lower friction coefficients. At high loads, there can be several fiber breaks, and so one additional break may not be critical. At low loads, a single fiber break cannot drive extensive additional damage; and so, again, the one break is not critical.

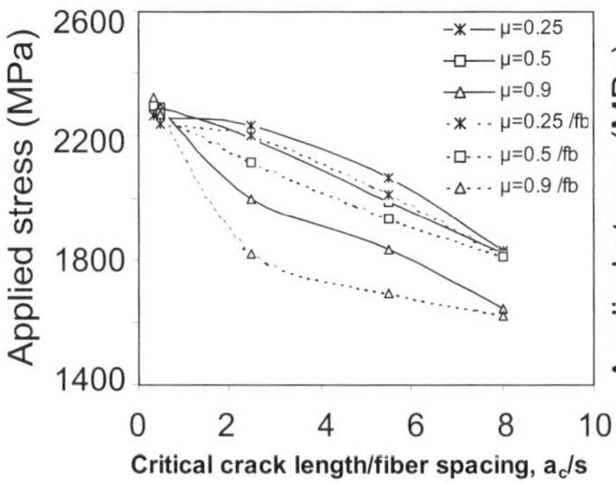

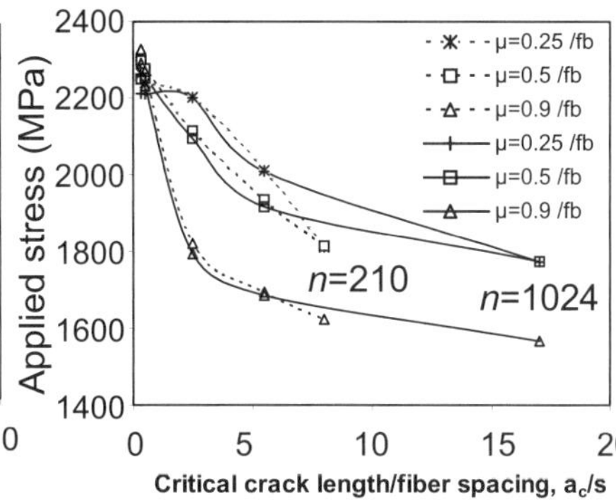

(a) (b)

Fig. 2.16. Composite strength vs. matrix crack length for different coefficients of friction: (**a**) composite size of 210 fibers, without (*solid lines*) and with (*dashed lines*) an initial central fiber break, (**b**) composite sizes of 210 (*dashed lines*) and 1,024 (*solid lines*) with a central fiber break. As-processed fiber strength parameters are listed in Table 2.5 (reprinted with permission from [34])

2.4.2 Fatigue Life Predictions

Fatigue life predictions require the knowledge of crack growth rate as well as the initial and critical crack lengths. The critical crack length (or composite strength) has been determined in Sect. 2.4.1. To complete the computation, we must compute the number of fatigue cycles needed to reach the critical size a_c at the corresponding critical applied strength (Fig. 2.16). Here, the crack growth rate in the matrix of TMC was calculated by using fatigue properties of "neat" matrix material. Dowling and Iyyer [8] have suggested that the low-cycle fatigue crack growth rate $\mathrm{d}a/\mathrm{d}N$ is associated with an effective stress intensity factor range ΔK_{eff} according to a Paris law

$$\frac{\mathrm{d}a}{\mathrm{d}N} = c\Delta K_{\mathrm{eff}}^{b}, \tag{2.13}$$

where $b = 8.22$ and $c = 7 \times 10^{-16}$ m per cycle are materials constants determined via fatigue crack growth experiments on IMI834 [17]. ΔK_{eff} is related to the effective strain energy release rate range, ΔJ_{eff}, consisting of an elastic strain component ΔJ_{e} and plastic strain component ΔJ_{p}. Following Dowling and Iyyer [8], an approximation valid for an internal circular crack is used in this study to calculate the plastic term ΔJ_{p}. The term ΔJ_{e} was calculated using the FE model as follows. In linear elastic fracture mechanics, J_e equals the strain energy release rate G_I. The strain energy release rate for matrix crack growth was determined with the modified crack closure integral technique [29], given by

$$G_I = \lim_{\Delta A \to 0} \left\{ \frac{1}{\Delta A} \sum_{i=1}^{M} F_z^i u_z^i \right\}, \tag{2.14}$$

where F_z^i is the axial force on node i at the crack tip position, u_z^i is the crack tip opening of the node i after permitting the crack to grow, M is the number of crack tip nodes, and ΔA is the area of crack propagation.

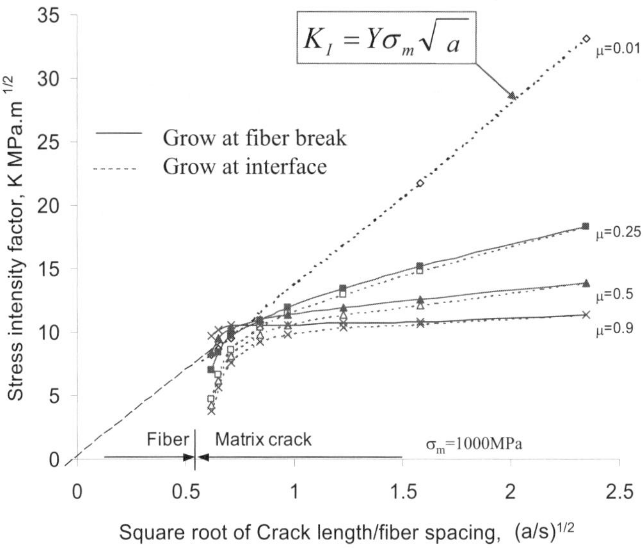

Fig. 2.17. Elastic contribution to the effective stress intensity factor K_{eff} vs. square root of the crack length a for different friction coefficients, both without (*open symbols*) and with (*solid symbols*) a central fiber break, at a matrix stress of 1,000 MPa (applied stress amplitude of 1,600 MPa, s = fiber spacing) (reprinted with permission from [34])

The elastic component of K_{eff} for a matrix stress of 1,000 MPa (corresponding to an applied stress amplitude of 1,600 MPa) is shown in Fig. 2.17 vs. crack length $a = a_m + R$, where a_m is the annular matrix crack length. The stress intensity factor for $\mu = 0.01$ follows the linear elastic fracture mechanics relation $K_I = Y\sigma_m\sqrt{\pi a}$ that is expected in the absence of fiber bridging. $Y = 0.536$ is a geometry factor whose value differs from the value $Y = 2/\pi = 0.637$ for a circular flaw in homogeneous elastic material

[33] because the higher-modulus fibers reduce the stress in the matrix. With increasing matrix crack length a ($a > 2.5s$), K_{eff} begins to approach a steady-state value independent of the crack length, and the bridging contribution to K_{eff} scales as $K_b \propto \tau^{1/3} a^{1/2}$, similar to predictions of McCartney [21]. When the initial crack includes an in-plane fiber break, the crack growth rate is greatly accelerated in the early stages, as shown in Fig. 2.17 and as expected due to the loss of fiber bridging. For larger fatigue cracks, the crack growth rate becomes largely independent of the initial crack details; the large number of bridging fibers establishes the approach to the steady-state regime.

The calculations also show that K_{eff} increases nearly linearly before the matrix exceeds the cyclic yield point (about 850 MPa), after which K_{eff} increases more rapidly. The interfacial friction has a strong influence on K_{eff} in both elastic and plastic stages, with increased friction leading to a reduction in K_{eff} because the fiber bridging is more effective when fiber sliding is more restrained.

With the composite strength and crack growth rate as a function of fatigue crack size, as shown in Figs. 2.16 and 2.17, it is straightforward to determine the number of cycles required, at that applied stress, to grow the fatigue crack from the initial size to the final critical size. Since the initial fiber breakage does significantly reduce the fatigue life, it is important in making predictions to determine the likelihood of fiber damage upon application of the initial load. A single initial fiber break anywhere in the entire composite can serve as the site for a fast-growing fatigue crack and, hence, early fatigue failure. The probability of failure of a single fiber somewhere in a composite specimen containing n fibers with a gauge length of L is, from the Weibull statistics of fiber failure, given by

$$P_{\text{f}}(\sigma_{\text{comp}}) = 1 - \exp\left[\frac{Ln}{L_0}\left(\frac{\sigma_{\text{comp}} - \sigma_{m\infty}(1-f)}{f\sigma_0}\right)^m\right]. \tag{2.15}$$

For small composites (e.g., $n = 210$, $L = 6$ mm), the 50% probability level for one fiber break is fairly high, at about 2,050 MPa. For a larger composite (e.g., $n = 1{,}024$, $L = 6$ mm), the 50% probability level is reduced to about 1,900 MPa. Hence, the typical stress level between having no initial fiber break and finding at least one initial fiber break is a function of composite size. To account for the possibility of initial fiber breakage, the average fatigue lifetime at any stress level should be weighted by the probability of obtaining such a fiber break. Thus, the lifetime N vs. applied stress is

$$N = N_b(\sigma_{comp})P_f(\sigma_{comp}) + N_{ub}(\sigma_{comp})[1 - P_f(\sigma_{app})], \quad (2.16)$$

where $N_b(\sigma_{comp})$ and $N_{nb}(\sigma_{comp})$ are the lifetimes for the cases of an initial fiber break and no initial fiber break at the applied stress level of interest.

The fatigue life (*S–N* curve) is calculated using the results shown in Fig. 2.18 for $N_b(\sigma_{comp})$ and $N_{nb}(\sigma_{comp})$, and (2.15) and (2.16) for composites with $n = 210$ and 1,024 fibers of gauge length 6 mm. The results, obtained with no adjustable parameters, are shown in Fig. 2.18 along with the experimental *S–N* data on SCS-6 fiber-reinforced IMI834 titanium alloy [32], for which the coefficient of friction is about 0.3 [16] and the sample size closely matches the simulated size. The low-cycle fatigue predictions for $\mu = 0.25$ are in very good agreement with the experimental results at stress levels higher than 1,800 MPa (lifetimes below 10^4 cycles). Below about 1,800 MPa for $\mu = 0.25$, the model predicts the aforementioned fatigue threshold whereas the actual composites continue to degrade, so that the life is overpredicted. However, the model uses only the pristine as-processed fiber strengths and explicit fatigue degradation of the fibers appears to be the cause of the reduced fatigue life for $N > 10^4$ cycles.

Guo et al. [12] found that after 10^4 cycles at a stress amplitude of 450 MPa, the extracted SiC (SCS-6) fiber surface shows a morphology similar to (uncoated) SCS-0 SiC fibers. The tensile strength of the SCS-6 fibers extracted from the fatigued specimens was reduced to a level nearly the same as that of SCS-0 fibers ($\sigma_0 = 2{,}300$ MPa, $m = 7.2$ at $L_0 = 25$ mm). Thus, it appears that low-cycle fatigue loading reduces the strength of SCS-6 fibers to that of SCS-0 fibers. We have used these fatigued fiber properties as relevant for $N > 10^4$ cycles (thereby assuming no further fatigue degradation of the fibers, consistent with the interpretation that the strength reduces only to that of the SCS-0 fibers) to calculate the composite strength vs. fatigue crack size and, subsequently, a fatigue life vs. stress. The initial fatigue crack at these low applied stresses is taken to be without a fiber break since the fibers only weaken after the $\approx 10^4$ cycles. The resulting fatigue prediction is also shown in Fig. 2.18 and contains no adjustable parameters. The predicted *S–N* curve agrees with the experimental data over the range from 1,200 to 1,500 MPa. Taken together, the two predicted results for fatigue life fall nearly along the same line, suggesting that if the fiber fatigue effect could be introduced in an appropriately gradual manner then the entire fatigue life curve would be very accurately predicted by our analysis. For the reduced fiber strengths, there is again a fatigue threshold predicted but it is at about 1,050 MPa and there is no experimental data at such low stresses. From the results shown in Fig. 2.18, interface friction is seen to play a more important, and varying, role in fatigue than in tensile strength. A higher friction coefficient is predicted to be beneficial for high-stress/low-cycle fatigue but to be detrimental at lower stresses or higher cycles.

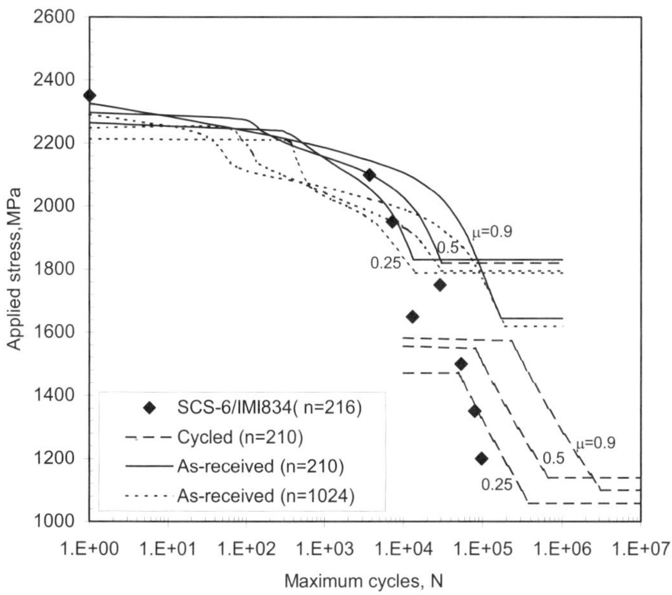

Fig. 2.18. Applied stress vs. fatigue cycles at failure (*S–N* curve, $R = 0.1$), as measured (*solid diamonds*) and as predicted (*lines*), for an SCS-6/IMI834 titanium matrix composite for different friction coefficients. Predictions use as-fabricated fiber strength ($\sigma_0 = 4{,}580$ MPa, $L_0 = 25$ mm, and $m = 17$) and fatigued/cycled fiber strengths ($\sigma_0 = 2{,}300$ MPa, $L_0 = 25$ mm, and $m = 7.2$) (reprinted with permission from [34])

2.5 Conclusions

An approach to the hierarchical modeling of composite failure has been presented. The proposed multiscale modeling involves the passing of key information from smaller to larger scales. The approach employs the FEM at the smallest scale to obtain detailed information on stress transfer from broken fibers to unbroken fibers as a function of elastic constants, fiber volume fraction, fiber/matrix interface conditions, and matrix deformation.

This detailed information is condensed into average axial SCFs on fibers around a break, which is then used as the Green's function in a larger-scale model of stochastic fiber damage evolution. In some materials, a SLM can replace the FEM/Green's function combination. Simulations of composite failure on small systems using the shear-lag or Green's function model are then performed. In the some cases, the simulated sizes can be comparable to actual test specimens so that direct comparison between model and experiment can be made; in general, this is not the case. Extensive simulations of tensile failure on small sample sizes are then used together with analytic size-scaling concepts to generate predictions for the strength vs. size and probability of failure of much larger (component size) specimens. Such large sizes could never be simulated directly, even using the highly efficient Green's function model.

Important features of the failure and deformation at each scale govern the ultimate macroscopic behavior. The interface friction coefficient and matrix-yielding behavior determine the load transfer. The shear-lag and Green's function simulations demonstrate how much damage must evolve, given the underlying load transfer, to drive tensile failure. The simulations then also provide statistical data on the size scaling. Each level of analysis is required for an accurate overall predictive methodology.

The proposed multiscale modeling approach has been used to predict the tensile strength of large-scale PMCs and MMCs. For PMCs, the predictions are in good agreement with the experimental results. In application of the method to MMCs, the success is mixed. The strength predictions for alumina fiber/Al matrix composites are comparable to the experimental values but not as accurate as SiC fiber/Ti matrix composites. The size scaling for alumina fiber/Al matrix composites is not accurately captured, with the experimental strengths decreasing faster than the predicted strengths. The absolute magnitude of the composite strength could be simply due to fiber degradation during processing, as noted above.

The present model does not consider several potential damage mechanisms that can occur in real as-fabricated coupons and components. Preexisting fiber breaks, which can occur during processing, have not been included. Given appropriate information on the break density, however, initial breaks can easily be incorporated into the current models. Spatial irregularity of the fibers has not been included, although work by Foster [10] shows that some spatial disorder has almost no effect on the distribution of strength in a composite. The possibility of touching fibers, wherein one fiber failure immediately precipitates the second fiber failure, may be detrimental to composite strength as well and has not been addressed here. Algorithms for introducing such correlated fiber breaking are easy to generate and incorporate into the shear-lag and Green's function model.

Time- and cycle-dependent deformation and failure are both important applications issues for fiber composites. Fatigue loading induces matrix fatigue cracks, which increase the stresses on the fibers and drives damage at loads well below the quasistatic tensile strength. The multiscale modeling method has been extended to low-cycle fatigue life predictions. A finite element model is coupled with a Green's function model to simulate the major damage mechanisms occurring under fatigue loading of TMC and have predicted the low-cycle fatigue life (*S–N* curve). A finite element model containing a matrix crack bridged by SiC fibers is used to calculate both the matrix crack tip stress intensity factor and the local fiber stress concentrations due to the matrix crack as a function of the crack size. The effective crack tip stress intensity factor, including the effect of the matrix plasticity, is then used to calculate the growth rate of the bridged matrix crack. A 3D Green's function method then uses three outputs from the FE model (1) the fiber stress states due to the matrix crack, (2) the average stress transfer from a broken fiber to unbroken fibers, and (3) the surface fiber stress concentrations, to simulate the fiber damage process and composite strength at any fixed fatigue crack size. The composite strength for a given fatigue crack size together with the number of cycles required to grow the fatigue crack to the given size at an applied stress equal to this composite strength then determines the fatigue life. Detailed application of this approach to the SCS-6/IMI834 shows very good agreement with no adjustable parameters when the appropriate fiber strengths are employed (as-processed value at low cycles, fatigued value at high cycles).

Creep deformation also influences local stress states, and the present multiscale approach may be useful as a means of transferring details of time-dependent deformation at small scales into larger-scale models of fiber damage evolution in a computationally efficient manner. Direct degradation of the reinforcing fibers, via fatigue crack growth or slow crack growth, can also occur. The present models can incorporate such degradation directly at the Green's function level by introducing time- or cycle-dependent fiber strengths. In some cases, the stresses acting on growing flaws in the fibers can, however, greatly exceed the average axial fiber stresses used in the Green's function models. Thus, additional multiscale models must be used to characterize fiber strength degradation at the fiber scale due to underlying flaw growth mechanics at much smaller scales.

The present modeling and methodology set the stage for the inclusion of other damage mechanisms that may be relevant to as-processed materials or in-service application. Within the framework of the model, one could study: the effects of processing damage on damage evaluation, fatigue growth and failure, and the quasistatic and creep failure of materials with small notches. Furthermore, the current models apply not only to metallic and polymeric matrix composites, but also to ceramic matrix composites. Future work remains to attack some of these key problems and applications to other materials systems.

Acknowledgments

This work was supported by the U.S. Air Force Office of Scientific Research within the Mechanics of Multifunctional Material and Microsystems program.

References

1. Bakuckas JG, Johnson WS (1994) NASA Technical Memorandum 109135
2. Bao G, McMeeking RM (1994) Fatigue crack growth in fiber-reinforced metal–matrix composites. Acta Metall. Mater. 42:2415–2425
3. Begley MR, McMeeking RM (1995) Numerical-analysis of fiber bridging and fatigue-crack growth in metal–matrix composite materials. Mater. Sci. Eng. A 200:12–20
4. Beyerlein IJ, Phoenix SL (1996) Stress concentrations around multiple fiber breaks in an elastic matrix with local yielding or debonding using quadratic influence superposition. J. Mech. Phys. Solids 44:1997–2039
5. Connell SJ, Zok FW (1997) Measurement of the cyclic bridging law in a titanium matrix composite and its application to simulating crack growth. Acta Mater. 45:5203–5211
6. Curtin WA (1991) Theory of mechanical-properties of ceramic–matrix composites. J. Am. Ceram. Soc. 74:2837–2845
7. Curtin WA (1999) Stochastic damage evolution and failure in fiber-reinforced composites. Adv. Appl. Mech. 36:163–253
8. Dowling NE, Iyyer NS (1987) Fatigue crack growth and closure at high cyclic strains. Mater. Sci. Eng. 96:99–107
9. Ewaids HL, Wanhill RJH (1984) Fracture Mechanics. Edward Arnold and Delftse Uitgevers Maatschappij, London, p 49
10. Foster G (1997) M.S. Thesis. Virginia Polytechnic Institute and State University, Blacksburg, VA
11. Gambone ML, Wawner FE (1997) The effect of surface flaws on SiC fiber strength in an SiC/Ti-alloy composite. J. Compos. Mater. 31:1062–1079
12. Guo SQ, Kagawa Y, Tanaka Y, Masuda C (1998) Microstructure and role of outermost coating for tensile strength of SiC fiber. Acta Metall. 46:4941–4954
13. Harlow DG, Phoenix SL (1981) Probability distribution for the strength of composite materials: a convergent sequence of tight. Int. J. Fracture 17:601–629
14. Hedgepeth JM (1961) NASA Technical Report D-822

15. Hedgepeth JM, Van Dyke PJ (1967) Local stress concentrations in imperfect filamentary composite materials. J. Compos. Mater. 1:294–304
16. Hemptenmacher J, Assler H, Peters PWM, Dudek HJ (1998) Prakt Metallogr-Pr (in German) 35:295–305
17. Kordisch T, Nowack H (1988) Mat.-wiss. u. Werkstofftech (in German) 29:215–228
18. Leucht R, Dudek HJ (1994) Properties of SiC-fiber reinforced titanium-alloys processed by fiber coating and hot isostatic pressing. Mater. Sci. Eng. A 188:201–210
19. Mahesh S, Phoenix SL (2004) Lifetime distributions for unidirectional fibrous composites under creep-rupture loading. Int. J. Fracture 127:303–360; Mahesh S, Beyerlein IJ, Phoenix SL (1999) Size and heterogeneity effects on the strength of fibrous composites. Physica D 133:371–389; Mahesh S, Phoenix SL (2004) Absence of a tough–brittle transition in the statistical fracture of unidirectional composite tapes under local load sharing. Phys. Rev. E 69:026102
20. Majumdar BS, Gundel DB, Dutton RE, Warrier SG, Pagano NJ (1998) Evaluation of the tensile interface strength in brittle-matrix composite systems. J. Am. Ceram. Soc. 81:1600–1610
21. McCartney LN (1987) Mechanics of matrix cracking in brittle-matrix fiber-reinforced composites. Proc. R. Soc. Lond. A 409:329–350
22. Oden JT, Vemaganti K, Moes N (1999) Hierarchical modeling of heterogeneous solids. Comput. Meth. Appl. Mech. Eng. 172:3–25
23. Okabe T, Nishikawa M, Takeda N, Sekine H (2006) Effect of matrix hardening on the tensile strength of alumina fiber-reinforced aluminum matrix composites. Acta Mater. 54:2557–2566
24. Okabe T, Takeda N (2002) Size effect on tensile strength of unidirectional CFRP composites – experiment and simulation. Compos. Sci. Technol. 62:2053–2064
25. Okabe T, Takeda N, Kamoshida Y, Shimizu M, Curtin WA (2001) A 3D shear-lag model considering micro-damage and statistical strength prediction of unidirectional fiber-reinforced composites. Compos. Sci. Technol. 61:1773–1787
26. Phoenix SL, Raj R (1992) Scalings in fracture probabilities for a brittle matrix fiber composite. Acta Metall. Mater. 40:2813–2828
27. Raghavan P, Ghosh S (2004) Adaptive multiscale computational modeling of composite materials. Comput. Mod. Eng. Sci. 5:151–170
28. Ramamurty U, Zok FW, Leckie FA, Deve HE (1997) Strength variability in alumina fiber-reinforced aluminum matrix composites. Acta Mater. 45:4603–4613
29. Rybicki EF, Kanninen MF (1997) A finite element calculation of stress intensity factors by a modified crack closure integral. Eng. Fract. Mech. 9:931–938
30. Sastry AM, Phoenix SL (1993) Load redistribution near nonaligned fiber breaks in a 2-dimensional unidirectional composite using break-influence superposition. J. Mater. Sci. Lett. 12:1596–1599

31. Telesman J, Ghosn LJ, Kantzos P (1993) Methodology for prediction of fiber-bridging in composites. J. Compos. Technol. Res. 15:234–241
32. Warrier SG, Majumdar BS (1999) Elastic shielding during fatigue-crack growth of titanium matrix composites. Metall. Mater. Trans. A 30:277–286
33. Xia ZH, Curtin WA (2000) Tough to brittle transitions in ceramic matrix composites with increasing interfacial shear stress. Acta Mater. 48:4879–4892
34. Xia ZH, Curtin WA (2001) Life prediction of titanium MMCs under low-cycle fatigue. Acta Mater. 49:1633–1646
35. Xia ZH, Curtin WA (2001) Multiscale modeling of fracture and failure in Al/Al_2O_3 composites. Compos. Sci. Technol. 61:2247–2257
36. Xia ZH, Curtin WA (2002) Green's function vs. shear-lag models of damage failure in fiber-reinforced composites. Compos. Sci. Technol. 62:1279–1288
37. Xia ZH, Curtin WA (2002) Shear-lag vs. finite element models for stress transfer in fiber-reinforced composites. Compos. Sci. Technol. 62:1141–1149
38. Xia ZH, Curtin WA, Peters PMW (2001) Multiscale modeling of failure in metal matrix composites. Acta Mater. 49:273–287

Chapter 3: Adaptive Concurrent Multilevel Model for Multiscale Analysis of Composite Materials Including Damage

Somnath Ghosh

John B. Nordholt Professor, Department of Mechanical Engineering,
The Ohio State University, Columbus, OH, USA

3.1 Introduction

The past few decades have seen rapid developments in the science and technology of a variety of advanced heterogeneous materials like polymer, ceramic, or metal matrix composite, functionally graded materials, and porous materials, as well as various alloy systems. Many of these engineered materials are designed to possess optimal properties for different functions, e.g., low weight, high strength, superior energy absorption and dissipation, high impact and penetration resistance, superior crashworthiness, better structural durability, etc. Tailoring their microstructures and properties to yield high structural efficiency has enabled these materials to provide enabling mission capabilities, which has been a key factor in their successful deployment in the aerospace, automotive, electronics, defense, and other industries.

Reinforced composites are constituted of stiff and strong fibers, whiskers or particulates of, e.g., glass, graphite, boron, or aluminum oxide, which are dispersed in primary phase matrix materials made of, e.g., epoxy, steel, titanium, or aluminum. Micrographs of a silicon particulate reinforced aluminum alloy (DRA) and an epoxy matrix composite (PMC), consisting of graphite fibers, are shown in Fig. 3.1. The presence of reinforcing phases generally enhances physical and mechanical properties like strength, thermal expansion coefficient, and wear resistance of the composite.

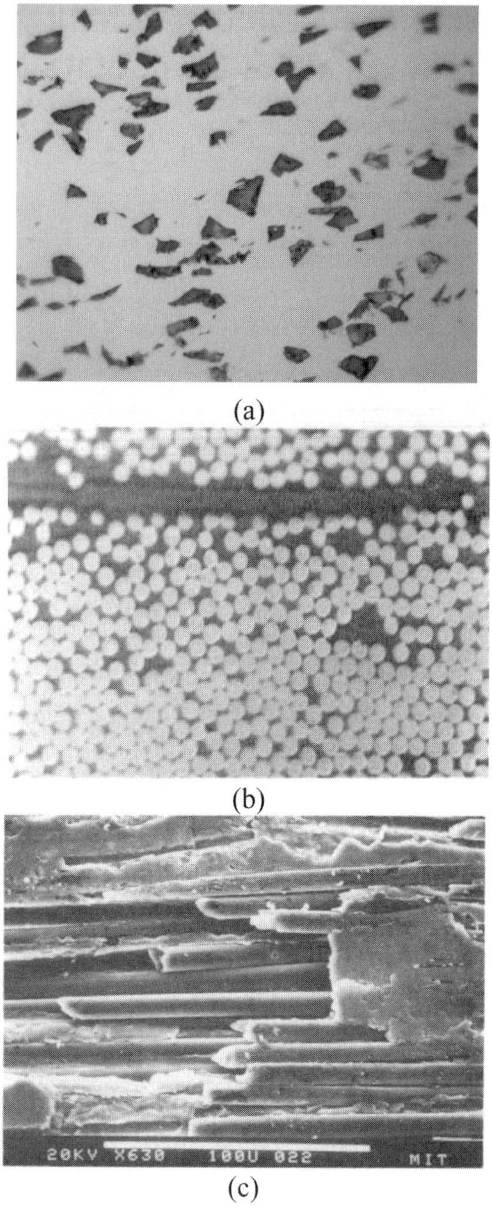

Fig. 3.1. Micrographs of (**a**) SiC particle-reinforced aluminum matrix composite showing particle cracking, (**b**) graphite-epoxy, fiber-reinforced polymer matrix composite, (**c**) fiber breakage in a polymer matrix composite

Processing methods, like powder metallurgy or resin transfer molding, often contribute to nonuniformities in microstructural morphology, e.g., in reinforcement spatial distribution, size or shape, or in the constituent material and interface properties. These nonuniformities can influence the degree of property enhancement. However, the presence of the nonuniform microstructural heterogeneities can have a strong adverse effect on their failure properties like fracture toughness, strain to failure, ductility, and fatigue resistance. Damage typically initiates at microstructural "weak spots" by inclusion (fiber or particle) fragmentation or decohesion at the inclusion-matrix interface. The cracks often bifurcate into the matrix and link up with other damage sites and cracks to evolve across larger scales and manifest as dominant cracks that cause structural failure. Structural failure of composite materials is thus inherently a multiple scale phenomenon. Microstructural damage mechanisms and structural failure properties are sensitive to the local variations in morphology, such as clustering, directionality, or connectivity and variations in reinforcement shape or size. Figure 3.1a shows particle and matrix cracking in a SiC-reinforced DRA microstructure, and Fig. 3.1c is the micrograph of a graphite-epoxy PMC showing failure by fiber breakage and matrix rupture. Experimental studies, e.g., in [5, 18], have established that particles in regions of clustering or alignment have a greater propensity toward fracture.

The need for robust design procedures for reliable and effective composite materials provides a compelling reason for the accelerated development of competent modeling methods that can account for the structure–material interaction and relate the microstructure to properties and failure characteristics. The models should accurately represent phenomena at different length scales and also optimize the computational efficiency through effective multiscale domain decomposition.

3.2 Homogenization and Multiscale Models

It is prudent to use the notion of multispatial scales in the analysis of composite materials and structures due to the inherent existence of various scales. Conventional methods of analysis have used effective properties obtained from homogenization of response at microscopic length scales. A number of analytical models have evolved within the framework of small deformation linear elasticity theory to predict homogenized macroscale constitutive response of heterogeneous materials, accounting for the characteristics of microstructural behavior. The underlying principle of these models is the Hill–Mandel condition of homogeneity [41], which states that for large differences in microscopic and macroscopic length scales, the volume averaged strain energy is obtained as the product of the volume averaged stresses and strains in the representative volume element or RVE, i.e.,

$$\int_{\Omega} \sigma_{ij}^{*}\varepsilon_{ij}^{*}\,\mathrm{d}\Omega = \left\langle \sigma_{ij}^{*}\varepsilon_{ij}^{*} \right\rangle = \left\langle \sigma_{ij}^{*} \right\rangle \left\langle \varepsilon_{ij}^{*} \right\rangle. \tag{3.1}$$

Here σ_{ij}^{*} and ε_{ij}^{*} are the general statically admissible stress field and kinematically admissible strain field in the microstructure, respectively, and Ω is a microstructural volume that is equal to or larger than the RVE. The representative volume element or RVE in (3.1) corresponds to a microstructural subregion that is representative of the entire microstructure in an average sense. For composites, it is assumed to contain a sufficient number of inclusions, which makes the effective moduli independent of assumed homogeneous tractions or displacements on the RVE boundary. The Hill–Mandel condition introduces the notion of a homogeneous material that is energetically equivalent to a heterogeneous material. Cogent reviews of various homogenization models are presented in Mura [9, 52]. Based on the eigenstrain formulation, an equivalent inclusion method has been introduced by Eshelby [22] for stress and strain distributions in an infinite elastic medium containing a homogeneous inclusion. Mori–Tanaka estimates, e.g., in [8], consider nondilute dispersions where inclusion interaction is assumed to perturb the mean stress and strain field. Self-consistent schemes introduced by Hill [40] provide an alternative iterative methodology for obtaining mean field estimates of thermoelastic properties by placing each heterogeneity in an effective medium. Notable among the various estimates and bounds on the elastic properties are the variational approach using extremum principles by Hashin et al. [39] and Nemat-Nasser et al. [53], the probabilistic approach by Chen and Acrivos [14], the self-consistent model by Budiansky [11], the generalized self-consistent models by Christensen and Lo [16], etc. These predominantly analytical models, however, do not offer adequate resolution to capture the fluctuations in microstructural variables that have significant effects on properties. Also, arbitrary morphologies, material nonlinearities, or large property mismatches in constituent phases cannot be adequately treated.

The use of computational micromechanical methods like the finite element method, boundary element method, spring lattice models, etc. has become increasingly popular for accurate prediction of stresses, strains, and other evolving variables in composite materials [9, 10, 83]. Within the framework of computational multispatial scale analyses of heterogeneous materials, two classes of methods have emerged, depending on the nature of coupling between the scales. The first group, known as “hierarchical models" [17, 23, 30, 31, 37, 43, 63, 77, 78] entails bottom-up coupling in which information is passed unidirectionally from lower to higher scales, usually in the form of effective material properties. A number of hierarchical models have incorporated the asymptotic homogenization theory developed by Benssousan [7], Sanchez-Palencia [68], and Lions [47] in conjunction with computational micromechanics models. Homogenization implicitly assumes uniformity of macroscopic field variables. Uncoupling of governing equations at different scales is achieved through incorporation of periodicity boundary conditions on the microscopic representative volume elements or RVEs, implying periodic repetition of a local microstructural region. Consequently, the models are used to predict evolution of variables at the macroscopic scale using homogenized constitutive relations, as well as in the periodic microstructural RVE. The latter analysis can be conducted as a postprocessor to the macroscopic analysis with macroscopic strain as the input. Hierarchical multiscale computational analyses of reinforced composites have been conducted by, e.g., Fish et al. [23], Kikuchi et al. [37], Terada et al. [78], Tamma and Chung [17, 77], and Ghosh et al. [30, 31, 43]. Hierarchical models involving homogenization for damage in composites have also been developed by Ghosh et al. in [63, 65] from the microstructural Voronoi cell FEM model, Lene et al. [21, 44], Fish et al. [25], and Allen et al. [2, 3, 20], among others.

While the "bottom-up" hierarchical models are efficient and can accurately predict macroscopic or averaged behavior, such as stiffness or strength, their predictive capabilities are limited with problems involving localization, failure, or instability. Macroscopic uniformity of response variables, like stresses or strains, is not a suitable assumption in regions of high gradients like free edges, interfaces, material discontinuities, or in regions of localized deformation and damage. On the other hand, RVE periodicity is unrealistic for nonuniform microstructures, e.g., in the presence of clustering of heterogeneities or localized microscopic damage. Even with a uniform phase distribution in the microstructure, the evolution of localized stresses, strains, or damage path can violate periodicity conditions. Such shortcomings for composite material modeling have been discussed for modeling heterogeneous materials by Pagano and Rybicki [58, 67], Oden and Zohdi [55, 84], Ghosh et al. [35, 62, 64], Fish et al. [24]. The solution of micromechanical problems in the vicinity of stress singularity was suggested in [58, 67] in the context of composite laminates with free edges. These problems have been effectively tackled by the second class of models known as "concurrent" multiscale modeling methods [24, 29, 35, 36, 51, 55, 56, 58, 61, 62, 64, 67, 71, 79, 82, 84].

Concurrent multiscale models differentiate between regions requiring different resolutions to invoke two-way (bottom-up and top-down) coupling of scales in the computational domain. These models provide effective means for analyzing heterogeneous materials and structures involving high solution gradients. Substructuring allows for macroscopic analysis using homogenized material properties in some parts of the domain while zooming in at selected regions for detailed micromechanical modeling. Macroscopic analysis, using bottom-up homogenization in regions of relatively benign deformation, enhances the efficiency of the computational analysis due to the reduced order models with limited information on the microstructural morphology. The top-down localization process, on the other hand, incorporates cascading down to the microstructure in critical regions of localized damage or instability. These regions need explicit representation of the local microstructure, and micromechanical analysis is conducted for accurately predicting localization or damage path. Microscopic computations involving complex microstructures are often intensive and computationally prohibitive. Selective microstructural analysis in the concurrent setting makes the overall computational analysis feasible, provided the "zoom-in" regions are kept to a minimum.

A variety of alternative methods have been explored for adaptive concurrent multiscale analysis in [51, 55, 56, 84, 79, 82]. Concurrent multiscale analysis using adaptive multilevel modeling with the microstructural Voronoi cell FEM model has been conducted by Ghosh et al. [29, 35, 36, 61, 62, 64] for modeling composites with free edges or with evolving damage resulting in dominant cracks. Guided by physical and mathematical considerations, the introduction of adaptive multiple scale modeling is a desirable feature for optimal selection of regions requiring different resolutions to minimize discretization and modeling errors. Ghosh and coworkers have also developed adaptive multilevel analysis using the microstructural Voronoi cell FEM model for modeling elastic–plastic composites with particle cracking and porosities in [35] and for elastic composites with debonding at the fiber–matrix interface in [29, 36]. This chapter is devoted to a discussion of adaptive concurrent multiple scale models developed by the author for composites with and without damage.

3.3 Multilevel Computational Model for Concurrent Multiscale Analysis of Composites Without Damage

A framework of an adaptive multilevel model is presented for macroscale to microscale analysis of composite materials in the absence of microstructural damage. The model consists of three levels of hierarchy, as shown in Fig. 3.2. These are:

(1) *Level-0* macroscopic computational domain of Fig. 3.2b using material properties that are obtained by homogenizing the material response in the microstructural RVE of Fig. 3.2a.
(2) *Level-1* computational domain of macroscopic analysis that is followed by a postprocessing operation of microscopic RVE analysis. This level, shown in Fig. 3.2c, is used to decipher whether RVE-based homogenization is justified in this region.
(3) *Level-2* computational domain of pure microscopic analysis, where the assumption of the microscopic RVE for homogenization is not valid.
(4) *Intermediate transition layer* sandwiched between the macroscopic (*level-0/level-1*) and microscopic (*level-2*) computational domains.

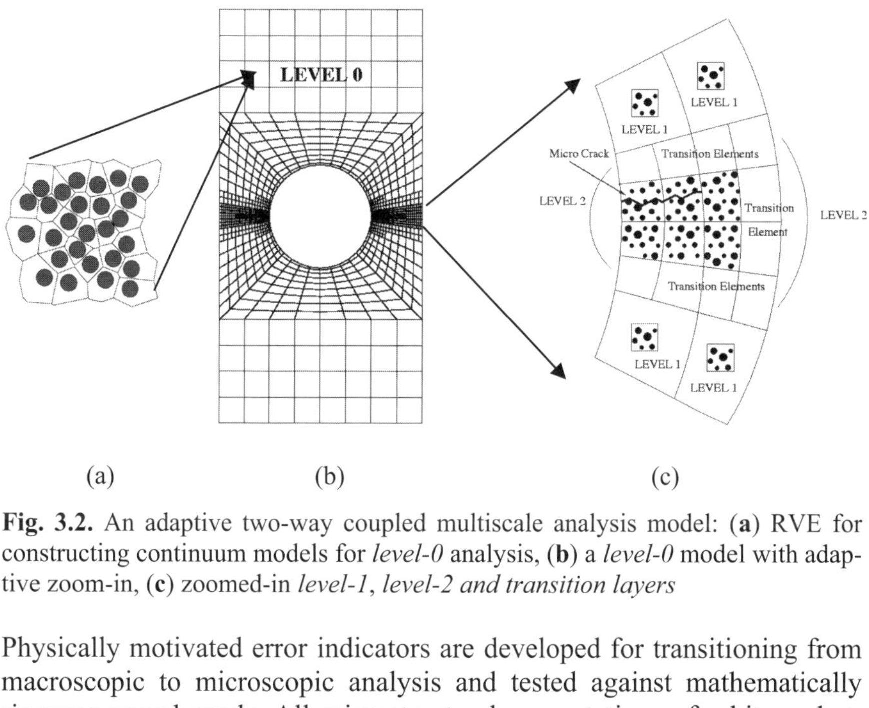

Fig. 3.2. An adaptive two-way coupled multiscale analysis model: (**a**) RVE for constructing continuum models for *level-0* analysis, (**b**) a *level-0* model with adaptive zoom-in, (**c**) zoomed-in *level-1, level-2 and transition layers*

Physically motivated error indicators are developed for transitioning from macroscopic to microscopic analysis and tested against mathematically rigorous error bounds. All microstructural computations of arbitrary heterogeneous domains are conducted using the adaptive Voronoi cell finite element model [26, 34, 48–50].

3.3.1 Hierarchy of Domains for Heterogeneous Materials

Consider a heterogeneous domain composed of multiple phases of linear elastic materials, which occupies an open bounded domain $\Omega_{\text{het}} \subset \mathfrak{R}^3$, with a Lipschitz boundary $\partial\Omega_{\text{het}} = \Gamma_u \cup \Gamma_t, \Gamma_u \cap \Gamma_t = \varnothing$. Γ_u and Γ_t corresponds to displacement and traction boundaries, respectively. The body forces $\boldsymbol{f} \in L^2(\Omega_{\text{het}})$ and surface tractions $\boldsymbol{t} \in L^2(\Omega_{\text{het}})$ are vector-valued functions. The multilevel computational model for this domain uses problem descriptions for two types of domains.

Micromechanics problem for the heterogeneous domain Ω_{het}

The micromechanics problem for the entire domain includes explicit consideration of multiple phases in Ω_{het} with the location dependent elasticity tensor $\boldsymbol{E}(\boldsymbol{x})$, which is a bounded function in $\mathfrak{R}^{9\times 9}$ that satisfies conventional conditions of ellipticity (positive strain energy for admissible strain fields) and symmetry. The displacement field **u** for the actual problem can be obtained as the solution to the conventional statement of principle of virtual work, expressed as

$$\text{Find}\,\mathbf{u}, \mathbf{u}|_{\Gamma_u} = \bar{\mathbf{u}},$$

such that

$$\int_{\Omega_{\text{het}}} \nabla\mathbf{v} : \mathbf{E} : \nabla\mathbf{u}\, \mathrm{d}\Omega = \int_{\Omega_{\text{het}}} \mathbf{f} \cdot \mathbf{v}\, \mathrm{d}\Omega + \int_{\Gamma_t} \mathbf{t} \cdot \mathbf{v}\, \mathrm{d}\Gamma \quad \forall \mathbf{v} \in \mathbf{V}(\Omega_{\text{het}}), \tag{3.2}$$

where $\mathbf{V}(\Omega)$ is a space of admissible functions defined as

$$\mathbf{V}(\Omega) = \{\mathbf{v} : \mathbf{v} \in \mathbf{H}^1(\Omega); \mathbf{v}|_{\Gamma_u} = 0\}. \tag{3.3}$$

For heterogeneous materials with a distribution of different phases, such as fibers, particles, or voids, the constituent material properties $\boldsymbol{E}(\boldsymbol{x})$ may vary considerably with spatial position. Consequently, conventional finite element models are likely to incorporate inordinately large meshes for accuracy, which results in expensive computations. A regularized version of the actual problem, using homogenization methods can be of significant value in reducing the computing efforts through reduced order models.

Regularized problem in a homogenized domain Ω_{hom}

A regularized solution $\mathbf{u}^H$ to the actual problem can be obtained by using a homogenized linear elasticity tensor $\boldsymbol{C}^H(\boldsymbol{x})$ in solving the boundary value problem, which is characterized by the principle of the virtual work:

$$\text{Find}\,\mathbf{u}^H, \mathbf{u}^H|_{\Gamma_u} = \bar{\mathbf{u}}$$

such that

$$\int_{\Omega_{\text{hom}}} \nabla \mathbf{v} : \boldsymbol{C}^H : \nabla \mathbf{u}^H \, \mathrm{d}\Omega = \int_{\Omega_{\text{hom}}} \mathbf{f} \cdot \mathbf{v} \, \mathrm{d}\Omega + \int_{\Gamma_t} \mathbf{t} \cdot \mathbf{v} \, \mathrm{d}\Gamma \quad \forall \mathbf{v} \in \mathbf{V}(\Omega_{\text{hom}}). \tag{3.4}$$

The homogenized elasticity tensor is assumed to satisfy symmetry and ellipticity conditions, and it is required to produce an admissible stress field $\boldsymbol{\sigma}^H (= \boldsymbol{C}^H : \nabla \mathbf{u}^H)$ satisfying the traction boundary condition: $\boldsymbol{n} \cdot \boldsymbol{\sigma}^H$ = $\mathbf{t}(\boldsymbol{x}) \; \forall \, \boldsymbol{x} \in \Gamma_t$. Determination of statistically homogeneous material parameters requires an isolated representative volume element or RVE $Y(\boldsymbol{x}) \subset \mathfrak{R}^3$, over which averaging can be performed. The resulting field variables like stresses and strains are also statistically homogeneous in the RVE and may be obtained from volumetric averaging as

$$\sigma^H = \frac{1}{|Y|} \int_Y \sigma(\mathbf{y}) \, \mathrm{d}Y, \quad \varepsilon^H = \frac{1}{|Y|} \int_Y \varepsilon(\mathbf{y}) \, \mathrm{d}Y, \qquad |Y| = \int_Y \mathrm{d}Y. \tag{3.5}$$

In classical methods of estimating homogenized elastic moduli $\boldsymbol{C}^H(\boldsymbol{x})$, the RVE is subjected to prescribed surface displacements or tractions, which in turn produce uniform stresses or strains in a homogenous medium. Various micromechanical theories have been proposed to predict the overall constitutive response by solving RVE-level boundary value problems, followed by volumetric averaging [9, 52]. The scale of the RVE $Y(\boldsymbol{x})$ is typically very small in comparison with the dimension L of the structure. The asymptotic homogenization theory, proposed in [7, 47, 68], is also effective in multiscale modeling of physical systems that contain multiple length scales. This method is based on asymptotic expansion of the solution fields, e.g., displacement and stress fields, in the microscopic spatial coordinates about their respective macroscopic values. The composite microstructure in the RVE is assumed to be locally Y-periodic. Correspondingly, any variable $\boldsymbol{f}^\varepsilon$ in the RVE is also assumed to be Y-periodic, i.e., $\boldsymbol{f}^\varepsilon(\boldsymbol{x},\boldsymbol{y}) = \boldsymbol{f}^\varepsilon(\boldsymbol{x},\boldsymbol{y}+\boldsymbol{k}\boldsymbol{Y})$. Here $\boldsymbol{y} = \boldsymbol{x}/\varepsilon$ corresponds to the microscopic coordinates in $Y(\boldsymbol{x})$. Here, $\varepsilon \ll 1$ is a small positive number representing the ratio of microscopic to macroscopic length scales, and $\boldsymbol{k}$ is a 3 × 3 array of integers. Superscript ε denotes the association with both length scales $(\boldsymbol{x},\boldsymbol{y})$. In homogenization theory, the displacement field is asymptotically expanded about $\boldsymbol{x}$ with respect to the parameter ε as

$$u_i^{\varepsilon}(\boldsymbol{x}) = u_i^0(\boldsymbol{x},\boldsymbol{y}) + \varepsilon u_i^1(\boldsymbol{x},\boldsymbol{y}) + \varepsilon^2 u_i^2(\boldsymbol{x},\boldsymbol{y}) + \cdots. \tag{3.6}$$

Since the stress tensor is obtained from the spatial derivative of $u_i^{\varepsilon}(\boldsymbol{x})$ as

$$\sigma_{ij}^{\varepsilon}(\boldsymbol{x},\boldsymbol{y}) = \frac{\mathbf{1}}{\varepsilon}\sigma_{ij}^0(\boldsymbol{x},\boldsymbol{y}) + \sigma_{ij}^1(\boldsymbol{x},\boldsymbol{y}) + \varepsilon\sigma_{ij}^2(\boldsymbol{x},\boldsymbol{y}) + \varepsilon^2\sigma_{ij}^3(\boldsymbol{x},\boldsymbol{y}) + \cdots, \tag{3.7}$$

where

$$\sigma_{ij}^0 = C_{ijkl}^{\varepsilon}\frac{\partial u_k^0}{\partial y_t}, \quad \sigma_{ij}^1 = C_{ijkl}^{\varepsilon}\left(\frac{\partial u_k^0}{\partial x_t} + \frac{\partial u_k^1}{\partial y_t}\right), \quad \sigma_{ij}^2 = C_{ijkl}^{\varepsilon}\left(\frac{\partial u_k^1}{\partial x_t} + \frac{\partial u_k^2}{\partial y_t}\right). \tag{3.8}$$

By applying periodicity conditions on the RVE boundary, i.e., $\int_{\partial Y}\sigma_{ij}n_j\,\mathrm{d}\partial Y = 0$, it is possible to decouple the governing equations into a set of microscopic and macroscopic problems, respectively. These are:

Microscopic equations

$$\frac{\partial\hat{\sigma}_{ij}^{kl}(\boldsymbol{y})}{\partial y_j} = 0 \qquad \text{(Equilibrium)},$$

$$\hat{\sigma}_{ij}^{kl}(\boldsymbol{y}) = C^{\varepsilon}{}_{ijpm}\left[\delta_{kp}\delta_{lm} + \frac{\partial\chi^{kl}{}_p}{\partial y_m}\right] \qquad \text{(Constitutive)}. \tag{3.9}$$

The superscripts k and l in (3.9) correspond to the components of the macroscopic strain that cause the microscopic stress components $\hat{\sigma}_{ij}^{kl}$. The subscripts i, j, etc. in this equation on the other hand correspond to microscopic tensor components.

Macroscopic equations

$$\frac{\partial\Sigma_{ij}(\boldsymbol{x})}{\partial x_j} + f_i = 0 \qquad \text{(Equilibrium)},$$

$$\Sigma_{ij}(\boldsymbol{x}) = \frac{1}{|Y|}\left[\int_Y C_{ijkl}^{\varepsilon}\left(\delta_{km}\delta_{lm} + \frac{\partial\chi_k^{mn}}{\partial y_l}\right)\mathrm{d}Y\right]\frac{\partial u_m^0}{\partial x_n} = C_{ijmn}^H e_{mn}(\boldsymbol{x}) \quad \text{(Constitutive)}. \tag{3.10}$$

The interscale transfer operators in these relations are defined as

$$\sigma^1_{ij} = \hat{\sigma}^{kl}_{ij}(\boldsymbol{y})\frac{\partial u^0_k}{\partial x_l}(\boldsymbol{x}) \quad \text{(Stress–strain)},$$

$$u^1_i = \chi^{kl}_i(\boldsymbol{y})\frac{\partial u^0_k}{\partial x_l}(\boldsymbol{x}) \quad \text{(Strain–displacement)}. \tag{3.11}$$

In (3.9)–(3.11), χ^{kl}_i is a Y-periodic function representing the characteristic modes of deformation in the RVE and

$$\Sigma_{ij}(\boldsymbol{x}) = \langle \sigma^{\varepsilon}{}_{ij}(\boldsymbol{x},\boldsymbol{y})\rangle_Y, \quad e_{ij}(\boldsymbol{x}) = \frac{1}{2}\left\{\frac{\partial u^0_i}{\partial x_j} + \frac{\partial u^0_j}{\partial x_i}\right\} = \langle e^{\varepsilon}{}_{ij}(\boldsymbol{x},\boldsymbol{y})\rangle_Y \tag{3.12}$$

are homogenized macroscopic stress and strain tensors, respectively, that are obtained by volumetric averaging. The asymptotic homogenization method provides good convergence characteristics with respect to certain norms, in addition to bounds on effective properties. Solutions of RVE boundary value problems with imposed unit macroscopic strains are used in the calculation of the anisotropic homogenized elasticity tensor $C^H{}_{ijkl}(\boldsymbol{x})$. The RVE boundaries are subjected to periodicity conditions, implying that all boundary nodes separated by the periods Y_1, Y_2, Y_3 along the three orthogonal coordinate directions will follow the displacement constraints:

$$u_i(x_1,x_2,x_3) = u_i(x_1 \pm k_1Y_1, x_2 \pm k_2Y_2, x_3 \pm k_3Y_3), \quad i = 1,2,3. \tag{3.13}$$

Following macroscopic analysis with the homogenized moduli $C_{ijkl}{}^H(\boldsymbol{x})$, numerical simulations of the RVE boundary value problems yield stresses and strains in the microstructural RVE.

Limitations of the regularized problem in Ω_{hom}

Limitations in solving the regularized problem for variables in heterogeneous microdomains arise from assumptions of relatively uniform macroscopic fields and periodicity of the RVE. In uncoupling the macroscopic problem in Ω_{hom} from the microscopic RVE problem in Y, it is assumed that the RVE has infinitesimal dimensions in comparison with the macroscopic scale, i.e., $\varepsilon \to 0$. While solutions of the problems in Ω_{hom} approach those for the actual domain Ω_{het} in this limit, considerable differences may result when the scale factor ε is finite and the RVE solutions are not periodic. The occurrence of such errors is significant in regions of high local gradients, free edges, or discontinuities.

3.3.2 Multiple Levels Coupling Multiple Scales

In the multilevel methodology developed in Ghosh et al. [35, 62, 64], the overall heterogeneous computational domain is adaptively decomposed into a set of nonintersecting open subdomains, which may each belong to one of Ω_{het} (domain for microscopic analysis) Ω_{hom} (regularized domain for macroscopic analysis), or to a combination thereof. The resulting computational domain $\bar{\Omega}_{\text{het}}$ may be expressed as the union of subdomains belonging to different levels expressed as

$$\begin{aligned}
&\bar{\Omega}_{\text{het}} = \bigcup_{k=1}^{N_0} \Omega_k^{l0} \cup \bigcup_{k=1}^{N_1} \Omega_k^{l1} \cup \bigcup_{k=1}^{N_2} \Omega_k^{l2} \cup \bigcup_{k=1}^{N_{\text{tr}}} \Omega_k^{\text{tr}}, \quad \text{where} \quad \Omega_k^{l0} \cap \Omega_l^{l0} = 0,\\
&\Omega_k^{l1} \cap \Omega_l^{l1} = 0, \quad \Omega_k^{l2} \cap \Omega_l^{l2} = 0, \quad \Omega_k^{\text{tr}} \cap \Omega_l^{\text{tr}} = 0 \quad \forall k \neq l,\\
&\text{and } \Omega_k^{l0} \cap \Omega_l^{l1} = 0, \quad \Omega_k^{l0} \cap \Omega_l^{l2} = 0, \quad \Omega_k^{l1} \cap \Omega_l^{l2} = 0,\\
&\Omega_k^{l1} \cap \Omega_l^{\text{tr}} = 0 \quad \forall k,l.
\end{aligned} \tag{3.14}$$

Here the superscripts $l0$, $l1$, and $l2$ correspond to *level-0*, *level-1*, or *level-2* subdomains in the computational hierarchy; and superscripts tr correspond to the transition region between *level-0/1* and *level-2* subdomains. Computations in different levels require different algorithmic treatments. The number of levels may not exactly correspond to the number of scales, even though they are connected to individual scales. The constituent subdomains, e.g., Ω_k^{l0} need not be contiguous and may occupy disjoint locations in $\bar{\Omega}_{\text{het}}$. However, certain restrictions apply with respect to sharing of contiguous subdomain boundaries. If $\partial\Omega_k^{l1}, \partial\Omega_k^{l1}, \partial\Omega_k^{l2}$ and $\partial\Omega_k^{\text{tr}}$ represent boundaries of the corresponding level subdomains, then,

- $\partial\Omega_k^{l0} \cap \partial\Omega_l^{l1} = \partial\Omega_k^{l0-l1} \quad \forall k,l.$ Also $\partial\Omega_{kl}^{l0-l1}$ has the same characteristics as $\partial\Omega_k^{l0}$ or $\partial\Omega_l^{l1}$, since $\partial\Omega_k^{l0}$ and $\partial\Omega_l^{l1}$ have compatible displacements.
- $\partial\Omega_k^{l0} \cap \partial\Omega_l^{l2} = 0 \quad \forall k,l$, i.e., $\partial\Omega_k^{l0}$ and $\partial\Omega_l^{l2}$ are not contiguous or may not share common edges.
- $\partial\Omega_k^{l2} \cap \partial\Omega_l^{\text{tr}} = \partial\Omega_{kl}^{l2-\text{tr}} \quad \forall k,l.$ Also $\partial\Omega_{kl}^{l2-\text{tr}}$ has the same characteristics as $\partial\Omega_k^{l2}$, since $\partial\Omega_k^{l2}$ and $\partial\Omega_l^{\text{tr}}$ have compatible displacements.
- $\partial\Omega_k^{l0/l1} \cap \partial\Omega_l^{\text{tr}} = \partial\Omega_{kl}^{l0/l1-\text{tr}} \quad \forall k,l$. Also the interfaces of $\partial\Omega_k^{l0/l1}$ and $\partial\Omega_l^{\text{tr}}$ are not compatible in general, and hence special constraint conditions need to be developed for $\partial\Omega_{kl}^{l0/l1-\text{tr}}$.

A few cycles in the iterative solution process are required to settle into an "optimal" distribution of the computational levels, even for linear elastic problems. The three levels of computational hierarchy, in the order of sequence of evolution are discussed next.

Computational subdomain level-0 Ω^{l0}

Macroscopic analysis with homogenized properties is performed in the *level-0* subdomain. Unless the microstructural morphology suggests strong nonperiodicity, the computational model can generally start with the assumption that $\bar{\Omega}_{\text{het}} = \Omega^{l0}\left(=\bigcup_{k=1}^{N_0}\Omega_k^{l0}\right)\subset\Omega_{\text{hom}}$, i.e., all elements belong to the *level-0* subdomain. This subdomain assumes relatively uniform deformation with "statistically" periodic microstructures, where the regularized problem formulation is nearly applicable. Upon establishing a representative volume element $Y(\boldsymbol{x})$ for the material at a point $\boldsymbol{x}$, the asymptotic expansion-based homogenization method is implemented to yield an assumed orthotropic homogenized elasticity tensor $C^H{}_{ijkl}(\mathbf{x})$ from (3.10). Components of $C^H{}_{ijkl}(\boldsymbol{x})$ for plane problems are calculated from the solution of three separate boundary value problems of the RVE with periodic boundary conditions and imposed unit macroscopic strains given as

$$\begin{Bmatrix} e_{11} \\ e_{22} \\ e_{12} \end{Bmatrix} = \begin{Bmatrix} 1 \\ 0 \\ 0 \end{Bmatrix}, \quad \begin{Bmatrix} e_{11} \\ e_{22} \\ e_{12} \end{Bmatrix} = \begin{Bmatrix} 0 \\ 1 \\ 0 \end{Bmatrix}, \quad \begin{Bmatrix} e_{11} \\ e_{22} \\ e_{12} \end{Bmatrix} = \begin{Bmatrix} 0 \\ 0 \\ 1 \end{Bmatrix}. \tag{3.15}$$

The homogenized elastic stiffness components $C^H{}_{1111}$, $C^H{}_{2222}$, $C^H{}_{1212}$, $C^H{}_{1133}$, $C^H{}_{2233}$, and $C^H{}_{1122}$ are calculated from the volume averaged stresses Σ_{ij} according to (3.10). In the event that the elastic coefficient $C^H{}_{3333}$ is needed, a fourth boundary value problem should be solved with $\left(e_{11,}e_{22,}e_{12,}e_{33}\right)^{\text{T}} = (0,0,0,1)^{\text{T}}$. Since the microstructure and the corresponding RVE can change from element to element $\left(E_{l0}\in\Omega^{l0}\right)$ in the computational domain, each element E_{l0} should be assigned its location specific RVE $Y(\boldsymbol{x})$. Drastically different moduli in adjacent elements could lead to nonphysical stress concentrations. Smoothing schemes may be needed for regularization in these regions for macroscopic analysis. However, switching levels can enable a smooth transition from one RVE to another through the introduction of intermediate *level-2* regions.

Level-0 mesh enrichment by h- and hp-adaptation

Computational models in the level-0 subdomains are enhanced adaptively by selective *h*- or *hp*-mesh refinement strategy based on suitably chosen "error" criteria. Local enrichment through successive mesh refinement or interpolation function augmentation serves a dual purpose in the multilevel computational strategy. The first goal is to identify regions of high discretization "error" and improve convergence through mesh enhancement in a finite element subspace $\mathbf{V}_{\text{adap}} \subset \mathbf{V}$ with the requirement

Find $\mathbf{u}^{\text{H}}_{\text{adap}}, \mathbf{u}^{\text{H}}_{\text{adap}}|_{\Gamma_{\mathbf{u}}} = \bar{\mathbf{u}}$, satisfying:

$$\int_{\Omega_{\text{hom}}} \nabla\mathbf{v} : \mathbf{E}^{\mathbf{H}} : \nabla\mathbf{u}^{\text{H}}_{\text{adap}} \mathrm{d}\Omega = \int_{\Omega_{\text{hom}}} \mathbf{f} \cdot \mathbf{v} \mathrm{d}\Omega + \int_{\Gamma_t} \mathbf{t} \cdot \mathbf{v} \mathrm{d}\Gamma \quad \forall \mathbf{v} \in \mathbf{V}_{\text{adap}} \tag{3.16}$$

such that $\left\| \mathbf{u}^{\text{H}} - \mathbf{u}^{\text{H}}_{\text{adap}} \right\| \leq$ preset tolerance.

The second is to identify regions of high modeling error due to limitations of the regularized problem in representing the heterogeneous domain and to zoom in on these regions to create higher resolution. These regions are generally characterized by large solution gradients and localization of regularized macroscopic variables. Element refinement is helpful in reducing the length-scale disparity between macroscopic elements in Ω_{hom} and the local microstructure Ω_{het}.

In [35, 64], the *h*-adaptation procedure has been used to subdivide macroscopic elements into smaller elements in regions of high stress or strain gradients, while keeping the order of interpolation fixed. The rate of convergence of this method for nonsmooth domains is quite limited, especially with solution singularities, e.g., in Szabo and Babuska [74]. As a remedy, the *hp*-version of finite element refinement has been established [1]. This method is capable of producing exponentially fast convergence in the finite element approximations to the energy norm for solutions of linear elliptic boundary value problems on nonsmooth domains, such as those with singularities. The rate of convergence of the *hp*-finite element model is estimated by the inequality

$$\left\| \mathbf{u} - \mathbf{u}^{hp}_{fe} \right\| \leq C h^{\mu} p^{-(m-1)} \left\| u \right\|, \tag{3.17}$$

where h is the mesh size, p is the order of interpolation polynomial, m corresponds to the regularity of the solution, C is a constant and $\mu = \min(p, m-1)$. The parameter m dictates the distribution and sequence of *h*- and *p*-refinements in the *hp*-adaptation scheme. Smaller m leads to algebraic rates, while large m for smooth solutions yield exponential rate of convergence with successive p-refinements. The adaptation scheme follows the criteria: Perform p-refinement if $p+2 \leq m$; and perform h-refinement if $p+2 > m$.

It is necessary to solve a sequence of element level regularized boundary value problems in Ω_{hom} to estimate the local regularity parameter m. If $\boldsymbol{\phi}_{p+q}(k)$ characterizes the error estimator in the FE space $Y_{p+q}(k)$ for the kth element, using polynomials of order $p+q$ (q is the enhancement), i.e.,

$$
\begin{aligned}
B_k(\boldsymbol{\phi}_{p+q}, \mathbf{v}) = & -\int_{\Omega^k_{\mathrm{hom}}} \nabla\mathbf{v} : (\mathbf{E}^{\mathrm{H}} \nabla \mathbf{u}^{\mathrm{H}}_{\mathrm{fem}}\, \mathrm{d}\Omega + \int_{\Omega^k_{\mathrm{hom}}} \mathbf{f}\cdot\mathbf{v}\,\mathrm{d}\Omega \\
& + \int_{\partial\Omega^k_{\mathrm{hom}}} \mathbf{g}_k \cdot \mathbf{v}\,\mathrm{d}\partial\Omega \quad \forall \mathbf{v} \in Y_{p+q}(k),
\end{aligned}
\tag{3.18}
$$

$$
\text{where}\quad \mathbf{g}_k = \frac{1}{2}[\boldsymbol{\sigma}_k \cdot \mathbf{n}_k + \boldsymbol{\sigma}_{k'} \cdot \mathbf{n}_k] \in \partial\Omega^k_{\mathrm{hom}} \bigcap \partial\Omega^{k'}_{\mathrm{hom}}
$$

$$
\text{and}\quad \mathbf{g}_k = \mathbf{t} \in \partial\Omega^k_{\mathrm{hom}} \bigcap \Gamma_t \text{ is the approximate traction on } \partial\Omega^k_{\mathrm{hom}}. \tag{3.19}
$$

Here $\boldsymbol{\phi}_{p+q}(k)$ is interpreted as the finite element approximation to the true error $\mathbf{e}(k) = \mathbf{u}^H - \mathbf{u}^H_{\mathrm{fem}}$ in element k, such that the total error is bounded by the sum of the element-wise error estimators $\|\mathbf{e}\|^2_k \leq \sum_k \|\boldsymbol{\phi}\|^2_k$. The parameter m is estimated by solving the local element boundary value problem in (3.18) for three successive values of q and solving for C_k, m and $\|\boldsymbol{\phi}\|_k$ from the approximate convergence criterion.

A numerical example of the regularized problem

Convergence of the hp-adaptive refinement is explored for a composite laminate (Fig. 3.3a) in this example. The top half of the laminate (above A-A) consists of 30.7% volume fraction of silicon carbide fibers in an epoxy matrix with homogenized orthotropic elasticity matrix (in GPa) as $C^H{}_{1111} = 9.1$, $C^H{}_{2222} = 9.1$, $C^H{}_{1212} = 2.3$, $C^H{}_{1133} = 3.7$, $C^H{}_{2233} = 4.1$, and $C^H{}_{1122} = 104.2$ .The bottom half consists of a monolithic matrix material with properties $E_{\mathrm{epoxy}} = 3.45$ GPa, $\nu_{\mathrm{epoxy}} = 0.35$. Due to symmetry in the xz and yz planes, only one quarter of the laminate is modeled. Symmetric boundary conditions are employed on the surfaces $x = 0$ and $y = 0$, and the top and right surfaces are assumed to be traction free.

The regularized laminate problem is subsequently analyzed using the *h*- and *hp*-adapted *level-0* finite element codes, subjected to constant axial strain $\varepsilon_{zz} = 1.0$ in the out-of-plane direction. While the analytical transverse stress σ_{yy} is approximately two orders lower compared to the leading order stress σ_{zz}, it exhibits a singularity of the form $\sigma_{yy} = C_s r^{\lambda+1}$ near the interface-free edge juncture A ($x/h = 4$). Here r is the distance from the singular point at the free edge and C_s is a constant along each radial line at a fixed angle θ, depending on material properties. The exponent λ has been evaluated to be 0.9629358 in [6] from traction and displacement continuity at the material interface and traction free conditions on edges. The initial mesh consists of 200 QUAD4 elements. Adaptations are performed in each element until the element error meets the criterion $\|\phi\|_k \leq 0.25\|\phi\|_{\max}$. The *h*- and *hp*-adapted mesh are shown in Figs. 3.2c and 3.3a, respectively. Following iterative cycles, the converged *h*-adapted mesh consists of 1,664 elements with 3,282 degrees of freedom, while the converged *hp*-adapted mesh consists of 344 elements with 1,834 degrees of freedom. The smallest element size in both cases is $0.0025h_e$, where h_e is the initial element size.

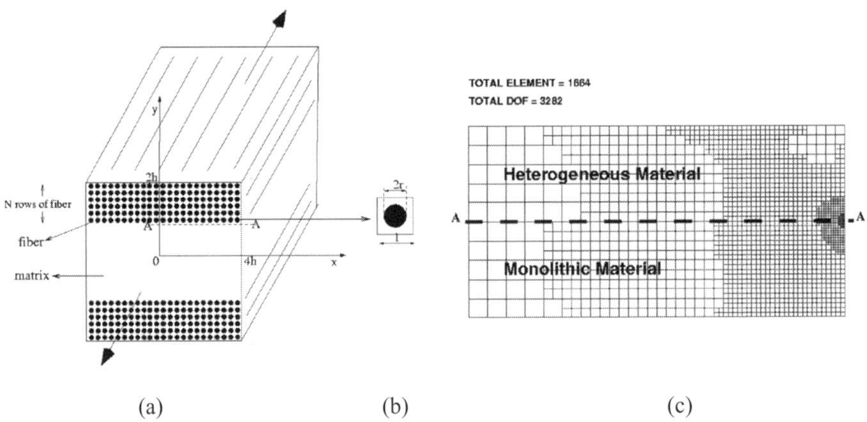

Fig. 3.3. (**a**) Unidirectional composite laminate subjected to out-of-plane loading; (**b**) a representative volume element of the microstructure, with a single fiber in a square matrix; (**c**) FE model with *h*-adapted mesh

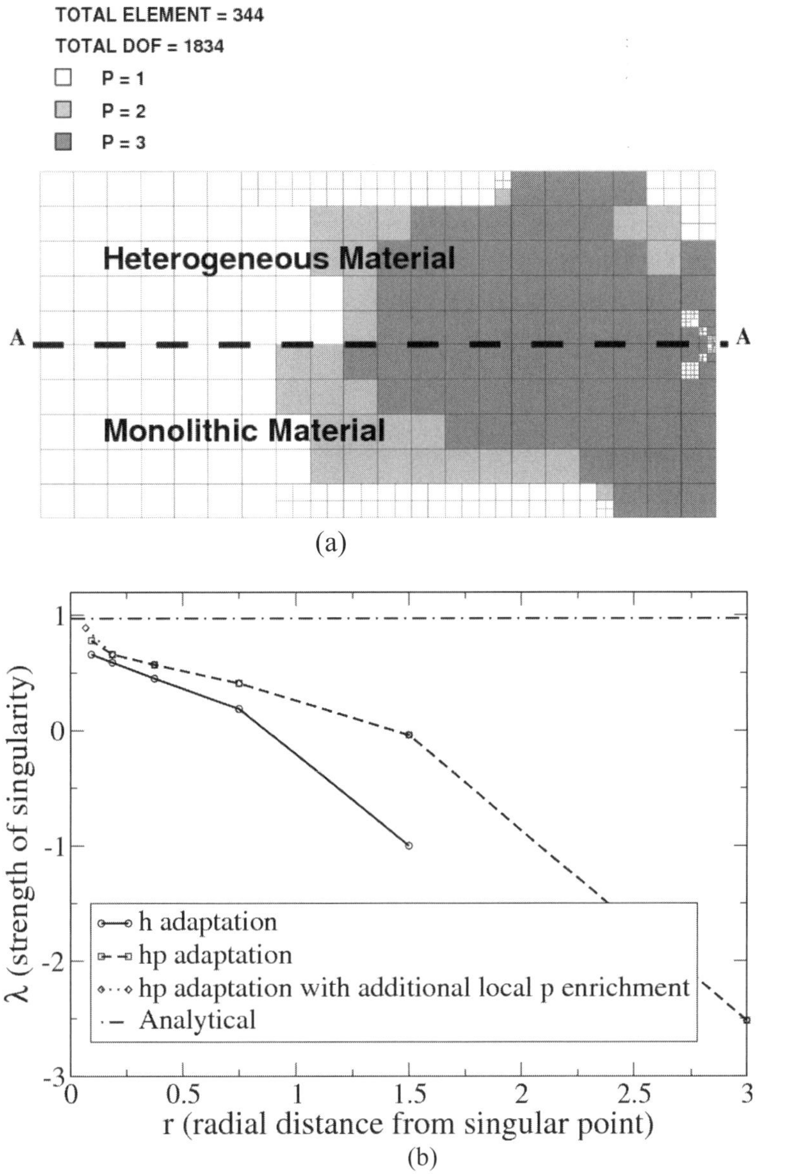

Fig. 3.4. (**a**) hp-adapted meshes in the regularized domain Ω^{l0}, (**b**) convergence of the strength of singularity for the h- and hp-adapted meshes

The strength of the singularity λ controls the rate of convergence and its value may be determined in the course of the adaptive refinements. The value of λ is obtained by evaluating σ_{yy} at two different values of r close to the singular point, and its convergence is shown in Fig. 3.4b. For the same smallest element size, the h-adapted mesh reaches up to a value of $\lambda = 0.66$, whereas the hp-adapted mesh goes up to $\lambda = 0.78$. Upon further enriching elements near the singular point by p-adaptation, λ reaches 0.89.

Local and pollution errors in the regularized problem

A posteriori error estimates based on elemental stresses or strain energy, e.g., jumps in variables, their gradients, or element residuals, are *local* in nature. Babuska [6] and Oden [54] have introduced element pollution error as one that is produced due to residual forces in other contiguous and noncontiguous elements in the mesh. Pollution error can be significant with uniform meshes in problems containing singularities, and local error estimation methods are incapable of detecting them. Consequently, in domains consisting of cracks, free edges, laminate interfaces, etc. accurate element error estimates in the energy norm may benefit from the addition of pollution errors to the local errors, i.e.,

$$\|er\|^{e} = \|er\|^{e}_{\text{local}} + \|er\|^{e}_{\text{pollution}} \tag{3.20}$$

with equidistribution of error estimates in the mesh, the pollution error is negligible. However in problems where the singularity exponent λ is less than half the order of interpolation p, i.e., $2\lambda < p$, the pollution error is significant. The basic algorithm develops an equivalent residual as the sum of element-wise local and pollution residuals as

$$B(\mathbf{er}_h, \mathbf{v}_h) = \sum_{k=1}^{N_E} [B(\mathbf{er}_{kh}^{\text{local}}, \mathbf{v}_h) + B(\mathbf{er}_{kh}^{\text{poll}}, \mathbf{v}_h)], \quad \text{where} \quad \sum_{k=1}^{N_E} B(\mathbf{er}_{kh}^{\text{local}}, \mathbf{v}_h)$$

$$= \sum_{k=1}^{N_E} \int_{\Omega^k} [\mathbf{f} + \nabla\cdot(\mathbf{E}\nabla\mathbf{u}_{\text{H}})]_k \mathbf{v}_h \mathrm{d}\Omega + \int_{\partial\Omega^k} \hat{\mathbf{g}}_k \cdot \mathbf{v}_h \,\mathrm{d}\partial\Omega \quad \forall \mathbf{v}_h \in V^h \tag{3.21}$$

$$\sum_{k=1}^{N_E} B(\mathbf{er}_{kh}^{\text{poll}}, \mathbf{v}_h) = \sum_{j=1, j\neq k}^{N_E} \int_{\Omega^j} [\mathbf{f} + \nabla(\mathbf{E} : \nabla\mathbf{u}_{\text{H}})]_j \mathbf{v}_h \,\mathrm{d}\Omega + \int_{\partial\Omega^j} \hat{\mathbf{g}}_j \mathbf{v}_h \,\mathrm{d}\partial\Omega \quad \forall \mathbf{v}_h \in V^h$$

$$\begin{aligned} \text{and} \quad \hat{\mathbf{g}}_k &= \left[\boldsymbol{\sigma}_k \cdot \mathbf{n}_k - \boldsymbol{\sigma}_{k'} \cdot \mathbf{n}_{k'}\right] \in \partial\Omega^k_{\text{hom}} \cap \partial\Omega^{k'}_{\text{hom}}, \\ \hat{\mathbf{g}}_k &= \mathrm{t} - \boldsymbol{\sigma}_{\text{k}} \cdot \mathbf{n}_k \in \partial\Omega^k_{\text{hom}} \cap \Gamma_{t'}. \end{aligned} \tag{3.22}$$

Here $\boldsymbol{V}^H$ is the polynomial subspace of $\boldsymbol{V}$, $\boldsymbol{V}^h$ is an enriched space approximation of $\boldsymbol{V}^H$. The major steps in the evaluation of the pollution error are given in [54, 62].

Composite laminate subjected to out-of-plane loading

The problem of composite laminate with free edge, similar to the one in Sect. 3.2.1 is studied to understand the effect of local and pollution errors. The top half of the laminate is a composite with 28.2% volume fraction of boron fiber in epoxy matrix with effective orthotropic homogenized properties:

E_{11} (psi)	E_{22} (psi)	E_{33} (psi)	G_{12} (psi)	ν_{12}	ν_{31}	ν_{23}
0.99×10^6	0.99×10^6	17.2×10^6	0.27×10^6	0.43	0.29	0.29

The bottom half is monolithic epoxy material with properties $E_{\text{epoxy}} = 0.5\times10^6$ psi and $\nu_{\text{epoxy}} = 0.34$. Out-of-plane loading is simulated using generalized plane strain condition with prescribed $\varepsilon_{zz} = 0.1\%$. Due to symmetry in the *xz* and *yz* planes, only one quarter of the laminate is modeled. Symmetric boundary conditions are employed on the surfaces $x = 0$ and $y = 0$, and the top and right surfaces are assumed to be traction free.

For a uniform mesh, the local error is concentrated near the intersection of the interface and the free edge region, whereas the pollution error is more diffused and occurs in bands, starting at points slightly away from the intersection point around the free edge. When *h*-adaptation is applied, the maximum local error reduces from 6.285×10^{-3} to 1.176×10^{-4}, while the pollution error reduces from 5.115×10^{-4} to 3.105×10^{-5}. This is shown in Fig. 3.5. For this problem, the inclusion of pollution error in the total element error estimate is found to add little to the criteria for *h*- and *hp*-adaptation and only local error is considered henceforth.

Micromechanical analysis with the Voronoi cell FEM

Accurate micromechanical modeling of deformation and damage in complex heterogeneous microstructures requires very high resolution models. Micromechanical analysis in the multilevel computational framework is conducted by the Voronoi cell finite element model (VCFEM) developed by Ghosh et al. in [26, 34, 46, 48–50] for accurate and efficient image-based modeling of nonuniform heterogeneous microstructures. Morphological arbitrariness in dispersion, shape, and size of heterogeneities, as acquired from actual micrographs are readily modeled by this method. The VCFEM computational mesh results from tessellating the microstructure

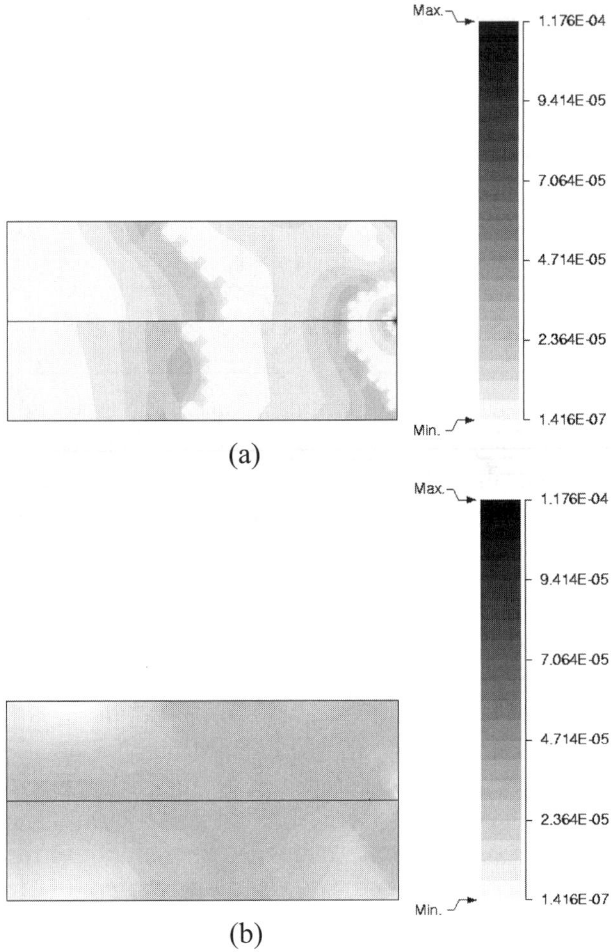

Fig. 3.5. Distribution of (**a**) local and (**b**) pollution error for the *h*-adapted mesh

$\Omega^{\mathrm{VCE}}_{\mathrm{micro}}$ with dispersed heterogeneities into a network of N_{VCE} multisided Voronoi polygon or cell elements, i.e., $\Omega^{\mathrm{VCE}}_{\mathrm{micro}} = \bigcup_{e=1}^{N_{\mathrm{VCE}}} \Omega^{\mathrm{VCE}}_{e}$, as shown in Fig. 3.8. Each Voronoi cell with embedded heterogeneities (particle, fiber, void, crack, etc.) represents the region of contiguity for the heterogeneity and is treated as an element in VCFEM. In this sense, a Voronoi cell element manifests the basic structure of the material microstructure and its evolution and is considerably larger than conventional FEM elements. Incorporation of known functional forms from analytical micromechanics substantially enhances its convergence. The VCFEM formulation is based on the assumed stress hybrid finite element method, and makes independent assumptions of equilibrated stress fields $\left(\sigma_{ij}^{m/c}\right)$ in the interior of each element $\Omega_{\text{micro}}^{\text{VCE}}$ for both the matrix and inclusion phases as well as compatible displacement fields $\left(u_i^e\right)$ on the element boundary $\partial\Omega_e^{\text{VCE}}$ and $\left(u_i^c\right)$ on the matrix–inclusion interface $\partial\Omega_c^{\text{VCE}}$. Considerable success has been achieved in modeling thermoelastic–plastic problems [48, 50], problems with microstructural damage by particle cracking [26, 49] and debonding [34, 45] by VCFEM. Recent extensions of VCFEM include multiple crack simulations by an extended VCFEM or X-VCFEM [46] and 3D VCFEM model in [27].

VCFEM is based on a hybrid formulation with independent assumptions on equilibrated stress fields σ_{ij} defined in the matrix phase Ω_m^{VCE} and the inclusion phase Ω_c^{VCE} of each Voronoi cell element. Special forms of stress functions are developed for equilibrated stress fields from known analytical micromechanics solutions. The stress functions are comprised of polynomials, shape-based reciprocal functions, and wavelet functions to facilitate accurate stress concentrations near the interface or crack tips. Compatible displacements are generated on the boundaries $\partial\Omega_e^{\text{VCE}}$ and interfaces $\partial\Omega_c^{\text{VCE}}$ (shown in Fig. 3.13b) by interpolating nodal displacements using polynomial shape functions. The stress and displacement interpolations may be expressed in matrix equation forms as

$$
\begin{aligned}
&\left\{\boldsymbol{\sigma}^m\right\}=\left[\mathbf{P}^m(x,y)\right]\left\{\boldsymbol{\beta}^m\right\}\subset\Omega_m^{\text{VCE}} \text{ and } \left\{\boldsymbol{\sigma}^c\right\}=\left[\mathbf{P}^c(x,y)\right]\left\{\boldsymbol{\beta}^c\right\}\subset\Omega_c^{\text{VCE}},\\
&\left\{\mathbf{u}^e\right\}=\left[\mathbf{L}^e\right]\left\{\mathbf{q}^e\right\} \text{ on } \partial\Omega_e^{\text{VCE}} \text{ and } \left\{\mathbf{u}^c\right\}=\left[\mathbf{L}^c\right]\left\{\mathbf{q}^c\right\} \text{ on } \partial\Omega_c^{\text{VCE}},
\end{aligned}
\tag{3.23}
$$

where $\{\boldsymbol{\beta}\ \},\{\boldsymbol{\beta}\ \},[\mathbf{P}\ (x,y)]$ and $[\mathbf{P}\ (x,y)]$ are the stress coefficients and interpolating functions, and $\{\mathbf{q}^e\}$, $\{\mathbf{q}^c\}$, $[\mathbf{L}^e]$ and $[\mathbf{L}^c]$ are the nodal displacement vectors and the corresponding interpolation matrices, respectively. A complementary energy functional for a Voronoi cell element may be defined as

$$
\begin{aligned}
&\Pi_e^{\text{VCE}}(\boldsymbol{\sigma}^m,\boldsymbol{\sigma}^c,\mathbf{u}^e,\mathbf{u}^c)=\int_{\Omega_m^{\text{VCE}}}\frac{1}{2}\boldsymbol{\sigma}^m:\mathbf{S}^m:\boldsymbol{\sigma}^m\,\mathrm{d}\Omega+\int_{\Omega_c^{\text{VCE}}}\frac{1}{2}\boldsymbol{\sigma}^c:\mathbf{S}^c:\boldsymbol{\sigma}^c\,\mathrm{d}\Omega\\
&-\int_{\partial\Omega_e^{\text{VCE}}}\boldsymbol{\sigma}^m\cdot\mathbf{n}^e\cdot\mathbf{u}^e\,\mathrm{d}\partial\Omega+\int_{\Gamma_t^{\text{VCE}}}\overline{\mathbf{t}}\cdot\mathbf{u}^e\,\mathrm{d}\Gamma+\int_{\partial\Omega_c^{\text{VCE}}}(\boldsymbol{\sigma}^m-\boldsymbol{\sigma}^c)\cdot\mathbf{n}^c\cdot\mathbf{u}^c\,\mathrm{d}\partial\Omega\\
&\forall\Omega_e^{\text{VCE}}=\Omega_m^{\text{VCE}}\cup\Omega_c^{\text{VCE}},
\end{aligned}
\tag{3.24}
$$

where $\mathbf{S}^m$ and $\mathbf{S}^c$ are the elastic compliance tensors in the matrix phase Ω_m^{VCE} and inclusion phase Ω_c^{VCE} of each element, $\mathbf{n}^e$ and $\mathbf{n}^c$ are the outward normals on $\partial\Omega_e^{\mathrm{VCE}}$ and interfaces $\partial\Omega_c^{\mathrm{VCE}}$, respectively, and $\bar{\mathbf{t}}$ is the prescribed traction on the boundary $\Gamma_{tm}^{\mathrm{VCE}}$. The corresponding total energy functional for the ensemble of all Voronoi cell elements in the domain is

$$\Pi_{\mathrm{Total}}^{\mathrm{VCE}} = \sum_{e=1}^{N_{\mathrm{VCE}}} \Pi_e^{\mathrm{VCE}} = \sum_{e=1}^{N_{\mathrm{VCE}}} \int_{\Omega_e^{\mathrm{VCE}}} \frac{1}{2}\boldsymbol{\sigma}:\mathbf{S}:\boldsymbol{\sigma}\,\mathrm{d}\Omega - \sum_{e=1}^{N_{\mathrm{VCE}}} \int_{\partial\Omega_e^{\mathrm{VCE}}} \boldsymbol{\sigma}\cdot\mathbf{n}^e\cdot\mathbf{u}^e\,\mathrm{d}\partial\Omega$$
$$+\sum_{e=1}^{N_\Gamma} \int_{\Gamma_t^{\mathrm{VCE}}} \bar{\mathbf{t}}\cdot\mathbf{u}^e\,\mathrm{d}\Gamma + \sum_{e=1}^{N_{\mathrm{VCE}}} \int_{\partial\Omega_c^{\mathrm{VCE}}} (\boldsymbol{\sigma}^m - \boldsymbol{\sigma}^c)\cdot\mathbf{n}^c\cdot\mathbf{u}^c\,\mathrm{d}\partial\Omega. \quad (3.25)$$

Setting the first variations of Π^{VCE} with respect to the element stresses $\boldsymbol{\sigma}^m, \boldsymbol{\sigma}^c$, and also the first variations of $\Pi_{\mathrm{Total}}^{\mathrm{VCE}}$ with respect to the displacements $\mathbf{u}^e$ and $\mathbf{u}^c$ to zero results in the following equations, respectively,

$$\int_{\Omega_m^{\mathrm{VCE}}} \boldsymbol{\sigma}^m:\mathbf{S}^m:\delta\boldsymbol{\sigma}^m\,\mathrm{d}\Omega - \int_{\partial\Omega_e^{\mathrm{VCE}}} \delta\boldsymbol{\sigma}^m\cdot\mathbf{n}^e\cdot\mathbf{u}^e\mathrm{d}\partial\Omega + \int_{\partial\Omega_c^{\mathrm{VCE}}} \delta\boldsymbol{\sigma}^m\cdot\mathbf{n}^c\cdot\mathbf{u}^c\mathrm{d}\partial\Omega = 0 \quad \text{(a)}$$
$$\int_{\Omega_m^{\mathrm{VCE}}} \boldsymbol{\sigma}^c:\mathbf{S}^c:\delta\boldsymbol{\sigma}^c\,\mathrm{d}\Omega - \int_{\partial\Omega_c^{\mathrm{VCE}}} \delta\boldsymbol{\sigma}^c\cdot\mathbf{n}^c\cdot\mathbf{u}^c\,\mathrm{d}\partial\Omega = 0 \quad \text{(b)}$$
$$-\sum_{e=1}^{N_{\mathrm{VCE}}} \int_{\partial\Omega_e^{\mathrm{VCE}}} \boldsymbol{\sigma}^m\cdot\mathbf{n}^e\cdot\delta\mathbf{u}^e\,\mathrm{d}\partial\Omega + \sum_{e=1}^{N_{\mathrm{VCE}}} \int_{\Gamma_t^{\mathrm{VCE}}} \bar{\mathbf{t}}\cdot\delta\mathbf{u}^e\,\mathrm{d}\Gamma = 0 \quad \text{(c)} \qquad (3.26)$$
$$\sum_{e=1}^{N_{\mathrm{VCE}}} \int_{\partial\Omega_c^{\mathrm{VCE}}} (\boldsymbol{\sigma}^m - \boldsymbol{\sigma}^c)\cdot\mathbf{n}^c\cdot\delta\mathbf{u}^c\,\mathrm{d}\partial\Omega = 0 \quad \text{(d)}$$

Equations (3.26a) and (3.26b) correspond to weak forms of the kinematic relations in the matrix and inclusion domains of each element, respectively. Solving these equations yields the relations between domain stresses and the boundary/interface displacements. Equations (3.26c) and (3.26d) correspond to weak forms of the traction reciprocity condition at the element boundary and the matrix–inclusion interface, respectively. The boundary and interface displacements can be determined by substituting the stress–displacement relations from (3.26a) and (3.26b), into (3.26c) and (3.26d) and solving them. The reader is referred to [26, 34, 46, 48–50] for details on VCFEM.

An adaptive VCFEM is developed in [50] to enhance solution accuracy and convergence of micromechanical solutions. Two error indicators are introduced to facilitate this adaptation:

1. *Traction reciprocity error on element boundaries and internal interfaces.* To estimate the quality of solution induced by the weak satisfaction of traction continuity on the Voronoi cell element boundary, an average traction continuity error (ATRE) is defined as

$$\text{ATRE}=\frac{\sum_{\hat{e}=1}^{\hat{N}_e} er_T^{\hat{e}}+\sum_{\hat{e}=1}^{\hat{N}_e} er_T^{\hat{c}}}{\hat{N}_e+\hat{N}_c},\ \text{where}\ \ er_T^{\hat{e}}=\frac{1}{\bar{\sigma}}\left[\frac{\int_{\partial Y_e^{\hat{e}}}\left([[\mathbf{t}]]\cdot[[\mathbf{t}]]\right)^{1/2}\mathrm{d}\partial Y}{n^*\int_{\partial Y_e^{\hat{e}}}\mathrm{d}\partial Y}\right]$$

$$\text{and}\quad er_T^{\hat{c}}=\frac{1}{\bar{\sigma}}\left[\frac{\int_{\partial Y_e^{\hat{c}}}\left([[\mathbf{t}]]\cdot[[\mathbf{t}]]\right)^{1/2}\mathrm{d}\partial Y}{n^*\int_{\partial Y_e^{\hat{c}}}\mathrm{d}\partial Y}\right]. \tag{3.27}$$

Equations (3.27), $\hat{N}_e$ and $\hat{N}_c$ are the total of all segments $\partial Y_e^{\hat{e}}$ and $\partial Y_e^{\hat{c}}$ on all element boundaries $\partial\Omega_e^{\mathrm{VCE}}$ and interfaces $\partial\Omega_c^{\mathrm{VCE}}$, respectively. The stress σ in the denominator is the absolute maximum principal value of the volume-averaged stress tensor in the microstructure, i.e.,

$$\bar{\sigma}_{ij}=\frac{\int_{\partial\Omega}\sigma_{ij}\,\mathrm{d}\partial\Omega}{\int_{\partial\Omega}\mathrm{d}\partial\Omega},$$

n^* is the number of degrees of freedom per node and $[[\mathbf{t}]]$ is the traction discontinuity along element boundaries and interfaces.

2. *Error in kinematic relations in the element matrix and reinforcement phases.* The source of this error is the weak satisfaction of kinematic or compatibility relations in the matrix and inclusion phases of each Voronoi cell element. To quantify this effect, an average strain energy error indicator (ASEE) related to the kinematic relation is defined in [50] as

$$\text{ASEE}=\frac{\sqrt{\sum_{\mathrm{e}=1}^{\mathrm{N}} er_{\mathrm{SE}}^2}}{\mathrm{N}},\quad \text{where}\quad er_{\mathrm{SE}}^2=\frac{1}{SE^2}\left[\int_{\Omega_m^{\mathrm{VCE}}}{}^*\boldsymbol{\sigma}^m\,{}^*\!:\boldsymbol{\varepsilon}^m\,\mathrm{d}\Omega\right.$$

$$+\int_{\Omega^{\mathrm{VCE}}}{}^*\boldsymbol{\sigma}^c:{}^*\boldsymbol{\varepsilon}^c\,\mathrm{d}\Omega-\int_{\partial\Omega_c^{\mathrm{VCE}}}{}^*\boldsymbol{\sigma}^m\cdot\mathbf{n}^e\cdot{}^*\mathbf{u}^e\,\mathrm{d}\partial\Omega+\int_{\partial\Omega_c^{\mathrm{VCE}}}{}^*\boldsymbol{\sigma}^m\cdot\mathbf{n}^c\cdot{}^*\mathbf{u}^c\,\mathrm{d}\partial\Omega$$

$$\left.-\int_{\partial\Omega_c^{\mathrm{VCE}}}{}^*\boldsymbol{\sigma}^c\cdot\mathbf{n}^c\cdot{}^*\mathbf{u}^c\,\mathrm{d}\partial\Omega\right]. \tag{3.28}$$

The variables with superscripts *, i.e., ${}^*\boldsymbol{\sigma}^m, {}^*\boldsymbol{\varepsilon}^m$, ${}^*\mathbf{u}^e$, and ${}^*\mathbf{u}^c$ correspond to the change in element stress, strain, and displacement fields that result from enrichments in the stress interpolations $\boldsymbol{\sigma}^m$ and $\boldsymbol{\sigma}^c$. SE in ASEE is the strain energy of the entire micromechanical domain that is expressed as

$$\mathrm{SE}^2 = \sum_{e=1}^{N} \frac{1}{2} \int_{\Omega_m^{\mathrm{VCE}}} \boldsymbol{\sigma}^m : \boldsymbol{\varepsilon}^m \, \mathrm{d}\Omega + \sum_{e=1}^{N} \frac{1}{2} \int_{\Omega_c^{\mathrm{VCE}}} \boldsymbol{\sigma}^c : \boldsymbol{\varepsilon}^c \, \mathrm{d}\Omega. \tag{3.29}$$

er_{SE} is the element level error estimator in strain energy. The latter estimator is a measure of the change in strain energy due to stress enrichment and, hence, is positive for positive definite stiffness matrices. Adaptation for enhancing the rate of convergence of VCFEM solutions is executed in two stages. In the first stage, the traction continuity error in (3.27) is minimized by selectively enhancing boundary and interface displacement degrees of freedom in the directions of optimal displacement enrichments. In the second stage, stress function enrichment with higher order polynomial terms (${}^{enr}p$-adaptation) of each element is performed for reducing the strain energy error [32].

The effectiveness of the adaptive Voronoi cell finite element model is tested by comparison with a well-known problem in micromechanics solved in [58, 67]. The boundary value problem, schematically illustrated in Fig. 3.8, demonstrates the limitations of the effective modulus theory in predicting stress states in laminated composites near a free edge. The composite has two rows of reinforcement, each consisting of eight aligned cylindrical boron fibers aligned in the z-direction, perpendicular to the plane of the paper. The fiber radius to edge dimension ratio is $r/l = 0.3$. Only a quarter of the cross section is analyzed due to symmetry about the xy and xz planes. The resulting mesh consists of four Voronoi cell elements as shown in Fig. 3.6a. The microstructure is subjected to a constant out-of-plane axial strain $\varepsilon_{zz} = 1$, which is modeled using generalized plane strain conditions. The material properties are given below:

Material	Young's modulus, E (psi)	Poisson ratio
Boron (fiber)	$E_{\mathrm{bo}} = 60 \times 10^6$	$\nu_{\mathrm{bo}} = 0.2$
Epoxy (matrix)	$E_{\mathrm{epoxy}} = 0.5 \times 10^6$	$\nu_{\mathrm{epoxy}} = 0.34$

The analysis considers two boundary conditions:

(a) Edges $x = 0$ and $y = 0$ are symmetry surfaces, while $x = 4h$ and $y = 2h$ are traction free

(b) Edges $x = 0$ and $y = 2h$ are symmetry surfaces and $x = 4h$ and $y = 0$ are traction free

The matrix stresses are constructed from an Airy's stress function consisting of a fourth-order polynomial and a 36 term reciprocal function.

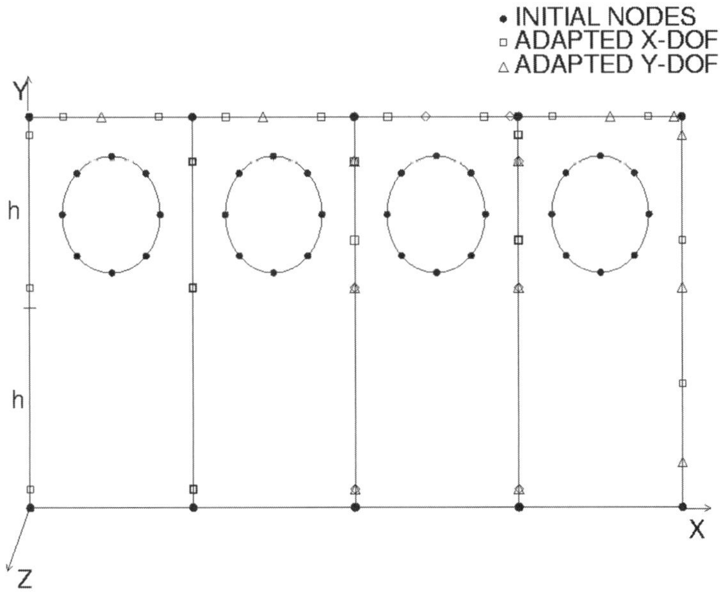

(a)

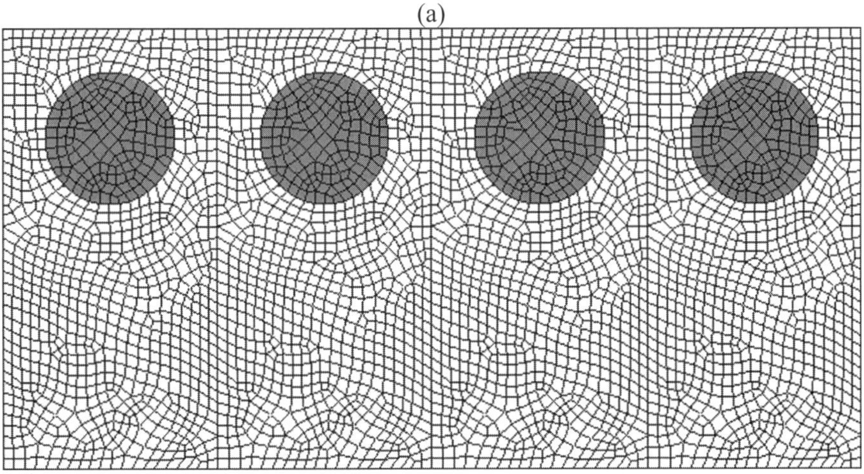

(b)

Fig. 3.6. (**a**) A microstructural hp-adapted VCFEM mesh showing locations of the initial and added nodes with x- and y-DOF – case (**a**); (**b**) ANSYS mesh

The inclusion stresses are constructed using a sixth-order polynomial function. Displacement fields are constructed with linear shape functions for element boundaries and quadratic shape functions for curved interface elements. The results of the adaptation cycle for the problem in case (a) are illustrated in Fig. 3.6 in terms of added displacement degrees of freedom. The preadaptation nodes are marked with a "filled circle." The x-direction nodal adaptations are marked with an "open square" while those in the y-direction are shown with "open triangle." The VCFEM solutions for both cases are compared with numerical results of micromechanical analysis provided in [67] and also with those from analysis by the finite element code ANSYS. The converged ANSYS mesh with 4230 QUAD4 elements and 4352 nodes is shown in Fig. 3.6b.

The transverse microscopic stress σ_{yy}, which is approximately two orders lower compared with the leading order stress σ_{zz}, is plotted along the horizontal section at $y = h$ in Fig. 3.7. Results are compared for the unadapted VCFEM, *hp*-adapted VCFEM, and the ANSYS model. The adapted VCFEM results agree very well with the ANSYS model. The figure also shows the singular stress solution near the free edge, which is obtained as a consequence of using the effective modulus theory by homogenization.

Statistically equivalent RVEs for nonuniform microstructures

Identification of the appropriate representative volume element or RVE is essential for estimating homogenized material properties needed in the computations of *level-0* or *level-1* elements. An RVE is not easily identifiable for material microstructures with nonuniform morphological distributions as shown in Fig. 3.8a. It is possible to identify an RVE only in a statistical sense, otherwise called a statistically equivalent RVE or SERVE. The SERVE is expected to exhibit a macroscopic behavior that is equivalent to the average behavior of the entire microstructural ensemble. A variety of statistical and computational tools are developed for identifying the SERVE for elastic composites with nonuniform dispersion of inclusions in [72]. The evolution of the SERVE with microstructural damage by interfacial debonding is examined in [73] using various metrics. As an example, the marked correlation function introduced in [60] is used in this chapter to delineate the SERVE size. This function characterizes the region of influence of a chosen heterogeneity on others in a domain with respect to chosen variables like stresses, strains, etc. The marked correlation function for a domain of an area A containing N inclusions is expressed as

$$M(r) = \frac{\mathrm{d}\dfrac{H(r)}{\mathrm{d}r}}{g(r)}, \quad \text{where} \quad H(r) = \frac{1}{m^2}\frac{A}{N^2}\sum_{i}^{N}\sum_{k=1}^{k_i} m_i m_k(r). \tag{3.30}$$

In (3.30), m_i is a mark associated with the ith inclusion, k_i is the number of inclusions which have their centers within a circle of radius r around the ith inclusion, and m is the mean of all the marks. Marks can be any field variable like the maximum principal stress or Von Mises stress or even a geometric feature associated with each inclusion. $H(r)$ is called the mark intensity function and $g(r)$ is the pair distribution function defined as

$$g(r) = \frac{1}{2\pi r}\frac{\mathrm{d}K(r)}{\mathrm{d}r}, \tag{3.31}$$

where $K(r)$ is a second-order intensity function explained in [32, 33, 72]. The radius of influence R_{inf} can be determined from a plot of $M(r)$ vs. r. R_{inf} corresponds to the value of r at which $M(r)$ stabilizes to a constant value. Upon determining R_{inf}, the SERVE may be constructed from the inclusions contained in a $R_{\text{inf}} \times R_{\text{inf}}$ square window of the micrograph. As shown in Fig. 3.8b, the local microstructure is first constructed by periodically repeating the set of inclusions that lie (wholly or partially) in the window in both the y_1 and y_2 directions for several period lengths. For each fiber at (y_1,y_2), periodically repetitive inclusions are placed at $(y_1 \pm k_1 Y_1, y_2)$, $(y_1, y_2 \pm k_2 Y_2)$, and $(y_1 \pm k_1 Y_1, y_2 \pm k_2 Y_2)$, where k_1, k_2 are integers. The resulting domain is then tessellated into a network of Voronoi cells as shown in Fig. 3.8b. The SERVE boundary, shown with bold lines in Fig. 3.8b, is the aggregate of all outside edges of Voronoi cells that are associated with the primary inclusions in the domain (shown in black). For nonuniform inclusion arrangements, the SERVE boundary will consist of multiple nonaligned edges. Nodes on the SERVE boundary created by this procedure are periodic. For every boundary node, a periodic pair, e.g., AA, BB, etc., can be identified on the boundary at a distance of one period along one or both of the coordinate directions. Periodicity constraint conditions on nodal displacements can then be easily imposed.

A numerical example is considered to demonstrate the effect of the SERVE size on macroscopic properties as well as on microscopic stresses. Maximum principal stress in the fiber and maximum Von Mises stress in matrix in each Voronoi cell are considered as *marks* in the correlation function, since they are good indicators of microstructural failure initiation.

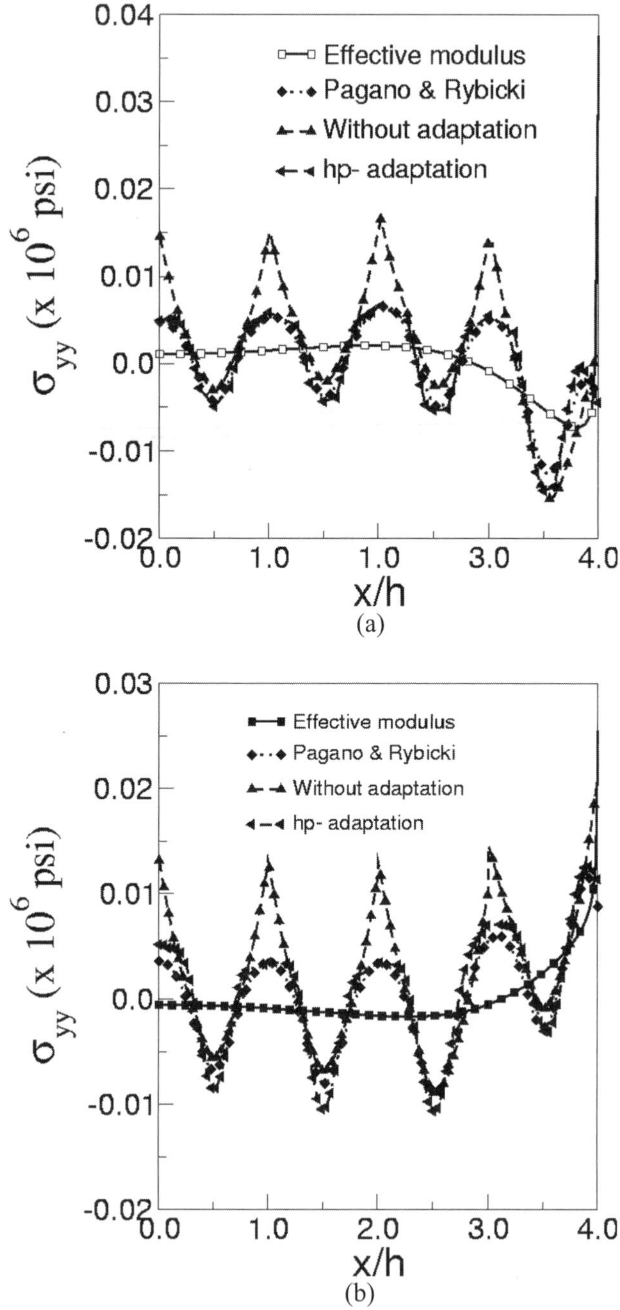

Fig. 3.7. σ_{yy} distribution at section A–A ($y = h$) for the composite section with one row of fiber for (**a**) case-a and (**b**) case-b

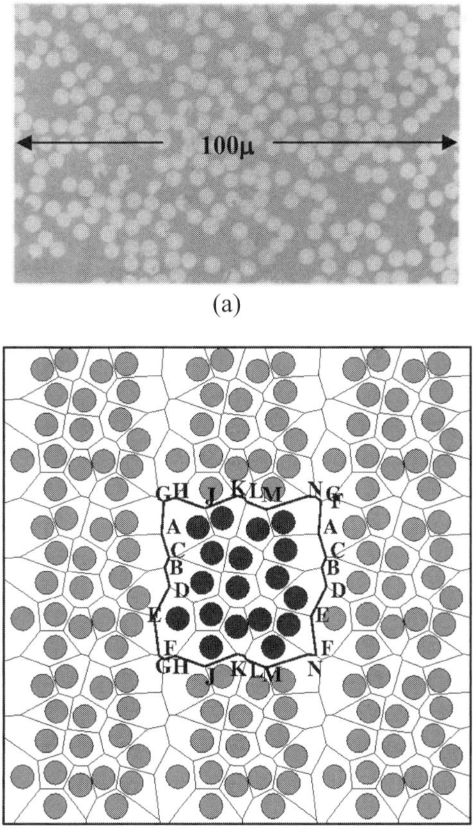

(a)

(b)

Fig. 3.8. (**a**) Optical micrograph of a polymer–matrix composite microstructure; (**b**) an RVE evolving from tessellation of microstructure with nonstraight edges

Plots of $M(r)$ for different marks are shown in Fig. 3.9a. The distance r is normalized with respect to the fiber radius $r_0 = 1.75\,\mu\text{m}$. $M(r)$ is high at distances less than $8r_0$ but stabilizes to a unit value at distances approximately greater than $8r_0$, corresponding to the region of influence of stress. It can also be seen that $M(r)$ plots for both Von Mises stress and principal stress are similar and stabilize approximately in the same radial range r. A similar behavior of $M(r)$ is also observed when the micrograph is loaded under biaxial tension as shown in Fig. 3.9b and, hence, $8r_0$ characterizes the size scale of the SERVE.

Convergence of macroscopic moduli and maximum microscopic stress with RVE size are studied for five different RVEs, consisting of 1, 8, 18, 35, and 55 fibers as shown in Fig. 3.10. The corresponding SERVE sizes

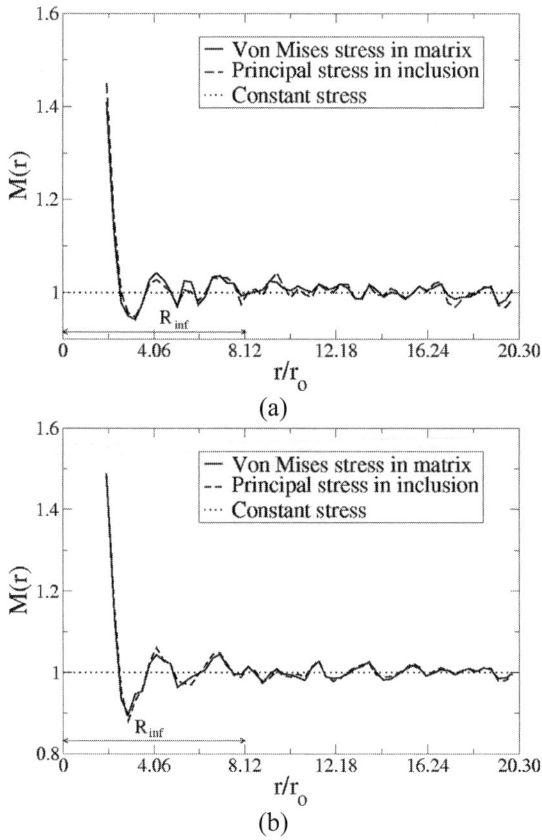

Fig. 3.9. Marked correlation functions $M(r)$ for (**a**) uniaxial and (**b**) biaxial load

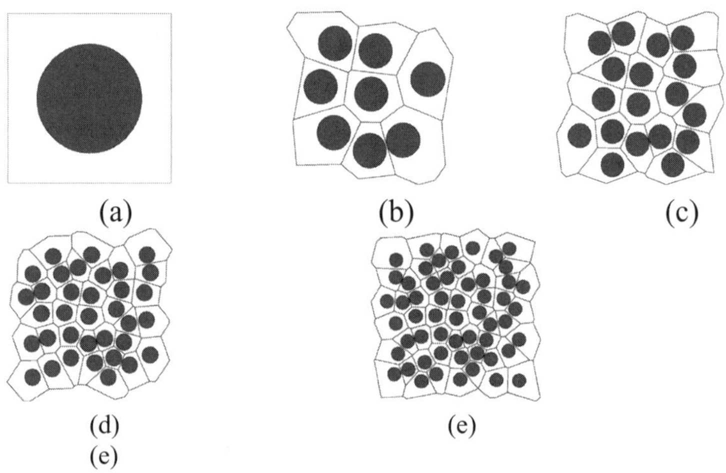

Fig. 3.10. SERVEs with 1, 8, 18, 35, and 55 fibers

are $r_0, 3r_0, 6r_0, 9r_0$, and $12r_0$, respectively. The morphology of each RVE is chosen from any arbitrary region in the micrograph. The Frobenius norm of the effective elastic modulus $\|E\|$ is plotted as a function of increasing RVE sizes at two different locations in the microstructure in Fig. 3.11a. The difference in the norm between the single fiber and 55 fibers is around 2%, while the difference between 18 fibers and 55 fibers is found to be less than 0.5%. A similar observation is also made when comparing the microscopic maximum Von Mises stress in the matrix and maximum principal stress in the fiber as functions of increasing RVE size in Fig. 3.11b.

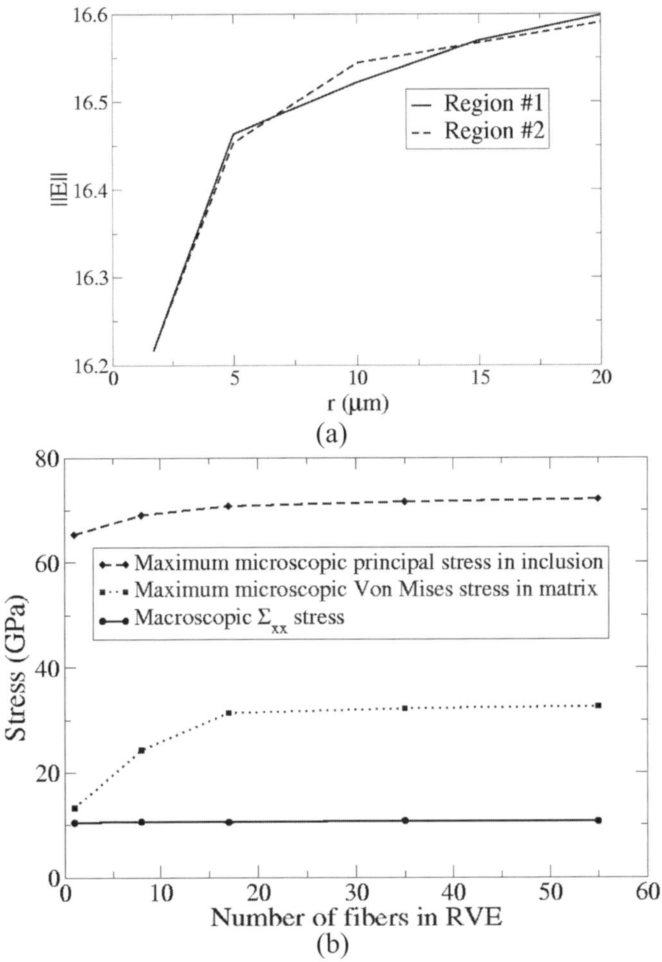

Fig. 3.11. Convergence of the (**a**) macroscopic stiffness and (**b**) microscopic stresses with increasing RVE sizes

The difference in maximum Von Mises stress in the matrix for the single fiber RVE and 55 fiber RVE is almost 60%, whereas the corresponding difference for the 18 fiber RVE and the 55 fiber RVE is less than 4%. Hence, a SERVE consisting of 18 fibers is deemed adequate for homogenization.

Computational subdomain level-1 Ω^{l1}

The *level-1* subdomains facilitate switchover from homogenization-based analysis in *level-0* subdomains to micromechanical analysis in *level-2* subdomains. They are seeded in regions of locally high gradients of macroscopic variables in *level-0* simulations. The formulation for Ω^{l1} is the same as for Ω_{hom} in Sect. 3.3.1. It serves as a "swing" region, where microscopic variables in the SERVE, as well as macroscopic gradients, are used to decide whether homogenization is valid in this region. Major steps in *level-1* element computations are:

1. Evaluate homogenized elastic stiffness C_{ijkl}^{H} for the macroscopic analysis using (3.10) with applied macroscopic unit strains corresponding to (3.15) together with periodicity.
2. Evaluate element stiffness and load vectors for elements E_{l0} and E_{l1} using C_{ijkl}^{H}, and solve the global FE equations for macroscopic displacements, stresses, and strains.
3. Perform RVE analysis in the postprocessing stage with macroscopic strains $e_{ij} = (1/2)((\partial u_i^0/\partial x_j) + (\partial u_j^0/\partial x_i))$ imposed from Step 2 and periodic boundary conditions. Microscopic stresses, strains, and other variables are computed in the RVE of every element $\left(E_{l1} \in \Omega^{l1}\right)$ for developing appropriate level switching criteria.

Macroscopic elements in $\Omega^{l1}\left(E_{l1}\right)$ are also adaptively enriched by *h*- and *hp*-refinement. No special treatment is required for displacement compatibility between E_{l0} and E_{l1} elements, since their boundaries are similar with identical displacement interpolation.

Criteria for level-0 to level-1 transition

Elements in the computational subdomain $\Omega^{l1}\left(E_{l1} \in \Omega^{l1}\right)$ are computationally much more expensive than *level-0* elements E_{l0}. Hence, the selection of appropriate criteria for switching from E_{l0} to E_{l1} elements is critical to enhance efficiency by optimally limiting the number of E_{l1} elements. These criteria depend on the important variables for the problem in question. Various switching criteria, based on gradients of physically significant stress measures, have been tested in [29, 36, 61]. As an example, element *k* will be required to undergo a *level* $0 \rightarrow 1$ transition if

$$E_k\left(\frac{\left(\Sigma_{\text{eqv}}\right)_k}{\left(\Sigma_{\text{eqv}}\right)_{\max}}\right) \geq C_1 E_{\text{avg}}, \quad \text{where} \quad E_{\text{avg}} = \left(\frac{\sum_{i=1}^{N_E} E_i^2}{NE}\right)^{1/2}. \tag{3.32}$$

Here C_1 is a prescribed tolerance, Σ_{eqv} is the equivalent stress in element *k*, and E_k can have one of the following forms:

$$\begin{aligned}
&\text{(a)} \quad E_k^2 = \frac{\int_{\partial\Omega^k_{\text{hom}}} [[\Sigma_{ij}]]^2 \, \mathrm{d}\partial\Omega}{\int_{\partial\Omega^k_{\text{hom}}} \mathrm{d}\partial\Omega}, \\
&\text{(b)} \quad E_k^2 = \frac{\int_{\partial\Omega^k_{\text{hom}}} [[\Sigma_{\text{pr}}]]^2 \, \mathrm{d}\partial\Omega}{\int_{\partial\Omega^k_{\text{hom}}} \mathrm{d}\partial\Omega}, \\
&\text{(c)} \quad E_k^2 = \frac{\int_{\partial\Omega^k_{\text{hom}}} \left([[T_x]]^2 + [[T_y]]^2\right) \mathrm{d}\partial\Omega}{\int_{\partial\Omega^k_{\text{hom}}} \mathrm{d}\partial\Omega}.
\end{aligned} \tag{3.33}$$

Σ_{ij} is a chosen macroscopic stress component, Σ_{pr} is the dominant principal stress, T_x, T_y are the element boundary traction components and $[[\cdots]]$ denotes the jump operator. The criterion in (3.32) reflects the fact that high gradients in regions of high stress levels are more relevant than those at low stress levels.

Computational subdomain level-2 Ω^{l2}

Level-2 subdomains of detailed microscopic analysis are characterized by the nonsatisfaction of homogenization conditions used in the *level-0* and *level-1* subdomains. Microstructural nonuniformities in the form of strongly nonperiodic, e.g., clustered dispersions or concentrated high stresses and strains with high gradients, occurring near a crack tip or free edge, necessitate the emergence of Ω^{l2}. Appropriate adaptation criteria are used to trigger switching from $E_{l1}\left(\in \Omega^{l1}\right)$ to $E_{l2}\left(\in \Omega^{l2}\right)$ elements for micromechanical analysis. It is expected that the local *h*- or *hp*-refinement in *level-0* or *level-1* elements will have reduced the size of these elements sufficiently prior to transition to the *level-2* elements such that a high spatial resolution is locally attained.

The elements E_{l2} are constructed by filling the *level-0/1* elements with the exact microstructure at that location. The region Ω_{l2}^{k} encompassed in the *k*th *level-2* element E_{l2}^{k} is obtained as the intersection of the local microstructural region Ω^{ε} with the *k*th *level-1* element Ω_{l1}^{k}, i.e., $\Omega_{l2}^{k} = \Omega^{\varepsilon} \cap \Omega_{l1}^{k}$. The steps in creating a *level-2* element are as follows:

- Identify a region $\Omega_{\hat{k}} \in \Omega_{\text{het}}$ that is located in the same region as Ω_{l2}^{k} and extends beyond it by at least two fiber lengths.
- Tessellate $\Omega_{\hat{k}}$ to generate a mesh of Voronoi cell elements as shown in Fig. 3.12a.
- Carve out the region Ω_{l2}^{k} by superposing the boundary of Ω_{l1}^{k} on $\Omega_{\hat{k}}$. This procedure will result in dissecting some of the fibers on the boundary of Ω_{l2}^{k}. When this happens, additional nodes are generated on the Voronoi cell boundary at locations where the fiber surface and Voronoi cell edges intersect the boundary of Ω_{l2}^{k}.

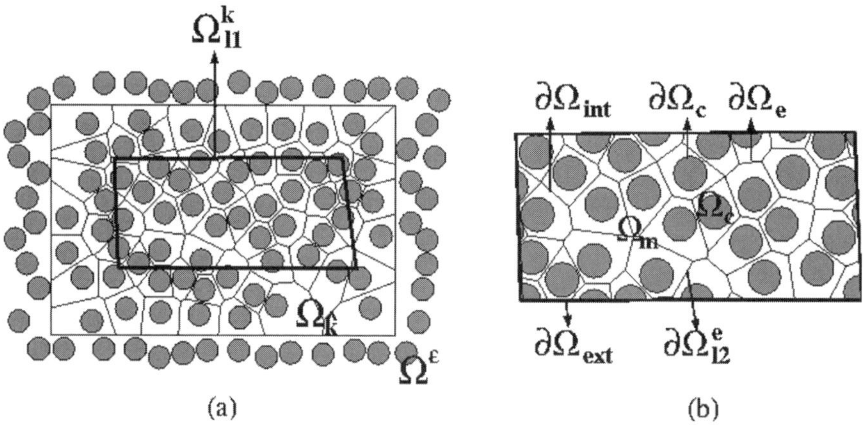

Fig. 3.12. (**a**) Carving out extended microstructural region in level-2 element; (**b**) level-2 element consisting of VC finite elements for microstructural modeling

Accurate, high-resolution modeling in these elements may require prohibitively high computing efforts with conventional finite element methods. The adaptive Voronoi cell finite element model [26, 34, 48–50] is, therefore, preferred for efficient micromechanical analysis.

Criteria for switching from level-1 to level-2

Departure from periodicity conditions in the microstructural SERVE is taken as an indicator for the *level-1→level-2* transition. This is in addition to the local gradients in macroscopic variables for *level-0→level-1* transition. A criterion for invoking the *level-1→level-2* change is defined as

$$\frac{\hat{\mathbf{F}}(\sigma_{ij},\varepsilon_{ij})^{l1}-\hat{\mathbf{F}}(\sigma_{ij},\varepsilon_{ij})^{\mathrm{RVE}}}{\hat{\mathbf{F}}(\sigma_{ij},\varepsilon_{ij})^{\mathrm{RVE}}}\geq C_2. \tag{3.34}$$

The function $\hat{\mathbf{F}}$ is a measure of a quantity of interest in terms of local variables $(\sigma_{ij},\varepsilon_{ij})$. In some of the numerical examples, $\hat{\mathbf{F}}$ is expressed as the average inclusion stress in the microstructure. The superscript $l1$ in (3.34) refers to element E^k_{l1}. The microstructural boundary value problem is solved with macroscopic displacement solutions from *level-0* imposed on E^k_{l1} boundary. The superscript RVE corresponds to the function being evaluated within each RVE only, by imposing macroscopic strains with periodic boundary conditions on the RVE.

Other criteria have also been used in this level transition. Among them are:

1. *Criterion based on strain energy density.* The ratio of local strain energy density to the average energy density in the RVE is important in the prediction of localization. The criterion suggests that if local strain energy density due to multiaxial straining significantly exceeds that due to uniaxial straining used in the evaluation of homogenization parameters, the onset of damage is likely. Hence *level-1 to level*-2 transition is made if

$$^{\mathrm{actual}}U^M_{\max}\geq R^M_{\max}\times {}^{\mathrm{actual}}U^M_{\mathrm{aver}}\quad \text{or}\quad {}^{\mathrm{actual}}U^I_{\max}\geq R^I_{\max}\times {}^{\mathrm{actual}}U^I_{\mathrm{aver}} \tag{3.35}$$

at more than 1% of all integration points. Here, $U^M=(1/2)S^M_{ijkl}\sigma^M_{ij}\sigma^M_{kl}$, $U^I=(1/2)S^I_{ijkl}\sigma^I_{ij}\sigma^I_{kl}$, and the energy density concentration factors are $R^M=U^M_{\max}/U^M_{\mathrm{aver}}$ and $R^I=U^I_{\max}/U^I_{\mathrm{aver}}$ for unit strain components. $U^M_{\max}$ and $U^I_{\max}$ are the maximum values of U^M and U^I at all integration points in the RVE, and U^M_{aver} and U^I_{aver} are the corresponding RVE-averaged energy densities. The maximum values for the four loading cases are noted as $R^M_{\max}$ and $R^I_{\max}$.

2. *Criterion based on equivalent stress*. In this criterion, *level-1→level-2* transition is made if the local equivalent stress exceeds the average, i.e.,

$$\left(\sigma^m_{\text{eqv}}\right)_{\max} > C_3 * \left(\sigma^m_{\text{eqv}}\right)_{\text{avg}}, \quad \left(\sigma^c_{\text{eqv}}\right)_{\max} > C_3 * \left(\sigma^c_{\text{eqv}}\right)_{\text{avg}} \tag{3.36}$$

at more than 1% of all integration points. $\left(\sigma^m_{\text{eqv}}\right)_{\max}, \left(\sigma^m_{\text{eqv}}\right)_{\text{avg}}$ and $\left(\sigma^c_{\text{eqv}}\right)_{\max}, \left(\sigma^c_{\text{eqv}}\right)_{\text{avg}}$ represent the maximum and average equivalent stresses in the matrix and inclusion phases.

3. *Criterion based on traction at the fiber–matrix interface*. Traction at the fiber–matrix interface is important for predicting failure by debonding. This criterion is postulated as *level-1→level-2* transition occurs if

$$|\hat{T}| > C_4 * |\hat{T}|_{\text{avg}} \quad \text{where} \quad |\hat{T}| = \left(\sqrt{T_n^2 + T_t^2}\right), \tag{3.37}$$

where $|\hat{T}|_{\text{avg}} = \dfrac{\sum_{i=1}^{N_I} |\hat{T}|}{N_I}$ is the average traction at the fiber–matrix interface and N_I is the total number of integration points on interface in the RVE. C_1, C_2, C_3, C_4 are chosen from numerical experiments.

Transition elements between elements in Ω_{l2} and $\Omega_{l1/l0}$

To facilitate gradual transition of scales across the element boundaries, a layer of transition elements $E_{\text{tr}} \left(\in \Omega^{\text{tr}}\right)$ is sandwiched between the macroscopic elements in $\Omega_{l1/l0}$ and microscopic elements in Ω_{l2}, as shown in Fig. 3.13. The elements E_{tr} are essentially *level-2* elements with compatibility and traction continuity constraints imposed at the interface with E_{l1} or E_{l0} elements. The transition elements are located beyond the *level-2* regions, away from critical hot spots.

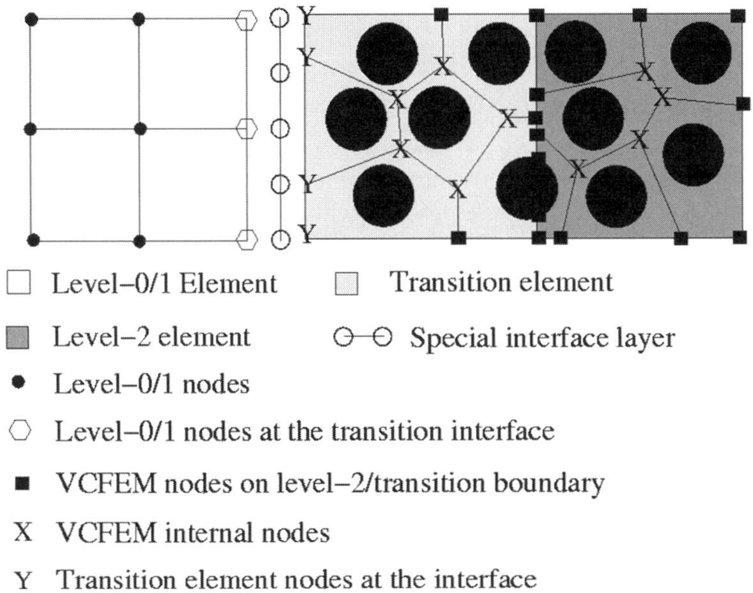

Fig. 3.13. Interface constraints between *level-0/1* and transition elements

A relaxed displacement constraint method is proposed in [36, 62], where a weak form of the interface displacement continuity is incorporated by using Lagrange multipliers, suggested in [4]. The total potential energy of the multilevel computational domain can then be expressed as

$$\Pi = \Pi_{\Omega_{l0}} + \Pi_{\Omega_{l1}} + \Pi_{\Omega_{l2}} + \Pi_{\Omega_{\mathrm{tr}}} + \int_{\Gamma_{\mathrm{int}}} \lambda_i^{l1}(v_i - u_i^{l1})\mathrm{d}\Gamma + \int_{\Gamma_{\mathrm{int}}} \lambda_i^{\mathrm{tr}}(v_i - u_i^{\mathrm{tr}})\mathrm{d}\Gamma, \quad (3.38)$$

where $\Pi_{\Omega_{l0}}, \Pi_{\Omega_{l1}}, \Pi_{\Omega_{l2}}$, and $\Pi_{\Omega_{\mathrm{tr}}}$ are the potential energies for elements in the respective subdomains, λ_i^{l1} and λ_i^{tr} are columns of Lagrange multipliers on the interfacial layer Γ_{int} belonging to Ω_{l1} and Ω_{tr}, respectively, for which the interfacial displacements are designated as u_i^{l1} and u_i^{tr}.

As shown in Fig. 3.13, an intermediate boundary segment is added with displacements v_i that may be interpolated with any order polynomial functions, independent of the interpolations for u_i^{l1} or u_i^{tr}. The Lagrange multipliers λ_i^{l1} and λ_i^{tr} correspond to the interface tractions on $\partial\Omega_{l1}$ and $\partial\Omega_{\mathrm{tr}}$, respectively. The displacements and the Lagrange multipliers on the intermediate boundary segment are interpolated from nodal values using suitably assumed shape functions

$$\{v\} = [L_{\text{int}}]\{\mathbf{q}_{\text{int}}\}, \quad \{\lambda^{l1}\} = [L_{\lambda^{l1}}]\{\boldsymbol{\Lambda}_{l1}\}, \quad \{\lambda^{\text{tr}}\} = [L_{\lambda^{\text{tr}}}]\{\boldsymbol{\Lambda}_{\text{tr}}\}. \tag{3.39}$$

To examine the effectiveness of the relaxed displacement constraint method, a composite laminate problem with two sandwiched lamina is solved. The top lamina consists of a uniform distribution of circular fibers of 30% volume fraction, while the bottom lamina has fibers of 10% volume fraction.

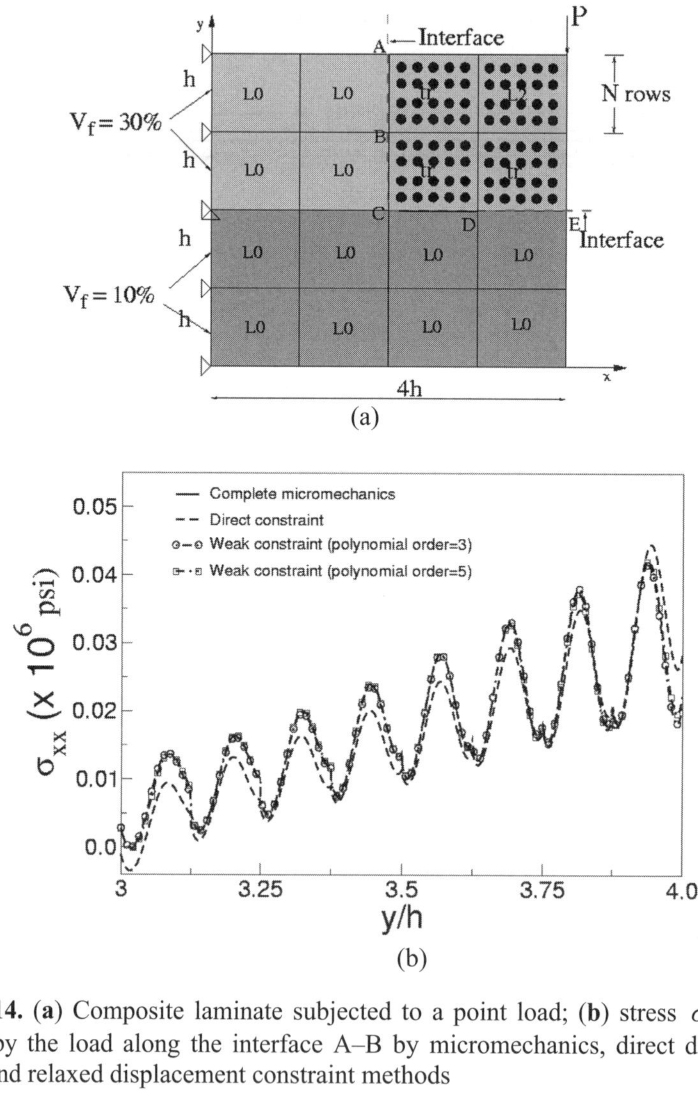

Fig. 3.14. (**a**) Composite laminate subjected to a point load; (**b**) stress σ_{xx} produced by the load along the interface A–B by micromechanics, direct displacement, and relaxed displacement constraint methods

A 10^6 lb point load is applied on the laminate as shown in Fig. 3.14. The fiber material has $E_{\text{fiber}} = 60 \times 10^6$ psi and $\nu_{\text{fiber}} = 0.2$, while the matrix material has $E_{\text{matrix}} = 0.5 \times 10^6$ psi and $\nu_{\text{matrix}} = 0.34$. As shown in the Fig. 3.14a, a portion of the top lamina is modeled using VCFEM-based *level-2* and transition elements.

This region consists of eight rows of fiber and, hence, each *level-2* or transition element may contain up to 64 fibers. The remaining elements are *level-0* with homogenized moduli. In the relaxed displacement constraint method, increasing order polynomials are considered for the displacement interpolation $\left[L_{\text{int}}\right]$ on the intermediate boundary segment between A and E. The shape functions $\left[L_{\lambda^{l1}}\right]$ and $\left[L_{\lambda^{\text{tr}}}\right]$ in (3.39) are assumed to be linear. The critical stress σ_{xx} distribution is shown along the section A–B for different interfacial conditions. The solutions of the multilevel models are compared with that of a fully micromechanical model analyzed by VCFEM. The plots in Fig. 3.14b show that relaxed displacement constraint method yields much better results compared with a direct constraint method done in [61, 62].

3.3.3 Coupling Levels in the Concurrent Multilevel FEM

The global stiffness matrix and load vectors are derived for the multilevel model consisting of *level-0*, *level-1*, *level-2*, and transition elements, i.e., $\Omega_{\text{het}} = \left\{\Omega_{l0} \cup \Omega_{l1} \cup \Omega_{\text{tr}} \cup \Omega_{l2} : \Omega_{l0} = \cup_{k=1}^{N_{l0}} E_{l0} ; \cup_{k=1}^{N_{l1}} E_{l1} ; \cup_{k=1}^{N_{tr}} E_{\text{tr}} ; \cup_{k=1}^{N_{l2}} E_{l2}\right\}$. The boundary is decomposed as $\Gamma_{\text{het}} = \left\{\Gamma_{l0} \cup \Gamma_{l1} \cup \Gamma_{l2} \cup \Gamma_{\text{tr}}\right\}$, where $\Gamma_{l0} = \partial\Omega_{l0} \cap \Gamma_{\text{het}}$, $\Gamma_{l1} = \partial\Omega_{l1} \cap \Gamma_{\text{het}}$, $\Gamma_{\text{tr}} = \partial\Omega_{\text{tr}} \cap \Gamma_{\text{het}}$ and $\Gamma_{l2} = \partial\Omega_{l2} \cap \Gamma_{\text{het}}$. The principle of virtual work equation for the entire multilevel computational domain for multiscale analysis is expressed as

$$
\begin{aligned}
&\int_{\Omega_{10}} \Sigma_{ij} \frac{\partial \delta u_i^{10}}{\partial x_j} \mathrm{d}\Omega - \int_{\Omega_{10}} f_i \delta u_i^{10} \,\mathrm{d}\Omega + \int_{\Omega_{11}} \Sigma_{ij} \frac{\partial \delta u_i^{11}}{\partial x_j} \mathrm{d}\Omega - \int_{\Omega_{11}} f_i \delta u_i^{11} \,\mathrm{d}\Omega \\
&+ \int_{\Omega_{12}} \sigma_{ij} \frac{\partial \delta u_i^{12}}{\partial x_j} \mathrm{d}\Omega - \int_{\Omega_{12}} f_i \delta u_i^{12} \,\mathrm{d}\Omega + \int_{\Omega_{tr}} \sigma_{ij} \frac{\partial \delta u_i^{\text{tr}}}{\partial x_j} \mathrm{d}\Omega - \int_{\Omega_{\text{tr}}} f_i \delta u_i^{\text{tr}} \,\mathrm{d}\Omega \\
&- \int_{\Gamma_{10}} t_i \delta u_i^{10} \,\mathrm{d}\Gamma - \int_{\Gamma_{11}} t_i \delta u_i^{11} \,\mathrm{d}\Gamma - \int_{\Gamma_{12}} t_i \delta u_i^{12} \,\mathrm{d}\Gamma - \int_{\Gamma_{\text{tr}}} t_i \delta u_i^{\text{tr}} \,\mathrm{d}\Gamma \\
&+ \delta \int_{\Gamma_{\text{int}}} \lambda_i^{10/11} (v_i - u_i^{10/11}) \mathrm{d}\Gamma + \delta \int_{\Gamma_{\text{int}}} \lambda_i^{\text{tr}} (v_i - u_i^{\text{tr}}) \mathrm{d}\Gamma = 0.
\end{aligned}
\tag{3.40}
$$

The traction continuity between *level-0* and *level-1*, as well as *level-2* and transition elements are satisfied in a weak sense. The boxed terms in (3.40) involve integration over the microstructural domains Ω_{l2} and Ω_{tr}, and are analyzed using VCFEM, described in Sect. 3.2.2. It is necessary to couple these terms with the other terms using homogenized properties, analyzed by conventional finite element models. To make the connection with the macroscopic elements in the model, the total energy in the ensemble of Voronoi cell elements in (3.25) is identified as the energy of the *level-2* or transition elements, i.e., $\Pi^{E_{l2}/E_{tr}} = \Pi^{VCE}_{Total}$. Furthermore, element boundaries of all Voronoi cell elements are split as $\sum_{e=1}^{N_{VCE}} \partial\Omega_e^{VCE} = \partial\Omega^{ext} \bigcup \partial\Omega^{int}$. $\partial\Omega^{ext}$ is the aggregate of all Voronoi element boundaries that coincides with *level-2* or transition element boundaries, shown with thicker lines in Fig. 3.12b, and $\partial\Omega^{int}$ corresponds to all other internal boundaries of VC elements. Substitution in (3.26c) yields

$$-\int_{\partial\Omega^{int}} \boldsymbol{\sigma}\cdot\mathbf{n}^e\cdot\delta\mathbf{u}^e\,\mathrm{d}\partial\Omega - \int_{\partial\Omega^{ext}} \boldsymbol{\sigma}\cdot\mathbf{n}^e\cdot\delta\mathbf{u}^e\,\mathrm{d}\partial\Omega + \sum_{e=1}^{N_\Gamma}\int_{\Gamma_t^{VCE}} \bar{\mathbf{t}}\cdot\delta\mathbf{u}^e\,\mathrm{d}\Gamma = 0. \quad (3.41)$$

In the absence of body forces, the boxed terms in (3.40) corresponding to the micromechanical energy in each *level-2 or transition* element can be restated by using divergence theorem as

$$\begin{aligned}
&\int_{E_{12}/E_{tr}} \boldsymbol{\sigma}\cdot\nabla\delta\mathbf{u}^{12/tr}\,\mathrm{d}\Omega - \int_{\Gamma_{12}/\Gamma_{tr}} \mathbf{t}\cdot\delta\mathbf{u}^{12/tr}\,\mathrm{d}\Gamma = \int_{\partial E_{12}/\partial E_{tr}} \boldsymbol{\sigma}\cdot\mathbf{n}^e\cdot\delta\mathbf{u}^{12/tr}\,\mathrm{d}\partial\Omega \\
&-\int_{E_{12}/E_{tr}} \nabla\boldsymbol{\sigma}\cdot\delta\mathbf{u}^{12/tr}\,\mathrm{d}\Omega - \int_{\Gamma_{12}/\Gamma_{tr}} \mathbf{t}\cdot\delta\mathbf{u}^{12/tr}\,\mathrm{d}\Gamma = \int_{\partial\Omega^{ext}} \boldsymbol{\sigma}\cdot\mathbf{n}^e\cdot\delta\mathbf{u}^{12/tr}\,\mathrm{d}\partial\Omega \qquad (3.42)\\
&-\sum_{e=1}^{N_{VCE}}\int_{\Omega_e^{VCE}} \nabla\boldsymbol{\sigma}\cdot\delta\mathbf{u}^{12/tr}\,\mathrm{d}\Omega - \sum_{e=1}^{N_\Gamma}\int_{\Gamma_t^{VCE}} \mathbf{t}\cdot\delta\mathbf{u}^{12/tr}\mathrm{d}\Gamma.
\end{aligned}$$

The term containing $\nabla\boldsymbol{\sigma}$ in (3.42) drops out, since equilibrated stress fields are used in VCFEM. It should be noted that the boundary vector $\mathbf{u}^{12/tr}$ is a subset of the VCFEM boundary displacements $\mathbf{u}^e$. The first term on the right-hand side is the contribution to the global stiffness and is obtained from VCFEM analysis by using static condensation in (3.41) to remove internal degrees of freedom on $\partial\Omega^{int}$ from the global stiffness. The displacement field along the edges of VCFEM elements is interpolated as

$$\{\mathbf{u}^e\} = [L_{VCE}]\{\mathbf{q}_{VCE}\}. \quad (3.43)$$

The degrees of freedom $\mathbf{q}_{\mathrm{VCE}}$ can be separated into $\mathbf{q}_{\mathrm{VCE}}^{\mathrm{ext}}$ and $\mathbf{q}_{\mathrm{VCE}}^{\mathrm{int}}$ depending on whether they belong to $\partial\Omega^{\mathrm{ext}}$ or $\partial\Omega^{\mathrm{int}}$, respectively. The stiffness matrix and the load vector of the ensemble of all Voronoi cell elements belonging to a *level-2* element or transition element can thus be partitioned as

$$\begin{pmatrix} K_{\mathrm{VCE}}^{\mathrm{ext,ext}} & K_{\mathrm{VCE}}^{\mathrm{ext,int}} \\ K_{\mathrm{VCE}}^{\mathrm{int,ext}} & K_{\mathrm{VCE}}^{\mathrm{int,int}} \end{pmatrix} \begin{Bmatrix} \mathbf{q}_{\mathrm{VCE}}^{\mathrm{ext}} \\ \mathbf{q}_{\mathrm{VCE}}^{\mathrm{int}} \end{Bmatrix} = \begin{Bmatrix} \mathbf{F}_{\mathrm{VCE}}^{\mathrm{ext}} \\ \mathbf{F}_{\mathrm{VCE}}^{\mathrm{int}} \end{Bmatrix}. \tag{3.44}$$

Static condensation of the internal degrees of freedom leads to

$$\begin{aligned} &\left[\left[K_{\mathrm{VCE}}^{\mathrm{ext,ext}}\right] - \left[K_{\mathrm{VCE}}^{\mathrm{ext,int}}\right]\left[K_{\mathrm{VCE}}^{\mathrm{int,int}}\right]^{-1}\left[K_{\mathrm{VCE}}^{\mathrm{int,ext}}\right]\right]\left\{\mathbf{q}_{\mathrm{VCE}}^{\mathrm{ext}}\right\} = \left\{\mathbf{F}_{\mathrm{VCE}}^{\mathrm{ext}}\right\} \\ &-\left[K_{\mathrm{VCE}}^{\mathrm{ext,int}}\right]\left[K_{\mathrm{VCE}}^{\mathrm{int,int}}\right]^{-1}\left\{\mathbf{F}_{\mathrm{VCE}}^{\mathrm{int}}\right\}. \end{aligned} \tag{3.45}$$

This form is used in global assembly. The displacements u_i^{l0} and u_i^{l1} in each *level-0* and *level-1* element are interpolated by the standard or hierarchical shape functions based on Legendre polynomials as

$$\{\mathbf{u}\}^{10} = [N_{10}]\{\mathbf{q}_{10}\} = \left[N_{10}^{\mathrm{I}} N_{10}^{\mathrm{O}}\right]\begin{Bmatrix} \mathbf{q}_{10}^{\mathrm{I}} \\ \mathbf{q}_{10}^{\mathrm{O}} \end{Bmatrix}, \{\mathbf{u}\}^{11} = [N_{11}]\{\mathbf{q}_{11}\} = \left[N_{11}^{\mathrm{I}} N_{11}^{\mathrm{O}}\right]\begin{Bmatrix} \mathbf{q}_{11}^{\mathrm{I}} \\ \mathbf{q}_{11}^{\mathrm{O}} \end{Bmatrix}, \tag{3.46}$$

where $\{\mathbf{q}_{10}^{\mathrm{I}}\}/\{\mathbf{q}_{11}^{\mathrm{I}}\}$ corresponds to the nodal degrees of freedom at the interface with transition elements and $\{\mathbf{q}_{10}^{\mathrm{O}}\}/\{\mathbf{q}_{11}^{\mathrm{O}}\}$ corresponds to the remaining degrees of freedom. A similar separation can also be done for nodal displacements of transition elements into displacements on this interface. The displacements and the Lagrange multipliers on the intermediate boundary segment between the *level-0/1* and transition elements are interpolated according to (3.39). Substituting (3.45), (3.46), and (3.39) in (3.40) results in a coupled set of matrix equations for the multilevel domain:

$$\begin{bmatrix} K_{10/11}^{\mathrm{I,I}} & K_{10/11}^{\mathrm{I,O}} & 0 & 0 & 0 & P_{10/11} & 0 \\ K_{10/11}^{\mathrm{O,I}} & K_{10/11}^{\mathrm{O,O}} & 0 & 0 & 0 & 0 & 0 \\ 0 & 0 & K_{\mathrm{tr}}^{\mathrm{I,I}} & K_{12/\mathrm{tr}}^{\mathrm{I,O}} & 0 & 0 & P_{\mathrm{tr}} \\ 0 & 0 & K_{12/\mathrm{tr}}^{\mathrm{O,I}} & K_{12/\mathrm{tr}}^{\mathrm{O,O}} & 0 & 0 & 0 \\ 0 & 0 & 0 & 0 & 0 & Q_{10/11} & Q_{\mathrm{tr}} \\ P_{10/11}^{\mathrm{T}} & 0 & 0 & 0 & Q_{10/11}^{\mathrm{T}} & 0 & 0 \\ 0 & 0 & P_{\mathrm{tr}}^{\mathrm{T}} & 0 & Q_{\mathrm{tr}}^{\mathrm{T}} & 0 & 0 \end{bmatrix} \begin{Bmatrix} \mathbf{q}_{10/11}^{\mathrm{I}} \\ \mathbf{q}_{10/11}^{\mathrm{O}} \\ \mathbf{q}_{\mathrm{tr}}^{\mathrm{I}} \\ \mathbf{q}_{12/\mathrm{tr}}^{\mathrm{O}} \\ \mathbf{q}_{\mathrm{int}} \\ \mathbf{\Lambda}_{10/11} \\ \mathbf{\Lambda}_{\mathrm{tr}} \end{Bmatrix} = \begin{Bmatrix} \mathbf{F}_{10/11}^{\mathrm{I}} \\ \mathbf{F}_{10/11}^{\mathrm{O}} \\ \mathbf{F}_{\mathrm{tr}}^{\mathrm{I}} \\ \mathbf{F}_{12/\mathrm{tr}}^{\mathrm{O}} \\ 0 \\ 0 \\ 0 \end{Bmatrix}. \tag{3.47}$$

Superscript I represents quantities on the interface with transition elements, while superscript O corresponds to other regions. The submatrices $K_{10/11}$, K_{12}, and K_{tr} and vectors $\mathbf{F}_{10/11}$, $\mathbf{F}_{12}$, and $\mathbf{F}_{tr}$ correspond to stiffness matrices and load vectors from the respective subdomains. The stiffness $[K_{12/tr}]$ and the load vectors $\{\mathbf{F}_{12/tr}\}$ are obtained from VCFEM analysis. The coupling between the *level-0/1* and transition elements is achieved through the $[P]$ and $[Q]$ matrices. The system of equations is solved using an iterative solver with the Lanczos method.

3.3.4 Numerical Examples with the Adaptive Multilevel Model

Three sets of numerical examples are solved to study the effectiveness of the multilevel computational model for heterogeneous materials.

Composite laminate with a free edge

A classical problem of a composite laminate with a free edge that was introduced by Pagano and Rybicki [58, 67] is solved by the multilevel adaptive computational model. The problem to be solved is illustrated in Fig. 3.3 with out-of-plane loading. The homogenized solution of this problem yields a singular stress field near the free edge between the composite ply and the monolithic material layer, due to the constraints imposed by the free edge and Poisson's effect. The stress singularity has been reported in [58] as (d^{-a}), where d is the radial distance from the edge and the exponent $a < 0.1$. However, the micromechanics solution does not show any singularity and hence the macroscopic solution is grossly misrepresented in this region. The material properties for the boron fiber and epoxy matrix are:

E_{bo} (psi)	ν_{bo}	E_{epoxy} (psi)	ν_{epoxy}
60×10^6	0.2	0.5×10^6	0.34

The ratio of fiber radius to edge length in the RVE is $r/l = 0.3$, corresponding to a local volume fraction of 28.2%. For 40 rows of fiber, the microstructural RVE is assumed to be a unit cell of size $l = h/40$. The homogenized orthotropic stiffness coefficients are obtained as: E_{xx} = E_{yy} = 0.99×10^6 psi, $E_{xx} = E_{yy} = 0.99\times10^6$ psi, $E_{zz} = 1.72\times10^6$ psi, G_{xy} = 0.27×10^6 psi, $\nu_{xy} = 0.43$, $\nu_{zx} = \nu_{zy} = 0.29$. Only a quarter of the laminate is modeled, accounting for symmetry about xz and yz planes by imposing symmetry boundary conditions on $x = 0$ and $y = 0$ surfaces. The top ($y = 2h$) and right ($x = 4h$) surfaces are assumed to be traction free. The out-of-plane loading is simulated using a generalized plane strain condition with prescribed $\varepsilon_{zz} = 1$. The problem solved is for the number of fiber rows ($n = 40$) corresponding to approximately 6,400 fibers. The initial mesh in the multilevel model consists of 200 QUAD4 *level-0* elements. The adaptive model consists of *hp*-adaptation, and the three level transitions for control of the discretization and modeling errors, respectively.

The *level-0* to *level-1* transition parameter $C_1 = 0.3$ in (3.32), E_k is based on traction discontinuity defined by (3.33c) and *level-1* to *level-2* transition takes place according to (3.34) with

$$\hat{\mathbf{F}} = \frac{1}{A_c}\int_{\omega} \sigma_{yy}(\mathbf{v})\,\mathrm{d}A_c. \tag{3.48}$$

$\hat{\mathbf{F}}$ is an inclusion area averaged stress and A_c is its cross-sectional area. Figure 3.15a shows the multilevel mesh consisting of 242 *level-0* elements, four *level-1* elements, six *transition* elements, and five *level-2* elements. Each *level-2* element in this model is assumed to contain a single unit cell or RVE. The same problem is also solved using the commercial code ANSYS with a mesh of 30,000 elements (50,000 nodes). A 3 × 3 array of nine fibers near the free edge and laminate interface are explicitly modeled using a highly refined mesh (see Fig. 3.15c) and coupled with the remaining macroscopic analysis mesh. The multilevel model has a significantly smaller size with DOF=2,000 (2 × no. of nodes in *level-0* + *level-1* + *level-2* elements + # of βs in *level-2* elements). Fig. 3.16a compares the stress σ_{yy} along the line $y/h = 1$ near the free edge $x/h = 4$ for two values of C_1 by (a) the homogenized material law (gives rise to a singularity), (b) the microscopic stress obtained by VCFEM, and (c) the microscopic stress from the ANSYS analysis. The singularity vanishes for the microscopic results and the ANSYS and multilevel model results compare very well.

In a second study, the effect of *level-1 to level-2* switch-over criteria in (3.35)–(3.37) are examined with $C_3 = 3.0$ and $C_4 = 1.5$. The criteria (a) and (b) in (3.35) and (3.36) lead to 242 *level-0* elements, four *level-1* elements, six transition elements, and five *level-2* elements. Criterion (c) in (3.37)

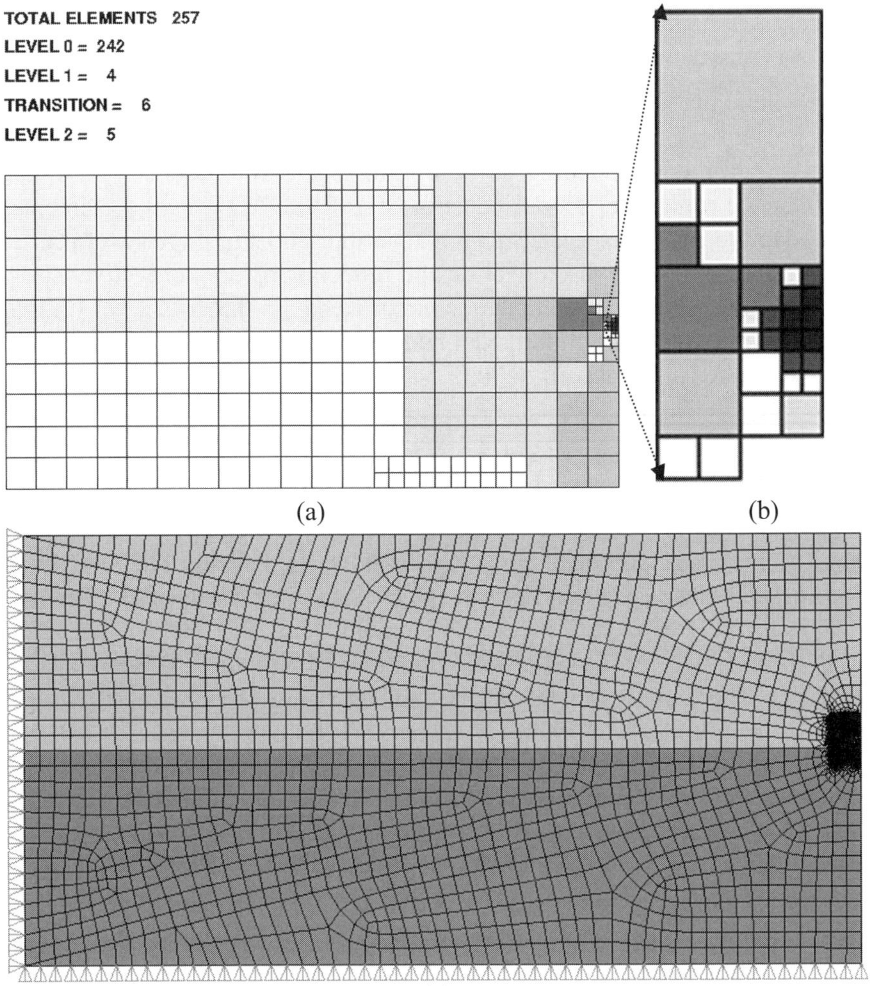

Fig. 3.15. (**a**) hp-Adapted multilevel mesh showing *level-0*, *level-1*, and *level-2* elements; (**b**) blow-up of the free-edge interface region mesh; and (**c**) ANSYS mesh with detailed microstructure modeled near the free edge

yields 245 *level-0* elements, four *level-1* elements, five transition elements, and three *level-2* elements, and criterion (d) in (3.34) yields 245 *level-0* elements, five *level-1* elements, five transition elements, and two *level-2* elements.

The stress σ_{yy} along $y/h=1$ near $x/h=4$ is compared in Fig. 3.16b. The agreement between the ANSYS and multilevel model results is excellent. Criteria (c,d) are more efficient due to a lesser number of *level-2* elements.

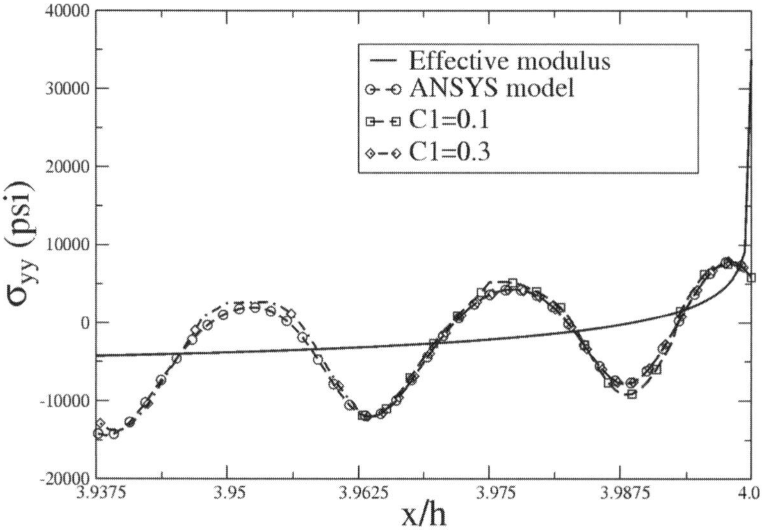

(a)

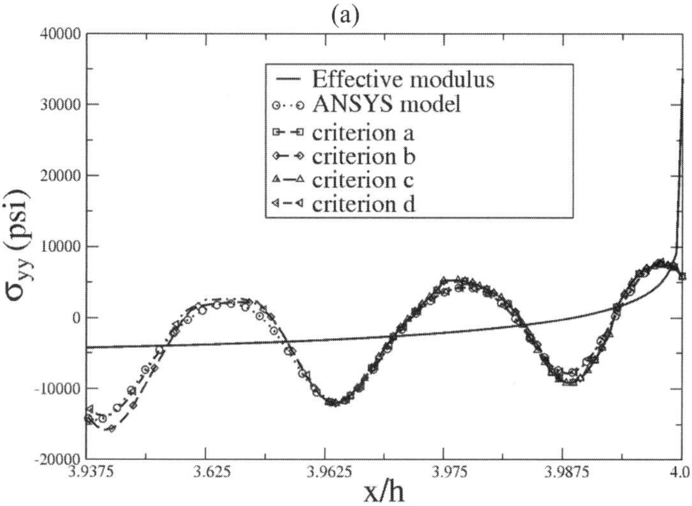

(b)

Fig. 3.16. Convergence of microscopic *level-2* stress σ_{yy} along section A–A near the critical free edge with different: (**a**) *level-0* to *level-1* and (**b**) *level-1* to *level-2* transition criteria

Comparison with GOALS algorithm-based multiscale modeling

In [56, 79], Oden et al. have introduced a theory of a posteriori modeling error estimates based on local quantities of interest, cast in terms of a linear functional $L(\mathbf{u})$. The goal-oriented adaptive local solution or GOALS algorithm is applied to the homogenized solution for estimating the local error in quantities of interest due to modeling a heterogeneous material as a homogenized medium. Subsequent to estimating the error, the algorithm adaptively adjusts the calculated quantities by adding microscale data until preset levels of accuracy are attained. The method entails solving an additional adjoint homogenized problem, in which $L(\mathbf{u})$ serves as the load vector. A measure β is defined as a local estimator of the modeling error as

$$|L(\mathbf{u}-\mathbf{u}^0)| \le \beta = \zeta_{\text{upp}}\bar{\zeta}_{\text{upp}} + \zeta_{\text{upp}} \|\mathbf{w}^0\|_{E(\Omega)}, \tag{3.49}$$

where ζ_{upp} and $\bar{\zeta}_{\text{upp}}$ are the upper bound of the energy norm-based modeling error in the primal problem and the adjoint problem respectively and $\|\mathbf{w}^0\|_{E(\Omega)}$ is the energy norm of the influence function. A domain of influence is determined as a local region or "cell" k for which the local error estimator β_k exceeds a prescribed tolerance. An *m-shaped* domain with randomly distributed cylindrical inclusions, having an average volume fraction 0.3 is depicted in Fig. 3.17a. The matrix material properties are $E = 100\,\text{MPa}$, $\nu = 0.2$, and inclusion properties are $E = 1{,}000\,\text{MPa}$, $\nu = 0.2$. The domain is subjected to a distributed load of $\mathbf{W} = 1\,\text{MN}\,\text{m}^{-1}$ under plane strain conditions. The domain is initially discretized into 42 *level-0* elements and homogenized properties for *level-0* and *level-1* elements are computed using a unit cell consisting of a single circular inclusion of 30% volume fraction. A local quantity of interest is ascertained in [79] as the inclusion area-averaged stress σ_{xx}:

$$L(\mathbf{v}) = \frac{1}{A_c}\int_{\omega} \sigma_{xx}(\mathbf{v})\mathrm{d}A_c, \tag{3.50}$$

where A_c is the inclusion cross-sectional area. The distribution of β_k is shown in the contour plot of Fig. 3.17b. In the application of the multilevel model, adaptation criteria in (3.32), (3.33a), and (3.34) are chosen. Consistent with (3.34), *the level-1 to level-2* switch takes place if

$$\hat{\sigma}_{xx}^{\,level\text{-}1} - \hat{\sigma}_{xx}^{\,\text{RVE}} \ge C_2 \hat{\sigma}_{xx}^{\,\text{RVE}}, \tag{3.51}$$

where

$$\hat{\sigma}_{xx} = \sum_{i}^{\#\,\text{inclusions}} \frac{1}{A_{ci}} \int_{\omega_i} \sigma_{xx}(\mathbf{v})\mathrm{d}A_{ci}.$$

The adaptation parameters are chosen as $C_1 = 0.1$ and $C_2 = 0.1$. The domain of influence has been calculated in [79] by the GOALS algorithm using the local quantity corresponding to an inclusion marked as ω, shown in Fig. 3.17a. The corresponding distribution of levels in the computational domain with the multilevel model are shown in Fig. 3.17c.

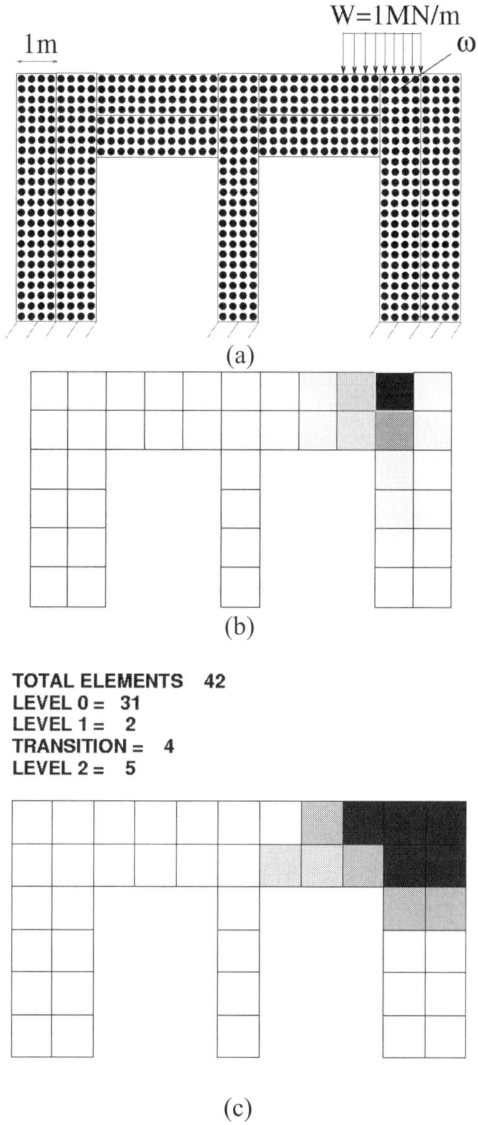

Fig. 3.17. (a) An *m*-shaped domain with uniformly distributed inclusions; (b) plot of β_k normalized with respect to its maximum for the quantity of interest L_2; (c) multilevel mesh with adaptation tolerances $C_1 = 0.1$, $C_2 = 0.1$

The GOALS algorithm [56, 79] conducts a microscopic analysis on six adjacent cells with high β_k (see Fig. 3.17b) to achieve 0.5% relative modeling error $(L(\mathbf{u}-\tilde{\mathbf{u}})/L(\mathbf{u}))\times 100\%$. In the present multilevel model, error is computed using $(L(\mathbf{u}-\mathbf{u}^0)/L(\mathbf{u}))= 0.743$ (as reported in [79]) and $L(\tilde{\mathbf{u}}) = 0.1554$ calculated from the solution of the *level-2* VCFEM in the same inclusion. The corresponding value of relative modeling error $(L(\mathbf{u}-\tilde{\mathbf{u}})/L(\mathbf{u}))\%$ is 0.45%, in comparison with the 0.5% in [56]. Thus the modeling error with multilevel adaptation is quite satisfactory.

A double lap aluminum-composite bonded joint

Adhesive bonded joints consisting of different materials are used to repair damaged structures in aircraft industries [66]. They can induce high stresses near the interface leading to failure initiation by fiber cracking, fiber–matrix interfacial debonding or interfacial delamination. A double-lap bonded joint with aluminum and boron-epoxy composite as the adherents is analyzed as shown in Fig. 3.18a. The dimension h is 64 mm with a total of 14 million fibers in the composite laminate. A perfect interface, corresponding to displacement continuity, is assumed between the aluminum and composite materials. Symmetry boundary conditions are employed in a quarter symmetry model with displacement component $u_y = 0$ on $y = 0$ and $u_x = 0$ along $x = 0$ as depicted in Fig. 3.18b. Displacement $u_x = 1$ is applied on the face $x = h$. The microstructure and RVE are the same as in Sect. 3.2.3. The material properties are

Material	Young's modulus, E (GPa)	Poisson ratio
Aluminum	$E_{\mathrm{al}} = 73.8$	$\nu_{\mathrm{al}} = 0.25$
Epoxy (matrix)	$E_{\mathrm{epoxy}} = 3.45$	$\nu_{\mathrm{epoxy}} = 0.3$
Boron (fiber)	$E_{\mathrm{boron}} = 413$	$\nu_{\mathrm{boron}} = 0.2$

The components of the homogenized elastic stiffness matrix for the composite are

E_{1111} (GPa)	E_{1122} (GPa)	E_{1133} (GPa)	E_{2222} (GPa)	E_{2233} (GPa)	E_{1212} (GPa)	E_{3333} (GPa)
9.93	4.39	4.14	10.59	4.27	2.58	137.32

The initial *level-0* mesh consists of 225 *level-0* QUAD4 elements. *Level-0* macroscopic stresses at the bonded interface $y = 0.5h$ are plotted in Fig. 3.19. In the composite, a high gradient of the tensile stress Σ_{xx} results near the interface A at $x = 0.25h$, with a high peak at A. Subsequently, Σ_{xx} drops to a very small value between $x = 0.25h$ and $x = 0.5h$. The composite stress Σ_{yy} (not shown) is compressive and exhibits a singular behavior at $x = 0.25h$ due to material mismatch and free edge constraints. The shear stress Σ_{xy} is generally zero in the composite along this line,

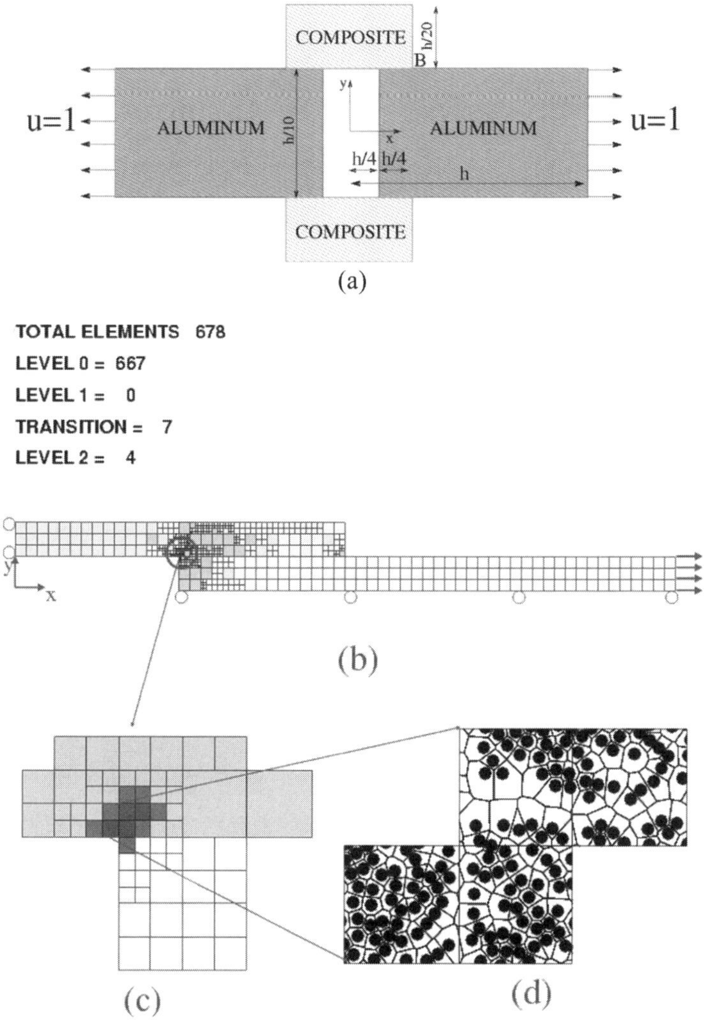

Fig 3.18. (**a**) Double-lap aluminum/boron-epoxy composite bonded joint, (**b**) macroscopic model of multilevel mesh, (**c**) zoomed in region of the macroscopic mesh undergoing level transition, (**d**) microscopic VCFEM analysis *level-2* regions

with the exception near A, where it exhibits a sharp gradient with a sign reversal. The small peaks at $x = 0.5h$ result from free edge conditions. In the aluminum panel, the stresses Σ_{xx} and Σ_{xy} start from zero at $x = 0.25h$ and reach a maximum with a very high gradient near the point A. Subsequently, they stabilize at lower values, satisfying the traction free boundary conditions on the top surface $y = 0.05h$. The stress Σ_{yy} is also compressive and very high near the interface $x = 0.25h$. These macroscopic results

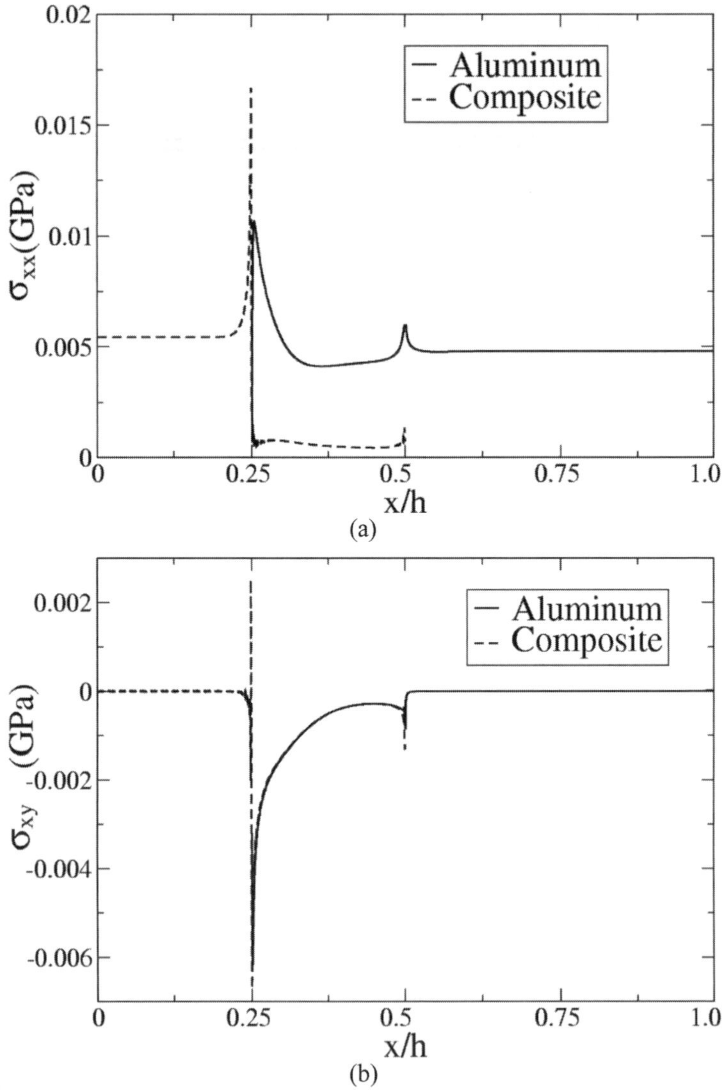

Fig. 3.19. Stresses (**a**) Σ_{xx}; (**b**) Σ_{xy} in aluminum and composite at $y = 0.05h$

qualitatively match the predictions of stresses made in [66]. The adapted multiple levels showing the microstructural region are depicted in Fig. 3.18. E_k in (3.32) is based on Σ_{xx} and *level-1* to *level-2* transition takes place according to (3.51). The evolved multilevel mesh in Fig. 3.18b, c has 667 *level-0* elements, seven transition elements, and four *level-2* elements. The *level-2* elements consist of a total of 203 microstructural Voronoi cell elements. Plots in Fig. 3.20 compares (a) the *level-0* macroscopic stress, (b) the *level-2* microscopic stress, and (c) the average microscopic stress, in the x-direction near the critical point A. The homogenized stresses do not match with the average microscopic stresses near A. However, they are

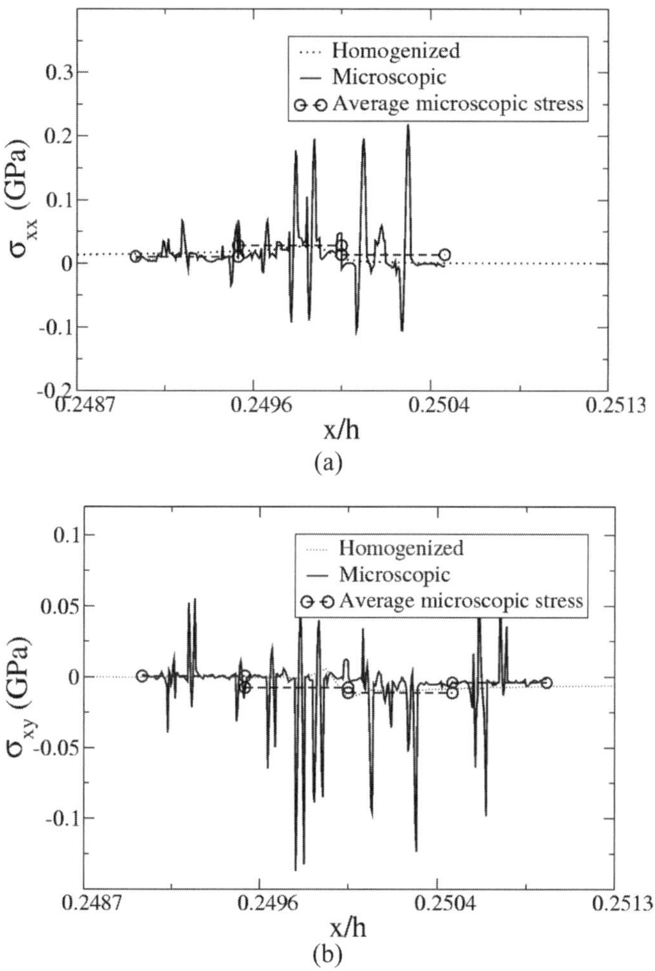

Fig. 3.20. *Level-2* stress (**a**) σ_{xx}; (**b**) σ_{xy} in aluminum and composite at $y = 0.05h$

the same away from the critical region, proving that homogenization is not effective at critical singular regions. The solution demonstrates the ability of the multilevel computational model in analyzing real problems with high efficiency and accuracy.

3.4 Multilevel Model for Damage Analysis in Composites

The adaptive concurrent multiscale modeling framework developed in Sect. 3.3 is extended to problems of composite structures undergoing damage initiation and growth due to microstructural damage induced by debonding at the fiber–matrix interface. Important changes in the three-level framework in the presence of damage are:

1. Incremental formulation is necessary to account for the history and path dependence of evolving damage.
2. The Voronoi cell FEM explicitly incorporates evolving damage (by interfacial debonding here) in the microstructure with inclusions.
3. An anisotropic continuum damage mechanics (CDM) model is developed for constitutive modeling of *level-0* elements to replace the constant homogenized stiffness in pure elastic problems.

The CDM model has been developed for unidirectional fiber-reinforced composites undergoing interfacial debonding by using homogenization theory in [36, 63]. It homogenizes the damage incurred through initiation and growth of interfacial debonding in a microstructural RVE and can effectively handle arbitrary loading conditions. An important assumption that is made in the derivation of this CDM is that the size of the RVE SERVE remains the same throughout the damage process. Extensive discussion of evolving damage in composites is provided in Talreja et al. [12, 76]. In [73], it has been shown that as the extent of damage increases with increasing strain, the SERVE size also increases. Continual increase in the SERVE size with evolving damage provides ground for its restricted use in homogenization schemes that use RVEs for evaluating continuum constitutive models. The breakdown of SERVE leads to the consideration of the *level-2* elements in these regions.

In [36], the CDM model of [63] is incorporated in an adaptive concurrent multilevel computational model to analyze multiscale evolution of damage in composites. Damage by fiber–matrix interface debonding is explicitly modeled over extended microstructural regions at critical locations using the Voronoi cell FEM developed in [34, 45], where a layer of cohesive springs model in the fiber–matrix interface. In this section, the adaptive multilevel modeling framework is discussed for composites with evolving damage with numerical examples demonstrating its effectiveness.

3.4.1 Voronoi Cell FEM with Microstructural Damage

The micromechanical Voronoi cell FEM outlined in Sect. 3.3.2 is extended for incorporating explicit damage evolution in the form of interfacial debonding as detailed in [34, 45]. Figure 3.21a shows a schematic of a typical Voronoi cell element with nonlinear cohesive zone springs characterizing the matrix–inclusion interface springs. Cohesive zone models are effective in depicting material failure as a separation process across an extended crack tip [36, 42, 63]. They introduce softening constitutive equations relating crack surface tractions to the material separation across the crack. The tractions across the interface reach a maximum, subsequently decrease, and eventually vanish with increasing interfacial separation. Motivated by interatomic potentials in atomistic modeling, many cohesive laws use a potential function ϕ to describe the traction–displacement relation during material separation. The traction–displacement relation of a bilinear model in [57] is depicted in Fig. 3.21b, c. The magnitude of traction t is expressed as a bilinear function of the interfacial separation as

$$t = \frac{\delta}{\delta_c}\sigma_{\max} \quad \forall \delta < \delta_c \quad \text{and} \quad t = \frac{\delta - \delta_e}{\delta_c - \delta_e}\sigma_{\max} \quad \forall \delta \geq \delta_c, \tag{3.52}$$

from which the normal and tangential traction components are derived as

$$T_n = \frac{\partial \phi}{\partial \delta_n} = \begin{cases} \sigma_{\max}\dfrac{\delta_n}{\delta_c} \\ \sigma_{\max}\dfrac{\delta - \delta_e}{\delta_c - \delta_e}\dfrac{\delta_n}{\delta} \\ 0 \end{cases} \quad \text{and} \quad T_t = \frac{\partial \phi}{\partial \delta_t} = \begin{cases} \beta^2\sigma_{\max}\dfrac{\delta_t}{\delta_c} & \forall \delta \leq \delta_c \\ \beta^2\sigma_{\max}\dfrac{\delta - \delta_e}{\delta_c - \delta_e}\dfrac{\delta_t}{\delta} & \forall \delta_c < \delta \leq \delta_e\,. \\ 0 \quad \forall \delta > \delta_e \end{cases} \tag{3.53}$$

When the normal displacement δ_n is positive, the traction at the interface increases linearly to a maximum value of $\sigma_{\max}$ (point A in Fig. 3.21b, c) corresponding to a value of δ_c before it starts decreasing to zero at a value of δ_e (point C). The unloading behavior in the hardening region is linear following the loading path. In the softening region, the unloading proceeds along a different linear path from the current position to the origin with a reduced stiffness given by the traction–displacement relation (line *BO*):

$$t = \sigma_{\max} \frac{\delta_{\max} - \delta_e}{\delta_c - \delta_e} \frac{\delta}{\delta_{\max}} \quad \forall \delta_c \le \delta_{\max} \le \delta_e \quad \text{and} \quad \delta \le \delta_{\max} \tag{3.54}$$

An irreversible damage path (*OBC*) is followed for reloading. Both normal and tangential tractions vanish when $\delta > \delta_e$. When the normal displacement is negative in compression, stiff penalty springs with high stiffness are introduced between the node pairs at the interface. The location of the separation at the debonding point is independent of the location of the peak of the curve for the bilinear model. This gives flexibility to adjust interfacial parameters for the peak and debonding locations to match the experimental observations as discussed in [34, 45].

In an incremental formulation, the complementary energy functional for each element in (3.24) is expressed in terms of the incremented stresses and displacements as

$$\begin{aligned}
&\Pi^{\mathrm{VCE}}_e\left(\sigma^m_{ij}, \Delta\sigma^m_{ij}, \sigma^c_{ij}, \Delta\sigma^c_{ij}, u^e_i, \Delta u^e_i, u^m_i, \Delta u^m_i, u^c_i, \Delta u^c_i\right) = \\
&-\int_{\Omega^{\mathrm{VCE}}_m \cup \Omega^{\mathrm{VCE}}_c} \frac{1}{2} S_{ijkl} \Delta\sigma_{ij} \Delta\sigma_{kl} \mathrm{d}\Omega - \int_{\Omega^{\mathrm{VCE}}_m \cup \Omega^{\mathrm{VCE}}_c} S_{ijkl} \Delta\sigma_{ij} \sigma_{kl} \mathrm{d}\Omega \\
&+ \int_{\partial\Omega^{\mathrm{VCE}}_e} \left(\sigma^m_{ij} + \Delta\sigma^m_{ij}\right) n^e_j \left(u^e_i + \Delta u^e_i\right) \mathrm{d}\partial\Omega - \int_{\partial\Omega^m_c} \left(\sigma^m_{ij} + \Delta\sigma^m_{ij}\right) n^c_j \left(u^m_i + \Delta u^m_i\right) \mathrm{d}\partial\Omega \\
&+ \int_{\partial\Omega^c_c} \left(\sigma^c_{ij} + \Delta\sigma^c_{ij}\right) n^c_j \left(u^c_i + \Delta u^c_i\right) \mathrm{d}\partial\Omega - \int_{\partial\Omega^{\mathrm{VCE}}_c} \int_{u^m_n - u^c_n}^{u^m_n + \Delta u^m_n - u^c_n - \Delta u^c_n} T^m_n \mathrm{d}\left(u^m_n - u^c_n\right) \mathrm{d}\partial\Omega \\
&- \int_{\partial\Omega^{\mathrm{VCE}}_c} \int_{u^m_t - u^c_t}^{u^m_t + \Delta u^m_t - u^c_t - \Delta u^c_t} T^m_t \mathrm{d}\left(u^m_t - u^c_t\right) \mathrm{d}\partial\Omega - \int_{\Gamma^{\mathrm{VCE}}_{tm}} \left(t_i + \Delta t_i\right)\left(u^e_i + \Delta u^e_i\right) \mathrm{d}\Gamma
\end{aligned} \tag{3.55}$$

Terms $\frac{1}{2}\sigma_{ij}\mathbf{S}_{ijkl}\sigma_{kl}$ and $\frac{1}{2}\mathbf{S}_{ijkl}\Delta\sigma_{ij}\Delta\sigma_{kl} + \mathbf{S}_{ijkl}\Delta\sigma_{ij}\sigma_{kl}$ are the complementary energy density and its increment, respectively. The prefix Δ corresponds to increments, and subscripts n and t correspond to the normal and tangential directions at the matrix–inclusion interface. Here $\mathbf{n}^e$ and $\mathbf{n}^c$ are the outward normal on $\partial\Omega_e$ and $\partial\Omega_c$, respectively. The two terms on the matrix–inclusion interface $\partial\Omega^{\mathrm{VCE}}_c$ provide the work done by the interfacial tractions $\mathbf{T}^m = T^m_n \mathbf{n}^m + T^m_t \mathbf{t}^m$ due to interfacial separation $\left(\mathbf{u}^m - \mathbf{u}^c\right)$. The integration over the incremental displacements at the interface $\partial\Omega_c$ is conducted by the backward Euler method. The total energy functional for each *level-2* or *transition* element containing N_{VCE} Voronoi cell elements is obtained by adding individual element contributions as in (3.25).

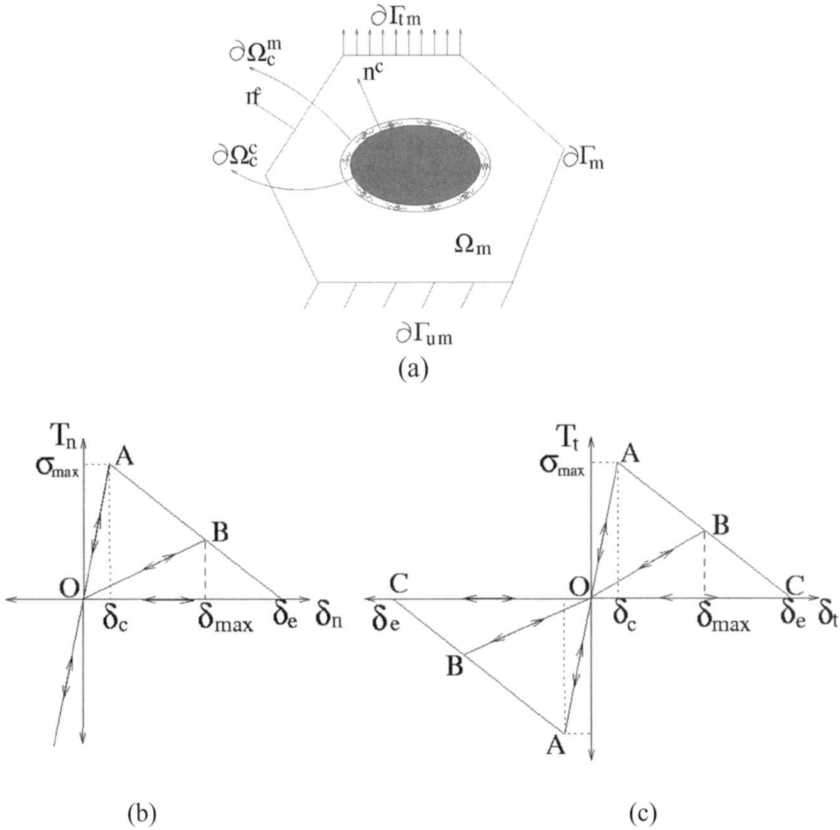

Fig. 3.21. (**a**) Voronoi cell FE with fiber–matrix interface using cohesive springs, (**b**) normal, and (**c**) tangential, traction–displacement behavior for bilinear cohesive zone model

Substituting stress and displacement increment interpolations of (3.26) in (3.71) and setting variations with respect to the stress coefficients $\Delta\boldsymbol{\beta}^m$ and $\Delta\boldsymbol{\beta}^m$, respectively, to zero, results in the weak form of the element kinematic relation, stated in a condensed matrix form as

$$\left[\mathbf{H}^e\right]\begin{Bmatrix}\boldsymbol{\beta}^m+\Delta\boldsymbol{\beta}^m\\ \boldsymbol{\beta}^c+\Delta\boldsymbol{\beta}^c\end{Bmatrix}=\left[\mathbf{G}^e\right]\begin{Bmatrix}\mathbf{q}^e+\Delta\mathbf{q}^e\\ \mathbf{q}^m+\Delta\mathbf{q}^m\\ \mathbf{q}^n+\Delta\mathbf{q}^n\end{Bmatrix}. \qquad (3.56)$$

The weak forms of the global traction continuity conditions are subsequently solved by setting the variation of the total energy function, with respect to $\Delta\mathbf{q}^e$, $\Delta\mathbf{q}^m$, and $\Delta\mathbf{q}^c$, to zero. This results in the weak form of the traction reciprocity conditions, stated in a condensed form as

$$\sum_{e=1}^{N_{vc}}\left[\mathbf{G}^e\right]\begin{Bmatrix}\boldsymbol{\beta}^m+\Delta\boldsymbol{\beta}^m\\ \boldsymbol{\beta}^c+\Delta\boldsymbol{\beta}^c\end{Bmatrix}=\sum_{e=1}^{N_{vc}}\left\{\mathbf{R}^e\right\}. \tag{3.57}$$

Substituting (3.56) into (3.57) yields

$$\sum_{e=1}^{N_{vc}}\left[\mathbf{G}^e\right]\left[\mathbf{H}^e\right]^{-1}\left[\mathbf{G}^e\right]\{\mathbf{q}+\Delta\mathbf{q}\}=\sum_{e=1}^{N_{vc}}\left\{\mathbf{R}^e\right\} \tag{3.58}$$

that is solved iteratively. Several numerical examples, validating this VCFE model, are solved in [34, 45].

As discussed in Sect. 3.3.2, the postprocessing phase for *level-1* elements requires the evaluation of different variables in the RVE from known values of macroscopic strains. A small variant of the formulation in (3.55) is needed for the energy functional of a SERVE (***Y***). The functional with *Y*-periodic displacements, *Y*-antiperiodic tractions on the boundary and imposed macroscopic strain $(e_{ij}+\Delta e_{ij})$ is written as

$$\begin{aligned}
\Pi_e = &-\int_{Y_m}\frac{1}{2}S^m_{ijkl}\Delta\sigma^m_{ij}\Delta\sigma^m_{kl}\mathrm{d}Y-\int_{Y_m}S^m_{ijkl}\sigma^m_{ij}\Delta\sigma^m_{kl}\mathrm{d}Y-\int_{Y_c}\frac{1}{2}S^c_{ijkl}\Delta\sigma^c_{ij}\Delta\sigma^c_{kl}\mathrm{d}Y\\
&-\int_{Y_c}S^c_{ijkl}\sigma^c_{ij}\Delta\sigma^c_{kl}\mathrm{d}Y+\int_{\partial Y_e}\left(\sigma^m_{ij}+\Delta\sigma^m_{ij}\right)n^e_j\left(u^e_i+\Delta u^e_i\right)\mathrm{d}\partial\Omega\\
&-\int_{\partial Y^m_c}\left(\sigma^m_{ij}+\Delta\sigma^m_{ij}\right)n^c_j\left(u^m_i+\Delta u^m_i\right)\mathrm{d}\partial\Omega+\int_{\partial Y^c_c}\left(\sigma^c_{ij}+\Delta\sigma^c_{ij}\right)n^c_j\left(u^c_i+\Delta u^c_i\right)\mathrm{d}\partial\Omega\\
&-\int_{\partial Y_c}\int_{u^m_n-u^c_n}^{u^m_n+\Delta u^m_n-u^c_n-\Delta u^c_n}T^m_n\mathrm{d}\left(u^m_n-u^c_n\right)\mathrm{d}\partial\Omega-\int_{\partial Y_c}\int_{u^m_t-u^c_t}^{u^m_t+\Delta u^m_t-u^c_t-\Delta u^c_t}T^m_t\mathrm{d}\left(u^m_t-u^c_t\right)\mathrm{d}\partial\Omega\\
&-\int_{Y_m}(e_{ij}+\Delta e_{ij})\Delta\sigma^m_{ij}\mathrm{d}\Gamma-\int_{Y_c}(e_{ij}+\Delta e_{ij})\Delta\sigma^c_{ij}\mathrm{d}\Gamma.
\end{aligned} \tag{3.59}$$

The boxed term corresponds to the additional energy due to the imposed macroscopic strain field on *Y*. Euler–Lagrange equations for this functional are the multiscale kinematic relations,

$$\begin{aligned}
&\varepsilon_{ij}\left(\mathbf{x},\mathbf{y}\right)+\Delta\varepsilon_{ij}\left(\mathbf{x},\mathbf{y}\right)=S_{ijkl}(\sigma_{ij}+\Delta\sigma_{ij})=(e_{ij}(\mathbf{x})+\Delta e_{ij}(\mathbf{x}))\\
&+\frac{1}{2}\left[\frac{\partial(u_i(\mathbf{y})+\Delta u_i(\mathbf{y}))}{\partial y_j}+\frac{\partial(u_j(\mathbf{y})+\Delta u_j(\mathbf{y}))}{\partial y_i}\right]\quad\forall\mathbf{y}\in Y_m,Y_c
\end{aligned} \tag{3.60}$$

3.4.2 Anisotropic CDM Model for *Level-0* Subdomain Ω^{l0}

For microstructures with randomly evolving microcracks or debonding causing diffused damage, the homogenized material behavior is best represented by a CDM law [2, 12, 20, 76]. An anisotropic CDM model with a fourth-order damage tensor has been developed from rigorous micromechanical analyses in [36, 63]. The general form of CDM models [2, 42, 70, 75] introduce a fictitious effective stress $\tilde{\Sigma}_{ij}$ acting on an effective resisting area $\tilde{A}$, which is caused by reduction of the original resisting area A due to material degradation from the presence of microcracks and stress concentration in the vicinity of cracks. The effective stress $\tilde{\Sigma}_{ij}$ is related to the actual Cauchy stress Σ_{ij} through the relation $\tilde{\Sigma}_{ij} = M_{ijkl}(\mathbf{D})\Sigma_{kl}$, where M_{ijkl} is a fourth-order damage effect tensor that is a function of the fourth-order damage tensor $\mathbf{D}(=D_{ijkl}\mathbf{e}_i \otimes \mathbf{e}_j \otimes \mathbf{e}_k \otimes \mathbf{e}_l)$. The hypothesis of equivalent elastic energy is used to evaluate M_{ijkl}, and hence establish a relation between the damaged and undamaged stiffnesses [15, 19, 80]. Equivalence is established by equating the elastic energy in the damaged state to that in a hypothetical undamaged state as

$$W(\mathbf{\Sigma},\mathbf{D}) = \frac{1}{2}\Sigma_{ij}(E_{ijkl}(\mathbf{D}))^{-1}\Sigma_{kl} = W(\tilde{\mathbf{\Sigma}},\mathbf{0}) = \frac{1}{2}\tilde{\Sigma}_{ij}(E^{\mathrm{o}}_{ijkl})^{-1}\tilde{\Sigma}_{kl}, \tag{3.61}$$

where E^{o}_{ijkl} is the elastic stiffness tensor in the undamaged state and $E_{ijkl}(\mathbf{D})$ is the stiffness in a damaged state. The relation between the damaged and undamaged stiffnesses is thus

$$E_{ijkl} = (M_{pqij})^{-1}E^{\mathrm{o}}_{pqrs}(M_{rskl})^{-1} \tag{3.62}$$

with an appropriate assumption of a function for M_{ijkl}, (3.62) can be used to formulate a damage evolution model using micromechanics and homogenization. A damage evolution surface is introduced to delineate the interface between damaged and undamaged domains in the strain e_{ij}-space as

$$F = \frac{1}{2}e_{ij}P_{ijkl}e_{kl} - \kappa(\alpha W_{\mathrm{d}}) = 0. \tag{3.63}$$

Here $W_{\mathrm{d}}\left(=\int \frac{1}{2}e_{ij}e_{kl}\mathrm{d}E_{ijkl}\right)$ corresponds to the dissipation of the strain energy density due to stiffness degradation for constant strain without an external work supply. Also called the degrading dissipation energy, it is an internal variable denoting the current state of damage. P_{ijkl} is a symmetric negative-definite fourth-order tensor that will be expressed as a function of the strain tensor e_{ij}, α is a scaling parameter and κ is a function of W_{d}. Assuming associativity rule in the stiffness space, the evolution of the fourth-order secant stiffness is obtained as

$$\dot{E}_{ijkl} = \dot{\lambda}\frac{\partial F}{\partial\left(\frac{1}{2}e_{ij}e_{kl}\right)} = \dot{\lambda}P_{ijkl}. \tag{3.64}$$

$P_{ijkl}(e_{mn})$ corresponds to the direction of the rate of stiffness degradation tensor $\dot{E}_{ijkl}$. For a composite material with interfacial debonding, the direction of rate of stiffness degradation varies with increasing damage and hence $P_{ijkl}(e_{mn})$ does not remain a constant throughout the loading process. The model requires the evaluation of κ, α, and P_{ijkl} in (3.63). These are determined from the results of VCFEM-based micromechanical simulations of an RVE with periodic boundary conditions. The function $\kappa(W_{\mathrm{d}})$ is evaluated for a reference loading path and all other strain paths are scaled with respect to this reference. Upon determination of the maximum value W_{d} for a reference loading condition, the value of α for any strain path can be obtained by simple scaling. To account for the variation of $P_{ijkl}(e_{mn})$, any macroscopic strain evolution path is discretized into a finite set of points. The values of P_{ijkl} are explicitly evaluated at these points from RVE-based simulations. Values of P_{ijkl} for any arbitrary macroscopic strain value can then be determined by interpolating between nodal values using shape functions of a 3D linear hexahedral element. Details of the parameter evaluation process in the macroscopic CDM model are discussed in [36, 63]. Gradients of important field variables are evaluated from macroscopic analysis using the CDM to assess the deviation of macroscopic uniformity. Such gradients may be the effect of microscopic nonhomogeneity in the form of highly localized stresses and strains or damage.

Numerical example with the anisotropic CDM

The macroscopic finite element model with its constitutive relations represented by the CDM model is validated by comparison of results with those obtained by homogenizing micromechanical solutions. The macroscopic model consists of a single QUAD4 element. For the microstructure, a nonuniform RVE with 20 circular fibers of volume fraction 21.78% (see Fig. 3.22a) is constructed with periodic boundary. As explained in [62, 64] and Sect. 3.3.2 (Fig. 3.8b), a local microstructure is first constructed by repeating the set of randomly distributed fibers that lie in a window in both the x and y directions for several period lengths. The multifiber domain is tessellated into a network of Voronoi cells for the entire region, and the boundary of the RVE is generated as the aggregate of the outside edges of Voronoi cells associated with the original set of fibers. Periodicity constraint conditions on nodal displacements can then be easily imposed by constraining the node pairs to move identically.

The material properties of the elastic matrix (m) and fibers (f) and the cohesive zone model parameters are:

E_{m} (GPa)	ν_{m}	E_{f} (GPa)	ν_{f}	δ_c (m)	δ_e (m)	$\sigma_{\max}$ (GPa)
4.6	0.4	210.0	0.3	5.0×10^{-5}	20.0×10^{-4}	0.2

Three different macroscopic strain paths are considered for loading conditions, viz.:

(L1) $\varepsilon_{xx} \neq 0, \quad \varepsilon_{yy} = \varepsilon_{xy} = 0$

(L2) $\varepsilon_{xy} \neq 0, \quad \varepsilon_{yy} = \varepsilon_{xx} = 0$

(L3) $\varepsilon_{xx} = \varepsilon_{yy} = -\varepsilon_{xy} \neq 0$

The parameter $\kappa(W_{\mathrm{d}})$ is evaluated from the reference loading (L1) corresponding to uniaxial tension. The strain state (L2) corresponds to shear loading condition, while the load (L3) represents a combination of all strain components. The macroscopic stress–strain plots by the CDM model are compared with the homogenized micromechanical analyses results in Figs. 3.21b–d. All the nonzero stress components are plotted for each of the loading conditions, and excellent agreement is observed. In the shear loaded case, while σ_{xx} and σ_{yy} are zero prior to the onset of damage, they continue to increase with softening and debonding of the interface. This is due to the different interface behavior in tension and compression. For the combined straining case, a more complex stress–strain behavior is observed. The debonding initiation and propagation is dispersed in the microstructure with 20 fibers and hence a very gradual reduction of stiffness is observed. The homogenized CDM model developed, predicts the true macroscopic damage behavior with high accuracy and efficiency.

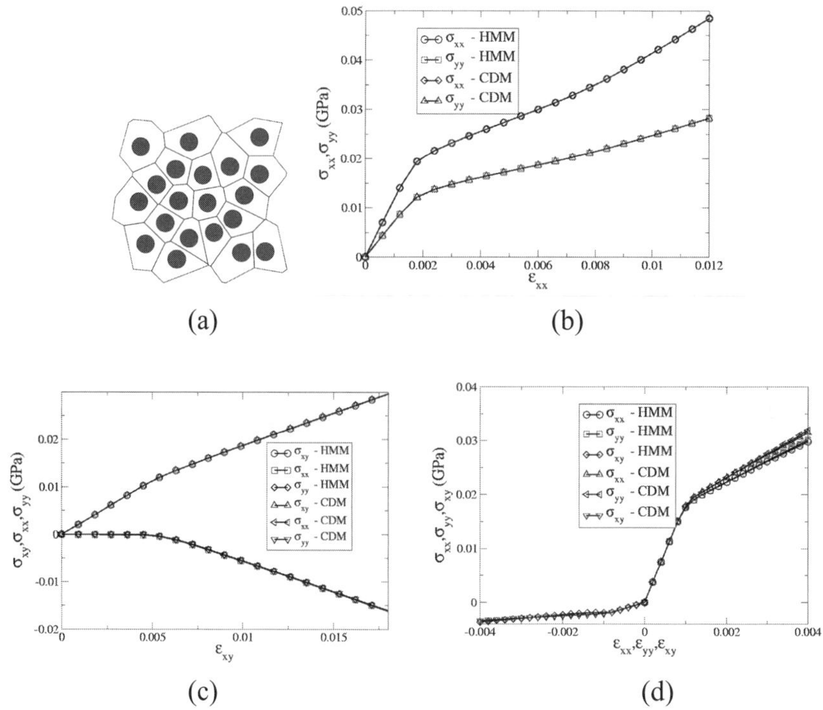

Fig. 3.22. Comparison of macroscopic stress–strain curves by CDM and homogenizing micromechanical solution: (**a**) a statistically equivalent RVE with 20 circular fibers, stress–strain plots for load cases, (**b**) L1, (**c**) L2, and (**d**) L3

3.4.3 Coupling Levels in the Concurrent Multiscale Algorithms

In a manner similar to Sect. 3.3.3, the incremental form of the equation of principle of virtual work equation for $\varOmega_{\text{het}}$ at the end of an increment can be written as the sum of contributions from each individual domain as

$$\begin{aligned}
&\int_{\varOmega_{10}} (\varSigma_{ij} + \Delta\varSigma_{ij}) \frac{\partial \delta u_i^{10}}{\partial x_j} \mathrm{d}\varOmega - \int_{\varGamma_{10}} (t_i + \Delta t_i) \delta u_i^{10} \mathrm{d}\varGamma + \int_{\varOmega_{11}} (\varSigma_{ij} + \Delta\varSigma_{ij}) \frac{\partial \delta u_i^{11}}{\partial x_j} \mathrm{d}\varOmega \\
&- \int_{\varGamma_{11}} (t_i + \Delta t_i) \delta u_i^{11} \mathrm{d}\varGamma + \int_{\varOmega_{12}} (\sigma_{ij} + \Delta\sigma_{ij}) \frac{\partial \delta u_i^{12}}{\partial x_j} \mathrm{d}\varOmega - \int_{\varGamma_{12}} \left(t_i + \Delta t_i\right) \delta u_i^{12} \mathrm{d}\varGamma \\
&+ \int_{\varOmega_{\text{tr}}} \sigma_{ij} \frac{\partial \delta u_i^{\text{tr}}}{\partial x_j} \mathrm{d}\varOmega - \int_{\varGamma_{\text{tr}}} (t_i + \Delta t_i) \delta u_i^{\text{tr}} \mathrm{d}\varGamma \\
&+ \delta \int_{\varGamma_{\text{int}}} \left(\lambda_i^{\text{tr}} + \Delta\lambda_i^{\text{tr}}\right)\left(v_i + \Delta v_i - u_i^{\text{tr}} - \Delta u_i^{\text{tr}}\right) \mathrm{d}\varGamma \\
&+ \delta \int_{\varGamma_{\text{int}}} \left(\lambda_i^{10/11} + \Delta\lambda_i^{10/11}\right)\left(v_i + \Delta v_i - u_i^{10/11} - \Delta u_i^{10/11}\right) \mathrm{d}\varGamma = 0.
\end{aligned} \tag{3.65}$$

The discretized algebraic form of (3.65) is solved by using the Newton–Raphson iterative solver. Setting up the tangent stiffness matrix requires consistent linearization by taking directional derivative of (3.65) along incremental displacement vectors $\Delta\mathbf{u}$ and $\Delta\mathbf{v}$ and the Lagrange multipliers $\Delta\boldsymbol{\lambda}$. For the *i*th iteration in the solution of the incremental variables, assembled matrix equations have the following structure.

$$\begin{bmatrix} K^{\mathrm{I,I}}_{10/11} & K^{\mathrm{I,O}}_{10/11} & 0 & 0 & 0 & P_{10/11} & 0 \\ K^{\mathrm{O,I}}_{10/11} & K^{\mathrm{O,O}}_{10/11} & 0 & 0 & 0 & 0 & 0 \\ 0 & 0 & K^{\mathrm{I,I}}_{\mathrm{tr}} & K^{\mathrm{I,O}}_{12/\mathrm{tr}} & 0 & 0 & P_{\mathrm{tr}} \\ 0 & 0 & K^{\mathrm{O,I}}_{12/\mathrm{tr}} & K^{\mathrm{O,O}}_{12/\mathrm{tr}} & 0 & 0 & 0 \\ 0 & 0 & 0 & 0 & 0 & Q_{10/11} & Q_{\mathrm{tr}} \\ P^{\mathrm{T}}_{10/11} & 0 & 0 & 0 & Q^{\mathrm{T}}_{10/11} & 0 & 0 \\ 0 & 0 & P^{\mathrm{T}}_{\mathrm{tr}} & 0 & Q^{\mathrm{T}}_{\mathrm{tr}} & 0 & 0 \end{bmatrix}^{i} \begin{Bmatrix} \Delta\mathbf{q}^{\mathrm{I}}_{10/11} \\ \Delta\mathbf{q}^{\mathrm{O}}_{10/11} \\ \Delta\mathbf{q}^{\mathrm{I}}_{\mathrm{tr}} \\ \Delta\mathbf{q}^{\mathrm{O}}_{12/\mathrm{tr}} \\ \Delta\mathbf{q}_{\mathrm{int}} \\ \Delta\boldsymbol{\Lambda}_{10/11} \\ \Delta\boldsymbol{\Lambda}_{\mathrm{tr}} \end{Bmatrix}^{i} = \begin{Bmatrix} \Delta\mathbf{F}^{\mathrm{I}}_{10/11} \\ \Delta\mathbf{F}^{\mathrm{O}}_{10/11} \\ \Delta\mathbf{F}^{\mathrm{I}}_{\mathrm{tr}} \\ \Delta\mathbf{F}^{\mathrm{O}}_{12/\mathrm{tr}} \\ \Delta\mathbf{F}_{\mathrm{int}} \\ \Delta\mathbf{F}_{\lambda^{10/11}} \\ \Delta\mathbf{F}_{\lambda^{12/\mathrm{tr}}} \end{Bmatrix}^{i}. \tag{3.66}$$

All the components have the same meaning as explained in Sect. 3.3.3. The stiffness $K_{12/\mathrm{tr}}$ and the load vector $\Delta\mathbf{F}_{12/\mathrm{tr}}$ for *level-2* and transition elements are obtained by VCFEM calculations followed by static condensation to represent the virtual work in terms of the boundary terms only.

3.4.4 Adaptation Criteria for Mesh Refinement and Level Change

The following criteria are used for mesh-refinement and level transitions due to discretization and modeling error, respectively, in the multilevel model. Many of these adaptation criteria are physically based, depending on the problem in consideration, since rigorous mathematical error bounds may not even exist for these nonlinear problems with damage. Consequently, other indicators may be used as appropriate.

Refinement of level-0 and level-1 meshes by h-adaptation

Computational models in *level-0* and *level-1* subdomains are enriched by *h-adaptation* to reduce discretization "error" and to identify regions of modeling error by zooming in on regions of localization with high gradients. For simulations using the CDM model, the adaptation criterion is formulated in terms of the traction jump across adjacent element boundaries, representing local stress gradients. The condition is stated as:

Refine element *"k,"* if the traction jump error satisfies the condition

$$E_k^{ij} \geq C_1 * E_{\text{avg}}^{ij}, \quad \text{where} \quad E_{\text{avg}}^{ij} = \left(\frac{\sum_{i=1}^{\text{NE}} \left(E_k^{ij} \right)^2}{\text{NE}} \right)^{1/2}$$

$$\text{and} \quad (E^{ij})_k^2 = \frac{\int_{\partial\Omega_e} \left([[T_x]]^2 + [[T_y]]^2 \right) \mathrm{d}\partial\Omega}{\int_{\partial\Omega_e} \mathrm{d}\partial\Omega}. \tag{3.67}$$

Here NE is the total number of *level-0* and *level-1* elements in the entire computational domain, T_x, T_y are the components of element boundary tractions in the x and y directions, and [[ ]] is the jump operator across element boundary $\partial\Omega_e$. $C_1 < 1$ is chosen from numerical experiments.

Criteria for switching from level-0 to level-1 elements

Level-0 to *level-1* element transition takes place according to criteria signaling the departure from conditions of homogenizability. The criteria are based on macroscopic variables in the CDM model of *level-0* elements. The degrading dissipation energy W_{d} is a strong indicator of localized damage evolution and hence, a criterion is formulated as:

Switch element "*k*" from *level-0* to *level-1* if:

$$E_k^{\text{gde}} * (W_{\mathrm{d}})_k > C_2 * E_{\max}^{\text{gde}} * (W_{\mathrm{d}})_{\max}, \tag{3.68}$$

where E_k^{gde} is the norm of the local gradient of $(W_{\mathrm{d}})_k$, expressed as $E_e^{\text{gde}} = \sqrt{\left(\frac{\partial (W_{\mathrm{d}})_e}{\partial x_1} \right)^2 + \left(\frac{\partial (W_{\mathrm{d}})_e}{\partial x_2} \right)^2}$, $E_{\max}^{\text{gde}}$ is the maximum value of E_k^{gde} in all elements and $C_2 (< 1)$ is a prescribed tolerance. The criterion (3.68) is helpful for seeking out regions with high gradients of W_{d} in regions of high W_{d} itself. The local gradient is accurately evaluated using the Zienkiewicz–Zhu (ZZ) gradient patch recovery method [81], where W_{d} is interpolated using a polynomial function over a patch of elements connected to a nodal point. The gradients of W_{d} in each element are calculated from the nodal values using element shape functions.

Criteria for switching from level-1 to level-2 elements

For elements in which macroscopic uniformity does not hold in the sense of (3.68), departure from RVE periodicity condition is used to trigger a switch from *level-1* to *level-2*. The switching criterion is developed in terms of evolving variables, e.g., the average strain at the fiber–matrix interface in the local microstructural RVE. The average strain is defined as

$$D_{ij} = \frac{1}{\int_{\cup\partial\Omega_c} \mathrm{d}\partial\Omega} \int_{\cup\partial\Omega_c} \varepsilon_{ij} \mathrm{d}\partial\Omega = \frac{1}{\int_{\cup\partial\Omega_c} \mathrm{d}\partial\Omega} \int_{\cup\partial\Omega_c} ([u_i]n_j + [u_j]n_i)\mathrm{d}\partial\Omega, \quad (3.69)$$

where the integral is evaluated over all fiber–matrix interfaces in the RVE. The jump in displacement across the fiber–matrix interface with a normal n_i is denoted by $[u_i]$. For perfect interfaces $[u_i]$ will be zero. Thus, D_{ij} corresponds to the contributions to the macroscopic strain due to damage only, since $D_{ij} = 0$ in the absence of damage. Departure from periodicity will result in a significantly altered averaged strain D_{ij} in response to different conditions on the boundary of the microstructural region. For example, let $D_{ij}^{e,l2}$ correspond to a solution of the boundary value problem of the local microstructure included in a *level-2* element (see Fig. 3.12), subject to boundary displacements that have been obtained from macroscopic *level-0/level-1* analysis. The microstructural scale is explicit in this analysis, since periodicity is not imposed on the boundary. On the other hand, let $D_{ij}^{e,\mathrm{RVE}}$ be from the solution of a boundary value problem of the local RVE with imposed macroscopic strains and subjected to periodic boundary displacements constraints. The difference in these two strains for a *level-1* element k is quantified as

$$E_k^{\mathrm{dper}} = \max\left(\left|D_{11}^{k,l2} - D_{11}^{k,\mathrm{RVE}}\right|, \left|D_{22}^{k,l2} - D_{22}^{k,\mathrm{RVE}}\right|, \left|D_{12}^{k,l2} - D_{12}^{k,\mathrm{RVE}}\right|\right). \quad (3.70)$$

For evaluating $D_{ij}^{e,l2}$ in the incremental solution, only the increments in the present step are calculated by the *level-1* macroscopic displacement boundary conditions. It is assumed that the RVE-based solution is valid all the way up to (but excluding) the present step. The departure from periodicity is measured in terms of the difference in averaged strains E_e^{dper}. The criterion thus reads:

Switch element "*k*" from *level-1* to *level-2* if:

$$E_e^{\mathrm{dper}} > C_3 D_{\max}^{\mathrm{RVE}}, \quad (3.71)$$

where $D_{\max}^{\mathrm{RVE}}$ is the maximum value of $D_{ij}^{k,\mathrm{RVE}}$ in all *level-1* elements.

Remark: Once the level-2 and transition elements have been identified, it is important to update the local states of stress, strain, and damage to the current state. This step should precede the coupled concurrent analysis. For this analysis, the history of the macroscopic displacement solution on the boundaries of the *level-0/level-1* elements is used. The local micromechanical (VCFEM) boundary value problem for the *level-2* element is incrementally solved from the beginning to obtain the stress, strain, and damage history in the microstructure from the macroscopic boundary displacement history.

3.4.5 Numerical Examples with the Adaptive Multilevel Model

Two numerical examples are solved to study the effectiveness of the multilevel computational model in analyzing damage in composite materials.

Multilevel model vs. micromechanical analysis

This example is aimed at establishing the effectiveness of the multilevel model in analyzing a nonuniform composite microstructure by comparing its predictions with those by pure micromechanical analysis. It is computationally intensive to conduct the reference micromechanical analysis with evolving damage for very large microstructural regions. Consequently, a computational domain with a small population of fibers, as shown in the optical micrograph of Fig. 3.23a, is considered. The polymer matrix composite micrograph has a random dispersion of uniaxial fibers. The dimensions of the micrograph analyzed are $100\,\mu\text{m} \times 70.09\,\mu\text{m}$, containing 264 circular fibers. Each fiber has a diameter of $1.645\,\mu\text{m}$ for a total volume fraction of 32%. Though the domain is not adequate for a clear separation between continuum and micromechanical regions (since relatively large regions are needed to realize an RVE), the results of this example show the effectiveness of the overall framework.

The optical micrograph is mapped onto a simulated microstructure with circular fibers that is tessellated into a mesh of 264 Voronoi cell elements, shown in Fig. 3.23b. The constituent materials are an epoxy resin matrix, stainless steel reinforcing fibers, and a very thin film of freekote ($< 0.1\,\mu\text{m}$) at the fiber–matrix interface. The freekote imparts weak strength to the steel–epoxy interface, which allows a stable growth of the debond crack for experimental observation. The experimental methods of material and interface characterization have been discussed in [34]. Both the matrix and fiber materials are characterized by isotropic elasticity properties, viz. $E_{\text{epoxy}} = 4.6\,\text{GPa}$, $\nu_{\text{epoxy}} = 0.4$, $E_{\text{steel}} = 210\,\text{GPa}$, $\nu_{\text{steel}} = 0.3$.

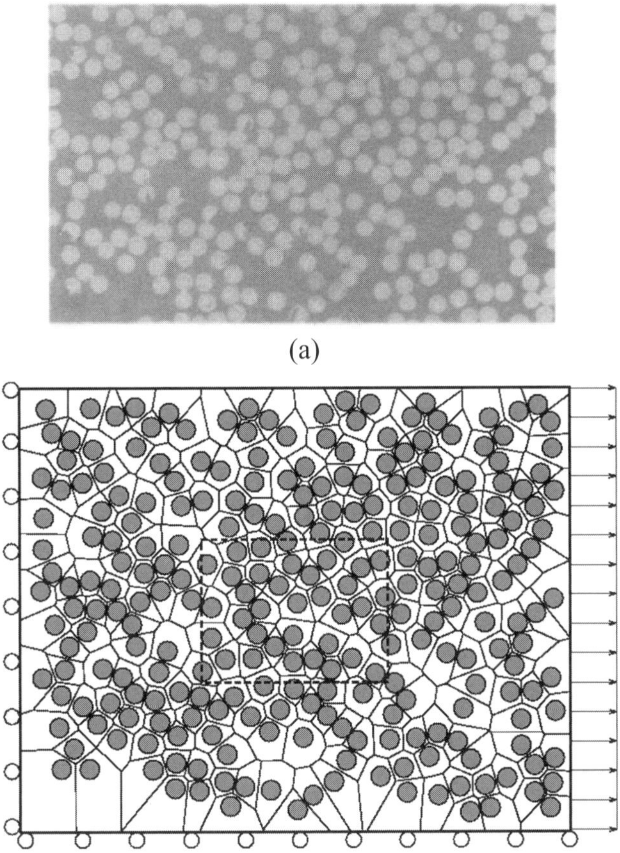

(a)

(b)

Fig. 3.23. (**a**) Optical micrograph of a steel fiber–epoxy matrix composite with 264 fibers and (**b**) the simulated computational model with a Voronoi cell mesh

The cohesive model properties are: $\delta_c = 5.1e-5m$, $\delta_e = 3.1e-4m$, and $\sigma_{\max} = 0.005\,\text{GPa}$. The micrograph is stretched in horizontal tension by 20 equal increments of $0.1\,\mu\text{m}$, to a total strain of $\varepsilon_{xx} = 0.1\%$. The displacement is imposed on the right edge, as shown in Fig. 3.23b.

The pure micromechanical VCFEM solution using the mesh of Fig. 3.23b is presented in [45] and is used here as reference solutions for the multiscale simulation. Figure 3.26a shows the contour plot of microscopic stress σ_{xx} at the final step of the micromechanical simulation, with a depiction of interfacial debonding. The right side of the microstructure shows significant localized damage. Debonding initiates at the top and percolates to the bottom of the microstructure along a narrow band. Multiscale analysis is performed by the concurrent multilevel model and the results are compared with those from the micromechanical VCFEM analysis. For the multilevel model, the entire computational region of 264 fibers is first divided into nine macroscopic finite elements as shown in Fig. 3.24a. For evaluating the homogenized constitutive properties for each element, the statistically equivalent representative volume element, or SERVE, for the microstructure underlying each macroscopic element is first identified. Statistical methods for identification of the SERVE have been discussed in Sect. 3.3.2.3 and also in [13, 32, 33, 60, 69, 72, 73]. However, since the number of fibers in the micrograph is limited in this exercise, the SERVE for each element is assumed to consist of all the fibers belonging to that element. For example, to generate the SERVE for an element window in the micrograph of Fig. 3.23b, all fibers whose centers are located within this window are first identified as constituents of the RVE and the SERVE is then created. The number of fibers and their distribution in the SERVE of each macroscopic element is shown in Fig. 3.24a.

The number of elements is only nine in this example. Consequently, *level-0* simulations with the CDM model are bypassed, and all elements are *level-1* at the start of the multilevel simulation. The factor C_3 in (3.71) is taken as 0.2. However the $D_{ij}^{k,l2} - D_{ij}^{k,\mathrm{RVE}}$ terms for each element in (3.70) are replaced by the difference in RVE-based averaged strains between adjacent elements, i.e., $D_{ij}^{e1,\mathrm{RVE}} - D_{ij}^{e2,\mathrm{RVE}}$. Also, instead of transition elements, a single layer of microscopic Voronoi cell elements is included for transitioning between the *level-1* and *level-2* elements. In Fig. 3.24b the Voronoi elements containing the grey fibers constitute the *transition* layer, while those containing the black fibers belong to *level-2*. An interface segment Γ_{int} is inserted between the *transition* and *level-1* elements at a distance $\mathbf{L}_{\mathrm{tr}/l2}$ from the right edge. Convergence properties of the multilevel model are studied for two cases, viz. $\mathbf{L}_{\mathrm{tr}/l2}/\mathbf{L} = 0.35$ and $\mathbf{L}_{\mathrm{tr}/l2}/\mathbf{L} = 0.45$. This is achieved by changing the initial *level-1* element size.

In Fig. 3.24b, only three elements (3, 6, and 9) on the right side of the initial mesh switch from *level-1* to *level-2*. A comparison of results by (a) VCFEM-based micromechanical analyses (all *level-2* elements), (b) homogenization-based macroscopic analysis (all *level-1* elements), and (c) concurrent multilevel analysis (*level-1* and *level-2* elements) is made.

Contour plots of σ_{11} (GPa) showing interfacial debonding at the end of the simulation are shown in Fig. 3.24b, c. The discrepancy between the

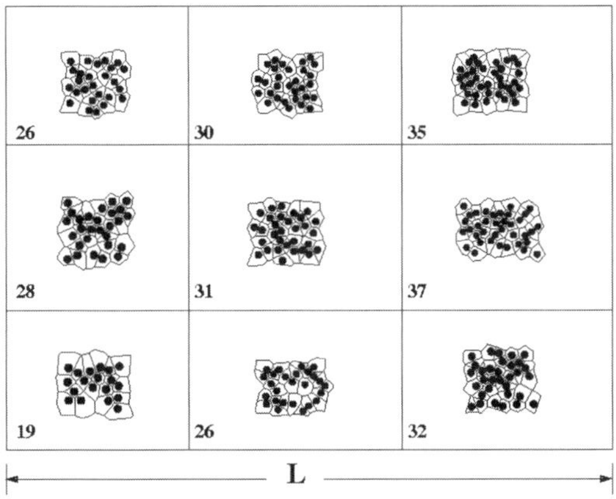

(a)

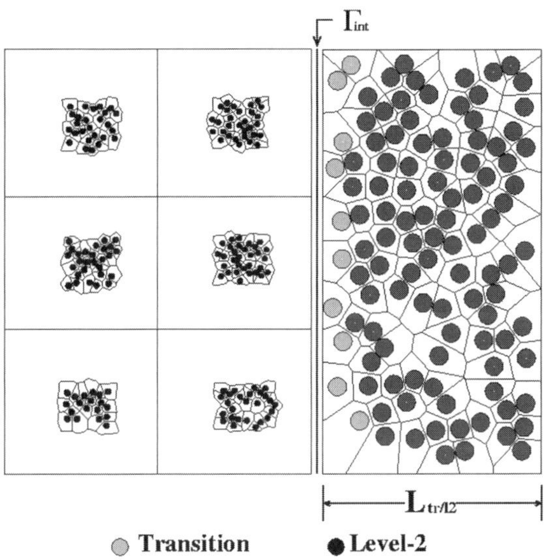

(b)

Fig. 3.24. Mesh for the computational domain: (**a**) macroscopic mesh with different RVE in every element and (**b**) multilevel model with the interface between macroscopic and microscopic VCFE elements damage paths predicted by the micromechanical and the multilevel analyses reduces sharply with increasing $\mathbf{L}_{tr/l2}/\mathbf{L}$ value. This can be attributed to the fact that the damage path is very sensitive to the macro–micro interface conditions. Since the sample size is small and there is no real periodicity in the microstructure, the proximity of the *level-1* boundary to the damage localization zone alters the local boundary conditions. However, as this distance is increased, the microscopic stress distribution, debonding pattern and damage zone replicates the micromechanical analysis results. This is due to the fact that the damage localization has little effect on the *level-1/level-2* boundary with increasing distance. The distribution of the micromechanical stresses σ_{11}, generated by pure micromechanical and multilevel analyses, is plotted along a line through the middle of the micrograph in Fig. 3.25. The micromechanical stresses show only minor oscillations about an average value of 0.005 GPa in the region to the left of the *level-1–level-2* interface. In the region to the right, where damage is predominant, there is clearly convergence of the stresses with increasing $\mathbf{L}_{tr/l2}/\mathbf{L}$ value. The macroscopic or averaged stress–strain response for element 1 (always *level-1*) and element 9 (changes levels) are plotted in Fig. 3.25. The volume averaged stresses and strains are evaluated by averaging the local fields over the microscopic domain as

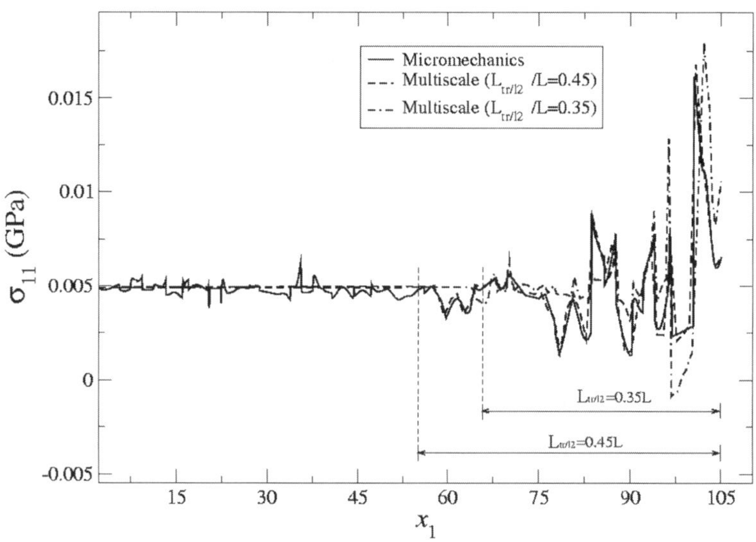

Fig. 3.25. Comparison of microscopic stress σ_{11} by different methods along a line through the middle of microstructure

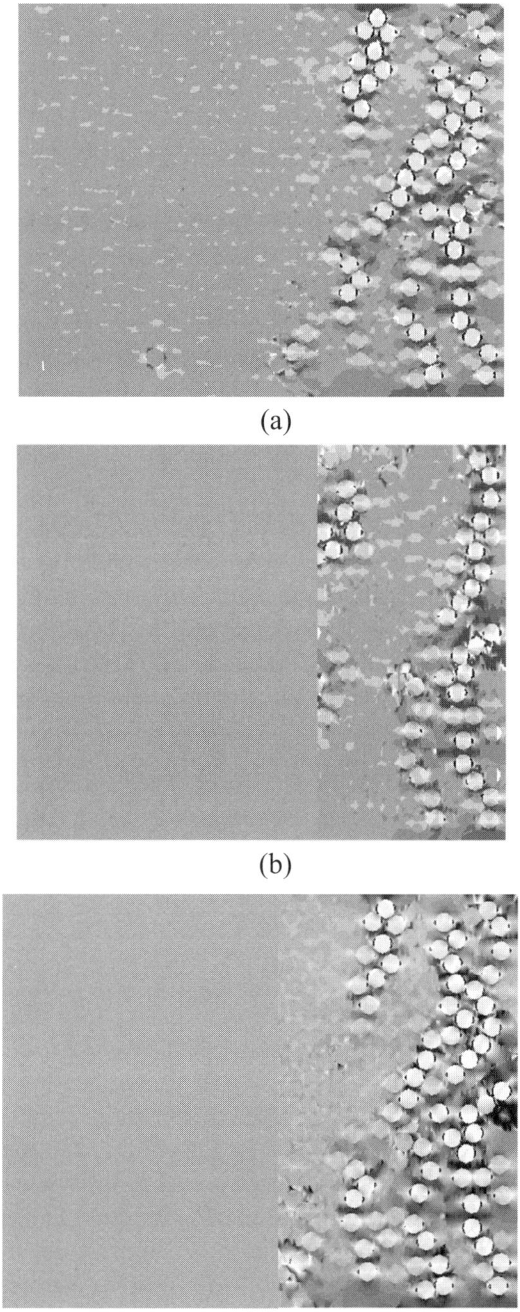

Fig. 3.26. Contour plot of σ_{11} showing interfacial debonding at the end of simulation for: (**a**) pure micromechanical analysis, (**b**) multiscale analysis with $\mathbf{L}_{tr/l2} / \mathbf{L} = 0.35$, and (**c**) multiscale analysis with $\mathbf{L}_{tr/l2} / \mathbf{L} = 0.45$

$$\Sigma_{ij} = \frac{1}{\Omega}\int_{\Omega}\sigma_{ij}(x_1,x_2)\mathrm{d}\Omega \quad \text{and} \quad e_{ij} = \frac{1}{\Omega}\int_{\Omega}\varepsilon_{ij}(x_1,x_2)\mathrm{d}\Omega - D_{ij}, \tag{3.72}$$

where D_{ij} is the strain jump defined in (3.68). The results for all the models are in good agreement for element 1, where there is no significant microstructural damage. The small difference is due to the periodicity constraints imposed on the microstructure. Also, there is a difference between the results of $\mathbf{L}_{tr/l2}/\mathbf{L} = 0.35$ and $\mathbf{L}_{tr/l2}/\mathbf{L} = 0.45$, due to the interface conditions at Γ_{int}. However, as is expected, the results are quite different for element 9, where significant damage is observed in Fig. 3.26. The *level-1* analysis shows significant deviation from the micromechanical analysis due to imposed periodicity in the damage zone. Once again, the results improve significantly with increasing $\mathbf{L}_{tr/l2}/\mathbf{L}$ ratio.

A composite double lap joint with microstructural debonding

The double lap bonded adhesive joint with boron–epoxy composites, discussed in Sect. 3.4.3, is again analyzed with interfacial damage. An adhesive, ABCD in Fig. 3.27a, is used to bond the two composite materials. Both plies above and below the adhesive are made of unidirectional boron fiber–epoxy matrix composite materials. The fibers are uniformly arranged in a square array, implying a square unit cell with a single circular fiber of $V_f = 20\%$. The epoxy matrix and boron fibers have the same properties as described in the previous section. The material properties of the isotropic adhesive and the bilinear cohesive law parameters for the matrix–fiber interface are:

$E_{adhesive}$ (GPa)	$\nu_{adhesive}$	δ_c (m)	δ_e (m)	σ_{max} (GPa)
4.6	0.4	5.0×10^{-5}	20.0×10^{-4}	0.2

Only a quarter of the joint is modeled from considerations of symmetry in boundary and loading conditions. In the model, the top ply above the adhesive is assumed to consist of ten rows of fiber, while the bottom row consists of five rows resulting in a total of 450 fibers. The number of fibers is kept low, so that a reference micromechanical analysis can be easily done for this example with a mesh of 450 Voronoi elements (square unit cell). The displacement component is $u_1 = 0$ along the face $x_2 = 0$ due to symmetry about the x_2 axis. The displacement components along the face $x_1 = 8h$ are $u_1 = 0$ and $u_2 = 0$ corresponding to a fixed edge. A total tensile displacement $u_1 = 1.2 \times 10^{-3} h$ is applied on the face of the lower ply at $x_1 = 0$ in 15 uniform increments. Three different approaches are used to solve this problem (1) a macroscopic model using the CDM model, (2) a detailed micromechanical VCFEM analysis, and (3) multiscale analysis by the multilevel model.

The starting mesh in the multilevel model of the bonded joint consists of a uniform grid of 470 QUAD4 elements for macroscopic analysis as shown in Fig. 3.27b. The constitutive relation for each element is a fourth-order anisotropic CDM model developed for this unit cell with interfacial cohesive zone in [63]. Figure 3.27d shows the gradient of the dissipation energy, i.e., ∇W_d, at the final stage of loading. Damage initiates near the bottom left corner A of the adhesive joint and propagates downward to span the entire region on the left of point A. Level transition parameters are $C_2 = 0.5$ and $C_3 = 0.1$. The corresponding evolution of various levels

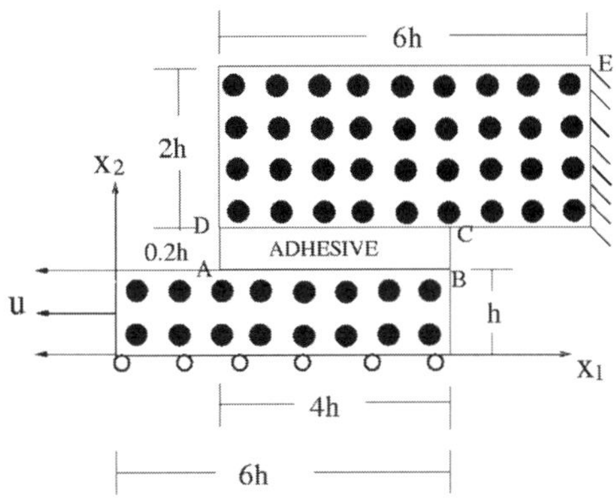

(a)

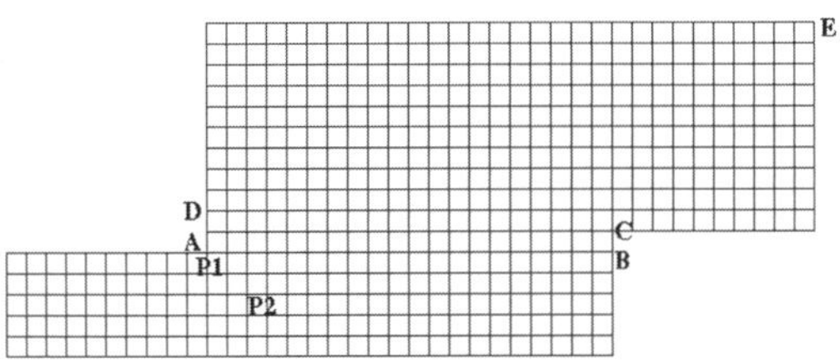

(b)

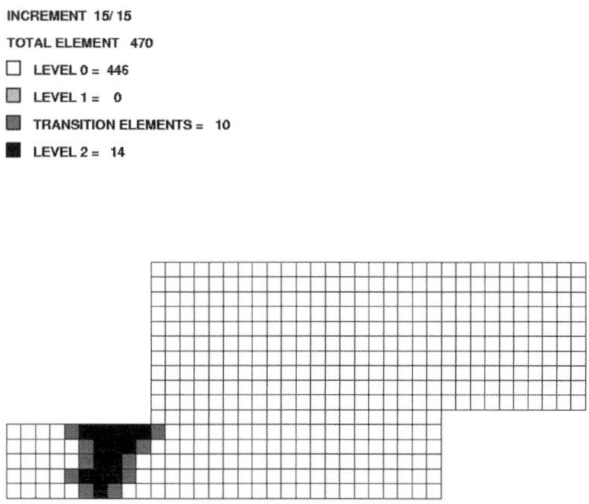

(c)

(d)

Fig. 3.27. (**a**) A composite double lap joint, (**b**) *level-0* computational mesh, (**c**) evolution of the multilevel computational model with level transition at the final loading stage, and (**d**) contour plot of dissipation energy gradient ∇W_d

in the model is depicted in Fig. 3.27c. At the final step, the multilevel mesh consists of 446 *level-0* elements, 0 *level-1* elements, 14 *level-2* elements, and 10 *transition* elements. All *level-2* elements emerge in critical regions where both ∇W_d and W_d are high.

Figure 3.28a, b depicts the contours of microscopic stress σ_{11} and the regions of debonding. The results of the multilevel model are in excellent agreement with the micromechanical analysis, both with respect to

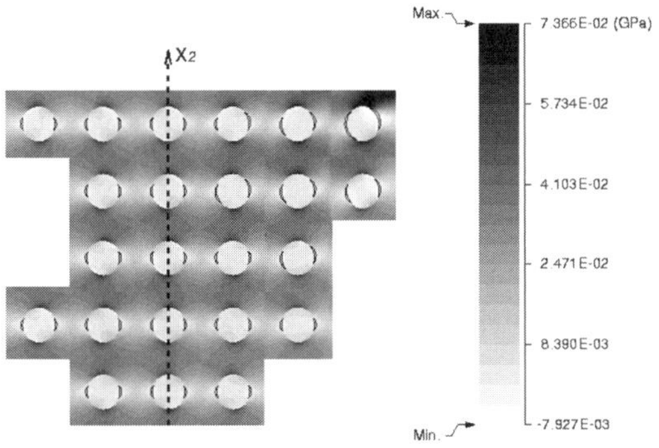

(a)

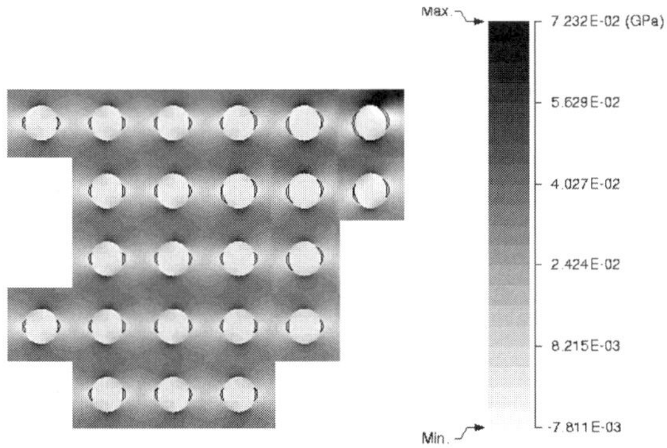

(b)

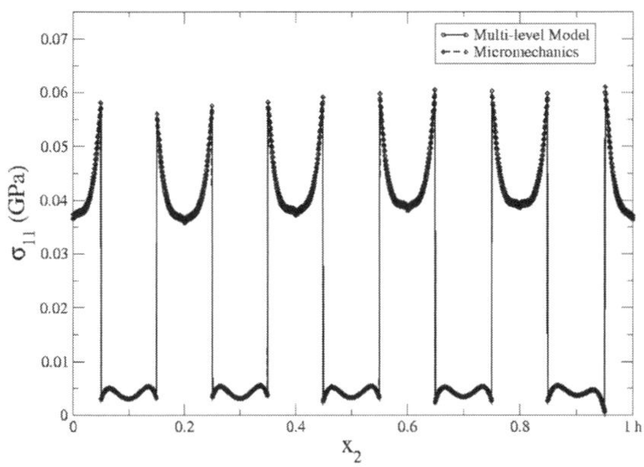

(c)

Fig. 3.28. *Level-2* solution near the corner A showing microscopic stress distribution (GPa) and interfacial debonding at the end of the analysis by: (**a**) pure micromechanical analysis, (**b**) multiscale analysis, and (**c**) comparison of σ_{11} along the vertical line through the microstructure by multilevel and micromechanical analysis debonding regions and evolving variables. The maximum error in σ_{11} is around 1%. The excellent agreement is further corroborated in the plot of σ_{11} along the vertical line through the microstructure in Fig. 3.28c. Figure 3.29a shows the macroscopic $\Sigma_{11} - e_{11}$ plots obtained from (a) macroscopic CDM-based analysis, (b) micromechanical analysis followed by

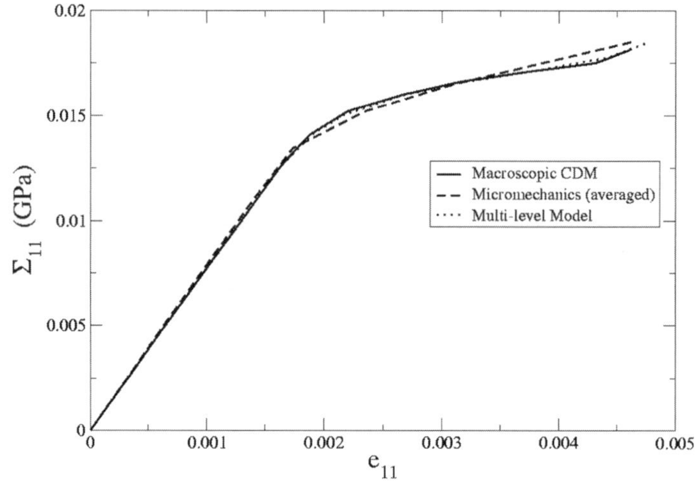

(a)

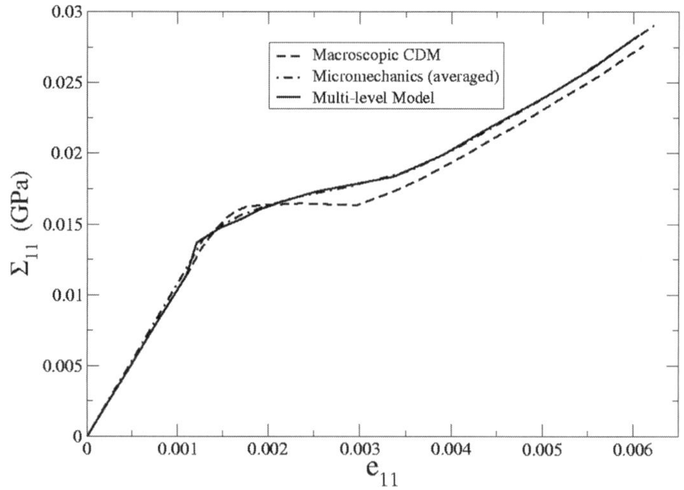

(b)

Fig. 3.29. Macroscopic $\Sigma_{11} - e_{11}$ plot by different methods at: (**a**) P2 and (**b**) P1 averaging, and (c) multiscale analysis with the multilevel model at two different locations, P1 and P2 shown in Fig. 3.27b. At P2, with low damage and its gradient, solutions by the CDM model and micromechanics are in relatively good agreement. At this point, the multiscale model uses the CDM constitutive law. However, the CDM results are quite different from the other two at P1, a hotspot where the damage and its gradient are high. The multilevel model and micromechanics results match quite well here.

Computational efficiency of the multilevel model is examined by a comparison of the CPU time on an IA32 computer cluster for the different models. The relative CPU times are as follows: (1) 71 s for all *level-0*, (2) 300,330 s for all *level-1*, (3) 300,310 s for all *level-2*, and (4) 42,260 s for the multilevel model. Although the macroscopic CDM analysis is faster, it can lead to significant errors. The complete *level-1* solution is even slower than the micromechanics solution. Accurate analysis with the multilevel model is at least seven times faster than the complete micromechanics and *level-1* solutions. The efficiency increases rapidly with increasing number of fibers in the analysis as shown in [36].

3.5 Conclusions

A comprehensive framework for adaptive concurrent multilevel computational analysis is developed in this chapter for multiscale analysis of fiber-reinforced composite materials with damage prediction. While, microstructural damage is manifested by fiber–matrix interfacial debonding here, any possible mode, e.g., multicrack evolution [46], can be incorporated into this framework. The multilevel model invokes two-way coupling of scales, viz. a bottom-up coupling with homogenization at lower scales to introduce reduced order continuum models and a top-down coupling at critical hot spots to transcend scales for following the microstructural damage evolution. Adaptive capabilities enable effective domain decomposition in the evolving problem with damage, keeping a balance between computational efficiency and accuracy. Numerical examples establish the accuracy and efficiency aspects of the model, as well as demonstrate its capability in handling problems involving large composite domains. Overall, this work lays an effective foundation for solving multiscale problems involving localization, damage, and crack evolution that may be impossible to achieve using any single scale model.

Acknowledgments

This work has been supported by the Air Force Office of Scientific Research (Program Director: Dr. B. L. Lee). This sponsorship is gratefully acknowledged. Computer support by the Ohio Supercomputer Center is also gratefully acknowledged. The author would like to thank his graduate students P. Raghavan, H. Bhatnagar, J. Bai, J. Jain, S. Manchiraju, G. Venkatramani, and K. Kennedy for their help with the results and editing of this document.

Figures 3.1–3.19 in this chapter are reprinted from*: Journal*: Computer Methods in Applied Mechanics and Engineering, *Vol.* 193(6–8), *Authors*: P. Raghavan and S. Ghosh, *Title of article*: Concurrent multiscale analysis of elastic composites by a multilevel computational model, *Copyright year* (2004) with permission from Elsevier. Figures 3.20 and 3.21 are reprinted from: *Journal*: Mechanics of Materials, *Vol.* 37 (9), *Authors*: P. Raghavan and S. Ghosh, *Title of article*: A continuum damage mechanics model for unidirectional composites undergoing interfacial debonding, *Copyright year* (2005) with permission from Elsevier.

References

1. Ainsworth M, Oden JT (1992) A procedure for a posteriori error estimation for hp-finite element methods. Comput Meth Appl Mech Engng 101:73–96
2. Allen DH (2002) Homogenization principles and their application to continuum damage mechanics. Comp Sci Tech 61:2223–2230
3. Allen DH, Yoon C (1998) Homogenization techniques for thermoviscoelastic solids containing cracks. Int J Solids Struct 35:4035–4054
4. Aminpour MA, Ransom JB, McCleary SL (1995) A coupled analysis method for structures with independently modeled finite element subdomains. Int J Num Meth Engng 38:3695–3718
5. Argon AS, Im J (1975) Separation of second phase particles in spheroidized 1045 steel, Cu-0.6Cr alloy, and maraging steel in plastic straining. Metall Trans 6A:839–851
6. Babuska I, Strouboulis T, Upadhyay CS, Gangaraj SK (1995) A-posteriori estimation and adaptive control of the pollution error in the h-version of the finite element method. Int J Numer Meth Engng 38:4208–4235
7. Benssousan A, Lions JL, Papanicoulau G (1978) Asymptotic Analysis for Periodic Structures. North Holland, Amsterdam
8. Benveniste Y (1987) A new approach to the application of Mori–Tanaka's theory in composite materials. Mech Mater 6:147–157

9. Boehm HJ (2004) A short introduction to continuum micromechanics. In: Boehm HJ (ed) Mechanics of Microstructured Materials, CISM Courses and Lectures, Vol 464. Int Centre for Mech Sci, Springer Wien, NY, pp 1–40
10. Boehm HJ, Rammerstoffer FG, Weissenbek E (1993) Some simple models for micromechanical investigations of fiber arrangement effects in MMC's. Comp Mater Sci 1:177–194
11. Budiansky B (1965) On the elastic moduli of some heterogeneous materials. J Mech Phys Solids 13:223–227
12. Bulsara VN, Talreja R, Qu J (1999) Damage in ceramic matrix composites with arbitrary microstructures. ASME MD 80:441–446
13. Bulsara VN, Talreja R, Qu J (1999) Damage initiation under transverse loading of unidirectional composites with arbitrarily distributed fibers. Comp Sci Tech 59:673–682
14. Chen HS, Acrivos A (1978) The effective elastic moduli of composite materials containing spherical inclusions at non-dilute concentrations. Int J Solids Struct 14:349–364
15. Chow CL, Wang J (1987) An anisotropic theory of elasticity for continuum damage mechanics. Int J Frac 20:381–390
16. Christensen RM, Lo KH (1979) Solutions for effective shear properties in three phase sphere and cylinder models. J Mech Phys Solids 27:315–330
17. Chung PW, Tamma KK, Namburu RR (2000) A finite element thermo-viscoelastic creep for heterogeneous structures with dissipative correctors. Finite Elem Anal Des 36:279–313
18. Clyne TW, Withers PJ (1993) An Introduction to Metal Matrix Composites. Cambridge University Press
19. Cordebois JP, Sidoroff F (1982) Anisotropic damage in elasticity and plasticity. J Mech Theor Appl 45–60
20. Costanzo F, Boyd JG, Allen DH (1996) Micromechanics and homogenization of inelastic composite materials with growing cracks. J Mech Phys Solids 44(3):333–370
21. Devries F, Dumontet H, Duvaut G, Lene F (1989) Homogenization and damage for composite structures. Int J Num Meth Engng 27:285–298
22. Eshelby JD (1957) The determination of the elastic field of an ellipsoidal inclusion and related problems. Proc R Soc (London) A 241:376–396
23. Fish J, Belsky V (1995) Multigrid method for periodic heterogeneous media, part II: multiscale modeling and quality control in multidimensional case. Comp Meth Appl Mech Engng 126:17–38
24. Fish J, Shek K (2000) Multiscale analysis of composite materials and structures. Comp Sci Tech 60:2547–2556
25. Fish J, Yu Q, Shek K (1999) Computational damage mechanics for composite materials based on mathematical homogenization. Int. J Num Meth Engng 45:1657–1679
26. Ghosh S, Moorthy S (1998) Particle fracture simulation in non-uniform microstructures of metal–matrix composites. Acta Mater 46:965–982

27. Ghosh S, Moorthy S (2004) Three dimensional Voronoi cell finite element model for modeling microstructures with ellipsoidal heterogeneities. Comput Mech 34 (6):510–531
28. Ghosh S, Mukhopadhyay SN (1991) A two dimensional automatic mesh generator for finite element analysis of random composites. Comp Struct 41:245–256
29. Ghosh S, Raghavan P (2004) Multiscale model for damage analysis in fiber reinforced composites with interfacial debonding materials. Int J Multiscale Comp Engng 2(4):621–645
30. Ghosh S, Lee K, Moorthy S (1995) Multiple scale analysis of heterogeneous elastic structures using homogenization theory and Voronoi cell finite element method. Int J Solids Struct 32:27–62
31. Ghosh S, Lee K, Moorthy S (1996) Two scale analysis of heterogeneous elastic-plastic materials with asymptotic homogenization and Voronoi cell finite element model. Comp Meth Appl Mech Engng 132:63–116
32. Ghosh S, Nowak Z, Lee K (1997) Quantitative characterization and modeling of composite microstructures by Voronoi cells. Acta Mater 45:2215–2234
33. Ghosh S, Nowak Z, Lee K (1997) Tessellation based computational methods for the characterization and analysis of heterogeneous microstructures. Comp Sci Tech 57:1187–1210
34. Ghosh S, Ling Y, Majumdar BS, Kim R (2000) Interfacial debonding in multiple fiber-reinforced composites. Mech Mater 32:561–591
35. Ghosh S, Lee K, Raghavan P (2001) A multi-level computational model for multiscale damage analysis in composite and porous materials. Int J Solids Struct 38(14):2335–2385
36. Ghosh S, Bai J, Raghavan P (2007) Concurrent multi-level model for damage evolution in microstructurally debonding composites. Mech Mater 39(3):241–266
37. Guedes JM, Kikuchi N (1991) Preprocessing and post processing for materials based on the homogenization method with adaptive finite element methods. Comp Meth Appl Mech Engng 83:143–198
38. Guo BQ, Babuska I (1986), The h–p version of the finite element method: Part I: The basic approximation results. Comp Mech 1:21–41
39. Hashin Z, Strikman S (1963) A variational approach to the theory of the elastic behavior of multiphase materials. J Mech Phys Solids 11:127–140
40. Hill R (1965) A self consistent mechanics of composite materials. J Mech Phys Solids 13:213–222.
41. Hill R (1967) The essential structure of constitutive laws for metal composites and polycrystals. J Mech Phys Solids 15:79–95
42. Kachanov LM (1987) Introduction to Continuum Damage Mechanics. Martinus Nijhoff, Dordrecht, Netherlands
43. Lee K, Ghosh S (1996) Small deformation multiscale analysis of heterogeneous materials with the Voronoi Cell finite element model and homogenization theory. Comp Mater Sci 7:131–146

44. Lene F (1986) Damage Constitutive Relations for Composite Materials. Engng Fract Mech 25(5):713–728
45. Li S, Ghosh S (2004) Debonding in composite microstructures with morphological variations. Int J Comput Meth 1(1):121–149
46. Li S, Ghosh S (2006) Extended Voronoi cell finite element model for multiple cohesive crack propagation in brittle materials. Int J Numer Meth Engng 65:1028–1067
47. Lions JL (1980) On some homogenization problems. Z Angew Math Mech 62:251–262
48. Moorthy S, Ghosh S (1996) A model for analysis of arbitrary composite and porous microstructures with Voronoi cell finite elements. Int J Numer Meth Engng 39:2363–2398
49. Moorthy S, Ghosh S (1998) A Voronoi cell finite element model for particle cracking in composite materials. Comp Meth Appl Mech Engng 151:377–400
50. Moorthy S, Ghosh S (2000) Adaptivity and convergence in the Voronoi cell finite element model for analyzing heterogeneous materials. Comp Meth Appl Mech Engng 185:37–74
51. Morgan RC, Babuska I (1991) An approach for constructing families of homogenized equations for periodic media, Part I: An integral representation and its consequences: SIAM J Math Anal 22:1–15
52. Mura T (1987) Micromechanics of Defects in Solids, 2nd edn. Kluwer Academic Publishers, Dordrecht
53. Nemat-Nasser S, Yu N, Hori M (1993) Bounds and estimates of overall moduli of composites with periodic microstructure. Mech Mater 15:163–181
54. Oden JT, Feng Y (1996) Local and pollution error estimation for finite element approximations of elliptic boundary value problems. J Comp Appl Math 74:245–293
55. Oden JT, Zohdi TI (1997) Analysis and adaptive modeling of highly heterogeneous elastic structures. Comput Meth Appl Mech Engng 148:367–391
56. Oden JT, Vemaganti K, Moes N (1999) Hierarchical modeling of heterogeneous solids. Comput Meth Appl Mech Engng 172:3–25
57. Ortiz M, Pandolfi (1999) A finite-deformation irreversible cohesive element for three-dimensional crack-propagation analysis. Int J Numer Meth Eng 44:1267–1282
58. Pagano NJ, Rybicki EF (1974) On the significance of effective modulus solutions for fibrous composites. J Comp Mater 8:214–228
59. Pagano NJ, Yuan FG (2000) Significance of effective modulus theory (homogenization) in composite laminate mechanics. Comp Sci Tech 60:2471–2488
60. Pyrz R (1994) Correlation of microstructure variability and local stress field in two-phase materials. Mat Sci Eng A 177:253–259
61. Raghavan P, Ghosh S (2004) Adaptive multiscale computational modeling of composite materials. Comp Model Engng Sci 5(2):151–170

62. Raghavan P, Ghosh S (2004) Concurrent multiscale analysis of elastic composites by a multi-level computational model. Comp Meth Appl Mech Engng 193(6–8):497–538
63. Raghavan P, Ghosh S (2005) A continuum damage mechanics model for unidirectional composites undergoing interfacial debonding. Mech Mater 37 (9):955–979
64. Raghavan P, Moorthy S, Ghosh S, Pagano NJ (2001), Revisiting the composite laminate problem with an adaptive multi-level computational model. Comp Sci Tech 61:1017–1040
65. Raghavan P, Li S, Ghosh S (2004) Two scale response and damage modeling of composite materials. Finite Elem Anal Des 40(12):1619–1640
66. Rastogi N, Soni SR, Nagar A (1998) Thermal stresses in aluminum-to-composite double-lap bonded joints. Adv Engng Software 29:273–281
67. Rybicki EF, Pagano NJ (1976) A study of the influence of microstructure on the modified effective modulus approach for composite laminates. In: Proceedings of the 1975 Int Conf Comp Mater 2, pp 198–207
68. Sanchez-Palencia E (1980) Non-homogeneous media and vibration theory. (Lecture notes in Physics, Vol 127, Springer, Berlin Heidelberg New York)
69. Shan Z, Gokhale AM (2002) Representative volume element for non-uniform micro-structure. Comp Mater Sci 24:361–379
70. Simo JC, Ju JW (1987) Strain and stress-based continuum damage models, Part I: Formulation. Int J Solids Struct 23(7):821–840
71. Smit RJM, Brekelmans WAM, Meijer HEH (1998) Prediction of the mechanical behavior of nonlinear heterogeneous systems by multi-level finite element modeling. Comp Meth Appl Mech Engng 155(1–2):181–192
72. Swaminathan S, Ghosh S, Pagano NJ (2006) Statistically equivalent representative volume elements for composite microstructures, Part I: Without damage. J Comp Mater 40(7):583–604
73. Swaminathan S, Ghosh S, Pagano NJ (2006) Statistically equivalent representative volume elements for composite microstructures, Part II: With evolving damage. J Comp Mater 40(7):605–621
74. Szabo B, Babuska I (1991) Finite Element Analysis. John Wiley & sons
75. Talreja R (1991) Continuum modeling of damage in ceramic matrix composites. Mech Mater 12:165–180
76. Talreja R (1994) Damage Mechanics of Composite Materials. Elsevier Science Ltd
77. Tamma KK, Chung PW (1999) Woven Fabric Composites: Developments in Engineering Bounds, Homogenization and Applications. Int J Numer Meth Engng 45:1757–1790
78. Terada K, Kikuchi N (2000) Simulation of the multiscale convergence in computational homogenization approaches. Int J Solids Struct 37:2285–2311
79. Vemaganti K, Oden JT (2001) Estimation of local modeling error and goal-oriented adaptive modeling of heterogeneous materials, Part II: a computational environment for adaptive modeling of heterogeneous elastic solids. Comp Meth Appl Mech Engng 190:6029–6214

80. Zhu YY, Cescotto S (1995) A fully coupled elasto-visco-plastic damage theory for anisotropic materials. Int J Solids Struct 32 (11):1607–1641
81. Zienkiewicz OC, Zhu JZ (1992) The super-convergence patch recovery and a posteriori error estimates, Part 1: The recovery technique. Inter J Numer Methods Engrg 33:1331–1364
82. Zohdi TI, Wriggers P (1999) A domain decomposition method for bodies with heterogeneous microstructure based on material regularization. Int J Solids Struct 36:2507–2525
83. Zohdi TI, Wriggers P (2004) Introduction to Computational Micromechanics. (Lecture Notes in Applied and Computational Mechanics, Vol 20)
84. Zohdi TI, Oden JT, Rodin GJ (1999) Hierarchical modeling of heterogeneous solids. Comput Meth Appl Mech Engng 172:3–25

Chapter 4: Multiscale and Multilevel Modeling of Composites

Young W. Kwon

Naval Postgraduate School, Monterey, CA, USA

4.1 Introduction

Composites have been used increasingly in various engineering applications which include, but are not limited to, the aerospace, automobile, sports, and leisure industries. To improve properties of composites so that they become stronger, stiffer, tougher, refractory, etc., it would be very useful to design the composite materials from the atomic levels. This requires proper multiscale and multilevel modeling techniques so that those techniques can be used for the design stage of new composites as well as the analysis of existing composites. This chapter presents multiscale and multilevel modeling techniques for different kinds of composite architectures which include particle-reinforced, fiber-reinforced, and woven fabric composites. The following sections describe these techniques.

4.2 Particulate Composites

4.2.1 Multiscale Analysis for Particulate Composites

A particle-reinforced composite, or particulate composite, is one of the simplest forms of composites. It has particles embedded in a matrix material. As a result, the multiscale analysis hierarchy is simple for the particulate composite, as illustrated in Fig. 4.1. The analysis connects the microscale, such as particles and matrix, to the mesoscale, such as the representative particulate composite, and finally to macroscale composites, such as a particulate composite structure [13, 23–32]. The multiscale analysis has two routes which complement each other for a complete cycle of analysis. The first route is the Stiffness Loop, and the other is the Stress Loop. For the Stiffness Loop, effective material properties are computed for an upper scale from material and geometric properties of the neighboring lower scale. For example, the effective particulate composite material properties are calculated from the particle and matrix material and geometric properties. The Particulate Module computes the effective properties, and it is described later. Then, the effective material properties are used for structural analysis of a composite as illustrated in Fig. 4.1.

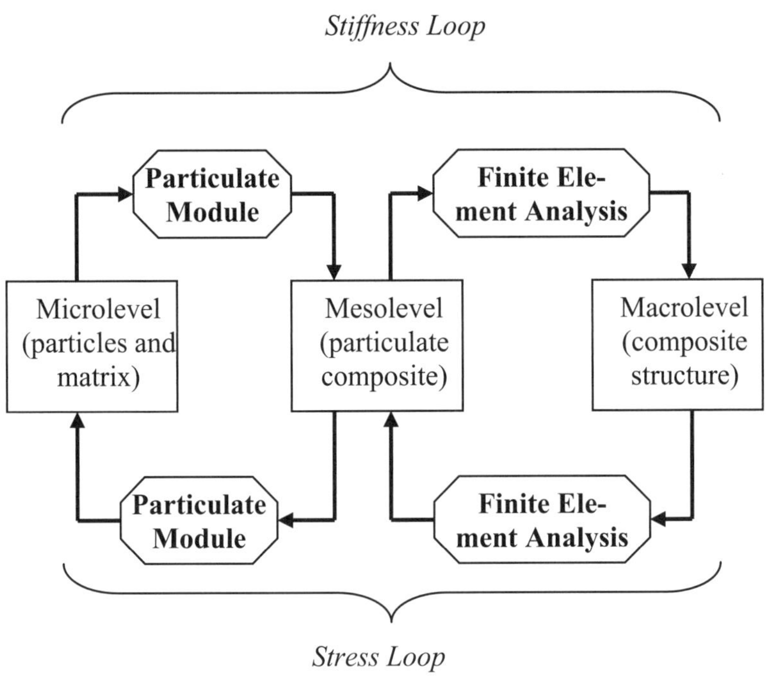

Fig. 4.1. Multiscale analysis hierarchy for a particulate composite

Structural analysis of the composite results in stresses, strains, and displacements at the macroscale. The stresses and strains are the composite level values, i.e., smeared values for the particles and the matrix. It is necessary to decompose the composite structural level stresses and strains into constituent level stresses and strains to apply the damage or failure criteria to the constituent materials, such as the particles and matrix. The same module used for the Stiffness Loop, i.e., Particulate Module is also used to compute the stresses and strains in the particle and matrix.

Sections 4.2.2 and 4.2.3 describe, respectively, the Particulate Module and the damage mechanics and crack initiation criterion used for the study.

4.2.2 Particulate Module

A representative unit cell is used for the present module. The purpose of this module is twofold. The first is to predict the effective stiffness of a particulate composite from the particle and matrix material properties as well as their geometric data. The second is to determine the microlevel stresses and strains occurring in the particle and matrix from the stresses and strains of the composite level. As a result, the module is used for both the Stiffness Loop and the Stress Loop.

To develop a representative unit cell for a particulate composite, a single representing particle surrounded by a matrix material is assumed. In general, every particle may have a different shape; however, the shape of the representative particle is simplified. A spherical shape would be a reasonable assumption. However, for mathematical simplicity, a cubic shape is assumed.

A microscale analysis of different shapes of particles using the boundary element method [20] showed that the effective material properties of the composite were insensitive to the actual shape of the particle. However, the same study indicated that the microscale stresses were rather sensitive to the particle shape. The sensitivity was mostly due to the stress concentration at sharp corners. For actual composites, each particle has a different shape and stress concentration. Practically, there is no way to consider all those different particle shapes and their stress concentration effects. A possible solution to this complex problem is using statistical mechanics. However, that approach is also very time-consuming. If it is assumed that the macro-level failure strength is more or less uniform for test coupons made out of the same particulate composite, the local stress concentration effects due to different shapes of particles may be smeared out in the composite test coupons. In this regard, a more regular shape of particle in the representative unit cell may be considered. Furthermore, the average values of stresses for the representative particle will be computed. This also makes the actual shape of particle less relevant for the unit cell.

Figure 4.2 shows the representative unit cell. With the assumption of symmetry, only one-eighth of the unit cell is shown in the figure, where the representative embedded particle is denoted by subcell *a*. The surrounding matrix material is represented by subcells *b* to *h* as illustrated in the figure. To clearly represent the relative positions of all subcells within the unit cell, the subcells are shown independently in the figure with lines and springs denoting their connections to neighboring subcells. The lines indicate continuous material between any two neighboring materials while the springs denote any potential interface effect between the particle and the matrix. The spring constant can be adjusted with either a strong or weak interface. The present model can have three different interface material properties along the three directions. However, for an isotropic material and damage behavior, all interface material properties, i.e., the spring constants, will be assumed to be the same. The size of the particle subcell a is $(V_p)^{1/3}$ where V_p is the particle volume fraction of the composite.

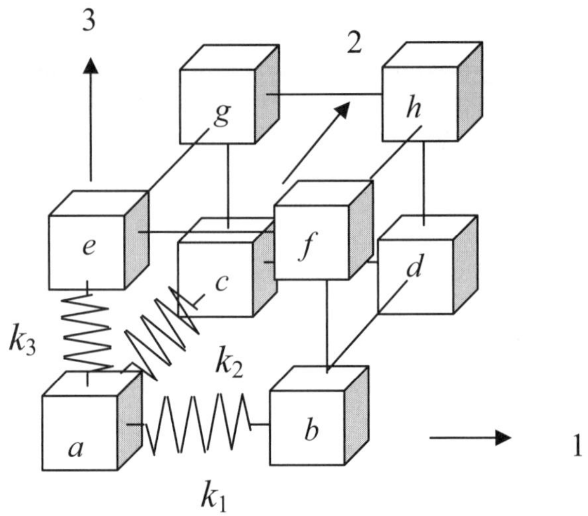

Fig. 4.2. Representative unit cell for a particulate composite

For each subcell, average stresses and strains are considered for the following derivation. Stresses must satisfy the equilibrium between any neighboring subcells as shown below

$$\sigma_{11}^a = \sigma_{11}^b, \quad \sigma_{11}^c = \sigma_{11}^d, \quad \sigma_{11}^e = \sigma_{11}^f, \quad \sigma_{11}^g = \sigma_{11}^h, \tag{4.1}$$

$$\sigma_{22}^a = \sigma_{22}^c, \quad \sigma_{22}^b = \sigma_{22}^d, \quad \sigma_{22}^e = \sigma_{22}^g, \quad \sigma_{22}^f = \sigma_{22}^h, \tag{4.2}$$

$$\sigma_{33}^a = \sigma_{33}^e, \quad \sigma_{33}^b = \sigma_{33}^f, \quad \sigma_{33}^c = \sigma_{33}^g, \quad \sigma_{33}^d = \sigma_{33}^h, \tag{4.3}$$

where superscript denotes the subcell identification as shown in Fig. 4.2 and subscript indicates the stress component. These equations are for normal components of stresses. Similar equations can be written for shear components, but they are omitted here to save space.

The subcell strains satisfy the following compatibility equations by assuming uniform deformation of the unit cell under periodic boundary conditions

$$l_{\rm p}\varepsilon_{11}^{a} + l_{\rm m}\varepsilon_{11}^{b} + \left(l_{\rm p}^{2}\sigma_{11}^{a} / k_{1}\right) = l_{\rm p}\varepsilon_{11}^{c} + l_{\rm m}\varepsilon_{11}^{d} = l_{\rm p}\varepsilon_{11}^{e} + l_{\rm m}\varepsilon_{11}^{f} = l_{\rm p}\varepsilon_{11}^{g} + l_{\rm m}\varepsilon_{11}^{h}, \quad (4.4)$$

$$l_{\rm p}\varepsilon_{22}^{a} + l_{\rm m}\varepsilon_{22}^{c} + \left(l_{\rm p}^{2}\sigma_{22}^{a} / k_{2}\right) = l_{\rm p}\varepsilon_{22}^{b} + l_{\rm m}\varepsilon_{22}^{d} = l_{\rm p}\varepsilon_{22}^{e} + l_{\rm m}\varepsilon_{22}^{g} = l_{\rm p}\varepsilon_{22}^{f} + l_{\rm m}\varepsilon_{22}^{h}, \quad (4.5)$$

$$l_{\rm p}\varepsilon_{33}^{a} + l_{\rm m}\varepsilon_{33}^{e} + \left(l_{\rm p}^{2}\sigma_{33}^{a} / k_{3}\right) = l_{\rm p}\varepsilon_{33}^{b} + l_{\rm m}\varepsilon_{33}^{f} = l_{\rm p}\varepsilon_{33}^{c} + l_{\rm m}\varepsilon_{33}^{g} = l_{\rm p}\varepsilon_{33}^{d} + l_{\rm m}\varepsilon_{33}^{h}, \quad (4.6)$$

where

$$l_{\rm p} = V_{\rm p}^{1/3}, \quad (4.7)$$

$$l_{\rm m} = 1 - l_{\rm p}. \quad (4.8)$$

Other necessary mathematical expressions are constitutive equations for the particle and matrix materials as well as for the composite. For the present particulate composite material, both constituent materials and the effective composite material are considered as isotropic materials. Furthermore, the unit cell stresses and strains are assumed to be the volume averages of subcell stresses and strains

$$\bar{\sigma}_{ij} = \sum_{n} V^{n}\sigma_{ij}^{n}, \quad (4.9)$$

$$\bar{\varepsilon}_{ij} = \sum_{n} V^{n}\varepsilon_{ij}^{n}, \quad (4.10)$$

where superimposed bar denotes the composite (unit cell level) stresses and strains, and V^n is the volume fraction of the nth subcell. The summation is over all subcells.

Algebraic manipulation of the previous equations finally yields the two main expressions as given below

$$[E^{\rm eff}] = [V][E][R], \quad (4.11)$$

$$\{\varepsilon\} = [R]\{\bar{\varepsilon}\}, \quad (4.12)$$

in which $[E^{\rm eff}]$ is the effective composite material property matrix, $[V]$ is the matrix composed of subcell volume fractions, $[E]$ is the matrix consisting of constituent material properties, and $[R]$ is the matrix relating the subcell strain vector consisting of particle and matrix strains $\{\varepsilon\}$ to the unit cell strain vector (effective composite strain) $\{\overline{\varepsilon}\}$.

Equation (4.11) is used for the Stiffness Loop while (4.12) is used for the Stress Loop. Once the microlevel strains are computed from (4.12), the constitutive equation of the particle and matrix material, respectively, is used at the microlevel to compute the microlevel stresses. Then, damage and failure criteria are applied to the microlevel stresses and strains.

4.2.3 Damage Mechanics and Crack Initiation Criterion

One of the advantages of applying damage and failure criteria at the microlevel is that even if the composite material has a different particle volume fraction, it is not necessary to obtain new composite material strength data from an experiment. Furthermore, all damage or failure modes can be simplified at the microlevel so that the damage or failure mechanism can be understood more clearly. For example, damage in a particulate composite can be classified into three categories as illustrated in Fig. 4.3: particle breakage, matrix cracking, and particle/matrix interface debonding. The interface debonding may be considered as matrix cracking at the boundary of particles. Different damage or failure criteria may be used for different damage or failure modes. For example, an isotropic damage theory can be applied to the matrix material if the material is isotropic and the damage progression is also assumed to be isotropic.

An experimental study of crack initiation and growth from a round notch tip in a composite showed that a crack initiated at the notch tip and

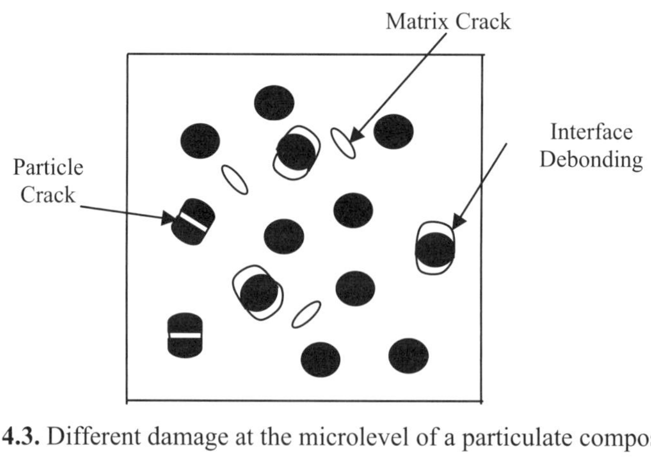

Fig. 4.3. Different damage at the microlevel of a particulate composite

grew until it reached a certain size. Then, the crack tip became blunted for a while until it propagated further. Using the computational model, it was hoped to predict the initial crack length before blunting and subsequent crack propagation. The crack size before the initial blunting is called the initial crack length. To predict such an initial crack size, the damage mechanics were used along with the criterion described below.

Let us consider a perforated plate under tension. Because of stress concentration, the stress very near the hole is much greater than the nominal value. Such high stress occurring very near the hole also results in damage at that location earlier than other locations. As the damage progresses very near the hole, the material at the same location becomes softer with greater damage. This means that even though the strain at very near the hole continues to grow with damage growth, the stress at the same location becomes lower with softer materials. Eventually, the stress very near the hole becomes lower than that in other locations until the stress at the tip of the hole goes down to nil. Such a process for stress reduction along with an increase of damage is illustrated in Fig. 4.4.

The case in Fig. 4.4d indicates damage saturation at the edge of the hole so that the stress there becomes nil. This means a crack can initiate from the hole edge at the onset of damage saturation. Then, the main question for the initial crack size is how far the crack will propagate from the initiation to form an initial crack before blunting. To answer this question, the material behavior near the hole edge is examined. This investigation shows that the material very near the hole edge has material softening. In other words, the slope of the stress–strain curve at the material softening zone becomes negative. This implies that the material softening zone is unstable. As a result, the crack initiated at the hole edge at the onset of damage saturation is expected to grow through the unstable material zone. This indicates that the initial crack size is equal to the length of the unstable material zone in front of the hole edge, as indicated by l_c in Fig. 4.4d. In summary, the criterion to predict the initial crack length is stated below:

At the onset of damage saturation at the edge of a hole, i.e., the stress becomes nil at that location, the length of the unstable material zone, i.e., material softening zone, is the initial crack size.

This criterion was tested against experimental data. The predicted results agreed very well with experimental results. For example, a particulate composite made of hard particles embedded in a very soft matrix material was studied. In this case, because particles are much stronger than the matrix material, a crack formed in the matrix material. Hence, the multiscale technique described in Sect. 4.2.1 was applied to the particulate composite, and the damage mechanics along with the proposed initial crack length criterion was applied to the matrix material level stresses and strains. The difference between the experimental and predicted results of the initial crack lengths formed at the edge of holes was almost uniformly between 5 and 10%. Figure 4.5 shows an initial crack formed in a particulate composite [30, 31].

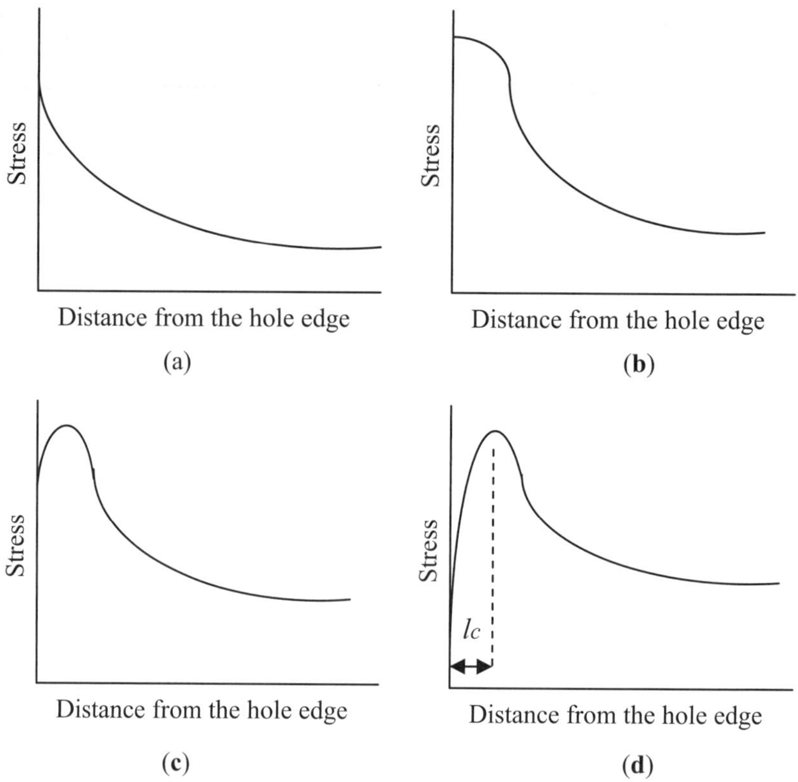

Fig. 4.4. Stress plots from the edge of a hole. Each stress plot from (**a**) to (**d**) is associated with increase of damage very near the hole along with load increase: (**a**) no damage state, (**b**) and (**c**) progressive damage states, and (**d**) saturated damage state

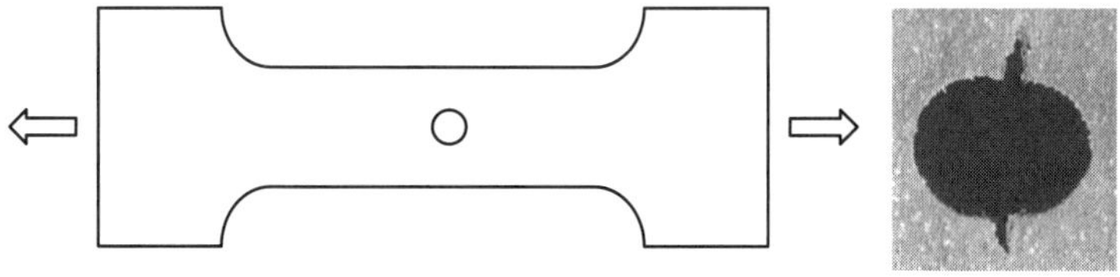

Fig. 4.5. Specimen under tensile loading. The figure on the *right* shows initial cracks formed at the edges of the hole

4.2.4 Study of Microstructural Inhomogeneity

Because the multiscale technique presented previously uses the material properties at the constituent material level, i.e., microscale level, it is easy to model inhomogeneous microstructure, such as a nonuniform particle distribution inside a composite. Even if the overall stiffness of a composite is much less dependent on the actual particle distribution, the effective strength of the composite depends on the particle distribution. For different particle distributions of the same amount of volume fraction, result in different local stresses which control failure at different load levels.

An experimental study was conducted for a particulate composite to determine the particle distribution. In this study, a large square specimen was cut into smaller sizes of square specimens subsequently. Then, at each level of cutting, the same sizes of specimens were examined using an X-ray technique to measure the particle volume fraction of every respective specimen. If the particle volume fraction were uniform in the original specimen, all smaller specimens would have the same particle volume fraction. However, as expected in a real specimen, there was a deviation of the particle volume fraction which is directly related to the mean intensity of the X-ray passing through each composite specimen. Therefore, the standard deviation was computed for each size of small samples. The study indicated that as the specimen size became smaller than a critical size, the standard deviation of the mean intensity of X-ray began to increase significantly. This result implies that an average particle distribution is quite uniform over a domain size greater than a critical size.

To model such an inhomogeneity of particle distribution, a composite specimen was divided into a number of domains with each domain size equal to the critical size. Then, the particle volume fraction was assumed to be the same for each domain. Every domain was further divided into much smaller subdomains. Particle volume fractions were varied randomly

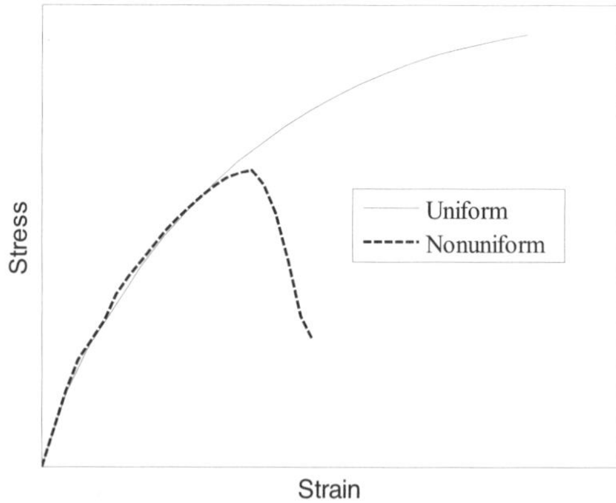

Fig. 4.6. Comparison of stress–strain curves for uniform and nonuniform particle distribution cases

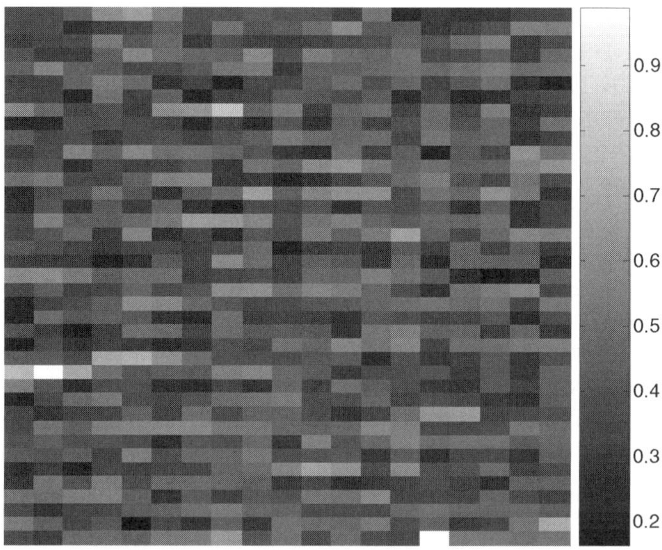

(a)

Fig. 4.7. (continue)

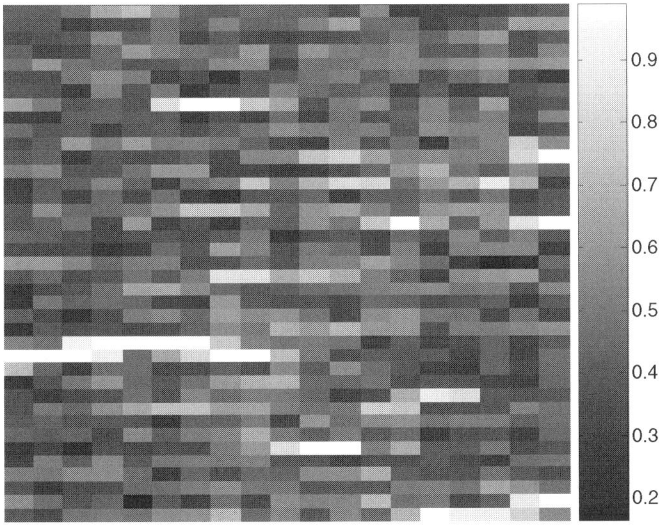

(**b**)

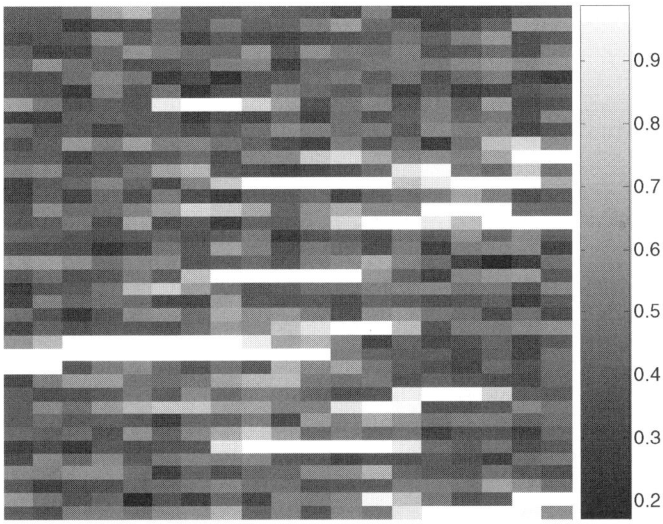

(**c**)

Fig. 4.7. Progressive damage plot from (**a**) lower damage state to (**c**) higher damage state. *White lines* indicate damage saturation, i.e., cracks among subdomains within each domain while the average particle volume fraction was equal to the domain particle volume fraction. By doing so, local stress effect due to inhomogeneous microstructure in a composite can be studied [28, 31].

Comparing the stress–strain curves between uniform and nonuniform particle distribution cases for the whole specimen, the nonuniform distribution case resulted in much lower failure strength and strain values as illustrated in Fig. 4.6. A nonuniform particle distribution resulted in locally higher stresses and strains than the uniform case. This caused crack initiations earlier for the nonuniform case and eventually earlier failure as shown in Fig. 4.7. The figure plots damage parameter distributions as damage increases. The lighter color in the gray scale indicates a greater damage state. The highest damage is represented by white, which denotes cracks.

This study also provided information regarding what would be a useful failure criterion for such a composite discussed above. Because the composite has much stronger and stiffer particles than the matrix, failure in the matrix did not affect much the effective stress at the composite level. Hence, the composite stresses would not be a good choice for a failure criterion at the macrolevel. On the other hand, the composite strains represent the damage very closely, and would be a good selection for a failure criterion at the macrolevel.

4.3 Fibrous Composites

4.3.1 Multiscale Analysis for Fibrous Composites

Fiber-reinforced composites can be constructed by multiple layers. Every layer has long unidirectional fibers embedded in a matrix material, and the fiber orientation of each layer is generally varied from layer to layer. For the fibrous composite, the multiscale analysis hierarchy is depicted in Fig. 4.8. Comparing to the particulate composite, the fibrous composite requires one more step for the multiscale analysis, which is the Lamination Module [7, 14–16, 19].

As illustrated in Fig. 4.8, the overall hierarchy has a Stiffness Loop and a Stress Loop for a complete cycle. The same modules are also used for both loops. First of all, the fiber and matrix material and geometric properties are used in the Fibrous Module to determine the effective

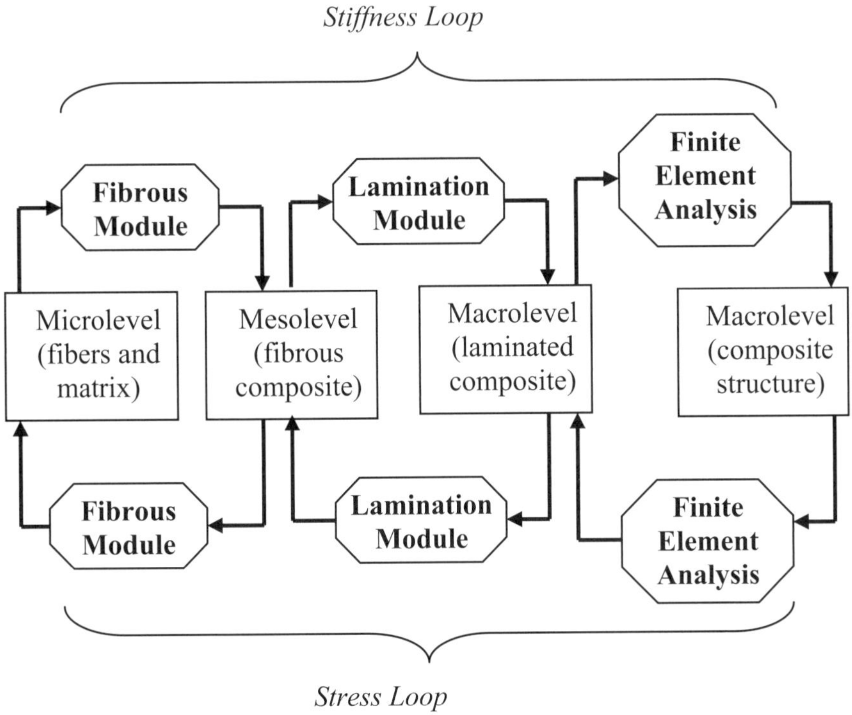

Fig. 4.8. Multiscale analysis hierarchy for a fibrous composite

material properties of a unidirectional fibrous composite. These composite properties are used for each lamina with its orientation of fibers with respect to the global coordinate system. The "Lamination Module" computes the effective properties of the laminated composite. Then, those properties are used for finite element analysis of a laminated composite structure. This completes the Stiffness Loop. Then, the reverse order is used to decompose stresses and strains at the macrolevel into those in the microlevel, i.e., stresses and strains in the fiber and matrix materials.

Once microlevel stresses and strains are computed, damage and/or failure criteria are applied to them. Because damage and failure are described at the constituent level, damage and failure modes are simplified and physics-based. At the microlevel, there are three potential damages and failures: fiber breakage, matrix cracking, and interface debonding. Different damage or failure criteria can be applied to those three different damage modes.

At the macrolevel, there are more complicated damage modes. For example, interlamina delamination, matrix cracking in a crossply lamina, and fiber splitting in a longitudinal layer are some of potential damage modes. All of these are associated with matrix cracking. Depending on the location of the matrix failure and its orientation to the layer and fiber orientations, the damage modes are differentiated as described above. For the macrolevel damage or failure criteria, each of those modes may require a different criterion and failure strength. However, when using the constituent level damage or failure criteria, all of those can be described using the same criterion associated with matrix damage. Simply, the damage or failure location and orientation will dictate the difference at the macroscale failure modes. As a result, damage and failure modes can be understood in unified and simplified concepts.

One of the advantages of the multiscale analysis technique is its flexibility for adaptation. For instance, if the fiber volume fraction was varied at the average sense or locally, the same multiscale technique can be used without any modification. It is not necessary to measure experimentally new strength and stiffness associated with the new volume fraction. Section 4.3.2 presents the modules used for the multiscale analysis of fibrous composites.

4.3.2 Fibrous Module

The Fibrous Module is based on a representative unit cell which contains a representative fiber surrounded by a matrix material. The fiber and matrix materials have, in general, significantly different properties. This is also true for the coefficients of thermal expansion. Due to this mismatch, thermal stresses arise at the micro level between the fiber and matrix. To minimize such thermal stresses, the fiber/matrix interface is designed to have a weak layer of bonding and slide each other as necessary. In this case, it would be beneficial to include the fiber/matrix interface layer in the representative unit cell model [10].

The unit cell model is shown in Fig. 4.9. The unit cell has a representative fiber along the x-axis, and cross-section containing the representative fiber, interface layer, and the matrix material in the y–z plane. Mostly, the fiber materials are elastic up to fracture while the matrix materials may be elastic, viscoelastic, or elastoplastic [3–6, 17]. As a result, an incremental formulation is presented here to incorporate inelastic deformation.

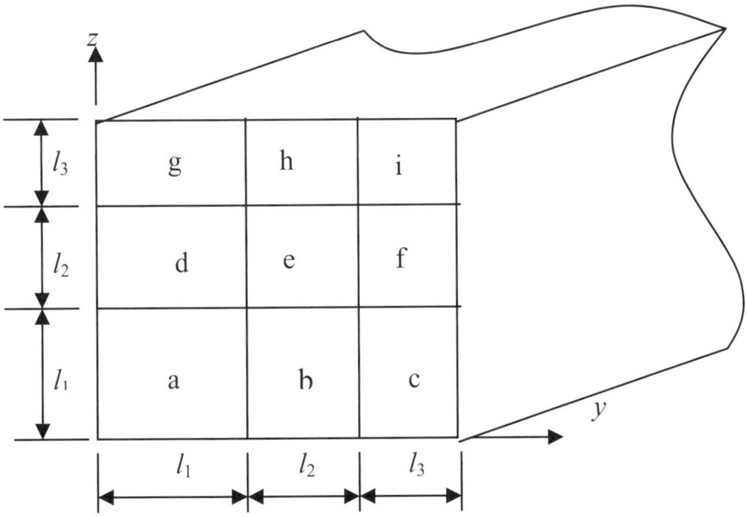

Fig. 4.9. Representative unit cell for a fibrous composite

To derive expressions to compute effective material properties for the Stiffness Loop and the stress decomposition for the Stress Loop, the following stress equilibrium of subcells is considered for the normal stress components

$$\begin{aligned} &\Delta\sigma_y^a = \Delta\sigma_y^b, \quad \Delta\sigma_y^b = \Delta\sigma_y^c, \quad \Delta\sigma_y^d = \Delta\sigma_y^e, \\ &\Delta\sigma_y^e = \Delta\sigma_y^f, \quad \Delta\sigma_y^g = \Delta\sigma_y^h, \quad \Delta\sigma_y^h = \Delta\sigma_y^i, \end{aligned} \tag{4.13}$$

$$\begin{aligned} &\Delta\sigma_z^a = \Delta\sigma_z^d, \quad \Delta\sigma_z^d = \Delta\sigma_z^g, \quad \Delta\sigma_z^b = \Delta\sigma_z^e, \\ &\Delta\sigma_z^e = \Delta\sigma_z^h, \quad \Delta\sigma_z^c = \Delta\sigma_z^f, \quad \Delta\sigma_z^f = \Delta\sigma_z^i, \end{aligned} \tag{4.14}$$

where superscripts are used to identify the subcells and stresses are assumed to be uniform within each subcell. Similar expressions can be developed for the shear stress components. The first set of equations is equilibrium in the y-axis and the second set is for the z-axis.

The deformation compatibilities among subcells are expressed as followed for the y- and z-axis, respectively

$$\begin{aligned} l_1\Delta\varepsilon_y^a + l_2\Delta\varepsilon_y^b + l_3\Delta\varepsilon_y^c &= l_1\Delta\varepsilon_y^d + l_2\Delta\varepsilon_y^e + l_3\Delta\varepsilon_y^f \\ &= l_1\Delta\varepsilon_y^g + l_2\Delta\varepsilon_y^h + l_3\Delta\varepsilon_y^i, \end{aligned} \tag{4.15}$$

$$\begin{aligned} l_1\Delta\varepsilon_z^a + l_2\Delta\varepsilon_z^d + l_3\Delta\varepsilon_z^g &= l_1\Delta\varepsilon_z^b + l_2\Delta\varepsilon_z^e + l_3\Delta\varepsilon_z^h \\ &= l_1\Delta\varepsilon_z^c + l_2\Delta\varepsilon_z^f + l_3\Delta\varepsilon_z^i, \end{aligned} \tag{4.16}$$

where

$$l_1 = \sqrt{v_\mathrm{f}}, \quad l_2 = \sqrt{l_1^2 + v_\mathrm{m}} - l_1, \quad l_3 = 1 - l_1 - l_2, \tag{4.17}$$

with v_f and v_m denoting the fiber and the matrix volume fractions. Here, the fiber/matrix interface layer is not included in the matrix volume fraction. The deformation compatibility in the fiber direction is expressed as

$$\Delta\varepsilon_x^a = \Delta\varepsilon_x^b, \quad \Delta\varepsilon_x^b = \Delta\varepsilon_x^c, \tag{4.18}$$

$$\Delta\varepsilon_x^b = \Delta\varepsilon_x^e, \quad \Delta\varepsilon_x^d = \Delta\varepsilon_x^e, \tag{4.19}$$

$$\Delta\varepsilon_x^c = \Delta\varepsilon_x^f, \quad \Delta\varepsilon_x^f = \Delta\varepsilon_x^i, \quad \Delta\varepsilon_x^g = \Delta\varepsilon_x^h, \quad \Delta\varepsilon_x^h = \Delta\varepsilon_x^i. \tag{4.20}$$

Each subcell satisfies its own constitutive equation depending on the material behavior, as given below

$$\Delta\sigma_{ij}^n = E_{ijkl}^n \left(\Delta\varepsilon_{kl}^n - \alpha_{kl}^n \Delta\theta\right), \tag{4.21}$$

where α_{kl}^n is the coefficient of thermal expansion tensor, and $\Delta\theta$ is the temperature change. Furthermore, for elastoplastic deformation, the material property tensor, E_{ijkl}^n may be expressed as

$$[E_\mathrm{ep}] = [E_\mathrm{e}] - \frac{[E_\mathrm{e}]\{q\}\{q\}^\mathrm{T}[E_\mathrm{e}]}{H' + \{q\}^\mathrm{T}[E_\mathrm{e}]\{q\}}, \tag{4.22}$$

in which $[E_\mathrm{ep}]$ and $[E_\mathrm{e}]$ are the elastoplastic and elastic material property matrices, respectively, $\{q\}$ is computed from

$$\{q\} = \frac{\partial F}{\partial \sigma_{ij}^n}, \tag{4.23}$$

where F is the yield function, and H' is the slope of the stress vs. plastic strain plot.

The unit cell stress and strain increments are the volume average of the subcell stress and strain increments as follows

$$\Delta\bar{\sigma}_{ij} = \sum V^k \Delta\sigma_{ij}^k \tag{4.24}$$

and

$$\Delta\bar{\varepsilon}_{ij} = \sum V^k \Delta\varepsilon_{ij}^k, \tag{4.25}$$

where V^k is the volume fraction of the kth subcell. The subcell stresses and strains are related by the effective constitutive equation as below:

$$\Delta\bar{\sigma}_{ij} = \bar{E}_{ijkl}(\Delta\bar{\varepsilon}_{kl} - \bar{\alpha}_{kl}\Delta\theta). \tag{4.26}$$

Putting together the above equations results in the following expressions

$$[T]\{\Delta\varepsilon\} = \{f\}, \tag{4.27}$$

in which $[T]$ is the matrix containing material and geometric properties, $\{\Delta\varepsilon\}$ is the vector containing strain increments of all subcells, and vector $\{f\}$ is expressed as

$$\{f\}^{\mathrm{T}} = \left\{\{0\}^{\mathrm{T}} \quad \{\Delta\bar{\varepsilon}\}^{\mathrm{T}}\right\}. \tag{4.28}$$

Solving for $\{\Delta\varepsilon\}$ yields

$$\{\Delta\varepsilon\} = [R]\{f\} = [R_2]\{\Delta\bar{\varepsilon}\}, \tag{4.29}$$

where matrix $[R]$ consists of two submatrices as shown below:

$$[R] = [T]^{-1} = \left[[R_1] \quad [R_2]\right]. \tag{4.30}$$

Eventually, substitution of (4.29) into (4.21), and the resulting expression into (4.24) yield

$$[\bar{E}] = [V][E][R_2], \tag{4.31}$$

$$\{\bar{\alpha}\} = -[\bar{E}]^{-1}[V]\frac{[E][R_1]\{\Delta\tau\} - \{\tau\}}{\Delta\theta}. \tag{4.32}$$

Equation (4.31) gives the effective stiffness of the unidirectional composite. Here $[V]$ and $[E]$ are the matrices constructed from the volume fractions and the material properties of subcells, respectively. On the other hand, (4.32) is used to compute the effective coefficients of thermal expansion. The details of the derivation and explanation are provided in [4]. Equations (4.31) and (4.32) are utilized for the Stiffness Loop while (4.32) is necessary for the Stress Loop

$$\{\sigma\} = [E]([R_1]\{\Delta\tau\} + [R_2]\{\alpha^{\text{unit}}\}\Delta\theta) - \{\tau\}. \tag{4.33}$$

The above equation computes the microlevel stresses at the constituent material level.

Once the effective material properties of each unidirectional layer are calculated as discussed previously, the lamination theories are used to determine the overall properties of the stacked layer. Either the classical lamination or higher-order lamination theory may be used. If the interlayer delamination is to be included in the model, each layer needs to be modeled individually; or a partial stack of layers may be modeled. The Lamination Module will vary on a case-by-case basis.

4.3.3 Examples of Fibrous Composites

In the example discussed below, the fiber was assumed to be elastic with an elastic modulus of 420 GPa and Poisson's ratio of 0.25. The matrix was considered to be elastoplastic with elastic modulus 70 GPa, Poisson's ratio 0.33, yield strength 500 MPa, and linear hardening modulus 1.2 GPa. The effective material properties of the composite were computed for the elastic and plastic ranges for different fiber volume fractions. The "Fibrous Module" described in Sect. 4.3.2 was used for the computation, and the finite element analysis of the microscale analysis was also conducted for comparison. The finite element model considered the fiber and matrix discretely in detail. Figure 4.10 shows the effective stress–strain curve along the fiber direction for various ratios of fiber volume fractions. Both finite element and analytical results compare very well as shown in the figure. In general, the stress–strain curve is bilinear because of linear hardening of the matrix material after yielding. When there is a low-volume fraction of fibers, the bilinear characteristics are more obvious than a higher fiber volume fraction because the matrix plays more roles for the former compared to the latter. In other words, the matrix plastic deformation is more noticeable as the strain increases for a lower fiber volume fraction. On the other hand, a higher volume fraction shows a very small effect of the matrix plastic deformation. For the latter case, the stress–strain curve is very close to linear.

Figure 4.11 shows the effective stiffness of the composite in the transverse direction. As expected, the matrix material's plastic deformation affects significantly the transverse effective stiffness for the plastic strain range. When the fiber volume fraction is not large, both analytical and finite element solutions are close to each other even for the plastic range. As the fiber volume fraction increases, there is discrepancy between the two solutions for the plastic strain rage. Figure 4.12 plots the effective longitudinal Poisson's ratio as a function of the applied strain for various fiber volume fractions. The figure shows that Poisson's ratios jump to near 0.5 for the range of plastic deformation.

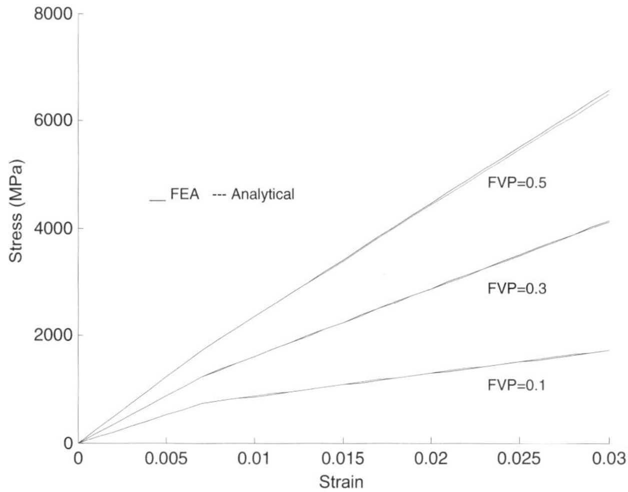

Fig. 4.10. Effective stress–strain plot in the longitudinal direction of a unidirectional fibrous composite for different fiber volume fractions

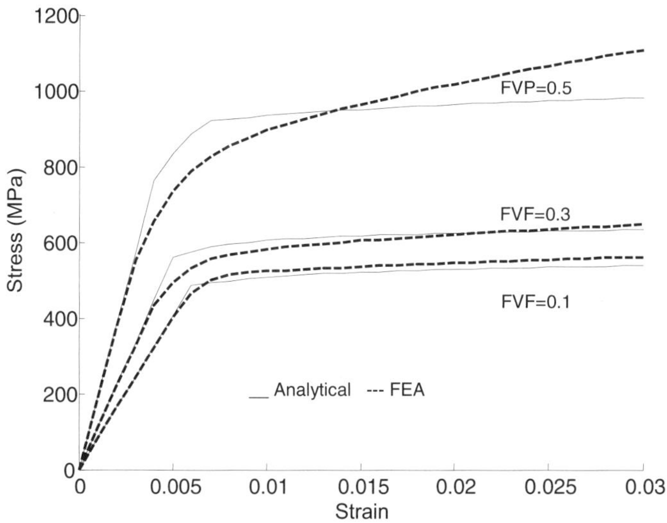

Fig. 4.11. Effective stress–strain plot in the transverse direction of a unidirectional fibrous composite for different fiber volume fractions

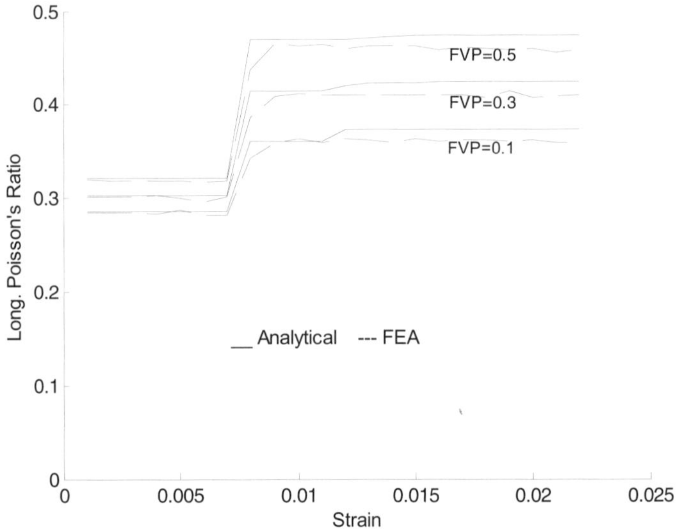

Fig. 4.12. Plot of strain vs. effective Poisson's ratio in the longitudinal direction of a unidirectional fibrous composite for different fiber volume fractions

4.4 Woven Fabric Composites

4.4.1 Multiscale Analysis for Woven Fabric Composites

Woven fabric composites have a more general analysis hierarchy than other kinds of composites. As a result, the analysis hierarchies for particulate and fibrous composites are special cases of those for woven fabric composites. Figure 4.13 illustrates the hierarchy for multiscale analysis of woven fabric composites. If the Fabric Module is omitted in Fig. 4.13, the procedure is the same as that for a fibrous composite. Because other modules were discussed previously, only the "Fabric Module" is presented here. The Fabric Module depends on the weaving pattern as sketched in Fig. 4.14. Because plain weave and twill weave are commonly used, Fabric Modules for those weave patterns are presented [8, 12, 18, 35].

4.4.2 Fabric Module for Plain Weave

The Fabric Module relates material properties of a unidirectional strand to the effective material properties of the woven fabric composite. Like the

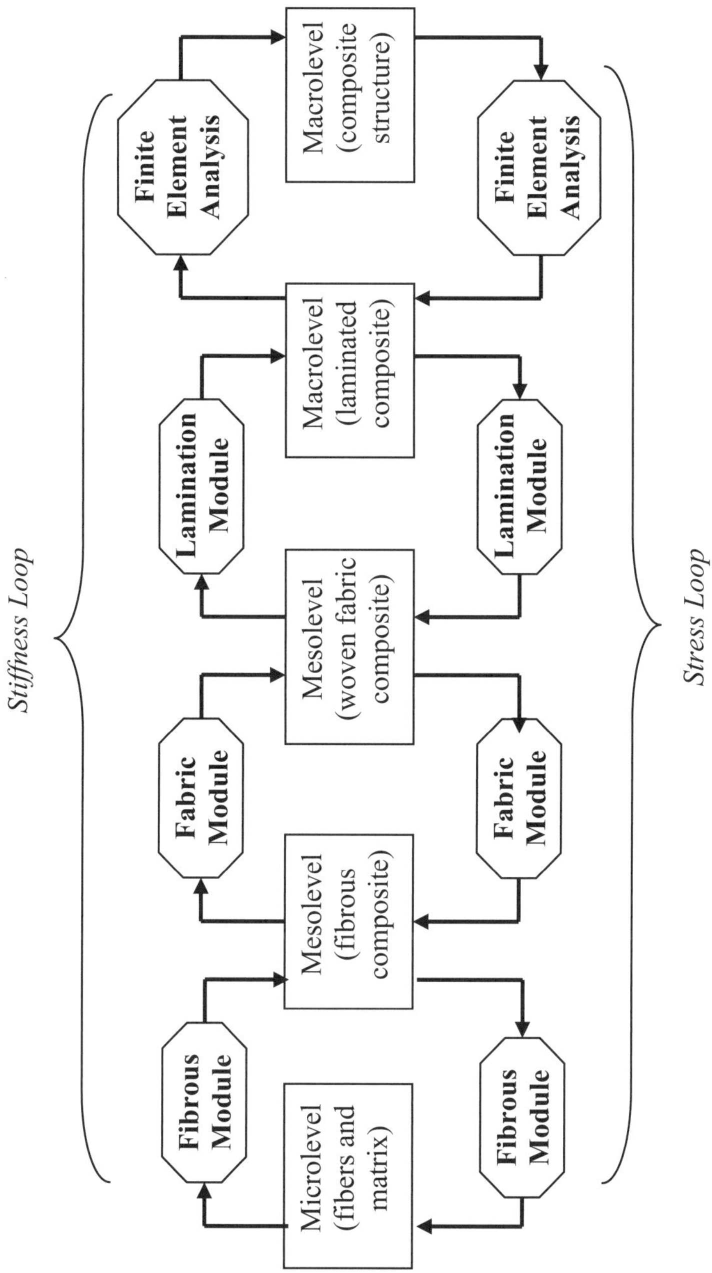

Fig. 4.13. Multiscale analysis hierarchy for a woven fabric composite

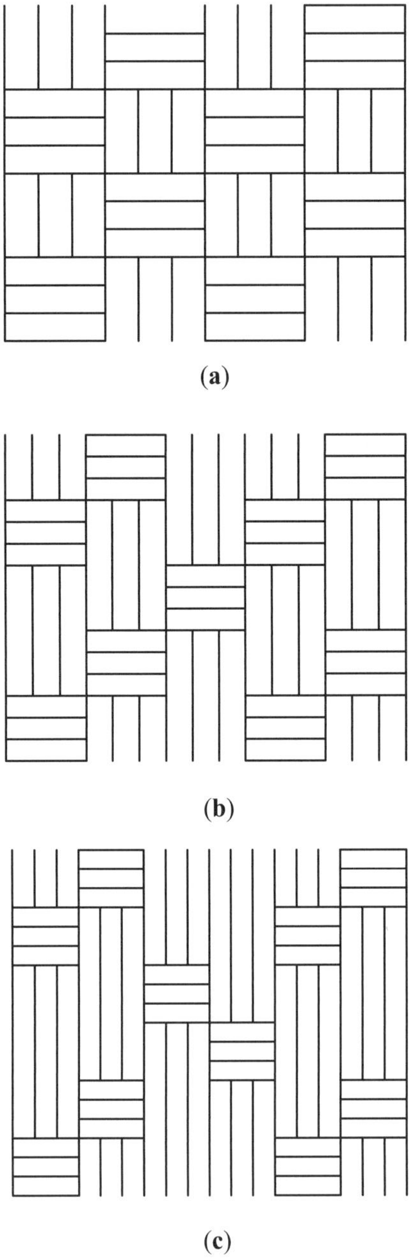

Fig. 4.14. Various weave patterns for woven fabric composites (**a**) Plain weave pattern (**b**) Twill weave pattern (**c**) Satin weave pattern

previous modules, this module has two functions: computation of the effective properties of the woven fabric using the strand materials and their weaving information, and decomposition of the woven strains and stresses into the strand strains and stresses. Plain weave is the simplest weave pattern of woven fabrics, as shown in Fig. 4.14. The representative unit cell for the plain weave is shown in Fig. 4.15.

The unit cell model for the plain weave has 13 subcells, and most of those subcells have fibers along the *x*- or *y*-axis. On the other hand, four subcells (*b*, *d*, *f*, and *h* in Fig. 4.15) have fibers in inclined orientations, and subcell *e* is filled with the matrix material.

First of all, the normal stress/strain components are discussed below. There are three normal strains for each subcell with the total of 39 strains $((\varepsilon_{kl}^{\text{str}})^n$, $n = a, b,\ldots,m)$. There are three normal strains at the unit cell level $(\varepsilon_{kl}^{\text{wf}})$. Those strains represent the average values of each subcell or the unit cell, respectively. To relate them, equilibrium and compatibility conditions are applied. At respective interfaces between any two neighboring subcells, the normal stresses on the interface plane should be in equilibrium.

$$(\sigma_x^{\text{str}})^a = (\sigma_x^{\text{str}})^b, \quad (\sigma_x^{\text{str}})^b = (\sigma_x^{\text{str}})^k, \tag{4.34}$$

$$(\sigma_x^{\text{str}})^c = (\sigma_x^{\text{str}})^j, \quad (\sigma_x^{\text{str}})^d = (\sigma_x^{\text{str}})^e, \tag{4.35}$$

$$(\sigma_x^{\text{str}})^e = (\sigma_x^{\text{str}})^f, \quad (\sigma_x^{\text{str}})^g = (\sigma_x^{\text{str}})^m, \tag{4.36}$$

$$(\sigma_x^{\text{str}})^h = (\sigma_x^{\text{str}})^i, \quad (\sigma_x^{\text{str}})^i = (\sigma_x^{\text{str}})^l, \tag{4.37}$$

$$(\sigma_y^{\text{str}})^a = (\sigma_y^{\text{str}})^l, \quad (\sigma_y^{\text{str}})^d = (\sigma_y^{\text{str}})^j, \tag{4.38}$$

$$(\sigma_y^{\text{str}})^d = (\sigma_y^{\text{str}})^g, \quad (\sigma_y^{\text{str}})^b = (\sigma_y^{\text{str}})^e, \tag{4.39}$$

$$(\sigma_y^{\text{str}})^e = (\sigma_y^{\text{str}})^h, \quad (\sigma_y^{\text{str}})^c = (\sigma_y^{\text{str}})^f, \tag{4.40}$$

$$(\sigma_y^{\text{str}})^f = (\sigma_y^{\text{str}})^m, \quad (\sigma_y^{\text{str}})^i = (\sigma_y^{\text{str}})^k, \tag{4.41}$$

$$(\sigma_z^{\text{str}})^a = (\sigma_z^{\text{str}})^j, \quad (\sigma_z^{\text{str}})^c = (\sigma_z^{\text{str}})^k, \tag{4.42}$$

$$(\sigma_z^{\text{str}})^g = (\sigma_z^{\text{str}})^l, \quad (\sigma_z^{\text{str}})^i = (\sigma_z^{\text{str}})^m, \tag{4.43}$$

where the superscript denotes the subcells in Fig. 4.15 and the subscript indicates the stress components. In addition, some strain compatibility conditions are assumed for a uniform deformation.

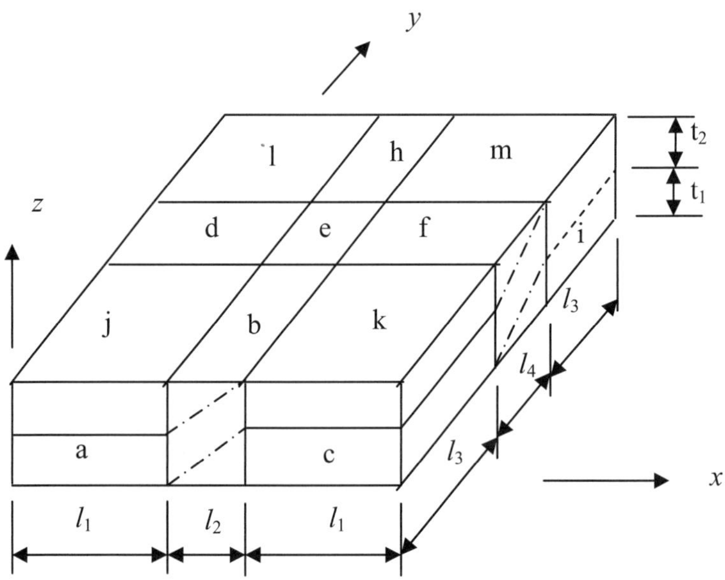

Fig. 4.15. Representative unit cell for the plain weave composite

$$(\varepsilon_x^{\text{str}})^a + (\varepsilon_x^{\text{str}})^k = (\varepsilon_x^{\text{str}})^c + (\varepsilon_x^{\text{str}})^j, \tag{4.44}$$

$$l_1(\varepsilon_x^{\text{str}})^j + l_2(\varepsilon_x^{\text{str}})^b + l_1(\varepsilon_x^{\text{str}})^c = l_1(\varepsilon_x^{\text{str}})^d + l_2(\varepsilon_x^{\text{str}})^e + l_1(\varepsilon_x^{\text{str}})^f, \tag{4.45}$$

$$l_1(\varepsilon_x^{\text{str}})^d + l_2(\varepsilon_x^{\text{str}})^e + l_1(\varepsilon_x^{\text{str}})^f = l_1(\varepsilon_x^{\text{str}})^g + l_2(\varepsilon_x^{\text{str}})^h + l_1(\varepsilon_x^{\text{str}})^m, \tag{4.46}$$

$$(\varepsilon_x^{\text{str}})^g + (\varepsilon_x^{\text{str}})^m = (\varepsilon_x^{\text{str}})^i + (\varepsilon_x^{\text{str}})^l, \tag{4.47}$$

$$(\varepsilon_y^{\text{str}})^a + (\varepsilon_y^{\text{str}})^l = (\varepsilon_y^{\text{str}})^g + (\varepsilon_y^{\text{str}})^j, \tag{4.48}$$

$$l_3(\varepsilon_y^{\text{str}})^j + l_4(\varepsilon_y^{\text{str}})^d + l_3(\varepsilon_y^{\text{str}})^g = l_3(\varepsilon_y^{\text{str}})^b + l_4(\varepsilon_y^{\text{str}})^e + l_3(\varepsilon_y^{\text{str}})^h, \tag{4.49}$$

$$l_3(\varepsilon_y^{\text{str}})^b + l_4(\varepsilon_y^{\text{str}})^e + l_3(\varepsilon_y^{\text{str}})^h = l_3(\varepsilon_y^{\text{str}})^c + l_4(\varepsilon_y^{\text{str}})^f + l_3(\varepsilon_y^{\text{str}})^m, \tag{4.50}$$

$$(\varepsilon_y^{\text{str}})^c + (\varepsilon_y^{\text{str}})^m = (\varepsilon_y^{\text{str}})^i + (\varepsilon_y^{\text{str}})^k, \tag{4.51}$$

$$t_1(\varepsilon_z^{\text{str}})^a + t_2(\varepsilon_z^{\text{str}})^j = (t_1 + t_2)(\varepsilon_z^{\text{str}})^b, \tag{4.52}$$

$$t_1(\varepsilon_z^{\text{str}})^c + t_2(\varepsilon_z^{\text{str}})^k = (t_1 + t_2)(\varepsilon_z^{\text{str}})^b, \tag{4.53}$$

$$t_1(\varepsilon_z^{\text{str}})^c + t_2(\varepsilon_z^{\text{str}})^k = (t_1 + t_2)(\varepsilon_z^{\text{str}})^d, \tag{4.54}$$

$$(\varepsilon_z^{\text{str}})^d = (\varepsilon_z^{\text{str}})^e, \tag{4.55}$$

$$(\varepsilon_z^{\text{str}})^e = (\varepsilon_z^{\text{str}})^f, \tag{4.56}$$

$$t_1(\varepsilon_z^{\text{str}})^g + t_2(\varepsilon_z^{\text{str}})^l = (t_1 + t_2)(\varepsilon_z^{\text{str}})^f, \tag{4.57}$$

$$t_1(\varepsilon_z^{\text{str}})^g + t_2(\varepsilon_z^{\text{str}})^l = (t_1 + t_2)(\varepsilon_z^{\text{str}})^h, \tag{4.58}$$

$$t_1(\varepsilon_z^{\text{str}})^i + t_2(\varepsilon_z^{\text{str}})^m = (t_1 + t_2)(\varepsilon_z^{\text{str}})^h. \tag{4.59}$$

Here, l_i and t_i are the dimensions of the plain weave construction as shown in Fig. 4.15. Finally, the unit cell strain and stress are assumed to be the volume average of the subcell strains and stresses, respectively. That is,

$$\varepsilon_{ij}^{\text{wf}} = \sum_{n=a,\ldots} V^n(\varepsilon_{ij}^{\text{str}})^n, \tag{4.60}$$

$$\sigma_{ij}^{\text{wf}} = \sum_{n=a\ldots} V^n(\sigma_{ij}^{\text{str}})^n. \tag{4.61}$$

Here, V^n is the volume fraction of the nth subcell. The constitutive equation for each subcell is expressed as

$$(\sigma_{ij}^{\text{str}})^n = (E_{ijkl}^{\text{str}})^n(\varepsilon_{kl}^{\text{str}})^n \quad (i,j,k,l = 1,2,3), \tag{4.62}$$

where the summation sign convention is applied only to the subscripts, k and l. For each subcell, $(E_{ijkl}^{\text{str}})^n$ should be determined based on the orientation of the strand inside the subcell. Algebraic manipulation of the previous equations, i.e., (4.34)–(4.62), yields the following relationship:

$$E_{ijkl}^{\text{wf}} = f\left((E_{ijkl}^{\text{str}})^n, l_i, t_i\right), \tag{4.63}$$

$$(\varepsilon_{ij}^{\text{str}})^n = g\left(\varepsilon_{ij}^{\text{wf}}, l_i, t_i\right). \tag{4.64}$$

Equations (4.63) and (4.64) are equivalent to (4.52) and (4.50) for the fiber-strand module, respectively. Equation (4.63) computes the effective material properties of the woven fabric composite based on the material and geometric properties of the strands while (4.64) calculates the strains at the strand level from the woven fabric strains. Once the strand strains are computed, strand stresses can be computed from the constituent equations.

A similar derivation can be made for shear components. For example, the shear component parallel to x–y plane can be derived as shown below. The shear stress equilibrium at the subcell interfaces is written as

$$t_1(\sigma_{xy}^{\text{str}})^a + t_2(\sigma_{xy}^{\text{str}})^j = (t_1 + t_2)(\sigma_{xy}^{\text{str}})^b, \tag{4.65}$$

$$t_1(\sigma_{xy}^{\text{str}})^c + t_2(\sigma_{xy}^{\text{str}})^k = (t_1 + t_2)(\sigma_{xy}^{\text{str}})^b, \tag{4.66}$$

$$t_1(\sigma_{xy}^{\text{str}})^a + t_2(\sigma_{xy}^{\text{str}})^j = (t_1 + t_2)(\sigma_{xy}^{\text{str}})^d, \tag{4.67}$$

$$t_1(\sigma_{xy}^{\text{str}})^g + t_2(\sigma_{xy}^{\text{str}})^l = (t_1 + t_2)(\sigma_{xy}^{\text{str}})^d, \tag{4.68}$$

$$t_1(\sigma_{xy}^{\text{str}})^g + t_2(\sigma_{xy}^{\text{str}})^l = (t_1 + t_2)(\sigma_{xy}^{\text{str}})^h, \tag{4.69}$$

$$t_1(\sigma_{xy}^{\text{str}})^i + t_2(\sigma_{xy}^{\text{str}})^m = (t_1 + t_2)(\sigma_{xy}^{\text{str}})^h, \tag{4.70}$$

$$t_1(\sigma_{xy}^{\text{str}})^i + t_2(\sigma_{xy}^{\text{str}})^m = (t_1 + t_2)(\sigma_{xy}^{\text{str}})^f, \tag{4.71}$$

$$(\sigma_{xy}^{\text{str}})^b = (\sigma_{xy}^{\text{str}})^e. \tag{4.72}$$

Assumed strain compatibility expressions are:

$$(\varepsilon_{xy}^{\text{str}})^a = (\varepsilon_{xy}^{\text{str}})^j, \tag{4.73}$$

$$(\varepsilon_{xy}^{\text{str}})^c = (\varepsilon_{xy}^{\text{str}})^k, \tag{4.74}$$

$$(\varepsilon_{xy}^{\text{str}})^g = (\varepsilon_{xy}^{\text{str}})^l, \tag{4.75}$$

$$(\varepsilon_{xy}^{\text{str}})^c = (\varepsilon_{xy}^{\text{str}})^i. \tag{4.76}$$

Use of (4.65)–(4.76) along with (4.60)–(4.62) for the shear component yields the shear component part of (4.63) and (4.64).

4.4.3 Fabric Module for 2/2-Twill Weave

The representative unit cell model for the 2/2-twill weave is sketched in Fig. 4.16. The unit cell was divided into 77 subcells. Of the 77 subcells, some were considered to have the same average stress states. The finite element model was run on several materials under a uniformed displacement to validate this assumption. Figure 4.17 is a schematic depiction of the subcells. Those having the same number designation are assumed to have the same average stress state. As a result, 17 independent subcells are used in the unit cell model. Forty-eight linearly independent equations were developed which relate the normal stresses and normal strains of the 17 uniquely stressed subcells. These 48 equations are the basis for the unit cell model. The average stress and strain of each subcell are used in all subsequent equations. The first set of equations represents the stress equilibrium at the subcell interfaces. Applying equilibrium to any two neighboring subcells results in the following normal stress equations, where the subscripts indicate stress components and the superscripts designate the subcell numbers.

The subsequent equations are a set of normal stress equilibriums in the 1-, 2-, and 3-directions, respectively

$$\begin{aligned}
&\sigma_{11}^{2}=\sigma_{11}^{8},\quad \sigma_{11}^{2}=\sigma_{11}^{15},\quad \sigma_{11}^{2}=\sigma_{11}^{14},\quad \sigma_{11}^{3}=\sigma_{11}^{7},\quad \sigma_{11}^{3}=\sigma_{11}^{15},\\
&\sigma_{11}^{3}=\sigma_{11}^{14},\quad \sigma_{11}^{5}=\sigma_{11}^{9},\quad \sigma_{11}^{4}=\sigma_{11}^{6},\quad \sigma_{11}^{1}=\sigma_{11}^{16},\quad \sigma_{11}^{1}=\sigma_{11}^{17},\\
&\sigma_{11}^{1}=\sigma_{11}^{10}+\sigma_{11}^{11},\quad \sigma_{11}^{1}=\sigma_{11}^{12}+\sigma_{11}^{13},
\end{aligned} \tag{4.77}$$

$$\begin{aligned}
&\sigma_{22}^{4}=\sigma_{22}^{16},\quad \sigma_{22}^{4}=\sigma_{22}^{17},\quad \sigma_{22}^{4}=\sigma_{22}^{12},\quad \sigma_{22}^{5}=\sigma_{22}^{17},\quad \sigma_{22}^{5}=\sigma_{22}^{17},\\
&\sigma_{22}^{5}=\sigma_{22}^{11},\quad \sigma_{22}^{3}=\sigma_{22}^{13},\quad \sigma_{22}^{2}=\sigma_{22}^{10},\quad \sigma_{22}^{1}=\sigma_{22}^{14},\quad \sigma_{22}^{1}=\sigma_{22}^{15},\\
&\sigma_{22}^{1}=\sigma_{22}^{6}+\sigma_{22}^{7},\quad \sigma_{22}^{1}=\sigma_{22}^{8}+\sigma_{22}^{9},
\end{aligned} \tag{4.78}$$

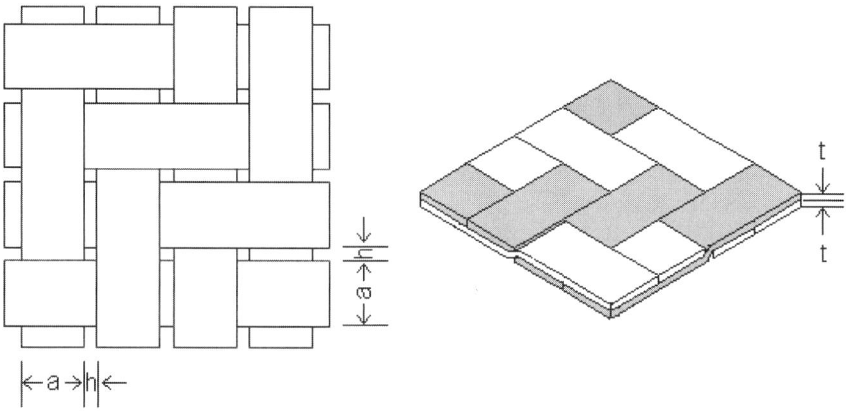

Fig. 4.16. Representative unit cell for 2/2-twill weave

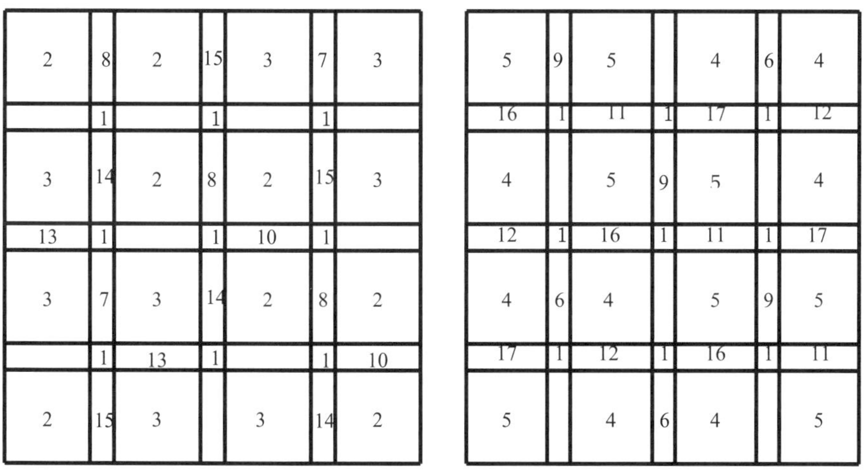

Horizontal strand direction Vertical strand direction

Fig. 4.17. Subcell numbering for the 2/2-twill unit cell

$$\sigma_{33}^{2}=\sigma_{33}^{5},\quad \sigma_{33}^{3}=\sigma_{33}^{4},\quad \sigma_{33}^{6}=\sigma_{33}^{7},$$
$$\sigma_{33}^{8}=\sigma_{33}^{9},\quad \sigma_{33}^{10}=\sigma_{33}^{11},\quad \sigma_{33}^{12}=\sigma_{33}^{13}. \tag{4.79}$$

The following equations represent the directional strain compatibility assuming a uniform deformation of the unit cell, where a and h are the dimensions of the fill and warp strands of the composite material, as shown in Fig. 4.17

$$\begin{aligned}
&2a(\varepsilon_{11}^{2}+\varepsilon_{11}^{3})+h(\varepsilon_{11}^{7}+\varepsilon_{11}^{8})=2a(\varepsilon_{11}^{4}+\varepsilon_{11}^{5})+h(\varepsilon_{11}^{6}+\varepsilon_{11}^{9}),\\
&2a(\varepsilon_{11}^{2}+\varepsilon_{11}^{3})+h(\varepsilon_{11}^{8})=2a(\varepsilon_{11}^{4}+\varepsilon_{11}^{5})+h(\varepsilon_{11}^{9}),\\
&2a(\varepsilon_{11}^{2}+\varepsilon_{11}^{3})+h(\varepsilon_{11}^{8}+\varepsilon_{11}^{14}+\varepsilon_{11}^{15})=a(\varepsilon_{11}^{10}+\varepsilon_{11}^{13}+\varepsilon_{11}^{16}+\varepsilon_{11}^{17})+3h(\varepsilon_{11}^{1}),\\
&\varepsilon_{11}^{10}+\varepsilon_{11}^{12}=\varepsilon_{11}^{11}+\varepsilon_{11}^{13},
\end{aligned} \tag{4.80}$$

$$\begin{aligned}
&2a(\varepsilon_{22}^{2}+\varepsilon_{22}^{3})+h(\varepsilon_{22}^{11}+\varepsilon_{22}^{12})=2a(\varepsilon_{22}^{4}+\varepsilon_{22}^{5})+h(\varepsilon_{22}^{10}+\varepsilon_{22}^{13}),\\
&2a(\varepsilon_{22}^{2}+\varepsilon_{22}^{3})+h(\varepsilon_{22}^{13})=2a(\varepsilon_{22}^{4}+\varepsilon_{22}^{5})+h(\varepsilon_{22}^{12}),\\
&2a(\varepsilon_{22}^{4}+\varepsilon_{22}^{5})+h(\varepsilon_{22}^{12}+\varepsilon_{22}^{16}+\varepsilon_{22}^{17})=a(\varepsilon_{22}^{6}+\varepsilon_{22}^{9}+\varepsilon_{22}^{14}+\varepsilon_{22}^{15})+3h(\varepsilon_{22}^{1}),\\
&\varepsilon_{22}^{6}+\varepsilon_{22}^{8}=\varepsilon_{22}^{7}+\varepsilon_{22}^{9},
\end{aligned} \tag{4.81}$$

$$\begin{aligned}
&\varepsilon_{33}^{2}+\varepsilon_{33}^{5}=\varepsilon_{33}^{16},\quad \varepsilon_{33}^{2}+\varepsilon_{33}^{5}=\varepsilon_{33}^{15},\quad \varepsilon_{33}^{2}+\varepsilon_{33}^{5}=\varepsilon_{33}^{17},\quad \varepsilon_{33}^{2}+\varepsilon_{33}^{5}=\varepsilon_{33}^{14},\\
&\varepsilon_{33}^{2}+\varepsilon_{33}^{5}=\varepsilon_{33}^{1},\quad \varepsilon_{33}^{2}+\varepsilon_{33}^{5}=\varepsilon_{33}^{6}+\varepsilon_{33}^{7},\quad \varepsilon_{33}^{2}+\varepsilon_{33}^{5}=\varepsilon_{33}^{12}+\varepsilon_{33}^{13}\quad ,\\
&\varepsilon_{33}^{2}+\varepsilon_{33}^{5}=\varepsilon_{33}^{6}+\varepsilon_{33}^{8},\quad \varepsilon_{33}^{2}+\varepsilon_{33}^{5}=\varepsilon_{33}^{10}+\varepsilon_{33}^{11},\quad \varepsilon_{33}^{2}+\varepsilon_{33}^{5}=\varepsilon_{33}^{3}+\varepsilon_{33}^{4}.
\end{aligned} \tag{4.82}$$

The constitutive equation for each subcell is given below in (4.83), where $(\sigma_{ij}^{\text{str}})^n$ and $(\varepsilon_{kl}^{\text{str}})^n$ are the nth subcell stresses and strains, and $(E_{ijkl}^{\text{str}})^n$ is the subcell material property matrix in terms of the unit cell axes

$$(\sigma_{ij}^{\text{str}})^n=(E_{ijkl}^{\text{str}})^n(\varepsilon_{kl}^{\text{str}})^n. \tag{4.83}$$

The stiffness matrix $(E_{ijkl}^{\text{str}})^n$ is determined from stress and strain equations in conjunction with the proper transformation matrices.

A result of weaving is that fibers must be undulated. The undulation angle is measured from the plane in which the majority of the fibers lay straight. The undulated portions of the strand must have their material properties adjusted to that of the global coordinate system. A brief explanation of stiffness transformation follows.

Let coordinate system x, y, and z be the global coordinate system, and x_1, y_1, and z_1 be the local coordinate system where the fiber in the strand of interest is aligned with the local axis. The stress and strain transformation from the local coordinates to global coordinates will be:

$$\begin{aligned} \{\sigma\}^{x_1y_1z_1} &= [T_\sigma]\{\sigma\}^{xyz}, \\ \{\varepsilon\}^{x_1y_1z_1} &= [T_\varepsilon]\{\varepsilon\}^{xyz}. \end{aligned} \tag{4.84}$$

The constitutive equations are:

$$\begin{aligned} \{\sigma\}^{xyz} &= [C]^{xyz}\{\varepsilon\}^{xyz}, \\ \{\sigma\}^{x_1y_1z_1} &= [C]^{x_1y_1z_1}\{\varepsilon\}^{x_1y_1z_1}. \end{aligned} \tag{4.85}$$

Mathematical manipulation of the above equations results in the following stiffness transformation equations

$$\begin{aligned} \{\sigma\}^{x_1y_1z_1} &= [C]^{x_1y_1z_1}[T_\varepsilon]\{\varepsilon\}^{xyz}, \\ [T_\sigma]\{\sigma\}_{kl}^{xyz} &= [C]^{x_1y_1z_1}[T_\varepsilon]\{\varepsilon\}^{xyz}, \\ \{\sigma\}^{xyz} &= [T_\sigma]^{-1}[C]^{x_1y_1z_1}[T_\varepsilon]\{\varepsilon\}^{xyz}, \\ \{\sigma\}^{xyz} &= [C]^{xyz}\{\varepsilon\}^{xyz}, \end{aligned} \tag{4.86}$$

where

$$[C]^{xyz} = [T_\sigma]^{-1}[C]^{x_1y_1z_1}[T_\varepsilon]. \tag{4.87}$$

Finally, the transformation matrices are as follows

$$[E_{ijkl}^{\text{str}}] = [T_\sigma]^{-1}[\bar{E}^{\text{str}}][T_\varepsilon], \tag{4.88}$$

where

$$[T_\varepsilon]^{-1} = \begin{bmatrix} m^2 & n^2 & 0 & 0 & 0 & mn \\ n^2 & m^2 & 0 & 0 & 0 & -mn \\ 0 & 0 & 1 & 0 & 0 & 0 \\ 0 & 0 & 0 & m & -n & 0 \\ 0 & 0 & 0 & n & m & 0 \\ -2mn & 2mn & 0 & 0 & 0 & m^2-n^2 \end{bmatrix}, \tag{4.89}$$

$$[T_\sigma]^{-1} = \begin{bmatrix} m^2 & n^2 & 0 & 0 & 0 & 2mn \\ n^2 & m^2 & 0 & 0 & 0 & -2mn \\ 0 & 0 & 1 & 0 & 0 & 0 \\ 0 & 0 & 0 & m & -n & 0 \\ 0 & 0 & 0 & n & m & 0 \\ -mn & mn & 0 & 0 & 0 & m^2-n^2 \end{bmatrix}. \tag{4.90}$$

The matrices $[T_\sigma]$ and $[T_\varepsilon]$ are the stress and strain transformation matrices. Transformation matrices account for a reduction or increase in the composite's material properties when the composite is rotated with respect to the coordinate axes from which the property was measured.

The final equations describe the relationship between the woven fabric unit cell stresses (or strains) and the strand subcell stresses (or strains). The woven fabric unit cell stresses and strains are computed as volumetric average of subcell stresses and strains

$$\sigma_{ij}^{\text{wf}} = \sum_{n=1}^{17} V^n (\sigma_{ij}^{\text{str}})^n, \tag{4.91}$$

$$\varepsilon_{ij}^{\text{wf}} = \sum_{n=1}^{17} V^n (\varepsilon_{ij}^{\text{str}})^n, \tag{4.92}$$

where V^n is the volume fraction of the nth subcell. After manipulating all the equations, the following relationships result:

$$E_{ijkl}^{\text{wf}} = f\left((E_{ijkl}^{\text{str}})^n, a, h, t\right), \tag{4.93}$$

$$(\varepsilon_{ij}^{\text{str}})^n = g(\varepsilon_{ij}^{\text{wf}}, a, h, t). \tag{4.94}$$

These two equations provide the bidirectional passage of the strand-fabric module. Equation (4.93) is used to compute the effective woven fabric stiffness E_{ijkl}^{wf} of a 2/2-twill composite from the stiffness $(E_{ijkl}^{\text{str}})^n$ of unidirectional strand and the weave geometric dimensions a, h, and t. Additionally, (4.94) decomposes the 2/2-twill composite strains into the subcell strains. The subcell stresses can then be calculated using (4.83).

4.4.4 Examples of Woven Fabric Composites

First of all, the Fibrous Module and Fabric Module were tested independently. The Fibrous Module was validated extensively in previous works [7, 15, 23], and it was tested again for a carbon/epoxy strand. The fiber and matrix materials are shown in Table 4.1. The effective stiffness and strength properties of the strand made of the fiber and matrix materials in Table 4.1 and a fiber volume fraction of 0.7 are compared with the results in [2] and in Table 4.2.

Plain weave composites were studied using the strand-fabric module. (Material properties of the strands of the woven-fabric composites are tabulated in Table 4.3.) The calculated effective material properties are compared to other results in Tables 4.4 and 4.5 for the carbon/epoxy and e-glass/vinylester woven composites, respectively. The present results compared well to other experimental data or analytical results.

Another plain weave composite made of the strand in Table 4.2 was also studied for its tensile strength. The predicted strength using the fiber-strand and strand-fabric modules is compared to the test result in [2]. As shown in Table 4.6, the predicted strength agrees well with the test result.

Table 4.1. Properties of fiber and matrix materials

	E_1 (GPa)	E_2 (GPa)	G_{12} (GPa)	G_{23} (GPa)	γ_{12}	γ_{23}	Tensile strength (MPa)	Shear strength (MPa)
Fiber	221	3.8	13.8	5.5	0.20	0.25	3,585	
Matrix	4.4	4.4	1.6	1.6	0.34	0.34	159	100

Table 4.2. Properties of strand material

	E_1 (GPa)	E_2 (GPa)	γ_{12}	γ_{23}	Long. strength (MPa)	Trans. strength (MPa)
[28]	151.0	10.1	0.24	0.50	2,550	152
Present	156.0	10.2	0.24	0.54	2,543	148

Table 4.3. Material properties of strands and resin

	E_1 (GPa)	E_2 (GPa)	G_{12} (GPa)	G_{23} (GPa)	γ_{12}	γ_{23}
E-glass/ vinylester	57.5	18.8	7.44	7.26	0.25	0.29
Carbon/epoxy	134	10.2	5.52	3.43	0.30	0.49
Epoxy	3.45	3.45	1.28	1.28	0.35	0.35

Table 4.4. Comparison of plain weave composite made of carbon/epoxy

	E_1 (GPa)	E_2 (GPa)	G_{12} (GPa)	G_{23} (GPa)	γ_{12}	γ_{23}
Exp. [28]	55.5	–	4.93	–	0.06	–
Result [29]	56.1	10.4	5.08	3.71	0.03	0.59
Present result	54.9	10.2	4.28	3.47	0.02	0.47

Table 4.5. Comparison of plain weave composite made of e-glass/vinylester

	E_1 (GPa)	E_2 (GPa)	G_{12} (GPa)	G_{23} (GPa)	γ_{12}	γ_{23}
Exp. [30]	24.8	8.5	6.5	4.2	0.1	0.28
Result [30]	25.3	13.4	5.19	5.24	0.12	0.29
Present result	24.4	12.4	5.92	4.52	0.12	0.28

Table 4.6. Plain weave and woven fabric composite

	Tensile strength (MPa)
Present	753
[28]	750

The next study was a plain weave composite plate subjected to a bending load. A quarter of the plate was modeled with two symmetric planes. The fiber and matrix properties are listed in Table 4.1. The longitudinal fiber stresses were computed for a square simply supported plate with a uniform unit pressure loading because they are the ultimate load carrying elements. The magnitudes of fiber-carrying stresses and their contour plots are shown in Fig. 4.18.

Laminated composite plates made of a 2/2-twill composite with a center hole were analyzed next. The objective was to simulate failure initiation and progression of 2/2-twill woven fabric composite plates with drilled holes at the center. Plates with three different hole sizes were considered. The hole diameters were 3, 6, and 9 mm. Figure 4.19 shows quarter of each plate with finite element meshes.

A finite element model of the plates contained 143 nodes and 240 two-dimensional triangular elements. Because of symmetry, only a quarter of each plate was modeled. Symmetric boundary conditions were applied to the bottom and left boundaries; the inner radius was left free; and a uniform displacement was applied along the right boundary of the mesh.

The applied load was incrementally increased until the failure criterion was realized in the element of interest. In this case, the elements of interest

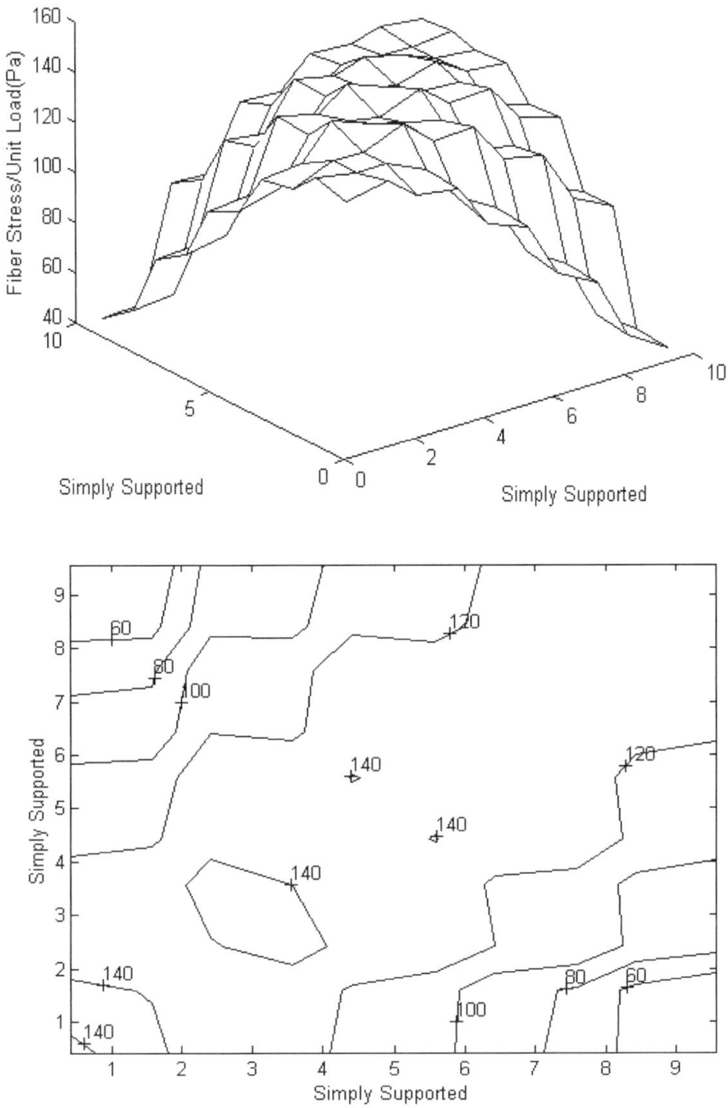

Fig. 4.18. Plot of fiber stresses for a square, simply supported plate made of a plain weave composite subjected to a uniform pressure loading

are those corresponding to the virtual strain gage in Fig. 4.19. Because of the high-volume fraction of the fibers, the ultimate strength of the fibers was set as the failure criterion for this analysis. With every incremental increase in applied strain, the multiscale model was used to determine the stresses in the fibers. Whenever a fiber reached its ultimate strength, the material property at the point degraded in the finite element model. The results are tabulated in Table 4.7. The ultimate failure strengths of the three composite plates with circular holes were predicted using the present analysis model and compared to the experimental data [36]. The predicted strengths are close to both experimental and previously predicted results.

Since a gross assumption of the unit cell model is that the strains are constant in each subcell, it may be of interest to determine how the subcell strains compare to the average strains in the subcells of a detailed finite element model. The following strains were applied to the woven fabric for the sake of analysis

$$\varepsilon_{ii}^{\mathrm{wf}} = \{0.0160 \quad -0.0009 \quad 0\}^{\mathrm{T}}.$$

The finite element model was divided into volumes which corresponded to the subcells of the unit cell model. The finite element volume strains were determined in two ways. First, the average of the strains for all elements contained in the finite element volume was calculated. As a second check, an average nodal displacement on two opposing faces was calculated. The difference in the average nodal displacement for each face was divided by the original face length to determine an average strain in the volume. Only the strains in the direction of applied stress were compared, the *y*-direction for this case.

Three subcells were chosen for comparison with the finite element model. Those comparisons are tabulated in Table 4.8. The subcell numbers correspond to the number in Fig. 4.17. Both models agreed very well.

Table 4.7. Predicted and experimental strength of 2/2-twill composites with holes

Hole diameter (mm)	Experimental strength (MPa) [31]	Predicted strength (MPa)
3	435	494
6	395	438
9	333	354

Table 4.8. Comparison of subcell strains between the present unit cell and detailed finite element models

Subcell no.	Unit cell model	Finite element model
1	0.0221	0.0277
4	0.0165	0.0142
11	0.0164	0.0150

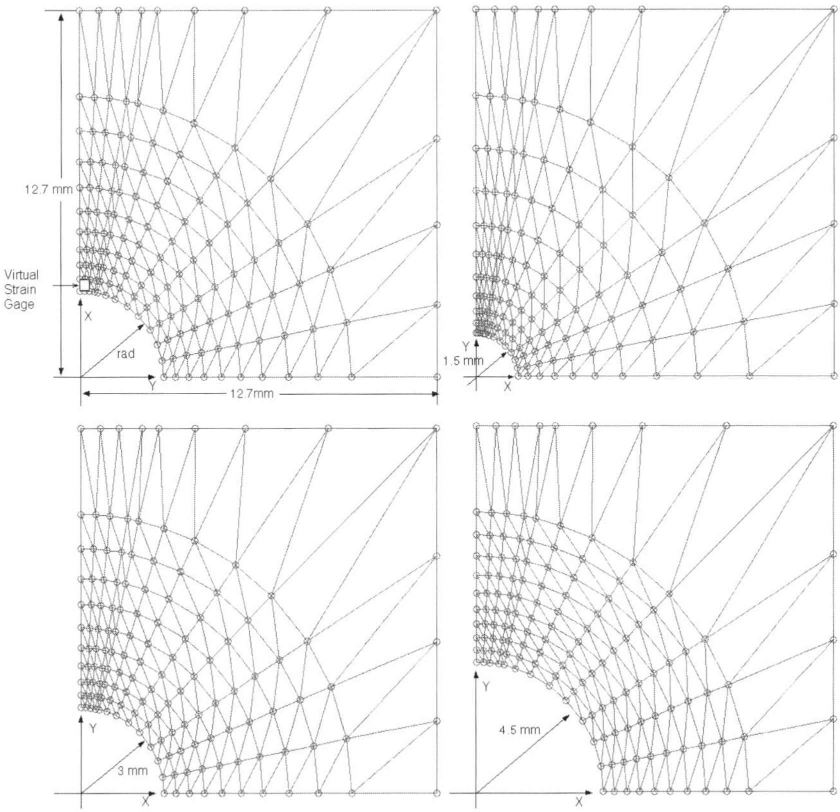

Fig. 4.19. Finite element mesh of quarters of plates with circular holes

4.5 Conclusions

Multiscale analysis techniques become increasingly important as we continuously strive to stretch composite material properties up to their ultimate limits. In that aspect, we may want to design and build composites starting from the atomic level [9, 11, 21, 22, 33, 34]. In this case, the multiscale techniques presented here should be extended to implement nanoscale characteristics. Otherwise, the present techniques are useful to design and analyze various composite structures from micro- to macroscale of engineering applications.

Acknowledgments

The author expresses his sincere gratitude to his graduate students and visiting scholars who have participated in the research projects presented in this manuscript.

References

1. Aitharaju VR, Averill RC (1999) Three-dimensional properties of woven-fabric composites. Composites Science and Technology 59: 1901–1911
2. Blackketter DM, Walrath DE, Hansen AC (1993) Modeling damage in a plain weave fabric-reinforced composite material. Journal of Composites Technology & Research 15: 136–142
3. Kwon YW (1991) Elasto-viscoplastic analysis of fiber-reinforced composites. Engineering Computations 8: 273–284
4. Kwon YW (1991) Material nonlinear analysis of composite plate bending using a new finite element formulation. Computers & Structures 41: 1111–1117
5. Kwon YW (1992) Finite element analysis of thermoelastoplastic stresses in composites. European Journal of Mechanical Engineering 37: 83–88
6. Kwon YW (1992) Thermo-elastoviscoplastic finite element plate bending analyses of composites. Engineering Computations 9: 595–607
7. Kwon YW (1993) Calculation of effective moduli of fibrous composites with micro-mechanical damages. Composite Structures 25: 187–192
8. Kwon YW (2001) Multi-level approach for failure in woven fabric composites. Advanced Engineering Materials 3: 713–717
9. Kwon YW (2003) Discrete atomic and smeared continuum modeling for static analysis. Engineering Computations 20: 964–978
10. Kwon YW (2005) Micromechanical, thermomechanical study of a refractory fiber/matrix/coating system. Journal of Thermal Stresses 28: 439–453
11. Kwon YW (2005) Multiscale modeling of mechanical behavior of polycrystalline materials. Journal of Computer-Aided Materials Design 11: 43–57
12. Kwon YW, Altekin A (2002) Multi-level, micro–macro approach for analysis of woven fabric composites. Journal of Composite Materials 36: 1005–1022
13. Kwon YW, Baron DT (1998) Numerical predictions of progressive damage evolution in particulate composites. Journal of Reinforced Plastics and Composites 17: 691–711

14. Kwon YW, Berner JM (1994) Analysis of matrix damage evolution in laminated composite plates. Engineering Fracture Mechanics 48: 811–817
15. Kwon YW, Berner JM (1995) Micromechanics model for damage and failure analyses of laminated fibrous composites. Engineering Fracture Mechanics 52: 231–242
16. Kwon YW, Berner JM (1997) Matrix damage analysis of fibrous composites: effects of thermal residual stresses and layer sequences. Computers & Structures 64: 375–382
17. Kwon YW, Byun KY (1990) Development of a new finite element formulation for the elasto-plastic analysis of fiber reinforced composites. Computers & Structures 35: 563–570
18. Kwon YW, Cho WM (2004) Multiscale thermal stress analysis of woven fabric composite. Journal of Thermal Stresses 27: 59–73
19. Kwon YW, Craugh LE (2001) Progressive failure modeling in notched cross-ply fibrous composites. Applied Composite Materials 8: 63–74
20. Kwon YW, Eren H (2000) Micromechanical study of interface stresses and failure in fibrous composites using boundary element method. Polymers & Polymer Composites 8: 369–386
21. Kwon YW, Harrell AF (2004) How many monomer repeat units are necessary for reliable molecular dynamics simulation. Polymers & Polymer Composites 12: 483–489
22. Kwon YW, Jung SH (2004) Atomic model and coupling with continuum model for static equilibrium problems. Computers & Structures 82: 1993–2000
23. Kwon YW, Kim C (1998) Micromechanical model for thermal analysis of particulate and fibrous composites. Journal of Thermal Stresses 21: 21–39
24. Kwon YW, Lee JH, Liu CT (1997) Modeling and simulation of crack initiation and growth in particulate composites. Transactions of the ASME: Journal of Pressure Vessel Technology 119: 319–324
25. Kwon YW, Lee JH, Liu CT (1998) Study of damage and crack in particulate composites. Composites Part B: Engineering 29: 443–450
26. Kwon YW, Liu CT (1997) Study of damage evolution in composites using damage mechanics and micromechanics. Composite Structures 38: 133–139
27. Kwon YW, Liu CT (1998) Damage growth in a particulate composite under a high strain rate loading. Mechanics Research Communications 25: 329–336
28. Kwon YW, Liu CT (1998) Effects of non-uniform particle distributions on damage evolution in pre-cracked, particulate composite specimens. Polymers & Polymer Composites 6: 387–397
29. Kwon YW, Liu CT (1999) Numerical study of damage growth in particulate composites. Transactions of the ASME: Journal of Engineering Materials and Technology 121: 476–482

30. Kwon YW, Liu CT (2000) Prediction of initial crack size in particulate composites with a circular hole. Mechanics Research Communications 27: 421–428
31. Kwon YW, Liu CT (2001) Effect of particle distribution on initial cracks forming from notch tips of composites with hard particles embedded in a soft matrix. Composites Part B: Engineering 32: 199–208
32. Kwon YW, Liu CT (2003) Microstructural effects on damage behavior in particle reinforced composites. Polymers & Polymer Composites 11: 1–8
33. Kwon YW, Manthena C (2006) Homogenization technique of discrete atoms into smeared continuum. Internal Journal of Mechanical Sciences 48: 1352–1359
34. Kwon YW, Manthena C, Oh JJ, Srivastava D (2005) Vibrational characteristics of carbon nanotubes as nanomechanical resonators. Journal of Nanoscience and Nanotechnology 5: 703–712
35. Kwon YW, Roach K (2004) Unit-cell model of 2/2-twill woven fabric composites for multiscale analysis. Computer Modeling in Engineering and Sciences 5: 63–72
36. Ng S, Tse P, Lau K (1998) Progressive failure analysis of 2/2-twill weave fabric composites with moulded-in circular hole. Composites Part B: Engineering 32: 139–152
37. Scida D, Aboura Z, Benzeggagh MM, Bocherens E (1999) A micromechanics model for 3d elasticity and failure of woven-fibre composite materials. Composites Science and Technology 59: 505–517

Chapter 5: A Micromechanics-Based Notion of Stress for Use in the Determination of Continuum-Level Mechanical Properties via Molecular Dynamics

Francesco Costanzo and Gary L. Gray

Department of Engineering Science and Mechanics, The Pennsylvania State University, University Park, PA 16802, USA

5.1 Introduction

By formulating a continuum homogenization problem that includes inertia effects, a link is established between continuum homogenization and the estimation of effective mechanical properties for particle ensembles whose interactions are governed by potentials (e.g., as is seen in molecular dynamics). The focus of this chapter is on showing that there is a fundamental consistency of ideas between continuum mechanics and the study of discrete particle systems, and that it is possible to define a notion of effective stress applicable to discrete systems that can be claimed to have the same meaning as it has in continuum mechanics.

5.2 Motivation, Objectives, and Organization

The last 15 years have seen an astonishing growth in nanomechanics-related research. During this time, experimental and theoretical mechanicians alike have had to adapt to a fast-evolving research landscape. Like many others, the authors of this chapter found themselves delving into specialized fields of study such as molecular dynamics (MD) and struggling to learn new languages and methodologies that were outside what they trained on during their graduate work. With this in mind, this chapter is in part the result of the authors' learning experience in how to use MD to compute mechanical properties of solids. In going through this learning process, the authors had to confront the fundamental issue of what it means to compute the stress response of a particle system and how this measure of stress is related to the continuum mechanical notion of stress. Clearly, this question is not new, since it dates back to the pioneering work by Cauchy who formalized the very notion of stress. However, we feel that we have added something new to the discussion in that we have approached the problem from the viewpoint of continuum homogenization and, in so doing, not only were we able to extend the continuum homogenization notion of effective stress to MD, but we were also able to construct a practical Lagrangian MD scheme that is rigorously based on classical mechanics.

From a conceptual viewpoint, the outcome of this work is that a good part of the MD that is used in nanomechanics can be comfortably understood with classical mechanics and homogenization ideas. In other words, it is possible to define an acceptable concept of stress for discrete systems without ever relying on ideas from statistical mechanics or a kinetic theory of matter. While this fact may be well understood by some researchers, we feel that it is not sufficiently known among classically trained engineers, and we hope that this chapter may reinforce the idea that there is a fundamental unity between the study of continuum and discrete systems.

The organization of this chapter is based on the idea that classical homogenization of heterogeneous systems is intimately related to MD, since both disciplines deal with the computation of effective properties of matter. Hence, we will start by reviewing some basic concepts of homogenization of linear elastic media. We will then discuss the extension of these concepts to the case of homogenization in the context of large deformation. Once this review is done, we will formulate a continuum homogenization problem that shares the basic properties of MD problems. We will show that the homogenization scheme in question can be turned into an MD scheme in which stress is defined such that it can be said to have the same meaning that it has in continuum homogenization. Finally, we will compare the continuum homogenization-based stress concept with the virial stress, the latter being the stress concept typically used in MD.

Before proceeding further, we wish to mention that some elements of this chapter have been presented in [1,2,8,9]. The main contribution of this chapter lies in a presentation that is intended to give a coherent vision of how continuum homogenization and MD are related. With this said, this chapter does contain some new results consisting of more general proofs, with respect to what had been previously published, on the equivalence between a continuum-based notion of effective stress and virial stress.

5.3 Notation

The material system under consideration will be denoted by Ω in its deformed configuration and will be denoted by Ω_κ in its reference configuration. Both Ω and Ω_κ are assumed to be regular subsets of a three-dimensional Euclidean point space. The boundaries of Ω and Ω_κ will be denoted by $\partial\Omega$ and $\partial\Omega_\kappa$, respectively. The volumes of Ω and Ω_κ will be denoted by $\mathrm{Vol}(\Omega)$ and $\mathrm{Vol}(\Omega_\kappa)$, respectively. The boundaries $\partial\Omega$ and $\partial\Omega_\kappa$ are oriented by the outward unit normal vector fields $\boldsymbol{n}$ and $\boldsymbol{n}_\kappa$, respectively. The position of points in the reference configuration will be denoted by χ and in the deformed configuration by $\boldsymbol{x}$.

The operators "Div" and "div" indicate the divergence operators with respect to χ and $\boldsymbol{x}$, respectively. Similarly, the operators "Grad" and "grad" indicate the gradient operators with respect to χ and $\boldsymbol{x}$, respectively.

We will use upper-case sans serif letters, such as A, to denote second-order tensors and lower-case bold italic letters, such as $\boldsymbol{a}$, to denote vectors. The notation $\boldsymbol{a} \otimes \boldsymbol{b}$ denotes the tensor product of the vectors $\boldsymbol{a}$ and $\boldsymbol{b}$. The symbol $\triangleq$ will indicate a definition.

5.4 Homogenization of Linear Elastic Heterogeneous Media: A Brief Review

To better illustrate how MD and continuum homogenization are related, it is useful to review some basic concepts from the theory of homogenization of linear elastic heterogeneous media. We will therefore review the essential objectives of homogenization theory and some basic definitions concerning *effective* mechanical properties. In subsequent sections, we will discuss how these definitions need to be adjusted to be useful in a fully nonlinear context in preparation for their application to discrete particle systems.

5.4.1 Homogenization Objectives

Referring to Fig. 5.1, consider a structural component made of a heterogeneous material with overall dimensions that are much larger than the characteristic length over which the material's constitutive properties vary. Conceptually, under the assumption that the material is linear elastic, in

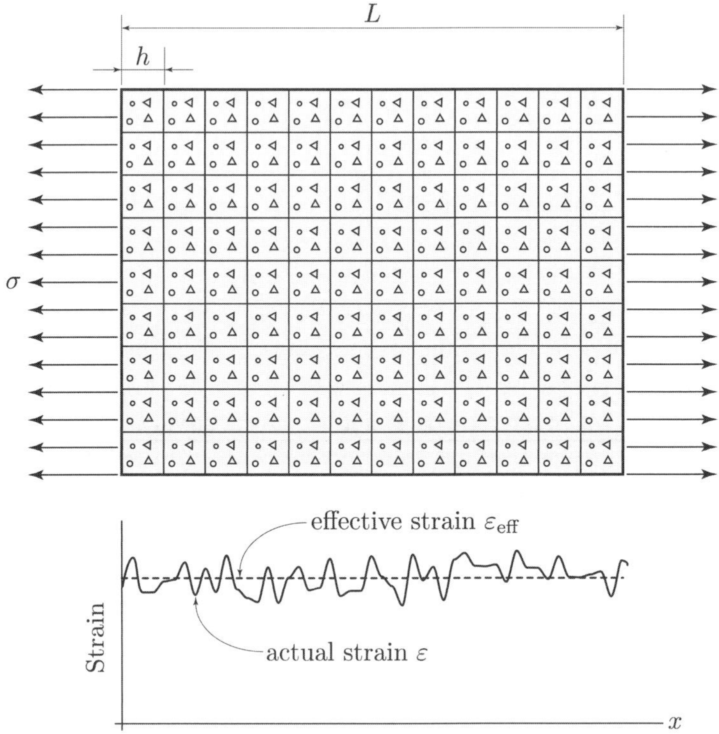

Fig. 5.1. A panel consisting of a heterogeneous material

quasistatic conditions, and in the absence of body forces, the prediction of the component's stress/strain response requires the solution of a boundary value problem (BVP) of the following type

$$\text{BVP}_{\text{exact}}: \quad \text{Div}(\mathsf{C}(\chi)[\varepsilon(\chi)]) = \mathbf{0} \quad \text{along with BCs,} \tag{5.1}$$

where χ denotes position, $\mathsf{C}(\chi)$ is the (fourth-order) tensor of elastic moduli, $\varepsilon(\chi)$ is the small strain tensor field, and the expression "BCs" stands for "boundary conditions." For convenience, we denote by $\boldsymbol{\sigma}(\chi)$ the stress field corresponding to $\varepsilon(\chi)$, i.e., $\boldsymbol{\sigma}(\chi) = \mathsf{C}(\chi)[\varepsilon(\chi)]$. Clearly, the structural component's stress/strain response to some applied loading will reflect the spatial variability of the elastic moduli, as schematically represented by the solid line in Fig. 5.1. Unfortunately, from a computational viewpoint, the spatial variability in question may make the solution of the problem in (5.1) difficult, if not impossible, to obtain. With this in mind, a practical way to approach the design of highly heterogeneous components is to construct a predictive capability that allows one to (1) model the material as homogeneous so as to more easily determine the system's "average" response (see the dashed line in Fig. 5.1) and (2) estimate the deviations from the "average" behavior since this information is essential in assessing failure conditions. The purpose of homogenization is to have both types of predictive capability, though we will only explore the first type here. Before doing so, it is important to recognize that, at this stage, we do not know whether or not what we have called the "average" response will in fact be an average in a strict mathematical sense. Hence, we will refer to the "average" strain and stress response as the *effective* strain and stress response and we will denote these quantities as $\varepsilon_{\rm eff}$ and $\boldsymbol{\sigma}_{\rm eff}$, respectively.

As suggested above, a fundamental objective of continuum homogenization is to use the knowledge of the material's microstructure to formulate a BVP whose solution is the system's effective response, i.e., homogenization theory delivers the possibility of predicting the effective system's response by solving the following BVP

$$\mathrm{BVP}_{\rm eff}: \quad \mathrm{Div}(\mathsf{C}_{\rm eff}[\varepsilon_{\rm eff}(\boldsymbol{\chi})]) = \mathbf{0} \quad \text{along with BCs,} \tag{5.2}$$

where it is essential to notice that, in the new BVP, the moduli $\mathsf{C}_{\rm eff}$, which are called the *material's effective moduli*, are not a function of position. Therefore, one way to interpret (5.2) is to say that homogenization theory takes information concerning the original *heterogeneous* material and maps it into the properties of an equivalent *homogenous* material. Finally, we will refer to the field $\mathsf{C}_{\rm eff}[\varepsilon_{\rm eff}(\boldsymbol{\chi})]$ as the effective stress field and we will denote it by $\boldsymbol{\sigma}_{\rm eff}(\boldsymbol{\chi})$, i.e.,

$$\boldsymbol{\sigma}_{\rm eff} = \mathsf{C}_{\rm eff}[\varepsilon_{\rm eff}(\boldsymbol{\chi})]. \tag{5.3}$$

So far, we have only sketched a conceptual map of what homogenization does without considering the important details needed to show that one can indeed go from the BVP in (5.1) to that in (5.2). Most of these "details" are outside the scope of this chapter and they can be easily found in the literature. For example, excellent references on the subject are the presentations in [18, 20, 28, 31]. For discussions that are more technical from a mathematical viewpoint, one can see the presentations in [3, 4, 15]. While we will stay away from the technical details of homogenization theory, a few important remarks are now needed for extending homogenization ideas to MD.

Remark 1 *(Representative Volume Element).* To solve the BVP in (5.2), one must first determine the effective moduli $\mathsf{C}_{\rm eff}$ and this can be done via several methods. Often, especially in engineering applications, the determination of the effective moduli is carried out by solving a special BVP defined over a portion of the material such that both the composition and the geometry of this portion are able to represent the material as a whole. This subset of material is called a *representative volume element* (RVE), which is schematically shown in Fig. 5.2. In general the determination of the RVE may not be

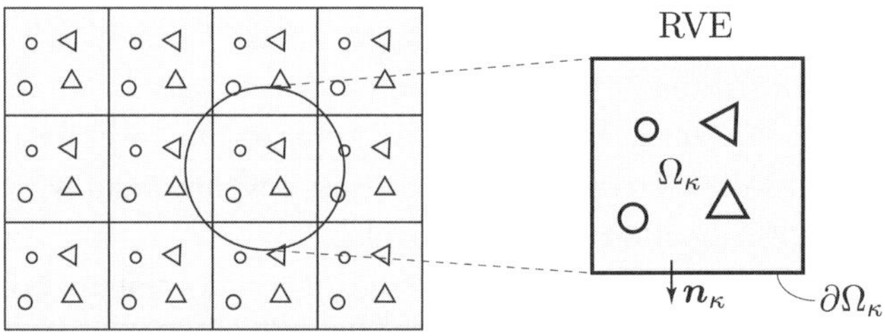

Fig. 5.2. A representative volume element for the special case of a periodic medium

straightforward, however, for periodic media, the RVE is readily identified with the periodic cell of the material. Furthermore, in the case of periodic media, there are rigorous proofs showing that the determination of C_{eff} by asymptotic expansion methods (see, e.g., [3, 4]) delivers the same result as the solution of the RVE BVP so long as the periodicity of the material is properly accounted for.

Remark 2 (Definition of Effective Quantities). Roughly speaking, in formal homogenization theory, $\boldsymbol{\varepsilon}_{\text{eff}}(\boldsymbol{\chi})$ and $\boldsymbol{\sigma}_{\text{eff}}(\boldsymbol{\chi})$ are defined as the leading terms of an asymptotic expansion of the fields $\boldsymbol{\varepsilon}(\boldsymbol{\chi})$ and $\boldsymbol{\sigma}(\boldsymbol{\chi})$ with respect to a scaling parameter, say λ, defined as the ratio between the length over which the moduli vary and the overall (large) dimension of the component (for example, referring to Fig. 5.1, one can set $\lambda = h/L$). With this in mind, and referring to Fig. 5.2, one can show that, when using the RVE as a way to determine the effective moduli, these definitions can be given the following form

$$\boldsymbol{\varepsilon}_{\text{eff}} = \frac{1}{2\,\mathrm{Vol}(\Omega_\kappa)} \int_{\partial\Omega_\kappa} (\boldsymbol{u} \otimes \boldsymbol{n}_\kappa + \boldsymbol{n}_\kappa \otimes \boldsymbol{u})\,\mathrm{d}A, \tag{5.4}$$

$$\boldsymbol{\sigma}_{\text{eff}} = \frac{1}{\mathrm{Vol}(\Omega_\kappa)} \int_{\partial\Omega_\kappa} (\boldsymbol{\sigma}\boldsymbol{n}_\kappa \otimes \boldsymbol{\chi})\,\mathrm{d}A, \tag{5.5}$$

where $\boldsymbol{u}$ denotes the displacement field. The essential feature of these definitions is that $\boldsymbol{\varepsilon}_{\text{eff}}(\boldsymbol{\chi})$ and $\boldsymbol{\sigma}_{\text{eff}}(\boldsymbol{\chi})$ are determined by gathering information on the boundary of the RVE rather than its interior. If the RVE is a simply connected regular domain, a straightforward application of the divergence theorem tells us that

$$\boldsymbol{\varepsilon}_{\text{eff}} = \frac{1}{\text{Vol}(\Omega_\kappa)} \int_{\Omega_\kappa} \boldsymbol{\varepsilon}\, \mathrm{d}V \quad \text{and} \quad \boldsymbol{\sigma}_{\text{eff}} = \frac{1}{\text{Vol}(\Omega_\kappa)} \int_{\Omega_\kappa} \boldsymbol{\sigma}\, \mathrm{d}V, \tag{5.6}$$

where the second of (5.6) requires that the pointwise[1] balance law is $\text{Div}(\mathsf{C}(\boldsymbol{\chi})[\boldsymbol{\varepsilon}(\boldsymbol{\chi})]) = \mathbf{0}$. Equation (5.6) implies that there are cases in which the word "effective" does mean "volume average," but, in general, effective strain and stress must be understood as given in (5.4) and (5.5) to be useful mathematical constructs. In addition, there is a strong physically based reason for defining effective quantities via boundary integrals, as eloquently remarked by Hill ([13]; see also [18, 27, 28]):

> Experimental determinations of mechanical behaviour rest ultimately on measured loads or mean displacements over pairs of opposite faces of a representative cube. Macro-variables intended for constitutive laws should thus be capable of definition in terms of surface data alone, either directly or indirectly. It is not necessary, by any means, that macro-variables so defined should be unweighed volume averages of their microscopic counterparts.

Remark 3 *(Basic Properties of $\boldsymbol{\varepsilon}_{\mathit{eff}}$ and $\boldsymbol{\sigma}_{\mathit{eff}}$).* In a small strain theory, one expects the strain and stress measures to be symmetric tensors. Referring to (5.4), it is easy to see that the effective strain is, by definition, a symmetric tensor. Furthermore, one can easily show that under most conditions $\boldsymbol{\sigma}_{\text{eff}}$ is symmetric. What needs to be observed here is that, at least at first glance, no special steps are needed to make sure that the above-defined effective quantities have the properties that one usually expects of the corresponding pointwise quantities. As we will see, this is certainly not the case when dealing with the definitions of the effective stress and deformation concepts in nonlinear homogenization.

5.4.2 Boundary Conditions for the RVE Problem

When relying on an RVE for the determination of the effective moduli, one must pose and solve a BVP over the RVE in question. This BVP is usually

[1] The "point" in pointwise refers to a *continuum material point*, by which we mean a point in a regular subset of $\mathbb{R}^3$. This is not to be confused with a material particle, by which we mean an abstract physical entity endowed with a given fixed mass.

called a *localization problem* and its governing partial differential equations are those in (5.1). As far as the BCs are concerned, these need to be carefully stated to match the particular nature of the problem. In fact, in a localization problem one is not interested in computing the solution's pointwise behavior. Rather, one needs to control the solution's *effective* behavior in such a way that the effective moduli can be calculated. With this in mind, it turns out that it is indeed possible to control the value of the effective strain or stress by specifying some specific sets of BCs. These BCs are as follows:

1. *Uniform strain*: $\boldsymbol{u} = \hat{\boldsymbol{\varepsilon}}\boldsymbol{\chi}$ on $\partial\Omega_{\kappa}$, with $\hat{\boldsymbol{\varepsilon}}$ a *given* symmetric second-order tensor.

2. *Uniform stress*: $\boldsymbol{\sigma}\boldsymbol{n}_{\kappa} = \hat{\boldsymbol{\sigma}}\boldsymbol{n}_{\kappa}$ on $\partial\Omega_{\kappa}$, with $\hat{\boldsymbol{\sigma}}$ a *given* symmetric second-order tensor.

3. *Periodic*: If the RVE is a periodic cell, then the displacement field is decomposed such that $\boldsymbol{u} = \hat{\boldsymbol{\varepsilon}}\boldsymbol{\chi} + \boldsymbol{u}^*$ *everywhere* in the RVE, with $\hat{\boldsymbol{\varepsilon}}$ a *given* symmetric second-order tensor and with $\boldsymbol{u}^*$ being an unknown vector field whose boundary values are constrained to be periodic, i.e., $\boldsymbol{u}^*$ is constrained to take on identical values on homologous points of the boundary. Furthermore, in addition to constraining the boundary values of the field $\boldsymbol{u}^*$, one must also constrain the behavior of the field $\boldsymbol{\sigma}\boldsymbol{n}_{\kappa}$ to be antiperiodic.

If one chooses BCs of type 1 or 3, it is relatively straightforward to prove (see, e.g., [18, 28]) that the controlled parameter $\hat{\boldsymbol{\varepsilon}}$ determines the value of the effective strain, i.e., $\boldsymbol{\varepsilon}_{\text{eff}} = \hat{\boldsymbol{\varepsilon}}$. If one chooses condition 2, then it is not difficult to show that the controlled parameter $\hat{\boldsymbol{\sigma}}$ determines the value of the effective stress, i.e., $\boldsymbol{\sigma}_{\text{eff}} = \hat{\boldsymbol{\sigma}}$.

From a conceptual viewpoint, the determination of the elastic moduli in RVE-based linear homogenization is carried out by the following procedure. Choosing uniform strain BCs for the sake of discussion, one can set $\hat{\varepsilon}_{11} = 1$ and all other components of $\hat{\boldsymbol{\varepsilon}}$ equal to zero. Then, one solves the RVE BVP and thus determines the $\boldsymbol{\sigma}$ component of the solution. Next, one uses the $\boldsymbol{\sigma}$ field in question, along with (5.5), to determine $\boldsymbol{\sigma}_{\text{eff}}$. Finally, due to the linearity of problem and referring to (5.3), the $\boldsymbol{\sigma}_{\text{eff}}$ just computed coincides with the "$ij11$" components of C_{eff} ($ij = 1, 2, 3$). This process is then repeated by selecting setting all components of $\hat{\boldsymbol{\varepsilon}}$ equal to zero, except say $\hat{\varepsilon}_{pq}$, which is set to unity so that the "$ijpq$" components of the elastic moduli can be found.

5.5 The RVE Problem and Large Deformations

In this section, we discuss the concepts of effective strain and stress in a context of large deformations. This discussion is again meant to properly setup a stage for the extension of continuum homogenization ideas to MD problems. Choosing to work in a large deformation context is motivated by the fact that we want the discussion to be as general as possible.[2] Before proceeding to the presentation of effective measures of deformation and stress, it is important to remark that the field of nonlinear homogenization is not as well developed as the corresponding linear theory. In particular, there are fewer theoretical results linking an asymptotic approach to homogenization to the RVE-based averaging procedures. With this in mind, as has been done by other authors (see, e.g., [12–14,25]), we will simply assume that the RVE problem is a valid way to compute effective properties. This assumption allows us to focus our attention on the RVE approach to homogenization, as opposed to considering the (more technically difficult) formal asymptotic approach.

5.5.1 Definition of Effective Deformation and Stress

In general, in a context of large deformation, one must take into consideration two measures of stress, namely the Cauchy stress and the first Piola–Kirchhoff stress, depending on whether one chooses the deformed or reference configurations, respectively, to write the system's equation of motion. The two notions of stress are related by the well-known relation (see, e.g., [11])

$$\mathsf{S} = \det(\mathsf{F})\mathsf{T}(\mathsf{F}^{-1})^{\mathrm{T}}, \tag{5.7}$$

where S denotes the first Piola–Kirchhoff stress tensor, F denotes the deformation gradient, T denotes the Cauchy stress tensor, and the superscript T denotes transposition. The relationship in (5.7) reminds us that, when we define effective deformation and stress in a context of large deformation, we need to (1) provide definitions that are based both on the reference and the deformed configurations and (2) discuss how these definitions relate to one another. With this in mind, we introduce two *independent* measures of effective deformation: the *effective deformation gradient tensor*, denoted by

[2] In choosing to work in a regime of large deformations, we assume that the kinematics at both the micro- and macroscales is fully nonlinear. Correspondingly, we do not assume that any aspect of the constitutive theory is linear.

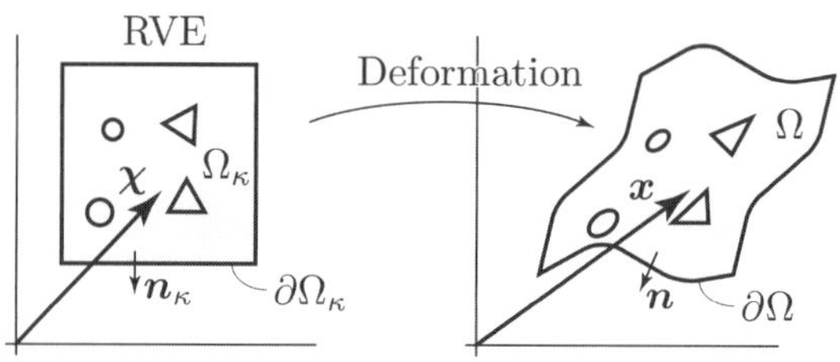

Fig. 5.3. The RVE in its reference (*left*) and deformed (*right*) configurations

$[\![\mathsf{F}]\!]$, and the *effective inverse deformation tensor*, denoted by $[\![\mathsf{F}^{-1}]\!]$. These quantities are defined as follows

$$[\![\mathsf{F}]\!] \triangleq \frac{1}{\mathrm{Vol}(\Omega_\kappa)} \int_{\partial\Omega_\kappa} \boldsymbol{x} \otimes \boldsymbol{n}_\kappa \, \mathrm{d}A, \quad [\![\mathsf{F}^{-1}]\!] \triangleq \frac{1}{\mathrm{Vol}(\Omega)} \int_{\partial\Omega} \boldsymbol{\chi} \otimes \boldsymbol{n} \, \mathrm{d}a, \quad (5.8)$$

where the RVE is subject to a motion $\boldsymbol{x} = \boldsymbol{x}(\boldsymbol{\chi}, t)$, and the symbols are defined in Fig. 5.3. As far as stress is concerned, we will define the *effective first Piola–Kirchhoff stress tensor* and the *effective Cauchy stress tensor*, denoted by $[\![\mathsf{S}]\!]$ and $[\![\mathsf{T}]\!]$, respectively, as follows:

$$[\![\mathsf{S}]\!] \triangleq \frac{1}{\mathrm{Vol}(\Omega_\kappa)} \int_{\partial\Omega_\kappa} (\mathsf{S}\boldsymbol{n}_\kappa) \otimes \boldsymbol{\chi} \, \mathrm{d}A, \quad [\![\mathsf{T}]\!] \triangleq \frac{1}{\mathrm{Vol}(\Omega)} \int_{\partial\Omega} (\mathsf{T}\boldsymbol{n}) \otimes \boldsymbol{x} \, \mathrm{d}a. \quad (5.9)$$

To the best of the authors' knowledge, the definitions of effective deformation and effective stress in a regime of large deformation were first systematically discussed in [13] (see also [14]). Important contributions to this subject also include the works in [12, 26]. More recent discussions have been given in [16, 25]. It should be pointed out that Hill [13, 14] does not include in his discussions the definition of $[\![\mathsf{F}^{-1}]\!]$. However, we feel that the definition of $[\![\mathsf{F}^{-1}]\!]$ is important because it has a bearing on the type of phenomena that we will choose as being *physically meaningful* when extending the above notions of effective deformation and stress to discrete systems.

Going back to (5.8) and (5.9), it is important to notice that we have once more defined effective quantities via boundary integrals rather than via volume averages. However, as was done in Sect. 5.4, under standard regularity and smoothness assumptions, a straightforward application of the divergence theorem yields (cf. [13])

$$[\![\mathsf{F}]\!] = \frac{1}{\mathrm{Vol}(\Omega_\kappa)} \int_{\Omega_\kappa} \mathsf{F} \, \mathrm{d}V \quad \text{and} \quad [\![\mathsf{F}^{-1}]\!] = \frac{1}{\mathrm{Vol}(\Omega)} \int_{\Omega} \mathsf{F}^{-1} \, \mathrm{d}v. \quad (5.10)$$

Furthermore, if to these assumptions one adds that the underlying deformation process is governed by $\operatorname{div}\mathsf{T} = \mathbf{0}$ (or $\operatorname{Div}\mathsf{S} = \mathbf{0}$), i.e., governed by a *quasistatic* form of the local balance of linear momentum without body forces, an application of the divergence theorem allows one to show that

$$[\![\mathsf{S}]\!] = \frac{1}{\mathrm{Vol}(\Omega_\kappa)} \int_{\Omega_\kappa} \mathsf{S}\,\mathrm{d}V \quad \text{and} \quad [\![\mathsf{T}]\!] = \frac{1}{\mathrm{Vol}(\Omega)} \int_{\Omega} \mathsf{T}\,\mathrm{d}v. \tag{5.11}$$

Before proceeding further, we should keep in mind that one of the objectives of this chapter is to extend continuum homogenization notions of effective stress and strain to the discrete systems analyzed via MD. Although often disregarded in continuum homogenization of elastic systems, time evolution and time averaging are central to MD calculations. Hence, we introduce here a time averaging operation that will be employed later in the chapter. This operation will be denoted by the use of angle brackets and defined as follows

$$\langle f \rangle \triangleq \lim_{\mathcal{T}\to\infty} \frac{1}{\mathcal{T}} \int_{t_0}^{t_0+\mathcal{T}} f(t)\mathrm{d}t, \tag{5.12}$$

where t denotes time, $f(t)$ is a generic function of time, and t_0 is the initial time. Without loss of generality, we will assume that $t_0 = 0$. Adopting concepts from statistical mechanics, we will view the time average operation defined in (5.12) as a way of translating the effects of fast dynamics into corresponding thermal effects.

5.5.2 Meaningful Deformation Processes

Now that we have introduced the definition of effective deformation and stress in a regime of large deformation, we need to address the problem of clearly identifying those RVE motions for which the definitions in question are useful in some sense. Specifically, it should be observed that each of the definitions we have given is independent of the others. Therefore, one cannot expect that, for example, $[\![\mathsf{F}^{-1}]\!] = [\![\mathsf{F}]\!]^{-1}$ for all possible RVE motions. By the same reasoning, in general, we cannot expect that the effective Cauchy and the first Piola–Kirchhoff stresses are related by a relationship such as (5.7), i.e., the relation satisfied the corresponding pointwise stress measures.

These observations indicate that there is a need for the establishment of conditions that guarantee the ability to attach physical meaning to the definitions given above. In fact, it can be argued that Hill's macrohomogeneity conditions [13, 14], i.e., those conditions under which certain products of effective quantities are equal to the volume average of the product of the corresponding local quantities, play the role of defining the set of physically meaningful averaging processes. Here, since we are interested in a rigorous extension of the continuum concepts of Cauchy stress and first Piola–Kirchhoff stress to discrete systems, we propose slightly more stringent requirements (with respect to Hill's macrohomogeneity conditions). We therefore introduce the following definition.

Definition 1 (Meaningful Deformation Processes). By a large deformation process with meaningful space averages, we mean a deformation process possessing all of the following properties:

1. $[\![\mathsf{F}]\!]^{-1} = [\![\mathsf{F}^{-1}]\!]$, with $\det([\![\mathsf{F}]\!]) > 0$.
2. $\mathrm{Vol}(\Omega) = \det([\![\mathsf{F}]\!])\,\mathrm{Vol}(\Omega_\kappa)$.
3. $[\![\mathsf{S}]\!] = \det([\![\mathsf{F}]\!])[\![\mathsf{T}]\!]([\![\mathsf{F}]\!]^{-1})^{\mathrm{T}}$.

This definition is motivated by a desire to have effective quantities that formally behave just like their local counterparts. Now, similarly to Hill's approach (cf. [14]), instead of attempting to derive necessary and sufficient conditions for satisfying Definition 1, we will only provide a list of *sufficient conditions*. These conditions are found by demanding that the RVE motions satisfy specific BCs. As observed in Sect. 5.4, the "right" choice of BCs is crucial for successfully solving the RVE problem that delivers the effective elastic moduli. In this section, we see that the right choice of BCs is crucial for establishing the very meaning of the definitions of effective quantities. In determining the type of BCs in question, one can start with analyzing the three "canonical" BCs we have discussed in Sect. 5.4.2. With this in mind, the first step is to properly redefine these BCs in a regime of large deformations. We do this next.

In the present context, we define *uniform strain BCs* as follows

$$\boldsymbol{x}(\boldsymbol{\chi}, t) = \hat{\mathsf{F}}(t)\boldsymbol{\chi} \quad \text{for } \boldsymbol{\chi} \in \partial\Omega_\kappa, \tag{5.13}$$

where, for all t of interest, $\hat{\mathsf{F}}(t)$ is a *prescribed* second-order tensor with positive determinant. The definition given here matches the definition given by Hill [13, 14].[3]

Uniform stress BCs are now defined as follows

$$\mathsf{T}(\boldsymbol{x}, t)\boldsymbol{n}(\boldsymbol{x}, t) = \hat{\Sigma}(t)\boldsymbol{n}(\boldsymbol{x}, t) \quad \text{for } \boldsymbol{x} \in \partial\Omega, \tag{5.14}$$

[3]Equation (5.13) is presented under the assumption that the origin of the coordinate system is at the mass center and that the total linear momentum of the system is zero.

where, for all t of interest, $\hat{\Sigma}(t)$ is a *prescribed* symmetric second-order tensor. In this case it is important to remark that, contrary to the case of the uniform strain BCs, the uniform stress BCs stated here do not match those discussed by Hill [13, 14].

As far as *periodic* BCs are concerned (for a very careful discussion of these BCs, see [9]), we redefine them as follows: We say that the motion $\boldsymbol{x} = \boldsymbol{x}(\chi, t)$ and the boundary traction field $\mathsf{S}(\chi, t)\boldsymbol{n}_\kappa(\chi)$ satisfy periodic BCs if:

1. $\boldsymbol{x}(\chi, t)$ can be given the form

$$\boldsymbol{x}(\chi, t) = \hat{\mathsf{F}}(t)\chi + \tilde{\boldsymbol{u}}(\chi, t), \tag{5.15}$$

   where, for all t of interest, $\hat{\mathsf{F}}(t)$ is a *prescribed* second-order tensor with positive determinant and $\tilde{\boldsymbol{u}}(\chi, t)$ is an Ω_κ-periodic displacement vector field.

2. $\mathsf{S}(\chi, t)\boldsymbol{n}_\kappa(\chi)$ is an Ω_κ-antiperiodic.

When adopting periodic BCs, it is important to keep in mind that the fields $\tilde{\boldsymbol{u}}$ and $\mathsf{S}\boldsymbol{n}_\kappa$ are unknown. In other words, the periodic BCs do not *prescribe* the boundary values of either the motion or the traction field. Rather, they only pose constraints on the class of functions to which both the displacement and the traction fields can belong.

Now that these definitions have been stated we present an important result.

Proposition 1. *For any regular, bounded, and simply connected RVE, a smooth deformation process complying with either uniform strain or periodic BC is one such that*

$$[\![\mathsf{F}]\!] = \hat{\mathsf{F}}, \tag{5.16}$$

i.e., one can control the effective deformation gradient tensor via the prescribed $\hat{\mathsf{F}}$. *Furthermore, under these BCs, the resulting effective quantities satisfy the conditions stated in Definition 1.*

This proposition combines a number of results whose rigorous proof has been given in [9]. Unfortunately, it is not possible to prove that the adoption of the uniform stress BCs allows one to satisfy Definition 1.

The consequence of the above result is that, on the one hand, there are sets of conditions under which our discussion is indeed meaningful. On the other hand, the uniform stress BCs, while perfectly acceptable in linear homogenization, are no longer usable in the present context. Although we have not discussed MD up to now, we will see that the loss of the uniform stress BCs as one of the admissible conditions poses possibly severe constraints on how one can define stress-controlled continuum homogenization-based MD schemes.

5.6 Continuum Homogenization and MD

5.6.1 Basic Ideas About MD

To construct a link between continuum homogenization and MD, we first make some cursory observations about MD. We start with observing that, from a conceptual viewpoint, in MD calculations one predicts the motion of a system of N particles using Newton's second law, i.e., $\boldsymbol{f}_i = m_i \ddot{\boldsymbol{r}}_i$, where $\boldsymbol{f}_i$, m_i, and $\boldsymbol{r}_i$ are the force acting on, the mass, and the position of the ith particle in the system, respectively, and where a dot over a quantity denotes the material time derivative, so that $\ddot{\boldsymbol{r}}_i$ is the acceleration of particle i (it is understood that the system's motion is being observed by an inertial observer). The particle ensemble under consideration is viewed as occupying a region of space called a *cell*, which is considered part of an infinite lattice of identical cells. Therefore, in MD computations, the cell under study is considered to be subjected to *periodic* BCs when the particles in the cells are allowed to interact with the particles in the cells that surround the main reference cell (see [9] for additional details). The particles in the cells surrounding the main cell are often called *image particles*. With this in mind, the force $\boldsymbol{f}_i$ acting on particle i is best viewed as follows

$$\boldsymbol{f}_i = \boldsymbol{f}_i^{\text{int}} + \boldsymbol{f}_i^{\text{ext}}, \tag{5.17}$$

where $\boldsymbol{f}_i^{\text{int}}$ is the force on particle i due to its interaction with the other $N-1$ particles in the cell. For this reason, $\boldsymbol{f}_i^{\text{int}}$ can be called an *internal* force, whereas $\boldsymbol{f}_i^{\text{ext}}$ is the force on particle i due to its interaction with the image particles, and therefore *external* to the ensemble. As far as the calculation of $\boldsymbol{f}_i^{\text{int}}$ is concerned, the internal force is derived as the gradient of the total potential energy of the system, this potential energy being the sum of all the potential energies that describe the bonds among the particles in the system. From a mathematical viewpoint, this means that given the N particles in the system and the knowledge of how these particles interact with one another, one can form a function $\Psi = \Psi(\boldsymbol{r}_1, \boldsymbol{r}_2, \dots, \boldsymbol{r}_N)$, namely the total potential energy of the system, such that

$$\boldsymbol{f}_i^{\text{int}} = -\frac{\partial \Psi(\boldsymbol{r}_1, \boldsymbol{r}_2, \dots, \boldsymbol{r}_N)}{\partial \boldsymbol{r}_i}. \tag{5.18}$$

As far as the calculation of $\boldsymbol{f}_i^{\text{ext}}$ is concerned, as it turns out, it is carried out in a way very similar to the calculation for the internal forces, i.e., by constructing the potential energy resulting from the interaction between a particle i in the cell and the image particles outside the cell. In view of how the force on a particle is calculated, and borrowing the language of the theory of elasticity, one could say that the particle ensemble is typically taken to be a *hyperelastic* material, i.e., a material whose internal response is completely characterized by a stored energy function (see, e.g., [21, p. 206] or [29, p. 302]).

Another crucial element of MD calculations is the idea, borrowed from statistical mechanics, that the system's total kinetic energy can be mapped via the equipartition theorem (see, e.g., [10]) into a measure of the system's temperature.

Almost without exception, the calculation of stress in particle systems studied via MD is done by computing the system's *virial stress*, which is defined as follows

$$\mathsf{P} \triangleq \frac{1}{\mathrm{Vol}(\Omega)} \sum_{i=1}^{N} \left(\frac{\partial \Psi}{\partial \boldsymbol{r}_i} \otimes \boldsymbol{r}_i - m_i \dot{\boldsymbol{r}}_i \otimes \dot{\boldsymbol{r}}_i \right), \tag{5.19}$$

where $\dot{\boldsymbol{r}}_i$ is the velocity of particle i. In general, the raw measure of stress provided by P is time averaged and treated as a measure of Cauchy stress (for a recent and detailed review of the concept of virial stress, see [32]). Therefore, the measure of effective stress typically done in MD is a time and volume average of a quantity that is related to the amount of potential and kinetic energy in the system.

The question to address is now as follows: Can one formulate a continuum homogenization problem that is formally identical to an MD-based measure of stress? The answer to this question is in the affirmative and to see how to use continuum homogenization to mimic MD, we need to make the following important remark: Traditionally, when applied to the characterization of the elastic response of a material, continuum homogenization is used to map the properties of a *heterogeneous* system *under quasistatic conditions* into those of a companion *homogeneous* system, again *under quasistatic conditions*. By contrast, an MD calculation maps the (discrete) properties of a system subject to the full form of Newton's second law, i.e., including inertial effects, and maps them into the properties of an equivalent thermomechanical system in equilibrium.[4] Therefore, going back to

[4]By contrast, *molecular statics* determines the properties of a material under static equilibrium conditions and therefore it is applicable only in the zero temperature limit.

continuum homogenization, what we need to consider is an RVE, consisting of a hyperelastic medium (heterogeneous or not), whose evolution is governed by a fully dynamic equation of motion, i.e.,

$$\mathrm{Div}(\mathsf{S}) = \rho_{\mathcal{K}} \dot{\boldsymbol{v}}, \tag{5.20}$$

where $\rho_{\mathcal{K}}$ is the density distribution in the reference configuration and $\boldsymbol{v}$ is the material velocity field. Next, the local properties of this RVE must then be mapped into the properties of an equivalent homogeneous *thermoelastic* solid under quasistatic conditions. Adopting this conceptual framework yields a method for bridging continuum homogenization and MD. This will be shown in detail in the remainder of the chapter.

5.6.2 Effective Cauchy Stress

As we noted earlier, the measure of effective stress in MD is obtained as a time and volume average. Hence, the first result we intend to illustrate concerns what happens to the continuum homogenization notion of effective Cauchy stress under (5.20) when computed as a volume average rather than through boundary integrals and when averaged over time.

For simplicity, we will assume that the RVE is a regular and simply connected domain. In addition, we will assume that there exist positive constants α and β such that, for all times $t \in [0, \infty)$, we have

$$\|\rho \boldsymbol{v} \otimes \boldsymbol{x}\| < \alpha \quad \text{and} \quad \beta < \mathrm{Vol}(\Omega), \tag{5.21}$$

where ρ is the RVE mass density distribution in the deformed configuration. Furthermore, we assume that the volume time rate of change is controllable in the following sense, namely that there exist a positive constant γ and a constant $\delta \in (0, 1)$ such that, for all $\tau \in [0, \infty)$, we have

$$\int_0^\infty \left| \frac{\mathrm{d}}{\mathrm{d}t} \mathrm{Vol}(\Omega) \right| \mathrm{d}t < \gamma \tau^\delta. \tag{5.22}$$

From a practical viewpoint, the above assumptions require that the RVE motion be controllable in such a way that the RVE volume neither shrinks to zero nor grows "too fast" and in such a way that the momentum of the system remains bounded. These assumptions are always achievable under uniform strain and periodic BCs since, from a purely kinematical viewpoint, these BCs imply the satisfaction of the relations in Definition 1, which, in turn, imply the full controllability of the system's volume by an appropriate choice of $\hat{\mathsf{F}}(t)$. In addition, the assumption concerning the system's momentum is easily achievable in that the BCs in question do not let the RVE to accelerate as a whole or to "spin out of control." With this in mind, we prove the following result.[5]

Proposition 2 (Effective Stress as a Volume Average). *Under the assumptions in* (5.21) *and* (5.22)*, and for a regular simply connected RVE governed by* (5.20)*, we have*

$$\langle [\![\mathsf{T}]\!] \rangle = \left\langle \frac{1}{\mathrm{Vol}(\Omega)} \int_\Omega (\mathsf{T} - \rho \boldsymbol{v} \otimes \boldsymbol{v}) \mathrm{d}v \right\rangle. \tag{5.23}$$

Before presenting the proof of this result, it is important to discuss its meaning. Equation (5.23) says that the time average of the effective stress has a structure that is very similar to that of the time average of the virial stress. In fact, for a hyperelastic material, T is completely derived from a potential, as is the first term on the right-hand side of (5.19). Furthermore, (5.23) displays the term $(\rho\, \mathrm{d}v)\boldsymbol{v} \otimes \boldsymbol{v}$, which is formally identical to the term $m_i \dot{\boldsymbol{r}}_i \otimes \dot{\boldsymbol{r}}_i$ appearing in (5.19). Clearly, we still have some work to do to show that indeed $\langle [\![\mathsf{T}]\!] \rangle$ is the same as $\langle \mathsf{P} \rangle$, since we do not have a discrete equivalent of $[\![\mathsf{T}]\!]$ yet. We will deal with this issue later in this chapter.

Proof of Proposition 2. Starting from the second of (5.9) and using the divergence theorem, we obtain

$$\mathrm{Vol}(\Omega)[\![\mathsf{T}]\!] = \int_\Omega (\mathsf{T}\, \mathrm{grad}\, \boldsymbol{x} + (\mathrm{div}\, \mathsf{T}) \otimes \boldsymbol{x}) \mathrm{d}v = \int_\Omega (\mathsf{T} + \rho \dot{\boldsymbol{v}} \otimes \boldsymbol{x}) \mathrm{d}v, \tag{5.24}$$

where we have used the fact that $\mathrm{grad}\, \boldsymbol{x} = \mathsf{I}$ along with (5.20).

Next, using the transport theorem (see, e.g., [11]), we can rewrite the last term in (5.24) as follows:

$$\int_\Omega \rho \dot{\boldsymbol{v}} \otimes \boldsymbol{x}\, \mathrm{d}v = \frac{\mathrm{d}}{\mathrm{d}t} \int_\Omega \rho \boldsymbol{v} \otimes \boldsymbol{x}\, \mathrm{d}v - \int_\Omega \rho \boldsymbol{v} \otimes \boldsymbol{v}\, \mathrm{d}v. \tag{5.25}$$

Substituting the above result into (5.24) and dividing by $\mathrm{Vol}(\Omega)$, we obtain

$$[\![\mathsf{T}]\!] = \frac{1}{\mathrm{Vol}(\Omega)} \int_\Omega (\mathsf{T} - \rho \boldsymbol{v} \otimes \boldsymbol{v}) \mathrm{d}v + \frac{1}{\mathrm{Vol}(\Omega)} \frac{\mathrm{d}}{\mathrm{d}t} \int_\Omega (\rho \boldsymbol{v} \otimes \boldsymbol{x}) \mathrm{d}v. \tag{5.26}$$

Next we consider the time integral of the last term in the above expression over the interval $(0, \tau)$

[5] A less general version of this result was presented in [9].

$$\begin{aligned}\int_0^{\mathcal{T}}\left\{\frac{1}{\mathrm{Vol}(\Omega)}\frac{\mathrm{d}}{\mathrm{d}t}\int_\Omega(\rho\boldsymbol{v}\otimes\boldsymbol{x})\mathrm{d}v\right\}\mathrm{d}t &= \left[\frac{1}{\mathrm{Vol}(\Omega)}\int_\Omega(\rho\boldsymbol{v}\otimes\boldsymbol{x})\mathrm{d}v\right]_0^{\mathcal{T}} \\ &+\int_0^{\mathcal{T}}\left\{\frac{1}{(\mathrm{Vol}(\Omega))^2}\left(\frac{\mathrm{d}}{}\,\mathrm{Vol}(\Omega)\right)\int_\Omega(\rho\boldsymbol{v}\otimes\boldsymbol{x})\mathrm{d}v\right\}\mathrm{d}t,\end{aligned}\tag{5.27}$$

where use has been made of integration by parts (with respect to t). From the above relation, we conclude that the norm of the term on the left-hand side is such that

$$\begin{aligned}&\left\|\int_0^{\mathcal{T}}\left\{\frac{1}{\mathrm{Vol}(\Omega)}\frac{\mathrm{d}}{\mathrm{d}t}\int_\Omega(\rho\boldsymbol{v}\otimes\boldsymbol{x})\mathrm{d}v\right\}\mathrm{d}t\right\| \\ &\qquad\leq\left\|\left[\frac{1}{\mathrm{Vol}(\Omega)}\int_\Omega(\rho\boldsymbol{v}\otimes\boldsymbol{x})\mathrm{d}v\right]_0^{\mathcal{T}}\right\| \\ &+\left\|\int_0^{\mathcal{T}}\left\{\frac{1}{(\mathrm{Vol}(\Omega))^2}\left(\frac{\mathrm{d}}{\mathrm{d}t}\,\mathrm{Vol}(\Omega)\right)\int_\Omega(\rho\boldsymbol{v}\otimes\boldsymbol{x})\mathrm{d}v\right\}\mathrm{d}t\right\|.\end{aligned}\tag{5.28}$$

In turn, taking advantage of the first of the assumptions in (5.21), for the first term on the right-hand side of the above expression, we can write

$$\begin{aligned}\left\|\left[\frac{1}{\mathrm{Vol}(\Omega)}\int_\Omega(\rho\boldsymbol{v}\otimes\boldsymbol{x})\mathrm{d}v\right]_0^{\mathcal{T}}\right\| &= \left\|\left[\frac{1}{\mathrm{Vol}(\Omega)}\int_\Omega(\rho\boldsymbol{v}\otimes\boldsymbol{x})\mathrm{d}v\right]_{t=\mathcal{T}}\right. \\ &\qquad\left.-\left[\frac{1}{\mathrm{Vol}(\Omega)}\int_\Omega(\rho\boldsymbol{v}\otimes\boldsymbol{x})\mathrm{d}v\right]_{t=0}\right\| \\ &\leq\left\|\left[\frac{1}{\mathrm{Vol}(\Omega)}\int_\Omega(\rho\boldsymbol{v}\otimes\boldsymbol{x})\mathrm{d}v\right]_{t=\mathcal{T}}\right\| \\ &\qquad+\left\|\left[\frac{1}{\mathrm{Vol}(\Omega)}\int_\Omega(\rho\boldsymbol{v}\otimes\boldsymbol{x})\mathrm{d}v\right]_{t=0}\right\| \\ &\leq\left[\frac{1}{\mathrm{Vol}(\Omega)}\int_\Omega\|\rho\boldsymbol{v}\otimes\boldsymbol{x}\|\mathrm{d}v\right]_{t=\mathcal{T}} \\ &\qquad+\left[\frac{1}{\mathrm{Vol}(\Omega)}\int_\Omega\|\rho\boldsymbol{v}\otimes\boldsymbol{x}\|\mathrm{d}v\right]_{t=0} \\ &<\left[\frac{\alpha\,\mathrm{Vol}(\Omega)}{\mathrm{Vol}(\Omega)}\right]_{t=\mathcal{T}}+\left[\frac{\alpha\,\mathrm{Vol}(\Omega)}{\mathrm{Vol}(\Omega)}\right]_{t=0} \\ &<2\alpha.\end{aligned}\tag{5.29}$$

As far as the second term on the right-hand side of the inequality in (5.28) is concerned, using the assumptions in (5.21) and (5.22), we have that

$$\begin{aligned}
&\left\| \int_0^{\mathcal{T}} \left\{ \frac{1}{(\mathrm{Vol}(\Omega))^2} \left(\frac{\mathrm{d}}{\mathrm{d}t} \mathrm{Vol}(\Omega) \right) \int_\Omega (\rho \boldsymbol{v} \otimes \boldsymbol{x}) \mathrm{d}v \right\} \mathrm{d}t \right\| \\
&\quad \le \int_0^{\mathcal{T}} \left\{ \frac{1}{(\mathrm{Vol}(\Omega))^2} \left| \frac{\mathrm{d}}{\mathrm{d}t} \mathrm{Vol}(\Omega) \right| \int_\Omega \| \rho \boldsymbol{v} \otimes \boldsymbol{x} \| \, \mathrm{d}v \right\} \mathrm{d}t \\
&\quad < \int_0^{\mathcal{T}} \frac{\alpha \, \mathrm{Vol}(\Omega)}{(\mathrm{Vol}(\Omega))^2} \left| \frac{\mathrm{d}}{\mathrm{d}t} \mathrm{Vol}(\Omega) \right| \mathrm{d}t \\
&\quad < \frac{\alpha}{\beta} \int_0^{\mathcal{T}} \left| \frac{\mathrm{d}}{\mathrm{d}t} \mathrm{Vol}(\Omega) \right| \mathrm{d}t < \frac{\alpha\gamma}{\beta} \mathcal{T}^\delta. \qquad (5.30)
\end{aligned}$$

Next, substituting the results in (5.29) and (5.30) into (5.28), we have that

$$\left\| \int_0^{\mathcal{T}} \left\{ \frac{1}{\mathrm{Vol}(\Omega)} \frac{\mathrm{d}}{\mathrm{d}t} \int_\Omega (\rho \boldsymbol{v} \otimes \boldsymbol{x}) \mathrm{d}v \right\} \mathrm{d}t \right\| < 2\alpha + \frac{\alpha\gamma}{\beta} \mathcal{T}^\delta. \qquad (5.31)$$

Finally, observing that

$$\lim_{\mathcal{T} \to \infty} \frac{1}{\mathcal{T}} \left(2\alpha + \frac{\alpha\gamma}{\beta} \mathcal{T}^\delta \right) = 0, \qquad (5.32)$$

and recalling the definition of the time average operation in (5.12), we conclude that we must have

$$\left\langle \frac{1}{\mathrm{Vol}(\Omega)} \frac{\mathrm{d}}{\mathrm{d}t} \int_\Omega (\rho \boldsymbol{v} \otimes \boldsymbol{x}) \mathrm{d}v \right\rangle = 0. \qquad (5.33)$$

Finally, taking the time average of (5.26) and taking advantage of the result in (5.33), we obtain (5.23). □

5.6.3 Application of BCs in the "Dynamic" RVE Problem

As discussed earlier in the section, to build a continuum homogenization problem that can mimic an MD calculation, one needs to adopt a fully dynamic form of the pointwise balance of linear momentum law, such as that in (5.20), which we repeat here for convenience

$$\mathrm{Div}(\mathsf{S}) = \rho_\kappa \dot{\boldsymbol{v}}. \qquad (5.34)$$

In addition to this balance law, the RVE problem needs to be accompanied by the material's constitutive equations and by a set of BCs. As far as constitutive equations are concerned, for a hyperelastic material, these are given by assigning a strain energy function $\psi_\kappa(\mathsf{F},\boldsymbol{\chi})$ so that

$$\mathsf{S}(\mathsf{F},\boldsymbol{\chi}) = \rho_\kappa \frac{\partial \psi_\kappa(\mathsf{F},\boldsymbol{\chi})}{\partial \mathsf{F}}. \tag{5.35}$$

As far as the BCs are concerned, as illustrated in Sect. 5.4.2, these are essential to the determination of the effective elastic properties of the material. In fact, the BCs allow one to control, say, the effective deformation and, at the same time, solve the governing equations for the resulting stress field, which can then be used to compute the effective stress response as a function of the specified value of the effective deformation. With this in mind, it is important to be aware of how the parameter controlled by the BCs actually appears in the problem.

Referring to (5.13), if one chooses uniform strain boundary conditions, then the control parameter $\hat{\mathsf{F}}(t)$ directly determines the location of the RVE boundary points in the deformed configuration. Hence, it is easy to see how the RVE "knows" about $\hat{\mathsf{F}}(t)$. However, in the case of periodic BCs, it is less obvious to see how the effective deformation of the RVE is directed to take on the prescribed value $\hat{\mathsf{F}}(t)$. To understand how this happens, one needs to carefully consider (5.15). The first thing to understand about (5.15) is that it is not an assumption: It is a convenient decomposition of the motion of the *entire* RVE. Specifically, it decomposes the entire RVE motion into a homogeneous deformation and an unknown displacement field $\tilde{\boldsymbol{u}}$. This decomposition is always admissible because the controlled parameter $\hat{\mathsf{F}}(t)$ is chosen to be invertible. Next, it must be understood that, under periodic BCs, the RVE "does not know" about $\hat{\mathsf{F}}(t)$ through boundary data, because the only fields whose boundary behavior is controlled are the unknown field $\tilde{\boldsymbol{u}}$ (constrained to be periodic) and the unknown field $\mathsf{S}\boldsymbol{n}_\kappa$ (constrained to be antiperiodic). We must, therefore, conclude that the effective deformation of the RVE is controlled via the governing equations. In fact, using the decomposition in (5.15) to compute the deformation gradient and the velocity of the RVE motion, we have

$$\mathsf{F} = \hat{\mathsf{F}}(\mathsf{I} + \tilde{\mathsf{H}}) \quad \text{and} \quad \dot{\boldsymbol{v}} = \ddot{\hat{\mathsf{F}}}(\boldsymbol{\chi} + \tilde{\boldsymbol{u}}) + 2\dot{\hat{\mathsf{F}}}\dot{\tilde{\boldsymbol{u}}} + \hat{\mathsf{F}}\ddot{\tilde{\boldsymbol{u}}}, \tag{5.36}$$

where $\tilde{\mathsf{H}} = \operatorname{Grad} \tilde{\boldsymbol{u}}$. With this in mind, one can then take F and $\boldsymbol{v}$ as given in (5.36) and combine them with (5.34) and (5.35) to formulate a BVP in the unknown field $\tilde{\boldsymbol{u}}$. By this substitution, we see that the control parameter $\hat{\mathsf{F}}$ will contribute to the governing partial differential equations as a forcing term.

To facilitate the discussion in Sect. 5.6.4, it is useful to notice that one could enforce even the uniform strain BCs as was done in the case of periodic BCs. In fact, no matter what BC set one chooses, given a nonsingular $\hat{\mathsf{F}}(t)$, one can *always* represent the RVE motion as

$$\boldsymbol{x}(\chi, t) = \hat{\mathsf{F}}(t)\boldsymbol{x}^*(\chi, t), \tag{5.37}$$

where $\boldsymbol{x}^*$ is an unknown vector field. Again, (5.37) is not an assumption, it is simply a decomposition that defines an unknown vector field $\boldsymbol{x}^*$. To determine this field, we first compute the RVE deformation gradient and the acceleration, which take on the form

$$\mathsf{F} = \hat{\mathsf{F}}\mathsf{F}^* \quad \text{and} \quad \dot{\boldsymbol{v}} = \ddot{\hat{\mathsf{F}}}\boldsymbol{x}^* + 2\dot{\hat{\mathsf{F}}}\dot{\boldsymbol{x}}^* + \hat{\mathsf{F}}\ddot{\boldsymbol{x}}^*, \tag{5.38}$$

where $\mathsf{F}^* = \operatorname{Grad} \boldsymbol{x}^*$. Next, the relations in (5.38) along with the constitutive relations in (5.35) are substituted into the momentum balance law to obtain a set of partial differential equations in the unknown field $\boldsymbol{x}^*$. Finally, under uniform strain BCs, the boundary value of $\boldsymbol{x}^*$ is such that $\boldsymbol{x}^* = \chi$, whereas in the case of periodic BCs, the boundary value of $\boldsymbol{x}^*$ is such that $\boldsymbol{x}^* = \chi + \boldsymbol{u}^*$, the field $\boldsymbol{u}^*$ being periodic. Clearly, in the case of periodic BCs, one still needs to also make sure that the field $\mathsf{S}\boldsymbol{n}_{\kappa}$ is antiperiodic.

5.6.4 A Lagrangian Continuum Homogenization Scheme

Now we can deal directly with the question concerning how the continuum homogenization notion of effective stress can be extended to discrete systems. The answer we provide lies in a reformulation of the RVE problem illustrated in Sect. 5.6.3 using the Lagrangian mechanics. Before delving into the details of this reformulation, it is essential to realize that the reformulation in question is *not* useful from the viewpoint of continuum homogenization because the latter requires the solution of a BVP, i.e., a set of partial differential equations and *boundary conditions*. The usefulness of the reformulation lies in the fact that it can be easily applied to discrete systems because, at least formally, the Lagrangian formulation we are about to illustrate does not require the evaluation of boundary information. Therefore, the proposed Lagrangian scheme can be applied to discrete systems without worrying about defining the boundary of such systems.

As observed in Sect. 5.6.3, we mimic what happens in MD by modeling the material as hyperelastic and subject to a fully dynamic version of the momentum balance law. Under these conditions, one can construct the

Lagrangian of the RVE, which, under uniform or periodic BCs, takes on the form (for a careful discussion on the effect of BCs on the Lagrangian of the RVE, see [1])

$$\mathscr{L}(\boldsymbol{x}, \boldsymbol{v}) = T - U, \tag{5.39}$$

where T and U denote the RVE's total kinetic and potential energies, respectively, and, in a continuum mechanics context, are defined in the usual way, i.e.,

$$T = \int_{\Omega_\kappa} \frac{1}{2} \rho_\kappa \boldsymbol{v} \cdot \boldsymbol{v} \, \mathrm{d}V \quad \text{and} \quad U = \int_{\Omega_\kappa} \rho_\kappa \psi_\kappa \, \mathrm{d}V, \tag{5.40}$$

In a continuum context, using the standard Lagrangian formalism, one can start from (5.39) and derive the governing equations of the RVE, namely (5.34) (see [1]). Again, we have no interest in pursuing this because we already have the governing equations of the RVE. Instead, noticing that the terms in (5.40) do not require an explicit knowledge of the boundary of the RVE, the focus of this discussion will be to illustrate how one can compute the effective stresses $[\![\mathsf{S}]\!]$ and $[\![\mathsf{T}]\!]$ *directly* from the Lagrangian function $\mathscr{L}$.

The calculation of $[\![\mathsf{S}]\!]$ and $[\![\mathsf{T}]\!]$ from $\mathscr{L}$ is done by first making sure that the parameter controlling the effective deformation of the RVE is explicitly embedded into the Lagrangian. In turn, this is done by adopting the strategy discussed at the end of Sect. 5.6.3, i.e., we adopt the decomposition in (5.37) and we turn the Lagrangian $\mathscr{L}$ into a function of the control parameter $\hat{\mathsf{F}}$, its time rate of change, the unknown motion $\boldsymbol{x}^*$, and the associated velocity field $\dot{\boldsymbol{x}}^*$

$$\mathscr{L}(\boldsymbol{x}, \boldsymbol{v}) = \mathscr{L}\left(\hat{\mathsf{F}}, \dot{\hat{\mathsf{F}}}, \boldsymbol{x}^*, \dot{\boldsymbol{x}}^*\right), \tag{5.41}$$

where, in view of (5.37), $\boldsymbol{v}$ takes on the form $\boldsymbol{v} = \dot{\hat{\mathsf{F}}}\boldsymbol{x}^* + \hat{\mathsf{F}}\dot{\boldsymbol{x}}^*$. Once the Lagrangian is given the above form, Andia et al. [1] have rigorously proven that

$$[\![\mathsf{S}]\!] = \frac{1}{\mathrm{Vol}(\Omega_\kappa)} \left[\frac{\mathrm{d}}{\mathrm{d}t} \left(\frac{\partial \mathscr{L}}{\partial \dot{\hat{\mathsf{F}}}} \right) - \frac{\partial \mathscr{L}}{\partial \hat{\mathsf{F}}} \right]. \tag{5.42}$$

Furthermore, since under uniform or periodic BCs one can rely on the relations in Definition 1, we have that

$$[\![\mathsf{T}]\!] = \frac{1}{\mathrm{Vol}(\Omega)} \left[\frac{\mathrm{d}}{\mathrm{d}t} \left(\frac{\partial \mathscr{L}}{\partial \dot{\hat{\mathsf{F}}}} \right) - \frac{\partial \mathscr{L}}{\partial \hat{\mathsf{F}}} \right] \hat{\mathsf{F}}^{\mathrm{T}}. \tag{5.43}$$

The importance of (5.42) and (5.43) lies in the fact that one can measure the effective stresses of the RVE without having to carry out the boundary integrations by which $[\![\mathsf{S}]\!]$ and $[\![\mathsf{T}]\!]$ are defined. As important, from a practical viewpoint, this means that one can use (5.42) and (5.43) to *define* $[\![\mathsf{S}]\!]$ and $[\![\mathsf{T}]\!]$ for a discrete system of particles as long as one can define the Lagrangian of the discrete system in question. With this in mind, constructing the Lagrangian of a system of particles interacting via a bond potential energy is a rather simple operation. In fact, denoting such a Lagrangian function by $\mathscr{L}_{\mathrm{MD}}$, we have that

$$\mathscr{L}_{\mathrm{MD}}(\boldsymbol{r}_1,\ldots,\boldsymbol{r}_N,\dot{\boldsymbol{r}}_1,\ldots,\dot{\boldsymbol{r}}_N)=\sum_{i=1}^{N}\frac{1}{2}m_i\dot{\boldsymbol{r}}_i\cdot\dot{\boldsymbol{r}}_i-\Psi(\boldsymbol{r}_1,\ldots,\boldsymbol{r}_N), \quad (5.44)$$

where the various quantities in the above equations have been introduced in Sect. 5.6.1. Next, we explicitly embed the control parameter $\hat{\mathsf{F}}$ into $\mathscr{L}_{\mathrm{MD}}$ similarly to what was done in the case of $\mathscr{L}$, i.e., by adopting the following decomposition of the system's motion:

$$\boldsymbol{r}_i(t)=\hat{\mathsf{F}}(t)\boldsymbol{r}_i^*(t), \quad i=1,\ldots,N. \quad (5.45)$$

The unknown functions $\boldsymbol{r}_i^*(t)$ are determined by solving the system's equations of motion. In turn, these equations are obtained by substituting (5.45) into (5.44) and by using the standard Lagrangian formalism, which, as demonstrated in [1], yields

$$m_i\hat{\mathsf{F}}\ddot{\boldsymbol{r}}_i^*+2m_i\dot{\hat{\mathsf{F}}}\dot{\boldsymbol{r}}_i^*+m_i\ddot{\hat{\mathsf{F}}}\boldsymbol{r}_i^*=-\hat{\mathsf{F}}^{-\mathrm{T}}\frac{\partial\Psi}{\partial\boldsymbol{r}_i^*}+\boldsymbol{f}_i^{\mathrm{e}}, \quad i=1,\ldots,N. \quad (5.46)$$

The result in (5.46) is important in that it shows that the somewhat abstract continuum homogenization scheme described earlier has been turned into a practical tool for carrying out MD simulations. In fact, by placing a particle ensemble in an initial reference volume Ω_κ, and by knowing the bond potentials that allow one to compute Ψ, one can assign any given effective deformation $\hat{\mathsf{F}}(t)$ and compute the resulting motion of the particle system by integrating the system of ordinary differential equations in (5.46). Once, this motion has been calculated, the solution can be postprocessed to compute the effective Cauchy and first Piola–Kirchhoff stresses for the discrete system which are now *defined* as follows

$$[\![\mathsf{S}]\!]_{\mathrm{MD}}\triangleq\frac{1}{\mathrm{Vol}(\Omega_\kappa)}\left[\frac{\mathrm{d}}{\mathrm{d}t}\left(\frac{\partial\mathscr{L}_{\mathrm{MD}}}{\partial\dot{\hat{\mathsf{F}}}}\right)-\frac{\partial\mathscr{L}_{\mathrm{MD}}}{\partial\hat{\mathsf{F}}}\right], \quad (5.47)$$

$$[\![\mathsf{T}]\!]_{\mathrm{MD}}\triangleq\frac{1}{\mathrm{Vol}(\Omega)}\left[\frac{\mathrm{d}}{\mathrm{d}t}\left(\frac{\partial\mathscr{L}_{\mathrm{MD}}}{\partial\dot{\hat{\mathsf{F}}}}\right)-\frac{\partial\mathscr{L}_{\mathrm{MD}}}{\partial\hat{\mathsf{F}}}\right]\hat{\mathsf{F}}^{\mathrm{T}}, \quad (5.48)$$

where we have used the subscript "MD" to underscore the fact that (5.47) and (5.48) are the definitions applicable in MD. In addition, it should be noted that the definition in (5.48) is obtained from that in (5.47) and the conditions stated in Definition 1. Therefore, in adopting the definitions in (5.47) and (5.48), it is crucial that the BCs enforced on the discrete system be compatible with Definition 1, e.g., periodic BCs, which are the BCs almost universally used in MD.

Carrying out the calculations implied by (5.47) and (5.48) and using (5.18), (5.45), and (5.46), one can show that $[\![\mathsf{S}]\!]_{\mathrm{MD}}$ and $[\![\mathsf{T}]\!]_{\mathrm{MD}}$ can be given the following simple forms (see [1]):

$$[\![\mathsf{S}]\!]_{\mathrm{MD}} = \frac{1}{\mathrm{Vol}(\Omega_\kappa)} \sum_{i=1}^{N} \boldsymbol{f}_i^{\mathrm{ext}} \otimes \boldsymbol{r}_i^* \quad \text{and} \quad [\![\mathsf{T}]\!]_{\mathrm{MD}} = \frac{1}{\mathrm{Vol}(\Omega)} \sum_{i=1}^{N} \boldsymbol{f}_i^{\mathrm{ext}} \otimes \boldsymbol{r}_i. \tag{5.49}$$

From a conceptual viewpoint, we consider the result in (5.47) and (5.48) to be extremely important in that we can claim that the effective stress measures we have introduced have exactly the same meaning in both a discrete context and a continuum context. For this reason, they demonstrate that one can indeed gain a great deal of understanding of MD methods for computing mechanical properties by relying just on classical mechanics concepts.

5.6.5 Virial Stress/Effective Cauchy Stress Equivalence

Now that we have obtained a continuum homogenization-based notion of effective Cauchy stress applicable to discrete systems, we are in a position to offer a meaningful comparison between such a notion and that of virial stress. This comparison will be expressed via the following.

Proposition 3 (Virial–Cauchy Stress Equivalence). *If there exist positive constants α and β such that, for all times $t \in [0, \infty)$ and all $i = 1, \ldots, N$,*

$$\|m_i \dot{\boldsymbol{r}}_i \otimes \boldsymbol{r}_i\| < \alpha \quad \textit{and} \quad \beta < \mathrm{Vol}(\Omega), \tag{5.50}$$

and assuming that the volume time rate of change is controllable in the sense that there exist a positive constant γ and a constant $\delta \in (0, 1)$ such that, for all $\tau \in [0, \infty)$,

$$\int_0^\infty \left| \frac{\mathrm{d}}{\mathrm{d}t} \mathrm{Vol}(\Omega) \right| \mathrm{d}t < \gamma \tau^\delta, \tag{5.51}$$

then

$$\langle [\![\mathsf{T}]\!]_{\mathrm{MD}} \rangle = \langle \mathsf{P} \rangle. \tag{5.52}$$

Proof of Proposition 3. We consider a reference cell consisting of an ensemble of N particles subject to periodic BCs. Next, using (5.17), we can write

$$\boldsymbol{f}_i^{\text{ext}} = \boldsymbol{f}_i - \boldsymbol{f}_i^{\text{int}}, \tag{5.53}$$

$\boldsymbol{f}_i$ being the total force acting on particle i and $\boldsymbol{f}_i^{\text{int}}$ being the force acting on particle i due to interactions with the other particles internal to the reference cell. Next, we recall that the total force $\boldsymbol{f}_i = m_i \ddot{\boldsymbol{r}}_i$ due to Newton's second law, so that, using (5.18), (5.53) can be rewritten as

$$\boldsymbol{f}_i^{\text{ext}} = m_i \ddot{\boldsymbol{r}}_i + \frac{\partial \Psi}{\partial \boldsymbol{r}_i}. \tag{5.54}$$

Substituting (5.54) into the second of (5.49), we have

$$[\![\mathsf{T}]\!]_{\text{MD}} = \frac{1}{\text{Vol}(\Omega)} \sum_{i=1}^{N} \left(m_i \ddot{\boldsymbol{r}}_i + \frac{\partial \Psi}{\partial \boldsymbol{r}_i} \right) \otimes \boldsymbol{r}_i. \tag{5.55}$$

Now, observe that the term $\ddot{\boldsymbol{r}}_i \otimes \boldsymbol{r}_i$ can be written as

$$\ddot{\boldsymbol{r}}_i \otimes \boldsymbol{r}_i = \frac{\mathrm{d}}{\mathrm{d}t}(\dot{\boldsymbol{r}}_i \otimes \boldsymbol{r}_i) - \dot{\boldsymbol{r}}_i \otimes \dot{\boldsymbol{r}}_i, \tag{5.56}$$

so that (5.55) can be written as

$$\begin{aligned} [\![\mathsf{T}]\!]_{\text{MD}} = {} & \frac{1}{\text{Vol}(\Omega)} \sum_{i=1}^{N} \left(\frac{\partial \Psi}{\partial \boldsymbol{r}_i} \otimes \boldsymbol{r}_i - m_i \dot{\boldsymbol{r}}_i \otimes \dot{\boldsymbol{r}}_i \right) \\ & + \frac{1}{\text{Vol}(\Omega)} \sum_{i=1}^{N} m_i \frac{\mathrm{d}}{\mathrm{d}t}(\dot{\boldsymbol{r}}_i \otimes \boldsymbol{r}_i). \end{aligned} \tag{5.57}$$

Next, recalling that the mass of each particle is a constant and substituting (5.19) into (5.57), we can give $[\![\mathsf{T}]\!]_{\text{MD}}$ the following form:

$$[\![\mathsf{T}]\!]_{\text{MD}} = \mathsf{P} + \frac{1}{\text{Vol}(\Omega)} \frac{\mathrm{d}}{\mathrm{d}t} \sum_{i=1}^{N} (m_i \dot{\boldsymbol{r}}_i \otimes \boldsymbol{r}_i). \tag{5.58}$$

Finally, observing that (5.50) and (5.51) are a restatement of (5.21) and (5.22), using the same strategy that we have used to derive (5.33), provided of course that the due adjustments are made to account for the fact that here we are in a discrete context, we have that

$$\left\langle \frac{1}{\text{Vol}(\Omega)} \frac{\mathrm{d}}{\mathrm{d}t} \sum_{i=1}^{N} (m_i \dot{\boldsymbol{r}}_i \otimes \boldsymbol{r}_i) \right\rangle = 0, \tag{5.59}$$

which implies that the time average of (5.58) yields (5.52). □

5.6.6 Remarks on the Difference Between $[\![\mathsf{T}]\!]_{\mathrm{MD}}$ and P

In the preceding sections, we have shown that the effective Cauchy stress developed by exclusively relying on continuum homogenization ideas does lead to a useful definition of effective stress for particle systems. In addition, we have shown that, when taken as a time average, the effective Cauchy stress is identical to the stress measure based on the virial stress, provided that the evolution of the RVE is bounded as indicated in Proposition 3. With this in mind, referring to (5.58), we see that, at every instant in time, the difference between the $[\![\mathsf{T}]\!]_{\mathrm{MD}}$ and P is given by the volume average of the time rate of change of the tensor

$$\mathsf{H} = \sum_{i=1}^{N} (m_i \dot{\boldsymbol{r}}_i \otimes \boldsymbol{r}_i). \tag{5.60}$$

The tensor H can be interpreted as the generalized moment of momentum (or angular momentum) of the system. In fact, taking advantage of the one-to-one mapping that exists between second-order skew-symmetric tensors and vectors (see, e.g., [11]), it is not difficult to show that the components of $\mathrm{skw}(\mathsf{H})$, the skew-symmetric part of H, are the components of the system's total angular momentum. Therefore, so long as the overall moment of the external forces acting on the reference cell (computed with respect to some fixed point in an inertial frame of reference) is equal to zero, as is normally the case in MD, then the balance of angular momentum for the reference cell demands that $(\mathrm{d}/\mathrm{d}t)\,\mathrm{skw}(\mathsf{H}) = 0$. In turn, this means that the difference between $[\![\mathsf{T}]\!]_{\mathrm{MD}}$ and P is given by the term

$$\frac{1}{\mathrm{Vol}(\Omega)}\,\mathrm{sym}(\dot{\mathsf{H}}), \tag{5.61}$$

where $\mathrm{sym}(\mathsf{H})$ is the symmetric part of H. As it turns out, the term in (5.61) does not vanish on a instant by instant basis. This can be seen in Fig. 5.4, which was obtained by conducting an MD simulation for an ensemble of 500 particles interacting via the Lennard–Jones potential and at constant volume (details about this simulation are reported in [2]). Keeping in mind that $[\![\mathsf{T}]\!]_{\mathrm{MD}}$ and P are indeed different notions of stress (for additional comments concerning this discussion, see [32]), thanks to Proposition 3, we know that the time average of their difference vanishes when averaging over long time periods. However, Andia et al. [2] have reported some surprising numerical results in which the time average of $([\![\mathsf{T}]\!]_{\mathrm{MD}} - \mathsf{P})$ becomes negligible even over relatively small timescales. However, a question that remains open to

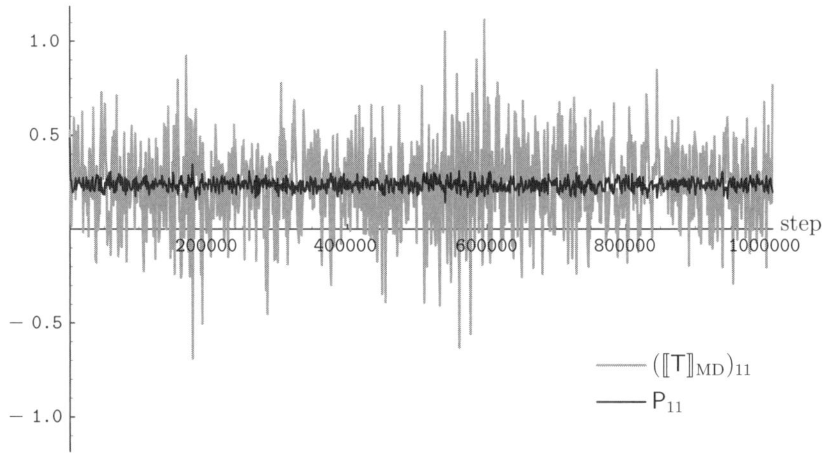

Fig. 5.4. Plot of the 11 components $[\![\mathsf{T}]\!]_{\mathrm{MD}}$ and P in nondimensional form obtained in a microcanonical ensemble simulation of a three-dimensional Lennard–Jones material. The *horizontal axis* shows the number of (identical) time integration steps during the MD simulation

investigation is whether or not there exists an intrinsic timescale over which the time average of $([\![\mathsf{T}]\!]_{\mathrm{MD}} - \mathsf{P})$ can be said to be negligible in some physically based way.

5.6.7 Is There a Continuum-Level Virial Stress?

From a conceptual viewpoint, this chapter is focused on how the concepts from a continuum context can be used to define equivalent concepts in a discrete context. What we have not dealt with are questions such as how to derive a continuum-level concept corresponding to the virial stress or, at a more fundamental level, how do we theoretically establish rigorous continuum limits of the mechanical response of a discrete system. This second question has been studied by many researchers (see, for example, [6, 17, 19]; new rigorous results have been presented in [5]) and is outside the scope of the present chapter. However, the first question can be answered within the framework presented here. In fact, carefully comparing the proofs of Propositions 2 and 3, there is strong indication that indeed one could define the continuum-level virial stress to be the following second-order tensor

$$\mathsf{P}_{\mathrm{CL}} \triangleq \mathsf{T} - \rho \boldsymbol{v} \otimes \boldsymbol{v}, \tag{5.62}$$

where the subscript CL stands for "continuum level." What is interesting about this result is that it is not at all original. In fact, although absent from most (if not all) of the continuum mechanics textbooks published in the last 20 years, in reality the stress tensor P_{CL} is well known in the fluid mechanics literature and has been discussed, although without being explicitly called "the virial stress," by Truesdell and Toupin [30, Article 207] and then recalled in their presentation of the virial theorem (cf. [7, Article 219]). Truesdell and Toupin [30] call the component $\rho \boldsymbol{v} \otimes \boldsymbol{v}$ of the tensor P_{CL} the "apparent stress due to transfer of momentum" and show that it appears naturally in a fully Eulerian restatement of the pointwise linear momentum balance law. Using the tensor P_{CL}, this restatement of the momentum balance law takes on the form

$$\operatorname{div} \mathsf{P}_{\mathrm{CL}} + \rho \boldsymbol{b} = \frac{\partial(\rho \boldsymbol{v})}{\partial t}, \tag{5.63}$$

where $\boldsymbol{b}$ is the (external) body force field per unit mass and $\partial(\rho \boldsymbol{v})/\partial t$ is often referred to as the "apparent" rate of change of linear momentum.

5.6.8 Remarks on the Proposed MD Scheme

As mentioned earlier, our continuum homogenization-based extension of the notion of effective stress to MD relied on a Lagrangian scheme which can be used in practice. This result is remarkable for at least two reasons (1) the derivation is based solely on classical mechanics ideas and (2) it is completely rigorous (again, within the confines of classical mechanics). With this in mind, it should be mentioned that there are various Lagrangian schemes for MD simulations of solid systems, most of which are variants of the scheme first proposed in [22–24]. Many of these Lagrangian schemes account for various effects that are not included in the Lagrangian scheme we have derived. However, these schemes are often based on ad hoc Lagrangian functions that cannot be reduced to the canonical form $T - U$ dictated by classical mechanics. Therefore, the Lagrangian scheme proposed herein offers the opportunity for rigorous comparisons with existing methods as has been done in the case of the original Parrinello–Rahman method in [1].

Another observation that can be made concerning the Lagrangian scheme derived herein is that, in view of (5.46), it is easily implementable as a "strain-control" MD method, i.e., a method in which $\hat{\mathsf{F}}$ dictates the simulation cell deformation.

With this in mind, it is possible to conceive of a stress-control variant of the method in which the governing equations are those in (5.46) *along with* (5.48), where the function $[\![\mathsf{T}]\!]_{\mathrm{MD}}$ is a given of the problem and where the effective deformation $\hat{\mathsf{F}}$ is found as part of the solution. The difficulty of this approach lies in the fact that one would have to solve a complex algebraic/differential system of nonlinear equations. To the authors' knowledge, the practical implementation of such an MD scheme has yet to be successfully done.

5.7 Conclusions

In this chapter, we have presented a strategy to use continuum homogenization ideas to determine the stress/deformation response of particle systems using MD. From a conceptual viewpoint, the main points of the chapter are that most of the MD techniques used for the determination of mechanical properties of solids can be understood by classical mechanics methods and, in fact, by methods often used by engineers for the study of composite materials. Specifically, we have proposed a rigorous extension of the concept of effective stress from continuum homogenization to the context of discrete particle systems. This extension has been carried out by formulating a practically implementable MD Lagrangian scheme, which, by being grounded in classical mechanics, offers a way of better understanding the sort of approximations that are done in the implementation of MD schemes based on ad hoc Lagrangian functions.

References

1. P. C. Andia, F. Costanzo, and G. L. Gray. A lagrangian-based continuum homogenization approach applicable to molecular dynamics simulations. *International Journal of Solids and Structures*, 42(24–25):6409–6432, 2005

2. P. C. Andia, F. Costanzo, and G. L. Gray. A classical mechanics approach to the determination of the stress–strain response of particle systems. *Modelling and Simulation in Materials Science and Engineering*, 14:741–757, 2006

3. N. Bakhvalov and G. Panasenko. *Homogenisation: Averaging Processes in Periodic Media*. Mathematical Problems in the Mechanics of Composite Materials. Kluwer, Dordrecht, 1989

4. A. Bensoussan, J.-L. Lions, and G. Papanicolaou. *Asymptotic Analysis for Periodic Structures*. North-Holland, Amsterdam, 1978

5. M. Berezhnyy and L. Berlyand. Continuum limit for three-dimensional mass-spring networks and discrete Korn's inequality. *Journal of the Mechanics and Physics of Solids*, 54(3):635–669, 2006
6. M. Born and K. Huang. *Dynamical Theory of Crystal Lattices*. Oxford Classic Texts in the Physical Sciences. Clarendon/Oxford University Press, Oxford/New York, 1988
7. R. J. E. Clausius. On a mechanical theorem applicable to heat. *Philosophical Magazine, Series 4*, 40:122–127, 1870
8. F. Costanzo, G. L. Gray, and P. C. Andia. On the notion of average mechanical properties in MD simulation via homogenization. *Modelling and Simulation in Materials Science and Engineering*, 12:S333–S345, 2004
9. F. Costanzo, G. L. Gray, and P. C. Andia. On the definitions of effective stress and deformation gradient for use in MD: Hill's macro-homogeneity and the virial theorem. *International Journal of Engineering Science*, 43(7):533–555, 2005
10. G. Gallavotti. *Statistical Mechanics: A Short Treatise*. Texts and Monographs in Physics. Springer, Berlin Heidelberg New York, 1999
11. M. E. Gurtin. *An Introduction to Continuum Mechanics*, Vol. 158 of *Mathematics in Science and Engineering*. Academic, San Diego, 1981
12. K. S. Havner. A discrete model for the prediction of subsequent yield surfaces in polycrystalline plasticity. *International Journal of Solids and Structures*, 7:719–730, 1971
13. R. Hill. On constitutive macro-variables for heterogeneous solids at finite strain. *Proceedings of the Royal Society of London A: Mathematical, Physical, and Engineering Sciences*, 326:131–147, 1972
14. R. Hill. On macroscopic effects of heterogeneity in elastoplastic media at finite strain. *Mathematical Proceedings of the Cambridge Philosophical Society*, 95:481–494, 1984
15. V. V. Jikov, S. M. Kozlov, and O. A. Oleinik. *Homogenization of Differential Operators and Integral Functionals*. Springer, Berlin Heidelberg New York, 1994
16. A. Krawietz. *Materialtheorie: Mathematische Beschreibung des Phänomenologischen Thermomechanischen Verhaltens*. Springer, Berlin Heidelberg New York, 1986
17. I. A. Kunin. *Elastic Media with Microstructure*. Springer Series in Solid-State Sciences, Vol. 26. Springer, Berlin Heidelberg New York, 1982

18. G. A. Maugin. *The Thermomechanics of Plasticity and Fracture*. Cambridge Texts in Applied Mathematics. Cambridge University Press, Cambridge, 1992
19. G. A. Maugin. *Nonlinear Waves in Elastic Crystals*. Oxford Mathematical Monographs. Oxford University Press, Oxford, 1999
20. T. Mura. *Micromechanics of Defects in Solids*. Mechanics of Elastic and Inelastic Solids, 2nd edition. Kluwer, Dordrecht, 1987
21. R. W. Ogden. *Non-Linear Elastic Deformations*. Dover, Mineola, 1997
22. M. Parrinello and A. Rahman. Crystal structure and pair potentials: a molecular-dynamics study. *Physical Review Letters*, 45(14):1196–1199, 1980
23. M. Parrinello and A. Rahman. Polymorphic transitions in single crystals: a new molecular dynamics method. *Journal of Applied Physics*, 52(12):7182–7190, 1981
24. M. Parrinello and A. Rahman. Strain fluctuations and elastic constants. *Journal of Chemical Physics*, 76:2662–2666, 1982
25. H. Petryk. Macroscopic rate-variables in solids undergoing phase transformation. *Journal of the Mechanics and Physics of Solids*, 46(5):873–894, 1998
26. J. R. Rice. Inelastic constitutive relations for solids: an internal-variable theory and its applications to metal plasticity. *Journal of the Mechanics and Physics of Solids*, 19:433–455, 1971
27. C. Stolz. General relationships between micro and macro scales for the non-linear behaviour of heterogeneous media. In: J. Gittus and J. Zarka, editors, *Modelling Small Deformations of Polycrystals*, pp 89–115. Elsevier Science, New York, 1986
28. P. M. Suquet. Elements of homogenization for inelastic solid mechanics. In: E. Sanchez-Palencia and A. Zaoui, editors, *Homogenization Techniques for Composite Media*, CISM Lecture Notes, pp 193–278. Springer, Berlin Heidelberg New York, 1987
29. C. A. Truesdell and W. Noll. *The Non-Linear Field Theories of Mechanics*, 3rd edition. Springer, Berlin Heidelberg New York, 2004
30. C. Truesdell and R. A. Toupin. The classical field theories. In: S. Flügge, editor, *Principles of Classical Mechanics and Field Theory*, Vol. III/1 of *Encyclopedia of Physics*, pp 226–858. Springer, Berlin Heidelberg New York, 1960

31. J. R. Willis. Variational and related methods for the overall properties of composites. In: *Advances in Applied Mechanics*, Vol. 21, pp 1–78. Academic, New York, 1981
32. M. Zhou. A new look at the atomic level virial stress: on continuum-molecular system equivalence. *Proceedings of the Royal Society of London A: Mathematical, Physical, and Engineering Sciences*, 459:2347–2392, 2003

Chapter 6: Multiscale Modeling and Simulation of Deformation in Nanoscale Metallic Multilayered Composites

F. Akasheh and H.M. Zbib

School of Mechanical and Materials Engineering
Washington State University, Pullman, WA, USA

6.1 Introduction

Nanoscale metallic multilayered (NMM) composites represent an important class of advanced engineering materials which have a great promise for high performance that can be tailored for different applications. Traditionally, NMM composites are made of bimetallic systems produced by vapor or electrodeposition. Careful experiments by several groups have clearly demonstrated that such materials exhibit a combination of several superior mechanical properties: ultrahigh strength reaching 1/3 to 1/2 of the theoretical strength of any of the constituent materials [28], high ductility [25], morphological stability under high temperatures and after large deformation [22], enhanced fatigue resistance of an order of magnitude higher than the values typically reported for the bulk form [35], and improved irradiation damage resistance [17, 27], again, as compared to the bulk. However, the basic understanding of the behavior of those materials is not yet at a level that allows them to be harnessed and designed for engineering applications.

The problem lies in the complexity and multiplicity of factors that govern their behavior. Although the concept of creating a stronger metal from two weaker ones by combining them in laminates has been proposed and understood by Koehler in 1970 [20], the nanometer scale introduces a new domain of complexity. At this length scale, the discrete nature of dislocations and their interactions becomes increasingly significant in dictating the response. Depending on the lattice structure and lattice parameters mismatch of the two materials, the layers can be under very high stress states; and interfaces may contain misfit dislocation structures. The miscibility of the materials and the chemical potential strongly affect the nature of interfaces and, hence, their interaction with dislocations. The fact that interfaces form an unusually high-volume fraction of the material makes them a major factor in governing the behavior. The combined complexity and interactions among all of the above-mentioned factors explains the deficiency in the theoretical understanding of the response of NMM composites.

The strong dependence of the mechanical behavior of NMM composites on unit dislocation processes and interfaces poses a challenge to modeling and simulating their behavior. Classical plasticity does not consider the physical mechanism underlying the deformation of the modeled continuum and fails to predict the dependence of the response of metallic structures on their size. Although classical crystal plasticity provides the correct physical framework for modeling dislocation-dependent plasticity, it fails to predict size effect and related phenomena because it does not accommodate geometrically necessary dislocations associated with gradients in plastic deformation. If any, it would be strain gradient plasticity theories that could provide the suitable framework for modeling NMM composites, although this remains a challenging problem and is far from being resolved at the present state of the field.

Multiscale modeling is one of the most promising modeling paradigms which appeared in the last decade for modeling macroscopic phenomena whose roots lie at a finer scale. The approach is based on the appropriate coupling of two models for each of the scales involved. In the case of NMM composites, such coupling involves the continuum mesoscale and dislocation microscale models, although a further coupling to the atomic scale is possible but practically very complex. Three-dimensional dislocation dynamics (DD) analysis is one of the most recent and powerful tool to model the behavior of metallic materials at the microscale in a more physical manner than existing plasticity models [8, 21, 33, 40, 41]. Since its development in the early 1990s, DD analysis has made significant advancement and proved useful in addressing several problems of interest in materials science and engineering. When coupled with the continuum level finite element (FE) analysis, the result is a multiscale model of elastoviscoplasticity which explicitly incorporates the physics of dislocation motion and interactions among themselves and with external loads, surfaces, and interfaces [37, 38]. Such a model provides a very useful tool perfectly suited to studying the behavior of micro- and nanosized metallic structures. The mechanical behavior of NMM composites is clearly one example of those problems.

Section 6.2 explores the subject of modeling and simulation of NMM composites using multiscale modeling. The basics of dislocation-based metal plasticity and its mathematical modeling through DD analysis are reviewed. Multiscale coupling of continuum mechanics and dislocation dynamics are then presented. Background on the mechanical behavior of NMM composites is presented in Sect. 6.3. Finally, the benefits of multiscale and other modeling tools for NMM composites are demonstrated using different examples.

6.2 Multiscale Modeling of Elastoviscoplasticity

Decades of research, since the existence of dislocations in crystal was first theorized, have established that metal plasticity is governed by the response of crystal defects, mainly dislocations, to external and internal loading. Macroscopically observed deformation of metals is the cumulative result of the motion of a very large number of dislocations. Although the theory of dislocations provides a complete description of the stress, strain, and displacement fields of a dislocation as well as of their motion under the effect of forces acting on them, the extension of this theoretical understanding to provide accurate physics-based prediction of the mechanical behavior of metals is practically impossible.

A typical density of dislocation in a moderately worked metal amounts to $10 \times 10^{12}\ m^{-2}$. A cubic millimeter of such metal contains about 1,000 m of curved dislocation lines. The huge computational demand in calculating the dynamics of such densities of dislocations, further complicated by the fact that dislocations have long-range interactions and can react with each other upon colliding to form intricate configurations with possibly new characteristics, is beyond the existing and near future computational capacities.

On the other hand, alternative continuum level modeling, although computationally feasible, remains phenomenological in nature. Even in the case of strain gradient plasticity and geometrically necessary dislocation-based theories, success of one theory in capturing certain aspects of size effects has been problem dependant; and it remains that no general framework is agreed upon. The status quo is mainly due to the complexity and multiplicity of dislocation interactions leading to size effects. For example, it is well known that a dislocation has a distortion field associated with it, which results in a long-range stress field that decays inversely proportional to the distance from the dislocation core. As the dimensions of the specimen become smaller, the interactions between these stress fields become increasingly significant, making the nonlocal effects increasingly pronounced. Furthermore, when the dimensions of a specimen become comparable to the range of the defect structure stress field, size effect arises due to the interaction of this field with the free surfaces (image stresses).

The Hall–Petch effect, which implies that strength is inversely proportional to the square root of a characteristic microstructural length scale, e.g., the grain size in microsized grains or the individual layer thickness in microscale multilayered structures, can be directly attributed to dislocation pileups at grain boundaries or layer interfaces, respectively. The stress needed to activate dislocation sources also depends on the grain size and their location within the grain, which reflects as a size effect in the early stages of deformation. Another size effect originates from low-energy dislocation structures, like cell structure or dislocation walls, which tend to form by dislocation patterning and reorganization. Capturing all this complexity is a formidable task for any phenomenology-based theory.

Plasticity in metals is an example of a problem that is multiscale in nature: The macroscopically observed behavior has its origin in the complex physics occurring at the microscale. A multiscale model for plasticity would implement a continuum level framework which avoids phenomenology by explicitly incorporating the physics of plasticity at the microscale through the DD analysis. The link between the two models is two-way: the DD model calculates and passes the plastic strain and the internal stress field due to dislocations at each material point (after proper homogenization), while the continuum model accounts for boundary conditions and internal surfaces and interfaces through the solution of an auxiliary boundary value problem and the superposition concept as detailed below.

In Sect. 6.2.1, we provide a brief background on dislocations in metals. The theoretical aspects of DD and their implementation in DD simulations are presented in Sect. 6.2.2. Then the multiscale dislocation dynamics plasticity model is presented in Sect. 6.2.3.

6.2.1 Basics of Dislocations in Metals

Dislocations are linear defects in crystals identified by their Burgers vector and line sense. Depending on the crystal structure, a dislocation can have one out of a finite set of Burgers vectors and can glide on one of a finite set of crystallographic planes. For example, in face-centered cubic (FCC) metals, there are six possible Burgers vectors, all of $a/2\langle 011\rangle$-type, a being the lattice parameter, and four $\{111\}$ slip planes. A combination of a

Burgers vector and a slip plane defines the slip system of a dislocation. The Burgers vector defines the direction of slip of the material, while the slip plane defines the plane on which the slip motion occurs. On its plane, the dislocation can have an arbitrary line sense, which can change as the dislocation glides. Although the Burgers vector is a characteristic of a dislocation, its slip plane is not because a dislocation can change its glide plane, a process known as *cross-slip*. Dislocations glide under the effect of shear stress resolved in the slip plane along the slip direction (direction of Burgers vector). Notice the difference between slip direction, which pertains to the direction of motion of the atoms, and the dislocation line motion. The macroscopically observed plastic deformation of a metallic continuum structure is the result of the irreversible glide motion of a large number of dislocations on multiple slip systems each with its own spatial orientation. The macroscopic plastic strain tensor ε^{p} is thus expressed by the following relation, which reflects the tensorial addition of several multiple contributions to slip each in a certain direction $\hat{s}^{(\beta)}$ on a particular $\hat{n}^{(\beta)}$

$$\dot{\varepsilon}^{\mathrm{p}} = \sum_{\beta} \dot{\gamma}^{(\beta)} (\hat{s}^{(\beta)} \otimes \hat{n}^{(\beta)})_{\mathrm{sym}}, \tag{6.1}$$

where $\dot{\varepsilon}^{\mathrm{p}}$ is the plastic strain increment, β is the slip system index, $\dot{\gamma}^{(\beta)}$ is the increment of slip on slip system β, $\hat{s}^{(\beta)}$ is the unit slip direction, and $\hat{n}^{(\beta)}$ is the slip plane normal.

Gliding dislocations can also collide with each other resulting in special types of interactions (short-range interactions) which are very complicated in nature and depend strongly on the interacting dislocations' slip systems, line senses, and approach trajectory. The main interactions include annihilation, jog formation, junction formation, and dipole formation. Furthermore, dislocations can also be trapped, ceasing to move either due to short-range interactions that leave them locked or due to long-range effects like pileups against obstacles or simply due to the occurrence of regions in the material where the stress field is not high enough to drive dislocations.

6.2.2 DD Simulations

The idea behind conducting DD simulations is to explicitly model the behavior of a dislocation population under applied load taking into consideration all the topological and kinematical characteristics of dislocations and their long- and short-range interactions as described above. Short-range interactions due to dislocation collision are accounted for through a set of physics-based rules learned from either atomic scale simulations or careful experimental observations. In short, DD analysis is the numerical implementation of the theory of dislocations to analyze the dynamics of a dislocation system in materials.

Generally, the simulation box in DD represents a representative volume element (RVE) of a larger specimen, although in some cases freestanding microsized components can make the simulation box. Unless a certain initial dislocation structure is desired, the simulation starts with a randomly generated dislocation structure. Dislocations are modeled as general curved lines in three-dimensional space made of an otherwise elastic medium characterized by its shear modulus, Poisson's ratio, and mass density. Dislocation lines are discritized into small segments, each associated with a dislocation node [39]. The nodes are the points at which forces on a dislocation from all dislocations in the system and from external loads are calculated. The governing equation for dislocation motion is then used to estimate the velocity, and hence the displacement, of each node in response to the net applied force. The node positions are updated accordingly, generating the new dislocation configuration and the process is repeated.

In this scheme, the analysis of the dynamics of continuous line objects reduces to those of a finite number of nodes. Typical to numerical algorithms, the mesh size (here the length of a segment) can be refined to obtain the desired accuracy in representing the topology of curved dislocation lines and their dynamics. The above sequence of calculations is repeated as time marches in appropriately chosen time steps, to the desired point of evolution of the dislocation system or the overall stress or strain levels. The details of the approach outlined above will be explored in the following section.

Dislocation equation of motion

The theory of dislocations provides the following governing equation for the motion of a straight dislocation segment *s* [11, 14, 18]:

$$m_s \dot{\upsilon}_s + \frac{1}{M_s}\upsilon = F_s. \tag{6.2}$$

Typical of a Newtonian-type equation of motion, it expresses the relation between the velocity of "an object" and the dislocation segment of effective mass m_s, moving in a viscous medium with a drag coefficient of

$1/M_s$ under the effect of a net force F_s. The effective mass per unit dislocation length $\tilde{m}$ has been given for the edge and screw components of a dislocation as follows [14]:

$$\tilde{m}_{\text{edge}} = \frac{W_o C^2}{\upsilon^4}\left(-16\gamma_l - 40\gamma_l^{-1} + 8\gamma_l^{-3} + 14\gamma + 50\gamma^{-1} - 22\gamma^{-3} + 6\gamma^{-5}\right) \quad (6.3a)$$

and

$$\tilde{m}_{\text{screw}} = \frac{W_o}{\upsilon^2}(-\gamma^{-1} + \gamma^{-3}) \quad (6.3b)$$

with $\gamma = \sqrt{1-(\upsilon/C)^2}$ and $\gamma_l = \sqrt{1-(\upsilon/C_l)^2}$. C and C_l are the transverse and longitudinal sound speeds in the elastic medium, υ is the dislocation speed, and W_o is the line energy of a dislocation per unit length given as $W_o = (Gb^2/4\pi)\ln(R/r_o)$ [13]. In the later expression, G is the shear modulus, b is the magnitude of the Burgers vector, and R and r_o are the external and internal cutoff radii, respectively. M_s is the dislocation mobility and it is typically a function of temperature and pressure. The net force F_s acting on a dislocation line can have several contributions to it depending on the problem. In general,

$$\begin{aligned} F_s = {} & F_{\text{Peierls}} + F_{\text{dislocation}} + F_{\text{self}} + F_{\text{external}} + F_{\text{obstacle}} \\ & + F_{\text{image}} + F_{\text{osmotic}} + F_{\text{thermal}}, \end{aligned} \quad (6.4)$$

where F_{Peierls} is the force from lattice friction opposing the motion of a dislocation, F_{self} is the force from the two neighboring dislocation segments directly connected to the segment under consideration, $F_{\text{dislocation}}$ is the net force from all other dislocation segments in the simulation domain, F_{external} is the force due to externally applied loads, F_{obstacle} is the interaction force between a dislocation and the stress field of an obstacle, F_{image} is the force experienced by a dislocation due to its presence near free surfaces or interfaces separating phases of different elastic properties, F_{osmotic} is the driving force in climb, and F_{thermal} is the force on the dislocation from thermal noise. In general, the force due to a general stress field σ is given by

$$F_s = l_s \sigma \cdot b_s \times \xi_s, \quad (6.5)$$

where l_s is the segment length and σ is the stress field "felt" by the dislocation segment, while b_s and ξ_s are the Burgers vector and the line sense, respectively, of the dislocation segment. For example, in the case of externally applied loads, the relevant stress field is σ^{a}, the net stress from all external loads along segment s and its force contribution will be $F_{\text{external}} = l_s \sigma^{a} \cdot b_s \times \xi$. The details of the calculation of $F_{\text{dislocation}}$ and F_{self} are not trivial and will be further detailed below.

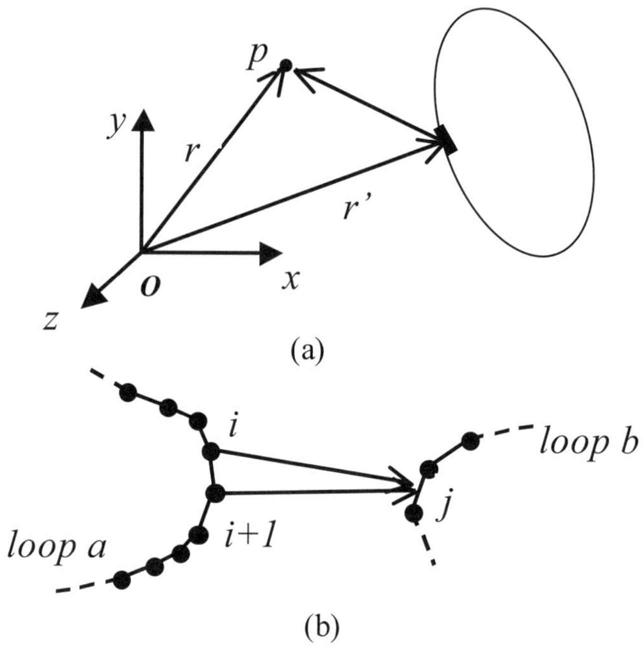

Fig. 6.1. (**a**) Integration of the stress field at a point p due to a dislocation loop and (**b**) the corresponding integration in the framework of DD by the linear element approximation

Evaluation of $F_{dislocation}$

As mentioned above, this force contribution comes from all of the dislocation segments in the system except for those two connected to the dislocation node under consideration. Dislocation theory provides the stress field of an arbitrary dislocation loop C at an arbitrary point p defined by the position vector $\boldsymbol{r}$ through the following expression [13] (see Fig. 6.1a).

$$\sigma_{\alpha\beta} = -\frac{G}{8\pi}\oint_C (b\times\nabla')\frac{1}{\mathrm{R}}\otimes \mathrm{d}l' + \frac{G}{4\pi}\oint_C \mathrm{d}l'\otimes(b\times\nabla')\frac{1}{\mathrm{R}} - \frac{G}{4\pi(1-\nu)}\oint_C \nabla'\cdot(b\times \mathrm{d}l')(\nabla\otimes\nabla - I\nabla^2)\mathrm{R}, \tag{6.6}$$

where $\boldsymbol{R}$ is position vector of p relative to the dislocation segment position $\boldsymbol{r}'$ and $I = e_1\otimes e_1 + e_2\otimes e_2 + e_3\otimes e_3$ is the unit dyadic. In the numerical implementation, dislocation curves are discretized into linear segments; and the above integrals over closed loops become sums over linear segments of length l_s; and the contribution from all segments is summed up to find the stress field at any desired point p

$$\sigma_{\alpha\beta} = \sum_{s=1}^{N_s-2}\left\{\begin{array}{l} -\dfrac{G}{8\pi}\displaystyle\int_s (b\times\nabla')\frac{1}{\mathrm{R}}\otimes \mathrm{d}l' + \frac{G}{4\pi}\int_s \mathrm{d}l'\otimes(b\times\nabla')\frac{1}{\mathrm{R}} \\ -\dfrac{G}{4\pi(1-\nu)}\displaystyle\int_s \nabla'\cdot(b\times \mathrm{d}l')(\nabla\otimes\nabla - I\nabla^2)\mathrm{R} \end{array}\right\}. \tag{6.7}$$

Furthermore, the integration over the segment length can be evaluated algebraically using the linear element approximation found in [3, 13]. According to this approach, the stress field at point p from a dislocation segment bound by nodes i and $i+1$ can be evaluated as [39] (see Fig. 6.1b)

$$\sigma_{\alpha\beta}(p) = \sigma_{\alpha\beta}^{i+1} - \sigma_{\alpha\beta}^{i}. \tag{6.8}$$

Evaluation of F_{self}

When applied to calculate the stress field at dislocation node j which belongs to the same dislocation segment whose stress contribution is being considered, the above procedure does not work due to the singular nature of the stress field at the dislocation core. To overcome this obstacle, a regularization scheme developed in [41] is implemented. Consider the dislocation bend consisting of a semi-infinite line and segment $(j, j+1)$, as shown in Fig. 6.2a. The glide force per unit length acting on a point on segment $(j, j+1)$ at a distance λ is explicitly given for the case where the adjacent segment is semi-infinite in length as [13]

$$\frac{F_g}{L} = \frac{G}{4\pi\lambda} f_g(\theta, b). \tag{6.9}$$

This expression can be used to find the average force per unit length on segment $(j, j+1)$ by integrating it over the length of the segment yielding

$$\left(\frac{F_{\mathrm{g}}}{L}\right)_{\mathrm{avg}} = \frac{G}{4\pi L} f_{\mathrm{g}}(\theta, b)\left(\ln\left(\frac{L}{b}\right) + \beta\right), \tag{6.10}$$

where β is an adjustable parameter that compensates for the energy contained in the dislocation core. Equation (6.10) is an equivalent expression to an alternative expression where an adjustable core cutoff radius r_o is used. To adapt the above solution to the case of a finite segment $(j-1, j)$, the superposition principle is used and the net glide component of the force on segment $(j, j+1)$ due to segment $(j-1, j)$ can be found by subtracting, from (6.10), the interaction force between additional semi-infinite segment and $(j, j+1)$ calculated using the standard procedure (Fig. 6.2b).

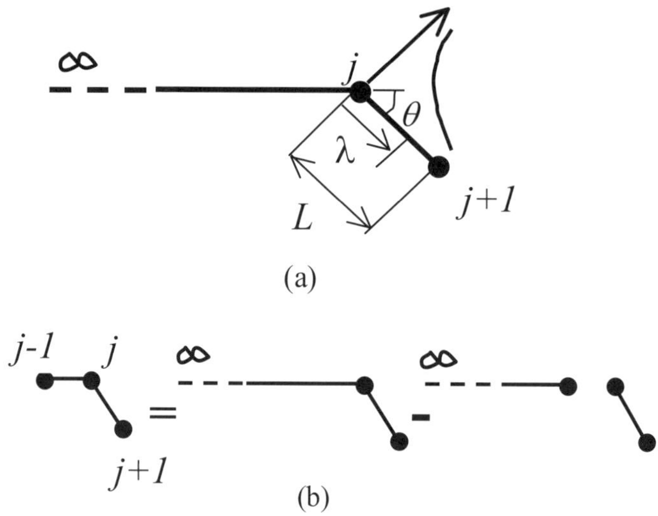

Fig. 6.2. Calculation of the Peach–Koehler force on a dislocation segment due to its direct neighboring segment

Treatment of boundary conditions

Typically, the simulation box used in DD analyses is an RVE representative of an infinite medium. To account for this model, special boundary conditions are needed. Two types of boundary conditions are applied in DD (1) reflection boundary conditions, which ensure the continuity of dislocation curves [41] and (2) periodic boundary conditions, which ensure both the conservation of the dislocation flux and the continuity of the dislocation curves [2].

As for the cases where the simulation box represents the complete specimen with finite domain and arbitrary loading conditions, the above boundary conditions are no longer valid; and a special treatment for the finite domain is needed. This treatment is implemented within the framework of the multiscale model and its discussion is presented in Sect. 6.2.3.

Evaluation of the macroscopic plastic strain

In metals, macroscopic deformation is the result of slip on different slip systems. The area swept by a gliding dislocation represents the area of the newly slipped region due to this motion. In the framework of DD, the increment of the plastic strain can be explicitly calculated from the area swept by the dislocation segments using this relation [39]

$$\dot{\varepsilon}^{\mathrm{p}} = \sum_{s=1}^{N_{\mathrm{s}}} \frac{l_{\mathrm{s}} \upsilon_{\mathrm{s}}}{2V} (n_{\mathrm{s}} \otimes b_{\mathrm{s}} + b_{\mathrm{s}} \otimes n_{\mathrm{s}}), \tag{6.11}$$

where N_{s} is the total number of dislocation segments, l_{s} is the segment length, υ_{s} is the segment glide velocity, b_{s} is the segment Burgers vector, n_{s} is the normal to the slip plane of the segment, and V is the volume of the RVE.

6.2.3 Multiscale DD Model

The coupling of continuum mechanics and DD calculations provides the physical link between the meso- and the microscales. At the continuum level, the typical laws governing an elastic continuum are implemented along with Hooke's law for the elastic regime, as usual. No constitutive law for the plastic behavior of the material is prescribed. Instead, the continuum level plastic strain is explicitly calculated from the actual motion of the underlying dislocations and homogenized at each material point.

Another quantity that is explicitly calculated in DD and passed to the continuum scale is the internal stress from dislocations (and any other defects exhibiting long-range, self-induced stress fields). In this manner, the continuum level back-stress concept and its direct effect on hardening are naturally incorporated. Furthermore, this framework allows the rigorous treatment of boundary conditions for free surfaces and interfaces separating heterogeneous media through the concept of image stresses and eigenstresses, respectively, as will be demonstrated below. This framework also facilitates the application of general loading conditions in DD simulations.

Treatment of finite domains

The stress fields employed in the DD calculations are those for a dislocation in infinite homogeneous media. In the case of finite domains, the stress fields are truncated at the boundaries and, thus, the dislocation can experience a force depending on its position relative to the free surfaces. The stress field calculations in this situation can be handled through the concept of superposition [5, 34, 39]. The elastic fields for the finite domain problem can be found by summing the elastic fields from two solutions: that for the dislocations as if they existed in an infinite medium, and the solution to a complementary problem where the domain is finite and tractions equal but opposite to those caused by the infinite stress fields at the finite domain boundary (Fig. 6.3)

$$\begin{aligned} u^{\mathrm{D}} &= u^{\mathrm{D}\infty} + u^{*}, \\ \varepsilon^{\mathrm{D}} &= \varepsilon^{\mathrm{D}\infty} + \varepsilon^{*}, \\ \sigma^{\mathrm{D}} &= \sigma^{\mathrm{D}\infty} + \sigma^{*}, \end{aligned} \tag{6.12}$$

where the superscript D∞ indicates a defect field quantity as if the defect existed in an infinite homogeneous medium, while the superscript *

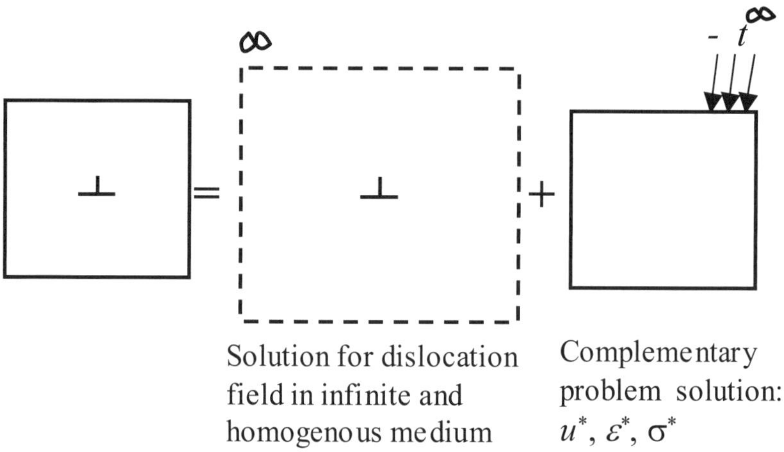

Fig. 6.3. Superposition principle application for the rigorous treatment of finite boundaries

indicates the solution to the complementary problem described above and satisfying the following boundary conditions:

$$\begin{aligned} t &= t^{a} - t^{\infty 1} - t^{\infty 2}, \quad \text{on } \partial\Omega, \\ u &= u^{a}, \qquad\qquad\quad\ \text{on } \partial\Omega_{u}. \end{aligned} \tag{6.13}$$

Here, t^a and u^a represent any externally applied tractions and any prescribed boundary displacements, respectively, on their corresponding parts of the boundary.

Treatment of heterogeneous media

Consider the infinite domain Ω consisting of two subdomains Ω_1 and Ω_2. Both Ω_1 and Ω_2 contain dislocations and other defects. The stress field in each medium $\sigma_{\Omega 1}$ and $\sigma_{\Omega 2}$ can be expressed as follows [39]

$$\begin{aligned} \sigma_{\Omega 1} &= \sigma^{D\infty 1} + \sigma^{D\infty 2} + \sigma^{\infty 12}, \\ \sigma_{\Omega 2} &= \sigma^{D\infty 2} + \sigma^{D\infty 1} + \sigma^{\infty 21}, \end{aligned} \tag{6.14}$$

where $\sigma^{D\infty 1}$ and $\sigma^{D\infty 2}$ are the stress fields due to the defect structure as if the whole domain were homogenous and made of the material of Ω_1 and Ω_2, respectively. $\sigma^{\infty 12} = [C_1 - C_2]\varepsilon^{D\infty 2}$ and $\sigma^{\infty 21} = [C_2 - C_1]\varepsilon^{D\infty 1}$ represent the image stress due to the difference in the elastic properties.

Treatment of the general case

Consider a finite domain Ω consisting of two subdomains Ω_1 and Ω_2. Each medium can have its own dislocation (and possibly other defects) structure which exhibits long-range effects. The total elastic fields can be expressed as the sum of four solutions (1) that for dislocations in Ω_1 as if they existed in an infinite medium made of the Ω_1 material, (2) that for the dislocations in Ω_2 as if they existed in an infinite medium made of the Ω_2 material, (3) the image fields due to the difference in elastic properties of the two media, and (4) the solution to a complementary problem where the domain is finite and traction equal but opposite to that caused by the infinite stress fields at the finite domain boundary, as described before. Furthermore, any externally applied loads can be included in the complementary problem [39]

$$
\begin{aligned}
\sigma_{\Omega 1} &= \sigma^{\mathrm{D}\infty 1} + \sigma^{\mathrm{D}\infty 2} + \sigma^{\infty 12} + \sigma^{*1}, \\
\sigma_{\Omega 2} &= \sigma^{\mathrm{D}\infty 2} + \sigma^{\mathrm{D}\infty 1} + \sigma^{\infty 21} + \sigma^{*2},
\end{aligned} \tag{6.15}
$$

where $\sigma^{*1} = [C_1]\varepsilon^{*}$ and $\sigma^{*2} = [C_2]\varepsilon^{*}$, and ε^* is the solution for the complementary problem where the finite domain is subjected to the following boundary conditions

$$
\begin{aligned}
t &= t^{\mathrm{a}} - t^{\infty 1} - t^{\infty 2}, && \text{applied to } \partial\Omega, \\
u &= u^{\mathrm{a}}, && \text{applied to } \partial\Omega_{\mathrm{u}},
\end{aligned}
$$

where t and u denote traction and displacement, the superscript a denotes externally applied quantities, and $\partial\Omega_{\mathrm{u}}$ is the part of the boundary to which external displacement boundary conditions are applied.

Finite element implementation

From the continuum point of view, the simulated material consists of an elastic medium with internal defect structure. The macroscopic behavior is governed by the basic laws of continuum mechanics. If a small strain deformation is assumed, then the total strain ε can be decomposed into an elastic part ε^{e} and a plastic part ε^{p} as follows:

$$
\begin{aligned}
\varepsilon &= \tfrac{1}{2}(\nabla u + u\nabla), \\
\varepsilon &= \varepsilon^{\mathrm{e}} + \varepsilon^{\mathrm{p}}.
\end{aligned} \tag{6.16}
$$

Furthermore, if Hooke's law is used as the constitutive law for the elastic regime, then one can write $\sigma = C[\varepsilon - \varepsilon^{\mathrm{p}}]$. Depending on the problem, the stress σ can have any of the contribution to stress mentioned above. In the most general case,

$$
\sigma_{\Omega i} = \sum_{i=1}^{N} \sigma^{\mathrm{D}\infty i} + \sum_{j=1}^{N} [C_i - C_j]\varepsilon^{\infty ij} + [C_i]\varepsilon^{*} \tag{6.17}
$$

with N being the number of subdomains making the structure and ε^* the solution for the complementary problem with boundary conditions:

$$
\begin{aligned}
t &= t^{\mathrm{a}} - \sum_{i=1}^{N} t^{\infty i}, && \text{on } \partial\Omega, \\
u &= u^{\mathrm{a}}, && \text{on } \partial\Omega_{\mathrm{u}}.
\end{aligned} \tag{6.18}
$$

The finite element formulation for the above problem results in the following [37, 39]

$$[M]\{\ddot{u}\}+[C]\{\dot{u}\}+[K]\{u\}=\{F^{a}\}+\{F^{\infty}\}+\{F^{B}\}+\{F^{p}\}, \qquad (6.19)$$

where $[M]$, $[C]$, and $[K]$ are the global mass, damping, and stiffness matrices, respectively. $\{F^{a}\}$ and $\{F^{\infty}\}$ represent, respectively, the externally applied force vector and the force vector from tractions on the free surfaces due to the truncations of the long-range dislocation fields. The body force vector $\{F^{B}\}$ results from the long-range stress fields in the dislocations (and other defects if present), i.e., the contribution of the $\sigma^{D\infty}$ terms. If present, $\{F^{B}\}$ also includes the contribution of the image forces due to the difference in the elastic moduli of the different subdomains, i.e., the second term on the right-hand side of (6.17). It is through $\{F^{B}\}$ that the long-range effect of dislocations belonging to any of the finite elements in the mesh on the dislocation in a certain element is considered. As for the interactions between the dislocations belonging to the same element, the interactions are calculated explicitly as described in Sect. 6.2.2. Finally, $\{F^{p}\}$ reflects the contribution of the stress term resulting from ε^{p}.

6.3 Nanoscale Metallic Multilayered Composites

NMM composites are typically produced by physical vapor deposition. The microstructure and mechanical properties of the composite are not only dependent on the absolute properties of the individual constituent metals, but also on their relative values (for example their lattice parameters and elastic properties mismatch) and on the individual layer thickness. Furthermore, the deposition process parameters sensitively affect the properties of the resulting structure.

A fundamental classification of the type of NMM systems, hence their behavior, is based on the compatibility of slip systems in the two phases and is commonly used. Coherent systems exhibit “almost” continuous slip systems across the interfaces such that dislocations gliding on a certain slip plane in one layer can continue to glide on the same plane in the neighboring layer. This condition requires the two metals to exhibit the same lattice structure, a small lattice parameter mismatch, and to be deposited epitaxially. Other terminology used to describe such systems is "transparent interface" systems [16]; transparent in the sense that dislocations can cross the interface into the neighboring material.

On the other hand, in incoherent systems, also known as "opaque interface" systems, slip systems in the two phases do not match, which means that dislocations cannot continue to glide in the neighboring layers. Instead, interfaces can act as slip barriers through different mechanisms. For example, in the Cu/Nb FCC/BCC system, interfaces act as sinks for interlayer dislocations as they enter the interface and their core spreads [15]. Incoherent systems involve metals with different crystal structure or ones with similar crystal structure but with large lattice parameters mismatch or large lattice misorientation.

Interface crossing is a critical plasticity propagation process in NMM composites because it marks the end of confined layer slip and the spread of dislocation activity in both phases of a coherent system, defining the limit of strength. In this work, the discussion will be limited to the strength of NMM composites made of coherent systems, in particular the FCC–FCC Cu/Ni system with $\{100\}$ interface. This orientation is also known as cube-on-cube. The lattice parameters of Cu and Ni are 0.3615 and 0.3524 amounting to a small mismatch of 2.6%, which allows for continuous slip between the Cu and Ni layers. Furthermore, we will restrict the discussion to the effect of the layer thickness on the mechanical and structural properties of NMM composites.

An investigation of the strength of NMM composites should start by studying the stress relaxation process of the as-deposited composite. The relaxation process dictates the nature of the initial dislocation structure, the interface properties, and the internal stress state in the composite. In turn, this as-deposited structure influences the ultimate mechanical response under loading. Due to its critical role, a brief discussion of stress relaxation in coherent multilayered systems should be in place at this point. The epitaxial coherent growth of two different materials forces the atomic positions of the materials to coincidence in spite of the slight difference in lattice parameters and atomic positions. For coherency to be maintained, the lattice with the larger size has to compress while that with the smaller size has to expand for them to match. The result is a strained multilayer structure having coherency strains and stresses of alternating sign (tension/compression). Coherency stresses can be extremely high, measuring about 2.6 GPa in the case of Cu/Ni system [16]. Consequently, as the deposition process proceeds, the structure's elastic energy increases linearly with the increase in thickness. At a certain point, it becomes energetically favorable for the system to reduce its elastic energy by injecting dislocations resulting in a semicoherent structure whose overall energy is less.

A critical layer thickness h_c is identified below which the structure is coherent and made of strained layers and above which the structure is semicoherent and has a network of interfacial dislocations and a reduced level of coherency stresses [6, 23, 24]. The relaxation process proceeds by the glide of what is known as threading dislocations. Threading dislocations are glide dislocations that originate from faults in atomic arrangement occurring during the deposition process. Once generated in one layer, they continue to replicate themselves in the layers deposited thereafter maintaining the same slip system. When the layer thickness is above the critical thickness, the resolved shear stress component of the coherency stresses drives the threading dislocations along their slip planes while the alternating stress state acts to confine them to their respective layers. This glide process is known as the *Orowan bowing* and results in the common hairpin dislocation structure (Fig. 6.4).

Another suggested source of dislocations is the nucleation of half-loops from free surfaces followed by their propagation within the layer [19]. Furthermore, in the process of threading, dislocations can interact with other threading dislocations or interfacial dislocations deposited by earlier threading events. Cross-slip is commonly observed in NMM composites and provides another dislocation multiplication mechanism. Threading dislocations can also react with each other to form Lomer locks which are commonly observed in an increasing number as the layer thickness increases [25, 29, 30].

The end result of all these interactions among dislocations and between dislocations and the evolving stress field is a structure with a complex network of interfacial dislocations (including glide dislocations, sessile Lomer-type dislocations, and dislocation bends), blocked threading dislocation, dislocation junctions, and jogs along with a remanent nonuniform distribution of internal stresses. The presence of "residual" internal stress in the structure is a consequence of the fact that full relaxation removing all coherency stresses is practically impossible due to several factors that impede the relaxation process. Such factors include dislocation–dislocation interactions, lattice friction, and the existence of other defects, all of which act to block the motion of threading dislocations. Even if the misfit dislocation network necessary to minimize the system energy is imagined to exist, there will still be a nonzero internal stress due to the stress fields of these dislocations.

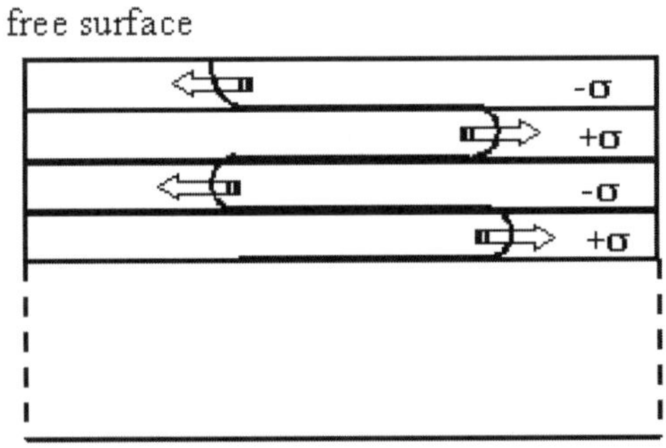

Fig. 6.4. The glide of a threading dislocation in different layers

The equilibrium semirelaxed structure described above becomes the initial structure for the consequent loading of the composite in service. If the applied load is high enough, it will provide the necessary driving force for dislocation activity to resume, with the softer layer flowing first. Meanwhile, the harder layer would be under elastic loading; and the laminate structure, as a whole, is still capable of supporting increased loading levels due to the hard phase still being in the elastic regime. For the overall structure to yield, both layers have to deform plastically. One scenario leading to this is the crossing of the dislocations from the soft layer to the harder one.

As outlined in the previous discussion, the basic model to describe the primary plasticity mechanism in NMM composites is that of Orowan bowing of a threading dislocation. Based on this, model expressions for the critical thickness and strength of layered composites were developed by [6, 23] and further elaborated and modified by several researchers. From a mechanistic point of view, the idea is based on the balance between the force exerted by the misfit stress on the threading dislocation and the tension force in the created dislocation lines. By equating those forces, one finds the channeling stress dependence on layer thickness to be proportional to $\ln(h)/h$, h being the layer thickness. The same results can be arrived at through an energetic approach.

This simple model, however, underestimates the measured strength of NMM composites as well as h_c. This is not surprising because the Orowan bowing model does not account for the effect that the presence of other dislocations in the system has on the critical condition for the stability of a threader. This effect can be due to long-range interactions as well as short-range interactions. Other proposed reasons for this discrepancy include barriers to dislocation nucleation, kinetic effects, Peirels friction, step formation at surfaces and interfaces, and Koehler forces resulting from elastic properties mismatch. This work will focus on the effect of the dislocation–dislocation interaction (both long and short range) on the strength of NMM composites.

Considering a network of interfacial dislocations rather than a single dislocation, several authors developed energy expressions for stable dislocation arrays in multilayered systems as a function of layer thickness [4, 9, 10, 12, 36]. Although these models are more realistic and resulted in improved h_c and strength predictions, they are based on simple configurations where the dislocations in the array are equally spaced and infinite in length. In real situations, however, those assumptions are rarely representative of the dislocation networks distribution.

To understand the strengthening effect of predeposited interfacial dislocations on the stress needed to drive a threading dislocation, hereafter referred to as the *channeling stress* [31], Freund [7] considered the effect of glide interfacial dislocations that intersect the path of a threading dislocation in an unpassivated film. It was concluded that the presence of an interfacial dislocation forces the threading dislocation to "squeeze" through a narrowed-down channel with effective thickness h^* as opposed to the actual layer thickness h. A blocking criterion was suggested based on this analysis, which predicts a significant hardening effect due to this interaction. For thicknesses below 100 nm, this effect was about 50%. Using Freund's approach, Nix [31] evaluated the effect of passivation and elastic properties mismatch between the film and the substrate on the channeling strength. For the case of unpassivated film with "very rigid" substrate, the channeling strength increased by 30% over the reference (no interfacial dislocations present) strength for the same film thickness. For the case of $10b$ (b: magnitude of Burgers vector) thick passivation layer and uniform elastic properties for the film, passivation, and substrate, the strength increase was about 50%.

In Freund's calculations, the interfacial dislocation was assumed to be a straight and pinned line not allowed to undergo short-range interactions. Furthermore, the threading dislocation was considered to be made of one straight segment, while the critical passing point, corresponding to the effective channel thickness h^*, was assumed to occur directly above the interfacial dislocation. As will be seen, these assumptions make this model far displaced from a real-life scenario. In fact, the blockage mechanism is completely absent in NMM composites when the interfacial dislocation is of the glide type and short-range interactions become the main factor that dictate the outcome of this process.

In an attempt to make more realistic predictions of the strength of nanoscale multilayered systems, Misra et al. [26] developed a model based on the Orowan bowing for a single threading dislocation in a layer having equally spaced interfacial dislocations and extended it to include the resistance to dislocation crossing from one layer to the next through the interface. Using an energy approach, the stress needed to propagate the hairpin was estimated for two bounding cases. In the upper bound case, the threading dislocation was assumed to propagate in the presence of an array of equally spaced interfacial Lomer-type dislocations as well as "left over" unrelaxed coherency stress inversely proportional to the layer thickness. In the lower bound case, the dislocation array was ignored. In both cases, the interfacial dislocation arrays were assumed to be the source of resistance to dislocation crossing into the neighboring layer. For yield to occur, the applied stress must be sufficiently large to propagate the threading dislocation and overcome the interface resistance. Lower bound estimates with the proper choice of the core size produced a good match to the experimental data for the strength of Cu–Ni systems.

The above discussion demonstrates the strong dependence of the mechanical response of NMM composites on the underlying dislocation structure and its interactions with interfaces and free surfaces. The discussion also pointed out that, in spite of their better predictions and insightfulness, analytical models developed to predict the behavior of NMM composites remain far from capturing the complexity of real systems. DD provides the right framework to address such problems because it treats dislocations and their interaction explicitly. Furthermore, the multi-scale coupling of DD with continuum mechanics allows the rigorous treatment of surfaces and interfaces at the macroscale.

Nevertheless, care should be exercised when interfaces are modeled in the DD framework. The physics of interfaces and their interactions with dislocation are very complicated and should be addressed at the atomic level. Another limitation to DD analysis in the modeling of NMM composites is the lower bound on layer thicknesses which can be accurately handled. As the layer thickness decreases, the length of the segments must also decrease for the accurate representation of dislocation curves. However, the segment length can become so small so as to permit the overlap of dislocation cores where the linear elasticity solution is not defined.

6.4 Modeling and Simulation of NMM Composites: Examples

The ultimate goal of modeling NMM composites is to provide accurate predictions for their mechanical response under various loading conditions. The achievement of this goal is important for the design of NMM composites for different engineering applications. Due to the complexity of the physics of NMM composites, this goal has been elusive to analytical approaches. The use of DD to study plasticity mechanisms in NMM composites and how they contribute to the macroscopically observed response will be demonstrated in this section.

We start by considering the modeling of four significant unit processes in isolation from other effects which inevitably exist in real systems (1) the basic Orowan bowing process, (2) the interaction between a threading dislocation and orthogonal interfacial dislocations intersecting its glide path, (3) the interaction between a threading dislocation and parallel interfacial dislocations, and (4) threading in a surface layer. Next, the collective behavior of a system of dislocations as in real-life scenarios will be discussed. In these simulations, all the above-mentioned unit processes, and others, naturally interplay and lead to the overall response of the composite.

6.4.1 Modeling of Unit Dislocation Process in NMM Composites

In complex systems where many mechanisms interact to produce the overall behavior, it seems plausible to disseminate the complexity by identifying a number of mechanisms thought to be significant and investigating them one at the time and in isolation of any other mechanisms. The next question becomes that of appropriately combining their behavior in a statistical manner to arrive at the overall behavior. This involves not only the statistics of each process individually, but also possible correlations among them. The benefits of investigating, within the framework of multiscale DD, a number of such significant and frequently occurring dislocation processes in NMM composites are demonstrated.

Orowan bowing

Figure 6.5 shows the crystallography of the problem and the setup used to model the Orowan bowing in a confined layer. It consists of a {001} orientation Cu layer with rigid walls. The physical properties of Cu used in this calculation and all the ones to follow in this work are listed in

Table 6.1. This layer represents a single buried layer in an infinite elastic medium of the same properties as that of the material of the layer, which implies that no image forces exist. Slip is confined to the layer by its rigid walls impenetrable by dislocations. An infinite $a/2\langle 011\rangle$ $\{111\}$-type dislocation resides on its slip plane with that portion contained in the layer representing the threading dislocation. If the stress applied to the layer is sufficiently high, the dislocation will bow out. Channeling stress defines the minimum stress needed to cause the dislocation to bow out indefinitely (i.e., become unstable).

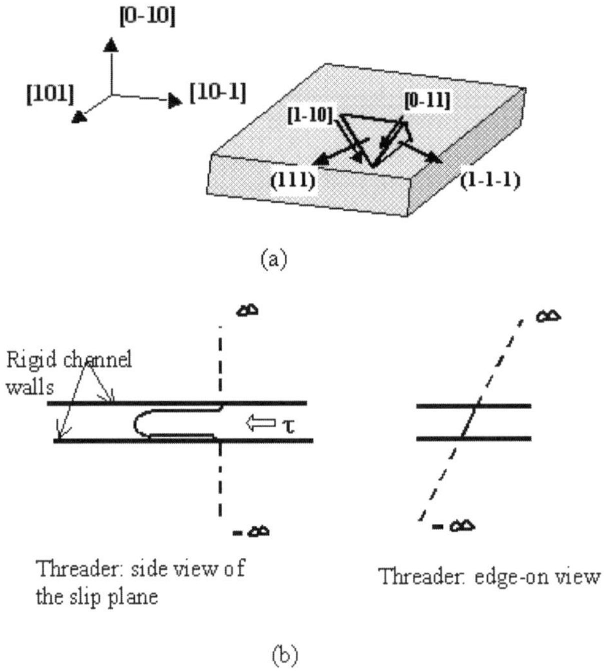

Fig. 6.5. (**a**) Crystallographic orientation of the layer and (**b**) setup for Orowan bowing simulation: infinite $a/2\langle 011\rangle$-type dislocation with a portion contained in a rigid channel subject to stress such that the dislocation threads in its slip plane

Table 6.1. Physical properties of Cu used in DD calculations

Density (Kg m^{-3})	8,980.0
Burgers vector magnitude ( )	2.556
Shear modulus (GPa)	38.46
Poisson's ratio	0.3
Core size (*b*: Burgers' vector magnitude)	1.0
Mobility (1 Pa^{-1} s^{-1})	1.0×10^4

Figure 6.6 shows the DD prediction of the channeling stress as a function of layer thickness. The numerical parameter β in (6.10) was determined to be 0.5 through a fitting procedure for the DD results so that they match the corresponding stress obtained from an analytical model based on dislocation interaction energies in the system (see Fig. 6.7). As expected, the channeling strength increases as the layer thickness decreases.

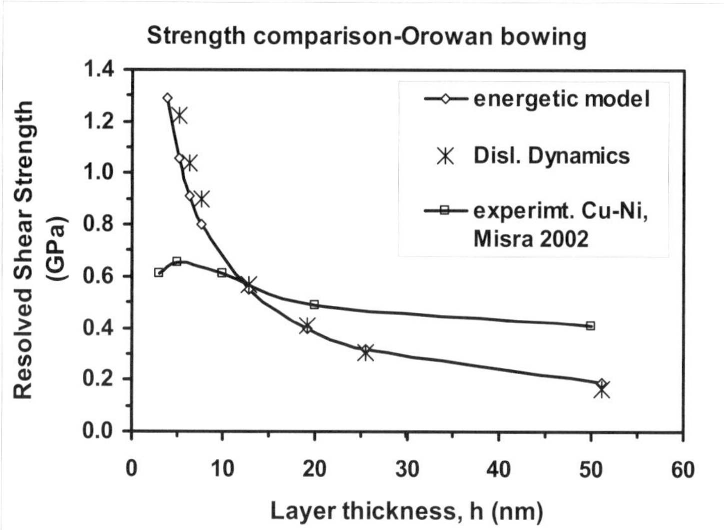

Fig. 6.6. Comparison of the strength due to Orowan bowing as predicted by DD and the energetic model (Fig. 6.7) with the measured strength of Cu/Ni multilayered structure

Also plotted in Fig. 6.6 is the measured strength of a real Cu/Ni system. Although the model bears little resemblance to the real system, the mild difference between the elastic moduli of Cu and Ni makes the comparison reasonable due to the minor effect of image force in such a system. This, however, does not apply for the few nanometer thickness range as the interface-related mechanism dominates over Orowan bowing [26]. This is why both DD and the analytical model results continue to predict increasing channeling strength in this range while the real system shows saturation followed by softening. Outside this range, both models underestimate the strength of a multilayer system. This should not be a surprise since a model based on a single bowing dislocation is not expected to capture the interactions between bowing dislocations and other bowing

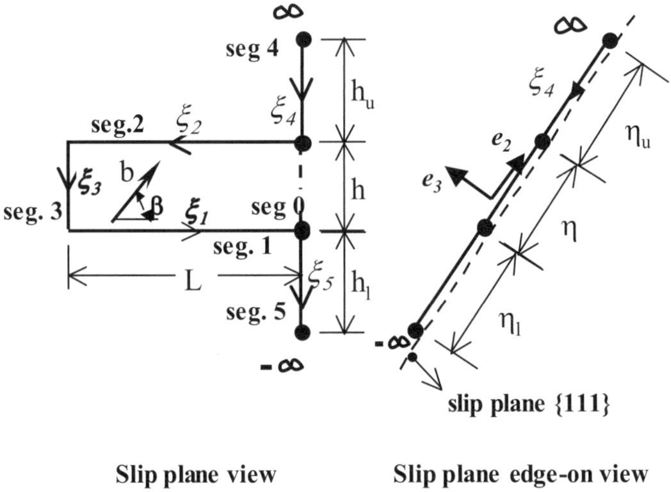

Fig. 6.7. Dislocation configuration modeling the threading process in a confined layer

dislocations and/or predeposited interfacial dislocations. This leads us to consider the DD simulation of the next two mechanisms.

Interaction between threading and orthogonal interfacial dislocations

The encounter between a threading dislocation and an interfacial dislocation orthogonally intersecting its path should be a very common event given the biaxial nature of loading in NMM composites. Freund [7] studied this interaction mechanism and developed the blockage model along with the associated concept of the effective channel thickness. As mentioned in the introductory part of this section, DD analysis indicates that the assumptions underlying this model suppress the possibility of short-range dislocation interactions making this model unrealistic. In fact, DD simulations [1] indicate that short-range interactions dictate the outcome of this type of interaction and that the blocking mechanism is not realistic all together in the case of a threader's encounter with glide-type interfacial dislocations.

In the case of an encounter between a threading dislocation and nonreacting Lomer-type dislocations, the blockage mechanism is realistic and has been observed. Nevertheless, the assumptions underlying the blockage mechanism are still coarse. In the process of bypassing an intersecting Lomer-type interfacial dislocation, the threader tends to adjust its configuration in a complex dynamic manner so as to minimize the overall system energy, rendering unclear the notion of h^* and a narrowed-down channel right above the obstacle through which the dislocation "squeezes-in" to overcome the stress field. Figure 6.8 shows simulation snapshots for the dynamic process of leading a threader to bypass a Lomer-type obstacle.

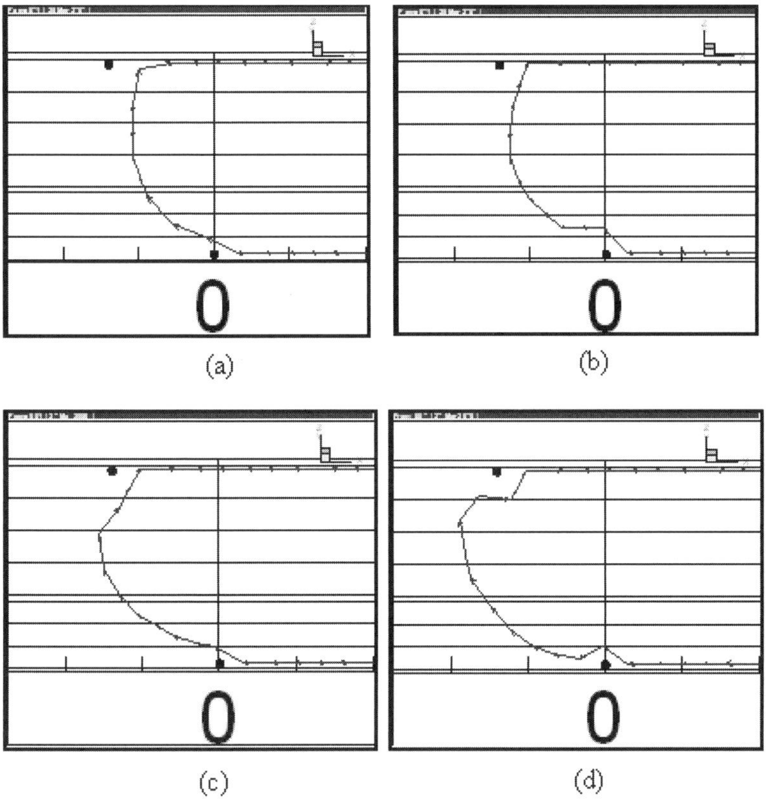

Fig. 6.8. Snapshots ((**a**)–(**d**) ordered in increasing simulation time) from DD simulations showing the complex sequence of configurations by which a threading dislocation bypasses a Lomer-type dislocation at the interface

As mentioned above, short-range dislocation interactions dominate the outcome of the encounter between threading and predeposited interfacial dislocations. Depending on the particular slip systems of the threader and the interfacial dislocation involved in the four representative encounters dictated by the crystallographic nature of the problem (Fig. 6.9), annihilation or jog formation can occur. Simulations indicate that the strongest effect on the strength is that of the annihilation interaction occurring when the Burgers vectors of both dislocations are collinear [1].

Figure 6.10 shows the channeling stress dependence on layer thickness as modified by this process. In the plot, the measured strength of Cu/Ni NMM composite and the reference strength due to the basic Orowan

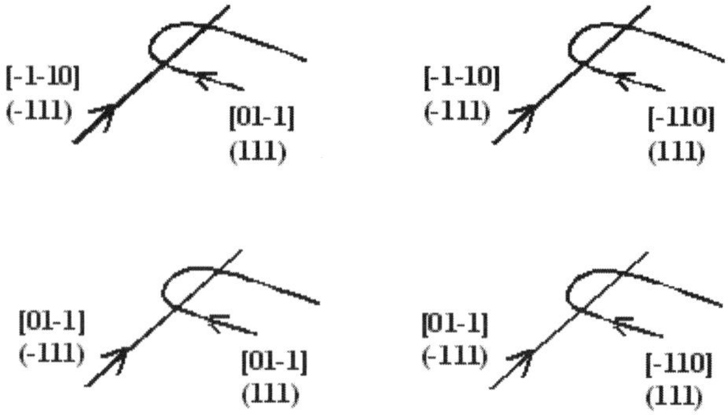

Fig. 6.9. The four encounters representing all possible intersections between threading dislocation and $a/2\langle 011\rangle$ -type interfacial dislocation dipoles (same crystallographic setup as that shown in Fig. 6.5)

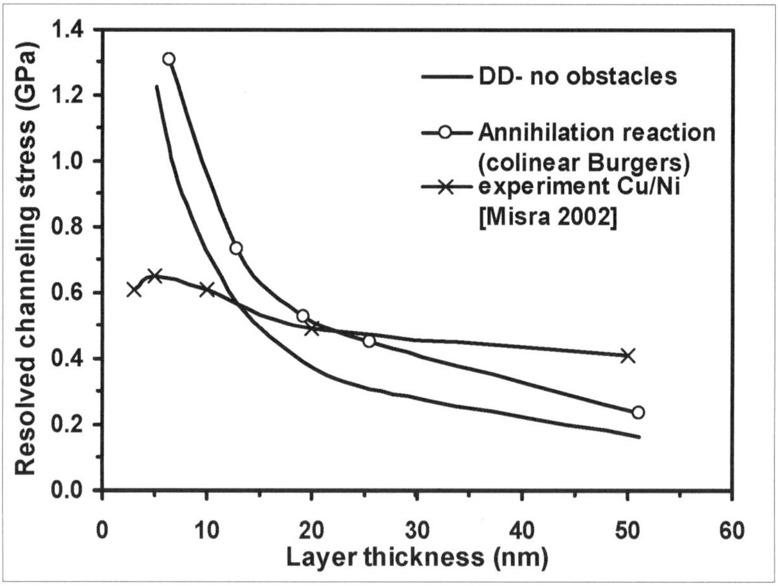

Fig. 6.10. Comparison between DD predictions of the channeling strength due to the basic Orowan bowing and that due to the orthogonal interaction between a threading and an interfacial dislocation of the same Burgers vector (also shown is the experimentally measured strength of Cu/Ni NMM composite)

model are also included. As can be seen from the figure, the strength predictions due to the strongest orthogonal dislocation interactions in NMM composites are significantly higher than those of the reference "plain" Orowan bowing process. Furthermore, they are in better agreement with the measured strength of the Cu/Ni composite system in the range of 22 nm, approximately, and above; while below this range the strength is overestimated.

Together, these observations indicate a model-based collinear orthogonal interaction results in better predictions of strength in NMM composites, although this is still incomplete. In the lower layer thickness structures, the results strongly suggest that softening mechanisms become more significant in governing the response of NMM composites.

Interaction between threading and parallel interfacial dislocations

Another frequent long-range dislocation interaction in NMM composites is the parallel threading of a dislocation in the vicinity of a predeposited interfacial dislocation and parallel to it. The closer the threading dislocation is to the interfacial dislocation, the stronger will be the effect of the later dislocation on the driving force for threading. Depending on the direction of the Burgers vectors of the two dislocations, the presence of the interfacial dislocation can impede or augment the threading process. Figure 6.11 shows DD results for the thickness dependence of the channeling stress as a function of the spacing (normalized with respect to layer thickness) between the threading and the parallel interfacial dislocation.

As the spacing between the threading dislocation and the predeposited dislocation dipole becomes smaller, the interaction gets stronger. The figure also demonstrates the effect of the relative sense of the Burgers vector of the two dislocations on the channeling strength. The cases identified by "parallel" in Fig. 6.11 refer to the case where the two dislocations have collinear Burgers vectors, while "antiparallel" refers to the case where the two dislocations have opposite Burgers vectors. In the former case, the interaction energy between the threader and the interfacial dislocation is positive, meaning that additional work has to be supplied to overcome the interaction. In the latter case, the interaction energy is negative implying that less work, relative to the "plain" Orowan bowing, needs to be spent. Finally, Fig. 6.11 overlays the measured strength of the Cu/Ni nanocomposite systems for comparison. Besides the significance of their effect on the channeling stress, parallel interactions have important implications on the minimum spacing between parallel interfacial dislocations. A new threader will need to be sufficiently far from a neighboring threader already threaded at an earlier stage in the relaxation process.

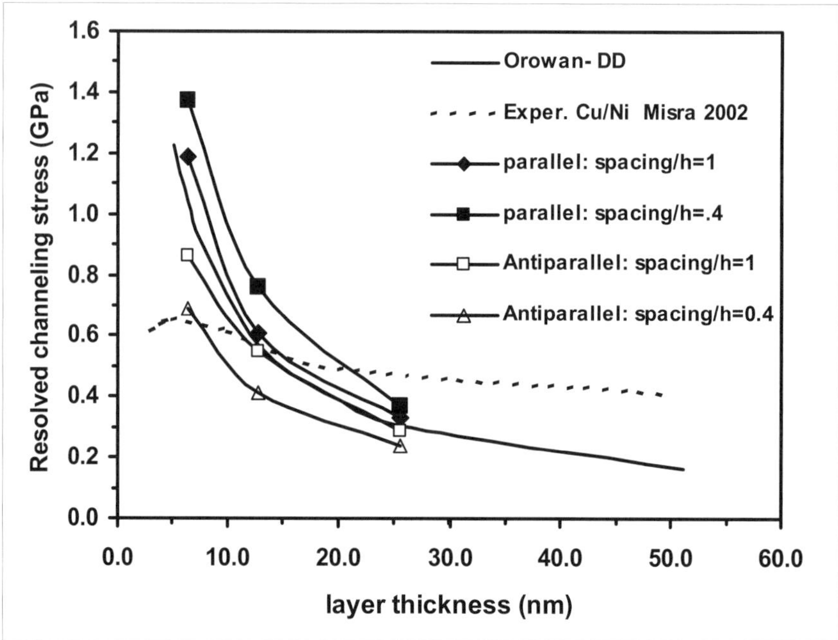

Fig. 6.11. Comparison between DD predictions of the channeling strength due to the basic Orowan bowing and that due to the parallel interaction between a threading and an interfacial dislocation of the same Burgers vector (also shown is the experimentally measured strength of Cu/Ni NMM composite)

Threading in surface layer: effect of free surface on threading strength

In the previous three examples, unit dislocation mechanisms were studied as they occur in a layer embedded in an infinite medium with homogenous elastic properties similar to that of the layer. In this example, the threading process in a surface layer of a large stack is considered. The fundamental difference is that the presence of a free surface will impose an additional stress (image stress) on the threading dislocation.

Image stresses induce an attractive force on dislocations close to the boundary pulling them toward the free surface and out of the medium. As pointed out in Sect. 6.2.3, the multiscale DD model, coupled with FE analysis, rigorously accounts for this effect through the superposition principle and the solution of the complementary elasticity problem, as explained there.

To assess the surface effect, the stress needed to propagate a threading dislocation in a surface layer was estimated with/without the multiscale coupling with FE analysis enabled. A layer thickness of 25 nm was considered, and the threading dislocation considered had the slip system *a*/2[01-1] (111), with the same crystallographic setup as that indicated in Fig. 6.5a. The surface layer and the rest of the domain were subjected to equal but opposite sign stresses as is the case in threading induced by coherency stress during the relaxation process.

It was found that a biaxial stress of 0.4 GPa was necessary to propagate the threader when surface treatment through the multiscale analysis was disabled, while in the case it was enabled, the corresponding stress was 0.2 GPa. This difference is significant and indicates that a free surface enhances the threading process in a surface layer. Further investigation of this effect for a different layer thickness is in progress and will be the subject of a future publication.

6.4.2 Modeling of Dislocation Systems in NMM Composites

Although the understanding of unit dislocation mechanisms is a crucial first step toward predicting the overall response of NMM composites, the complexity of dislocation interactions in real systems makes the task of extending an understanding of isolated units to overall behavior nontrivial, if not impossible. As can be concluded from comparing the measured strength of NMM composites and predicted strength based on parallel and orthogonal interactions between threading and interfacial dislocations (Figs. 6.10 and 6.11), no single mechanism can be claimed to be dominant over the whole range of layer thicknesses. Furthermore, the evolution of the dislocation structure under an extremely complex and continuously evolving stress field can significantly change the dynamics of unit process. For example, DD simulations performed by Pant et al. [32] showed that, while in isolation from other dislocations, two threading dislocations gliding in opposite directions on parallel planes can form a dipole and get stuck. However, a third threading dislocation approaching this dipole on a close parallel plane can "free" one of the dislocations in the existing dipole and form a new dipole. The high density of dislocations in a real system makes it impossible to track the effect on the overall response of the possibly large number of similar interactions between the background dislocations and the unit mechanisms.

The above complexity of the dislocation behavior in NMM composites can be handled using DD analysis because dislocation motion and interactions are treated explicitly. Their motion is directly calculated from the net stress field due to all other dislocations and to any applied loads, while boundary conditions can be rigorously treated through the coupling to FE. Short-range interactions are also accounted for through physical constitutive rules. Thus, capturing the natural evolution of a real system is possible and feasible in spite of the heavy computational expense. Nevertheless, the question of the validity of such massive simulations hinges on the validity of the initial dislocation structure from which loading starts.

Arriving at a valid initial structure should also be the natural outcome of the process leading to it, i.e., the relaxation process of the as-deposited composite. DD simulations for the relaxation process of an as-deposited NMM structure and its consequent loading are presented in the following sections. The relaxation simulation starts from an initial random distribution of threading dislocations driven by an alternating system of coherency stresses, as suggested by the physics of the deposition and relaxation process in real systems, and ends at the point when the dynamics evolution of the systems ceases, marking the arrival at a relaxed equilibrium structure. The loading simulation starts from this relaxed structure by applying strain rate-controlled uniform biaxial loading. From this simulation, stress–strain curves can be collected from which the strength of the composite is captured.

Modeling of the relaxation process and as-deposited structure

The goal of these simulations is to arrive at a representative relaxed structure for NMM composites with physical initial dislocation content and residual coherency stresses. The simulation box is made of four layers, and represents an infinite domain made of alternating layers of the bimetal system. To ensure this, periodic boundary conditions are applied to the simulation box in all three directions. Each layer is 12.7 nm thick, and its elastic properties are those of Cu (Table 6.1). All the layers have the same elastic properties; however, they are under a stress field of equal magnitude but alternating sign in alternating layers. The stress field is due to coherency stress resulting from lattice parameters mismatch between the two materials making the alternating layers. Strictly speaking, this setup represents a bimetal system with matching elastic properties but different lattice parameters. However, as was explained earlier, this idealization is still reasonable for the Cu/Ni system.

The coherency stress used in the simulation is 2.6 GPa [16] and is biaxial. The simulation begins with a random distribution of dislocation loops spanning the four layers. If the layer thickness is larger that the critical thickness h_c, dislocations glide in opposite directions in the alternating layers due to the alternating stress state. In the process, dislocations naturally interact with the stress field and among themselves result in the relaxation of the coherency stress and at the expense of generating a new dislocation structure.

The process continues until dislocation motion ceases indicating the attainment of an equilibrium state between the dislocations and the final distribution of internal stress. Figure 6.12 shows DD simulation of the final structure after the relaxation process has ended. As can be seen, the structure consists dominantly of a network of orthogonal interfacial dislocations, dislocation bends resulting from cross-slip, and blocked threading dislocations. Another dislocation mechanism that was observed during the relaxation process but leaves no traces of its occurrence in the final structure is annihilation of threaders of opposite signs. As they do so, the threading segments get eliminated leaving only the trailing hairpin arm at the interfaces.

Modeling of the biaxial loading process

As mentioned previously, the resulting structure at the end of the relaxation process is considered to be representative of that of an as-deposited structure that has relaxed, although incompletely as discussed previously, its coherency stresses through the threading process. To study the strength of NMM composites, this structure serves as the starting point for loading. The same periodic four-layer system used in the relaxation simulation is used here. Tensile biaxial loading at a controlled strain rate is applied.

Figure 6.13a is a side snapshot showing dislocation activity under loading just after the yield point has been reached. It shows that the top and the third-from-top layers, which had compressive coherency stresses in the relaxation phase, are now showing reversal of dislocation activity due to the opposite sign loading. This recovery process is limited to those dislocations which have not undergone irreversible short-range interactions and those which threaded to some length that is short of their threading stability point. The other two layers, which were under tensile coherency stresses in the relaxation process, are showing a resumption of dislocation activity. Figure 6.13b shows the stress evolution curve captured by the simulation. As can be seen, the yield point occurs at about 2.2 GPa. The difference between this value and the 2.6 GPa coherency stress experienced in the relaxation phase reflects the fact that relaxation is not complete due to the impediments to relaxation expected to exist in real systems, as discussed earlier.

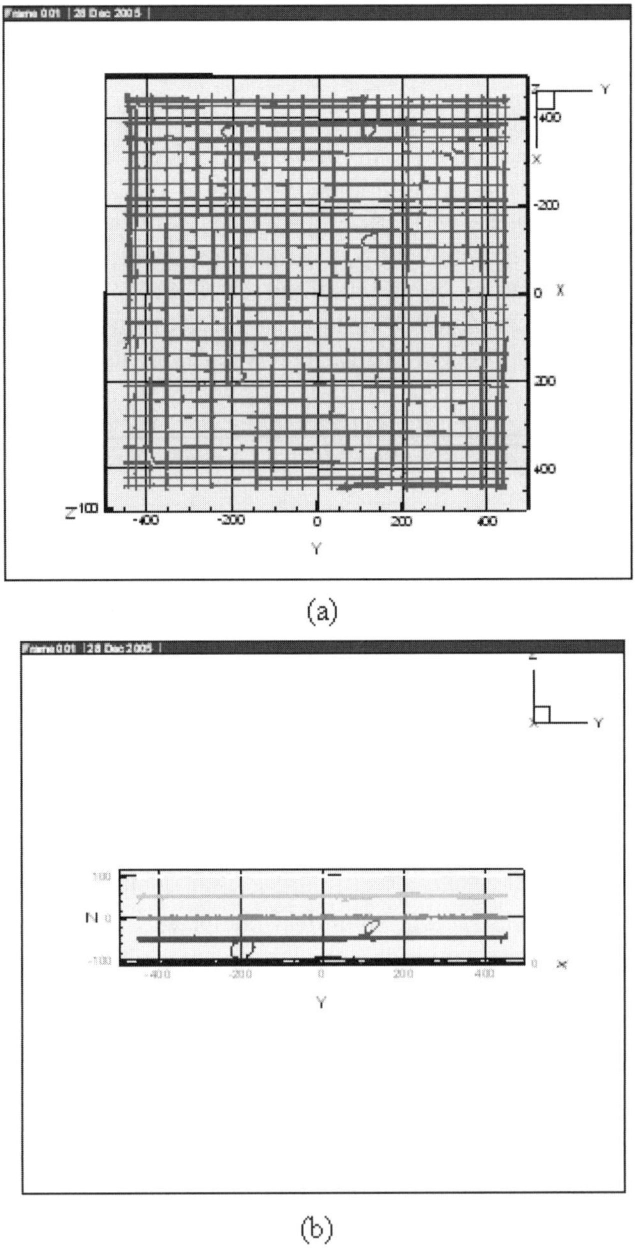

Fig. 6.12. Relaxed dislocation configuration due to coherency stress in a four-layered structure representative of a bimetal multilayered structure

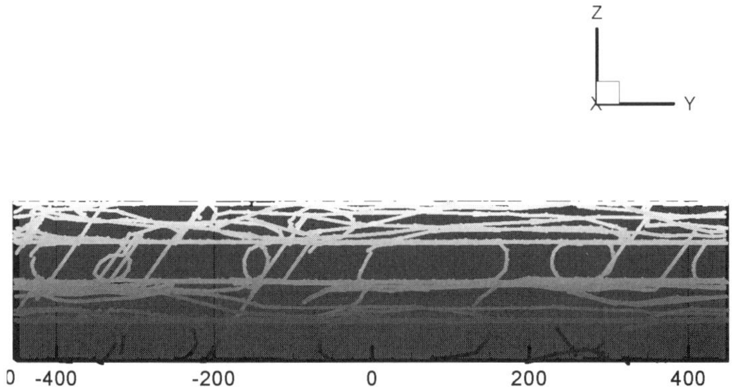

(a)

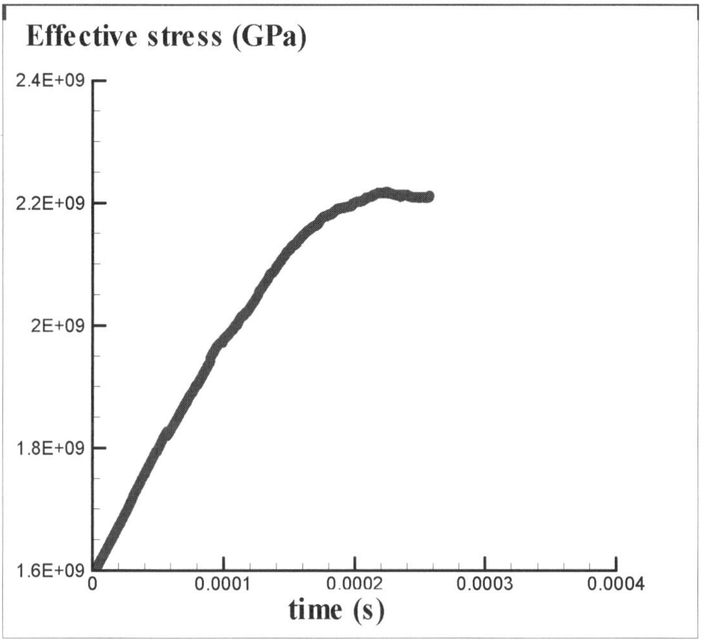

(b)

Fig. 6.13. Biaxial loading of the relaxed multilayer structure shown in Fig. 6.12: (**a**) side view of the dislocation activity after yielding and (**b**) stress evolution as captured by the simulation

References

1. Akasheh F, Zbib HM, Hirth JP, Hoagland RG, Misra A. 2007. Dislocation dynamics analysis of dislocation intersections in nanoscale metallic multilayered composites. Journal of Applied Physics 101:084314
2. Bulatov VV, Rhee M, Cai W. 2000. Periodic Boundary Conditions for Dislocation Dynamics Simulations in Three Dimensions. Proceedings of the MRS meeting
3. DeWit R. 1960. The continuum theory of stationary dislocations. Solid State Physics 10:249–292
4. Feng X, Hirth JP. 1992. Critical layer thickness for inclined dislocation stability in multilayer structures. Journal of Applied Physics 72:1386–1393
5. Fivel MC, Roberston CF, Canova G, Bonlanger L. 1998. Three-dimensional modeling of indent-induced plastic zone at a mesocale. Acta Materialia 46:6183–6194
6. Frank FC, van der Merwe JH. 1949. One-dimensional dislocations. I. Static theory. Proceedings of the Royal Society of London A 198:205
7. Freund LB. 1990. A criterion for arrest of a threading dislocation in a strained epitaxial layer due to an interface misfit dislocation in its path. Journal of Applied Physics 68:2073–2080
8. Ghoniem NM, Amodeo RJ. 1990. Numerical simulation of dislocation patterns during plastic deformation. In: Walgraef D, Ghoniem NM, editors. Patterns, Defects and Material Instabilities. Kluwer, Dordecht, pp 303–329
9. Gosling TJ, Bullough R, Jain SC, Willis JR. 1993. Misfit dislocation distributions in capped (buried) strained semiconductor layers. Journal of Applied Physics 73:8267–8278
10. Gosling TJ, Willis JR, Bullough R, Jain SC. 1993. The energetics of dislocation array stability in strained epitaxial layers. Journal of Applied Physics 73:8297–8303
11. Hirth JP. 1992. Injection of dislocations into strained multilayer structures. In: Semiconductors and Semimetals. Academic, London, pp 267–292
12. Hirth JP, Feng X. 1990. Critical layer thickness for misfit dislocation stability in multilayer structures. Journal of Applied Physics 67:3343–3349
13. Hirth JP, Lothe J. 1982. Theory of Dislocations. Wiley, New York, p 857
14. Hirth JP, Zbib HM, Lothe J. 1998. Forces on high velocity dislocations. Modelling and Simulation in Materials Science and Engineering 6:165–169
15. Hoagland RG, Hirth JP, Misra A. 2006. On the role of weak interfaces in blocking slip in nanoscale layered composites. Philosophical Magazine 86:3537–3558
16. Hoagland RG, Mitchell TE, Hirth JP, Kung H. 2002. On the strengthening effects of interfaces in multilayer fcc metallic composites. Philosophical Magazine A 82:643–664
17. Hochbauer T, Misra A, Hattar K, Hoagland RG. 2005. Influence of interfaces on the storage of ion-implanted He in multilayered metallic composites. Journal of Applied Physics 98:123516

18. Huang H, Ghoniem N, Diaz de la Rubia T, Rhee M, Zbib HM, Hirth JP. 1999. Development of physical rules for short range interactions in BCC crystals. ASME-JEMT 121:143–150
19. Kamat SV, Hirth JP. 1990. Dislocation injection in strained multilayer structures. Journal of Applied Physics 67:6844–6850
20. Koehler JS. 1970. Attempt to design a strong solid. Physical Review B 2:547–551
21. Kubin LP, Canova G. 1992. The modelling of dislocation patterns. Scripta Metallurgica 27:957–962
22. Mara N, Sergueeva A, Misra A, Mukherjee AK. 2004. Structure and high-temperature mechanical behavior relationship in nano-scaled multilayered materials. Scripta Materialia 50:803–806
23. Matthews JW, Blakeslee AE. 1974. Defects in epitaxial multilayers. I. Misfit dislocations. Journal of Crystal Growth 27:118–125
24. Matthews JW, Mader S, Light TB. 1970. Accommodation of misfit across the interface between crystals of semiconducting elements or compounds. Journal of Applied Physics 41:3800–3804
25. Misra A, Hirth JP, Hoagland RG, Embury DJ, Kung H. 2004. Dislocation mechanisms and symmetric slip in rolled nano-scaled metallic multilayers. Acta Materialia 52:2387–2394
26. Misra A, Hirth JP, Kung H. 2002. Single-dislocation-based strengthening mechanisms in nanoscale metallic multilayers. Philosophical Magazine A 82:2935–2951
27. Misra A, Hoagland RG. 2005. Effects of elevated temperature annealing on the structure and hardness of copper/niobium nanolayered films. Journal of Materials Research 20:2046–2054
28. Misra A, Kung H. 2001. Deformation behavior of nanostructured metallic multilayers. Advanced Engineering Materials 3:217–222
29. Mitlin D, Misra A, Mitchell TE, Hoagland RG, Hirth JP. 2004. Influence of overlayer thickness on the density of Lomer dislocations in nanoscale Ni–Cu bilayer thin films. Applied Physics Letters 85:1686–1688
30. Mitlin D, Misra A, Radmilovic V, Nastasi M, Hoagland RG, Embury D, Hirth JP, Mitchell T. 2004. Formation of misfit dislocations in nanoscale Ni–Cu bilayer films. Philosophical Magazine 84:719–736
31. Nix WD. 1998. Yielding and strain hardening of thin metal films on substrates. Scripta Materialia 39:545–554
32. Pant P, Schwarz KW, Baker SP. 2003. Dislocation interactions in thin FCC metal films. Acta Materialia 51:3243–3258
33. Schwarz KW, Tersoff J. 1996. Interaction of threading and misfit dislocations in a strained epitaxial layer. Applied Physics Letters 69:1220–1222
34. Van der Giessen E, Needleman A. 1995. Discrete dislocation plasticity: a simple planar model. Materials Science and Engineering 3:689–735
35. Wang YC, Misra A, Hoagland RG. 2006. Fatigue properties of nanoscale Cu/Nb multilayers. Scripta Materialia 54:1593–1598

36. Willis JR, Jain SC, Bullough R. 1990. The energy of an array of dislocations: implications for strain relaxation in semiconductor heterostructures. Philosophical Magazine A 62:115–129
37. Yasin H, Zbib HM, Khaleel MA. 2001. Size and boundary effects in discrete dislocation dynamics: coupling with continuum finite element. Materials Science and Engineering A 309–310:294–299
38. Zbib HM, Diaz de la Rubia TA. 2001. Multiscale Model of Plasticity: Patterning and Localization. Material Science for the 21st Century, The Society of Materials Science, Japan, Vol. A, pp 341–347
39. Zbib HM, Diaz de la Rubia T. 2002. A multiscale model of plasticity. International Journal of Plasticity 18:1133–1163
40. Zbib HM, Rhee M, Hirth JP. 1996. 3D simulation of curved dislocations: discretization and long range interactions. In: Abe T, Tsuta T, editors. Advances in Engineering Plasticity and Its Applications. Pergamon, Amsterdam, pp 15–20
41. Zbib HM, Rhee M, Hirth JP. 1998. On plastic deformation and the dynamics of 3D dislocations. International Journal of Mechanical Science 40:113–127

Chapter 7: Multiscale Modeling of Composites Using Analytical Methods

L.N. McCartney

NPL Materials Centre
National Physical Laboratory
Middlesex TW11 0LW, UK

7.1 Introduction

Both fiber and particulate composite materials provide applications in materials science where the multiscale microstructure leads to the need for multiscale modeling. The length scales encountered range from the fiber and particle sizes whose dimensions are measured in microns, to the individual plies in laminates whose thicknesses are measured in fractions of millimeters, to the laminates themselves whose thicknesses in the laboratory are measured in millimeters, e.g., 40–50 mm. The laminates then form parts of composite structures whose sizes are measured in meters, although modeling at this scale will not form part of this chapter. While conventional composites are based on essentially homogeneous matrices, which can be polymeric, metallic, or ceramic, advanced composites are also being considered to have matrices, which are themselves composites reinforced by submicron particles or whiskers, e.g., carbon nanotubes. Such developments lead to the need to be able to estimate the properties of composite laminates that have multiscale reinforcements, e.g., fibers in particulate/whisker-reinforced matrices. Also, there is a need to predict the onset of damage in the materials when they are operating in service conditions.

This chapter will focus on two aspects of the problem that complement work already published dealing with multiscale analytical modeling for damaged composite laminates [7]. The first aspect, considered in Sects. 7.2–7.4, is the development of new understanding with regard to the prediction of the properties of undamaged particulate and unidirectional fiber-reinforced composites. Recent work [8] has involved the study of a methodology first published in 1873 that was developed by Maxwell [3] when considering the effective electrical conductivity of a conducting medium in which a dilute distribution of particles having a different conductivity was dispersed. It has now been shown [8] that this methodology can be applied much more widely with the result that many other effective properties of both particulate and fiber-reinforced composites can be estimated. A description of the key results of this investigation will first be presented in this chapter together with a discussion of the relationship of the new results to existing formulae that can be used to estimate effective composite properties. The results given will be very useful when estimating undamaged properties for matrices if reinforced by submicron particles, and the undamaged properties of the individual plies of laminates that will become damaged when loaded in service conditions. Recent work to be reported in [8] has reconsidered the bounds on properties that arise from the use of variational methods [1, 2, 10]; and in Sect. 7.4, sets of conditions are given that identify whether the extreme values of properties are upper or lower bounds.

The onset of microstructural damage in the form of fiber and interface fracture for unidirectional composites, and of ply cracking and delamination in laminated composites, leads to a deterioration of thermoelastic properties. For structural applications of composites, such as in plates with bolt holes, stress concentrations lead to localized damage and to localized changes in modulus, Poisson's ratios, and thermal expansion coefficients that cause load to be transferred to other parts of the structure. Damage development in structures is, thus, a gradual inhomogeneous process of material deterioration that eventually culminates in the catastrophic failure of the structure. The local damage-induced load transfer can lead to composite components out performing their expected performance on the basis of laboratory coupon data. Sections 7.5 and 7.6 of this chapter are concerned with laminated composites for which ply cracking is the only damage mode, although results are expected to be valid more generally. The damage model for laminates will require the properties of undamaged plies, and use can be made of the results presented in Sects. 7.2–7.4 that relate to the fiber/matrix length scale rather than the ply/laminate length scales.

While a methodology for the prediction of damage formation has been described in [7], it involves the use of various interrelationships between the effective properties of damaged laminates. These relationships were derived from the development of an accurate stress-transfer model [4–6] that estimated the values of the effective thermoelastic properties of the laminate. The objective here is to show how the interrelationships developed can be derived independently of the stress-transfer model. In addition, the analysis is extended to demonstrate that the interrelationships are valid also for laminates having orthogonal sets of ply cracks.

Ply cracking damage is usually the first significant form of damage that occurs when a laminate is loaded. The occurrence of ply cracks leads to a degradation of laminate properties; and such property degradation needs to be taken into account when assessing the integrity of composite structures using numerical methods, such as finite element analysis (FEA). The use of FEA in a structural setting demands that the three-dimensional properties of laminates be available to the software. When considering the structural integrity of a component, account needs to be taken of the degradation of all local properties when damage occurs in the form of ply cracking. Such phenomena, leading to the redistribution of stress in the structure that will affect the onset of component failure during loading, need to be modeled realistically. The results given in Sects. 7.5 and 7.6 provide most of the necessary property degradation relationships. The prediction of damage formation in laminates is, thus, set on a firm basis; and the results of these sections provide a rigorous framework of general validity that can be used with confidence in design methodologies.

7.2 Application of Maxwell's Method to Particulate Composites

In the field of electricity and magnetism, Maxwell [3] (as early as 1873) developed a method of estimating the electrical conductivity of an isotropic cluster of spherical particles of the same size embedded in a matrix having different conduction properties. The method was based on the exact solution for an isolated sphere embedded in an infinite matrix subject to a uniform gradient of electrostatic potential. This solution is applied both to the individual particles in the cluster (which were assumed to be noninteracting), and to the effective composite material that can be used to replace the particle cluster without affecting the potential distribution in the matrix at large distances from the cluster.

The effective composite medium is taken to have a radius such that the matrix and particles enclosed have the same particle volume fraction as the composite for which properties are required. When observed at large distances from an isolated particle, the electrostatic potential has the form of the sum of the unperturbed potential distribution (that arises when the particle is not present) plus a term that is inversely proportional to the square of the radial distance from the center of the particle. The coefficient of the perturbation term depends on the particle radius and the electrical conductivities of both the particle and the matrix.

The isolated particle solution is applied both to the single sphere of effective composite material representing the cluster and to each particle in the cluster. Maxwell states that an assumption is made that the particles do not interact, in which case at large distances from a cluster of particles, the perturbing effect of the particles can be expressed as the sum of the perturbations that each particle would cause if isolated in the matrix at a given point. This assumption implies that results are likely to be valid only for low volume fractions of reinforcement, although evidence is presented suggesting much wider applicability.

The purpose of this section is to report results that have recently been derived [8] where Maxwell's methodology has been applied to the estimation of many other properties for both isotropic particulate composites and anisotropic fiber-reinforced composites.

7.2.1 Applying Maxwell's Approach to Multiphase Particulate Composites

Because of the use of the far field in Maxwell's methodology for estimating the properties of particulate composites, it is possible to consider multiple spherical reinforcements. Suppose, in a cluster of particulate reinforcements embedded in an infinite matrix that there are N different types such that for $i = 1,\ldots,N$ there are n_i spherical particles of radius a_i. The properties of the particles of type i are denoted by a superscript i and subscript p, where k will denote bulk moduli, μ will denote shear moduli, and α will denote thermal expansion coefficients. The cluster is assumed to be homogeneous regarding the distribution of particles and leads to isotropic effective properties. A suffix m will be used to denote matrix properties.

The cluster of all types of particle is now considered to be enclosed in a sphere of radius b such that the volume fraction of particles of type i within the sphere of radius b is given by $V_{\mathrm{p}}^{i} = n_i a_i^3 / b^3$. The volume fractions must satisfy the relation

$$V_{\mathrm{m}} + \sum_{i=1}^{N} V_{\mathrm{p}}^{i} = 1, \tag{7.1}$$

where V_{m} is the volume fraction of matrix material. For the case of multiple phases, it has been shown that the effective bulk modulus k_{eff}, shear modulus μ_{eff}, and thermal expansion coefficient α_{eff} are given by [8]

$$\frac{1}{k_{\text{eff}}} = \frac{1}{1+\Lambda}\left(\frac{1}{k_{\text{m}}} - \frac{3\Lambda}{4\mu_{\text{m}}}\right), \tag{7.2}$$

$$\mu_{\text{eff}} = \mu_{\text{m}} \frac{1-(7-5\nu_{\text{m}})\Gamma}{1+2(4-5\nu_{\text{m}})\Gamma}, \tag{7.3}$$

$$\alpha_{\text{eff}} = \alpha_{\text{m}} + \Omega\left(\frac{1}{k_{\text{eff}}} + \frac{3}{4\mu_{\text{m}}}\right), \tag{7.4}$$

where

$$\Lambda = \sum_{i=1}^{N} \frac{\dfrac{1}{k_{\text{m}}} - \dfrac{1}{k_{\text{p}}^{i}}}{\dfrac{1}{k_{\text{p}}^{i}} + \dfrac{3}{4\mu_{\text{m}}}} V_{\text{p}}^{i}, \tag{7.5}$$

$$\Gamma = \sum_{i=1}^{N} \frac{(\mu_{\text{m}} - \mu_{\text{p}}^{i})V_{\text{p}}^{i}}{2(4-5\nu_{\text{m}})\mu_{\text{p}}^{i} + (7-5\nu_{\text{m}})\mu_{\text{m}}}, \tag{7.6}$$

$$\Omega = \sum_{i=1}^{N} \frac{\alpha_{\text{p}}^{i} - \alpha_{\text{m}}}{\dfrac{1}{k_{\text{p}}^{i}} + \dfrac{3}{4\mu_{\text{m}}}} V_{\text{p}}^{i}. \tag{7.7}$$

Application to a two-phase system

Consider a cluster of n spherical particles, having the same properties and the same radius a, embedded in an infinite matrix of different properties. The cluster is just enclosed by a sphere of radius b and the particle distribution is sufficiently homogeneous for it to lead to isotropic properties for the composite formed by the cluster and the matrix lying within this sphere. If the particulate volume fraction of the composite is denoted by V_{p}, then

$$V_{\text{p}} = \frac{na^3}{b^3} = 1 - V_{\text{m}}, \tag{7.8}$$

where V_m is the volume fraction of the matrix. The suffices p and m will be used to refer properties k, μ, and α to the particles and matrix, respectively.

It can be shown from (7.2) to (7.7) that the effective bulk modulus, shear modulus, and thermal expansion coefficient are given by

$$\frac{1}{k_{\rm eff}} = \frac{\dfrac{4\mu_{\rm m}}{k_{\rm p}k_{\rm m}} + \dfrac{3V_{\rm m}}{k_{\rm m}} + \dfrac{3V_{\rm p}}{k_{\rm p}}}{4\mu_{\rm m}\left(\dfrac{V_{\rm p}}{k_{\rm m}} + \dfrac{V_{\rm m}}{k_{\rm p}} + \dfrac{3}{4\mu_{\rm m}}\right)}, \tag{7.9}$$

$$\mu_{\rm eff} = \mu_{\rm m}\left[1 + \frac{15(1-\nu_{\rm m})(\mu_{\rm p}-\mu_{\rm m})V_{\rm p}}{2(4-5\nu_{\rm m})(V_{\rm m}\mu_{\rm p}+V_{\rm p}\mu_{\rm m}) + (7-5\nu_{\rm m})\mu_{\rm m}}\right], \tag{7.10}$$

$$\alpha_{\rm eff} = \alpha_{\rm m} + \frac{\dfrac{1}{k_{\rm eff}} + \dfrac{3}{4\mu_{\rm m}}}{\dfrac{1}{k_{\rm p}} + \dfrac{3}{4\mu_{\rm m}}} V_{\rm p}(\alpha_{\rm p}-\alpha_{\rm m}). \tag{7.11}$$

It has been shown [5] that it is possible to express these relations as the sum of mixtures estimates plus correction terms so that

$$\frac{1}{k_{\rm eff}} = \frac{V_{\rm p}}{k_{\rm p}} + \frac{V_{\rm m}}{k_{\rm m}} - \frac{\left(\dfrac{1}{k_{\rm p}} - \dfrac{1}{k_{\rm m}}\right)^2 V_{\rm p}V_{\rm m}}{\dfrac{V_{\rm p}}{k_{\rm m}} + \dfrac{V_{\rm m}}{k_{\rm p}} + \dfrac{3}{4\mu_{\rm m}}}, \tag{7.12}$$

$$\mu_{\rm eff} = V_{\rm p}\mu_{\rm p} + V_{\rm m}\mu_{\rm m} - \frac{(\mu_{\rm p}-\mu_{\rm m})^2}{V_{\rm p}\mu_{\rm m} + V_{\rm m}\mu_{\rm p} + \dfrac{9k_{\rm m}+8\mu_{\rm m}}{6(k_{\rm m}+2\mu_{\rm m})}\mu_{\rm m}} V_{\rm p}V_{\rm m}, \tag{7.13}$$

$$\alpha_{\rm eff} = V_{\rm p}\alpha_{\rm p} + V_{\rm m}\alpha_{\rm m} - \frac{\left(\dfrac{1}{k_{\rm p}} - \dfrac{1}{k_{\rm m}}\right)(\alpha_{\rm p}-\alpha_{\rm m})V_{\rm p}V_{\rm m}}{\dfrac{V_{\rm p}}{k_{\rm m}} + \dfrac{V_{\rm m}}{k_{\rm p}} + \dfrac{3}{4\mu_{\rm m}}}. \tag{7.14}$$

The mixtures estimates are given by the first two terms on the right-hand side of (7.12)–(7.14), while the third term is the correction term that must be applied to the mixtures rule. The results (7.12) and (7.14) are identical to those derived by applying the spherical shell model of the particulate composite to a representative volume element comprising just one particle and a matrix region that is consistent with the volume fraction of the composite. The values of the results (7.12)–(7.14) are identical to one of the bounds obtained when using variational methods [1, 2, 10]. These results suggest that the assumption by Maxwell [3] of low volume fractions is not necessary; an issue that will be discussed in [5].

The formulae (7.2)–(7.4) and (7.12)–(7.14) completely characterize the properties of an isotropic particulate composite and are expressed in the form of a mixtures estimate and a correction term. These formulae can be used to estimate the effective properties of a microreinforced matrix where the reinforcing phase is particulate in nature.

7.3 Application of Maxwell's Method to Fiber Composites

Because of the use of the far field in Maxwell's methodology for estimating the properties of composites, it is now possible to consider multiple fiber, rather than particulate reinforcements. Suppose in a cluster of fibers that there are N different types such that for $i = 1,\ldots,N$ there are n_i fibers of radius a_i. The properties of the fibers of type i are denoted by a superscript i. Poisson's ratios are to be denoted by ν, and axial and transverse properties will be denoted by suffices A and T, respectively. The cluster is assumed to be homogeneous regarding the distribution of fibers and leads to transverse isotropic effective properties. The suffix or superscript m will be used to denote matrix properties.

The cluster of all types of fiber is now considered to be enclosed in a cylinder of radius b such that the volume fraction of fibers of type i within the cylinder of radius b is given by $V_\mathrm{f}^i = n_i a_i^2 / b^2$. The volume fractions must satisfy the relation

$$V_\mathrm{m} + \sum_{i=1}^{N} V_\mathrm{f}^i = 1. \qquad (7.15)$$

7.3.1 Properties Derived from the Lamé Solution

By making use of Maxwell's methodology in conjunction with the Lamé solution for two bonded concentric cylinders, several properties of a fiber-reinforced composite can be estimated. It has been shown [8] that the following effective properties for the multiphase fiber-reinforced composite

Transverse bulk modulus: $k_{\mathrm{T}}^{\mathrm{eff}}$
Axial Poisson's ratio: $\nu_{\mathrm{A}}^{\mathrm{eff}}$
Axial thermal expansion coefficient: $\alpha_{\mathrm{A}}^{\mathrm{eff}}$
Transverse thermal expansion coefficient: $\alpha_{\mathrm{T}}^{\mathrm{eff}}$

may be estimated using the formulae

$$\frac{1}{k_{\mathrm{T}}^{\mathrm{eff}}} = \frac{1}{1-\Lambda_1}\left(\frac{1}{k_{\mathrm{T}}^{\mathrm{m}}} + \frac{\Lambda_1}{\mu_{\mathrm{T}}^{\mathrm{m}}}\right), \tag{7.16}$$

$$\nu_{\mathrm{A}}^{\mathrm{eff}} = \nu_{\mathrm{A}}^{\mathrm{m}} - \Lambda_2\left(\frac{1}{\mu_{\mathrm{T}}^{\mathrm{m}}} + \frac{1}{k_{\mathrm{T}}^{\mathrm{eff}}}\right), \tag{7.17}$$

$$(\alpha_{\mathrm{T}}^{\mathrm{eff}} + \nu_{\mathrm{A}}^{\mathrm{eff}}\alpha_{\mathrm{A}}^{\mathrm{eff}}) = (\alpha_{\mathrm{T}}^{\mathrm{m}} + \nu_{\mathrm{A}}^{\mathrm{m}}\alpha_{\mathrm{A}}^{\mathrm{m}}) + \Lambda_3\left(\frac{1}{\mu_{\mathrm{T}}^{\mathrm{m}}} + \frac{1}{k_{\mathrm{T}}^{\mathrm{eff}}}\right), \tag{7.18}$$

where

$$\Lambda_1 = \sum_{i=1}^{N} V_{\mathrm{f}}^{i}\,\frac{\dfrac{1}{k_{\mathrm{T}}^{i}} - \dfrac{1}{k_{\mathrm{T}}^{\mathrm{m}}}}{\dfrac{1}{\mu_{\mathrm{T}}^{\mathrm{m}}} + \dfrac{1}{k_{\mathrm{T}}^{i}}} = \frac{\dfrac{1}{k_{\mathrm{T}}^{\mathrm{eff}}} - \dfrac{1}{k_{\mathrm{T}}^{\mathrm{m}}}}{\dfrac{1}{\mu_{\mathrm{T}}^{\mathrm{m}}} + \dfrac{1}{k_{\mathrm{T}}^{\mathrm{eff}}}}, \tag{7.19}$$

$$\Lambda_2 = \sum_{i=1}^{N} V_{\mathrm{f}}^{i}\,\frac{\nu_{\mathrm{A}}^{\mathrm{m}} - \nu_{\mathrm{A}}^{i}}{\dfrac{1}{\mu_{\mathrm{T}}^{\mathrm{m}}} + \dfrac{1}{k_{\mathrm{T}}^{i}}} = \frac{\nu_{\mathrm{A}}^{\mathrm{m}} - \nu_{\mathrm{A}}^{\mathrm{eff}}}{\dfrac{1}{\mu_{\mathrm{T}}^{\mathrm{m}}} + \dfrac{1}{k_{\mathrm{T}}^{\mathrm{eff}}}}, \tag{7.20}$$

$$\Lambda_3 = \sum_{i=1}^{N} V_f^i \frac{(\alpha_T^i + \nu_A^i \alpha_A^i) - (\alpha_T^m + \nu_A^m \alpha_A^m)}{\dfrac{1}{\mu_T^m} + \dfrac{1}{k_T^i}} = \frac{(\alpha_T^{eff} + \nu_A^{eff} \alpha_A^{eff}) - (\alpha_T^m + \nu_A^m \alpha_A^m)}{\dfrac{1}{\mu_T^m} + \dfrac{1}{k_T^{eff}}}. \tag{7.21}$$

Application to a two-phase system

It follows from (7.16) to (7.21) that

$$\frac{1}{k_T^{eff}} \equiv \frac{2(1-\nu_T^{eff})}{E_T^{eff}} - \frac{4(\nu_A^{eff})^2}{E_A^{eff}} = \frac{\dfrac{V_m}{k_T^m \mu_T^m} + \dfrac{1}{k_T^f k_T^m} + \dfrac{V_f}{k_T^f \mu_T^m}}{\dfrac{1}{\mu_T^m} + \dfrac{V_m}{k_T^f} + \dfrac{V_f}{k_T^m}}, \tag{7.22}$$

$$\nu_A^{eff} = \nu_A^m + V_f \frac{\nu_A^f - \nu_A^m}{\dfrac{1}{\mu_T^m} + \dfrac{1}{k_T^f}} \left(\frac{1}{\mu_T^m} + \frac{1}{k_T^{eff}} \right), \tag{7.23}$$

$$\alpha_T^{eff} + \nu_A^{eff} \alpha_A^{eff} = \alpha_T^m + \nu_A^m \alpha_A^m + V_f \frac{(\alpha_T^f + \nu_A^f \alpha_A^f) - (\alpha_T^m + \nu_A^m \alpha_A^m)}{\dfrac{1}{\mu_T^m} + \dfrac{1}{k_T^f}} \left(\frac{1}{\mu_T^m} + \frac{1}{k_T^{eff}} \right), \tag{7.24}$$

where $V_f = 1 - V_m = na^2 / b^2$ is the volume fraction of n fibers of radius a embedded in matrix within a cylinder of radius b.

The relations (7.22)–(7.24) are now written as the sum of a mixtures term plus a correction term so that

$$\frac{1}{k_{\mathrm{T}}^{\mathrm{eff}}}=\frac{V_{\mathrm{f}}}{k_{\mathrm{T}}^{\mathrm{f}}}+\frac{V_{\mathrm{m}}}{k_{\mathrm{T}}^{\mathrm{m}}}-\frac{\left(\dfrac{1}{k_{\mathrm{T}}^{\mathrm{f}}}-\dfrac{1}{k_{\mathrm{T}}^{\mathrm{m}}}\right)^2}{\dfrac{V_{\mathrm{f}}}{k_{\mathrm{T}}^{\mathrm{m}}}+\dfrac{V_{\mathrm{m}}}{k_{\mathrm{T}}^{\mathrm{f}}}+\dfrac{1}{\mu_{\mathrm{T}}^{\mathrm{m}}}}V_{\mathrm{f}}V_{\mathrm{m}}, \tag{7.25}$$

$$\nu_{\mathrm{A}}^{\mathrm{eff}}=V_{\mathrm{f}}\nu_{\mathrm{A}}^{\mathrm{f}}+V_{\mathrm{m}}\nu_{\mathrm{A}}^{\mathrm{m}}-\frac{(\nu_{\mathrm{A}}^{\mathrm{f}}-\nu_{\mathrm{A}}^{\mathrm{m}})\left(\dfrac{1}{k_{\mathrm{T}}^{\mathrm{f}}}-\dfrac{1}{k_{\mathrm{T}}^{\mathrm{m}}}\right)}{\dfrac{V_{\mathrm{f}}}{k_{\mathrm{T}}^{\mathrm{m}}}+\dfrac{V_{\mathrm{m}}}{k_{\mathrm{T}}^{\mathrm{f}}}+\dfrac{1}{\mu_{\mathrm{T}}^{\mathrm{m}}}}V_{\mathrm{f}}V_{\mathrm{m}}, \tag{7.26}$$

$$\begin{aligned}\alpha_{\mathrm{T}}^{\mathrm{eff}}+\nu_{\mathrm{A}}^{\mathrm{eff}}\alpha_{\mathrm{A}}^{\mathrm{eff}}&=V_{\mathrm{f}}(\alpha_{\mathrm{T}}^{\mathrm{f}}+\nu_{\mathrm{A}}^{\mathrm{f}}\alpha_{\mathrm{A}}^{\mathrm{f}})+V_{\mathrm{m}}(\alpha_{\mathrm{T}}^{\mathrm{m}}+\nu_{\mathrm{A}}^{\mathrm{m}}\alpha_{\mathrm{A}}^{\mathrm{m}})\\&\quad-\frac{\dfrac{1}{k_{\mathrm{T}}^{\mathrm{f}}}-\dfrac{1}{k_{\mathrm{T}}^{\mathrm{m}}}}{\dfrac{V_{\mathrm{f}}}{k_{\mathrm{T}}^{\mathrm{m}}}+\dfrac{V_{\mathrm{m}}}{k_{\mathrm{T}}^{\mathrm{f}}}+\dfrac{1}{\mu_{\mathrm{T}}^{\mathrm{m}}}}[(\alpha_{\mathrm{T}}^{\mathrm{f}}+\nu_{\mathrm{A}}^{\mathrm{f}}\alpha_{\mathrm{A}}^{\mathrm{f}})-(\alpha_{\mathrm{T}}^{\mathrm{m}}+\nu_{\mathrm{A}}^{\mathrm{m}}\alpha_{\mathrm{A}}^{\mathrm{m}})]V_{\mathrm{f}}V_{\mathrm{m}}.\end{aligned} \tag{7.27}$$

These results correspond to one of the bounds derived using variational methods [1, 2, 10] which are identical to those obtained using the concentric cylinder model for a unidirectional, fiber-reinforced composite.

7.3.2 Axial Shear

It has been shown [8] that the effective transverse shear modulus for the multiphase fiber-reinforced composite is given by

$$\mu_{\mathrm{A}}^{\mathrm{eff}}=\mu_{\mathrm{A}}^{\mathrm{m}}\frac{1-\Lambda}{1+\Lambda}, \tag{7.28}$$

where

$$\Lambda\equiv\sum_{i=1}^{N}V_{\mathrm{f}}^{i}\frac{\mu_{\mathrm{A}}^{\mathrm{m}}-\mu_{\mathrm{A}}^{i}}{\mu_{\mathrm{A}}^{i}+\mu_{\mathrm{A}}^{\mathrm{m}}}=\frac{\mu_{\mathrm{A}}^{\mathrm{m}}-\mu_{\mathrm{A}}^{\mathrm{eff}}}{\mu_{\mathrm{A}}^{\mathrm{eff}}+\mu_{\mathrm{A}}^{\mathrm{m}}}. \tag{7.29}$$

Application to a two-phase system

When just one type of fiber is present, it follows from (7.28) and (7.29) that

$$\mu_{\mathrm{A}}^{\mathrm{eff}} = \mu_{\mathrm{A}}^{\mathrm{m}} \frac{(1+V_{\mathrm{f}})\mu_{\mathrm{A}}^{\mathrm{f}} + V_{\mathrm{m}}\mu_{\mathrm{A}}^{\mathrm{m}}}{V_{\mathrm{m}}\mu_{\mathrm{A}}^{\mathrm{f}} + (1+V_{\mathrm{f}})\mu_{\mathrm{A}}^{\mathrm{m}}}. \tag{7.30}$$

The result (7.30) is now written in the form of the mixtures estimate plus a correction term as follows:

$$\mu_{\mathrm{A}}^{\mathrm{eff}} = V_{\mathrm{f}}\mu_{\mathrm{A}}^{\mathrm{f}} + V_{\mathrm{m}}\mu_{\mathrm{A}}^{\mathrm{m}} - \frac{(\mu_{\mathrm{A}}^{\mathrm{f}} - \mu_{\mathrm{A}}^{\mathrm{m}})^2 V_{\mathrm{f}} V_{\mathrm{m}}}{V_{\mathrm{f}}\mu_{\mathrm{A}}^{\mathrm{m}} + V_{\mathrm{m}}\mu_{\mathrm{A}}^{\mathrm{f}} + \mu_{\mathrm{A}}^{\mathrm{m}}}. \tag{7.31}$$

This result corresponds to one of the bounds derived using variational methods [1, 2, 10] and is identical to the result obtained when using the concentric cylinder model for a unidirectional, fiber-reinforced composite.

7.3.3 Transverse Shear

It has been shown [8] that the effective transverse shear modulus for the multiphase composite is given by

$$\frac{1}{\mu_{\mathrm{T}}^{\mathrm{eff}}} = \frac{1}{1+\Lambda}\left[\frac{1}{\mu_{\mathrm{T}}^{\mathrm{m}}} - \Lambda\left(\frac{1}{\mu_{\mathrm{T}}^{\mathrm{m}}} + \frac{2}{k_{\mathrm{T}}^{\mathrm{m}}}\right)\right], \tag{7.32}$$

where

$$\Lambda \equiv \sum_{i=1}^{N} V_{\mathrm{f}}^{i} \frac{\dfrac{1}{\mu_{\mathrm{T}}^{\mathrm{m}}} - \dfrac{1}{\mu_{\mathrm{T}}^{i}}}{\dfrac{1}{\mu_{\mathrm{T}}^{\mathrm{m}}} + \dfrac{2}{k_{\mathrm{T}}^{\mathrm{m}}} + \dfrac{1}{\mu_{\mathrm{T}}^{i}}} = \frac{\dfrac{1}{\mu_{\mathrm{T}}^{\mathrm{m}}} - \dfrac{1}{\mu_{\mathrm{T}}^{\mathrm{eff}}}}{\dfrac{1}{\mu_{\mathrm{T}}^{\mathrm{m}}} + \dfrac{2}{k_{\mathrm{T}}^{\mathrm{m}}} + \dfrac{1}{\mu_{\mathrm{T}}^{\mathrm{eff}}}}. \tag{7.33}$$

Application to a two-phase system

When just one type of fiber is present, it follows from (7.32) and (7.33) that

$$\mu_{\mathrm{T}}^{\mathrm{eff}} = \mu_{\mathrm{T}}^{\mathrm{m}} \frac{2\mu_{\mathrm{T}}^{\mathrm{f}}\mu_{\mathrm{T}}^{\mathrm{m}} + V_{\mathrm{m}}k_{\mathrm{T}}^{\mathrm{m}}\mu_{\mathrm{T}}^{\mathrm{m}} + (1+V_{\mathrm{f}})\mu_{\mathrm{T}}^{\mathrm{f}}k_{\mathrm{T}}^{\mathrm{m}}}{V_{\mathrm{m}}(k_{\mathrm{T}}^{\mathrm{m}} + 2\mu_{\mathrm{T}}^{\mathrm{m}})\mu_{\mathrm{T}}^{\mathrm{f}} + (1+V_{\mathrm{f}})k_{\mathrm{T}}^{\mathrm{m}}\mu_{\mathrm{T}}^{\mathrm{m}} + 2V_{\mathrm{f}}(\mu_{\mathrm{T}}^{\mathrm{m}})^2}. \tag{7.34}$$

The result (7.34) is now written as a mixtures estimate plus a correction term so that

$$\mu_{\mathrm{T}}^{\mathrm{eff}} = V_{\mathrm{f}}\mu_{\mathrm{T}}^{\mathrm{f}} + V_{\mathrm{m}}\mu_{\mathrm{T}}^{\mathrm{m}} - \frac{(\mu_{\mathrm{T}}^{\mathrm{f}} - \mu_{\mathrm{T}}^{\mathrm{m}})^2 V_{\mathrm{f}} V_{\mathrm{m}}}{V_{\mathrm{f}}\mu_{\mathrm{T}}^{\mathrm{m}} + V_{\mathrm{m}}\mu_{\mathrm{T}}^{\mathrm{f}} + \dfrac{k_{\mathrm{T}}^{\mathrm{m}}\mu_{\mathrm{T}}^{\mathrm{m}}}{k_{\mathrm{T}}^{\mathrm{m}} + 2\mu_{\mathrm{T}}^{\mathrm{m}}}}. \tag{7.35}$$

This result corresponds to one of the bounds derived using variational methods [1, 2, 10].

As results derived using Maxwell's methodology can correspond exactly to bounds obtained using variational methods, and to results obtained using the concentric cylinders model, it is clear that Maxwell's approach is not necessarily restricted to low fibre volume fractions where fibre interactions are negligible.

To date, it has not been possible to find a method of using Maxwell's methodology to estimate the axial modulus and axial thermal expansion coefficient for a fiber-reinforced composite. However, the concentric cylinder model for a composite [6] leads to the following relations, which are expressed in the form of a mixtures estimate plus a correction term,

$$E_{\mathrm{A}}^{\mathrm{eff}} = V_{\mathrm{f}} E_{\mathrm{A}}^{\mathrm{f}} + V_{\mathrm{m}} E_{\mathrm{A}}^{\mathrm{m}} + \frac{4(\nu_{\mathrm{A}}^{\mathrm{f}} - \nu_{\mathrm{A}}^{\mathrm{m}})^2}{\dfrac{V_{\mathrm{f}}}{k_{\mathrm{T}}^{\mathrm{m}}} + \dfrac{V_{\mathrm{m}}}{k_{\mathrm{T}}^{\mathrm{f}}} + \dfrac{1}{\mu_{\mathrm{T}}^{\mathrm{m}}}} V_{\mathrm{f}} V_{\mathrm{m}}, \tag{7.36}$$

$$\begin{aligned} E_{\mathrm{A}}^{\mathrm{eff}}\alpha_{\mathrm{A}}^{\mathrm{eff}} &= V_{\mathrm{f}} E_{\mathrm{A}}^{\mathrm{f}}\alpha_{\mathrm{A}}^{\mathrm{f}} + V_{\mathrm{m}} E_{\mathrm{A}}^{\mathrm{m}}\alpha_{\mathrm{A}}^{\mathrm{m}} \\ &+ \frac{4(\nu_{\mathrm{A}}^{\mathrm{f}} - \nu_{\mathrm{A}}^{\mathrm{m}})}{\dfrac{V_{\mathrm{f}}}{k_{\mathrm{T}}^{\mathrm{m}}} + \dfrac{V_{\mathrm{m}}}{k_{\mathrm{T}}^{\mathrm{f}}} + \dfrac{1}{\mu_{\mathrm{T}}^{\mathrm{m}}}} [\alpha_{\mathrm{T}}^{\mathrm{f}} + \nu_{\mathrm{A}}^{\mathrm{f}}\alpha_{\mathrm{A}}^{\mathrm{f}} - \alpha_{\mathrm{T}}^{\mathrm{m}} - \nu_{\mathrm{A}}^{\mathrm{m}}\alpha_{\mathrm{A}}^{\mathrm{m}}] V_{\mathrm{f}} V_{\mathrm{m}}. \end{aligned} \tag{7.37}$$

The statement is now complete of all the analytical formulae that can be used to predict and completely characterize the thermoelastic properties of an undamaged unidirectional, fiber-reinforced composite. It only remains to state, in Sect. 7.4, formulae for the extreme values that arise from variational calculations [1, 2, 10] and to provide conditions that determine whether the extreme values are upper or lower bounds.

7.4 Bounds for Composite Properties

As already mentioned above, variational methods have been used [1, 2, 10] to estimate upper and lower bounds for the effective properties of both isotropic particulate and anisotropic fiber-reinforced composites. In many cases it is well known that the expressions for the bounds are such that interchanging fiber and matrix parameters merely interchanges the upper and lower bounds. One of the bounds corresponds to estimates that can be derived from spherical shell or concentric cylinder models, as described in [1]. An examination of the literature regarding the conditions that determine the type of bound (either upper or lower) has revealed that there is a good deal of ambiguity. The objective of this section is to state the bounds for a particular property and to identify unambiguously the conditions that determine whether the extreme value of the property is an upper or lower bound. It should be noted that the required bounds are most easily derived by considering formulae for properties that have (as in Sect. 7.3) been expressed as a mixtures estimate plus a correction term. It is easily seen that interchanging fiber and matrix parameters affects only one term in the denominator of the correction term. Many of the following conditions, determining the upper and lower bounds to be given, do not seem to appear in the literature. It should be noted that the correction terms all have a common structure involving proportionality to the product, $V_\mathrm{p}V_\mathrm{m}$ or $V_\mathrm{f}V_\mathrm{m}$, and to the product of two reinforcement and matrix property differences, and that interchanging the reinforcement and matrix properties does not change the value of these products. It is emphasized that although the expressions given for the bounds differ in form to those that are quoted in [1, 2, 10], their values are in fact identical to the corresponding published expressions.

7.4.1 Bounds for Properties of Particulate Composites

Bulk modulus

Bounds for the bulk modulus of a particulate composite may be written in the following form

$$\overline{\frac{1}{k}}-\frac{\left(\dfrac{1}{k_\mathrm{p}}-\dfrac{1}{k_\mathrm{m}}\right)^2 V_\mathrm{p}V_\mathrm{m}}{\dfrac{V_\mathrm{m}}{k_\mathrm{p}}+\dfrac{V_\mathrm{p}}{k_\mathrm{m}}+\dfrac{3}{4\mu_\mathrm{m}}}\le\frac{1}{k_\mathrm{eff}}\le\overline{\frac{1}{k}}-\frac{\left(\dfrac{1}{k_\mathrm{p}}-\dfrac{1}{k_\mathrm{m}}\right)^2 V_\mathrm{p}V_\mathrm{m}}{\dfrac{V_\mathrm{m}}{k_\mathrm{p}}+\dfrac{V_\mathrm{p}}{k_\mathrm{m}}+\dfrac{3}{4\mu_\mathrm{p}}},\tag{7.38}$$

where

$$\overline{\frac{1}{k}}=\frac{V_\mathrm{p}}{k_\mathrm{p}}+\frac{V_\mathrm{m}}{k_\mathrm{m}}.\tag{7.39}$$

These bounds are valid only if

$$\mu_{\mathrm{p}} \leq \mu_{\mathrm{m}}, \tag{7.40}$$

and the bounds are reversed if $\mu_{\mathrm{p}} \geq \mu_{\mathrm{m}}$.

Shear modulus

Bounds for the shear modulus of a particulate composite may be written in the following form

$$\begin{aligned}
&\bar{\mu} - \frac{(\mu_{\mathrm{p}} - \mu_{\mathrm{m}})^2 V_{\mathrm{p}} V_{\mathrm{m}}}{V_{\mathrm{p}}\mu_{\mathrm{m}} + V_{\mathrm{m}}\mu_{\mathrm{p}} + \dfrac{(9k_{\mathrm{m}} + 8\mu_{\mathrm{m}})\mu_{\mathrm{m}}}{6(k_{\mathrm{m}} + 2\mu_{\mathrm{m}})}} \leq \mu_{\mathrm{eff}} \\
&\leq \bar{\mu} - \frac{(\mu_{\mathrm{p}} - \mu_{\mathrm{m}})^2 V_{\mathrm{p}} V_{\mathrm{m}}}{V_{\mathrm{p}}\mu_{\mathrm{m}} + V_{\mathrm{m}}\mu_{\mathrm{p}} + \dfrac{(9k_{\mathrm{p}} + 8\mu_{\mathrm{p}})\mu_{\mathrm{p}}}{6(k_{\mathrm{p}} + 2\mu_{\mathrm{p}})}},
\end{aligned} \tag{7.41}$$

where

$$\bar{\mu} = V_{\mathrm{p}}\mu_{\mathrm{p}} + V_{\mathrm{m}}\mu_{\mathrm{m}}. \tag{7.42}$$

These bounds are valid for the practically important case

$$k_{\mathrm{p}} \geq k_{\mathrm{m}}, \quad \mu_{\mathrm{p}} \geq \mu_{\mathrm{m}}, \tag{7.43}$$

and the bounds are reversed if

$$k_{\mathrm{p}} \leq k_{\mathrm{m}}, \quad \mu_{\mathrm{p}} \leq \mu_{\mathrm{m}}.$$

Other cases can arise that are not considered here.

Thermal expansion

Bounds for the thermal expansion coefficient of a particulate composite may be written in the following form

$$\bar{\alpha} + \frac{(k_p - k_m)(\alpha_p - \alpha_m)V_p V_m}{k_p V_p + k_m V_m + \dfrac{3k_m k_p}{4\mu_m}} \le \alpha_{eff} \le \bar{\alpha} + \frac{(k_p - k_m)(\alpha_p - \alpha_m)V_p V_m}{k_p V_p + k_m V_m + \dfrac{3k_m k_p}{4\mu_p}}, \tag{7.44}$$

where

$$\bar{\alpha} = V_p \alpha_p + V_m \alpha_m. \tag{7.45}$$

It should be noted that Rosen and Hashin [10, Eq. (2.27)] do in fact express their result in the form of a mixtures term plus a correction term. It follows that the bounds apply only if the following condition is satisfied

$$(k_p - k_m)(\mu_p - \mu_m)(\alpha_p - \alpha_m) \ge 0, \tag{7.46}$$

and the bounds are reversed if $(k_p - k_m)(\mu_p - \mu_m)(\alpha_p - \alpha_m) \le 0$.

7.4.2 Bounds for Properties of Fiber-Reinforced Composites

Axial modulus

The bounds are given by

$$\overline{E_A} + \frac{4(\nu_A^f - \nu_A^m)^2 V_f V_m}{\dfrac{V_m}{k_T^f} + \dfrac{V_f}{k_T^m} + \dfrac{1}{\mu_T^m}} \le E_A^{eff} \le \overline{E_A} + \frac{4(\nu_A^f - \nu_A^m)^2 V_f V_m}{\dfrac{V_m}{k_T^f} + \dfrac{V_f}{k_T^m} + \dfrac{1}{\mu_T^f}}, \tag{7.47}$$

where

$$\overline{E_A} = V_f E_A^f + V_m E_A^m. \tag{7.48}$$

These bounds are valid only if

$$\mu_T^f \ge \mu_T^m, \tag{7.49}$$

and the bounds are reversed if $\mu_T^f \le \mu_T^m$.

Poisson's ratio

The bounds are given by

$$\overline{\nu_A} - \frac{(\nu_A^f - \nu_A^m)\left(\dfrac{1}{k_T^f} - \dfrac{1}{k_T^m}\right)V_f V_m}{\dfrac{V_m}{k_T^f} + \dfrac{V_f}{k_T^m} + \dfrac{1}{\mu_T^m}} \le \nu_A^{eff} \le \overline{\nu_A} - \frac{(\nu_A^f - \nu_A^m)\left(\dfrac{1}{k_T^f} - \dfrac{1}{k_T^m}\right)V_f V_m}{\dfrac{V_m}{k_T^f} + \dfrac{V_f}{k_T^m} + \dfrac{1}{\mu_T^f}}, \tag{7.50}$$

where

$$\overline{\nu_A} = V_f \nu_A^f + V_m \nu_A^m. \tag{7.51}$$

These bounds are valid only if

$$(\nu_A^f - \nu_A^m)(k_T^f - k_T^m)(\mu_T^f - \mu_T^m) \ge 0, \tag{7.52}$$

and the bounds are reversed if $(\nu_A^f - \nu_A^m)(k_T^f - k_T^m)(\mu_T^f - \mu_T^m) \le 0$.

Transverse bulk modulus

The bounds are given by

$$\overline{\frac{1}{k_T}} - \frac{\left(\dfrac{1}{k_T^f} - \dfrac{1}{k_T^m}\right)^2 V_f V_m}{\dfrac{V_m}{k_T^f} + \dfrac{V_f}{k_T^m} + \dfrac{1}{\mu_T^m}} \le \frac{1}{k_T^{eff}} \le \overline{\frac{1}{k_T}} - \frac{\left(\dfrac{1}{k_T^f} - \dfrac{1}{k_T^m}\right)^2 V_f V_m}{\dfrac{V_m}{k_T^f} + \dfrac{V_f}{k_T^m} + \dfrac{1}{\mu_T^f}}, \tag{7.53}$$

where

$$\overline{\frac{1}{k_T}} = \frac{V_f}{k_T^f} + \frac{V_m}{k_T^m}. \tag{7.54}$$

These bounds are valid only if

$$\mu_T^f \le \mu_T^m, \tag{7.55}$$

and the bounds are reversed if $\mu_T^f \ge \mu_T^m$.

Transverse shear modulus

The bounds are given by

$$\overline{\mu_{\mathrm{T}}} - \frac{(\mu_{\mathrm{T}}^{\mathrm{f}} - \mu_{\mathrm{T}}^{\mathrm{m}})^2 V_{\mathrm{f}} V_{\mathrm{m}}}{V_{\mathrm{m}}\mu_{\mathrm{T}}^{\mathrm{f}} + V_{\mathrm{f}}\mu_{\mathrm{T}}^{\mathrm{m}} + \dfrac{k_{\mathrm{T}}^{\mathrm{m}}\mu_{\mathrm{T}}^{\mathrm{m}}}{k_{\mathrm{T}}^{\mathrm{m}} + 2\mu_{\mathrm{T}}^{\mathrm{m}}}} \le \mu_{\mathrm{T}}^{\mathrm{eff}} \le \overline{\mu_{\mathrm{T}}} - \frac{(\mu_{\mathrm{T}}^{\mathrm{f}} - \mu_{\mathrm{T}}^{\mathrm{m}})^2 V_{\mathrm{f}} V_{\mathrm{m}}}{V_{\mathrm{m}}\mu_{\mathrm{T}}^{\mathrm{f}} + V_{\mathrm{f}}\mu_{\mathrm{T}}^{\mathrm{m}} + \dfrac{k_{\mathrm{T}}^{\mathrm{f}}\mu_{\mathrm{T}}^{\mathrm{f}}}{k_{\mathrm{T}}^{\mathrm{f}} + 2\mu_{\mathrm{T}}^{\mathrm{f}}}}, \tag{7.56}$$

where

$$\overline{\mu_{\mathrm{T}}} = V_{\mathrm{f}}\mu_{\mathrm{T}}^{\mathrm{f}} + V_{\mathrm{m}}\mu_{\mathrm{T}}^{\mathrm{m}}. \tag{7.57}$$

These bounds are valid for the practically important case

$$k_{\mathrm{T}}^{\mathrm{f}} \ge k_{\mathrm{T}}^{\mathrm{m}}, \quad \mu_{\mathrm{T}}^{\mathrm{f}} \ge \mu_{\mathrm{T}}^{\mathrm{m}}, \tag{7.58}$$

and the bounds are reversed if

$$k_{\mathrm{T}}^{\mathrm{f}} \le k_{\mathrm{T}}^{\mathrm{m}}, \quad \mu_{\mathrm{T}}^{\mathrm{f}} \le \mu_{\mathrm{T}}^{\mathrm{m}}.$$

Other cases can arise that are not considered here.

Axial shear modulus

The bounds are given by

$$\overline{\mu_{\mathrm{A}}} - \frac{(\mu_{\mathrm{A}}^{\mathrm{f}} - \mu_{\mathrm{A}}^{\mathrm{m}})^2 V_{\mathrm{f}} V_{\mathrm{m}}}{V_{\mathrm{m}}\mu_{\mathrm{A}}^{\mathrm{f}} + V_{\mathrm{f}}\mu_{\mathrm{A}}^{\mathrm{m}} + \mu_{\mathrm{A}}^{\mathrm{m}}} \le \mu_{\mathrm{A}}^{\mathrm{eff}} \le \overline{\mu_{\mathrm{A}}} - \frac{(\mu_{\mathrm{A}}^{\mathrm{f}} - \mu_{\mathrm{A}}^{\mathrm{m}})^2 V_{\mathrm{f}} V_{\mathrm{m}}}{V_{\mathrm{m}}\mu_{\mathrm{A}}^{\mathrm{f}} + V_{\mathrm{f}}\mu_{\mathrm{A}}^{\mathrm{m}} + \mu_{\mathrm{A}}^{\mathrm{f}}}, \tag{7.59}$$

where

$$\overline{\mu_{\mathrm{A}}} = V_{\mathrm{f}}\mu_{\mathrm{A}}^{\mathrm{f}} + V_{\mathrm{m}}\mu_{\mathrm{A}}^{\mathrm{m}}. \tag{7.60}$$

These bounds are valid only if

$$\mu_{\mathrm{A}}^{\mathrm{m}} \le \mu_{\mathrm{A}}^{\mathrm{f}}, \tag{7.61}$$

and the bounds are reversed if $\mu_{\mathrm{A}}^{\mathrm{m}} \ge \mu_{\mathrm{A}}^{\mathrm{f}}$.

Axial thermal expansion

The bounds are given by

$$\overline{E_{\mathrm{A}}\alpha_{\mathrm{A}}}+\frac{4(\nu_{\mathrm{A}}^{\mathrm{f}}-\nu_{\mathrm{A}}^{\mathrm{m}})(\hat{\alpha}_{\mathrm{T}}^{\mathrm{f}}-\hat{\alpha}_{\mathrm{T}}^{\mathrm{m}})V_{\mathrm{f}}V_{\mathrm{m}}}{\dfrac{V_{\mathrm{m}}}{k_{\mathrm{T}}^{\mathrm{f}}}+\dfrac{V_{\mathrm{f}}}{k_{\mathrm{T}}^{\mathrm{m}}}+\dfrac{1}{\mu_{\mathrm{T}}^{\mathrm{m}}}}\le E_{\mathrm{A}}^{\mathrm{eff}}\alpha_{\mathrm{A}}^{\mathrm{eff}}$$
$$\le\overline{E_{\mathrm{A}}\alpha_{\mathrm{A}}}+\frac{4(\nu_{\mathrm{A}}^{\mathrm{f}}-\nu_{\mathrm{A}}^{\mathrm{m}})(\hat{\alpha}_{\mathrm{T}}^{\mathrm{f}}-\hat{\alpha}_{\mathrm{T}}^{\mathrm{m}})V_{\mathrm{f}}V_{\mathrm{m}}}{\dfrac{V_{\mathrm{m}}}{k_{\mathrm{T}}^{\mathrm{f}}}+\dfrac{V_{\mathrm{f}}}{k_{\mathrm{T}}^{\mathrm{m}}}+\dfrac{1}{\mu_{\mathrm{T}}^{\mathrm{f}}}}, \tag{7.62}$$

where

$$\overline{E_{\mathrm{A}}\alpha_{\mathrm{A}}}=V_{\mathrm{f}}E_{\mathrm{A}}^{\mathrm{f}}\alpha_{\mathrm{A}}^{\mathrm{f}}+V_{\mathrm{m}}E_{\mathrm{A}}^{\mathrm{m}}\alpha_{\mathrm{A}}^{\mathrm{m}},\quad \hat{\alpha}_{\mathrm{T}}=\alpha_{\mathrm{T}}+\nu_{\mathrm{A}}\alpha_{\mathrm{A}}. \tag{7.63}$$

These bounds are valid only if

$$(\nu_{\mathrm{A}}^{\mathrm{f}}-\nu_{\mathrm{A}}^{\mathrm{m}})(\hat{\alpha}_{\mathrm{T}}^{\mathrm{f}}-\hat{\alpha}_{\mathrm{T}}^{\mathrm{m}})(\mu_{\mathrm{T}}^{\mathrm{f}}-\mu_{\mathrm{T}}^{\mathrm{m}})\ge 0, \tag{7.64}$$

and the bounds are reversed if $(\nu_{\mathrm{A}}^{\mathrm{f}}-\nu_{\mathrm{A}}^{\mathrm{m}})(\hat{\alpha}_{\mathrm{T}}^{\mathrm{f}}-\hat{\alpha}_{\mathrm{T}}^{\mathrm{m}})(\mu_{\mathrm{T}}^{\mathrm{f}}-\mu_{\mathrm{T}}^{\mathrm{m}})\le 0$.

Transverse thermal expansion

The bounds are given by

$$\overline{\hat{\alpha}_{\mathrm{T}}}-\frac{\left(\dfrac{1}{k_{\mathrm{T}}^{\mathrm{f}}}-\dfrac{1}{k_{\mathrm{T}}^{\mathrm{m}}}\right)(\hat{\alpha}_{\mathrm{T}}^{\mathrm{f}}-\hat{\alpha}_{\mathrm{T}}^{\mathrm{m}})V_{\mathrm{f}}V_{\mathrm{m}}}{\dfrac{V_{\mathrm{m}}}{k_{\mathrm{T}}^{\mathrm{f}}}+\dfrac{V_{\mathrm{f}}}{k_{\mathrm{T}}^{\mathrm{m}}}+\dfrac{1}{\mu_{\mathrm{T}}^{\mathrm{m}}}}\le\hat{\alpha}_{\mathrm{T}}^{\mathrm{eff}}$$
$$\le\overline{\hat{\alpha}_{\mathrm{T}}}-\frac{\left(\dfrac{1}{k_{\mathrm{T}}^{\mathrm{f}}}-\dfrac{1}{k_{\mathrm{T}}^{\mathrm{m}}}\right)(\hat{\alpha}_{\mathrm{T}}^{\mathrm{f}}-\hat{\alpha}_{\mathrm{T}}^{\mathrm{m}})V_{\mathrm{f}}V_{\mathrm{m}}}{\dfrac{V_{\mathrm{m}}}{k_{\mathrm{T}}^{\mathrm{f}}}+\dfrac{V_{\mathrm{f}}}{k_{\mathrm{T}}^{\mathrm{m}}}+\dfrac{1}{\mu_{\mathrm{T}}^{\mathrm{f}}}}, \tag{7.65}$$

where

$$\overline{\hat{\alpha}_{\mathrm{T}}}=V_{\mathrm{f}}\hat{\alpha}_{\mathrm{T}}^{\mathrm{f}}+V_{\mathrm{m}}\hat{\alpha}_{\mathrm{T}}^{\mathrm{m}}. \tag{7.66}$$

These bounds are valid only if

$$(k_{\mathrm{T}}^{\mathrm{f}}-k_{\mathrm{T}}^{\mathrm{m}})(\hat{\alpha}_{\mathrm{T}}^{\mathrm{f}}-\hat{\alpha}_{\mathrm{T}}^{\mathrm{m}})(\mu_{\mathrm{T}}^{\mathrm{f}}-\mu_{\mathrm{T}}^{\mathrm{m}})\leq 0, \tag{7.67}$$

and the bounds are reversed if $(k_{\mathrm{T}}^{\mathrm{f}}-k_{\mathrm{T}}^{\mathrm{m}})(\hat{\alpha}_{\mathrm{T}}^{\mathrm{f}}-\hat{\alpha}_{\mathrm{T}}^{\mathrm{m}})(\mu_{\mathrm{T}}^{\mathrm{f}}-\mu_{\mathrm{T}}^{\mathrm{m}})\geq 0$.

7.5 General Framework for Prediction of Ply Cracking in Laminates

Having considered, in Sects. 7.2–7.4, methods of estimating the effective properties of both particulate and fiber-reinforced composites, and showing how the expression of the results in a special form enables the determination of the conditions for the extreme values being upper and lower bounds, it is now appropriate to consider laminated composites where results already derived can be used to estimate the properties of undamaged laminates. This section is concerned with the development of a generalized theoretical framework that enables the prediction of microstructural damage formation in any symmetric crossply laminate subject to in-plane loading involving both biaxial and through-thickness loading modes.

In McCartney [4–6], the details are given for a high-quality stress-transfer model that has been shown to be capable of deriving accurate stress and displacement distributions at all points within any symmetrical multiple-ply laminate containing an equally spaced array of cracks in some or all of the 90° plies. The analysis accounts for the presence of residual stresses arising from thermal expansion mismatch effects. On formation, cracks are assumed (when applying stress-transfer models) to be fully developed so that each ply crack intersects with two edges of the laminate. It is assumed that the crack faces are always stress-free. This latter assumption clearly restricts the validity of the analysis to certain loading states for which cracks are either open or just closed such that no compressive or shear stresses (possibly arising from friction) are transmitted across the crack surfaces.

Having developed a stress-transfer model that can predict both stress and displacement distributions at all points in a cracked symmetric crossply laminate, together with the effective thermoelastic constants, it is necessary to be able to use such results in a way that the conditions for ply crack initiation and progressive formation can be determined, as described in [7]. The theoretical framework to be developed in this chapter will, however, put aside the important aspect of stress-transfer modeling and address the problem of predicting ply crack formation using other extremely useful principles. It is found that a great deal of progress can be made without having to resort to the use of a stress-transfer model. In addition, the approach is likely to be valid for more general damage distributions that cannot, as yet, be analyzed using a stress-transfer model. The analysis proposed here is based upon an assumption of damage homogeneity at the macroscopic laminate level. This means that bending modes arising from nonuniform damage formation are not taken into account.

The important results that are derived progressively during the development of the general framework have been checked by making use of the detailed solutions that can be obtained from the application of the stress-transfer model [4, 5]. This reinforces the validity of the proposed framework and confirms that the stress-transfer model is indeed one of high quality. The results presented here, while developed for crossply laminates subject to general in-plane loading, can be extended [5, 6] to general symmetric laminates. In such cases, in-plane biaxial loading and in-plane shear loading effects are not then separable.

7.5.1 Definition of Effective Stresses and Strains for Damaged Laminates

Consider a multiple-ply crossply laminate subject to multiaxial loading without shear. During deformation, cracks parallel to the fibers in both 0° and 90° plies in the laminate may form progressively at locations that will lead to nonuniform crack distributions because of the statistical variability of fiber distribution and of the fiber, matrix, and interface properties. The laminate is assumed to be uncracked at the commencement of loading. The state of deformation in the laminate at any stage of loading and cracking is assumed to be governed by the field equations of linear elasticity theory. When the cracks that form are open during loading, the tractions on crack surfaces are zero. The outer faces of the laminate are assumed to be subject to the same uniform applied stress denoted by σ_t.

Boundary conditions need to be imposed on the external edges of the laminate. It is assumed that cracks in plies never form on the outer edges of the laminate and that in-plane loading is applied to the laminate by imposing uniform axial and transverse displacements on the outer edges. The laminate is assumed to have length $2L$ in the y-direction and width $2W$ in the z-direction. The normal displacement components in the y- and z-directions on the edges $y = \pm L$ and $z = \pm W$ are uniform and are denoted by $\pm v_0$ and $\pm w_0$, respectively, defining, as in [4–7], effective axial and transverse applied in-plane strains as follows:

$$\varepsilon = \frac{v_0}{L}, \quad \varepsilon_{\mathrm{T}} = \frac{w_0}{W}. \tag{7.68}$$

Because uniform displacements have been imposed on the edges of the laminate, the stresses σ_{yy} and σ_{zz} resulting on the edges are nonuniform while the shear stress σ_{yz} is zero. Effective applied axial and transverse stresses are defined, as in [4–7], by

$$\begin{aligned} \sigma &= \frac{1}{4hW}\int_{-W}^{W}\int_{-h}^{h}\sigma_{yy}(x,L,z)\mathrm{d}x\mathrm{d}z, \\ \sigma_{\mathrm{T}} &= \frac{1}{4hL}\int_{-L}^{L}\int_{-h}^{h}\sigma_{zz}(x,y,W)\mathrm{d}x\mathrm{d}y, \end{aligned} \tag{7.69}$$

where

$$h = \sum_{i=1}^{N+1} h_i, \tag{7.70}$$

so that $2h$ is the total thickness of the laminate. The effective out-of-plane transverse strain is defined by

$$\varepsilon_{\mathrm{t}} = \frac{1}{4hLW}\int_{-L}^{L}\int_{-W}^{W} u(h,y,z)\mathrm{d}y\mathrm{d}z. \tag{7.71}$$

7.5.2 Stress–Strain Relations for Damaged Laminates

The following theoretical developments are intended to apply to any symmetric crossply laminate that may be subject to microcracking involving the formation of fully developed cracks that traverse the entire width or length of some or all of the individual 90° plies in the laminate. The approach to be followed, while it will be developed for nonuniform distributions of fully developed ply cracks in some or all of the 90° plies, is in fact valid also for much more general damage patterns where small ply cracks (not traversing the full width of the laminate) are present. The approach to be described will apply to these damage modes provided that all cracks are free of compressive or frictional shear loading and that the damage formed is effectively homogeneous when viewed at the macroscopic scale. For such cases, the prediction of the effective thermoelastic constants of the damaged laminate will be much more complex than is the case for fully developed ply cracks.

Consider a multiple-ply crossply laminate whose damage is in the form of stress-free ply cracks. It is *assumed* that the distribution of damage is uniform on average such that the nonshear effective stress/strain relations of the damaged laminate may be expressed in the form

$$\varepsilon_{\mathrm{t}} = \frac{\sigma_{\mathrm{t}}}{E_{\mathrm{t}}} - \frac{\nu_{\mathrm{a}}}{E_{\mathrm{A}}}\sigma - \frac{\nu_{\mathrm{t}}}{E_{\mathrm{T}}}\sigma_{\mathrm{T}} + \alpha_{\mathrm{t}}\Delta T, \tag{7.72}$$

$$\varepsilon = -\frac{\nu_{\mathrm{a}}}{E_{\mathrm{A}}}\sigma_{\mathrm{t}} + \frac{\sigma}{E_{\mathrm{A}}} - \frac{\nu_{\mathrm{A}}}{E_{\mathrm{A}}}\sigma_{\mathrm{T}} + \alpha_{\mathrm{A}}\Delta T, \tag{7.73}$$

$$\varepsilon_{\mathrm{T}} = -\frac{\nu_{\mathrm{t}}}{E_{\mathrm{T}}}\sigma_{\mathrm{t}} - \frac{\nu_{\mathrm{A}}}{E_{\mathrm{A}}}\sigma + \frac{\sigma_{\mathrm{T}}}{E_{\mathrm{T}}} + \alpha_{\mathrm{T}}\Delta T, \tag{7.74}$$

where, for specified values of σ_{t}, σ, σ_{T}, and ΔT, the parameters ε_{t}, ε, and ε_{T} are, respectively, the effective out-of-plane, axial, and in-plane transverse strains of the damaged laminate, and where E_{A}, E_{T}, E_{t}, ν_{A}, ν_{a}, ν_{t}, α_{A}, α_{T}, and α_{t} denote the effective thermoelastic constants of the cracked laminate. It is assumed that the values of the thermoelastic constants in (7.72)–(7.74) may be determined by detailed stress analysis for any distribution of ply cracks. In McCartney [4–6], it is shown that the form (7.72)–(7.74) assumed for the stress–strain relations is in fact obtained from the use of stress-transfer models when ply cracks having only a single orientation are fully developed in each cracked ply. It is also shown how the thermoelastic constants may be determined for the special case of generalized plane strain conditions where ply cracking is uniformly distributed in the cracked 90° plies of the laminate. In McCartney and Schoeppner [9], it is shown how to account for nonuniform ply cracking in a laminate, assuming the same crack pattern occurs in each cracked ply.

Corresponding to (7.72)–(7.74), the stress/strain relations for an uncracked laminate subject to the same applied stresses σ_{t}, σ, σ_{T} and temperature difference ΔT are written as

$$\varepsilon_{\mathrm{t}}^{\mathrm{o}} = \frac{\sigma_{\mathrm{t}}}{E_{\mathrm{t}}^{\mathrm{o}}} - \frac{\nu_{\mathrm{a}}^{\mathrm{o}}}{E_{\mathrm{A}}^{\mathrm{o}}}\sigma - \frac{\nu_{\mathrm{t}}^{\mathrm{o}}}{E_{\mathrm{T}}^{\mathrm{o}}}\sigma_{\mathrm{T}} + \alpha_{\mathrm{t}}^{\mathrm{o}}\Delta T, \tag{7.75}$$

$$\varepsilon^{\mathrm{o}} = -\frac{\nu_{\mathrm{a}}^{\mathrm{o}}}{E_{\mathrm{A}}^{\mathrm{o}}}\sigma_{\mathrm{t}} + \frac{\sigma}{E_{\mathrm{A}}^{\mathrm{o}}} - \frac{\nu_{\mathrm{A}}^{\mathrm{o}}}{E_{\mathrm{A}}^{\mathrm{o}}}\sigma_{\mathrm{T}} + \alpha_{\mathrm{A}}^{\mathrm{o}}\Delta T, \tag{7.76}$$

$$\varepsilon_{\mathrm{T}}^{\mathrm{o}} = -\frac{\nu_{\mathrm{t}}^{\mathrm{o}}}{E_{\mathrm{T}}^{\mathrm{o}}}\sigma_{\mathrm{t}} - \frac{\nu_{\mathrm{A}}^{\mathrm{o}}}{E_{\mathrm{A}}^{\mathrm{o}}}\sigma + \frac{\sigma_{\mathrm{T}}}{E_{\mathrm{T}}^{\mathrm{o}}} + \alpha_{\mathrm{T}}^{\mathrm{o}}\Delta T, \tag{7.77}$$

where the parameters $\varepsilon_{\mathrm{t}}^{\mathrm{o}}$, ε^{o}, and $\varepsilon_{\mathrm{T}}^{\mathrm{o}}$ are, respectively, the effective out-of-plane, axial, and in-plane transverse strains of the undamaged laminate, where $E_{\mathrm{A}}^{\mathrm{o}}$ is the axial Young's modulus of an uncracked laminate and, similarly, for the other thermoelastic constants. It is shown in [6], for example, how the effective thermoelastic constants for an undamaged crossply laminate may be calculated.

7.5.3 Case of One Damage Mode: Cracking in 90° Plies

Consider now the special case when ply cracks can form only in the 90° plies of the laminate. When the applied loading is such that cracks are *just* closed, the stress and displacement distributions in the laminate correspond to those in an undamaged laminate subject to the same loading and uniform temperature. The stress–strain relations for undamaged 0° plies are given by

$$\varepsilon_{\mathrm{t}}^{\mathrm{o}} = \frac{\sigma_{\mathrm{t}}}{E_{\mathrm{t}}^{(0)}} - \frac{\nu_{\mathrm{a}}^{(0)}}{E_{\mathrm{A}}^{(0)}}\sigma - \frac{\nu_{\mathrm{t}}^{(0)}}{E_{\mathrm{T}}^{(0)}}\sigma_{\mathrm{T}} + \alpha_{\mathrm{t}}^{(0)}\Delta T, \tag{7.78}$$

$$\varepsilon^{\mathrm{o}} = -\frac{\nu_{\mathrm{a}}^{(0)}}{E_{\mathrm{A}}^{(0)}}\sigma_{\mathrm{t}} + \frac{\sigma}{E_{\mathrm{A}}^{(0)}} - \frac{\nu_{\mathrm{A}}^{(0)}}{E_{\mathrm{A}}^{(0)}}\sigma_{\mathrm{T}} + \alpha_{\mathrm{A}}^{(0)}\Delta T, \tag{7.79}$$

$$\varepsilon_{\mathrm{T}}^{\mathrm{o}} = -\frac{\nu_{\mathrm{t}}^{(0)}}{E_{\mathrm{T}}^{(0)}}\sigma_{\mathrm{t}} - \frac{\nu_{\mathrm{A}}^{(0)}}{E_{\mathrm{A}}^{(0)}}\sigma + \frac{\sigma_{\mathrm{T}}}{E_{\mathrm{T}}^{(0)}} + \alpha_{\mathrm{T}}^{(0)}\Delta T, \tag{7.80}$$

while the stress–strain relations for undamaged 90° plies are given by

$$\varepsilon_{\mathrm{t}}^{\mathrm{o}} = \frac{\sigma_{\mathrm{t}}}{E_{\mathrm{t}}^{(90)}} - \frac{\nu_{\mathrm{a}}^{(90)}}{E_{\mathrm{A}}^{(90)}}\sigma - \frac{\nu_{\mathrm{t}}^{(90)}}{E_{\mathrm{T}}^{(90)}}\sigma_{\mathrm{T}} + \alpha_{\mathrm{t}}^{(90)}\Delta T, \tag{7.81}$$

$$\varepsilon^{\mathrm{o}} = -\frac{\nu_{\mathrm{a}}^{(90)}}{E_{\mathrm{A}}^{(90)}}\sigma_{\mathrm{t}} + \frac{\sigma}{E_{\mathrm{A}}^{(90)}} - \frac{\nu_{\mathrm{A}}^{(90)}}{E_{\mathrm{A}}^{(90)}}\sigma_{\mathrm{T}} + \alpha_{\mathrm{A}}^{(90)}\Delta T, \tag{7.82}$$

$$\varepsilon_{\mathrm{T}}^{\mathrm{o}} = -\frac{\nu_{\mathrm{t}}^{(90)}}{E_{\mathrm{T}}^{(90)}}\sigma_{\mathrm{t}} - \frac{\nu_{\mathrm{A}}^{(90)}}{E_{\mathrm{A}}^{(90)}}\sigma + \frac{\sigma_{\mathrm{T}}}{E_{\mathrm{T}}^{(90)}} + \alpha_{\mathrm{T}}^{(90)}\Delta T. \tag{7.83}$$

The stress–strain relations (7.81)–(7.83) are based on the selection of an axial direction that is the same as that of the 0° plies, i.e., a global axial direction has been defined rather than local directions that depend on the fiber directions in a given ply. Most crossply laminates use the same material for both 0° and 90° plies in which case the stress–strain relations for both 0° and 90° plies of an undamaged laminate have the form

$$\varepsilon_{\mathrm{t}}^{\mathrm{o}} = \frac{\sigma_{\mathrm{t}}}{E_{\mathrm{t}}^{*}} - \frac{\nu_{\mathrm{a}}^{*}}{E_{\mathrm{A}}^{*}}\sigma - \frac{\nu_{\mathrm{t}}^{*}}{E_{\mathrm{T}}^{*}}\sigma_{\mathrm{T}} + \alpha_{\mathrm{t}}^{*}\Delta T, \tag{7.84}$$

$$\varepsilon^{\mathrm{o}} = -\frac{\nu_{\mathrm{a}}^{*}}{E_{\mathrm{A}}^{*}}\sigma_{\mathrm{t}} + \frac{\sigma}{E_{\mathrm{A}}^{*}} - \frac{\nu_{\mathrm{A}}^{*}}{E_{\mathrm{A}}^{*}}\sigma_{\mathrm{T}} + \alpha_{\mathrm{A}}^{*}\Delta T, \tag{7.85}$$

$$\varepsilon_{\mathrm{T}}^{\mathrm{o}} = -\frac{\nu_{\mathrm{t}}^{*}}{E_{\mathrm{T}}^{*}}\sigma_{\mathrm{t}} - \frac{\nu_{\mathrm{A}}^{*}}{E_{\mathrm{A}}^{*}}\sigma + \frac{\sigma_{\mathrm{T}}}{E_{\mathrm{T}}^{*}} + \alpha_{\mathrm{T}}^{*}\Delta T, \tag{7.86}$$

where the axial direction for starred properties is defined as being in the fiber direction. It can then be shown that

$$\begin{aligned} &E_{\mathrm{t}}^{(0)} = E_{\mathrm{t}}^{*}, \quad E_{\mathrm{A}}^{(0)} = E_{\mathrm{A}}^{*}, \quad E_{\mathrm{T}}^{(0)} = E_{\mathrm{T}}^{*}, \\ &\nu_{\mathrm{t}}^{(0)} = \nu_{\mathrm{t}}^{*}, \quad \nu_{\mathrm{A}}^{(0)} = \nu_{\mathrm{A}}^{*}, \quad \nu_{\mathrm{a}}^{(0)} = \nu_{\mathrm{a}}^{*}, \\ &\alpha_{\mathrm{t}}^{(0)} = \alpha_{\mathrm{t}}^{*}, \quad \alpha_{\mathrm{A}}^{(0)} = \alpha_{\mathrm{A}}^{*}, \quad \alpha_{\mathrm{T}}^{(0)} = \alpha_{\mathrm{T}}^{*}, \end{aligned} \tag{7.87}$$

and that

$$\begin{aligned} E_{\mathrm{t}}^{(90)} &= E_{\mathrm{t}}^{*}, \quad E_{\mathrm{A}}^{(90)} = E_{\mathrm{T}}^{*}, \qquad E_{\mathrm{T}}^{(90)} = E_{\mathrm{A}}^{*}, \\ \nu_{\mathrm{t}}^{(90)} &= \nu_{\mathrm{a}}^{*}, \quad \nu_{\mathrm{A}}^{(90)} = \nu_{\mathrm{A}}^{*}\frac{E_{\mathrm{T}}^{*}}{E_{\mathrm{A}}^{*}}, \quad \nu_{\mathrm{a}}^{(90)} = \nu_{\mathrm{T}}^{*}, \\ \alpha_{\mathrm{t}}^{(90)} &= \alpha_{\mathrm{t}}^{*}, \quad \alpha_{\mathrm{A}}^{(90)} = \alpha_{\mathrm{T}}^{*}, \qquad \alpha_{\mathrm{T}}^{(90)} = \alpha_{\mathrm{A}}^{*}. \end{aligned} \tag{7.88}$$

The results of Sect. 7.3 can be used to provide estimates of the various ply properties if the fiber volume fraction and fiber and matrix properties are known. The analysis given in this chapter will not use the identifications (7.87) and (7.88) in order that results can be applied to hybrid laminates where the 0° and 90° plies are made of different materials.

Since the stress $\sigma^{(90)}$ in the 90° plies, in the direction of the axial loading of the laminate, is everywhere zero at the point of closure, it then follows that the in-plane strains ε^{c} and $\varepsilon_{\mathrm{T}}^{\mathrm{c}}$ at the point of closure satisfy the following relations for any value of the through-thickness stress at closure, denoted by σ_{t}, and for any temperature difference ΔT,

$$\varepsilon^{\mathrm{c}} = -\frac{\nu_{\mathrm{a}}^{(90)}}{E_{\mathrm{A}}^{(90)}}\sigma_{\mathrm{t}} - \frac{\nu_{\mathrm{A}}^{(90)}}{E_{\mathrm{A}}^{(90)}}\sigma_{\mathrm{T}}^{(90)} + \alpha_{\mathrm{A}}^{(90)}\Delta T, \tag{7.89}$$

$$\varepsilon_{\mathrm{T}}^{\mathrm{c}} = -\frac{\nu_{\mathrm{t}}^{(90)}}{E_{\mathrm{T}}^{(90)}}\sigma_{\mathrm{t}} + \frac{\sigma_{\mathrm{T}}^{(90)}}{E_{\mathrm{T}}^{(90)}} + \alpha_{\mathrm{T}}^{(90)}\Delta T, \tag{7.90}$$

where $\sigma_{\mathrm{T}}^{(90)}$ is the in-plane transverse stress in the 90° plies. It follows from (7.90) that

$$\sigma_{\mathrm{T}}^{(90)} = E_{\mathrm{T}}^{(90)}\varepsilon_{\mathrm{T}}^{\mathrm{c}} + \nu_{\mathrm{t}}^{(90)}\sigma_{\mathrm{t}} - E_{\mathrm{T}}^{(90)}\alpha_{\mathrm{T}}^{(90)}\Delta T. \tag{7.91}$$

On substituting (7.91) into (7.89) it can be shown that

$$\varepsilon^{\mathrm{c}} = -A\sigma_{\mathrm{t}} - B\varepsilon_{\mathrm{T}}^{\mathrm{c}} + C\Delta T, \tag{7.92}$$

where

$$A = \frac{\nu_{\mathrm{a}}^{(90)} + \nu_{\mathrm{t}}^{(90)}\nu_{\mathrm{A}}^{(90)}}{E_{\mathrm{A}}^{(90)}} = \frac{\nu_{\mathrm{t}}^{*}}{E_{\mathrm{T}}^{*}} + \frac{\nu_{\mathrm{a}}^{*}\nu_{\mathrm{A}}^{*}}{E_{\mathrm{A}}^{*}},$$

$$B = \nu_{\mathrm{A}}^{(90)} \frac{E_{\mathrm{T}}^{(90)}}{E_{\mathrm{A}}^{(90)}} = \nu_{\mathrm{A}}^{*}, \tag{7.93}$$

$$C = \alpha_{\mathrm{A}}^{(90)} + \nu_{\mathrm{A}}^{(90)} \frac{E_{\mathrm{T}}^{(90)}}{E_{\mathrm{A}}^{(90)}} \alpha_{\mathrm{T}}^{(90)} = \alpha_{\mathrm{T}}^{*} + \nu_{\mathrm{A}}^{*}\alpha_{\mathrm{A}}^{*},$$

where the starred quantities denote the ply properties for a laminate made of just one type of ply material (see (7.87) and (7.88)). The relation (7.92) implies that the axial closure strain depends on the transverse strain $\varepsilon_{\mathrm{T}}^{\mathrm{c}}$ in addition to the through-thickness stress σ_{t} and the temperature difference ΔT.

7.5.4 Ply Crack Closure for Uniaxial Loading in Axial Direction

Consider now the critical point for the first closure of the ply cracks for the case of uniaxial loading such that $\sigma_{\mathrm{t}} = \sigma_{\mathrm{T}} = 0$. It follows from (7.72) to (7.74), on setting $\varepsilon_{\mathrm{t}} = \bar{\varepsilon}_{\mathrm{t}}^{\mathrm{c}}$, $\varepsilon = \bar{\varepsilon}^{\mathrm{c}}$, $\varepsilon_{\mathrm{T}} = \bar{\varepsilon}_{\mathrm{T}}^{\mathrm{c}}$, $\sigma = \bar{\sigma}^{\mathrm{c}}$, that

$$\bar{\varepsilon}_{\mathrm{t}}^{\mathrm{c}} = -\frac{\nu_{\mathrm{a}}}{E_{\mathrm{A}}}\bar{\sigma}^{\mathrm{c}} + \alpha_{\mathrm{t}}\Delta T, \tag{7.94}$$

$$\bar{\varepsilon}^{\mathrm{c}} = \frac{\bar{\sigma}^{\mathrm{c}}}{E_{\mathrm{A}}} + \alpha_{\mathrm{A}}\Delta T, \tag{7.95}$$

$$\bar{\varepsilon}_{\mathrm{T}}^{\mathrm{c}} = -\frac{\nu_{\mathrm{A}}}{E_{\mathrm{A}}}\bar{\sigma}^{\mathrm{c}} + \alpha_{\mathrm{T}}\Delta T. \tag{7.96}$$

For undamaged laminates, it follows that at the critical closure stress $\bar{\sigma}^{\mathrm{c}}$ the laminate is subject to the same strains as in the corresponding cracked laminate, with the result that from (7.75) to (7.77)

$$\bar{\varepsilon}_{\mathrm{t}}^{\mathrm{c}} = -\frac{\nu_{\mathrm{a}}^{\mathrm{o}}}{E_{\mathrm{A}}^{\mathrm{o}}}\bar{\sigma}^{\mathrm{c}} + \alpha_{\mathrm{t}}^{\mathrm{o}}\Delta T, \tag{7.97}$$

$$\overline{\varepsilon}^{\mathrm{c}} = \frac{\overline{\sigma}^{\mathrm{c}}}{E_{\mathrm{A}}^{\mathrm{o}}} + \alpha_{\mathrm{A}}^{\mathrm{o}} \Delta T, \tag{7.98}$$

$$\overline{\varepsilon}_{\mathrm{T}}^{\mathrm{c}} = -\frac{\nu_{\mathrm{A}}^{\mathrm{o}}}{E_{\mathrm{A}}^{\mathrm{o}}} \overline{\sigma}^{\mathrm{c}} + \alpha_{\mathrm{T}}^{\mathrm{o}} \Delta T. \tag{7.99}$$

On subtracting (7.97) to (7.99) from (7.94) to (7.96) to eliminate the closure strains $\overline{\varepsilon}_{\mathrm{t}}^{\mathrm{c}}$, $\overline{\varepsilon}^{\mathrm{c}}$, and $\overline{\varepsilon}_{\mathrm{T}}^{\mathrm{c}}$, it then follows that

$$\overline{\sigma}^{\mathrm{c}} = -\frac{\alpha_{\mathrm{t}} - \alpha_{\mathrm{t}}^{\mathrm{o}}}{\dfrac{\nu_{\mathrm{a}}}{E_{\mathrm{A}}} - \dfrac{\nu_{\mathrm{a}}^{\mathrm{o}}}{E_{\mathrm{A}}^{\mathrm{o}}}} \Delta T = -\frac{\alpha_{\mathrm{A}} - \alpha_{\mathrm{A}}^{\mathrm{o}}}{\dfrac{1}{E_{\mathrm{A}}} - \dfrac{1}{E_{\mathrm{A}}^{\mathrm{o}}}} \Delta T = -\frac{\alpha_{\mathrm{T}} - \alpha_{\mathrm{T}}^{\mathrm{o}}}{\dfrac{\nu_{\mathrm{A}}^{\mathrm{o}}}{E_{\mathrm{A}}^{\mathrm{o}}} - \dfrac{\nu_{\mathrm{A}}}{E_{\mathrm{A}}}} \Delta T. \tag{7.100}$$

It follows from (7.92) that for the case of uniaxial loading under discussion

$$\overline{\varepsilon}^{\mathrm{c}} = -B \overline{\varepsilon}_{\mathrm{T}}^{\mathrm{c}} + C \Delta T. \tag{7.101}$$

Thus on substituting (7.98) and (7.99) into (7.101) and solving for $\overline{\sigma}^{\mathrm{c}}$, the following alternative expression for the crack closure stress is obtained

$$\overline{\sigma}^{\mathrm{c}} = -\frac{E_{\mathrm{A}}^{\mathrm{o}} [\alpha_{\mathrm{A}}^{\mathrm{o}} + B \alpha_{\mathrm{T}}^{\mathrm{o}} - C]}{1 - \nu_{\mathrm{A}}^{\mathrm{o}} B} \Delta T. \tag{7.102}$$

Thus, it follows from (7.100) and (7.102) that the following inter-relationships are valid

$$\frac{\alpha_{\mathrm{t}} - \alpha_{\mathrm{t}}^{\mathrm{o}}}{\dfrac{\nu_{\mathrm{a}}}{E_{\mathrm{A}}} - \dfrac{\nu_{\mathrm{a}}^{\mathrm{o}}}{E_{\mathrm{A}}^{\mathrm{o}}}} = \frac{\alpha_{\mathrm{A}} - \alpha_{\mathrm{A}}^{\mathrm{o}}}{\dfrac{1}{E_{\mathrm{A}}} - \dfrac{1}{E_{\mathrm{A}}^{\mathrm{o}}}} = \frac{\alpha_{\mathrm{T}} - \alpha_{\mathrm{T}}^{\mathrm{o}}}{\dfrac{\nu_{\mathrm{A}}^{\mathrm{o}}}{E_{\mathrm{A}}^{\mathrm{o}}} - \dfrac{\nu_{\mathrm{A}}}{E_{\mathrm{A}}}} = k_1, \tag{7.103}$$

where

$$k_1 = \frac{E_{\mathrm{A}}^{\mathrm{o}} [\alpha_{\mathrm{A}}^{\mathrm{o}} + B \alpha_{\mathrm{T}}^{\mathrm{o}} - C]}{1 - \nu_{\mathrm{A}}^{\mathrm{o}} B}, \quad \overline{\sigma}^{\mathrm{c}} = -k_1 \Delta T. \tag{7.104}$$

From (7.93) it is clear that the parameter k_1 is a constant for an undamaged laminate.

7.5.5 Ply Crack Closure for Uniaxial Loading in In-Plane Transverse Direction

Consider now ply crack closure for the case of uniaxial loading such that $\sigma_{\mathrm{t}} = \sigma = 0$. By following the approach of Sect. 7.5.4, it can be shown, on setting $\varepsilon_{\mathrm{t}} = \hat{\varepsilon}_{\mathrm{t}}^{\mathrm{c}}$, $\varepsilon = \hat{\varepsilon}^{\mathrm{c}}$, $\varepsilon_{\mathrm{T}} = \hat{\varepsilon}_{\mathrm{T}}^{\mathrm{c}}$, $\sigma_{\mathrm{T}} = \hat{\sigma}_{\mathrm{T}}^{\mathrm{c}}$, that

$$\hat{\sigma}^{\mathrm{c}} = -\frac{\alpha_{\mathrm{t}} - \alpha_{\mathrm{t}}^{\mathrm{o}}}{\dfrac{\nu_{\mathrm{t}}^{\mathrm{o}}}{E_{\mathrm{T}}^{\mathrm{o}}} - \dfrac{\nu_{\mathrm{t}}}{E_{\mathrm{T}}}} \Delta T = -\frac{\alpha_{\mathrm{A}} - \alpha_{\mathrm{A}}^{\mathrm{o}}}{\dfrac{\nu_{\mathrm{A}}^{\mathrm{o}}}{E_{\mathrm{A}}^{\mathrm{o}}} - \dfrac{\nu_{\mathrm{A}}}{E_{\mathrm{A}}}} \Delta T = -\frac{\alpha_{\mathrm{T}} - \alpha_{\mathrm{T}}^{\mathrm{o}}}{\dfrac{1}{E_{\mathrm{T}}} - \dfrac{1}{E_{\mathrm{T}}^{\mathrm{o}}}} \Delta T. \tag{7.105}$$

It follows from (7.92) that for the case of transverse uniaxial loading under discussion

$$\hat{\varepsilon}^{\mathrm{c}} = -B\hat{\varepsilon}_{\mathrm{T}}^{\mathrm{c}} + C\Delta T. \tag{7.106}$$

Thus, on using the stress–strain relations (7.75)–(7.77) for an undamaged laminate applied to uniaxial transverse loading, the following alternative expression for the crack closure stress is obtained

$$\hat{\sigma}^{\mathrm{c}} = -\frac{E_{\mathrm{T}}^{\mathrm{o}}[\alpha_{\mathrm{A}}^{\mathrm{o}} + B\alpha_{\mathrm{T}}^{\mathrm{o}} - C]}{B - \nu_{\mathrm{A}}^{\mathrm{o}} E_{\mathrm{T}}^{\mathrm{o}} / E_{\mathrm{A}}^{\mathrm{o}}} \Delta T. \tag{7.107}$$

It then follows from (7.105) and (7.107) that

$$\frac{\alpha_{\mathrm{t}} - \alpha_{\mathrm{t}}^{\mathrm{o}}}{\dfrac{\nu_{\mathrm{t}}^{\mathrm{o}}}{E_{\mathrm{T}}^{\mathrm{o}}} - \dfrac{\nu_{\mathrm{t}}}{E_{\mathrm{T}}}} = \frac{\alpha_{\mathrm{A}} - \alpha_{\mathrm{A}}^{\mathrm{o}}}{\dfrac{\nu_{\mathrm{A}}^{\mathrm{o}}}{E_{\mathrm{A}}^{\mathrm{o}}} - \dfrac{\nu_{\mathrm{A}}}{E_{\mathrm{A}}}} = \frac{\alpha_{\mathrm{T}} - \alpha_{\mathrm{T}}^{\mathrm{o}}}{\dfrac{1}{E_{\mathrm{T}}} - \dfrac{1}{E_{\mathrm{T}}^{\mathrm{o}}}} = k_2, \tag{7.108}$$

where

$$k_2 = \frac{E_{\mathrm{T}}^{\mathrm{o}}[\alpha_{\mathrm{A}}^{\mathrm{o}} + B\alpha_{\mathrm{T}}^{\mathrm{o}} - C]}{B - \nu_{\mathrm{A}}^{\mathrm{o}} E_{\mathrm{T}}^{\mathrm{o}} / E_{\mathrm{A}}^{\mathrm{o}}}. \tag{7.109}$$

The constant k_2 is thus another laminate constant for an undamaged laminate. It follows from (7.104) and (7.109) that

$$k = \frac{k_1}{k_2} = \frac{E_{\mathrm{A}}^{\mathrm{o}}}{E_{\mathrm{T}}^{\mathrm{o}}} \frac{B - \nu_{\mathrm{A}}^{\mathrm{o}} E_{\mathrm{T}}^{\mathrm{o}} / E_{\mathrm{A}}^{\mathrm{o}}}{1 - \nu_{\mathrm{A}}^{\mathrm{o}} B}, \tag{7.110}$$

where the constant k is independent of the thermal expansion coefficients. It then follows from (7.103), (7.108), and (7.110) that the following interrelationships must also be valid

$$\frac{\dfrac{\nu_{\mathrm{t}}^{\mathrm{o}}}{E_{\mathrm{T}}^{\mathrm{o}}}-\dfrac{\nu_{\mathrm{t}}}{E_{\mathrm{T}}}}{\dfrac{\nu_{\mathrm{a}}^{\mathrm{o}}}{E_{\mathrm{A}}^{\mathrm{o}}}-\dfrac{\nu_{\mathrm{a}}}{E_{\mathrm{A}}}}=\frac{\dfrac{\nu_{\mathrm{A}}^{\mathrm{o}}}{E_{\mathrm{A}}^{\mathrm{o}}}-\dfrac{\nu_{\mathrm{A}}}{E_{\mathrm{A}}}}{\dfrac{1}{E_{\mathrm{A}}}-\dfrac{1}{E_{\mathrm{A}}^{\mathrm{o}}}}=\frac{\dfrac{1}{E_{\mathrm{T}}}-\dfrac{1}{E_{\mathrm{T}}^{\mathrm{o}}}}{\dfrac{\nu_{\mathrm{A}}^{\mathrm{o}}}{E_{\mathrm{A}}^{\mathrm{o}}}-\dfrac{\nu_{\mathrm{A}}}{E_{\mathrm{A}}}}=\frac{\alpha_{\mathrm{T}}-\alpha_{\mathrm{T}}^{\mathrm{o}}}{\alpha_{\mathrm{A}}-\alpha_{\mathrm{A}}^{\mathrm{o}}}=k, \qquad (7.111)$$

where k is a constant for an undamaged laminate and is independent of the thermal expansion coefficients.

7.5.6 Ply Crack Closure for Uniaxial Loading in Out-of-Plane Direction

Consider now ply crack closure for the case of uniaxial out-of-plane loading such that $\sigma=\sigma_{\mathrm{T}}=0$. Again, following the approach of Sect. 7.5.4, it follows from (7.72) to (7.74), on setting $\varepsilon_{\mathrm{t}}=\tilde{\varepsilon}_{\mathrm{t}}^{\mathrm{c}}$, $\varepsilon=\tilde{\varepsilon}^{\mathrm{c}}$, $\varepsilon_{\mathrm{T}}=\tilde{\varepsilon}_{\mathrm{T}}^{\mathrm{c}}$, and $\sigma_{\mathrm{T}}=\tilde{\sigma}_{\mathrm{T}}^{\mathrm{c}}$, that

$$\tilde{\sigma}_{\mathrm{t}}^{\mathrm{c}}=-\frac{\alpha_{\mathrm{t}}-\alpha_{\mathrm{t}}^{\mathrm{o}}}{\dfrac{1}{E_{\mathrm{t}}}-\dfrac{1}{E_{\mathrm{t}}^{\mathrm{o}}}}\Delta T=-\frac{\alpha_{\mathrm{A}}-\alpha_{\mathrm{A}}^{\mathrm{o}}}{\dfrac{\nu_{\mathrm{a}}^{\mathrm{o}}}{E_{\mathrm{A}}^{\mathrm{o}}}-\dfrac{\nu_{\mathrm{a}}}{E_{\mathrm{A}}}}\Delta T=-\frac{\alpha_{\mathrm{T}}-\alpha_{\mathrm{T}}^{\mathrm{o}}}{\dfrac{\nu_{\mathrm{t}}^{\mathrm{o}}}{E_{\mathrm{T}}^{\mathrm{o}}}-\dfrac{\nu_{\mathrm{t}}}{E_{\mathrm{T}}}}\Delta T. \qquad (7.112)$$

It follows from (7.92) that for the case of uniaxial loading under discussion

$$\tilde{\varepsilon}^{\mathrm{c}}=-A\tilde{\sigma}_{\mathrm{t}}^{\mathrm{c}}-B\tilde{\varepsilon}_{\mathrm{T}}^{\mathrm{c}}+C\Delta T. \qquad (7.113)$$

Thus, on using the stress–strain relations (7.75)–(7.77) for an undamaged laminate subject to a uniaxial through-thickness loading, the following alternative expression for the crack closure stress is obtained

$$\tilde{\sigma}_{\mathrm{t}}^{\mathrm{c}}=-\frac{\alpha_{\mathrm{A}}^{\mathrm{o}}+B\alpha_{\mathrm{T}}^{\mathrm{o}}-C}{A-\dfrac{\nu_{\mathrm{a}}^{\mathrm{o}}}{E_{\mathrm{A}}^{\mathrm{o}}}-\dfrac{\nu_{\mathrm{t}}^{\mathrm{o}}}{E_{\mathrm{T}}^{\mathrm{o}}}B}\Delta T. \qquad (7.114)$$

It then follows from (7.112) and (7.114) that

$$\frac{\alpha_t - \alpha_t^o}{\dfrac{1}{E_t} - \dfrac{1}{E_t^o}} = \frac{\alpha_A - \alpha_A^o}{\dfrac{\nu_a^o}{E_A^o} - \dfrac{\nu_a}{E_A}} = \frac{\alpha_T - \alpha_T^o}{\dfrac{\nu_t^o}{E_T^o} - \dfrac{\nu_t}{E_T}} = k_3, \tag{7.115}$$

where

$$k_3 = \frac{\alpha_A^o + B\alpha_T^o - C}{A - \dfrac{\nu_a^o}{E_A^o} - \dfrac{\nu_t^o}{E_T^o} B}. \tag{7.116}$$

k_3 is another constant for an undamaged laminate.

It follows from (7.104) and (7.116) that

$$k' = \frac{k_1}{k_3} = \frac{E_A^o A - \nu_a^o - \nu_t^o \dfrac{E_A^o}{E_T^o} B}{1 - \nu_A^o B}, \tag{7.117}$$

where the constant k' is independent of the thermal expansion coefficients.

It then follows from (7.103), (7.115), and (7.117) that the following interrelationships must also be valid

$$\frac{\dfrac{1}{E_t} - \dfrac{1}{E_t^o}}{\dfrac{\nu_a^o}{E_A^o} - \dfrac{\nu_a}{E_A}} = \frac{\dfrac{\nu_a^o}{E_A^o} - \dfrac{\nu_a}{E_A}}{\dfrac{1}{E_A} - \dfrac{1}{E_A^o}} = \frac{\dfrac{\nu_t^o}{E_T^o} - \dfrac{\nu_t}{E_T}}{\dfrac{\nu_A^o}{E_A^o} - \dfrac{\nu_A}{E_A}} = \frac{\alpha_t - \alpha_t^o}{\alpha_A - \alpha_A^o} = k', \tag{7.118}$$

where k' is another constant for an undamaged laminate and is independent of the thermal expansion coefficients.

7.5.7 Useful Independent Interrelationships

The relations (7.103), (7.108), (7.111), (7.115), and (7.118) are not, of course, all independent. It will, therefore, be very useful to identify a set of independent interrelationships. To achieve this objective, it is convenient to introduce the parameter D defined by

$$D = \frac{1}{E_A} - \frac{1}{E_A^o}. \tag{7.119}$$

The quantity D can be regarded as a definition of a damage parameter arising in the field of continuum damage mechanics [6]. It should be noted that as the axial modulus of a damaged laminate is nearly always measured as a function of crack density, the parameter D may be estimated from experimental data. Also, the dependence of the axial modulus on crack density is nearly always predicted by mathematical models of stress transfer, especially the elementary models based on shear lag theory. It can be shown, from (7.103), (7.108), and (7.118), that the following relations form the required set of independent interrelationships:

$$\frac{\nu_{\mathrm{A}}^{\mathrm{o}}}{E_{\mathrm{A}}^{\mathrm{o}}} - \frac{\nu_{\mathrm{A}}}{E_{\mathrm{A}}} = kD, \tag{7.120}$$

$$\frac{1}{E_{\mathrm{T}}} - \frac{1}{E_{\mathrm{T}}^{\mathrm{o}}} = k^2 D, \tag{7.121}$$

$$\frac{\nu_{\mathrm{a}}^{\mathrm{o}}}{E_{\mathrm{A}}^{\mathrm{o}}} - \frac{\nu_{\mathrm{a}}}{E_{\mathrm{A}}} = k' D, \tag{7.122}$$

$$\frac{1}{E_{\mathrm{t}}} - \frac{1}{E_{\mathrm{t}}^{\mathrm{o}}} = (k')^2 D, \tag{7.123}$$

$$\frac{\nu_{\mathrm{t}}^{\mathrm{o}}}{E_{\mathrm{T}}^{\mathrm{o}}} - \frac{\nu_{\mathrm{t}}}{E_{\mathrm{T}}} = kk' D, \tag{7.124}$$

$$\alpha_{\mathrm{A}} - \alpha_{\mathrm{A}}^{\mathrm{o}} = k_1 D, \tag{7.125}$$

$$\alpha_{\mathrm{T}} - \alpha_{\mathrm{T}}^{\mathrm{o}} = kk_1 D, \tag{7.126}$$

$$\alpha_{\mathrm{t}} - \alpha_{\mathrm{t}}^{\mathrm{o}} = k' k_1 D. \tag{7.127}$$

In the above interrelationships, the parameters k, k', and k_1 are constants for undamaged laminates defined by the relations (7.110), (7.117), and (7.104), respectively. Thus, the values of these constants are readily calculated using (7.87), (7.88), and (7.93) and the results given in [6] for undamaged laminate properties. On using the relations (7.120)–(7.127), the effective thermoelastic constants of a damaged laminate can be calculated in terms of the damage-dependent parameter D defined by (7.119).

The importance of these relationships must be emphasized. First of all, they enable an exact and simple means of fully characterizing the effects of microstructural damage on the effective thermoelastic constants of multiple-ply crossply laminates, requiring only the value of the parameter D which depends upon the axial modulus that is frequently measured in practice or can be calculated from a suitable model. A reasonable estimate of the axial modulus of a damaged laminate can be obtained using a wide variety of models. The relation (7.119) can then be used to estimate the damage parameter D, and the relations (7.120)–(7.127) provide a rigorous method of extending such results so that estimates can be made of many other properties of damaged laminates that are needed by finite element analyses of damaged composite structures. Secondly, the relationships enable an efficient method of calculating the effective thermoelastic constants of damaged laminates using stress-transfer models, avoiding a great deal of time-consuming numerical computation if using state-of-the-art stress-transfer models that are capable of predicting highly accurate results.

7.6 Ply Crack Closure for Orthogonal Ply Cracking

In Sect. 7.5, the special case was considered where cracks were allowed to form only in the 90° plies of a crossply laminate. A consequence of this restriction is the validity of the very useful relations (7.119)–(7.127). The more general case of orthogonal cracking is now addressed. Consider the special loading case when all the ply cracks in the 0° and 90° plies of the laminate *just* close during multiaxial loading, where the through-thickness stress may have any value σ_t. For these crack closure conditions, let $\sigma = \sigma^c$ and $\sigma_T = \sigma_T^c$ so that from (7.72) to (7.74)

$$\varepsilon_t^c = \frac{\sigma_t}{E_t} - \frac{\nu_a}{E_A}\sigma^c - \frac{\nu_t}{E_T}\sigma_T^c + \alpha_t \Delta T, \tag{7.128}$$

$$\varepsilon^c = -\frac{\nu_a}{E_A}\sigma_t + \frac{\sigma^c}{E_A} - \frac{\nu_A}{E_A}\sigma_T^c + \alpha_A \Delta T, \tag{7.129}$$

$$\varepsilon_{\mathrm{T}}^{\mathrm{c}} = -\frac{\nu_{\mathrm{t}}}{E_{\mathrm{T}}}\sigma_{\mathrm{t}} - \frac{\nu_{\mathrm{A}}}{E_{\mathrm{A}}}\sigma^{\mathrm{c}} + \frac{\sigma_{\mathrm{T}}^{\mathrm{c}}}{E_{\mathrm{T}}} + \alpha_{\mathrm{T}}\Delta T, \tag{7.130}$$

and from (7.75) to (7.77) that

$$\varepsilon_{\mathrm{t}}^{\mathrm{c}} = \frac{\sigma_{\mathrm{t}}}{E_{\mathrm{t}}^{\mathrm{o}}} - \frac{\nu_{\mathrm{a}}^{\mathrm{o}}}{E_{\mathrm{A}}^{\mathrm{o}}}\sigma^{\mathrm{c}} - \frac{\nu_{\mathrm{t}}^{\mathrm{o}}}{E_{\mathrm{T}}^{\mathrm{o}}}\sigma_{\mathrm{T}}^{\mathrm{c}} + \alpha_{\mathrm{t}}^{\mathrm{o}}\Delta T, \tag{7.131}$$

$$\varepsilon^{\mathrm{c}} = -\frac{\nu_{\mathrm{a}}^{\mathrm{o}}}{E_{\mathrm{A}}^{\mathrm{o}}}\sigma_{\mathrm{t}} + \frac{\sigma^{\mathrm{c}}}{E_{\mathrm{A}}^{\mathrm{o}}} - \frac{\nu_{\mathrm{A}}^{\mathrm{o}}}{E_{\mathrm{A}}^{\mathrm{o}}}\sigma_{\mathrm{T}}^{\mathrm{c}} + \alpha_{\mathrm{A}}^{\mathrm{o}}\Delta T, \tag{7.132}$$

$$\varepsilon_{\mathrm{T}}^{\mathrm{c}} = -\frac{\nu_{\mathrm{t}}^{\mathrm{o}}}{E_{\mathrm{T}}^{\mathrm{o}}}\sigma_{\mathrm{t}} - \frac{\nu_{\mathrm{A}}^{\mathrm{o}}}{E_{\mathrm{A}}^{\mathrm{o}}}\sigma^{\mathrm{c}} + \frac{\sigma_{\mathrm{T}}^{\mathrm{c}}}{E_{\mathrm{T}}^{\mathrm{o}}} + \alpha_{\mathrm{T}}^{\mathrm{o}}\Delta T, \tag{7.133}$$

where $\varepsilon_{\mathrm{t}}^{\mathrm{c}}$, ε^{c}, and $\varepsilon_{\mathrm{T}}^{\mathrm{c}}$ are the laminate strains when cracks in the 0° and 90° plies just close for any values of σ_{t} and ΔT. At the point of closure, both cracked and uncracked laminates will have the same stress and strain distributions. It is shown in Appendix A how the values of the crack closure stresses and strains appearing in (7.128)–(7.133) may be calculated from the properties of the 0° and 90° plies. On subtracting (7.131)–(7.133) from (7.128)–(7.130), it follows that

$$\begin{aligned}&\left[\frac{1}{E_{\mathrm{t}}} - \frac{1}{E_{\mathrm{t}}^{\mathrm{o}}}\right]\sigma_{\mathrm{t}} + \left[\frac{\nu_{\mathrm{a}}^{\mathrm{o}}}{E_{\mathrm{A}}^{\mathrm{o}}} - \frac{\nu_{\mathrm{a}}}{E_{\mathrm{A}}}\right]\sigma^{\mathrm{c}} \\ &\quad + \left[\frac{\nu_{\mathrm{t}}^{\mathrm{o}}}{E_{\mathrm{T}}^{\mathrm{o}}} - \frac{\nu_{\mathrm{t}}}{E_{\mathrm{T}}}\right]\sigma_{\mathrm{T}}^{\mathrm{c}} + [\alpha_{\mathrm{t}} - \alpha_{\mathrm{t}}^{\mathrm{o}}]\Delta T = 0,\end{aligned} \tag{7.134}$$

$$\begin{aligned}&\left[\frac{\nu_{\mathrm{a}}^{\mathrm{o}}}{E_{\mathrm{A}}^{\mathrm{o}}} - \frac{\nu_{\mathrm{a}}}{E_{\mathrm{A}}}\right]\sigma_{\mathrm{t}} + \left[\frac{1}{E_{\mathrm{A}}} - \frac{1}{E_{\mathrm{A}}^{\mathrm{o}}}\right]\sigma^{\mathrm{c}} \\ &\quad + \left[\frac{\nu_{\mathrm{A}}^{\mathrm{o}}}{E_{\mathrm{A}}^{\mathrm{o}}} - \frac{\nu_{\mathrm{A}}}{E_{\mathrm{A}}}\right]\sigma_{\mathrm{T}}^{\mathrm{c}} + [\alpha_{\mathrm{A}} - \alpha_{\mathrm{A}}^{\mathrm{o}}]\Delta T = 0,\end{aligned} \tag{7.135}$$

$$\left[\frac{\nu_{\rm t}^{\rm o}}{E_{\rm T}^{\rm o}}-\frac{\nu_{\rm t}}{E_{\rm T}}\right]\sigma_{\rm t}+\left[\frac{\nu_{\rm A}^{\rm o}}{E_{\rm A}^{\rm o}}-\frac{\nu_{\rm A}}{E_{\rm A}}\right]\sigma^{\rm c}$$
$$+\left[\frac{1}{E_{\rm T}}-\frac{1}{E_{\rm T}^{\rm o}}\right]\sigma_{\rm T}^{\rm c}+[\alpha_{\rm T}-\alpha_{\rm T}^{\rm o}]\Delta T=0. \qquad (7.136)$$

These relations must be satisfied for all possible damage states in the form of orthogonal ply cracks and for all values of $\sigma_{\rm t}$ and ΔT. When there are only ply cracks in the 90° plies, use can be made of the relations (7.119)–(7.127) in which case the relations (7.134)–(7.136) all reduce to the following single relationship

$$k'\sigma_{\rm t}+\sigma^{\rm c}+k\sigma_{\rm T}^{\rm c}+k_1\Delta T=0. \qquad (7.137)$$

From (7.104) it follows that (7.137) may be written as

$$k'\sigma_{\rm t}+\sigma^{\rm c}+k\sigma_{\rm T}^{\rm c}=\bar{\sigma}^{\rm c}, \qquad (7.138)$$

showing how the stresses $\sigma_{\rm t}$, $\sigma^{\rm c}$, and $\sigma_{\rm T}^{\rm c}$ are related to the closure stress $\bar{\sigma}^{\rm c}$ for a uniaxially loaded laminate for the special case when ply cracks form only in the 90° plies.

For orthogonal cracking, it can be shown on using (A16) that (7.134)–(7.136) can be satisfied for all damage states, and for all values of $\sigma_{\rm t}$ and ΔT, only if relations of the following type are satisfied

$$\left[\frac{\nu_{\rm a}^{\rm o}}{E_{\rm A}^{\rm o}}-\frac{\nu_{\rm a}}{E_{\rm A}}\right]R+\left[\frac{\nu_{\rm t}^{\rm o}}{E_{\rm T}^{\rm o}}-\frac{\nu_{\rm t}}{E_{\rm T}}\right]R_{\rm T}+\left[\frac{1}{E_{\rm t}}-\frac{1}{E_{\rm t}^{\rm o}}\right]=0, \qquad (7.139)$$

$$\left[\frac{1}{E_{\rm A}}-\frac{1}{E_{\rm A}^{\rm o}}\right]R+\left[\frac{\nu_{\rm A}^{\rm o}}{E_{\rm A}^{\rm o}}-\frac{\nu_{\rm A}}{E_{\rm A}}\right]R_{\rm T}+\left[\frac{\nu_{\rm a}^{\rm o}}{E_{\rm A}^{\rm o}}-\frac{\nu_{\rm a}}{E_{\rm A}}\right]=0, \qquad (7.140)$$

$$\left[\frac{\nu_{\rm A}^{\rm o}}{E_{\rm A}^{\rm o}}-\frac{\nu_{\rm A}}{E_{\rm A}}\right]R+\left[\frac{1}{E_{\rm T}}-\frac{1}{E_{\rm T}^{\rm o}}\right]R_{\rm T}+\left[\frac{\nu_{\rm t}^{\rm o}}{E_{\rm T}^{\rm o}}-\frac{\nu_{\rm t}}{E_{\rm T}}\right]=0, \qquad (7.141)$$

$$\left[\frac{\nu_{\mathrm{a}}^{\mathrm{o}}}{E_{\mathrm{A}}^{\mathrm{o}}}-\frac{\nu_{\mathrm{a}}}{E_{\mathrm{A}}}\right]P+\left[\frac{\nu_{\mathrm{t}}^{\mathrm{o}}}{E_{\mathrm{T}}^{\mathrm{o}}}-\frac{\nu_{\mathrm{t}}}{E_{\mathrm{T}}}\right]P_{\mathrm{T}}+[\alpha_{\mathrm{t}}-\alpha_{\mathrm{t}}^{\mathrm{o}}]=0, \tag{7.142}$$

$$\left[\frac{1}{E_{\mathrm{A}}}-\frac{1}{E_{\mathrm{A}}^{\mathrm{o}}}\right]P+\left[\frac{\nu_{\mathrm{A}}^{\mathrm{o}}}{E_{\mathrm{A}}^{\mathrm{o}}}-\frac{\nu_{\mathrm{A}}}{E_{\mathrm{A}}}\right]P_{\mathrm{T}}+[\alpha_{\mathrm{A}}-\alpha_{\mathrm{A}}^{\mathrm{o}}]=0, \tag{7.143}$$

$$\left[\frac{\nu_{\mathrm{A}}^{\mathrm{o}}}{E_{\mathrm{A}}^{\mathrm{o}}}-\frac{\nu_{\mathrm{A}}}{E_{\mathrm{A}}}\right]P+\left[\frac{1}{E_{\mathrm{T}}}-\frac{1}{E_{\mathrm{T}}^{\mathrm{o}}}\right]P_{\mathrm{T}}+[\alpha_{\mathrm{T}}-\alpha_{\mathrm{T}}^{\mathrm{o}}]=0, \tag{7.144}$$

where the quantities P, P_{T}, R, and R_{T} are laminate constants defined in Appendix A. The relations (7.139)–(7.144) must be satisfied for all possible damage states and this includes the case where there are only ply cracks in the 90° plies of the laminate for which the interrelationships (7.119)–(7.127) are valid. On substituting them in (7.139)–(7.144), it follows that the relations (7.139)–(7.141) reduce to the single relation

$$k' + R + kR_{\mathrm{T}} = 0, \tag{7.145}$$

and the relations (7.142)–(7.144) reduce to the single relation

$$P + kP_{\mathrm{T}} + k_1 = 0. \tag{7.146}$$

The validity of these relations has been verified numerically and this implies that the interrelationships (7.119)–(7.127), derived for laminates having ply cracking damage only in the 90° plies, are valid also for the case of orthogonal cracking where ply cracks are present in both 0° and 90° plies of a laminate. Appendix B provides further analysis to support this statement. Although the analysis in this section has been derived for sets of orthogonal fully developed ply cracks in a crossply laminate, the approach described should be valid for more general damage states involving partially formed ply cracks, i.e., cracks that do not traverse the entire section of a ply. The assumption for such damage states is that biaxial effective stresses can be applied to the laminate, for a given through-thickness stress σ_{t} and temperature difference ΔT that lead to the precise closure of any cracks that have formed. This condition must be satisfied for the analysis given in this section to have validity.

As described in [4, 6, 7], the interrelationships (7.119)–(7.127) enable a very simple expression for the change in the Gibbs energy (or complementary energy) per unit volume as a result of ply cracking in the laminate, namely

$$\overline{g} - \overline{g}_0 = -\frac{1}{2}\left(\frac{1}{E_{\mathrm{A}}} - \frac{1}{E_{\mathrm{A}}^{\mathrm{o}}}\right)[s - \overline{\sigma}^{\mathrm{c}}]^2 = -\frac{1}{2} D[s - \overline{\sigma}^{\mathrm{c}}]^2, \tag{7.147}$$

where s is an effective stress that is defined by

$$s = k'\sigma_{\mathrm{t}} + \sigma + k\sigma_{\mathrm{T}}, \tag{7.148}$$

and where use has been made of the relation (7.119). This is the stress that controls ply crack formation and takes full account of the combined effects of biaxial loading and through-thickness loading of the laminate. Thermal residual stresses are accounted for through the axial ply crack closure stress $\overline{\sigma}^{\mathrm{c}}$, which, from (7.104), is proportional to the temperature difference ΔT. It is remarkable that such a simple relationship can be derived that takes account of the complex stress-transfer mechanisms arising from multiaxial loads and thermal residual stresses. The relations (7.147) and (7.148) have been used to derive [6, 7] simple expressions for the ply cracking stresses that have very good potential for use in design methodologies.

7.7 Conclusions

The following conclusions are drawn from the theoretical analysis presented in this chapter.

7.7.1 Undamaged Properties of Composites (at Ply Level)

1. A methodology developed by Maxwell in 1873 for estimating the electrical conductivity of isotropic particulate composites has been shown in another publication to be of much wider applicability. Results are presented in this chapter giving formulae for many thermoelastic constants for both isotropic particulate composites and fiber-reinforced composites.

2. The results are shown to be identical with results that can be derived using spherical shell and concentric cylinder models, which, in turn, are identical to one of the bounds derived using variational methods, implying that Maxwell's methodology is not necessarily restricted to low volume fractions.

3. The estimates of the various thermoelastic constants for both particulate and fiber-reinforced composites derived using Maxwell's method, or other nonvariational methods, can all be expressed in the form of a mixtures estimate of the property plus a correction term. These results have a common structure; and they correspond to one of the bounds derived using variational methods, where the other bound can be found simply by interchanging reinforcement and matrix properties and volume fractions.

4. Definitive conditions for an extreme property value being an upper bound, and those for it being a lower bound, have easily been determined (based on results described in [8]). The conditions given do not seem to have been given before in the literature.

7.7.2 Damaged Properties of Composites (at Laminate Level)

1. By making a limited number of reasonable assumptions, it is possible to develop, without the use of a stress-transfer model, a detailed theoretical framework that can be used to characterize the properties of laminates when ply cracks form in crossply laminates subject to thermal residual stresses and to multiaxial loading involving combined in-plane biaxial and through-thickness applied stresses. Results derived have been confirmed by an accurate stress-transfer model.

2. By considering ply crack closure, it is possible to develop a set of interrelationships that must be satisfied by the effective thermoelastic constants of cracked crossply laminates and which are derived without the use of a stress-transfer model. The interrelationships indicate that the effective constants of a cracked laminate can be characterized by just one damage function D that is defined in terms of the axial modulus of the laminate for both damaged and undamaged states. The interrelationships provide a rational and rigorous method of degrading many laminate properties due to damage formation. The interrelationships have been shown to be valid also for laminates with orthogonal ply cracking. The interrelationships should be applied to structural analyses when modeling the local stress redistribution that can occur due to the local formation of laminate damage in the form of ply cracking.

3. The interrelationships enable the change of Gibbs free energy (or complementary energy) resulting from ply crack formation to be written in a very simple form that involves macroscopic quantities and which is exact within the assumptions made when developing the theoretical framework. One key feature is the identification of an effective applied stress that takes account of the combined effects on ply crack formation of biaxial and through-thickness loading. Another key feature is the demonstration that the effects of thermal residual stresses can be taken into account by introducing the uniaxial crack closure stress.

4. Stress-transfer models predicting the stress and displacement distributions in cracked laminates are needed only to predict the actual values of the effective thermoelastic constants of cracked laminates. All other aspects of the prediction of ply crack formation can be accounted for using the derived theoretical framework.

Acknowledgment

The research described in this report was carried out as part of the Materials Measurement Programme, a program of underpinning research financed by the UK Department of Trade and Industry.

References

1. Hashin Z, Analysis of composite materials – a survey, *J. Appl. Mech.* 1983; 50: 481–505
2. Hashin Z and S Shtrikman, A variational approach to the theory of the elastic behaviour of multiphase materials, *J. Mech. Phys. Solids* 1963; 11: 127–140
3. Maxwell JC, A Treatise on Electricity and Magnetism, Vol. 1, First edition 1873 (Third edition 1904), Clarendon, Oxford
4. McCartney LN, Predicting transverse crack formation in cross-ply laminates resulting from micro-cracking, *Compos. Sci. Technol.* 1998; 58: 1069–1081
5. McCartney LN, Model to predict effects of triaxial loading on ply cracking in general symmetric laminates, *Compos. Sci. Technol.* 2000; 60: 2255–2279 (see Errata in *Compos. Sci. Technol.* 2002; 62: 1273–1274)

6. McCartney LN, Physically based damage models for laminated composites, *Proc. Inst. Mech. Eng. L, J. Mater.: Des. Appl.* 2003; 217: 163–199
7. McCartney LN, Multiscale predictive modeling of cracking in laminate composites, Chapter 3 in Multiscale modeling of composite material systems, ed. Soutis C and PWR Beaumont, Woodhead, Cambridge, 2005
8. McCartney LN and Kelly A, work on Maxwell's methodology that is to be published.
9. McCartney LN and GA Schoeppner, Predicting the effect of non-uniform ply cracking on the thermoelastic properties of cross-ply laminates, *Compos. Sci. Technol.* 2002; 62: 1841–1856
10. Rosen WB and Z Hashin, Effective thermal expansion coefficients and specific heats of composite materials, *Int. J. Eng. Sci.* 1970; 8: 157–173

Appendix A: Crack Closure Stresses for Orthogonal Cracking

Consider an undamaged crossply laminate where the 0° plies are made of the same material and the 90° plies are made of the same material that could differ from that of the 0° plies. The stresses in each ply, made of the same material and having the same orientation, must have the same values. This arises because each ply experiences the same values ε and ε_T for the axial and transverse in-plane strains, respectively, and the same value for the through-thickness stress σ_t. It is useful to denote the axial and transverse in-plane stresses in the 0° plies by $\sigma^{(0)}$ and $\sigma_T^{(0)}$, respectively, and to denote the axial and transverse in-plane stresses in the 90° plies by $\sigma^{(90)}$ and $\sigma_T^{(90)}$, respectively.

For a laminate containing cracks in both 0° and 90° plies, the crack closure stresses are such that $\sigma^{(90)} = 0$ and $\sigma_T^{(0)} = 0$ for a given value of the through-thickness stress σ_t and for a given value of the temperature difference ΔT. The resulting stress state in the laminate corresponds to the case when orthogonal cracks in the laminate just close. For this special case, mechanical equilibrium asserts that

$$\sigma^{(0)} = \frac{h\sigma^{c}}{h^{(0)}}, \quad \sigma_T^{(90)} = \frac{h\sigma_T^{c}}{h^{(90)}}, \tag{A1}$$

where $h^{(0)}$ and $h^{(90)}$ denote the total thicknesses in the laminate of all 0° plies and all 90° plies, respectively, and where σ^{c} and σ_T^{c} are, respectively, the effective applied axial stress and the in-plane transverse stress at the point of crack closure. Substitution of (A1) into the stress–strain equations for both 0° and 90° plies leads to the relations

$$\begin{aligned} \varepsilon^{\mathrm{c}} &= -\frac{\nu_{\mathrm{a}}^{(0)}}{E_{\mathrm{A}}^{(0)}}\sigma_{\mathrm{t}} + \frac{h}{h^{(0)}E_{\mathrm{A}}^{(0)}}\sigma^{\mathrm{c}} + \alpha_{\mathrm{A}}^{(0)}\Delta T \\ &= -\frac{\nu_{\mathrm{a}}^{(90)}}{E_{\mathrm{A}}^{(90)}}\sigma_{\mathrm{t}} - \frac{h\nu_{\mathrm{A}}^{(90)}}{h^{(90)}E_{\mathrm{A}}^{(90)}}\sigma_{\mathrm{T}}^{\mathrm{c}} + \alpha_{\mathrm{A}}^{(90)}\Delta T, \end{aligned} \tag{A2}$$

$$\begin{aligned} \varepsilon_{\mathrm{T}}^{\mathrm{c}} &= -\frac{\nu_{\mathrm{t}}^{(0)}}{E_{\mathrm{T}}^{(0)}}\sigma_{\mathrm{t}} - \frac{h\nu_{\mathrm{A}}^{(0)}}{h^{(0)}E_{\mathrm{A}}^{(0)}}\sigma^{\mathrm{c}} + \alpha_{\mathrm{T}}^{(0)}\Delta T \\ &= -\frac{\nu_{\mathrm{t}}^{(90)}}{E_{\mathrm{T}}^{(90)}}\sigma_{\mathrm{t}} + \frac{h}{h^{(90)}E_{\mathrm{T}}^{(90)}}\sigma_{\mathrm{T}}^{\mathrm{c}} + \alpha_{\mathrm{T}}^{(90)}\Delta T, \end{aligned} \tag{A3}$$

where the superscripts 0 and 90 refer the thermoelastic constants to the properties of the 0° and 90° plies, respectively. The parameters ε^{c} and $\varepsilon_{\mathrm{T}}^{\mathrm{c}}$ are, respectively, the uniform axial and transverse in-plane strains in the laminate when the cracks are just closed.

Regarding (A2) and (A3) as simultaneous algebraic equations for the unknowns σ^{c} and $\sigma_{\mathrm{T}}^{\mathrm{c}}$, it can be shown when $\sigma_{\mathrm{t}} = 0$ that

$$\sigma^{\mathrm{c}} = P\Delta T, \quad \sigma_{\mathrm{T}}^{\mathrm{c}} = P_{\mathrm{T}}\Delta T, \tag{A4}$$

where

$$P = \frac{h^{(0)}E_{\mathrm{A}}^{(0)}}{h} \frac{\alpha_{\mathrm{A}}^{(90)} - \alpha_{\mathrm{A}}^{(0)} + \nu_{\mathrm{A}}^{(90)}\dfrac{E_{\mathrm{T}}^{(90)}}{E_{\mathrm{A}}^{(90)}}(\alpha_{\mathrm{T}}^{(90)} - \alpha_{\mathrm{T}}^{(0)})}{1 - \nu_{\mathrm{A}}^{(0)}\nu_{\mathrm{A}}^{(90)}\dfrac{E_{\mathrm{T}}^{(90)}}{E_{\mathrm{A}}^{(90)}}}, \tag{A5}$$

$$P_{\mathrm{T}} = \frac{h^{(90)}E_{\mathrm{T}}^{(90)}}{h} \frac{\alpha_{\mathrm{T}}^{(0)} - \alpha_{\mathrm{T}}^{(90)} + \nu_{\mathrm{A}}^{(0)}(\alpha_{\mathrm{A}}^{(0)} - \alpha_{\mathrm{A}}^{(90)})}{1 - \nu_{\mathrm{A}}^{(0)}\nu_{\mathrm{A}}^{(90)}\dfrac{E_{\mathrm{T}}^{(90)}}{E_{\mathrm{A}}^{(90)}}}. \tag{A6}$$

The corresponding values of the ply strains given by (A2) and (A3) are

$$\varepsilon^{\mathrm{c}} = Q\Delta T, \quad \varepsilon_{\mathrm{T}}^{\mathrm{c}} = Q_{\mathrm{T}}\Delta T, \tag{A7}$$

where

$$Q = \frac{\alpha_{\mathrm{A}}^{(90)} + \nu_{\mathrm{A}}^{(90)} \dfrac{E_{\mathrm{T}}^{(90)}}{E_{\mathrm{A}}^{(90)}} (\alpha_{\mathrm{T}}^{(90)} - \alpha_{\mathrm{T}}^{(0)} - \nu_{\mathrm{A}}^{(0)} \alpha_{\mathrm{A}}^{(0)})}{1 - \nu_{\mathrm{A}}^{(0)} \nu_{\mathrm{A}}^{(90)} \dfrac{E_{\mathrm{T}}^{(90)}}{E_{\mathrm{A}}^{(90)}}}, \tag{A8}$$

$$Q_{\mathrm{T}} = \frac{\alpha_{\mathrm{T}}^{(0)} + \nu_{\mathrm{A}}^{(0)} (\alpha_{\mathrm{A}}^{(0)} - \alpha_{\mathrm{A}}^{(90)}) - \nu_{\mathrm{A}}^{(0)} \nu_{\mathrm{A}}^{(90)} \dfrac{E_{\mathrm{T}}^{(90)}}{E_{\mathrm{A}}^{(90)}} \alpha_{\mathrm{T}}^{(90)}}{1 - \nu_{\mathrm{A}}^{(0)} \nu_{\mathrm{A}}^{(90)} \dfrac{E_{\mathrm{T}}^{(90)}}{E_{\mathrm{A}}^{(90)}}}. \tag{A9}$$

Similarly, when $\Delta T = 0$, it can be shown that

$$\sigma^{\mathrm{c}} = R \sigma_{\mathrm{t}}, \quad \sigma_{\mathrm{T}}^{\mathrm{c}} = R_{\mathrm{T}} \sigma_{\mathrm{t}}, \tag{A10}$$

where

$$R = \frac{h^{(0)} E_{\mathrm{A}}^{(0)}}{h} \frac{\dfrac{\nu_{\mathrm{a}}^{(0)}}{E_{\mathrm{A}}^{(0)}} - \dfrac{\nu_{\mathrm{a}}^{(90)}}{E_{\mathrm{A}}^{(90)}} + \nu_{\mathrm{A}}^{(90)} \dfrac{E_{\mathrm{T}}^{(90)}}{E_{\mathrm{A}}^{(90)}} \left(\dfrac{\nu_{\mathrm{t}}^{(0)}}{E_{\mathrm{T}}^{(0)}} - \dfrac{\nu_{\mathrm{t}}^{(90)}}{E_{\mathrm{T}}^{(90)}} \right)}{1 - \nu_{\mathrm{A}}^{(0)} \nu_{\mathrm{A}}^{(90)} \dfrac{E_{\mathrm{T}}^{(90)}}{E_{\mathrm{A}}^{(90)}}}, \tag{A11}$$

$$R_{\mathrm{T}} = \frac{h^{(90)} E_{\mathrm{T}}^{(90)}}{h} \frac{\dfrac{\nu_{\mathrm{t}}^{(90)}}{E_{\mathrm{T}}^{(90)}} - \dfrac{\nu_{\mathrm{t}}^{(0)}}{E_{\mathrm{T}}^{(0)}} + \nu_{\mathrm{A}}^{(0)} \left(\dfrac{\nu_{\mathrm{a}}^{(90)}}{E_{\mathrm{A}}^{(90)}} - \dfrac{\nu_{\mathrm{a}}^{(0)}}{E_{\mathrm{A}}^{(0)}} \right)}{1 - \nu_{\mathrm{A}}^{(0)} \nu_{\mathrm{A}}^{(90)} \dfrac{E_{\mathrm{T}}^{(90)}}{E_{\mathrm{A}}^{(90)}}}. \tag{A12}$$

The corresponding values of the ply strains given by (A2) and (A3) are

$$\varepsilon^{\mathrm{c}} = S \sigma_{\mathrm{t}}, \quad \varepsilon_{\mathrm{T}}^{\mathrm{c}} = S_{\mathrm{T}} \sigma_{\mathrm{t}}, \tag{A13}$$

where

$$S = \frac{\nu_{\mathrm{A}}^{(90)} \dfrac{E_{\mathrm{T}}^{(90)}}{E_{\mathrm{A}}^{(90)}} \left(\dfrac{\nu_{\mathrm{t}}^{(0)}}{E_{\mathrm{T}}^{(0)}} - \dfrac{\nu_{\mathrm{t}}^{(90)}}{E_{\mathrm{T}}^{(90)}} + \dfrac{\nu_{\mathrm{A}}^{(0)} \nu_{\mathrm{a}}^{(0)}}{E_{\mathrm{A}}^{(0)}} \right) - \dfrac{\nu_{\mathrm{a}}^{(90)}}{E_{\mathrm{A}}^{(90)}}}{1 - \nu_{\mathrm{A}}^{(0)} \nu_{\mathrm{A}}^{(90)} \dfrac{E_{\mathrm{T}}^{(90)}}{E_{\mathrm{A}}^{(90)}}}, \tag{A14}$$

$$S_{\mathrm{T}} = \frac{\nu_{\mathrm{A}}^{(0)}\left(\dfrac{\nu_{\mathrm{a}}^{(90)}}{E_{\mathrm{A}}^{(90)}} - \dfrac{\nu_{\mathrm{a}}^{(0)}}{E_{\mathrm{A}}^{(0)}}\right) - \dfrac{\nu_{\mathrm{t}}^{(0)}}{E_{\mathrm{T}}^{(0)}} + \nu_{\mathrm{A}}^{(0)}\nu_{\mathrm{A}}^{(90)}\dfrac{E_{\mathrm{T}}^{(90)}}{E_{\mathrm{A}}^{(90)}}\dfrac{\nu_{\mathrm{t}}^{(90)}}{E_{\mathrm{T}}^{(90)}}}{1 - \nu_{\mathrm{A}}^{(0)}\nu_{\mathrm{A}}^{(90)}\dfrac{E_{\mathrm{T}}^{(90)}}{E_{\mathrm{A}}^{(90)}}}. \qquad \text{(A15)}$$

In general, it follows from (A2–A4), (A7), (A10), and (A13) that

$$\sigma^{\mathrm{c}} = P\Delta T + R\sigma_{\mathrm{t}}, \quad \sigma_{\mathrm{T}}^{\mathrm{c}} = P_{\mathrm{T}}\Delta T + R_{\mathrm{T}}\sigma_{\mathrm{t}}, \qquad \text{(A16)}$$

$$\varepsilon^{\mathrm{c}} = Q\Delta T + S\sigma_{\mathrm{t}}, \quad \varepsilon_{\mathrm{T}}^{\mathrm{c}} = Q_{\mathrm{T}}\Delta T + S_{\mathrm{T}}\sigma_{\mathrm{t}}. \qquad \text{(A17)}$$

Thus, having specified the values of σ_{t} and ΔT, the results (A16) and (A17) show how the ply crack closure stresses and strains may be calculated.

Appendix B: Interrelationships for Orthogonal Ply Cracks

On using the analysis given in Appendix A, the relations (7.139)–(7.144) are now expressed in the form

$$Ra + R_{\mathrm{T}}b + f = 0, \tag{B1}$$

$$Rc + R_{\mathrm{T}}d + a = 0, \tag{B2}$$

$$Rd + R_{\mathrm{T}}e + b = 0, \tag{B3}$$

$$Pa + P_{\mathrm{T}}b + x = 0, \tag{B4}$$

$$Pc + P_{\mathrm{T}}d + y = 0, \tag{B5}$$

$$Pd + P_{\mathrm{T}}e + z = 0, \tag{B6}$$

where

$$\begin{aligned} a &= \frac{\nu_{\mathrm{a}}^{\mathrm{o}}}{E_{\mathrm{A}}^{\mathrm{o}}} - \frac{\nu_{\mathrm{a}}}{E_{\mathrm{A}}}, \quad b = \frac{\nu_{\mathrm{t}}^{\mathrm{o}}}{E_{\mathrm{T}}^{\mathrm{o}}} - \frac{\nu_{\mathrm{t}}}{E_{\mathrm{T}}}, \quad f = \frac{1}{E_{\mathrm{t}}} - \frac{1}{E_{\mathrm{t}}^{\mathrm{o}}}, \\ c &= \frac{1}{E_{\mathrm{A}}} - \frac{1}{E_{\mathrm{A}}^{\mathrm{o}}}, \quad d = \frac{\nu_{\mathrm{A}}^{\mathrm{o}}}{E_{\mathrm{A}}^{\mathrm{o}}} - \frac{\nu_{\mathrm{A}}}{E_{\mathrm{A}}}, \quad e = \frac{1}{E_{\mathrm{T}}} - \frac{1}{E_{\mathrm{T}}^{\mathrm{o}}}, \\ x &= \alpha_{\mathrm{t}} - \alpha_{\mathrm{t}}^{\mathrm{o}}, \quad y = \alpha_{\mathrm{A}} - \alpha_{\mathrm{A}}^{\mathrm{o}}, \quad z = \alpha_{\mathrm{T}} - \alpha_{\mathrm{T}}^{\mathrm{o}}. \end{aligned} \tag{B7}$$

It is useful to define the following parameters:

$$P' = \frac{P}{PR_{\mathrm{T}} - P_{\mathrm{T}}R}, \quad P_{\mathrm{T}}' = \frac{P_{\mathrm{T}}}{PR_{\mathrm{T}} - P_{\mathrm{T}}R},$$
$$R' = \frac{R}{PR_{\mathrm{T}} - P_{\mathrm{T}}R}, \quad R_{\mathrm{T}}' = \frac{R_{\mathrm{T}}}{PR_{\mathrm{T}} - P_{\mathrm{T}}R}. \tag{B8}$$

Solving (B1) and (B4) for the parameters a and b leads to

$$a = P_{\mathrm{T}}' f - R_{\mathrm{T}}' x, \quad b = R' x - P' f. \tag{B9}$$

Solving (B2) and (B5) for the parameters c and d leads to

$$c = P_{\mathrm{T}}' a - R_{\mathrm{T}}' y, \quad d = R' y - P' a. \tag{B10}$$

Solving (B1) and (B4) for the parameters d and e leads to

$$d = P_{\mathrm{T}}' b - R_{\mathrm{T}}' z, \quad e = R' z - P' b. \tag{B11}$$

On eliminating x, y, and z using (B9)–(B11), respectively, it can be shown that

$$R + R_{\mathrm{T}} \frac{b}{a} + \frac{f}{a} = 0,$$
$$R + R_{\mathrm{T}} \frac{d}{c} + \frac{a}{c} = 0, \tag{B12}$$
$$R + R_{\mathrm{T}} \frac{e}{d} + \frac{b}{d} = 0.$$

The relations (B12) must be satisfied for all possible damage states, where R and R_{T} are known laminate constants (see (A11) and (A12)); and in view of the relation (7.145), it is deduced that

$$\frac{b}{a} = \frac{d}{c} = \frac{e}{d} = k, \quad \frac{f}{a} = \frac{a}{c} = \frac{b}{d} = k', \tag{B13}$$

where the parameters k and k' are given by (7.110) and (7.117), respectively.

Since from (B13) $b = ka$, it follows from (B9) that

$$R'x - P'f = kP_{\mathrm{T}}'f - kR_{\mathrm{T}}'x, \tag{B14}$$

so that on using (B8)

$$(P + kP_{\mathrm{T}})f = (R + kR_{\mathrm{T}})x. \tag{B15}$$

It then follows from (7.145) and (7.146) that

$$x = \frac{k_1}{k'}f, \tag{B16}$$

where k_1 is a laminate constant defined by (7.104). Similarly, it can be shown, using (B10), (B11), and (B13), that

$$y = \frac{k_1}{k'}a, \tag{B17}$$

$$z = \frac{k_1}{k'}b. \tag{B18}$$

It follows from (B13) and (B16–B18) that

$$\frac{b}{a} = \frac{d}{c} = \frac{e}{d} = \frac{z}{y} = k, \quad \frac{f}{a} = \frac{a}{c} = \frac{b}{d} = \frac{x}{y} = k', \tag{B19}$$

and it then follows that

$$\frac{x}{a} = \frac{y}{c} = \frac{z}{d} = k_1. \tag{B20}$$

On using (B7), it is clear that the relations (B19) and (B20) are precisely the relations (7.103), (7.111), and (7.118) that were derived for the case of ply cracking only in the 90° plies. The analysis of this appendix thus demonstrates that the interrelationships (7.103), (7.111), and (7.118) are valid also for orthogonal systems of ply cracks in crossply laminates.

Chapter 8: Nested Nonlinear Multiscale Frameworks for the Analysis of Thick-Section Composite Materials and Structures

Rami Haj-Ali

Georgia Institute of Technology, Atlanta, GA 30332-0355, USA
rami.haj-ali@ce.gatech.edu

8.1 Introduction

This chapter presents nonlinear and time-dependent multiscale frameworks for the analysis of thick-section and multilayered composite materials and structures. Nested and hierarchical three-dimensional (3D) micromechanical models are formulated within the nonlinear analysis framework. The constitutive framework is composed of nonlinear material models for the matrix behavior, micromodels for the unidirectional lamina, and a sublaminate model for a repeating ply-stacking sequence. A unified development of a class of constant deformation cell (CDC) micromodels is presented to generate the effective nonlinear response of a unidirectional lamina from the response of its matrix and fiber constituents (subcells). Two structural modeling approaches for nonlinear analysis of laminated composites are proposed using 3D and shell nonlinear finite element (FE) analysis. The first, for the analysis of multilayered and thick-section composites, uses the 3D sublaminate model coupled with 3D FE structural models. The sublaminate represents the nonlinear effective continuum response of a through-thickness repeating stacking sequence at the FE material points (Gaussian integration points). The CDC micromodels can be employed for the different layers within the sublaminate model. The second structural approach is used for the analysis of thin-section laminated composite plates and shells in the form of a ply-by-ply. In this case, the micromodels are used to represent the effective response of each layer. New stress-update solution algorithms are developed for the micromodels and the sublaminate model; they are well suited for non-linear displacement-based FE. Different applications are presented and comparisons are made with reported experimental results. The proposed micromodels are shown to be very capable of predicting the response of different composite materials and structural systems, such as multilayered laminated composites and thick-section pultruded composites. The numerical stress-update algorithms are shown to be well behaved and robust. Applications presented using the proposed frameworks indicate their suitability as practical, general material, and structural analysis tools.

Unlike traditional structural materials, such as metals, composite materials add a new and exciting dimension to the engineering design process. Their effective material properties and strengths can be controlled based on the choice of the matrix and fiber materials, volume fractions, and multiaxial reinforcements, along with several other material, geometry and manufacturing parameters. Proper selection of these parameters in the design process can lead to an optimal structural design, such as a structure with minimum weight and a maximum resistance to the applied forces.

Composite materials are widely used in high-performance structures where high stiffness and strength combined with low weight are required. Today, many structural components are made from composite materials, especially in the aviation industry. However, it is still rare to find a complete structure that is fully made of composite materials. This indicates that the analysis, design, and manufacturing of composite structures have not yet fully reached a satisfactory level of reliability. Therefore, there is still a need to improve and introduce new analysis and design approaches that can predict the nonlinear and damage behavior of composites.

Recently, the use of composite technology in civil and infrastructure applications, such as bridges and construction joints, has been advocated. However, there are two major obstacles standing in the way: the relatively high manufacturing cost and the lack of sufficient predictive models to provide information on the behavior of such structures over their lifespan. Nevertheless, in some cases, the relatively high cost of using composite materials can be justified. For example, the use of composite materials in bridges can eliminate the need to reinforce the concrete with steel bars that are subject to corrosion, thereby prolonging the lifespan of the bridge which can compensate for the higher cost of the composite structure. In addition, it is evident that advances in mass manufacturing of composite materials will drive costs down. This provides additional incentive to continue the research on the behavior of composite structures in civil and infrastructure applications.

The tremendous advances in computer technology that have taken place over the last two decades have made possible the development of computational tools that routinely employ nonlinear analysis for practical engineering applications. The use of nonlinear stress–strain relations, such as those provided by plasticity and other inelastic models, is now considered a standard engineering practice. However, nonlinear structural modeling approaches that use 3D analysis are not widespread for laminated composites. This is due to many factors. Laminated composites are often considered as brittle materials without accounting for their nonlinear behavior. Therefore, elastic structural analysis and design are often considered sufficient. Furthermore, many laminated structures are thin shell structures that can be idealized using plane-stress constitutive models. However, nonlinear 3D structural analyses may be needed to produce reliable structural designs. Even in the case of thin shell structures, a realistic nonlinear 3D constitutive model is needed to depict accurately the structural response in the presence of edge effects and structural discontinuities. These discontinuities, such as crack tips, holes, and cutouts, usually have a significant impact on the response of the structure, because damage will typically initiate at and propagate from these locations. Therefore, it is important to develop nonlinear and three-dimensional material models to properly simulate the structural behavior with local nonlinear and damage responses.

Macroscale nonlinear constitutive models can be formulated directly at the lamina level. On the other hand, micromechanical models of nonlinear lamina behavior, which explicitly recognize the fiber and matrix constituents, are appealing because they can provide more detailed response information than macromechanical models. They are also potentially simpler to formulate because they operate at a more fundamental level than macromechanical models. However, the direct use of micromechanical models in practical nonlinear analysis of laminated structures requires compromise between accuracy and computational effort.

This chapter reviews multiscale material and structural frameworks that allow the application of several micromechanical models while performing nonlinear structural analysis. A class of simple 3D micromodels that strike a reasonable balance between accuracy and simplicity is reviewed. These nonlinear micromechanical models, e.g., for a unidirectional lamina, are incorporated into a hierarchical framework that is suitable for FE analysis. The structural analysis includes both nonlinear material and geometric effects. The hierarchical nature of this framework allows the use of several alternative combinations of material and structural modeling approaches. The nonlinear material behavior can arise from different sources: matrix nonlinear constitutive behavior, microfailure effects, e.g., matrix microcracking, fiber–matrix debonding, and fiber failure, e.g., fiber buckling. Several examples of structural analyses are presented and compared with experimental results where possible.

8.2 Multiscale Analysis of Laminated Composite Structures

A general 3D multiscale framework is proposed for the nonlinear analysis of laminated composite structures. Figure 8.1 illustrates the proposed analysis framework for multilayered structures using 3D or shell-based structural FE models. In the case of a 3D FE structural model, a sublaminate model is formulated to represent the nonlinear effective continuum response at each material point (Gaussian point) [24, 25, 33, 34]. The sublaminate model is used to generate a 3D through-thickness effective response of a representative stacking sequence.

In the case of shell elements, Fig. 8.1 illustrates that each layer is explicitly modeled with one or more integration points under plane-stress condition; and the sublaminate model is reduced to the classical lamination theory in this case. Constant transverse shear, cross-sectional stiffness is assumed for the shell elements. This assumption is valid where the transverse stresses in the different layers are very small compared to the in-plane stresses. The 3D micromechanical models provide for the effective nonlinear constitutive behavior for each Gaussian point. The shell element's effective through-thickness response is generated at select integration points on its reference surface by integrating the effective micromechanical response over all Gaussian points, as shown in Fig. 8.1.

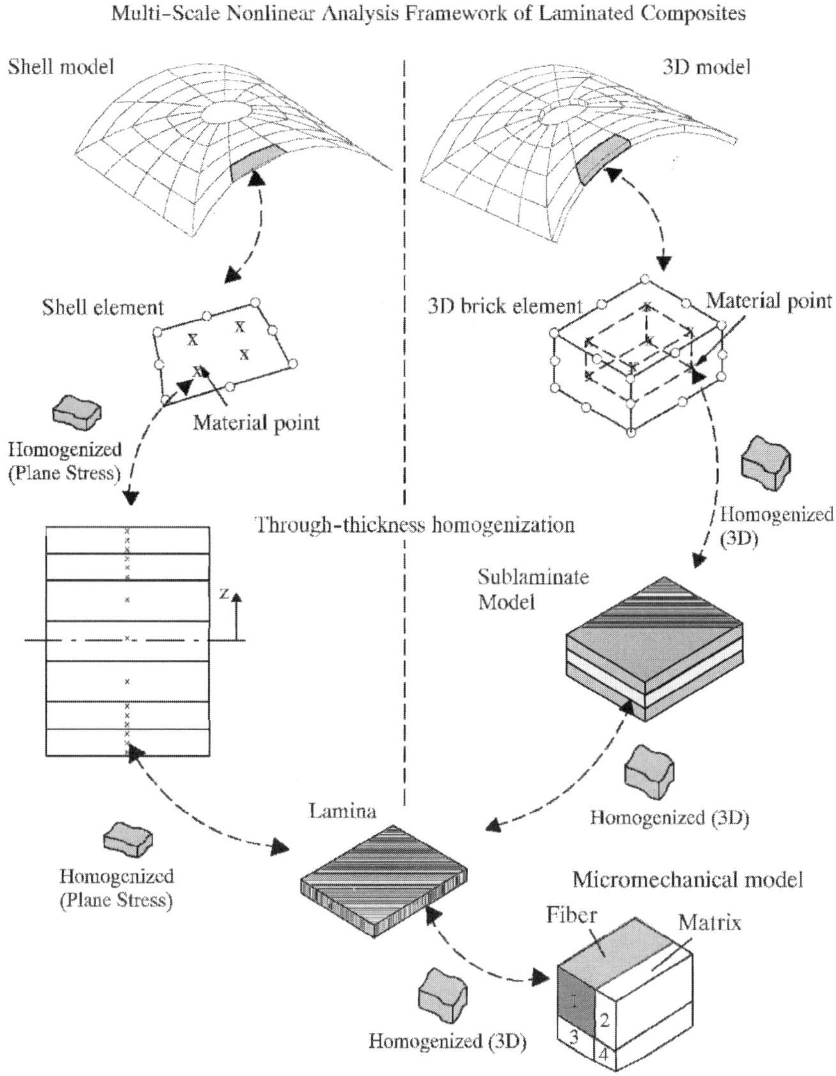

Fig. 8.1. A multiscale micromechanical–structural framework for nonlinear and viscoelastic analysis of laminated composite structures (adapted from [23])

8.3 A Simplified Class of Micromechanical Constitutive Models

A unified approach for defining and characterizing a simple and phenomenological class of nonlinear micromechanical models for fiber composites is presented in this section. A unified development of a class of CDC micromodels, or unit cell (UC), is presented to generate the effective nonlinear continuum response from the average response of its matrix and fiber constituents (subcells). The main advantage of these simple multicell models lies in their ability to generate the full 3D effective stress–strain response of fiber composites in a form that is suitable for integration into finite element structural analysis. The first part of this section sets out some general definitions and relations that are valid for all the micromodels in this class. Specific micromodels are presented in the later part of this section and through this chapter.

The objective of the CDC models is to generate the nonlinear effective stress–strain relations by employing a simple geometrical representation of the unit cell geometry and satisfy traction and displacement continuity between the cells in an average sense. Few assumptions are made at this stage regarding the fiber and matrix constitutive relations; specific material nonlinear constitutive behavior is characterized only at the more fundamental subcell level. The resulting unit cell effective stress–strain relations can be viewed, from a global/structural perspective, as a material model with microstructural constraints.

It is assumed that, for a given heterogeneous periodic medium, it is possible to define a basic unit cell that represents the medium's geometrical and material characteristics. Each unit cell is divided into a number of subcells. Within each subcell, the spatial variation of the displacement field is assumed such that the stresses and deformations are *spatially* uniform in each subcell. Traction continuity at an interface between subcells can, therefore, be satisfied only in an average sense. Some general definitions and linearized formulations are established that are applicable to any CDC micromodel.

The volume average stress over the unit cell is defined as

$$\bar{\sigma}_{ij} = \frac{1}{V}\int_V \sigma_{ij}(x)\mathrm{d}V = \frac{1}{V}\sum_{\alpha=1}^{N}\int_{V_{(\alpha)}} \sigma_{ij}^{(\alpha)}(x^{(\alpha)})\mathrm{d}V_{(\alpha)} = \frac{1}{V}\sum_{\alpha=1}^{N} V_{(\alpha)}\sigma_{ij}^{(\alpha)}, \quad (8.1)$$

where N is the number of subcells and V is the unit cell volume. A similar definition applies for volume average strain $\bar{\varepsilon}_{ij}$. The superscript α denotes the subcell number. An overbar denotes a unit cell average quantity. The variables x and $x^{(\alpha)}$ are the unit cell global and the subcell local coordinates, respectively. Stress and strain are uniform within each subcell by definition. Therefore, using matrix notation

$$\bar{\sigma} = \frac{1}{V}\sum_{\alpha=1}^{N} V_{(\alpha)}\sigma^{(\alpha)}, \quad \bar{\varepsilon} = \frac{1}{V}\sum_{\alpha=1}^{N} V_{(\alpha)}\varepsilon^{(\alpha)}, \quad V = \sum_{\alpha=1}^{N} V_{(\alpha)}, \tag{8.2}$$

where the stresses and strains are now written as vectors.

Next, a strain-concentration or strain-interaction fourth rank tensor B is defined for each subcell, which relates the subcell strain increment to the unit cell average strain increment

$$\mathrm{d}\varepsilon_{ij}^{(\alpha)} = B_{ijkl}^{(\alpha)}\mathrm{d}\bar{\varepsilon}_{kl}. \tag{8.3}$$

It is important to emphasize that the interaction matrices are unknown at this stage; they will be determined later in this section by solution of the unit cell governing equations. It can be easily shown that a subcell strain-interaction matrix is usually a function of the tangent stiffness and the relative volumes of all subcells.

Using the incremental form of (8.2) with (8.3), expressed in matrix notation, the average strain increment of the unit cell is:

$$\mathrm{d}\bar{\varepsilon} = \frac{1}{V}\sum_{\alpha=1}^{N} V_{(\alpha)}\mathrm{d}\varepsilon^{(\alpha)} = \frac{1}{V}\sum_{\alpha=1}^{N} V_{(\alpha)}B^{(\alpha)}\mathrm{d}\bar{\varepsilon}. \tag{8.4}$$

Since (8.4) must hold for an arbitrary average strain increment $\mathrm{d}\varepsilon$, the following relations must be satisfied

$$\frac{1}{V}\sum_{\alpha=1}^{N} V_{(\alpha)}B^{(\alpha)} = I \quad \text{and} \quad \sum_{\alpha=1}^{N} V_{(\alpha)}(B^{(\alpha)} - I) = 0, \tag{8.5}$$

where I is a unit matrix. The second relation in (8.5) follows from the first relation due to the volume sum relation expressed in (8.2). The matrix representation of the strain-concentration tensor is not symmetric. Next, the incremental stress–strain relations are used to express the stress increment in each of the subcells

$$\mathrm{d}\sigma^{(\alpha)} = C^{(\alpha)}\mathrm{d}\varepsilon^{(\alpha)} = C^{(\alpha)}B^{(\alpha)}\mathrm{d}\bar{\varepsilon}, \tag{8.6}$$

where $C^{(\alpha)}$ is the current tangent stiffness matrix of the subcell. The incremental form of the average stress can be expressed, using (8.6), as:

$$\mathrm{d}\bar{\sigma} = \frac{1}{V}\sum_{\alpha=1}^{N} V_{(\alpha)}\sigma^{(\alpha)} = \frac{1}{V}\sum_{\alpha=1}^{N} V_{(\alpha)}C^{(\alpha)}B^{(\alpha)}\mathrm{d}\bar{\varepsilon}. \tag{8.7}$$

Equation (8.7) can be expressed as

$$\mathrm{d}\bar{\sigma} = C^{*}\mathrm{d}\bar{\varepsilon}, \tag{8.8}$$

where C^* is the unit cell effective tangent stiffness matrix defined by:

$$C^{*} = \frac{1}{V}\sum_{\alpha=1}^{N} V_{(\alpha)}C^{(\alpha)}B^{(\alpha)}. \tag{8.9}$$

An alternative for deriving the stiffness matrix is to use the second variation of the strain energy density. This is demonstrated by the following relations:

$$\mathrm{d}\bar{\varepsilon}^{\mathrm{T}}\mathrm{d}\bar{\sigma} = \frac{1}{V}\sum_{\alpha=1}^{N} V_{(\alpha)}\mathrm{d}\varepsilon^{(\alpha)^{\mathrm{T}}}\mathrm{d}\sigma^{(\alpha)} = \mathrm{d}\bar{\varepsilon}^{\mathrm{T}}\left[\frac{1}{V}\sum_{\alpha=1}^{N} V_{(\alpha)}B^{(\alpha)^{\mathrm{T}}}C^{(\alpha)}B^{(\alpha)}\right]\mathrm{d}\bar{\varepsilon}. \tag{8.10}$$

Substituting (8.8) into the left-hand side of (8.10), the unit cell stiffness matrix is expressed as:

$$C^* = \frac{1}{V}\sum_{\alpha=1}^{N} V_{(\alpha)} B^{(\alpha)^{\mathrm{T}}} C^{(\alpha)} B^{(\alpha)}. \tag{8.11}$$

The two expressions for the effective stiffness matrix in (8.9) and (8.11) must be identical. It can be easily verified that, since the strain-concentration matrices $B^{(\alpha)}$ satisfy the relations in (8.5), the two stiffness expressions are, in fact, identical. Equation (8.11) shows that the unit cell stiffness matrix C^* is symmetric provided that the stiffness matrix of each of the subcells $C^{(\alpha)}$ is also symmetric. However, it is interesting to note that this property is not explicitly apparent by a first examination of the expression in (8.9).

Up to this stage, the properties of the strain-interaction matrices and the expression for the unit cell effective stiffness matrix have been dealt with. The only assumption that was made is that the subcells have uniform stress and strain. Therefore, these linearized relations are general for any CDC micromodel. To derive the strain-interaction matrices for a unit cell, the traction and displacement continuity conditions must be imposed, and stress–strain relations must be invoked. The fact that the strains and stresses are uniform in every subcell makes it possible to express the traction and displacement continuity conditions directly in terms of the average stress and strain vectors. The term *strain compatibility* will be used here to describe the relations between the strains in the subcells which satisfy displacement continuity in an average fashion. The combined set of equations that describe the strain compatibility and the traction continuity equations (micromechanical constraints) can ultimately be written in a general incremental form as:

$$\mathrm{d}R_\sigma = \overline{C}_i(\mathrm{d}\sigma^{(\alpha)}, \mathrm{d}\varepsilon^{(\alpha)}, \mathrm{d}\overline{\varepsilon}, V_{(\alpha)}, \alpha = 1,2,\ldots,N) = 0, \quad i = 1,\ldots,n. \tag{8.12}$$

Equation (8.12) is used to generate the strain-interaction matrices for the subcells. The incremental form of the stress–strain relations in the subcells (8.6) is used to express the constraints in terms of the incremental strains

$$\mathrm{d}R_\varepsilon = E_j(C^{(\alpha)}, \mathrm{d}\varepsilon^{(\alpha)}, \mathrm{d}\overline{\varepsilon}, V_{(\alpha)}, \alpha = 1,2,\ldots,N) = 0, \quad j = 1,\ldots,m. \tag{8.13}$$

The subset of (8.13) that represents the strain compatibility constraints satisfies (8.4). Equation (8.13) forms a set of linear equations in terms of the unknown incremental strain vectors for each of the subcells. The current state of the linearized micromechanical equations can be arranged in terms of these unknowns and the known values, the current tangent stiffness matrices, and the unit cell strain vector, and represented, in a general matrix form, as:

$$\underset{6N\times 6N}{\left[\begin{array}{c} \\ A \\ \\ \end{array}\right]} \underset{6N\times 1}{\begin{Bmatrix} \mathrm{d}\varepsilon^{(1)} \\ \mathrm{d}\varepsilon^{(2)} \\ \vdots \\ \mathrm{d}\varepsilon^{(N)} \end{Bmatrix}} = \underset{6N\times 6}{\left[\begin{array}{c} \\ D \\ \\ \end{array}\right]} \underset{6\times 1}{\{\mathrm{d}\bar{\varepsilon}\}}. \tag{8.14}$$

Equation (8.14) can be rearranged by dividing the subcells' strain components into two dependent groups with (m) and (n) number of components, respectively, to yield a new compact form that can be solved numerically in an efficient manner. The general structure of the linearized micromechanical equations for the CDC class of micromodels is:

$$\begin{Bmatrix} \underset{(m\times 1)}{\mathrm{d}R_{\varepsilon}} \\ \underset{(n\times 1)}{\mathrm{d}R_{\sigma}} \end{Bmatrix} = \begin{bmatrix} \underset{(m\times m)}{I} & \underset{(m\times n)}{\bar{A}_{ab}} \\ \underset{(n\times m)}{\bar{A}_{ba}} & \underset{(n\times n)}{\bar{A}_{bb}} \end{bmatrix} \begin{Bmatrix} \underset{(m\times 1)}{\mathrm{d}\varepsilon_{a}} \\ \underset{(n\times 1)}{\mathrm{d}\varepsilon_{b}} \end{Bmatrix} = \begin{bmatrix} \underset{(m\times 6)}{\bar{D}_{a}} \\ \underset{(n\times 6)}{0} \end{bmatrix} \left\{ \underset{(6\times 1)}{\mathrm{d}\bar{\varepsilon}} \right\}. \tag{8.15}$$

The bar notation over the components of the (A) matrix denotes the new arrangement of the terms of the original matrix. Once (8.14) or (8.15) is solved, the incremental stress in each of the subcells and the average stress of the unit cell can be back-calculated using the incremental stress–strain relations. The incremental strain-concentration matrices are expressed, using (8.3) and (8.14), by

$$\begin{bmatrix} B^{(1)} \\ B^{(2)} \\ \vdots \\ B^{(N)} \end{bmatrix} = \begin{bmatrix} & & \\ & A^{-1} & \\ & & \end{bmatrix} \begin{bmatrix} & & \\ & D & \\ & & \end{bmatrix}. \tag{8.16}$$

The stress analysis of a micromechanical unit cell becomes a straightforward procedure as a result of this formulation. Given an average strain increment and the history of deformations in the subcells, the strain-interaction matrices are formed using (8.16). The strain increments are subsequently formed in each of the subcells followed by the corresponding stress increments. This procedure is a linearized incremental stress analysis and will be referred to as the *trial state*. If only this linearized trial analysis is used, two types of error will result at each trial increment and will accumulate during the analysis. It is important to mention, however, that the strain compatibility and traction continuity constraints are exactly satisfied by the trial state which is composed of tangential approximations. The first error occurs in the strain increments because the strain-interaction matrices are derived using the tangent stiffness matrices of the subcells at the beginning of the increment. The second error occurs as a result of using the tangent stiffness to compute the stress increment. Therefore, a *correction scheme* must be used to accurately account for the nonlinear constitutive (with or without damage) material behavior (prediction) and its associated error in the incremental micromechanical equations. New general correction algorithms have been derived for different CDC type micromechanical models with nonlinear and time-dependent behavior, e.g., [16, 18, 20, 22–24].

A four-cell micromodel is formulated next using the previous tangential and stress-update formulations. This model was originally formulated using the method of cells (MOCs), e.g., [2–6]. Aboudi's MOC or GMOC has been shown to be well suited for highly nonlinear matrix response, such as that exhibited by metal matrix composites. However, integration of the MOC formulation in general 3D analysis of composite structures can be tremendously enhanced using the proposed numerical formulation because of the large computational effort that is needed to be performed at each material point (Gaussian point) of the FEA. Therefore, it is important to employ the above efficient stress-update and stress-correction formulations for this model that are suitable for nonlinear

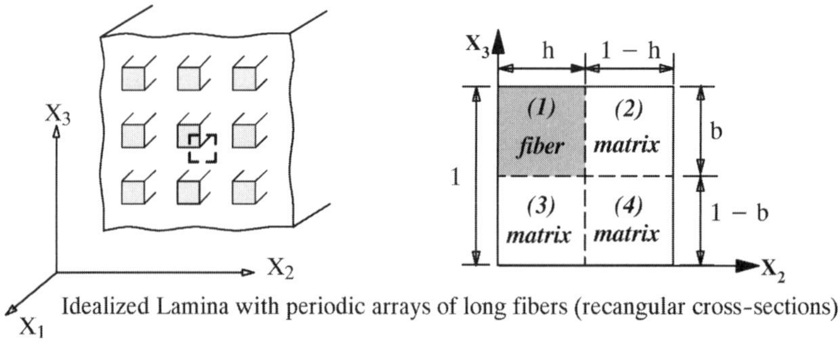

Fig. 8.2. Unit cell micromodel for unidirectional reinforced composites

structural analysis. Next, an incremental formulation of the four-cell model is presented in terms of the average stresses and strains in the subcells. New stress-update and stress-correction algorithms are developed which significantly reduce the computational effort that is needed. The new algorithms are formulated given a constant average strain rate for each time step, which makes them suitable for integration with FE constitutive framework.

The micromechanical model is shown in Fig. 8.2. The unidirectional composite, which consists of long fibers arranged unidirectionally in the matrix system, is idealized as doubly periodic array of fibers with rectangular cross-section. A quarter UC that consists of four subcells is modeled due to symmetry. The first subcell is a fiber constituent, while subcells 2–4 represent the matrix constituents. The long fibers are aligned in the x_1-direction. The other cross-section directions are referred to as the *transverse directions*. The x_3-direction is called the out-of-plane axis or lamina thickness direction. The total volume of the UC is taken to be equal to one. The volumes of the four subcells are:

$$V_1 = bh, \quad V_2 = h(1-b), \quad V_3 = b(1-h), \quad V_4 = (1-h)(1-b). \qquad (8.17)$$

The notations used for the stress and strain vectors are:

$$\begin{aligned} \mathrm{d}\sigma_k^{(\alpha)} &= \{\mathrm{d}\sigma_{11}, \mathrm{d}\sigma_{22}, \mathrm{d}\sigma_{33}, \mathrm{d}\tau_{12}, \mathrm{d}\tau_{13}, \mathrm{d}\tau_{23}\}, \quad \alpha = 1,\ldots,4, \\ \mathrm{d}\varepsilon_k^{(\alpha)} &= \{\mathrm{d}\varepsilon_{11}, \mathrm{d}\varepsilon_{22}, \mathrm{d}\varepsilon_{33}, \mathrm{d}\gamma_{12}, \mathrm{d}\gamma_{13}, \mathrm{d}\gamma_{23}\}, \quad k = 1,\ldots,6. \end{aligned} \qquad (8.18)$$

The 3D nonlinear constitutive integration for the fiber and matrix constituents is performed separately for each subcell. The fiber is linear elastic and transversely isotropic, while the matrix medium is viscoelastic.

The homogenization of the micromodel should satisfy displacement and traction continuity. Perfect bond is assumed along the interfaces of the subcells. In the fiber direction, the four subcells satisfy the same strain continuity relation. The axial average stress definition is used as a second independent relation to relate the effective axial stress to the stresses in the subcells. The following equations summarize the relations in the axial mode

$$\begin{aligned} &\mathrm{d}\varepsilon_1^{(1)} = \mathrm{d}\varepsilon_1^{(2)} = \mathrm{d}\varepsilon_1^{(3)} = \mathrm{d}\varepsilon_1^{(4)} = \mathrm{d}\bar{\varepsilon}_1, \\ &V_1 \mathrm{d}\sigma_1^{(1)} + V_2 \mathrm{d}\sigma_1^{(2)} + V_3 \mathrm{d}\sigma_1^{(3)} + V_4 \mathrm{d}\sigma_1^{(4)} = \mathrm{d}\bar{\sigma}_1, \end{aligned} \tag{8.19}$$

where overbar denotes an overall average quantity over the unit cell.

Along the interfaces between the subcells with normal in the x_2-direction, the in-plane stress components σ_{22} and τ_{12} must satisfy traction continuity conditions. The total strain components ε_{22} and γ_{12} from subcells 1 and 2 and subcells 3 and 4, respectively, should also satisfy strain compatibility conditions. These relations are written in an incremental form as:

$$\begin{aligned} &\mathrm{d}\sigma_2^{(1)} = \mathrm{d}\sigma_2^{(2)}, \\ &\mathrm{d}\sigma_2^{(3)} = \mathrm{d}\sigma_2^{(4)}, \\ &\frac{V_1}{V_1+V_2}\mathrm{d}\varepsilon_2^{(1)} + \frac{V_2}{V_1+V_2}\mathrm{d}\varepsilon_2^{(2)} = \mathrm{d}\bar{\varepsilon}_2, \\ &\frac{V_3}{V_3+V_4}\mathrm{d}\varepsilon_2^{(3)} + \frac{V_4}{V_3+V_4}\mathrm{d}\varepsilon_2^{(4)} = \mathrm{d}\bar{\varepsilon}_2, \end{aligned} \tag{8.20}$$

$$\begin{aligned} &\mathrm{d}\sigma_4^{(1)} = \mathrm{d}\sigma_4^{(2)}, \\ &\mathrm{d}\sigma_4^{(3)} = \mathrm{d}\sigma_4^{(4)}, \\ &\frac{V_1}{V_1+V_2}\mathrm{d}\varepsilon_4^{(1)} + \frac{V_2}{V_1+V_2}\mathrm{d}\varepsilon_4^{(2)} = \mathrm{d}\bar{\varepsilon}_4, \\ &\frac{V_3}{V_3+V_4}\mathrm{d}\varepsilon_4^{(3)} + \frac{V_4}{V_3+V_4}\mathrm{d}\varepsilon_4^{(4)} = \mathrm{d}\bar{\varepsilon}_4. \end{aligned} \tag{8.21}$$

Considering interfaces between subcells with normal in the x_3-direction, the out-of-plane stress components σ_{33} and τ_{13} must satisfy traction continuity conditions. The total strain components ε_{33} and γ_{13} from subcells 1 and 3 and subcells 2 and 4, respectively, should also satisfy strain compatibility conditions. These relations are expressed in incremental form as:

$$\begin{aligned}
&\mathrm{d}\sigma_3^{(1)} = \mathrm{d}\sigma_3^{(3)},\\
&\mathrm{d}\sigma_3^{(2)} = \mathrm{d}\sigma_3^{(4)},\\
&\frac{V_1}{V_1+V_3}\mathrm{d}\varepsilon_3^{(1)} + \frac{V_3}{V_1+V_3}\mathrm{d}\varepsilon_3^{(3)} = \mathrm{d}\bar{\varepsilon}_3,\\
&\frac{V_2}{V_2+V_4}\mathrm{d}\varepsilon_3^{(2)} + \frac{V_4}{V_2+V_4}\mathrm{d}\varepsilon_3^{(4)} = \mathrm{d}\bar{\varepsilon}_3,
\end{aligned} \tag{8.22}$$

$$\begin{aligned}
&\mathrm{d}\sigma_5^{(1)} = \mathrm{d}\sigma_5^{(3)},\\
&\mathrm{d}\sigma_5^{(2)} = \mathrm{d}\sigma_5^{(4)},\\
&\frac{V_1}{V_1+V_3}\mathrm{d}\varepsilon_5^{(1)} + \frac{V_3}{V_1+V_3}\mathrm{d}\varepsilon_5^{(3)} = \mathrm{d}\bar{\varepsilon}_5,\\
&\frac{V_2}{V_2+V_4}\mathrm{d}\varepsilon_5^{(3)} + \frac{V_4}{V_2+V_4}\mathrm{d}\varepsilon_5^{(4)} = \mathrm{d}\bar{\varepsilon}_5.
\end{aligned} \tag{8.23}$$

Finally, both types of interfaces should satisfy transverse shear stress continuity. Therefore, the transverse shear stresses in the four subcells are equal to the effective transverse shear stress. The transverse shear strains from the four subcells in the average strain definition are used to express the relations with the effective transverse shear strain of the UC. The transverse shear relations are summarized as:

$$\begin{aligned}
&\mathrm{d}\sigma_6^{(1)} = \mathrm{d}\sigma_6^{(2)} = \mathrm{d}\sigma_6^{(3)} = \mathrm{d}\sigma_6^{(4)} = \mathrm{d}\bar{\sigma}_6,\\
&V_1\mathrm{d}\varepsilon_6^{(1)} + V_2\mathrm{d}\varepsilon_6^{(2)} + V_3\mathrm{d}\varepsilon_6^{(3)} + V_4\mathrm{d}\varepsilon_6^{(4)} = \mathrm{d}\bar{\varepsilon}_6.
\end{aligned} \tag{8.24}$$

Equations (8.19)–(8.24) along with the stress–strain relations within each fiber and matrix subcells complete the micromechanical formulation of the unidirectional lamina. These relations are used in incremental (rate) form due to the nonlinear constitutive relations in the matrix subcells. Next, the strain components in the subcells are grouped into two parts (a) and (b). The first part corresponds to the incremental compatibility equations and the second part is the traction continuity relations (homogeneous equations). The two groups of strain vectors are defined by:

$$\underset{(1\times13)}{\mathrm{d}\varepsilon_{\mathrm{a}}^{\mathrm{T}}} = \left\{ \begin{array}{l} \mathrm{d}\varepsilon_1^{(1)}, \mathrm{d}\varepsilon_1^{(2)}, \mathrm{d}\varepsilon_1^{(3)}, \mathrm{d}\varepsilon_1^{(4)}, \mathrm{d}\varepsilon_2^{(1)}, \mathrm{d}\varepsilon_2^{(3)}, \mathrm{d}\varepsilon_4^{(1)}, \\ \qquad \mathrm{d}\varepsilon_4^{(3)}, \mathrm{d}\varepsilon_3^{(1)}, \mathrm{d}\varepsilon_3^{(2)}, \mathrm{d}\varepsilon_5^{(1)}, \mathrm{d}\varepsilon_5^{(2)}, \mathrm{d}\varepsilon_6^{(1)} \end{array} \right\}, \quad (8.25)$$

$$\underset{(1\times11)}{\mathrm{d}\varepsilon_{\mathrm{b}}^{\mathrm{T}}} = \left\{ \begin{array}{l} \mathrm{d}\varepsilon_2^{(2)}, \mathrm{d}\varepsilon_2^{(4)}, \mathrm{d}\varepsilon_4^{(2)}, \mathrm{d}\varepsilon_4^{(4)}, \mathrm{d}\varepsilon_3^{(3)}, \mathrm{d}\varepsilon_3^{(4)} \\ \qquad \mathrm{d}\varepsilon_5^{(3)}, \mathrm{d}\varepsilon_5^{(4)}, \mathrm{d}\varepsilon_6^{(2)}, \mathrm{d}\varepsilon_6^{(3)}, \mathrm{d}\varepsilon_6^{(4)} \end{array} \right\}. \quad (8.26)$$

Equations (8.19)–(8.24) can be expressed in terms of the strain increments in the subcells after substituting the incremental stress–strain relations. The rearrangement of the strain increments allows this set of equations to be transformed into the previous general CDC form presented in (8.15), where $\mathrm{d}R_\sigma$ is the residual form of the stress relations (traction continuity) expressed incrementally in terms of the strains in the subcells. The matrices that appear in (8.15) for this UC micromodel are listed below and can be identified by examining (8.17)–(8.24). The nonzero terms of $\bar{A}_{\mathrm{ab}}$ are:

$$\begin{aligned} &\bar{A}_{\mathrm{ab}}(5,1) = \bar{A}_{\mathrm{ab}}(6,2) = \bar{A}_{\mathrm{ab}}(7,3) = \bar{A}_{\mathrm{ab}}(8,4) = \bar{A}_{\mathrm{ab}}(13,9) = \frac{1-h}{h}, \\ &\bar{A}_{\mathrm{ab}}(9,5) = \bar{A}_{\mathrm{ab}}(10,6) = \bar{A}_{\mathrm{ab}}(11,7) = \bar{A}_{\mathrm{ab}}(12,8) = \bar{A}_{\mathrm{ab}}(13,10) = \frac{1-b}{b}, \quad (8.27) \\ &\bar{A}_{\mathrm{ab}}(13,11) = \frac{(1-b)(1-h)}{bh}. \end{aligned}$$

The nonzero terms of $\bar{D}_{\mathrm{a}}$ are:

$$
\begin{aligned}
&\bar{D}_{\mathrm{a}}(1,1)=\bar{D}_{\mathrm{a}}(2,1)=\bar{D}_{\mathrm{a}}(3,1)=\bar{D}_{\mathrm{a}}(4,1)=1,\\
&\bar{D}_{\mathrm{a}}(5,2)=\bar{D}_{\mathrm{a}}(6,2)=\bar{D}_{\mathrm{a}}(7,4)=\bar{D}_{\mathrm{a}}(8,4)=\frac{1}{h},\\
&\bar{D}_{\mathrm{a}}(9,3)=\bar{D}_{\mathrm{a}}(10,3)=\bar{D}_{\mathrm{a}}(11,5)=\bar{D}_{\mathrm{a}}(12,5)=\frac{1}{b},\\
&\bar{D}_{\mathrm{a}}(13,6)=\frac{1}{bh}.
\end{aligned}
\tag{8.28}
$$

The terms of $\bar{A}_{\mathrm{ba}}$ and $\bar{A}_{\mathrm{bb}}$ matrices are listed in (8.29) and (8.30), respectively. Only the inverse of the (11×11) submatrix in (8.15) is needed to solve for $\mathrm{d}\varepsilon_{\mathrm{a}}$ and $\mathrm{d}\varepsilon_{\mathrm{b}}$. The strain-concentration matrices are determined by solving $\mathrm{d}R_{\sigma}=0$ and $\mathrm{d}R_{\varepsilon}=0$ equations as previously outlined

$$
\bar{\mathrm{A}}_{\mathrm{ba}} = \left[\begin{array}{c|c|c|c|c|c|c|c|c|c|c|c|c}
-C^{(1)}_{12} & C^{(2)}_{12} & 0 & 0 & -C^{(1)}_{22} & 0 & 0 & 0 & -C^{(1)}_{23} & C^{(2)}_{23} & 0 & C^{(2)}_{25} & 0 \\ \hline
0 & 0 & -C^{(3)}_{12} & C^{(4)}_{12} & 0 & -C^{(3)}_{22} & 0 & -C^{(3)}_{24} & 0 & 0 & 0 & 0 & 0 \\ \hline
0 & C^{(2)}_{14} & 0 & 0 & 0 & 0 & -C^{(1)}_{44} & 0 & 0 & C^{(2)}_{34} & 0 & C^{(2)}_{45} & 0 \\ \hline
0 & 0 & -C^{(3)}_{14} & C^{(4)}_{14} & 0 & -C^{(3)}_{24} & 0 & -C^{(3)}_{44} & 0 & 0 & 0 & 0 & 0 \\ \hline
-C^{(1)}_{12} & 0 & C^{(3)}_{13} & 0 & -C^{(1)}_{23} & C^{(3)}_{23} & 0 & C^{(3)}_{34} & -C^{(1)}_{22} & 0 & 0 & 0 & 0 \\ \hline
0 & -C^{(2)}_{13} & 0 & C^{(4)}_{13} & 0 & 0 & 0 & 0 & 0 & -C^{(2)}_{33} & 0 & -C^{(2)}_{35} & 0 \\ \hline
0 & 0 & C^{(3)}_{15} & 0 & 0 & C^{(3)}_{25} & 0 & C^{(3)}_{45} & 0 & 0 & -C^{(1)}_{44} & 0 & 0 \\ \hline
0 & -C^{(2)}_{15} & 0 & C^{(4)}_{15} & 0 & 0 & 0 & 0 & 0 & -C^{(2)}_{35} & 0 & -C^{(2)}_{55} & 0 \\ \hline
0 & C^{(2)}_{16} & 0 & 0 & 0 & 0 & 0 & 0 & 0 & C^{(2)}_{36} & 0 & C^{(2)}_{56} & -C^{(1)}_{66} \\ \hline
0 & 0 & C^{(3)}_{16} & 0 & 0 & C^{(3)}_{26} & 0 & C^{(3)}_{46} & 0 & 0 & 0 & 0 & -C^{(1)}_{66} \\ \hline
0 & -C^{(2)}_{16} & 0 & C^{(4)}_{16} & 0 & 0 & 0 & 0 & 0 & -C^{(2)}_{36} & 0 & -C^{(2)}_{56} & 0
\end{array}\right]
\tag{8.29}
$$

$$
\overline{A}_{bb} = \left[\begin{array}{c|c|c|c|c|c|c|c|c|c|c}
C^{(2)}_{22} & 0 & C^{(2)}_{24} & 0 & 0 & 0 & 0 & 0 & C^{(2)}_{26} & 0 & 0 \\ \hline
0 & C^{(4)}_{22} & 0 & C^{(4)}_{24} & -C^{(3)}_{23} & C^{(4)}_{23} & -C^{(3)}_{25} & C^{(4)}_{25} & 0 & -C^{(3)}_{26} & C^{(4)}_{26} \\ \hline
C^{(2)}_{24} & 0 & C^{(2)}_{44} & 0 & 0 & 0 & 0 & 0 & C^{(2)}_{46} & 0 & 0 \\ \hline
0 & C^{(4)}_{24} & 0 & C^{(4)}_{44} & -C^{(3)}_{34} & C^{(4)}_{34} & -C^{(3)}_{45} & C^{(4)}_{45} & 0 & -C^{(3)}_{46} & C^{(4)}_{46} \\ \hline
0 & 0 & 0 & 0 & C^{(3)}_{33} & 0 & C^{(3)}_{35} & 0 & 0 & C^{(3)}_{36} & 0 \\ \hline
-C^{(2)}_{23} & C^{(4)}_{23} & -C^{(2)}_{34} & C^{(4)}_{34} & 0 & C^{(4)}_{33} & 0 & C^{(4)}_{35} & -C^{(2)}_{36} & 0 & C^{(4)}_{36} \\ \hline
0 & 0 & 0 & 0 & C^{(3)}_{35} & 0 & C^{(3)}_{55} & 0 & 0 & C^{(3)}_{56} & 0 \\ \hline
-C^{(2)}_{25} & C^{(4)}_{25} & -C^{(2)}_{45} & C^{(4)}_{45} & 0 & C^{(4)}_{35} & 0 & C^{(4)}_{55} & -C^{(2)}_{56} & 0 & C^{(4)}_{56} \\ \hline
C^{(2)}_{26} & 0 & C^{(2)}_{46} & 0 & 0 & 0 & 0 & 0 & C^{(2)}_{66} & 0 & 0 \\ \hline
0 & 0 & 0 & 0 & C^{(3)}_{36} & 0 & C^{(3)}_{56} & 0 & 0 & C^{(3)}_{66} & 0 \\ \hline
-C^{(2)}_{26} & C^{(4)}_{26} & -C^{(2)}_{46} & C^{(4)}_{46} & 0 & C^{(4)}_{36} & 0 & C^{(4)}_{56} & -C^{(2)}_{66} & 0 & C^{(4)}_{66}
\end{array}\right] \qquad (8.30)
$$

The micromechanical relations are exact only in the case of linear stress–strain relations in the fiber and matrix subcells. Due to the nonlinear response in one or more of the subcells, the incremental relations will usually violate the constitutive equations. Thus, an iterative correction scheme is needed to satisfy both the micromechanical constraints and the constitutive equations. The tasks for the micromechanical algorithm can be stated as: Given history variables in the subcells from previous converged solution and a constant average strain rate for the unit cell within the current time increment, update the effective stress, the effective stiffness, and the history variables at the end of the increment.

8.4 The Sublaminate Model

The term *thick-section composite laminate* is herein defined as a multi-layered laminate with a thickness greater than approximately 1/4 in. in which it is possible to identify a repeating sublaminate. The sublaminate consists of the smallest repeating stacking sequence of laminae (Fig. 8.1). Unlike the terminology used in the theory of plates and shells for thick-section structures, the term *thick-section*, used herein, does not always imply the existence of relatively large interlaminar stress and strain distributions. For example, a composite with thickness $t = 1.0$ in., stacking sequence of $[90_2/0]_{30S}$, and a radius of curvature of 80 in. and above is not normally considered to be a thick-section in the context of plate or shell theory; however, it does fall under the current definition for thick-section and multilayered composites.

A thick-section composite structure may exhibit nearly linear overall structural behavior almost up to failure. However, nonlinear structural response can also arise locally especially in the presence of edge effects and structural discontinuities, such as crack tips, holes, and cutouts. These stress concentrations can have a significant impact on damage and the overall behavior close to and postultimate. Therefore, it is important to develop analysis methods for thick-section composites that include both 3D and nonlinear capabilities.

In cases where plate or shell structural modeling is appropriate, and where the material response can be considered linear, the cross-sectional stiffness and flexural rigidities can be calculated using the classical laminate theory (CLT). The stress and strain distributions for the individual laminae can be back-calculated from central plane strains and curvatures obtained from the structural analysis. However, the existence of hundreds of individual plies in a typical thick-section composite structure makes ply-by-ply nonlinear analysis impractical. In this case, where nonlinear response and damage and/or interlaminar stress effects are important, the integration of the stress–strain relations must be performed numerically for all plies during the analysis. As a result, the CLT method is very difficult to apply. On the other hand, the large number of repeating plies in thick-section composites produces a structure which is, in effect, much more homogeneous than sections with a small number of plies. This allows for a through-thickness homogenization or "smearing" procedure to be used effectively at the laminate level.

A sublaminate is constructed from the smallest through-thickness repeating stacking sequence. The effective response of the sublaminate is used to define an equivalent nonlinear homogeneous continuum. The response of a nonlinear response of the composite structure is determined by the effective homogeneous behavior of material points with the few layers of generated by the sublaminate at each integration point in the FE structural mesh.

Figure 8.1 describes a 3D framework for the nonlinear structural analysis of thick-section laminated composites. This general approach involves a material model with a two-level hierarchy: a micromechanical model of a unidirectional lamina and a sublaminate model. The CDC class of micromodels, developed in the previous sections, can be used to represent the nonlinear response of each layer in the sublaminate. The 3D lamination theory is used in a nonlinear formulation to synthesize the effective continuum response of the sublaminate model. The sublaminate material model is designed to function through a standard interface with structural analysis packages; the micromodels and sublaminate model, which constitute a two-level material model, are hidden from element and structural-level processes. The main features of the nonlinear sublaminate model are formulated in this section using the 3D lamination theory to derive thermomechanical effective stress–strain relations for the sub-laminate. The instantaneous effective stiffness and thermal coefficients of the sublaminate are then derived.

The 3D lamination theory was used by Pagano [30, 31] to derive the cross-sectional properties of anisotropic laminates. In Pagano's work, the extensional, flexural, and coupling stiffness were generated for the entire section of the laminate. The formulation of Pagano is based on a linear elasticity solution which satisfies the external boundary conditions as well as the interface traction and displacement continuity between the layers. Therefore, the treatment is limited to the case where all force and moment resultants, and surface forces, are spatially constant. Sun and Li [39] employed the same deformation field and derived the extensional effective stiffness properties for a repeating sublaminate. Pecknold [32, 33] presented a simpler approach for deriving the effective properties of anisotropic sub-laminate using the same fundamental patterns of deformations. They proposed using a nonlinear micromechanical model for the individual layers in the sublaminate, which is integrated within a 3D FE analysis of thick-section composites. The cross-section is homogenized at a selected number of integration points, and the nonlinear response of the equivalent homogeneous material is obtained using the 3D lamination theory. The key idea is to replace the sublaminate by a well-defined equivalent homo-geneous material. The properties of this material are determined by requiring that the two, sublaminate and equivalent continuum, respond identically when subjected to certain fundamental patterns of stress or strain. To satisfy displacement and traction continuity conditions between the layers, homogeneous in-plane strain and homogeneous out-of-plane stress patterns are used. The conjugate stress and strain components are determined as through-thickness weighted averages of the corresponding quantities in each layer of the sublaminate.

Effective thermoelastic moduli of the sublaminate are derived considering each layer to be a general anisotropic material. The equivalent response of the repeating sublaminate is assumed to represent the average response of a multilayered and thick-section laminate in some local region. In this equivalent continuum approach, certain selected patterns of stress and strain are used to define equivalence between the actual sublaminate and an equivalent homogeneous continuum. This derivation is an exact solution when the applied in-plane strain and the out-of-plane surface tractions are spatially constant. Under these conditions, the interlaminar or out-of-plane stresses are also constant. The field equations in the form of equilibrium, interface traction, and displacement continuity are reviewed next.

It is useful to use the following notations for the stress and strain vectors:

$$\{\sigma\} = \begin{Bmatrix} \sigma_x \\ \sigma_y \\ \tau_{xy} \\ -- \\ \sigma_z \\ \tau_{xz} \\ \tau_{yz} \end{Bmatrix} \equiv \begin{Bmatrix} \sigma_\mathrm{i} \\ -- \\ \sigma_\mathrm{o} \end{Bmatrix}, \quad \{\varepsilon\} = \begin{Bmatrix} \varepsilon_x \\ \varepsilon_y \\ \gamma_{xy} \\ -- \\ \varepsilon_z \\ \gamma_{xz} \\ \gamma_{yz} \end{Bmatrix} \equiv \begin{Bmatrix} \varepsilon_\mathrm{i} \\ -- \\ \varepsilon_\mathrm{o} \end{Bmatrix}. \tag{8.31}$$

In (8.31), all stress and strain vectors are partitioned into in-plane and out-of-plane components. Perfect bond interface conditions are assumed between the layers. Therefore, the displacement continuity conditions at the interfaces are expressed as

$$u_\mathrm{i}^{(k)}\left(x_1, x_2, x_3^{(k)} = \frac{t^{(k)}}{2}\right) = u_\mathrm{i}^{(k+1)}\left(x_1, x_2, x_3^{(k+1)} = -\frac{t^{(k+1)}}{2}\right), \tag{8.32}$$

where (k) denotes a layer number and is the through-thickness direction. It can be easily verified using (8.32) that the interface displacement continuity equations require that

$$\varepsilon_{\mathrm{i}}^{(k)}\left(x_1, x_2, x_3^{(k)}=\frac{t^{(k)}}{2}\right)=\varepsilon_{\mathrm{i}}^{(k+1)}\left(x_1, x_2, x_3^{(k+1)}=-\frac{t^{(k+1)}}{2}\right). \tag{8.33}$$

The equilibrium and traction continuity at the interface are expressed by:

$$\sigma_{\mathrm{o}}^{(k)}\left(x_1, x_2, x_3^{(k)}=\frac{t^{(k)}}{2}\right)=\sigma_{\mathrm{o}}^{(k+1)}\left(x_1, x_2, x_3^{(k+1)}=-\frac{t^{(k+1)}}{2}\right). \tag{8.34}$$

The interface equations expressed in (8.33) and (8.34), along with the external boundary conditions and the stress–strain relations of the laminae, define the governing field equations.

The effective thermoelastic moduli of the anisotropic sublaminate are defined next. The formulation is expressed in terms of total stress and strain vectors. This formulation also applies in an incremental form for the effective tangent stiffness matrix of the laminate in the case where nonlinear stress–strain response in a layer is considered.

The through-thickness effective stress and strain vectors are defined as

$$\begin{Bmatrix} \bar{\varepsilon} \\ -- \\ \bar{\sigma} \end{Bmatrix}=\frac{1}{t}\sum_{k=1}^{N}\int_{-t^{(k)}/2}^{t^{(k)}/2}\begin{Bmatrix} \varepsilon \\ -- \\ \sigma \end{Bmatrix}^{(k)} \mathrm{d}x_3^{(k)}, \tag{8.35}$$

where t is the sublaminate thickness, $t^{(k)}$ is the kth lamina thickness, and N is the number of laminae in the sublaminate. An upper bar is used to denote a sublaminate effective (global) quantity.

The fundamental patterns of applied stress and strain should include spatially homogeneous patterns to generate the effective response of the multilayered laminate in the form of an equivalent continuum. Examining the previous field equations of the 3D lamination theory, it can be seen that

$\varepsilon_{\mathrm{i}}^{(k)} = \bar{\varepsilon}_{\mathrm{i}}$ and $\sigma_{\mathrm{o}}^{(k)} = \bar{\sigma}_{\mathrm{o}}$ satisfy the interface conditions (8.33) and (8.34), respectively. Therefore, the applied spatially homogeneous patterns are:

$$\begin{Bmatrix} \bar{\varepsilon}_{\mathrm{i}} \\ -- \\ \bar{\sigma}_{\mathrm{o}} \end{Bmatrix}. \tag{8.36}$$

Using the interface conditions (8.33) and (8.34), along with the applied homogeneous patterns (8.36), the stress and strain for each lamina (*k*) are

$$\begin{Bmatrix} \varepsilon_{\mathrm{i}} \\ -- \\ \sigma_{\mathrm{o}} \end{Bmatrix}^{(k)} = \begin{Bmatrix} \bar{\varepsilon}_{\mathrm{i}} \\ -- \\ \bar{\sigma}_{\mathrm{o}} \end{Bmatrix}, \quad k = 1,\ldots,N, \tag{8.37}$$

where (8.37) identically satisfies (8.35). The conjugate effective stresses and strains are expressed, using (8.35), by:

$$\begin{Bmatrix} \bar{\sigma}_{\mathrm{i}} \\ -- \\ \bar{\varepsilon}_{\mathrm{o}} \end{Bmatrix} = \sum_{k=1}^{N} \frac{t^{(k)}}{t} \begin{Bmatrix} \sigma_{\mathrm{i}} \\ -- \\ \varepsilon_{\mathrm{o}} \end{Bmatrix}^{(k)}. \tag{8.38}$$

Next, the displacement and traction continuity relations (8.37) are used, along with the stress–strain relations for the laminae, to form the sublaminate effective thermoelastic properties. To this end, the stress–strain relations for each lamina are expressed in the sublaminate global coordinate system. Using the notation of (8.31), the stress–strain relations for lamina (*k*) are expressed in the global coordinate system as

$$\begin{Bmatrix} \sigma_{\mathrm{i}} \\ \sigma_{\mathrm{o}} \end{Bmatrix} = \begin{bmatrix} C_{\mathrm{ii}} & C_{\mathrm{io}} \\ C_{\mathrm{oi}} & C_{\mathrm{oo}} \end{bmatrix} \begin{Bmatrix} \varepsilon_{\mathrm{i}} - \Delta T \alpha_{\mathrm{i}} \\ \varepsilon_{\mathrm{o}} - \Delta T \alpha_{\mathrm{o}} \end{Bmatrix}, \tag{8.39}$$

where ΔT is the change in temperature and α denotes the thermal expansion coefficients of the lamina. Next, (8.39) is partially inverted to obtain the form

$$\begin{Bmatrix} \sigma_{\mathrm{i}} \\ \varepsilon_{\mathrm{o}} - \Delta T \alpha_{\mathrm{o}} \end{Bmatrix} = \begin{bmatrix} A & B \\ -B^{\mathrm{T}} & D \end{bmatrix} \begin{Bmatrix} \varepsilon_{\mathrm{i}} - \Delta T \alpha_{\mathrm{i}} \\ \sigma_{\mathrm{o}} \end{Bmatrix},$$

$$D = C_{\mathrm{oo}}^{-1}, \qquad (8.40)$$

$$\text{in which} \quad B = C_{\mathrm{io}} C_{\mathrm{oo}}^{-1},$$

$$A = C_{\mathrm{ii}} - C_{\mathrm{io}} C_{\mathrm{oo}}^{-1} C_{\mathrm{oi}}.$$

Equation (8.40) is merely a general and convenient representation of the stress–strain relations that apply in general to any material with an invertible tangent stiffness matrix. However, this compact representation is an instrumental precursor to the simplified formulation for the effective stiffness of anisotropic laminates. The matrices B and D, defined in (8.40), should not be confused with the coupling and flexural stiffness matrices of the CLT, which are often denoted by the same symbols. Equation (8.40) is further simplified and written as:

$$\begin{Bmatrix} \sigma_{\mathrm{i}} \\ \varepsilon_{\mathrm{o}} \end{Bmatrix} = \begin{bmatrix} A & B \\ -B^{\mathrm{T}} & D \end{bmatrix} \begin{Bmatrix} \varepsilon_{\mathrm{i}} \\ \sigma_{\mathrm{o}} \end{Bmatrix} + \Delta T \begin{bmatrix} -A & 0 \\ B^{\mathrm{T}} & I \end{bmatrix} \begin{Bmatrix} \alpha_{\mathrm{i}} \\ \alpha_{\mathrm{o}} \end{Bmatrix}. \qquad (8.41)$$

Substituting (8.37) and (8.41), the stress–strain relation for the kth lamina is:

$$\begin{Bmatrix} \sigma_{\mathrm{i}} \\ \varepsilon_{\mathrm{o}} \end{Bmatrix}^{(k)} = \begin{bmatrix} A & B \\ -B^{\mathrm{T}} & D \end{bmatrix}^{(k)} \begin{Bmatrix} \bar{\varepsilon}_{\mathrm{i}} \\ \bar{\sigma}_{\mathrm{o}} \end{Bmatrix} + \Delta T \begin{bmatrix} -A & 0 \\ B^{\mathrm{T}} & I \end{bmatrix}^{(k)} \begin{Bmatrix} \alpha_{\mathrm{i}} \\ \alpha_{\mathrm{o}} \end{Bmatrix}^{(k)}. \qquad (8.42)$$

Equations (8.38) and (8.42) are used to express the effective in-plane stress and out-of-plane strain of the sublaminate as:

$$\begin{Bmatrix} \bar{\sigma}_\mathrm{i} \\ \bar{\varepsilon}_\mathrm{o} \end{Bmatrix} = \sum_{k=1}^{N} \frac{t^{(k)}}{t} \begin{bmatrix} A & B \\ -B^\mathrm{T} & D \end{bmatrix}^{(k)} \begin{Bmatrix} \bar{\varepsilon}_\mathrm{i} \\ \bar{\sigma}_\mathrm{o} \end{Bmatrix} + \Delta T \sum_{k=1}^{N} \frac{t^{(k)}}{t} \begin{bmatrix} -A & 0 \\ B^\mathrm{T} & I \end{bmatrix}^{(k)} \begin{Bmatrix} \alpha_\mathrm{i} \\ \alpha_\mathrm{o} \end{Bmatrix}^{(k)}. \quad (8.43)$$

It is desired to write the effective stress–strain relations for the sublaminate, similar to the laminae stress–strain relations in (8.39) and (8.41), as:

$$\begin{Bmatrix} \bar{\sigma}_\mathrm{i} \\ \bar{\sigma}_\mathrm{o} \end{Bmatrix} = \begin{bmatrix} \bar{C}_\mathrm{ii} & \bar{C}_\mathrm{io} \\ \bar{C}_\mathrm{oi} & \bar{C}_\mathrm{oo} \end{bmatrix}^{(k)} \begin{Bmatrix} \bar{\varepsilon}_\mathrm{i} - \Delta T \bar{\alpha}_\mathrm{i} \\ \bar{\varepsilon}_\mathrm{o} - \Delta T \bar{\alpha}_\mathrm{o} \end{Bmatrix}. \quad (8.44)$$

To determine the effective stiffnesses in (8.44), it is first partially inverted in the form

$$\begin{Bmatrix} \bar{\sigma}_\mathrm{i} \\ \bar{\varepsilon}_\mathrm{o} \end{Bmatrix} = \begin{bmatrix} \bar{A} & \bar{B} \\ -\bar{B}^\mathrm{T} & \bar{D} \end{bmatrix} \begin{Bmatrix} \bar{\varepsilon}_\mathrm{i} \\ \bar{\sigma}_\mathrm{o} \end{Bmatrix} + \Delta T \begin{bmatrix} -\bar{A} & 0 \\ \bar{B}^\mathrm{T} & I \end{bmatrix} \begin{Bmatrix} \bar{\alpha}_\mathrm{i} \\ \bar{\alpha}_\mathrm{o} \end{Bmatrix},$$

$$\text{where} \quad \begin{aligned} \bar{D} &= \bar{C}_\mathrm{oo}^{-1}, \\ \bar{B} &= \bar{C}_\mathrm{io} \bar{C}_\mathrm{oo}^{-1}, \\ \bar{A} &= \bar{C}_\mathrm{ii} - \bar{C}_\mathrm{io} \bar{C}_\mathrm{oo}^{-1} \bar{C}_\mathrm{oi}, \end{aligned} \quad (8.45)$$

and then compared with (8.43) to yield

$$\begin{bmatrix} \bar{A} & \bar{B} \\ -\bar{B}^\mathrm{T} & \bar{D} \end{bmatrix} = \sum_{k=1}^{N} \frac{t^{(k)}}{t} \begin{bmatrix} A & B \\ -B^\mathrm{T} & D \end{bmatrix}^{(k)} \quad (8.46)$$

and

$$\begin{bmatrix} -\bar{A} & 0 \\ \bar{B}^\mathrm{T} & I \end{bmatrix} \begin{Bmatrix} \bar{\alpha}_\mathrm{i} \\ \bar{\alpha}_\mathrm{o} \end{Bmatrix} = \sum_{k=1}^{N} \frac{t^{(k)}}{t} \begin{bmatrix} -A & 0 \\ B^\mathrm{T} & I \end{bmatrix}^{(k)} \begin{Bmatrix} \alpha_\mathrm{i} \\ \alpha_\mathrm{o} \end{Bmatrix}^{(k)}. \quad (8.47)$$

Therefore, the effective tangent stiffness matrix of the sublaminate can be calculated from (8.46) as:

$$\bar{C} = \begin{bmatrix} \bar{C}_{\mathrm{ii}} & \bar{C}_{\mathrm{io}} \\ \bar{C}_{\mathrm{oi}} & \bar{C}_{\mathrm{oo}} \end{bmatrix}, \quad \text{where} \quad \begin{aligned} \bar{C}_{\mathrm{oo}} &= \bar{D}^{-1}, \\ \bar{C}_{\mathrm{io}} &= \bar{B}\bar{D}^{-1}, \\ \bar{C}_{\mathrm{oi}} &= \bar{C}_{\mathrm{io}}^{\mathrm{T}}, \\ \bar{C}_{\mathrm{ii}} &= \bar{A} + \bar{B}\bar{D}^{-1}\bar{B}^{T}. \end{aligned} \tag{8.48}$$

The effective thermal expansion coefficients of the sublaminate are derived from (8.47) as:

$$\begin{aligned} \bar{\alpha}_{\mathrm{i}} &= \bar{A}^{-1}\sum_{k=1}^{N}\frac{t^{(k)}}{t}A^{(k)}\alpha_{\mathrm{i}}^{(k)}, \\ \bar{\alpha}_{\mathrm{o}} &= -\bar{B}^{\mathrm{T}}\bar{\alpha}_{\mathrm{i}} + \sum_{k=1}^{N}\frac{t^{(k)}}{t}\left(B^{(k)^{\mathrm{T}}}\alpha_{\mathrm{i}}^{(k)} + \alpha_{\mathrm{o}}^{(k)}\right). \end{aligned} \tag{8.49}$$

8.5 Multiscale Analysis of Thick-Section Pultruded Composites

Thick-section structural components can be manufactured by the pultrusion process using several reinforcement layers with a polymeric resin system. The result is a long prismatic structural component that can have shapes similar to the standard steel shapes, such as wide flange, channels, and angle sections. Layers with continuous filament mat (CFM) reinforcement are heavily used in this process to provide large volume and bind the unidirectional roving layers in transition areas of the prismatic cross-section. The relatively fast production of pultruded structural components allows for mass manufacturing and cost competitive composite materials. Pultruded composites can have a thickness ranging from 1/16 to 1 in. and an overall fiber volume fraction (FVF) of 0.3–0.5.

Due to a relatively low FVF, large thickness, and existence of the polymeric matrix, a nonlinear and time-dependent mechanical response is present. The nonlinear responses coupled with the time-dependent responses affect the overall behavior of the structure. It is important to form a combined material and structural framework that can simultaneously provide effective material and structural analyses.

Several experimental and analytical studies on time-independent buckling and postbuckling of thick-section composite (pultruded) columns have been performed. Vakanier et al. [41] performed linearized buckling analysis of columns with stocky wide-flange (WF) cross-sections using FE models. Most of geometries studied have relatively small slenderness ratio and allow for local flange buckling. Barbero and Tomblin [12] investigated global buckling loads for I-shape long columns. As expected, the results for the long columns were well predicted by the Euler buckling theory. Barbero et al. [10] studied interaction between the local and global buckling modes on intermediate length composite WF columns. Zureick and Scott [42] presented design guidelines for fiber-reinforced polymer (FRP) slender structural members under axial compression load based on global buckling limit states. Axially compression tests on box and I-shape cross-sections of E-glass/vinylester composite specimens were performed. Bank and Yin [8] investigated the postbuckling regime of composite I-beams, focusing on the web-flange junction failure. They performed a FE analysis using a node separation technique to simulate the local separation of the flange from the web following a local buckling of the flange.

Micromechanical modeling approaches in thick-section pultruded composite materials have been studied. Barbero [9, 37] proposed a linear micromechanical modeling approach to generate the overall effective stiffness of thick-section (pultruded) composite material systems. Their micromodel, which employed the periodic microstructure formulae of [11], was combined with the classical lamination theory and mechanics of laminated beams approach to determine the overall effective stiffness of composite beams.

Haj-Ali et al. [16, 19, 20, 22, 23] proposed a multiscale modeling approach for the nonlinear elastic responses of thick-section composite systems with roving and CFM reinforcements. This multiscale framework is schematically illustrated in Fig. 8.3.

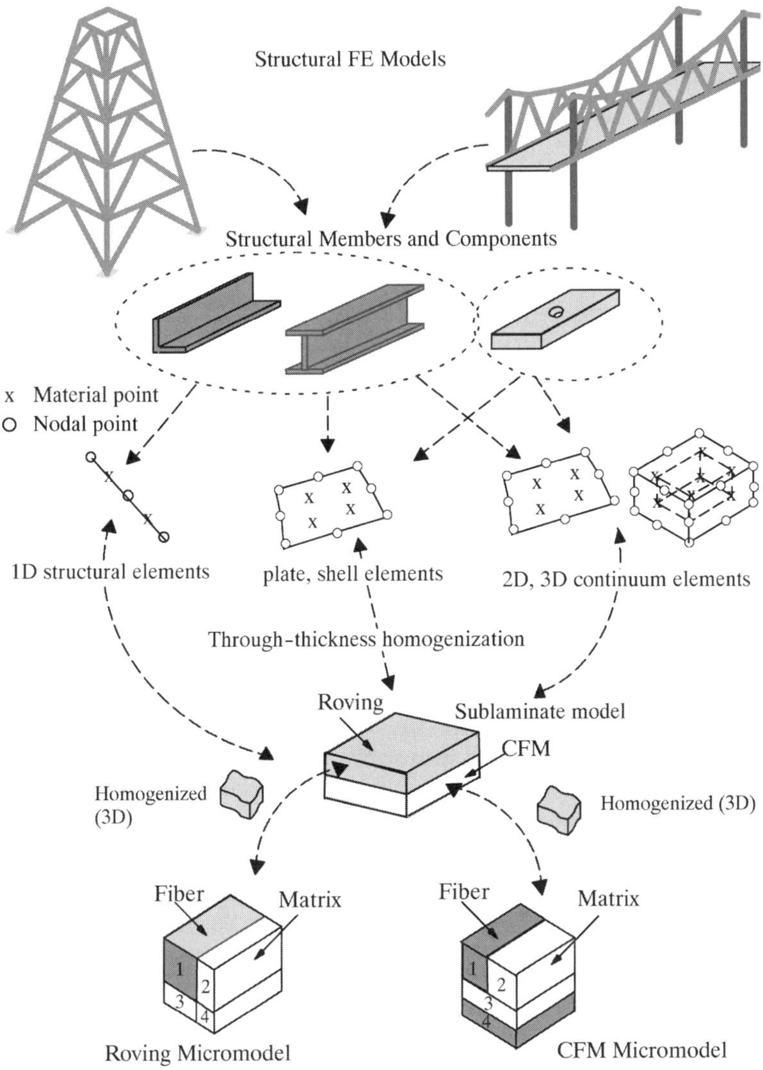

Fig. 8.3. Multiscale structural and micromechanical framework for the analysis of pultruded composite materials and structures (adapted from [21])

Both structural and continuum finite elements can be used. Different micromechanical models are employed for the reinforcement systems in the pultruded layers. Haj-Ali et al. [19, 20] were the first to introduce this combined nonlinear 3D micromechanical modeling approach for pultruded composites. Their framework was time independent and focused mainly on applying continuum elements. The structural level of this framework, shown in Fig. 8.3, represents FE models for pultruded structures using 1D (beam, truss), 2D (plane, shell), and 3D (brick) elements. A sublaminate model is used at each Gauss point in every element to generate a 3D effective anisotropic nonlinear viscoelastic response of the combined roving and CFM layers. Different stress–strain constraints are imposed on the sublaminate model to properly interface with the 1D, 2D, or 3D elements. In the case where beam and shell elements are used, a uniaxial stress–strain relation and a plane-stress condition must be imposed, respectively. In the lower level, two previously developed 3D micromechanical material models are then used for the roving and CFM layers [20, 21, 23].

Deformation plasticity with a Ramberg–Osgood strain–stress curve was used to describe the nonlinear static response in the matrix system. Tension and compression tests on off-axis coupons made of E-glass fiber and vinylester matrix were conducted to calibrate linear and nonlinear elastic material properties and to validate their proposed modeling approach. It was shown that the nonlinearity in the material response is significant for the off-axis specimens even for relatively low load levels. Haj-Ali and Muliana [21, 23] proposed a new *nonlinear* and *time-dependent* formulation for multilayered composite systems. The Schapery constitutive model [22, 35] was used for the time-dependent behavior of the isotropic matrix. Short-term off-axis creep compression tests on E-glass/vinylester coupons were performed to calibrate the nonlinear time-dependent material properties and to verify their proposed micromodels. Haj-Ali and Muliana [22] performed an integrated micromechanical–structural modeling approach to analyze creep behaviors of composite structures. The previously developed time-dependent micromechanical models of roving and CFM were used within continuum typed elements in general FE analyses. Creep responses of notched plate composite plates were used to validate the integrated micromechanical–structural modeling approach.

Studies on long-term behaviors of thick-section layered FRP composites are limited. Several experimental and analytical works have been focused on the uniaxial viscoelastic responses on composite specimens and structural components [7, 27, 28, 36, 38]. Spence [38] performed tests on unidirectional glass/epoxy pultruded coupons under compression creep for 840 h. The creep response was pronounced for applied loads higher than 30% of the compressive strength. Bank and Mosallam [7] conducted long-term creep tests for E-glass/vinylester structural members with continuous strand mat and roving layers. A plane portal frame, 6 ft. high × 9 ft. wide, was tested under a load level of 25% of the ultimate failure for 10,000 h.

Scott and Zureick [36] conducted compression creep tests on pultruded E-glass/vinylester coupons cut from I-shape pultruded sections in their longitudinal direction. The samples were subjected to three different stress levels for the duration, up to 16 months. Findley power law model was used with a constant exponent to calibrate their uniaxial time-dependent behavior. The stress-dependent coefficients were calibrated from the short duration tests (1,000 h).

Long-term behavior, such as creep buckling of thick-section composites, has not been widely addressed, especially using a multiscale modeling approach. This study presents a multiscale nonlinear viscoelastic framework for the time-dependent behavior of composite materials and structural systems. The multiscale modeling approach is a local–global structural framework that can integrate different constitutive material models at the lowest material scales, namely the fiber and polymeric matrix constituents. It can also generate the effective nonlinear anisotropic continuum response that is needed at the structural level. The composite system studied is reinforced with roving and CFM layers. The constitutive characterization for the fiber and matrix constituents is performed at the lowest level of the multiscale modeling framework. The Schapery's nonlinear single integral model is applied for the polymeric matrix system. It is assumed that both matrix constituents in the roving and CFM unit cells have the same isotropic nonlinear viscoelastic properties. An overall

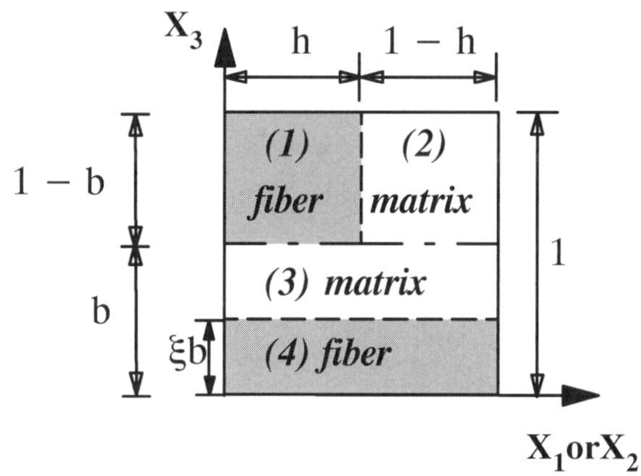

Fig. 8.4. Unit cell for the CFM layers in a pultruded composite

numerical integration method of the Schapery's nonlinear viscoelastic model is formulated at the matrix level. Linearized micromechanical formulations and a stress-correction algorithm for roving, CFM, and sublaminate systems were employed. Static postbuckling and creep collapse analyses to demonstrate the capability of the proposed framework were conducted by Haj-Ali and Muliana [21–23].

Micromechanical formulations for the two-layer sublaminate, CFM, and roving systems have been formulated and described in Haj-Ali and Muliana [21, 23]. These relations are used in incremental (rate) form due to the nonlinear constitutive relations in the matrix subcells. Equation (8.15) can be used to define the micromechanical relations for the sublaminate, CFM, and roving unit cells. The roving micromodel follows a square, four-cell unit cell model previously formulated for the unidirectional composites. The strain vectors corresponding to displacement compatibility and traction continuity in the two-layer sublaminate model are

$$\underset{(1\times 9)}{\mathrm{d}\varepsilon_{\mathrm{a}}^{\mathrm{T}}} = \{\mathrm{d}\varepsilon_{\mathrm{i}}^{(\mathrm{R})}, \mathrm{d}\varepsilon_{\mathrm{i}}^{(\mathrm{C})}, \mathrm{d}\varepsilon_{\mathrm{o}}^{(\mathrm{C})}\}, \quad \underset{(1\times 3)}{\mathrm{d}\varepsilon_{\mathrm{b}}^{\mathrm{T}}} = \{\mathrm{d}\varepsilon_{\mathrm{o}}^{(\mathrm{R})}\}, \tag{8.50}$$

where the subscript i and o indicate in-plane and out-of-plane strain components, respectively. The displacement compatibility and traction continuity strain vectors in the CFM model, shown in Fig. 8.4, are:

$$\begin{aligned} \underset{(1\times 12)}{\mathrm{d}\varepsilon_{\mathrm{a}}^{\mathrm{T}}} &= \{\mathrm{d}\varepsilon_{\mathrm{i}}^{(1)}, \mathrm{d}\varepsilon_{\mathrm{o}}^{(1)}, \mathrm{d}\varepsilon_{\mathrm{i}}^{(3)}, \mathrm{d}\varepsilon_{\mathrm{i}}^{(4)}\}, \\ \underset{(1\times 12)}{\mathrm{d}\varepsilon_{\mathrm{b}}^{\mathrm{T}}} &= \{\mathrm{d}\varepsilon_{\mathrm{i}}^{(2)}, \mathrm{d}\varepsilon_{\mathrm{o}}^{(2)}, \mathrm{d}\varepsilon_{\mathrm{o}}^{(3)}, \mathrm{d}\varepsilon_{\mathrm{o}}^{(4)}\}. \end{aligned} \tag{8.51}$$

The $\bar{A}$ and $\bar{D}$ matrices in (8.15) are determined for sublaminate and CFM micromodels. After some algebraic manipulations, the $\bar{A}$ and $\bar{D}$ matrices for the sublaminate model are

$$\bar{A} = \begin{bmatrix} \underset{(3\times 3)}{I} & \underset{(3\times 3)}{0} & \underset{(3\times 3)}{0} & \underset{(3\times 3)}{0} \\ \underset{(3\times 3)}{0} & \underset{(3\times 3)}{I} & \underset{(3\times 3)}{0} & \underset{(3\times 3)}{0} \\ \underset{(3\times 3)}{0} & \underset{(3\times 3)}{0} & \underset{(3\times 3)}{I} & \underset{(3\times 3)}{\frac{t_{\mathrm{R}}}{t_{\mathrm{C}}} I} \\ \underset{(3\times 3)}{-C_{\mathrm{oi}}^{(\mathrm{C})}} & \underset{(3\times 3)}{C_{\mathrm{oi}}^{(\mathrm{R})}} & \underset{(3\times 3)}{-C_{\mathrm{oo}}^{(\mathrm{C})}} & \underset{(3\times 3)}{C_{\mathrm{oo}}^{(\mathrm{R})}} \end{bmatrix}, \tag{8.52}$$

$$\bar{D}^{\mathrm{T}} = \begin{bmatrix} \underset{(3\times3)}{I} & \underset{(3\times3)}{I} & \underset{(3\times3)}{0} & \underset{(3\times3)}{0} \\ \underset{(3\times3)}{0} & \underset{(3\times3)}{0} & \underset{(3\times3)}{\frac{t}{t_C} I} & \underset{(3\times3)}{0} \end{bmatrix}, \tag{8.53}$$

where C is the subcell stiffness matrix and the subscripts "i" and "o" in the stiffness matrices indicate in-plane and out-of-plane components, respectively. The variables t, t_C, and t_R refer to sublaminate, CFM, and roving thicknesses, respectively. The $\bar{A}$ and $\bar{D}$ matrices for the CFM micromodel are expressed as

$$\bar{\mathrm{A}} = \left[\begin{array}{cccc|cccc}
\underset{(3\,\mathrm{x}\,3)}{\mathrm{I}} & \underset{(3\,\mathrm{x}\,3)}{0} & \underset{(3\,\mathrm{x}\,3)}{0} & \underset{(3\,\mathrm{x}\,3)}{0} & \underset{(3\,\mathrm{x}\,3)}{0} & \underset{(3\,\mathrm{x}\,3)}{0} & \underset{(3\,\mathrm{x}\,3)}{0} & \underset{(3\,\mathrm{x}\,3)}{0} \\
\underset{(3\,\mathrm{x}\,3)}{0} & \underset{(3\,\mathrm{x}\,3)}{\mathrm{I}} & \underset{(3\,\mathrm{x}\,3)}{0} & \underset{(3\,\mathrm{x}\,3)}{0} & \underset{(3\,\mathrm{x}\,3)}{0} & \underset{(3\,\mathrm{x}\,3)}{0} & \underset{(3\,\mathrm{x}\,3)}{0} & \underset{(3\,\mathrm{x}\,3)}{0} \\
\underset{(3\,\mathrm{x}\,3)}{0} & \underset{(3\,\mathrm{x}\,3)}{0} & \underset{(3\,\mathrm{x}\,3)}{\mathrm{I}} & \underset{(3\,\mathrm{x}\,3)}{0} & \underset{(3\,\mathrm{x}\,3)}{\frac{\mathrm{V}_2}{\mathrm{V}_1}\mathrm{I}} & \underset{(3\,\mathrm{x}\,3)}{0} & \underset{(3\,\mathrm{x}\,3)}{0} & \underset{(3\,\mathrm{x}\,3)}{0} \\
\underset{(3\,\mathrm{x}\,3)}{0} & \underset{(3\,\mathrm{x}\,3)}{0} & \underset{(3\,\mathrm{x}\,3)}{0} & \underset{(3\,\mathrm{x}\,3)}{\mathrm{I}} & \underset{(3\,\mathrm{x}\,3)}{0} & \underset{(3\,\mathrm{x}\,3)}{\frac{\mathrm{V}_2}{\mathrm{V}_1}\mathrm{I}} & \underset{(3\,\mathrm{x}\,3)}{\frac{\mathrm{V}_3}{\mathrm{V}_1}\mathrm{I}} & \underset{(3\,\mathrm{x}\,3)}{\frac{\mathrm{V}_4}{\mathrm{V}_1}\mathrm{I}} \\
\hline
\underset{(3\,\mathrm{x}\,3)}{0} & \underset{(3\,\mathrm{x}\,3)}{0} & \underset{(3\,\mathrm{x}\,3)}{-\mathrm{C}^{(1)}_{\mathrm{oi}}} & \underset{(3\,\mathrm{x}\,3)}{-\mathrm{C}^{(1)}_{\mathrm{oo}}} & \underset{(3\,\mathrm{x}\,3)}{\mathrm{C}^{(2)}_{\mathrm{oi}}} & \underset{(3\,\mathrm{x}\,3)}{\mathrm{C}^{(2)}_{\mathrm{oo}}} & \underset{(3\,\mathrm{x}\,3)}{0} & \underset{(3\,\mathrm{x}\,3)}{0} \\
\underset{(3\,\mathrm{x}\,3)}{0} & \underset{(3\,\mathrm{x}\,3)}{0} & \underset{(3\,\mathrm{x}\,3)}{-\mathrm{C}^{(1)}_{\mathrm{ii}}} & \underset{(3\,\mathrm{x}\,3)}{-\mathrm{C}^{(1)}_{\mathrm{io}}} & \underset{(3\,\mathrm{x}\,3)}{\mathrm{C}^{(2)}_{\mathrm{ii}}} & \underset{(3\,\mathrm{x}\,3)}{\mathrm{C}^{(2)}_{\mathrm{io}}} & \underset{(3\,\mathrm{x}\,3)}{0} & \underset{(3\,\mathrm{x}\,3)}{0} \\
\underset{(3\,\mathrm{x}\,3)}{\mathrm{C}^{(3)}_{\mathrm{oi}}} & \underset{(3\,\mathrm{x}\,3)}{0} & \underset{(3\,\mathrm{x}\,3)}{-\mathrm{C}^{(1)}_{\mathrm{oi}}} & \underset{(3\,\mathrm{x}\,3)}{-\mathrm{C}^{(1)}_{\mathrm{oo}}} & \underset{(3\,\mathrm{x}\,3)}{0} & \underset{(3\,\mathrm{x}\,3)}{0} & \underset{(3\,\mathrm{x}\,3)}{\mathrm{C}^{(3)}_{\mathrm{oo}}} & \underset{(3\,\mathrm{x}\,3)}{0} \\
\underset{(3\,\mathrm{x}\,3)}{0} & \underset{(3\,\mathrm{x}\,3)}{\mathrm{C}^{(4)}_{\mathrm{oi}}} & \underset{(3\,\mathrm{x}\,3)}{-\mathrm{C}^{(1)}_{\mathrm{oi}}} & \underset{(3\,\mathrm{x}\,3)}{-\mathrm{C}^{(1)}_{\mathrm{oo}}} & \underset{(3\,\mathrm{x}\,3)}{0} & \underset{(3\,\mathrm{x}\,3)}{0} & \underset{(3\,\mathrm{x}\,3)}{0} & \underset{(3\,\mathrm{x}\,3)}{\mathrm{C}^{(4)}_{\mathrm{oo}}}
\end{array}\right] \tag{8.54}$$

$$\overline{D}^T = \left[\begin{array}{cccc|cccc} \underset{(3\,x\,3)}{I} & \underset{(3\,x\,3)}{I} & \underset{(3\,x\,3)}{\frac{V_1 + V_2}{V_1} I} & \underset{(3\,x\,3)}{0} & \underset{(3\,x\,3)}{0} & \underset{(3\,x\,3)}{0} & \underset{(3\,x\,3)}{0} & \underset{(3\,x\,3)}{0} \\ \hline \underset{(3\,x\,3)}{0} & \underset{(3\,x\,3)}{0} & \underset{(3\,x\,3)}{0} & \underset{(3\,x\,3)}{\frac{V}{V_1} I} & \underset{(3\,x\,3)}{0} & \underset{(3\,x\,3)}{0} & \underset{(3\,x\,3)}{0} & \underset{(3\,x\,3)}{0} \end{array}\right] \tag{8.55}$$

where the variable V_α indicates unit volume fraction of the subcell number α.

The linearized micromechanical relations, derived in incremental formulation, will usually violate the constitutive equations because of the nonlinear and time-dependent response in the matrix subcells. An iterative correction scheme is needed to satisfy both the micromechanical constraints and the constitutive equations. The linearized micromechanical relations with tangential material matrices are used to generate trial incremental stresses and strains for the subcells (trial solution). The total micromechanical relations are then used to define a residual error for each micromodel. This residual is then used to correct the trial solution. This process is repeated until a converged solution that satisfies both micromechanical and nonlinear equations is reached.

A correction algorithm is needed in every nested micromodel. The input is in the form of an applied incremental strain. The stress–strain states from the previous step and the history variables are the known variables for each nested micromodel and scale. Inside the sublaminate model, the strain increment is distributed to the roving and CFM micromodels. Iterative correction schemes are developed separately for the roving and CFM models to minimize the errors and satisfy the micromechanical and constitutive relations. At the roving and CFM systems, the current stress–strain states, the tangent stiffness, together with the history variables are updated and sent to the sublaminate level. An iterative procedure is also performed inside the sublaminate system until the actual stress–strain relations, as well as homogenization constraints, are satisfied. Any iteration at the sublaminate level requires the full calculation procedure from the lower levels of the framework. Once all levels of error are satisfied, the sublaminate effective nonlinear continuum state is defined and communicated to the FE structural level.

8.6 Applications

Haj-Ali and Muliana [22, 23] examined the ability of the multiscale formulation to predict the nonlinear viscoelastic behavior of composite materials and structures. The effective response is generated from calibrated in situ properties of the matrix and fiber constituents. To that end, different creep tests available in the literature are used. Off-axis test results are available for glass/epoxy [26] and T300/5208 graphite/epoxy [40] composites. Prediction of the calibrated model is examined against test results not used in the calibration process. The current approach is similar but employs a refined 3D micromodel that can ultimately be used in both 2D and 3D structural models. Creep test results on glass/epoxy off-axis composite specimens reported by Lou and Schapery were used for validation of the current modeling approach. Linear viscoelastic calibration was performed using results from the 45° off-axis specimen under the lowest applied axial stress (1.382 ksi). Overall, the nonlinear calibration strikes a balance between all nonlinear curves as seen in Figs. 8.5 and 8.6.

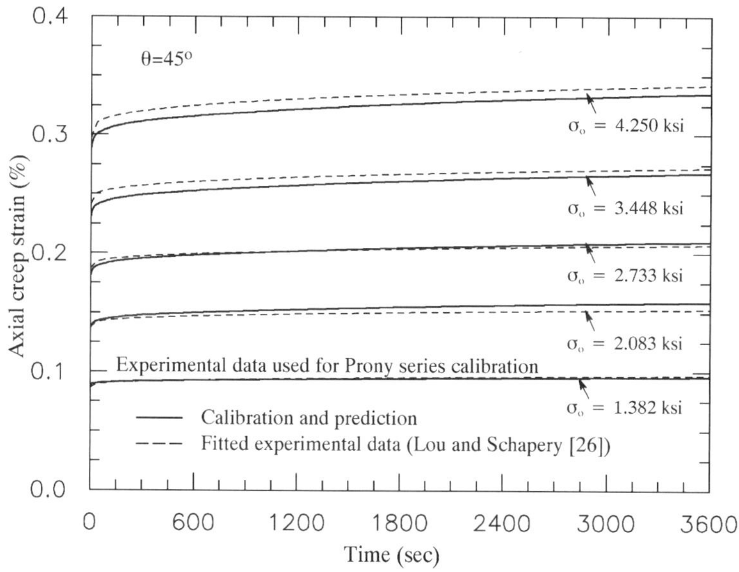

Fig. 8.5. Axial creep strain for 45° off-axis glass-epoxy laminate (adapted from [23])

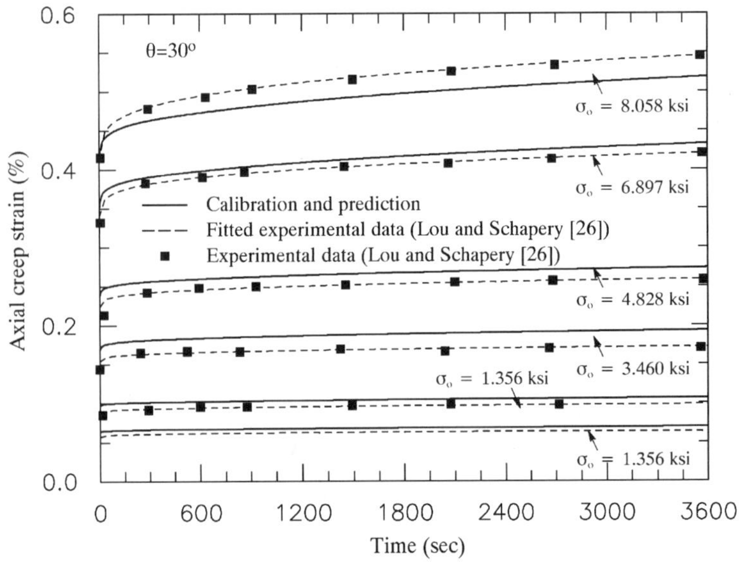

Fig. 8.6. Axial creep strain for 30° off-axis glass-epoxy laminate (adapted from [23])

Haj-Ali and Kilic [19, 20] applied the multiscale nonlinear models to predict the nonlinear response of thick-section pultruded plates under multiaxial stress–strain states. To this end, a series of tests were performed with off-axis notched pultruded plates subject to uniaxial compression. Four uniaxial strain gages were attached on the back and front surfaces of each plate. The off-axis angles used in the notched plates were 0°, 15°, 30°, 45°, 60°, and 90°. The multiscale framework was implemented in the [1] general FE code.

Figure 8.7 shows the off-axis tests and FE results in the form of the fourth strain gage (G4) that is the closest to the edge of the hole. A consistent softer response is shown from the G4 curves for all off-axis tests. As expected, the level of nonlinear response varies from the different plates, increasing with the increased off-axis angle. Compression buckling was monitored by examining the difference in the back-to-back strains. It was concluded that no significant out-of-plane bending occurred in these tests prior to reaching their ultimate loading states. Overall, very good prediction is shown by the FE models compared with the experimental results. This confirms the ability of the micromodels to predict a nonlinear multiaxial state of deformation with stress concentration due to the notched holes.

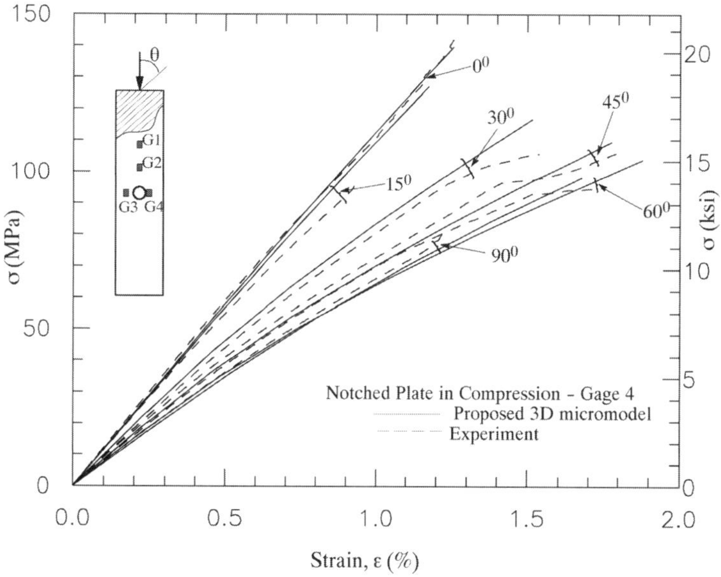

Fig. 8.7. Experimental and FE remote compression stress vs. axial strain measured from G4 for different off-axis orientations of notched pultruded composite plate (adapted from [20])

A nested 3D micromechanical and structural framework is presented for the nonlinear elastic, viscoelastic, and crack growth analysis of thick-section and multilayered composite materials. Micromodels for a unidirectional lamina, an in-plane random medium phenomenological model, and a sublaminate model for effective continuum response of a through-thickness repeating stacking sequence are formulated and integrated in the material–structural modeling approach. Numerical predictor–corrector stress-update algorithms are used and nested in all hierarchies. The analysis framework is general and can be used with a displacement-based nonlinear FE code. It can incorporate time, temperature, and moisture effects. Several structural simulations are demonstrated to illustrate the effectiveness of the framework.

Haj-Ali and El-Hajjar [13–15, 17, 18] performed experimental and numerical analyses to determine the translayer Mode I and Mode II fracture toughness and crack growth of a thick-section FRP pultruded composite using the proposed multiscale material and structural framework.

Crack growth tests with crack tip opening displacement (CTOD) measurements were conducted for different crack length to width ratios (*a*/*W*) as shown in Fig. 8.8. Figure 8.9 shows that the computational cohesive models were calibrated and used to predict Mode I and Mode II crack growth for eccentrically loaded single-edge-notch (tension), ESE(T), and notched butterfly specimens using a modified Arcan fixture. Figure 8.10 illustrates the predicted capability of the combined cohesive FE modeling with the proposed multiscale constitutive framework compared to experimental results in the form of load vs. CTOD and in the form of crack length vs. CTOD. The validity of the multiscale modeling approach before the onset of crack growth was also investigated using a new infrared thermography (IR) method [13–15]. The multiscale constitutive framework, combined with cohesive models, was shown to be effective in predicting the failure load and crack growth behavior in thick-section composites.

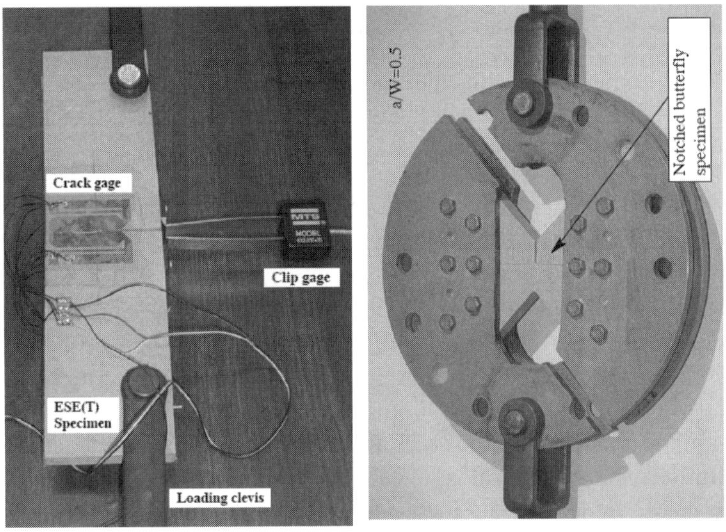

(a) (b)

Fig. 8.8. Two test setups: (**a**) fracture test with crack monitoring gage bonded to ESE(T) specimen and (**b**) modified Arcan fixture used for Mode II fracture testing (adapted from [18])

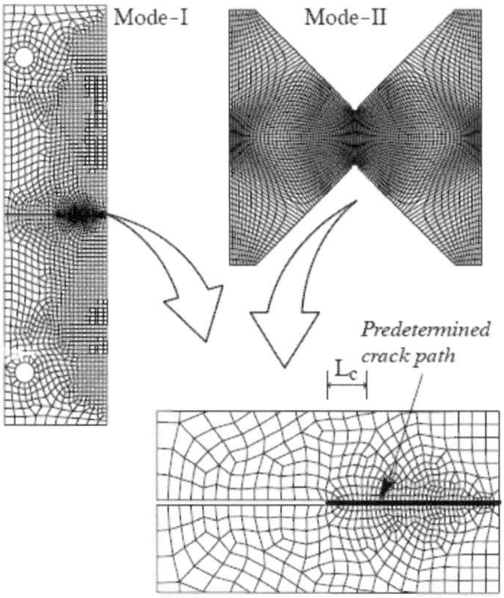

Fig. 8.9. Cohesive fracture modeling of through-crack in thick-section composites using FE cohesive models along with the multiscale material and structural modeling framework (adapted from [18])

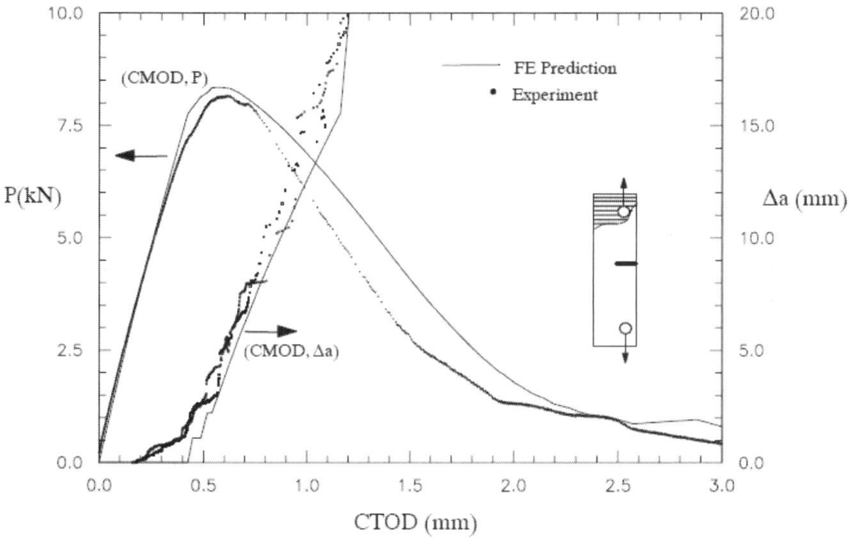

Fig. 8.10. Predicted crack growth for transversely oriented reinforcement of an ESE(T) composite specimen with $a/W = 0.3$ using cohesive FE model with the multiscale material framework (adapted from [18])

8.7 Conclusions

Effective multiscale micromechanical and structural frameworks are formulated for the nonlinear analysis of multilayered and thick-section FRP composite materials. The micromechanical models explicitly recognize the response of the layered composite systems and their fiber–matrix constituents. A unified numerical formulation is presented for a class of simplified micromodels. The formulation allows efficient application of the micromodels in large-scale nonlinear FE analysis of composite structures, where very large numbers of micromechanical calculations are carried out. Therefore, it is crucial that the proposed nonlinear stress-update and stress-correction algorithms are efficient and robust. The proposed material and structural framework were shown to be effective in predicting the nonlinear and time-dependent responses of laminated and pultruded FRP composites.

Acknowledgments

The author wishes to acknowledge past and continued support by the National Science Foundation (NSF) through the Civil and Mechanical Systems (CMS) Division under grant numbers 0409514 and 9876080. Previous partial grants from Lockheed Martin Co. under the Composites Affordability Initiative are also appreciated.

References

1. ABAQUS, Hibbitt, Karlsson, and Sorensen, Inc., User's Manual, Version 6.3 (2006)
2. Aboudi J (1987) Closed form constitutive equations for metal matrix composites. Int. J. Eng. Sci., 25, 1229
3. Aboudi J (1989) Micromechanical analysis of composites by the method of cells. Appl. Mech. Rev., 42(7), 193–221
4. Aboudi J (1990) The nonlinear behavior of unidirectional and laminated composites – a micromechanical approach. J. Reinf. Plast. Compos. 9, 13–32
5. Aboudi J (1991) Mechanics of Composite Materials – A Unified Micromechanical Approach. Elsevier, New York
6. Aboudi J (1993) Constitutive behavior of multiphase metal matrix composites with interfacial damage by the generalized cells model. In: Damage in Composite Materials, ed. Voyiadjis GZ, Elsevier, New York

7. Bank LC and Mosallam AS (1990) Creep and failure of a fill-size fiber reinforced plastic pultruded frame. ASME Petrol. Div. Pub.: Compos. Mater. Technol., 32, 49–56
8. Bank LC and Yin J (1999) Failure of web-flange junction in postbuckled pultruded i-beams. J. Compos. Constr., 3(4), 177–184
9. Barbero EJ (1991) Pultruded structural shape – from the constituents to the structural behavior. SAMPE J., 27(1), 25–30
10. Barbero EJ, Dede EK, and Jones S (2000) Experimental verification of buckling-mode interaction in intermediate-length composite columns. Int. J. Solids Struct., 37(29), 3919–3934
11. Barbero EJ and Luciano R (1995) Micromechanical formulas for the relaxation tensor of linear viscoelastic composites with transversely isotropic fibers. Int. J. Solids Struct., 32(13), 1859–1872
12. Barbero EJ and Tomblin J (1993) Euler buckling of thin-walled composite columns. Thin-Walled Struct., 17(4), 237–258
13. El-Hajjar RF and Haj-Ali RM (2003) A quantitative thermoelastic stress analysis method for pultruded composites. Compos. Sci. Technol. J., 63(7), 967–978
14. El-Hajjar RF and Haj-Ali RM (2004) In-plane shear testing of thick-section pultruded frp composites using a modified arcan fixture. Compos. Part B: Eng., 35(5), 421–428
15. El-Hajjar RF and Haj-Ali RM (2004) Infrared (IR) thermography for strain analysis in fiber reinforced plastics. Exp. Tech.: Soc. Exp. Mech., 28(2), 19–22
16. El-Hajjar RF and Haj-Ali RM (2005) Mode-I fracture toughness testing of thick section frp composites using the ese(t) specimen. Eng. Fract. Mech., 72(4), 631–643
17. Haj-Ali RM and El-Hajjar RF (2003) Crack propagation analysis of mode-i fracture in pultruded composites using micromechanical constitutive models. Mech. Mater., 35(9), 885–902
18. Haj-Ali RM, El-Hajjar RF, and Muliana HA (2006) Cohesive fracture modeling in thick-section composites. Eng. Fract. Mech. (in press)
19. Haj-Ali RM and Kilic MH (2002) Nonlinear behavior of pultruded FRP composites. Compos. Part B: Eng., 33(3), 173
20. Haj-Ali RM, Kilic MH, and Zureick AH (2001) Three-dimensional micromechanics based constitutive framework for analysis of pultruded composite structures. ASCE J. Eng. Mech., 127(7), 653–660
21. Haj-Ali RM and Muliana AH (2003) Micromechanical constitutive framework for the nonlinear viscoelastic behavior of pultruded composite materials. Int. J. Solids Struct., 40(5), 1037–1057
22. Haj-Ali RM and Muliana AH (2004) Numerical finite element formulation of the Schapery nonlinear viscoelastic material model. Int. J. Numer. Meth. Eng., 59(1), 25–45
23. Haj-Ali RM and Muliana AH (2004) A multiscale constitutive framework for the nonlinear analysis of laminated composite materials and structures. Int. J. Solids Struct., 41(13), 3461–3490

24. Haj-Ali RM and Pecknold DA (1996) Hierarchical material models with microstructure for nonlinear analysis of progressive damage in laminated composite structures. Structural Research Series No. 611, UILU-ENG-96-2007, Department of Civil Engineering, University of Illinois at Urbana-Champaign
25. Haj-Ali RM, Pecknold DA, and Ahmad MF (1993) Combined micro-mechanical and structural finite element analysis of laminated composites. ABAQUS Users' Conference, Achen, Germany, pp 233–247
26. Lou YC and Schapery RA (1971) Viscoelastic characterization of a nonlinear fiber-reinforced plastic. J. Compos. Mater., 5, 208–234
27. McClure G and Mohammadi Y (1995) Compression creep of pultruded e-glass reinforced plastic angles. J. Mater. Civil Eng., 7(4), 269–276
28. Mottram JT (1993) Short and long-term structural properties of pultruded beam assemblies fabricated using adhesive bonding. Compos. Struct., 25(1–4), 387–395
29. Muliana AH and Haj-Ali RM (2004) Nested nonlinear viscoelastic and micromechanical models for the analysis of pultruded composite materials and structures. Mech. Mater., 36(11), 1087–1110
30. Pagano NJ (1971) Stress gradients in laminated composite cylinders. J. Compos. Mater., 5, 260–265
31. Pagano NJ (1974) Exact moduli of anisotropic laminates. In: Mechanics of Composite Materials, ed. Sendeckyj GP, Academic, New York, pp 23–44
32. Pecknold DA (1990) A framework for 3-d nonlinear modeling of thick-section composites. DTRC-SME-90/92, David Taylor Research Center, Bethesda, MD
33. Pecknold DA and Haj-Ali RM (1993) Integrated micromechanical/structural analysis of laminated composites. In: Mechanics of Composite Materials – Nonlinear Effects, ed. Hyer MW, SES/ASME/ASCE Joint meeting, Charlottesville, VA, AMD, Vol. 159, pp 197–206
34. Pecknold DA and Rahman S (1994) Micromechanics-based structural analysis of thick laminated composites. Comput. Struct., 51(2), 163–179
35. Schapery RA (1969) On the characterization of nonlinear viscoelastic materials. Polym. Eng. Sci., 9(4), 295
36. Scott DW and Zureick AH (1998) Compression creep of a pultruded e-glass/vinylester composite. Compos. Sci. Technol., 58(8), 1361–1369
37. Sonti SS and Barbero E (1996) Material characterization of pultruded laminates and shapes. J. Reinf. Plast. Compos., 15(7), 701–717
38. Spence BR (1990) Compressive viscoelastic effects (creep) of a unidirectional glass/epoxy composite material. Proceedings of 35th International SAMPE Symposium, April 2–5, Vol. 35(2), pp 1490–1493
39. Sun CT and Li S (1988) Three-dimensional effective elastic constants for thick laminates. J. Compos. Mater., 22, 629–639
40. Tuttle ME and Brinson HF (1986) Prediction of the long-term creep compliance of general composite laminates. Exp. Mech., 26, 89–102

41. Vakanier AR, Zureick A, and Will KM (1991) Predictions of local flange buckling in pultruded shapes by finite element analysis. In: Proceedings of the ASCE Specialty Conference on Advanced Composite Materials in Civil Engineering and Structure, ASCE Material Engineering Division, eds. Iyer SL and Sen R, New York, pp 302–312
42. Zureick A and Scott D (1997) Short-term behavior and design of fiber reinforced polymeric slender members under axial compression. J. Compos. Constr., ASCE, 1(1), 140–149

Chapter 9: Predicting Thermooxidative Degradation and Performance of High-Temperature Polymer Matrix Composites

G.A. Schoeppner[1], G.P. Tandon[2] and K.V. Pochiraju [3]

[1]US Air Force Research Laboratory, Wright-Patterson Air Force Base, OH, USA
[2]University of Dayton Research Institute, Dayton, OH, USA
[3]Stevens Institute of Technology, Hoboken, NJ, USA

9.1 Introduction

Polymer matrix composites (PMCs) used in aerospace high-temperature applications, such as turbine engines and engine-exhaust-washed structures, are known to have limited life due to environmental degradation. Predicting the extended service life of composite structures subjected to mechanical, high temperature, moisture, and corrosive conditions is challenging due to the complex physical, chemical, and thermomechanical mechanisms involved. Additionally, the constituent phases of the material, in particular the matrix phase, continuously evolve with aging time. It is the aging-dependent evolution of the constituent properties that makes prediction of the long-term performance of PMCs in high-temperature environments so challenging. A comprehensive prediction methodology must deal with several complications presented by the highly coupled material aging, damage evolution, and thermooxidation processes.

While carbon fibers may be more resistant to oxidation and have longer relaxation times than the polymer matrix, the mechanical performance of the fiber–matrix interface at high temperatures may be critical to the evolution of damage and composite failure. The properties of the fiber–matrix interface or interphase region[1] are a function of the interaction of the fiber and matrix during the manufacturing process, the chemical compatibility of the fiber, and the sizing (if applicable) on the fiber.

To deal with the aforementioned complex interactions, modeling is expected to incorporate mechanisms active at multiple scales (constituent, microstructural, lamina, and laminate scales) and multiple domains of response (chemical, thermal, and mechanical). The motivation is to develop robust, life-performance, prediction modeling capabilities for designers who currently use cost prohibitive experimentally based design allowables and use weight loss to evaluate the thermal-oxidative stability of material systems.

In the face of such complexity and lack of knowledge regarding material behavior, industry currently relies on the use of environmental "knockdown factors" for the design of composite aerospace structures which operate in aggressive environments. One of the current methodologies used for determining environmental knockdown factors for designing composite aerospace structures is to determine reduced material allowables for each of the lamina properties (laminate ultimate strengths, notched strengths, fatigue, flexure properties, bearing strengths, etc.) for each of the material forms used in the structure. The material forms may include unitape, woven cloth, fiber preforms with mold filling, etc.

The large variety of tests performed on a statistically sufficient number of replicates for the various environmental conditions adds tremendous cost to the materials qualification and design allowables process. In addition, the worst case environmental extreme conditions are typically used for determining reduced materials allowables to help circumvent risk associated with the variability and inaccurate prediction of service environments. In addition to the environmental knockdown factors, damage tolerance knockdown factors are also applied to the designs. The damage tolerance factors may be determined through damaging full-scale components and testing them for a predetermined number of life cycles in a hot/wet environment. Using this approach, the end-of-life properties are measured and used in the allowables determination process for designing components.

The aerospace engine community (where high-temperature polymer matrix composites (HTPMCs) are most prevalent) often designs to end-of-life properties, and it is accepted that at the end of life for certain lightly

---

[1] Although fiber–matrix interface and fiber–matrix interphase are often used interchangeably, we formally define the fiber–matrix interface as the two-dimensional surface defined by the common fiber and matrix surfaces. We define the fiber–matrix interphase as the three-dimensional matrix region directly around the fiber that may have properties distinct from the bulk matrix properties.

loaded structures, the composites may have cracks. On the other hand, for highly loaded or critical structures, the presence of cracks, even at the end of life, is not accepted. This traditional approach for determining material design allowables, taking into account extremes in-service environments, requires significant testing (cost and schedule). For cost savings and risk mitigation, designers often resort to previously qualified materials, thereby negating the potential benefits of new material advancements. By developing a life prediction/performance modeling capability, the cost and time associated with developing material design allowables can provide opportunities for the insertion of new advanced materials.

This chapter focuses on predicting the performance of PMCs in high-temperature environments where thermooxidative conditions are extreme and the material is used near its tolerance limits. For the current design methodology used by aerospace structure designers, composites are considered to be "chemically static," ignoring the evolution of the chemical/environmental degradation and nonelastic mechanical response. However, the empirical design allowables are based on the expected end-of life properties by testing specimens that have been aged under representative service environments for the design life.

This chapter outlines the current state of the art, the shortcomings of existing capabilities, and future challenges for addressing modeling needs for PMCs in high-temperature oxidative environments. It is believed that the approach presented provides the methodology to accurately model both short- and long-term environmental effects on composite laminates for facilitating design allowables generation (including history-dependent failure prediction) and for predicting life expectancy (degradation state) for fielded composite structures.

A valuable resource for understanding the high-temperature behavior of PMCs is the tremendous volume of long-term high-temperature aging data generated in the NASA High-Speed Research (HSR) program. The HSR effort was a national effort to develop the next-generation supersonic passenger jet designed for a 60,000-h life with temperatures approaching 177°C. To meet the vehicle requirements, PMCs with high glass transition temperatures T_g and, thus, high-temperature use capabilities were thermomechanically loaded for very long-aging times.

The long-term testing conducted in this program demonstrates the challenge of using PMCs in high-temperature environments. An important contribution to understanding the high-temperature performance of the polymer composites evaluated in the program is the use of viscoelastic formulations to model long-term behavior [21, 42, 49, 68, 115, 116]. The assumption inherent in the use of these viscoelastic formulations is that the material is chemically static or the original chemical structure is thermally recoverable [98]. Although these formulations are an important aspect of the overall methodology to model the performance of HTPMCs, a comprehensive review of such is beyond the scope of the current discussion. The discussion here is limited to modeling the thermooxidative aging of HTPMCs and the role of the various constituents. Although hygrothermal degradation, which is a significant concern for HTPMCs, is not specifically addressed here, the methodology described herein can be applied to modeling hygrothermal degradation behavior. As with oxidation, hygrothermal degradation involves transport and chemical reactions of the polymer with the diffused media. The primary challenge for modeling the coupling of oxidative and hygrothermal degradation of HTPMCs is an understanding of the numerous degradation mechanisms.

Several major aging mechanisms that lead to weight loss and damage growth and, hence, degradation of the performance of the polymer resin, fibers, and their composite can be identified. They are:

- *Physical aging*. The thermodynamically reversible volumetric response due to slow evolution toward thermodynamic equilibrium is identified as physical aging. The decreased molecular mobility and free-volume reduction lead to strain and damage development in the material.
- *Chemical aging*. The nonreversible volumetric response due to chain scission reactions and/or additional crosslinking, hydrolysis, deploymerization, and plasticization is classified as chemical aging. A dominant chemical aging process for HTPMCs is thermooxidative aging – which is the nonreversible surface diffusion response and chemical changes occurring during oxidation of a polymer (hence a modality of chemical aging). The oxidative aging may lead to either the reduction in molecular weight as a result of chemical bond breakage and loss in weight from outgassing of low molecular weight gaseous species, or chain scission and formation of dangling chains in polymer networks.
- *Mechanical stress-induced aging*. Mechanical and thermal fatigue loading cause micromechanical damage growth within the material. The damage evolution, in turn, exacerbates the physical aging and thermooxidative response of the material. This aging mechanism may be the least understood and least modeled by researchers.

The capability to predict the performance of HTPMCs for both primary and secondary structural applications is elusive. The highly coupled physical, chemical, and mechanical response of these materials to the extreme hygrothermal environments provides formidable challenges to the composite mechanics community. Polymers (in particular amorphous polymers) have been studied from the standpoint of physical aging [98], chemical aging [66], and strain-dependent aging [118]. Although models to predict the aging response of polymers are available and the fiber response can be assumed to be elementary, the ability to model the thermooxidative aging of the fiber–matrix system (presence of fiber–matrix interface/interphase region complicates the issue with a coupling or sizing agent) is lacking.

While the time-dependent physical, chemical, and damage-induced degradation mechanisms have been studied for some resin systems, polymer composite thermal oxidation studies from a mechanistic perspective are nascent. Notable exceptions to this are the recent works of the groups of Colin and Verdu [27], Colin et al. [32], Skontorp et al. [97], Wang et al. [121], and Pochiraju and Tandon [73, 74], Tandon et al. [104], and past work by Wise et al. [125], Celina et al. [22], and McManus et al. [62]. Equally important are the experimental characterization efforts by such groups as Bowles et al. [14, 19], Tsuji et al. [113], Abdeljaoued [1], Johnson and Gates [49], and Schoeppner et al. [91]. However, most literature is confined to thermal oxidation of neat polymer systems and there are limited studies that correlate the thermochemical decomposition of high-temperature polymers to the resulting mechanical performance. Such works include Meador et al. [63–65] and Thorp [107] for PMR-15 high-temperature composites and Lincoln [56] for 5250-4 bismaleimide (BMI) composites.

Current emphasis is on the implementation and extension of multiscale models to represent the polymer behavior/properties as a function of the degradation state to include chemical degradation kinetics, micromechanical models with time-dependent polymer constitutive relationships, and ply-level and laminate level models to predict structural behavior. The behavior of the composite will, thus, be dependent on its current chemical, physical, and mechanical state as well as its service history. This multidisciplinary modeling approach will provide generalizable analytical and design tools for realistic prediction of performance, durability, and use life of HTPMCs.

The following requirements are identified for the formulation of the predictive thermooxidation models:

– *Appropriate mechanistic and kinetic modeling of polymer environmental degradation.* For the highly crosslinked polyimide composite of interest, the reactivity of the end cap is often a primary concern [63, 107]. Although the primary, and in some cases secondary, oxidation and hydrolytic degradation mechanisms can be identified, determination of mechanisms up to the final state of degradation is difficult if not impossible. Predicting thermal, physical, and mechanical performance based on the chemical state of the polymer is currently impractical for all but the very simplest of polymer systems. In the absence of this predictive capability, empirical correlation of the chemical state (if known) to mechanical properties is used to help define the constitutive models.

- *Polymer constitutive models that incorporate the chemical, thermal, and deformation state and history dependence.* Linear viscoelastic constitutive models have been successfully used to model the physical aging of the polymer phase of polymer composites used in high-temperature environments. Beyond physical aging, accounting for the effects of chemical aging (oxidative and nonoxidative) in a constitutive model requires path/history-dependent relationships. Constitutive models and/or empirically derived correlations that account for the dominant behavior of the material provide alternatives to models that can predict performance based on the chemical degradation state.
- *Integrated mechanistic models that explicitly represent fiber–matrix phases and interfaces to predict lamina properties.* The highly critical fiber–matrix interface/interphase region may govern the oxidative behavior of HTPMCs, but significant challenges exist in measuring properties of the interphase region. Prediction of strength/failure requires knowledge of the constituent properties, including strength and toughness, as a function of the degradation state or aging history. It is beyond current modeling capabilities to predict mechanical properties based on the chemical state of the polymer. For lack of a robust constitutive model, simplifying assumptions regarding the history-dependent properties can be made as a first-order approximation to predict the material behavior.
- *Structural laminate models with lamina property descriptions and discrete ply representations for laminate and composite failure modeling under history-dependent environmental service loads.* As for the micromechanical scale in which the fiber, matrix, and interphase regions are represented explicitly, the residual stresses on the ply-level and laminate level scales play a critical role in the thermal oxidation process. Accurate representation of the free-edge interlaminar stresses in multidirectional composites, taking into account stress-assisted diffusion, is essential for accurately modeling the oxidation-susceptible free surfaces.
- *Experimental validation of the simulation and transition into designer assistance tools and property databases.* To transfer the prediction methodology from scientists to designers, a new generation of tools that incorporate the simulation methods and experimental databases needs to be developed. To facilitate adoption of the tools by designers, any such development must be in collaboration with the practitioners from the airframe and propulsion segments of the aerospace industry. This methodology of tool development in collaboration with design practitioners was used in the DARPA Accelerated Insertion of Materials – Composite (AIM-C) program [82].

For those steps described above where prediction capability is lacking, experimental/empirical means must be used to represent the behavior of the material. It is these empirical relationships that preserve the capability to make predictions on how materials will behave when mechanistic models are not available. Experimental validation of predictive models for a given geometric scale provides a basis for building models at higher geometric scales. As an example, polymer constitutive models can be validated by testing neat resin polymers but validation of such models on the lamina or laminate scale is not viable. This is duc to the fact that fibers mask the polymer degradation behavior particularly for the fiber-dominated composite properties. Additionally, fibers can exacerbate some degradation mechanisms of polymer due to the introduction of fiber–matrix residual stresses and the introduction of fiber–matrix interface/interphase.

The multiscale modeling levels and the vital links between the various model levels are illustrated in Fig. 9.1. This description of the hierarchical scheme is an idealized representation of the methodology needed to mechanistically model the behavior of PMCs in aggressive environments. In particular, a kinetic description of the polymer phase of the material is necessary when the materials are subjected to or susceptible to chemical changes or degradation.

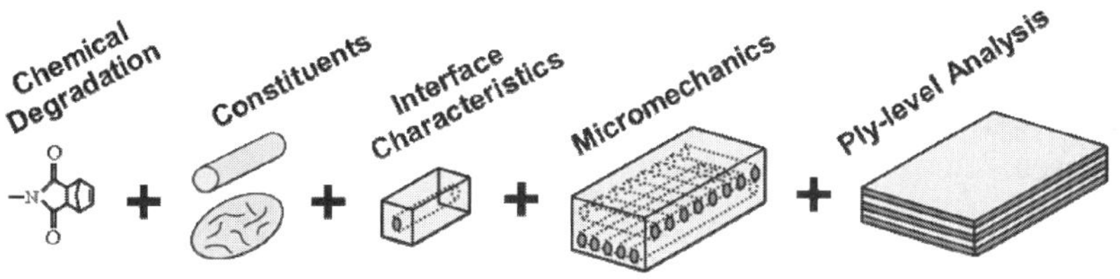

Fig. 9.1. Multiscale modeling levels

The primary focus of the discussion in this work is on predicting isothermal-oxidative aging of unitape laminates with particular focus on the composites using PMR-15 high-temperature polymer. Additionally, the reinforcement is limited to polyacrylonitrile (PAN)-based carbon fibers. PMR-15 is a widely used addition polyimide with a maximum service temperature of approximately 288°C. Among the class of high-temperature polymers are the bismaleimides, Avimid-N, thermosetting polyimides (AFR700B, LARC RP46), and phenylethynyl-terminated polyimides (PETI-5, AFR-PE-N) resin systems. Each of these material systems has unique degradation reactions, mechanisms, and kinetics particular to their chemical structure. Although experimental observations and predictions of the behavior of PMR-15 neat resin and composites are not necessarily representative of the behavior of these other high-temperature polymer material systems, the modeling methodology and procedures for determining material parameters may be directly applicable to predicting their behavior. On the other hand, the observed behavior of other high-temperature material systems can deviate significantly from PMR-15 and such deviations will be noted when appropriate. Current polymer and polymer matrix composite modeling approaches for chemical and physical aging, as well as the modeling process for predicting thermooxidative degradation, are discussed in the following sections.

9.1.1 Physical Aging of Polymers and PMCs

At temperatures below a polymer's glass transition temperature (T_g), the material is in a nonequilibrium state and undergoes a time-dependent rearrangement toward thermodynamic equilibrium. During this rearrangement (relaxation) toward the equilibrium state, there are time-dependent changes in volume, enthalpy, and entropy, as well as mechanical properties and this is known as *physical aging*. Physical aging is characterized by changes in stiffness, yield stress, density, viscosity, diffusivity, and fracture energy (toughness) as well as embrittlement in some polymeric materials. The physical aging rate depends on the distance the aging temperature is from the material's T_g. Therefore, the closer the aging temperature to the T_g, the greater the polymer molecular mobility and the greater the relaxation rate, while the driving force defined as the entire path to the equilibrium state, decreases.

Practical considerations of physical aging only become important when aging temperatures are near the T_g. Physical aging is a reversible process known to be easily altered with stress and temperature. For high-temperature PMCs that are used at temperatures near the material's T_g, physical aging may dramatically affect the time-dependent mechanical properties (creep and stress relaxation) and rate-dependent failure processes [56]. For highly crosslinked polyimide systems, it is difficult to separate the physical aging from chemical aging effects when conducting tests near the T_g because the aging effects are coupled. Whereas in some simple polymer systems, physical aging can be reversed by heating the polymer above the T_g and quenching, heating highly crosslinked polymers above the T_g induces chemical aging, thereby, altering the thermodynamic equilibrium state.

Earlier studies on the effects of physical aging on polymer composite behavior were conducted by Sullivan [99], McKenna [59], and Waldron and McKenna [118]. The majority of the reported work on predicting the long-term performance of PMCs in high-temperature environments is limited to physical aging models and the use of linear viscoelastic and time–temperature superposition models [21]. Inherent in the use of physical aging models is the assumption that the aging process is thermoreversible, therefore nonreversible chemical aging cannot be properly modeled using this approach. That is, the effects of irreversible processes, e.g., chemical decomposition, hydrolytic degradation, oxidation, etc., are neglected in these models or are assumed to be implicitly accounted for in determining the physical aging parameters. However, chemical aging can be the primary life-limiting degradation process for HTPMCs used at temperatures near their T_g. Neglecting thc modeling of the chemical degradation for composites in an oxidative environment will likely result in inaccurate and nonconservative predictions of performance.

9.1.2 Chemical Aging of Polymers and PMCs

Chemical aging, unlike physical aging, is typically not thermoreversible. Hydrolysis (the chemical reaction of the polymer with water) and oxidation (the chemical reaction of the polymer with oxygen) are the primary forms of chemical degradation in HTPMCs. The chemical changes occurring during oxidation include chemical bond breaks that result in a reduction in molecular weight, mechanical response changes, and a local loss of mass associated with outgassing of oxidation byproducts. Although the rate of oxidative chemical reactions is, in part, governed by the availability of reactive polymer, the oxygen concentration, and the reaction temperature for high T_g glassy polymers, the rate of oxidation can be greater than the rate of diffusion of oxygen into the polymer. For such circumstances, the oxidative process is diffusion rate limited. Diffusion rate-limited oxidation of neat polymer specimens typically results in the development of an oxidative layer or graded oxidative properties near the free surfaces of the specimen. Within the oxidized region of the polymer, it is typical that the tensile strength, strain to failure, flexural strength, density, and toughness decrease while the modulus increases. The effect of oxidation on changes in the T_g is dependent on the specific polymer system. Some polymers initially have a decrease and then an increase in T_g, others may have only a decrease, and still others may only have an increase in T_g. This may be due to competing chemical and physical aging phenomenon [56] or differences in the oxidation reaction mechanisms. Since HTPMCs typically operate at temperatures near their initial design T_g, any changes in the local or global T_g can have detrimental effects on performance.

Determination of the primary, secondary, and tertiary chemical degradation mechanisms in high-temperature polyimide composites is an extremely challenging task [92] but can yield substantial benefits for designing new polyimides that are less susceptible to degradation. The primary and secondary oxidation mechanisms for PMR-15 were determined by Meador et al. [63–65] in a series of tests on model compounds. It was determined that the nadic end group, in which the aliphatic carbons are consumed during oxidation, is the primary rapid mechanism for weight loss. The second, longer-term mechanism is the oxidation of the diamine bridging methylene to the carbonyl group. As part of a separate effort, Thorp [107] determined that the nadic end group is also responsible for the primary hydrolytic degradation mechanism. Similar types of studies have led to the development of polyimides with a more thermally stable phenylethynyl end group as a replacement for the degradation-susceptible nadic end group. These new material developments include the family of phenylethynyl-terminated polyimides such as NASA's PETI resins [47] and the Air Force-developed AFR-PE-N resins that are more thermally stable than PMR-15 [123].

9.1.3 Mechanical Stress-Induced Aging

Elevated temperatures can have a dominating effect on the strength and stiffness of HTPMCs [76], and the strength and stiffness at the operating temperature are primary considerations for selecting suitable materials for a given application. Primary consideration is also given to the long-term durability of the material in the elevated temperature environment. The effects of long-term mechanical loading on PMCs at elevated temperatures are manifested in the creep–relaxation response (viscoelastic–plastic behavior) and the load-induced damage developed under thermomechanical cyclic loading. Often a linear elastic representation of the fiber-dominated composite properties is sufficient, while various time-dependent linear and nonlinear viscoelastic–plastic models [85, 86] may be needed to represent the resin- or matrix-dominated properties for high-temperature applications. Additionally, there is a strong coupling between chemical aging and the damage development due to the changes in stiffness, strength, and toughness that occur during long-term aging.

For the purpose of the discussion here, mechanical stress-induced aging is associated with stress-assisted diffusion and aging as well as thermo-mechanical-induced damage or cracking. The development of damage in HTPMCs exacerbates the chemical and physical aging by introduction of stress concentrations that accelerate physical aging effects and exacerbates the chemical aging by introducing pathways for oxidants and other agents to advance deeper into the material. Damage typically takes the form of matrix cracks and fiber–matrix interface debonds, with the micromechanical cracks and transverse ply cracks coalescing to form larger ply-level cracks. For this reason, the resin- or matrix-dominated properties are of primary interest for modeling the high-temperature aging response.

The mechanistic approach of modeling discrete damage, such as transverse ply cracks [58, 89], provides accurate assessments of the effective response for a given damage state but has limitations on the number of discrete cracks or amount of discrete damage that can be represented. Therefore, mechanistic damage modeling is often reserved for analyzing local details and small components, or it can be used as a local model in a global analysis. Ultimately, for modeling large structures, phenomenological approaches of continuum damage mechanics [4, 100] can be used to represent the evolution of effective local damaged properties through the constitutive relationships, whereby widespread dispersed damage can be more easily represented.

9.2 Experimental Characterization and Observation

The research presented here focuses on PMR-15 resin, carbon fibers, and PMR-15/carbon fiber composites. Although other HTPMC material systems are of interest, the open literature contains very little data for all but the PMR-15 material. Other materials may behave significantly different from the PMR-15 material system, however the modeling methodology given here should be applicable to modeling thermal oxidation in other polymer systems. Thermooxidative aging experiments, which include the effects of both physical and chemical aging, were conducted on the neat polymer and composite specimens. Model development and experiments focus on thermo-oxidative degradation while hydrolytic degradation will be addressed in subsequent research efforts. The spatial and temporal variability of the oxidized material and the ensuing mechanical properties were monitored to determine material parameters for the implemented models.

Determination of model parameters for the multiscale modeling effort entails experimental characterization of the composite constituents, namely the fiber and the matrix polymer. However, characterization of the fiber and matrix is not sufficient to predict the behavior of the composite due to the critical fiber–matrix interface/interphase that develops during the cure process. Although a direct measure of the properties in the interface/interphase region is challenging, indirect measures of the influence of the fiber–matrix interphase based on experimental observation of the composite behavior can be used to predict the model parameters for the interphase.

9.2.1 Fiber Oxidation Degradation

The effects of service temperatures and oxidation on the mechanical properties of high-temperature polymer composites are primarily manifested in the polymer-dominated properties, namely the transverse properties (perpendicular to the fiber direction) and the shear properties. However, the fiber-dominated properties may be affected by degradation of the fiber and deterioration of the fiber–matrix interface. This may be particularly true for composites with glass fibers [44] in which magnesium and sodium can leach out from the fiber and possibly contribute to polymer/interface degradation. It is reported that graphite fibers containing significant amounts of sodium and potassium as contaminants are less thermooxidatively stable than graphite fibers with very low alkali metal contents [39, 43]. Studies conducted by Bowles and Nowak [12] indicate that extreme oxidative erosion of the Celion 6000 graphite fiber occurs at elevated temperatures in the presence of the polyimide matrix.

Bowles [8] has investigated the effects of different fiber reinforcements on thermooxidative stability of various fiber-reinforced PMR-15 composites. The ceramic Nicalon and Nextel fibers were found to drastically accelerate thermal oxidation of the corresponding composites because of the active fiber–matrix interface. Compared to polyimide matrices, reinforcing carbon fibers [60] are usually far more stable at the elevated temperatures considered. Studies conducted by Wong et al. [126] indicate that IM6 carbon fibers are more stable than the G30-500 fibers within the first 600 h during thermal oxidation at 371°C. The oxidation rate of the IM6 fibers then increases substantially, leading to complete decomposition of the fibers during the final stage. The sudden increase in the oxidation rate in the IM6 fibers implies the possibility of change in degradation mechanisms.

The three PAN-based carbon fibers of interest to the present body of work are two low modulus carbon fibers T650-35 and G30-500 that are typically used in HTPMCs, and one intermediate modulus carbon fiber, IM7. Isothermal aging studies were recently conducted [105] on these three types of fibers to study their oxidation behavior. Both sized and unsized fibers were exposed to different elevated temperatures for varying time periods in an attempt to understand the influence of the fiber sizing/coupling agent on their thermooxidative stability. Note that aging of the bare fibers may not necessarily be representative of the behavior of the in situ fibers embedded in the matrix, because the exposed surface area of the fibers in the composite is only a very small percentage of the total surface area of the fibers. Thermal degradation was quantified by the amount of weight loss measured, while degradation of mechanical properties was measured using single-filament tests on unaged and aged fibers. Finally, surface morphology changes were monitored in an attempt to relate physical and surface changes to the decreases in mechanical performance as a function of aging.

Weight loss studies

Figure 9.2 shows that the thermal degradation (as measured by weight loss) of unsized T650-35 carbon fibers is substantial for moderate temperatures above 316°C. It is seen that there is a large (~40%) weight reduction of T650-35 carbon fibers after 2,000 h of aging at 343°C. At the lower temperature of 232°C, the weight loss is not significant and the weight loss as a function of time follows approximately a linear relationship. However, as the aging temperature is increased to 343°C, the T650-35 fibers are observed to lose weight rapidly. There seems to be a change in slope in the weight loss curve at around 750 h of aging, possibly signifying a change of degradation mechanism. Typically, the weight loss data for neat polymer specimens are normalized with respect to the specimen surface area, since oxidative weight loss occurs from losses on exposed surfaces and edges of the specimen. Note that carbon fibers have an extremely large surface area due to their small diameter (typically 4–8 μm) and large aspect ratio (typically >1,000).

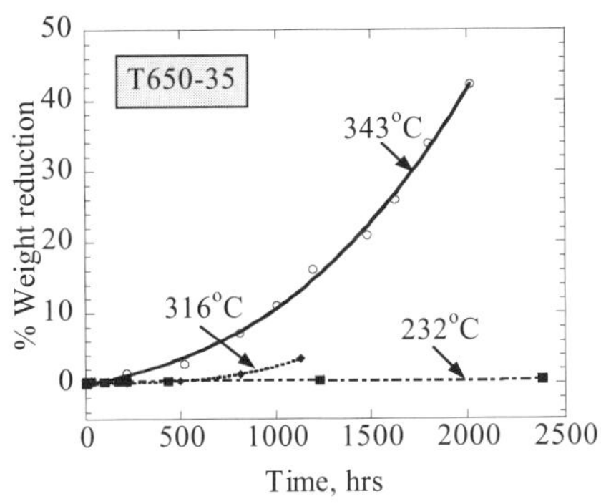

Fig. 9.2. Weight loss of T650-35 carbon fiber as a function of aging temperature

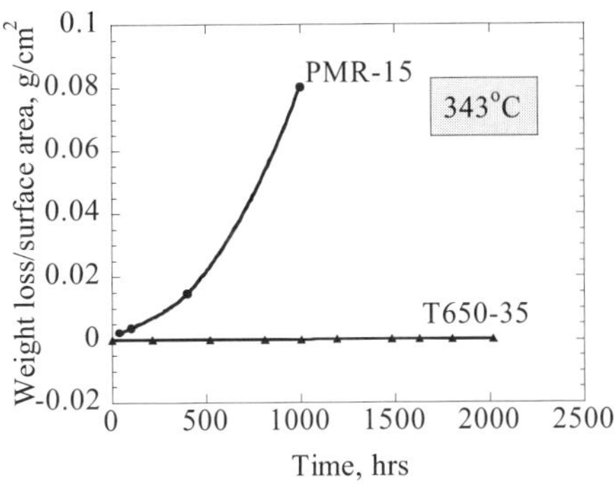

Fig. 9.3. Normalized weight loss of T650-35 fiber and PMR-15 resin

Figure 9.3 shows a comparison of the normalized weight loss of T650-35 carbon fiber and PMR-15 neat resin at 343°C. Note that the normalized fiber weight loss is almost negligible compared to that of neat resin over the entire 2,000 h of aging time. Thus, even though the carbon fiber loses a significant weight fraction with aging at elevated temperatures, the weight loss normalized by the fiber surface area is negligible compared to that of the resin. Weight loss studies on sized fibers [105] indicate that the sizing/coupling agent is released within a short time period (~24 h) of aging beyond which the weight loss trend is similar to the corresponding unsized fibers.

Mechanical properties

Single fiber specimens were tested at room temperature in tension using the single-filament test. Figures 9.4 and 9.5 show the normalized failure strength and failure strain, respectively, of unsized T650-35 carbon fibers aged at 343°C. The strengths and failure strains are normalized with respect to their corresponding unaged values, such that the observed decreases are a reflection of the reduction resulting from isothermal aging. A minimum of ten fibers were tested for each condition considered, and the standard deviation is shown as the error bar across the measured mean values. There is some scatter in the strength and failure strain data which is typically encountered in single fiber testing as failure is sensitive to the presence of flaws over the fiber gage length. Fibers aged at the elevated temperature of 343°C show a large decrease in the strength after 1,000 h of aging. The significant decrease in mechanical strength signifies that the carbon fibers should not be treated as static entities when composites containing these fibers are aged at this temperature for extended periods. However, the reported data are a worst case scenario in which all surfaces of the fibers are exposed during aging. In situ fibers of the composite have only a small fraction of their total surface area exposed to the oxidizing environment and should, therefore, suffer less degradation. Further, test data from single-filament testing [105] indicate that application of fiber sizing may

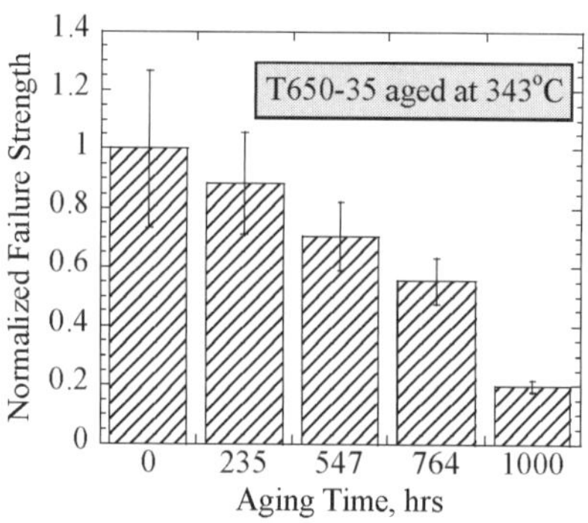

Fig. 9.4. Normalized strength of unsized T650-35 carbon fibers aged at 343°C

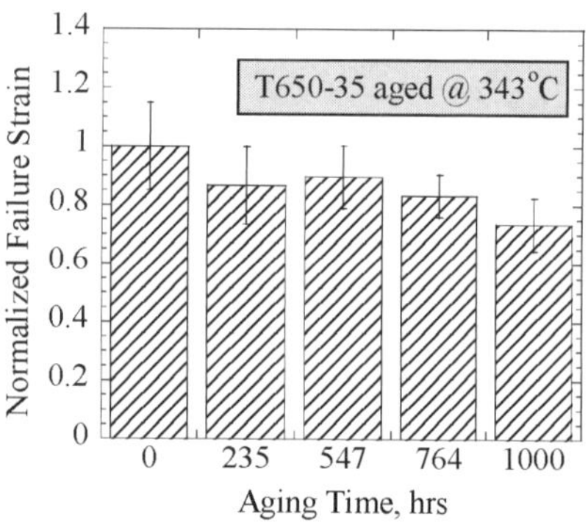

Fig. 9.5. Normalized failure strain of unsized T650-35 fibers aged 343°C

result in an improved performance of the carbon fiber in the unaged condition but could result in some loss of mechanical performance after aging at elevated temperatures. Failure is sensitive to the presence of flaws, and it is likely that the sizing/coupling agent on the fiber helps to decrease the influence of the fiber surface flaws resulting in a slightly better performance in the unaged condition.

Surface characterization

Scanning electron microscope (SEM) microstructural studies were conducted to determine the carbon fiber surface morphology changes during long-term isothermal aging. Figure 9.6a compares the SEMs of unsized T650-35 carbon fiber in the unaged condition with that of the fiber aged for 1,052 h at 343°C in Fig. 9.6b. No significant visible damage or surface morphology changes are observed, although some minimal amount of pitting is visible on the previously smooth fiber surfaces after 1,052 h of aging. Thus, the fiber surfaces provide little indication of the deterioration in the mechanical performance of the aged fiber.

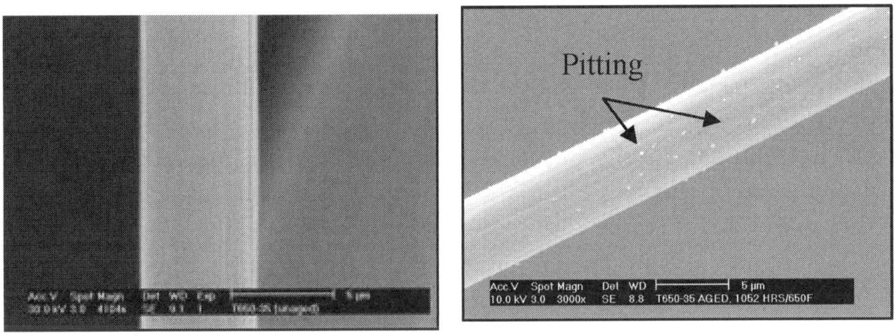

(**a**) (**b**)

Fig. 9.6. SEMs showing surface morphology of unsized T650-35 carbon fibers in (**a**) the unaged condition and (**b**) aged for 1,052 h at 343°C

Anisotropic diffusivity

It has been established that the fibers account for only a minimal amount of the total weight loss for HTPMCs. However, the role of fibers in the composite oxidation process, in particular their role in transporting oxygen into the composite, is not fully understood. Extruded PAN carbon fibers exhibit a preferred orientation in which the graphitic layers may extend for thousands of angstroms and extend straight for hundreds of angstroms parallel to the fiber axis providing high-strength and high-stiffness along the fiber axis direction. The mechanical (stiffness and strength) and thermal (coefficient of thermal expansion [52] and thermal conductivity [51]) properties of these oriented PAN-based fibers are highly anisotropic in nature. Typical density values of these fibers range between 1.75 and 1.86 g cm^{-3} and a porosity of approximately 15% with most of the porosity appearing in needle-shaped forms with axes of the needles parallel to the fiber axis [77]. Therefore, it follows that the diffusivity of the fibers will likely be anisotropic in nature. Unfortunately, experimental measures of carbon fiber diffusivity have not been obtained.

Numerous researchers have attributed the anisotropic diffusivity of unidirectional laminates to the preferential oxidation along the fiber–matrix interface or along the interphase region. However, the interphase region is a very small-volume percentage of the composite and a very small area fraction of the surface area. It is questionable that the diffusion of oxygen through and subsequent oxidation of the interphase region by itself can have such a significant effect on the composite oxidation process. More likely, the rapid oxidation rate along the fiber length indicates that, in addition to diffusion of oxygen from the specimen surface to the oxidation front, the interphase region has a supplemental oxygen path or source. Two possible mechanisms or scenarios for the supplemental oxygen source to the interphase are being investigated. Firstly, if oxidation in the interphase region leads to early fiber–matrix interface debonds, the debonds provide a pathway for oxygen to penetrate deeper into the composite. Knowing that there are residual curing stresses at the fiber–matrix interface, it is likely that such a scenario would entail the interface debond propagating along with the oxidation front. The second scenario is that the oxygen diffuses into the composite along the fiber and to the fiber–matrix interface at a rate much greater than through the neat resin or interphase. Based on the anisotropic oxidation behavior of unidirectional composites, the axial diffusivity of the fibers would have to be much greater than the diffusivity of the resin. Unfortunately, quantitative measures of the diffusivity of carbon fibers are lacking. Although one might anticipate that the bulk diffusivity of carbon is representative of that of the PAN fibers, this would only be the case if the morphology of the fibers and bulk carbon match, which in general is not the case.

For unit cell modeling of the diffusion behavior of unidirectional composites, the diffusivity of the resin, the interphase, and the fiber constituents must be defined. Only the diffusivity of the resin phase can be defined with any certainty due to lack of experimental data on the diffusivity of the fiber and interphase. Therefore, the role and the relative contribution of the fiber and the interphase to the effective diffusivity cannot be rigorously determined.

9.2.2 Neat Resin Behavior

The attributes of the oxidation process of high-temperature polymers and polymer composites vary depending on the chemistry of the polymers being tested. The high-temperature aging behavior of several neat polymer resins and their composites has been studied [1, 10, 13, 14, 26, 37, 38, 53, 64, 65, 95, 113]. In PMR-15, the oxidized material near the specimen's free surfaces is observed to have a darker appearance than the unoxidized interior using bright-field light microscopy Recent studies [19, 28, 63, 78, 90, 119] document the growth of the thermooxidative layer and the changes in elastic moduli and chemical composition resulting from isothermal aging. It has been observed for PMR-15 that the polymer degradation occurs mainly within a thin surface layer that develops and grows during thermal aging. Ripberger et al. [78] reported an oxidation layer thickness of approximately 55 μm after 50 h aging and a layer thickness of between 107 and 129 μm after 342 h of aging for PMR-15. Additionally, the interior core of the specimens (specimens greater than 2 mm thick) is generally protected from oxidative degradation during thermal aging and is relatively unchanged after aging.

In this work, for both the neat polymer and composite specimens that are used to monitor the propagation of oxidation, samples are dry-sectioned from aged larger specimens, as shown in Fig. 9.7, potted in 828-D230 epoxy resin, and cured at room temperature for 3 days. This sectioning procedure allows monitoring of four of the exposed free surfaces of the large specimen. The large specimens are subsequently placed back in the oven to continue the aging process. The diamond blade used to section these

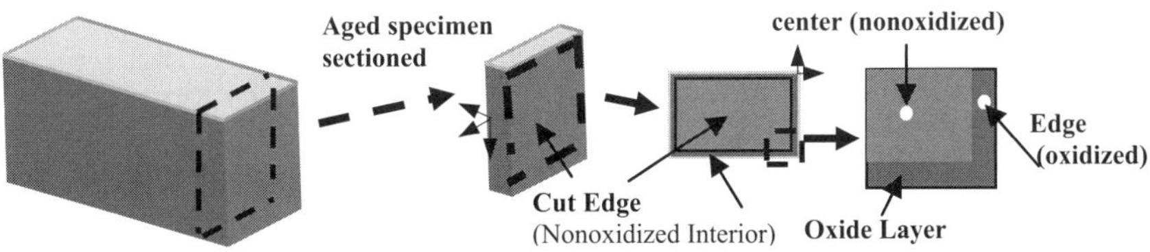

Fig. 9.7. Oxidation measurement procedures

samples is washed with acetone and wiped clean with paper towels prior to cutting to minimize the amount of contamination from the cutting wheel. The specimens are then wet-sanded with 600-grit sandpaper and distilled water and polished using a 0.3-μm alumina polishing media. Since the oxidized layer forms on all exposed free surfaces and propagates to the interior of the sample, the oxidized region is easily seen in the cross section. These potted specimens are used for both nanoindentation testing (MTS Nanoindenter XP) and optical and dark-field microscopy (Nikon Microphot-FXL, Model F84006) measurements. The MTS Nanoindenter has an X–Y positional stage with a resolution of 0.5 μm and a force resolution of 50.0 nN. The reported transitional resolution for the Nikon positioning stage is 1.0 μm. A Philips XL30 ESEM with energy dispersive spectroscopy (EDS) analysis capabilities is also used to examine the damage and the chemical state of the aged composite specimens.

Figure 9.8 shows a photomicrograph of a PMR-15 neat resin specimen isothermally aged in ambient air at 343°C for a period of 196 h. The figure clearly shows the oxidized region (much like a picture frame) on the two adjacent exposed free surfaces of the specimen. Between the outer oxidized layer and interior unoxidized region is a transition region, which is the active "reaction" or "process" zone. These observations allow easy measuring and characterization of the oxidized layer and the transition

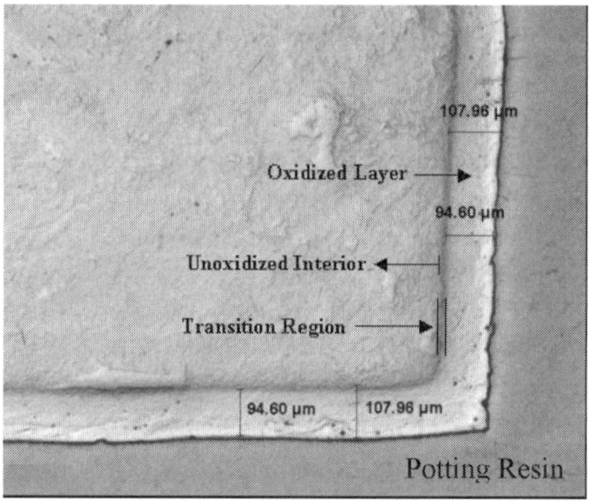

Fig. 9.8. Photomicrograph of PMR-15 resin after 196 h of aging at 343°C showing oxide layer formation, transition region, and unoxidized interior

regions as a function of aging temperature and time. Figure 9.9 shows the evolution of the oxidized layer and the active "reaction" zone thicknesses as a function of aging time. Within the first hour, an oxidized layer 11.0 μm thick forms on the exposed specimen surfaces. The thickness of the oxidized layer is seen to approach a plateau value as the oxidation growth rate reduces considerably for longer aging time periods, whereas the thickness of the active "reaction" zone remains nearly constant for the aging times considered.

PMR-15 samples aged in air were also evaluated using a nanoindenter by scanning across the oxidized layer, through the active "process" or "reaction" zone, and into the unoxidized central area of the specimen. The indentations were spaced so as to give a minimum of three data points within each region. These measurements were repeated at several locations around the perimeter of the oxidized sample.

Figure 9.10 shows the variation of elastic moduli from three individual scans as a function of distance from the specimen's edge for a specimen aged in air for 196 h at 343°C. The data are normalized to the modulus of the unaged material as determined by nanoindentation. Scatter exists in the data as a result of sampling through different regions of the specimen. This figure illustrates the spatial variability and the heterogeneous nature of the oxidation formation. In general, the modulus of the material in the oxidized region is higher toward the outer exposed surface than it is near the inner reaction zone. The modulus increase in the oxidation zone is consistent with embrittlement of the oxidized region.

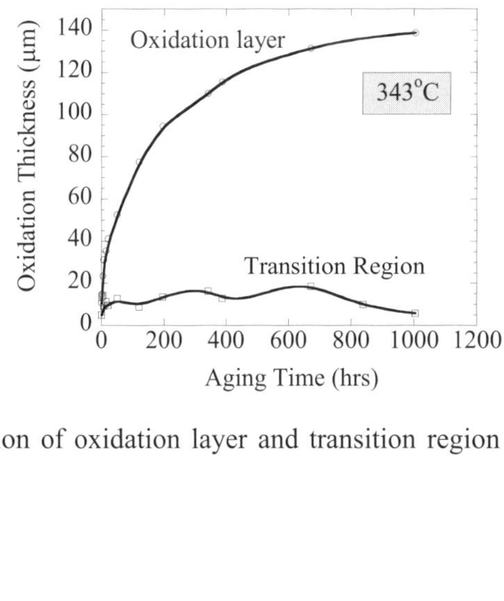

Fig. 9.9. Evolution of oxidation layer and transition region thickness with aging time

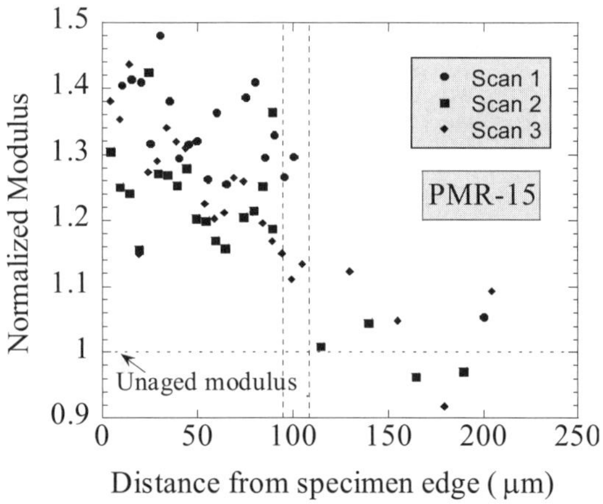

Fig. 9.10. Normalized modulus of PMR-15 resin after 196 h of aging at 343°C

Based on optical observations, the three distinct specimen regions are identified in Fig. 9.10 as the higher modulus oxidized surface layer, the lower modulus unoxidized interior, and the reaction zone or transition region in which the modulus reduces from the oxidized to unoxidized values. The average thickness measurements of oxidized zones and the transition zone, as obtained from optical measurements, are plotted as dotted vertical lines in Fig. 9.10. It is observed that the average thickness of the oxidation layer and active "reactive" zone measured by optical methods is in good agreement with the boundaries of three regions suggested by the nanoindentation data. Similar observations were made by Johnson et al. [50] using atomic force microscopy (AFM). They summarized that the outer "plateau" region is a homogeneous oxidized layer, which is a result of a zero-order reaction. The transition reaction zone is a diffusion-controlled oxidation zone, which is a result of a first-order reaction, and the third region in the specimen interior is the unoxidized PMR-15.

Figure 9.11 shows the average elastic moduli of the oxidized region and the unoxidized interior of PMR-15 specimens aged at 343°C as a function of aging time [78]. In addition, the average elastic moduli of a PMR-15 specimen thermally aged in a nonoxidizing environment is shown in the figure. The reported average values of all the moduli measurements made within their respective regions include data from multiple scans. There is only a marginal increase in the modulus of the oxidized layer with aging time, as seen in Fig. 9.11. Thus, once the material oxidizes, little change in material modulus occurs with aging time.

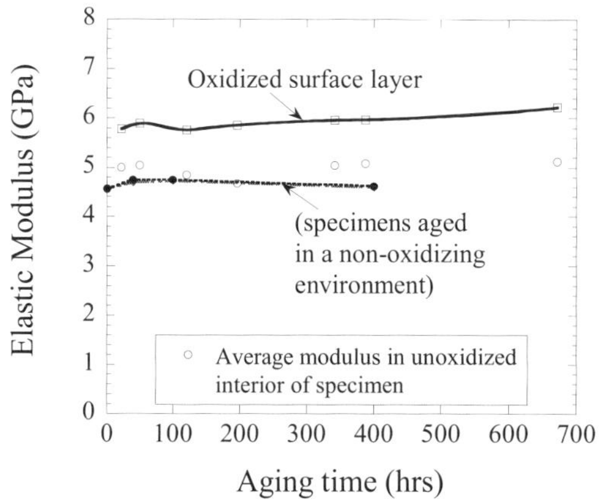

Fig. 9.11. Variation of elastic moduli with aging time

The open circle data points in Fig. 9.11 are the average modulus values in the interior, i.e., in the unoxidized region, for the specimens aged in air. These measurements are in good agreement with the elastic modulus of neat resin specimens aged in a nonoxidizing environment. Because the oxidation process is diffusion limited, the interior of the aged specimens is not oxidized and has properties similar to that of the specimens aged in a nonoxidizing or inert environment. These results, therefore, validate the assumptions of Tsuji et al. [113] that the properties of the specimens aged in a nonoxidizing environment can be taken to be similar to the properties of the unoxidized core material in the air-aged specimens.

Oxidation of the surface layer typically leads to a decrease in local density and a weight loss (due to outgassing of oxidation byproducts) both of which contribute to the shrinkage of the oxidized layer generating tensile stresses and possibly "spontaneous" cracks [14, 29, 38, 63, 78]. For PMR-15 neat resin specimens aged in air at 343°C, surface cracks in the outer oxidized layer begin to appear after 200 h of aging. These crack faces provide additional diffusion surfaces, create pathways for oxidants to penetrate deeper into the material, and accelerate the material degradation and growth of the oxidation layer [78].

Figure 9.12 shows the measured crack density as a function of aging time. The crack density is computed by dividing the total number of cracks formed on the sample edges at a given cross section by the specimen's perimeter. As one would expect, crack density increases with aging time. Beyond 800 h of aging, the severe degradation makes the crack density too complex to quantify. Figure 9.13 shows that, after 342 h of aging, crack depths have already reached 263 μm into the surface of the polymer. The

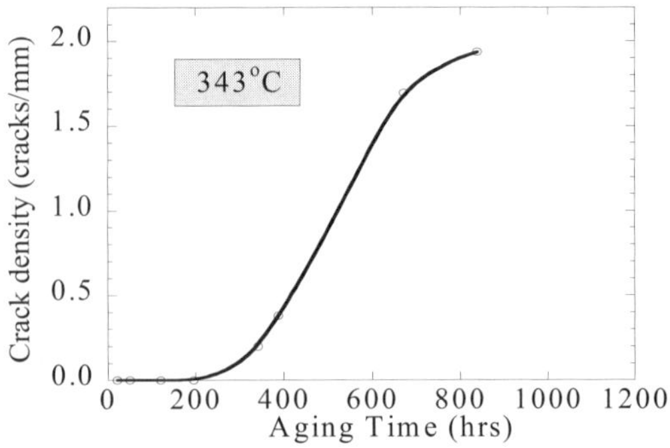

Fig. 9.12. Measurement of surface crack density for specimens aged in air

photomicrograph suggests that the cracks formed in several steps (primary, secondary, and tertiary cracks). As the cracks propagate into the surface and widen, new oxidative surfaces form around the crack front and subsequent cracking occurs. Thus, oxidants are able to penetrate further into the sample through extensive cracking.

Figure 9.14 illustrates severe oxidation damage observed in PMR-15 neat resin specimens after 670 h of aging in air at 343°C. Note the appearance of voids in the oxidized layer with a larger concentration near the surface. The voids form because inward diffusion of oxygen is slower than outward diffusion of degradation byproducts [63].

While optical microscopy techniques are successfully used to monitor oxidation in PMR-15 resin, the same is not true for other polyimide systems, such as the recently developed AFR-PE-4 resin, because the optical characteristics of the polyimide do not change when it is oxidized. Other techniques, such as dark-field imaging [95], polarized light microscopy,

Fig. 9.13. Crack penetration depth after 342 h of aging in air at 343°C

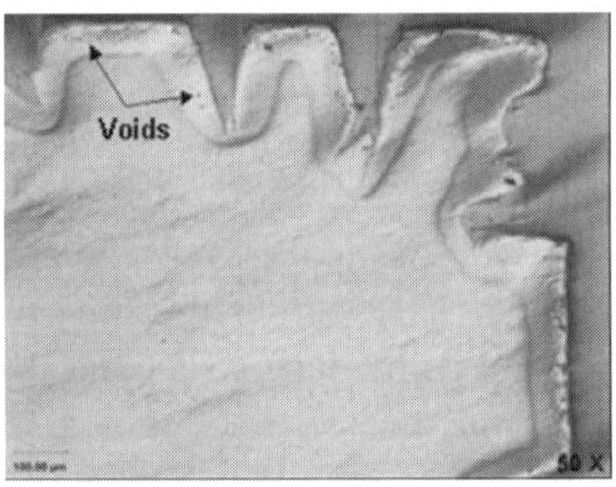

Fig. 9.14. Damaged PMR-15 from isothermal aging at 343°C for 670 h

and scanning electron microscopy in backscatter mode [63], were also used without successfully being able to optically view the oxidation layer [79]. This challenging problem motivated the use of surface analysis techniques, such as XPS and FTIR, in an attempt to detect changes in surface chemistry as a means to monitor oxidation development. Although these surface analysis techniques are able to detect increased concentrations of oxygen in the surface regions of aged specimens, the techniques are not sensitive enough to provide a quantitative measure of the extent of oxidation. Therefore, changes in the moduli, as measured by nanoindentation scans, are used to provide a quantitative measure of the degree of oxidation and monitor the propagation of the oxidation.

9.2.3 Composite Behavior

The anisotropic oxidative response of PMCs was first documented by Nelson [67] when he observed that the oxidation process is sensitive to the surface area for the different test specimen geometries that he investigated. He found that the dominant degradation mechanism for the graphite/polyimides is oxidation of the matrix at the laminate edges. Additionally, the materials degraded preferentially at the specimen surface perpendicular to the fiber (axial surface) and the rate of oxidation is hastened by microcracks opening on the axial surface increasing the surface area for oxidation. The anisotropic nature of oxidation in HTPMCs has also been observed by numerous other investigators including the works of Nam and Seferis [66] and Skontorp [95]. Although it is well documented that unidirectional composites preferentially oxidize in the fiber direction [11, 66, 67], details of the rate of oxidation propagation in the orthogonal directions of unidirectional composites are not well documented. There is also substantial evidence that, for woven composites, the oxidation preferentially advances in the in-plane direction along the fiber paths [20]. The oxidation in the transverse direction (normal to the specimen's top and bottom or tool surfaces) is typically constrained by the presence of the fibers.

When testing a T650-35/PMR-15, 8-harness, satin-weave graphite fiber material, Bowles [9] observed two types of surface degradations. Higher temperature aging (288–316°C) results in the formation of a light-colored surface layer that propagates into the material causing voids and microcracks to initiate and grow within the surface layer. At temperatures lower than 288°C, specimens show the same advancement of voids and microcracks into the surface but the oxidized light band of matrix material is not visible. That is, at lower temperatures, oxidation does not cause a visible change in the color of the polymer.

Composite oxidation is primarily a surface reaction phenomenon controlled by the diffusion and rate of reaction of oxygen with the material so that surfaces with different microstructural characteristics are expected to exhibit different oxidation behavior due to differences in diffusivity. Resin cure shrinkage and mismatches in the coefficient of thermal expansion of the fibers and matrix during the composite cure process give rise to localized micromechanical residual stresses in the fiber–matrix interphase region. The diffusion process in materials has been shown to increase with increasing stress levels even becoming nonlinear, as a function of stress, for materials with linear mechanical behavior [122]. Therefore, the highly stressed fiber–matrix interphase regions tend to oxidize at an accelerated rate compared to the lower stressed matrix phase of the composite.

Additionally, the local stoichiometry of the resin may be altered in the fiber–matrix interphase region by the presence of glass fiber-reinforced coupling agents or graphite fiber-reinforced sizing agents. Woven and braided preforms require the use of a fiber coupling or sizing agents to protect the fibers from damage during the weaving and braiding process. Owing to the significant use of woven fiber prepreg and preforms in HTPMC applications, the proper selection of fiber sizing is of paramount concern. The presence of the sizing can have a strong influence on the fiber–matrix interphase or interfacial properties [17, 70, 114] and can ultimately affect the local diffusivity and/or thermal-oxidative stability.

Using the notation of Bowles and Nowak [12], three different types of composite surfaces can be defined for unidirectional composite specimens as S_1 = area of nonmachined resin-rich surfaces (top and bottom or tool surfaces), S_2 = area of surfaces cut parallel to fibers, and S_3 = area of surfaces cut perpendicular to fibers. For woven composites, three different surface types can be similarly defined as Σ_1 = area of nonmachined resin-rich surfaces, Σ_2 = area of surfaces cut perpendicular to warp fibers, and Σ_3 = area of surfaces cut perpendicular to fill fibers. Figure 9.15 illustrates the surface area types for both the unidirectional and woven composites.

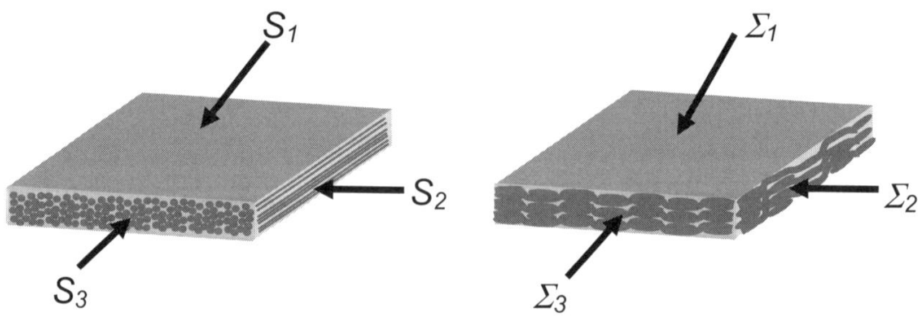

Fig. 9.15. Types of surface areas for unidirectional and woven composites

Dark-field microscopy [95] is used to monitor the oxidation propagation rates in both the axial direction (along the fiber) and the transverse direction (transverse to the fibers) of unidirectional G30-500/PMR-15 composites aged in air at 288°C. Test specimens are removed from the aging oven at specified times, and a small cross section of the specimen is cut off and mounted in an epoxy plug for polishing, as illustrated in Fig. 9.7. The original specimen is then placed back into the oven until the next specified aging time.

Figure 9.16 shows (a) the original stitched micrograph and (b) an enhanced micrograph that clearly distinguishes the white oxidized material from the black unoxidized material. The oxidation layer appears as a frame around the composite specimen just as seen in aged neat resin PMR-15 samples in Fig. 9.8. The specimen aged for 197 h is shown to have only minimal oxidation transverse to the fibers, but has moderate oxidation development in the axial direction. The method of enhancing the micrograph consists of constructing (using Adobe Photoshop® 7.0) a complete image of the entire composite by stitching together individual micrographs using standard light microscopy in the grayscale mode. Once the image is constructed, the apparent light-oxidized region is best fit in the lab mode to a pure white, specified as having a lightness value of 100, while the remaining unoxidized regions of the image are given a lightness value of zero. Thus, this image processing creates exactly two distinct grayscale colors: black and white.

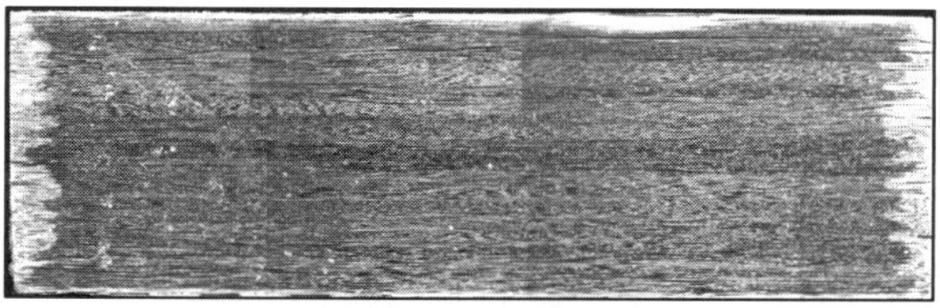

(a)

(b)

Fig. 9.16. (**a**) Array of eight stitched photomicrographs of oxidized specimen cross section and (**b**) enhanced micrograph of the specimen cross section

Figure 9.17 shows enhanced micrographs of G30-500/PMR-15 unidirectional composites after 407, 1,200, and 2,092 h of aging at 288°C that clearly show that oxidation substantially increases in both the axial and transverse directions with aging time. Next, the overall level of oxidation is quantified by measuring the percentage of the specimen cross-sectional surface area that is oxidized. Using the constructed enhanced micrograph image, a histogram of the image is used to determine the ratio of white to black pixels and the ratio is used to quantify the amount of surface area of the composite cross section that is oxidized. Quantification of the oxidation area using this process is a two-dimensional computation and does not account for stochastic variations of the oxidized cross section through the thickness of the specimen.

Figure 9.18 shows the average of the extent of the axial oxidation along the fiber direction as a function of aging time at 288°C. The figure also includes the average of the extent of the transverse oxidation from the enhanced micrographs. Figure 9.18 shows the dominance of the axial oxidation degradation as compared to the transverse degradation in composites which is attributed to the higher effective diffusivity on the S_3 surfaces. Figure 9.19 compares the oxidation growth in the transverse direction (from S_1 surface) of the composite to the oxidation growth in neat resin PMR-15. It appears that the presence of fibers may initially retard growth of oxidation in the transverse directions. Figures 9.18 and 9.19 show the significant differences between the oxidative behavior of the constituents and the composite owing to the fiber–matrix microstructure that may cause increases in diffusion and reaction rates.

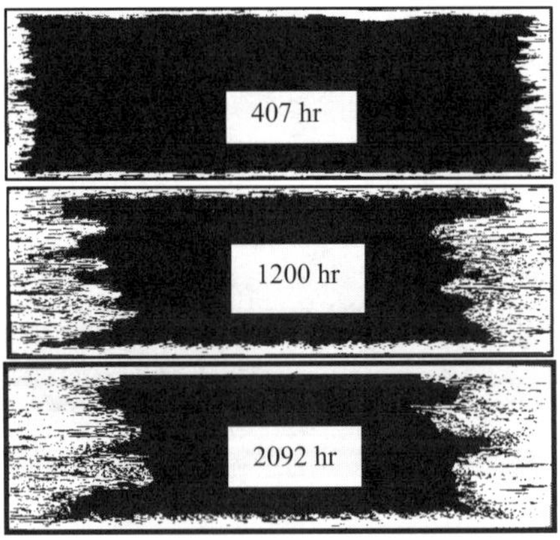

Fig. 9.17. Unidirectional G30-500/PMR-15 at 407, 1,200 and 2,092 h of oxidation at 288°C

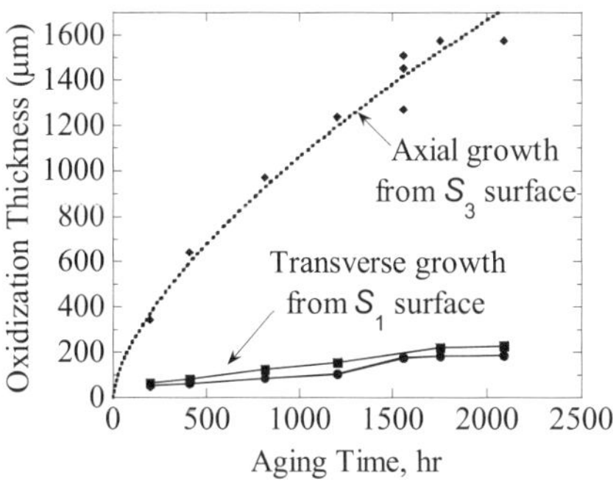

Fig. 9.18. Comparison of oxidation growth in S_1 and S_3 directions of composite

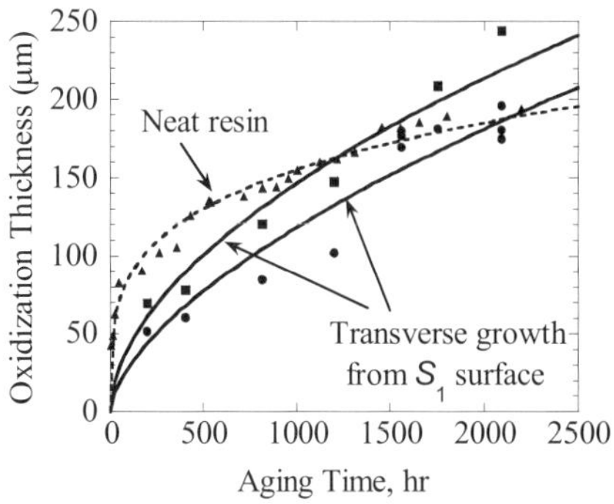

Fig. 9.19. Comparison of oxidation growth in S_1 direction of composite and in neat resin PMR-15

A Philips XL30 environmental scanning electron microscope (ESEM) is used to document and characterize the surfaces of the aged specimens. Figure 9.20 shows the photomicrographs of polished cross sections of axial surfaces S_3 for both nonoxidized and oxidized specimens of the PMR-15 unidirectional composites. It is commonly known that, when composite specimens are polished for imaging, the stiffer fibers wear at a slower rate than the parent matrix, leading to topographic differences between the fibers and the surrounding matrix. For the oxidized specimen (Fig. 9.20b), oxidative degradation of the fibers exacerbates the uneven polishing or rounding of the fibers as compared to the nonoxidized specimen shown in Fig. 9.20a. The PAN-based G30-500 fiber has a typical skin–core structure in which crystalline sheets are oriented radially in the skin but form a random granular-like structure in the core [71]. This construction of the fiber's cross section is not expected to polish evenly.

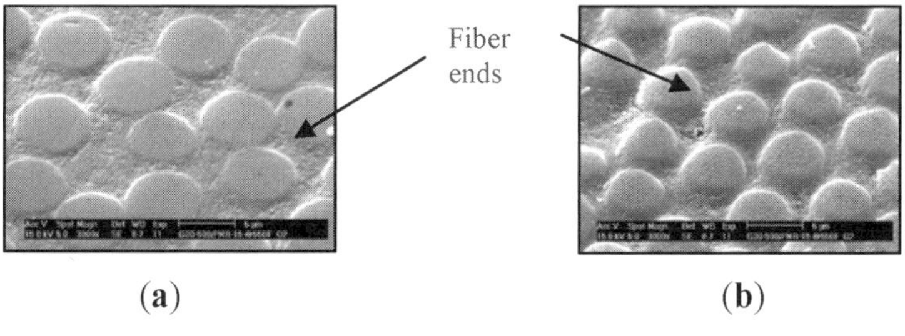

Fig. 9.20. SEM micrograph of the specimen cross section in (**a**) nonoxidized and (**b**) oxidized region of the composite

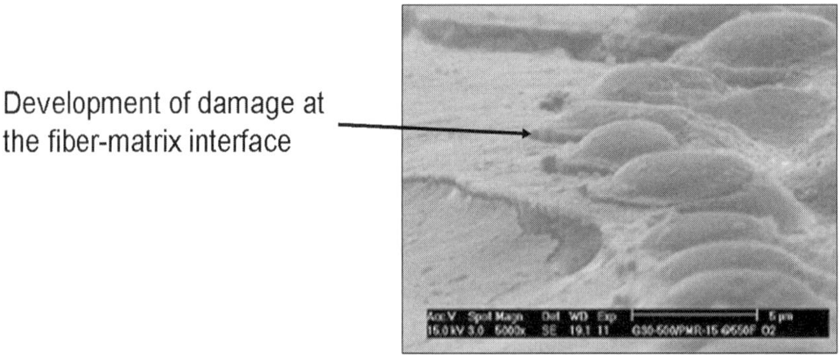

Fig. 9.21. Debonding at fiber–matrix interface at the oxidized fiber end

Higher magnification of an oxidized axial surface, as seen in Fig. 9.21, shows details of the rounding of the oxidized fibers and the opened interfacial fiber–matrix cracks [80].

In Fig. 9.22, a high magnification view of a transverse cross section S_2 for a specimen aged in air at 288°C for 2,092 h shows significant fiber–matrix debonding indicative of that shown in Fig. 9.21. Additionally during sectioning and polishing, the PMR-15 matrix resin fragmented in the highly oxidized regions. Figure 9.23 shows cracking on the axial S_3 surface that included both fiber–matrix debonding and matrix cracking that progressively open during the aging process. The fiber–matrix debonds that are evident on both the axial and transverse sections are a result of mass loss, shrinkage, and increased brittleness of the resin due to oxidation. The cracks provide added surfaces and pathways for oxidants to penetrate deeper into the composite during aging. Therefore, once cracking initiates on the axial surface at the fiber–matrix interface, an acceleration of the degradation occurs.

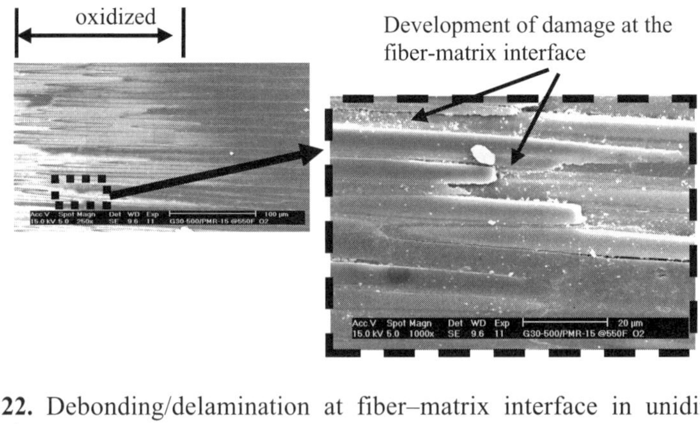

Fig. 9.22. Debonding/delamination at fiber–matrix interface in unidirectional composite

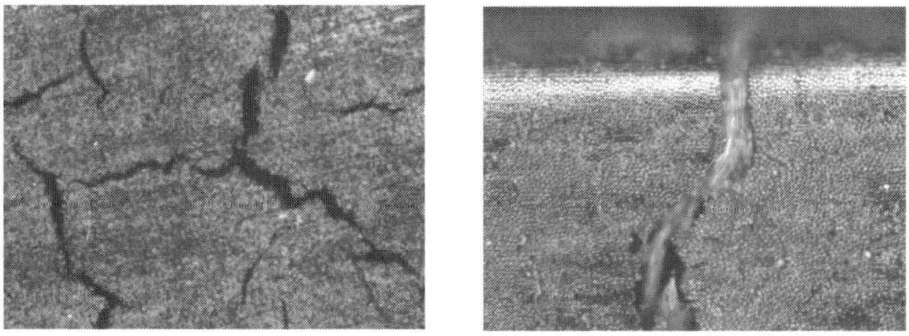

(a) (b)

Fig. 9.23. Photomicrographs of axial surface S_3 of a unidirectional specimen aged for 1,864 h at 288°C showing (**a**) matrix cracking perpendicular to fiber ends and (**b**) close-up view of a matrix crack providing pathway for enhanced diffusion

9.3 Accelerated Aging/Oxidation

Accelerated aging methods are needed to evaluate materials which are to be used under long-term exposure to elevated temperature in oxidative environments. The need for accelerated test methods for HTPMCs is manifested in the requirement to characterize these materials for their expected service life which can be several thousands of hours. The cost of aging these materials for this extended period is often prohibitive. A good accelerated test method neither introduces extraneous damage/degradation mechanisms nor omits any actual mechanisms at its use temperature. Moreover, the method must be validated by comparing mechanical properties, damage modes, and physical aging parameters, such as weight loss, with those from specimens tested under real-time conditions. Since the use temperature of HTPMCs is often near the glass transition temperature of the material, the ability to accelerate aging by increasing the temperature is limited. The use of pressure to accelerate the oxidative aging process is used in the aircraft engine community based in part on the fact that engine parts experience elevated pressures during use. Other conditioning factors reported in the literature include aging in different gaseous environments, aging under load, exposure to ultraviolet radiation, and exposure to moisture. An extensive assessment of accelerated test methods conducted for the HSR program is found in the HSR Materials Durability Guide [2, 41].

The only way to know if an aging mechanism can be properly accelerated is to identify and understand the mechanism and how it will be affected by acceleration [41]. Although we do not fully understand the oxidation mechanism in HTPMCs, we can determine some of the functional relationships that govern aging. For example, elevated pressure aging should not change the oxidative chemical mechanism but should accelerate it.

The acceleration of aging would be a result of increased transport of the oxygen to the interior of the specimen. Since the oxidation process is not reaction rate limited but diffusion rate limited, the elevated pressure would effectively increase diffusion rate, thereby accelerating the oxidation process. Specimens subjected to accelerated aging conditions will not necessarily have the same degradation state as specimens aged in standard conditions. This is evidenced by the fact that the elevated pressure aging specimens have a significantly greater oxidation layer thickness [103] than the ambient pressure-aged specimens. The accelerated aging test must be accompanied by analysis methods to relate the nonaccelerated aging data to the accelerated aging data. These analysis methods and tools will provide capabilities to predict the performance of the material in more general service environments.

9.3.1 Elevated Temperature Aging

Thermally activated rate processes are typically accelerated by increasing temperature. Acceleration by temperature occurs by reducing the activation energy of chemical bond rupture in the polymer macromolecule. Unfortunately, elevated temperature may promote degradation processes that do not occur at application temperatures. Temperature can also affect the rate of degradation by increasing the thermal stress in polymer composites caused by differences in the thermal expansion coefficient of the constituents.

A series of tests was conducted on two carbon/epoxy systems by Tsotsis [108, 109] and Tsotsis and Lee [110] to assess key factors in the characterization of the thermooxidative stability of composite materials. Mechanical properties were determined from specimens aged in air at 177°C up to several thousand hours. After 1,000 h at different temperatures, differential scanning calorimetry (DSC) as well as glass transition temperature T_g measurements suggest that higher aging temperatures lead to more highly crosslinked microstructures which may not be representative of actual aging at use temperatures. Aging below, but near the T_g, gives degradation rates that are nonlinear with respect to temperatures, thereby making estimates of useful lifetimes difficult at best.

Recent work by Ripberger et al. [78] and Tandon et al. [103] also examines the use of elevated temperature to accelerate the rate of thermo-oxidative degradation in PMR-15 resin. While the oxidized layer thickness is similar for all of the aging temperatures up to approximately 1,000 h, as shown in Fig. 9.24, the effect of elevated temperature on the weight loss

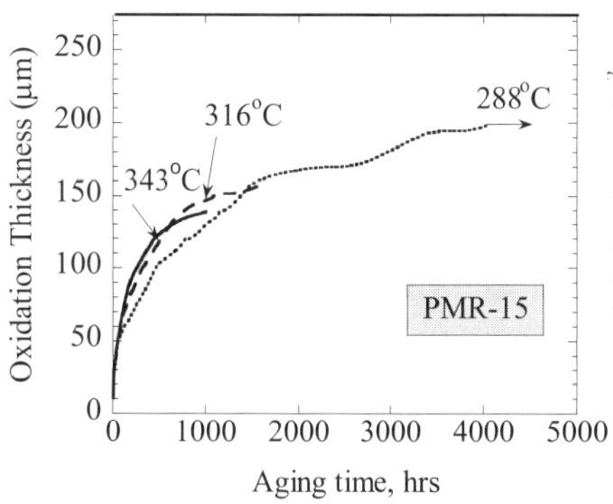

Fig. 9.24. Influence of aging temperature on the oxidation layer growth

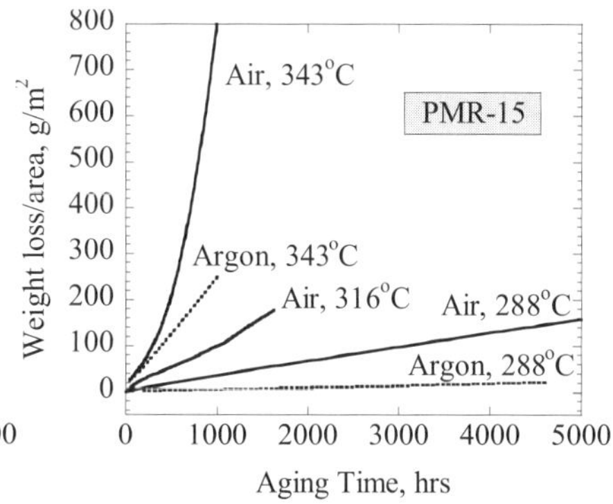

Fig. 9.25. Oxidation weight loss for air- and argon-aged neat resin PMR-15

is not similar. For the specimens aged at 288°C, which is near the use temperature, the majority of the weight loss is due to oxidation since the weight loss in an inert argon environment at 288°C is small. However, for specimens aged at 343°C, a substantial weight loss percentage is attributed to the nonoxidizing thermal aging (aging in an inert argon environment), as shown in Fig. 9.25. The aggressive 343°C aging temperature is near the glass transition temperature of the material. Thermal aging in an inert environment is believed to involve chemical changes associated with chain scission reactions, additional crosslinking, or reduction of crosslink density, etc., that can result in changes to the molecular weight of the polymer, altering the physical and mechanical properties. Thus, there is likely a change in the thermal aging mechanism between the specimens aged at 288 and 343°C. Therefore, one needs to be extremely careful in acelerated aging by elevating the temperature above the use temperature since the materials already operate near the T_g and one must be certain that the aging mechanisms do not change at the elevated temperature or be able to account for any changing mechanisms.

9.3.2 Elevated Pressure Aging

The rate of oxidation is sensitive to the partial pressure of oxygen at the composite surface and acceleration can be achieved by increasing the partial pressure of oxygen within the aging chamber. Increasing the partial pressure of oxygen can be achieved in one or two ways. First, the partial pressure of oxygen can be increased by increasing the mole fraction of oxygen in the aging chamber. For example, pure oxygen (O_2) can be used instead of ambient air. However, the use of pure oxygen in a high-temperature aging chamber may be a safety concern. Second, the partial pressure of oxygen in air can be increased by increasing the total pressure since the partial pressure of oxygen is directly proportional to the total air pressure. By studying the dependence of the flexural strength of glass-reinforced epoxy resin on temperature and oxygen pressure, Ciutacu et al. [25] demonstrated the importance of oxygen pressure as an accelerating factor in thermo-oxidative degradation. The results showed that the same thermooxidative degradation mechanisms for glass-reinforced epoxy resin occurred both in air and oxygen, at the pressure and temperatures used.

Subsequent studies conducted by Tsotsis et al. [111, 112] demonstrated that higher pressures of air or oxygen tend to increase the rate of degradation of polymeric composites. The methodology proposes augmenting elevated temperature aging with elevated pressure to accelerate the rate of thermooxidative degradation. The test results showed a distinct accelerating effect with the use of elevated pressures especially for the tensile shear AS4/3501-6 coupons. The results also suggested that elevated pressure may be a good tool for significantly reducing screening times for materials that will be subjected to long exposure in oxygen-containing environments at elevated temperatures.

Recent work by Tandon et al. [103] examined the use of elevated pressure in conjunction with a realistic use temperature to accelerate the rate of thermooxidative degradation in PMR-15 resin. Neat resin specimens were isothermally aged over a range of time periods in a pressure chamber (see Fig. 9.26), with a continuous supply of pressurized air, and placed inside an air-circulating oven at an elevated temperature. Unlike a sealed closed chamber in which the oxygen supply is depleted for long-aging

Fig. 9.26. Elevated pressure aging chamber placed inside the oven

times, the chamber maintains a target pressure while replenishing the air. Steel tubing attached to regulated house air (with two in-line desiccators) delivers a continuous supply of dry air and maintains a constant pressure of 0.414 MPa (60 psi). The second desiccator contains a purifier that removes any contaminants and impurities with an effective molecular diameter of more than 5 from the dry air before it enters the pressure chamber.

It is observed that the elevated pressure aging of the PMR-15 neat resin presumably has a significant effect on the rate of diffusion of oxygen into the specimen, accelerating the oxidation process and allowing the oxygen to diffuse deeper into the interior of the specimens. This results in greater oxidation layer thicknesses than are achievable in ambient air pressure environments as shown in Fig. 9.27. The effect of aging in 0.414 MPa (60 psi) pressured air further results in nearly a twofold increase in the rate of volume change and increases the weight loss rate of neat resin specimens by approximately a factor of 2 (see Fig. 9.28). Further, mechanical testing reveals that the specimens aged in the pressurized air environment have by far the lowest failure strain and that the strength reduction rate is large for short-aging times, as shown in Fig. 9.29. Since the oxidation process in PMR-15 resin is diffusion limited vs. reaction rate limited, the oxidation process is accelerated by the pressure leading to significant increases in mechanical property degradation.

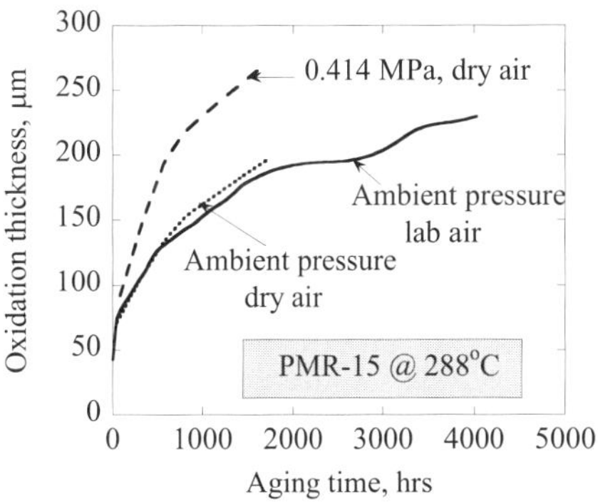

Fig. 9.27. Oxidation layer thickness for ambient and elevated pressure aging

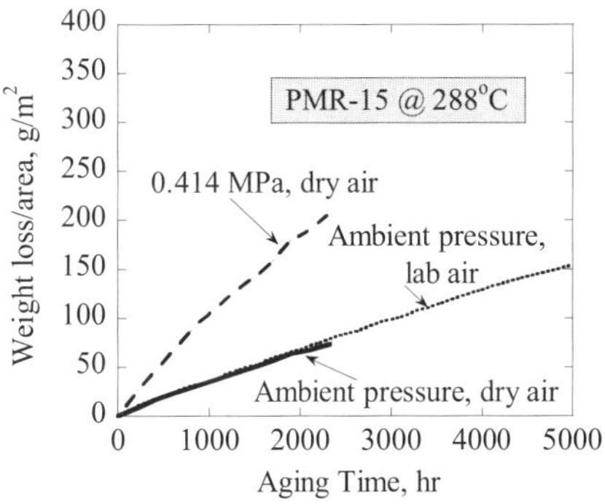

Fig. 9.28. Weight loss of ambient and elevated pressure-aged PMR-15 resin

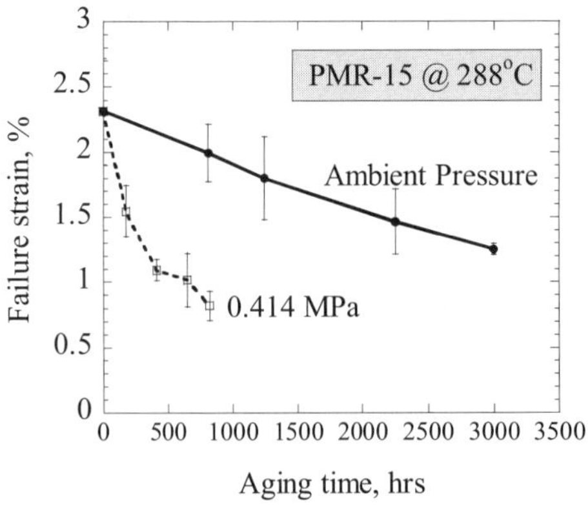

Fig. 9.29. Failure strain for specimens aged at ambient and elevated pressures

9.3.3 Stress-Assisted Aging

In a study by Bowles et al. [15], both unidirectional and crossply (±45) PMR-15 laminates were isothermally aged at 288°C for up to 1,000 h. They found that the weight loss of the crossply laminates is much greater than that of the unidirectional laminates. The primary reason given for the greater weight loss in the crossply laminates is that free-edge cracks, resulting from interlaminar residual stresses, in the crossply laminates are much more prominent and result in more extensive advancement of oxidation into the interior of the composite. Therefore residual stresses, both at the fiber–matrix micromechanical level as previously discussed and at the ply level, play an important role in the degradation process. Residual curing stresses and mechanical load-induced stresses contribute to aging in several ways that include influencing chemical reactions and enhancing diffusion [81]. Stresses can accelerate chemical reactions that cause bond rupture, accelerate chain scission reactions, and alter activation energy for chemical reactions. Popov et al. [75] showed that residual stress and externally applied stress have exactly the same effect on oxidation rate. They further observed that oxidation may induce stresses locally and that these stresses may subsequently influence reactions.

The effect of mechanical stress on long-term thermal aging was investigated in NASA's HSR program [2]. It was observed that the addition of mechanical stress has an accelerating effect on changes in the glass transition temperature in IM7/K3B composites. However, after 10,000 h of aging at 177°C under a constant axial load (loaded to 3,000 $\mu m/m$ strain level), the axial stress applied during aging had little or no effect on aged, unnotched tensile properties of quasi-isotropic layups for both IM7/K3B and IM7/PETI-5 composites. However, these findings cannot be generalized to other properties of the composites. Thermal aging primarily influences the matrix-dominated properties of composites, and the unnotched tensile strength of quasi-isotropic laminates is greatly influenced by the properties of the fibers aligned in the load direction.

To investigate the coupling effects of aging time, temperature, and stress, a low-cost pretensing fixture (inspired by the preload fixture design under the HSR program) was recently developed that allows thermal aging of neat resin specimens under applied load at elevated temperatures. A photograph of the test fixture is shown in Fig. 9.30. The specimen is securely tightened in between the two crossheads with the lower end kept fixed while the upper end is spring loaded to the desired stress level. The fixture assembly is then placed in the oven which is heated to the specified aging temperature, and the specimen is allowed to age for the specified time.

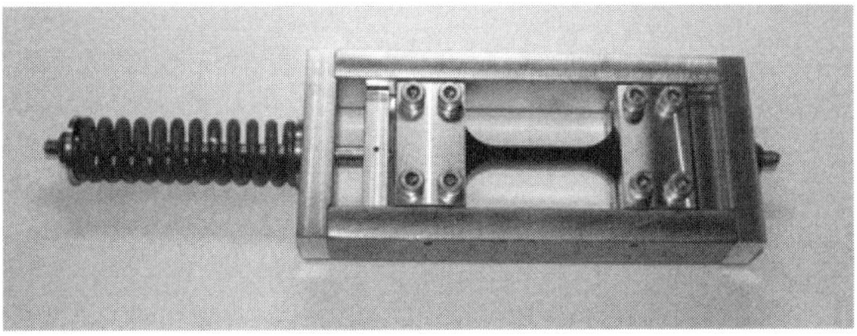

Fig. 9.30. Preload aging fixture for neat resin specimens

Figure 9.31 compares the total thickness of the oxidized region measured in toughened bismaleimide (5250-4) neat resin at 177°C for accelerated aging environments measured using the optical methods, as discussed earlier. Similar to the case of PMR-15 resin, aging under pressure (0.414 MPa (60 psi)) results in far greater oxidation zone thicknesses than are achievable in ambient air pressure environments. Additionally, using the tension aging test fixture shown in Fig. 9.30 (loaded to a stress level of 13.79 MPa (2 ksi)), results in a small increase in oxidation zone thickness at longer aging times compared to an unloaded specimen in an ambient lab air environment. The stress level of 13.79 MPa is chosen based on uniaxially tension-loaded specimens tested at 177°C to ensure that the mechanical response of the neat resin is limited to elastic behavior at the elevated temperature. The tensile strength of the resin reduces considerably under accelerated pressure aging, while only minor decreases are observed under stress-assisted aging when comparing the performance to ambient air pressure-aged specimens.

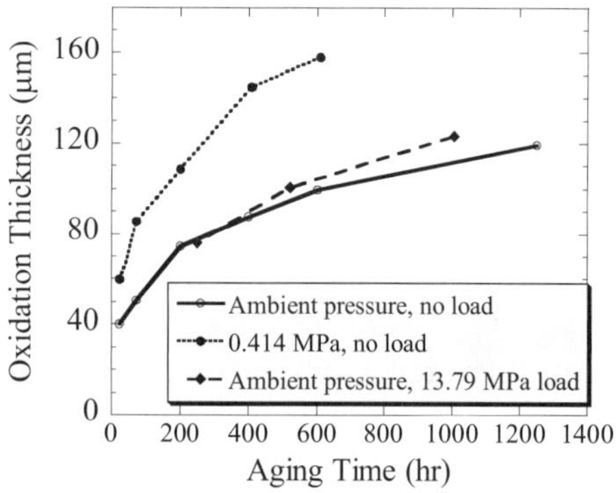

Fig. 9.31. Oxidation layer thickness for 5250-4 for accelerated aging conditions

9.4 Hierarchical Modeling

For effective design of aerospace components with HTPMCs, long-term stiffness and strength prediction and failure mechanism modeling is critical. A framework for hierarchical modeling of composite behavior from its constituent scale to its laminate scale, comprised of predictive models for the fiber(s), matrix, and interface/interphase long-term performance under aging conditions, is shown in Fig. 9.32. Although unit cell models have been used for the prediction of stiffness and strength of the composites, comprehensive unit cell scale constitutive behavior modeling of the HTPMCs under thermooxidation- and stress-induced damage evolutions is lacking.

As discussed previously, three major mechanisms drive age-related degradation of the polymeric matrix composite materials. They are (1) physical aging: creep and relaxation predominantly of the matrix and interphase, (2) chemical aging dominated by the thermooxidative degradation of the resin and the fiber–matrix interphase, and (3) micromechanical damage evolution and its effects on the constitutive and failure behavior of the material. While these effects have been studied for the neat resin in reasonable detail, the studies on composite behavior are nascent. Prediction of the mechanical behavior of HTPMCs under long-term thermomechanical fatigue and harsh environmental conditions requires rigorous simulation of physical aging and its effect on the viscoelastic behavior, the thermo-oxidative aging processes, and the damage evolution in the material constitutive behavior descriptions [53, 95, 120]. Several researchers, e.g., McManus and Cunningham [61], Wang and Chen [119], McManus et al.

[62], Schieffer et al. [87], Schoeppner and Curliss [88], and Colin and Verdu [27], have developed or are developing model-based design capabilities to predict the service life of the HTPMCs. Although mechanism-based models are preferred for predictive capability, when the primitive mechanisms are not understood, phenomenological models can be used. In this context, the relationship between the thermooxidative state of the polymers of interest and their mechanical properties is beyond current mechanistic modeling capabilities. Therefore, researchers rely on empirical correlations to describe the relationship between mechanical properties and aging history for HTPMCs.

The mechanism-based modeling approach, in which individual roles of constitutive materials and the fiber–matrix interface/interphase are recognized, is required to accurately represent the anisotropy of the physical aging, the diffusion and reaction kinetics, the stress-aging coupling, etc., to the level of fidelity required for predicting the performance and failure of HTPMC structures. Mechanism-based models whose aim is to model the relevant physics of the oxidation processes and degradation may be better suited than phenomenological models to predict the local spatial variation of the degradation state in structural components. In the same spirit as the hierarchical process shown in Fig. 9.32, McManus

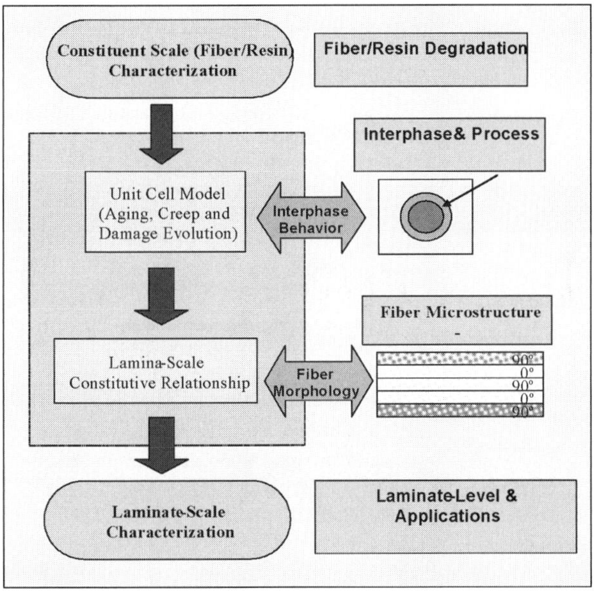

Fig. 9.32. Hierarchical modeling for long-term composite performance

et al. [62] and Schieffer et al. [87] presented approaches that model the various transport and degradation mechanisms at the appropriate geometric level. Using a micromechanical analysis, Schieffer et al. [87] predicted both the oxygen concentration profiles and the oxygen consumption through the thickness of unidirectional specimen by coupling diffusion kinetics and oxygen chemical reactions in the polymer. The model describes an oxygen concentration gradient in the oxidized layer that decreases from its maximum at the free surfaces of the composite laminate toward the interior of the specimen.

Additionally, simulations show that the volume variation of resin shrinkage in the oxidized layer is sufficient to initiate fiber–matrix interface debonding after only a few hours of aging. Micromechanical models, such as described by Schieffer et al. [87], can be coupled to three-dimensional ply-level analyses, e.g., [48], that are capable of modeling the spatial variability of degradation and accounting for the ply-level residual stresses that influence the oxidative degradation.

As shown in Fig. 9.33, a model that depicts the three essential aging mechanisms characterized by scalar state variables can be conceived. The physical aging, including its effect on creep/relaxation behavior, has been characterized by appropriate time–temperature–age shifting and is denoted by the ζ-axis. The thermooxidative development is defined by the state variable ϕ which typically denotes the oxidized material fraction at any

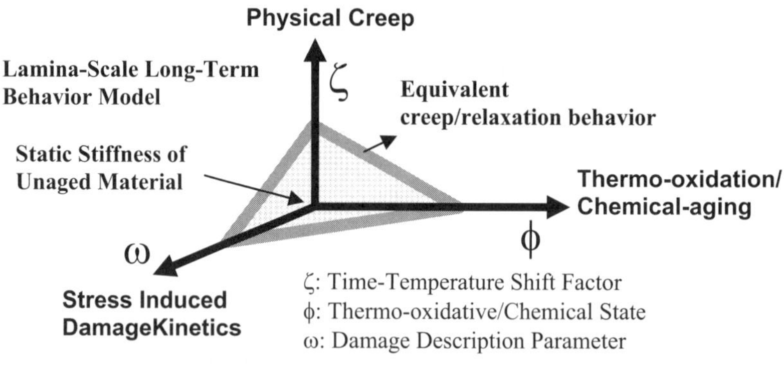

Fig. 9.33. The long-term lamina scale behavior model with various aging mechanisms. The *shaded area* illustrates the aging conditions with equivalent creep and relaxation behavior

material point in the structure. Although, the oxidation growth is being studied using a diffusion–reaction model and experimental characterization, the effect of oxidation on the mechanical behavior remains to be discussed. HTPMC structures subjected to extreme thermal loading are expected to undergo micromechanical damage evolution during their lifetime. The damage state in the composite is represented using a state variable ω. The damage growth dramatically influences the thermooxidative processes as well as the creep/relaxation behavior. Damage evolution and its influence on the oxidative and creep behavior must be characterized and simulated in the constitutive behavior model. The other coupled effects that include the effect of the oxidative layer on the creep behavior and the effect of the creep/relaxation behavior on the damage growth can contribute significantly to stiffness and strength degradation. Formulation of a failure or damage evolution law based on the stress/strain or fatigue state as well as the aging state of the material (for example, using parameters representing the three aging mechanisms) is a crucial step in designing structures with HTPMCs.

In this work, constituent level modeling of the fiber and the matrix is utilized in the unit cell analysis. The diffusion–reaction model for the polymer is developed using material parameters based on the evaluation of neat resin properties, and the models are compared to the oxidation development in the neat resin material. As discussed previously, the analysis is primarily focused on the PMR-15/carbon composite material system. Therefore, the polymer oxidation modeling utilizes PMR-15 neat resin material data from the literature and from experiments conducted for this modeling effort. Although it is recognized that the PAN-based carbon fibers of interest will degrade at temperatures above approximately 300°C (570°F), it is assumed that the fibers have static properties for the purpose of modeling the unit cell oxidation behavior.

9.4.1 Polymer Oxidation Modeling

The focus of the polymer modeling is twofold (1) to understand changes in the polymer properties with service environment exposure and (2) to accurately model these changes in the appropriate constitutive model that will be integrated in the micromechanics representation. The chemical mechanisms responsible for the initiation of thermal and/or oxidative degradation in high-temperature polyimide resins have been the subject of several extensive studies [63–65]. However, understanding and tracking these chemical processes beyond their initial stages is daunting and currently impractical for the class of high-performance polyimides used for aerospace applications.

If the primary, secondary, and tertiary chemical degradation mechanisms become known, their use in a multiscale modeling methodology to predict service performance will depend on the ability to predict mechanical properties based on the chemical state, using techniques such as molecular dynamics. To date, such capability does not exist for highly crosslinked polyimide materials. However, the effect of thermooxidative degradation can be approximated and modeled at a continuum level by representing the effective behavior of the material using suitable semiempirical or parametric models.

Figure 9.34 shows the various mechanisms of the thermooxidative degradation of the polymer. Note that the illustration shows the mechanisms as individual events. However, in reality the mechanisms are coupled, some may be dominant, and the sequence of their occurrence can be a function of the material and the aging environment, i.e., temperature, pressure, etc.

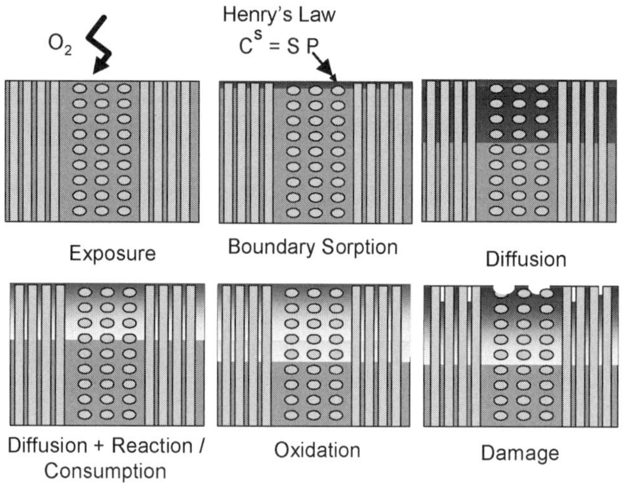

Fig. 9.34. Six phases of thermooxidative damage evolution which include exposure to oxygen environment, sorption at the boundary, diffusion/reaction, oxidative layer growth, and material damage

The oxidative behavior of a resin system is primarily controlled by the diffusivity of oxygen into the material and the kinetics of reaction of oxygen with the composite. The kinetics of reaction entails descriptions of both the oxygen consumption rate (or generally termed as the oxidation reaction rate) and the polymer (substrate) conversion rate (often overlooked in analyses). The substrate conversion rate is generally related to the oxygen consumption rate. While the oxygen availability is controlled by the diffusivity, the polymer availability is controlled by the chemistry and molecular composition of the polymer. As oxygen diffusivity is a non-terminal process, the reaction cessation is typically governed by the polymer availability. Materials differ in the nature of the oxidation behavior observed during experimentation depending on the relative dominance of the diffusive behavior and polymer availability or the reaction kinetics.

Experimental observations suggest that the behavior of PMR-15 is diffusion controlled, and the reaction rate is high enough to consume the available polymer in observable time periods (tens of hours). Therefore, growth of oxidation into the polymer is observed with a clear oxidized, active reaction, and unoxidized regions and the growth of the oxidized region with time. Other materials may behave significantly different from the PMR-15 material, however the modeling methodology given here should be applicable. Materials such as AFR-PE-4 and BMI do not show this clear demarcation indicating either slow reaction rates or conversely large polymer availability for reaction.

As an idealized representation of oxidized neat polymer PMR-15, three distinct material regions, representing different levels of oxidation, as illustrated in Fig. 9.35 can be used. Using this representation, the material phases are the oxidized polymer phase (typically near the surface, Region I), the active reaction zone (where a mix of oxidized and unoxidized polymers exists, Region II), and an unoxidized polymer (typically in the interior of the specimen, Region III). The region composed of the oxidized (I) and active oxidation zones (II) will be referred to as the *oxidative surface layer*. Although this three-phase description is analogous to the two-phase models (or skin/core models) described by Nam and Seferis [66] and Salin and Seferis [83], the description of three phases enables extension to situations with high conversion ratios in which the constancy of the substrate

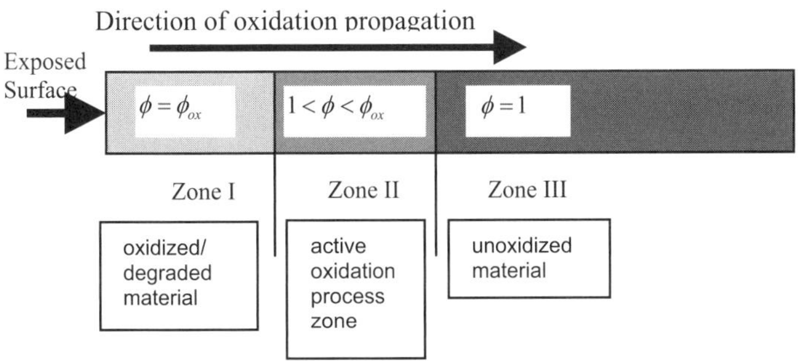

Fig. 9.35. The three zones in thermooxidation. The oxidized region is followed by active zone separating the oxidized and unoxidized regions

concentration is no longer valid and the substrate depletion leads to an autoretardation of the oxidation kinetics. A practical consequence is a decrease of the oxidation rate in the superficial (Region I) layers of the specimen, where the high conversion ratios are reached first, which leads to a moving of the oxidation front toward the specimen core. Such an extension was proposed by Colin and Verdu [27] and Colin et al. [33] and successfully applied to the thermal oxidation of amine-crosslinked epoxy and bismaleimide resins.

Sorption and diffusion modeling

For an aging process involving the diffusion and consumption of small molecules like O_2 by reaction with the polymer, there are critical conditions of reaction rate and/or thickness above which the process becomes kinetically controlled by the diffusion of the small molecules into the polymer [5]. Suitable lifetime prediction models must involve the thickness distribution of the oxidation reaction products. The latter can be predicted from diffusion laws, such as Fick's second law modified by a term relating the rate of consumption of the diffusing oxygen to the chemical reaction. In this analysis, $C(x,y,z,t)$ is defined as the concentration field of oxygen at any time within a domain with a diffusivity of D^*_{ij} and consumptive reaction with a rate, $R^*(C)$. The oxygen concentration is tracked through the Fickian diffusion equation with a reaction term and orthotropic diffusivity, as given by (9.1)

$$\frac{\partial C}{\partial t} = \left(D^*_{11} \frac{\partial^2 C}{\partial x^2} + D^*_{22} \frac{\partial^2 C}{\partial y^2} + D^*_{33} \frac{\partial^2 C}{\partial z^2} \right) - R^*(C) \tag{9.1}$$

subject to the boundary conditions:

$$C = C^{\mathrm{S}} \text{ on the exposed boundaries,}$$

$$\mathrm{d}C / \mathrm{d}t = 0 \text{ on symmetry boundaries.}$$

The sorption on the exposed boundaries is given by Henry's equation

$$C^{\mathrm{S}} = SP^{\mathrm{O}}, \tag{9.2}$$

where S is the solubility (which is the maximum quantity of solute that can dissolve in a certain quantity of solution at a specified temperature) and P^{O} is the partial pressure of the oxygen in the environment. The solubility S is assumed to be temperature independent in this analysis. Therefore, increasing the pressure of the gas in the system (or increasing the molar volume V/n in $PV = nRT$) increases the boundary concentration C^{S}. In (9.1), the superscript * is used to denote additional dependencies of these parameters on the substrate temperature and oxidation state. These dependencies are further detailed in the ensuing subsections.

When the polymer reacts with oxygen at a given point, the sites on the polymer available for oxidation become depleted and available oxygen molecules will continue to diffuse into the interior of the polymer to react. Therefore, a polymer availability state variable ϕ is defined that quantifies the availability of active polymer sites for reaction. For the sake of simplicity, a reactive site depletion parameter ϕ_{ox} is defined at which the oxygen will predominantly diffuse rather than react with the polymer. In reality, the material will continue to oxidize. However, the rate of conversion of the polymer to oxides in this region may be much slower. The state variable ϕ is parameterized to have the range $\phi_{\mathrm{ox}} < \phi < 1$, where ϕ_{ox} denotes completely oxidized polymer (idealized assumption) and $\phi = 1$ denotes unoxidized polymer. The introduction of the polymer oxidization state parameter ϕ, therefore, enables quantitative tracking of the reaction within the polymer. The reaction rate function $R^*(C)$ is influenced by the temperature T and available oxygen concentration C and is also a function of the polymer substrate oxidation state variable ϕ as shown in (9.3)

$$R^*(C) = g(\phi)R(C,T), \tag{9.3}$$

where $g(\phi)$ describes the reaction rate dependence on the oxidation state variable and $R(C, T)$ defines the reaction rate dependence on the concentration and temperature. The function $R(C, T)$ is defined as the reaction rate when there is an abundance of polymer reaction sites.

In isothermally aged PMR-15 specimens in an ambient air environment, modulus measurements using nanoindentation reveal that the oxidized region (I) has an approximately uniform modulus over the entire region, whereas the modulus of the active oxidation zone (II) decreased monotonically from Zone I stiffness levels to Zone III stiffness levels. As an idealization for PMR-15, it is therefore assumed that all of the reactions take place in the active reaction zone (II). In some other polymer systems, the diffusion–reaction relationship is such that the active reaction zone (II) encompasses the entire oxidative surface layer.

Figure 9.36 is an illustration of elastic modulus measurements made across the thickness of neat resin specimens of the polyimide AFR-PE-4 using nanoindentation. The data have been normalized with respect to the modulus of the unaged resin and show that there is no definitive region (I) in which the polymer is considered to be completely oxidized. Rather, the modulus decreases monotonically from the surface of the specimen to the unoxidized interior region (III). While the active zone (II) modeling describes the reaction rate and its dependence on oxygen availability (concentration fields), the diffusion–reaction system will be stationary (will not propagate into the polymer) if enough polymer in the reaction zone (II) is available for reaction. However, for longer aging times of tens to hundreds of hours, the active reaction zone (II) will propagate into the interior of the polymer.

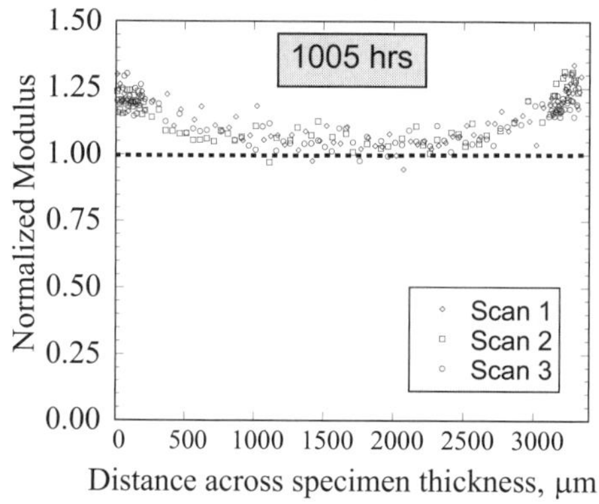

Fig. 9.36. Normalized oxidation modulus profile through entire thickness of AFR-PE-4 neat resin specimen

Oxidative aging leads to a reduction in molecular weight as a result of chemical bond breakage and loss in weight from outgassing of low molecular weight gaseous species. The most common assumption relating the reaction rate and weight loss is to assume that the rate of change in weight ($\mathrm{d}W/\mathrm{d}t$) is proportional to the reaction rate $R(C, T)$, i.e.,

$$\mathrm{d}W / \mathrm{d}t \propto -R(C,T), \tag{9.4}$$

where the negative sign is introduced in (9.4) since the rate of change in weight is negative and the oxygen consumption rate is positive. The oxidation state parameter at any point in the material is determined as the current weight of the material over its original (unoxidized) weight, i.e.,

$$\mathrm{d}\phi / \mathrm{d}t \propto \mathrm{d}W / \mathrm{d}t. \tag{9.5}$$

Combining (9.4) and (9.5), we obtain

$$d\phi(t)/dt = -\alpha R(C,T), \tag{9.6}$$

where α is a proportionality parameter that, in general, is time and temperature dependent. The initial state of the polymer is taken to be $\phi = 1$ and the final state (where the reaction terminates after complete conversion of the polymer) is taken to be $\phi = \phi_{ox}$. The solution to the system of (9.1)–(9.6) provides the transient evolution of oxygen concentration profiles and the oxidation state in the polymer domain. The system of differential equations constituting the kinetic model, in which is added an equation describing the substrate depletion, is solved numerically without any simplifying assumptions (concerning the long kinetic chains, the stationary state, or the existence of a relationship between the termination rate constants). This model accurately predicts the continual increase of the thickness of oxidized layer in the domain of high conversion ratios.

Temperature, pressure, and oxidation state dependence

The diffusivity of oxygen in the polymer depends on the temperature and the substrate oxidation state ϕ of the polymer material. It is assumed that the polymer is stress free. The temperature dependence of the diffusivities is expressed in its usual Arrhenius form as

$$D_{ij} = D_{ij}^{o} \exp(-E_{ij}^{a} / RT), \tag{9.7}$$

where the gas constant R should not be confused with the concentration-dependent consumption reaction rate $R(C, T)$. If it is assumed that the neat resin is isotropic, (9.7) can be simply expressed as $D_{ij} = D^{o} \exp(-E^{a} / RT)$. Using permeability tests at lower temperatures, Abdeljaoued [1] determined the preexponent (D^{o}) and the activation energy parameters (E^{a}) of PMR-15 neat resin in its virgin state, i.e., unoxidized condition, assuming isotropic diffusivity of the neat polymer. Direct experimental determination of the diffusivity parameters is usually nontrivial due to thin specimens required for observing gas permeability in reasonable timescales. Anisotropic diffusivity measurements are considerably more complicated due to the number of parameters that are needed. A diffusivity value of 53.6×10^{-6} mm^2 min^{-1} is calculated for PMR-15 resin at 288°C using (9.7) in conjunction with the unoxidized preexponent D^{o} and the activation energy parameters E^{a} listed in Table 9.1.

In general, the diffusivities for each of the material regions will be different, while the diffusivity of the oxidized polymer layer (Region I) is the controlling parameter since the diffusion of oxygen through the oxidized layer is the source of oxygen supply to the active reaction zone (Region II). For lack of experimental measurements of the diffusivities of

Table 9.1. PMR-15 polyimide resin diffusivity parameters

	Unoxidized, $\phi = 1$ (Abdeljaoued [1])	Oxidized, $\phi = \phi_{ox}$ (estimated using model [74])
D^o	6.10×10^{-11} m^2 s^{-1}	8.90×10^{-11} m^2 s^{-1}
E^a	19,700 J mol^{-1}	
R	8.31447 J (mol K)$^{-1}$	

the oxidized (I) and active reaction (II) regions for the PMR-15 material, it is initially assumed that the diffusivities for the three regions are equivalent and equal to the unoxidized region (III) diffusivity. Later, simulations are presented where the distribution of the diffusivity D_{ij} in the active reaction zone (Region II) is assumed to be a linear interpolation of the diffusivities of the oxidized (Region I) and unoxidized (Region III) materials, denoted by D_{ij}^{ox} and D_{ij}^{un}, respectively. The oxidation state variable ϕ dependence on the diffusivity is introduced using a rule of mixtures, namely

$$D_{ij}^{*} = D_{ij}(T,\phi) = \langle D_{ij}^{o} \rangle_{un} \exp(-\langle E_{ij}^{a} \rangle_{un} / RT) \frac{\phi - \phi_{ox}}{1 - \phi_{ox}} + \langle D_{ij}^{o} \rangle_{ox} \exp(-\langle E_{ij}^{a} \rangle_{ox} / RT) \frac{1 - \phi}{1 - \phi_{ox}}. \tag{9.8}$$

The $\langle \rangle_{un}$ notation is used to denote the quantities for the unoxidized ($\phi = 1$) substrate state and the $\langle \rangle_{ox}$ is used to denote those of oxidized state ($\phi = \phi_{ox}$). Therefore, the temperature-dependent diffusivity needs to be characterized for oxidized and unoxidized states, and for polymer states $\phi_{ox} < \phi < 1$ such that the diffusivity is given by linear interpolation using (9.8). For the remainder of this manuscript, superscripts and subscripts "ox" and "un" will be used with model parameters whenever needed for clarity. For clarification, unoxidized material does not imply unaged material.

The reaction rate term $R(C, T)$ in (9.3) models the reaction of the oxygen with the polymer for the case that there is an excess of polymer reaction sites. The reaction modeling presented here assumes that the reaction products (water and other volatiles) leave the polymer instantaneously, and modeling of the outgassing is not considered. The basic autooxidation scheme (BAS) that describes a simple kinetics oxidation scheme for polymers [45] is used. It is further assumed that the expression for the oxygen consumption rate determined in the stationary state remains valid in the domain of high conversion ratios.

Typically, the reaction rate is modeled with the Arrhenius-type kinetics model with a saturated reaction rate R_O as given by $R_O(T) = R_O' \cdot \exp(-R^a / RT)$ for capturing thc temperature dependence of the reaction rates [37, 62] or using mechanistic reaction models [1, 31, 33, 87, 117] for capturing the oxygen concentration dependence. The rate constant R_O' and the activation parameter R^a determine the temperature dependence of the reaction rate. The reaction rate function is reduced when the oxygen availability is reduced or when the substrate is fully converted, as given by (9.9)

$$R^*(C) = g(\phi)R(C,T) = \left[\frac{\phi - \phi_{ox}}{1 - \phi_{ox}}\right] R_O(T) f(C). \tag{9.9}$$

The models by Colin et al. [31, 33] and Abdeljaoued [1] differ on how the concentration dependence is expressed. To determine the saturation reaction rate R_O, an active reaction zone (II) analogy is employed. Considering the diffusion–reaction equation (9.1), the proportion of the diffusivity vs. reaction rate determines the zone over which the oxygen diffuses in and oxygen gets consumed by the reaction. This zone is indicative of the saturation reaction rate R_O in (9.9). The lower the reaction rate, the larger the active reaction zone (II) for a constant diffusivity. The function $f(C)$ in (9.9) models the situation in which the amount of oxygen available for reaction is lower than that required for the maximum reaction rate under saturation conditions. The reduction in reaction rate can be modeled following Abdeljaoued [1] using (9.10) or Colin et al. [31, 33] using (9.11):

$$f(C) = \frac{\beta C}{1 + \beta C}, \tag{9.10}$$

$$f(C) = \frac{2\beta C}{1 + \beta C}\left[1 - \frac{\beta C}{2(1 + \beta C)}\right]. \tag{9.11}$$

The former model (9.10) has been used for epoxy and bismaleimide matrices, and the latter model (9.11) has been used for the PMR-15 polyimide resin systems. Figure 9.37 shows the dependence of the reaction rate on the concentration as modeled by (9.10) and (9.11). The abscissa has a normalized concentration parameter βC in which the parameter β non-dimensionalizes the concentration field. According to (9.10), the saturation reaction rate is reached more slowly and at considerably higher concentration βC values than for (9.11). The expression given by (9.11) allows the saturation reaction rate to be reached quicker and at a finite βC value. For βC values of greater than approximately 3 (corresponding to $R(C,T)/R_0 \sim 90\%$), the reaction rate remains relatively uniform. For this analysis, the concentration dependence, as modeled in (9.11), was chosen for the simulations. The oxygen concentration at which the reaction behavior changes from an accelerated rate to approximately a uniform reaction rate can be easily controlled by varying the value of β (with a constant βC equal to 3).

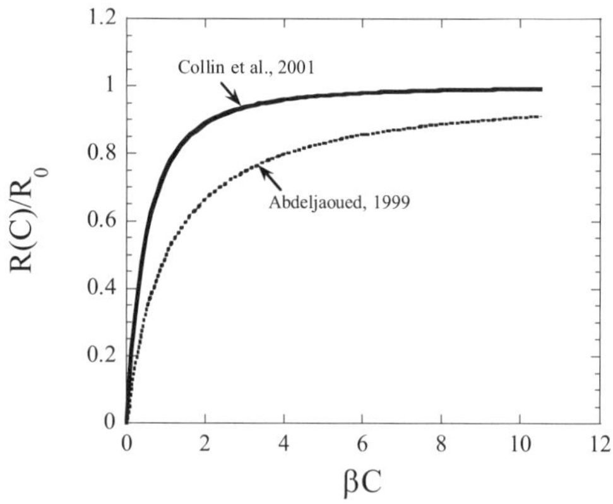

Fig. 9.37. Typical model of reaction rate dependence on concentration

The value of β can be determined from weight loss data obtained from specimens aged at two different oxygen partial pressures, typically in pure O_2 and in air. This formulation leads to the reaction rate approaching the saturation reaction rate $R(C,T) \rightarrow R_0(T)$ when the concentration approaches infinity $C \rightarrow \infty$. That is, when the reaction is not oxygen deprived the reaction rate reaches a maximum saturation rate, $R_0(T)$. To determine this value, weight loss measurements at two concentrations which translate to the reaction rates at those two concentration values are needed. Since the weight loss is taken to be proportional to the oxygen consumption rate, the ratio of the weight loss determined at two concentrations is the same as $R_1(C_1, 288°\text{C})/R_2(C_2, 288°\text{C})$. For example, from Abdeljaoued [1], the ratio of the weight loss between pure O_2 ($C_2 = 3.74$ mol m^{-3}) and air ($C_1 = 0.79$ mol m^{-3}) is about 0.7 at a temperature of 288°C. Therefore,

$$\frac{2\beta C_1}{1+\beta C_1}\left[1-\frac{\beta C_1}{2(1+\beta C_1)}\right] = 0.7\frac{2\beta C_2}{1+\beta C_2}\left[1-\frac{\beta C_2}{2(1+\beta C_2)}\right]. \tag{9.12}$$

Solving (9.12) produces three roots [–3.5593, –0.42657, 0.91947] with only one physically feasible value for PMR-15 resin to be $\beta = 0.919$.

The oxygen consumption expression contains two key parameters of the oxidation kinetics (1) the maximal oxidation rate of the polymer when excess oxygen is available $R_O = k_3^2[\text{PH}]^2/k_6$ for a given temperature T and (2) the reciprocal of the critical oxygen concentration beyond which oxygen excess is reached $\beta = k_2k_6/(2k_5k_3[\text{PH}])$. The reaction rate parameters k_2, k_3, k_5, and k_6 for the BAS are not determined explicitly to calculate the model parameters but describe mechanistically how R_O and β are related to the kinetics of oxidation [32]. It is clear that both R_O and β are dependent on the substrate concentration or the availability of polymer reaction sites [PH]. However, it is assumed that only R_O is especially affected by the substrate depletion and that the macroscopic weight loss measurements can be used to determine the reaction and polymer consumption parameters.

Correlation with weight loss observations

Experimental weight loss data are used to determine the value of ϕ_{ox} for PMR-15 neat resin. To determine this value, a relation between the oxidation layer size and the weight loss must first be established. In an inert gaseous environment, polymers undergo physical and nonoxidative chemical aging only, whereas in an oxidizing environment, specimens will undergo physical, nonoxidative chemical aging, and oxidative aging. The differences in the aging behavior of the specimens aged in an inert gaseous environment and those aged in an oxidizing gaseous environment can then be attributed to oxidative aging. Thermal aging experiments were conducted in both an inert atmosphere (argon) and an oxidative atmosphere (air), and the weight loss was recorded at various time intervals to determine the effect of oxidation on weight loss.

Let the initial dimensions of a specimen be given by length L, width W, and thickness T. For a particular aging time, if the weight loss fraction (weight loss/original weight) in the inert atmosphere is given as γ and weight loss fraction after the same time period in air is given by ε, then the weight loss fraction due to oxidation alone is $(\varepsilon - \gamma)$. Let the observed thickness of the completely oxidized region (I) and the active oxidation layer (II) be denoted by t_o and t_a, respectively, as shown in Fig. 9.38. The initial volume of the specimen V, the volume of the fully oxidized region V_o, and the volume of the active oxidation zone V_a are, therefore, given by:

$$V = LWT, \tag{9.13}$$

$$V_o = \{LWT - (L - 2t_o)(W - 2t_o)(T - 2t_o)\}, \tag{9.14}$$

$$V_a = \{LWT - V_o - (L - 2(t_a + t_o))(W - 2(t_a + t_o))(T - 2(t_a + t_o))\}. \tag{9.15}$$

Assuming that ϕ varies linearly in the active oxidation zone (II) from ϕ_{ox} to 1, the total weight loss is the sum of the weight loss in the oxidized zone (I) and the active zone (II) and is equal to the weight loss in the specimen, i.e.,

$$\left[(1-\phi_{ox})V_o + \left(\frac{1-\phi_{ox}}{2}\right)V_a\right]\rho = (LWT)[\varepsilon - \gamma]\rho. \tag{9.16}$$

Solving for ϕ_{ox},

$$\phi_{ox} = 1 - \frac{V(\varepsilon - \gamma)}{V_o + \frac{1}{2}V_a}. \tag{9.17}$$

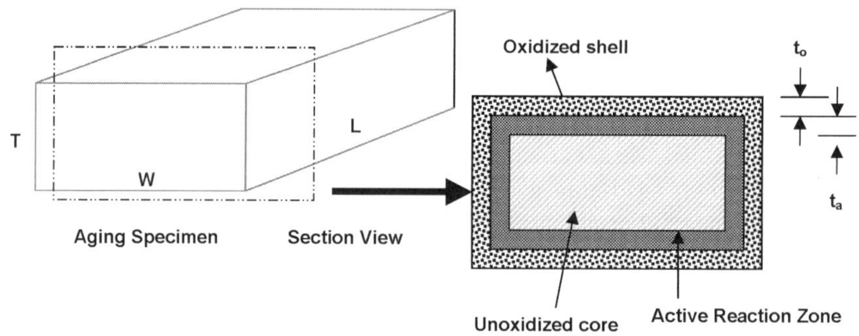

Fig. 9.38. Geometry of the specimen used for aging with all boundaries exposed to oxygen. The oxidized layer, active reaction zone, and the unoxidized regions are illustrated in the sectional view

Equations (9.13)–(9.17) are used to determine the value of ϕ_{ox} from oxidation layer size observations and weight loss data from aging specimens in inert (γ) and oxidative (ε) environments. The values of ϕ_{ox} as determined from the experimental data from PMR-15 aging tests are observed to decrease with aging time. This observation indicates that the oxidized region (I) material continues to degrade and lose weight with increasing aging time. Material loss from the specimen free surfaces and volumetric shrinkage are both observed during oxidative aging experiments and these effects arc predominant for PMR-15 neat resin specimens beyond 100 h of aging, especially at 343°C. Therefore, for PMR-15 resin, the ϕ_{ox} value is determined by averaging the predictions during the initial time period of 100 h at 288°C. The average ϕ_{ox} value as determined by (9.17) using experimental weight loss data is 0.187, which implies that the completely oxidized polymer weighs about 18.7% of the unoxidized polymer. Recognizing that the model represents an idealization of the oxidation process, in reality the nonzero ϕ_{ox} value may simply represent a state of the polymer at which the reaction rate is very slow compared to oxidation of unaged polymer. Unless otherwise mentioned, all of the simulations presented herein are performed using the average ϕ_{ox} value of 0.187. Determination of the oxidation parameter ϕ_{ox} by (9.17) is applicable up to the temperature-dependent aging time that surface cracks appear in the oxidized zone (I) of the specimen. Once surface cracks appear, the relationship is no longer valid.

Oxidation-dependent modulus change

Aging of the polymer at elevated temperatures results in changes of the mechanical properties of the material. Changes in the stiffness of the neat resin can occur for aging in both oxidizing and nonoxidizing environments. For oxidized regions (I) of PMR-15 neat resin specimens, the stiffness is greater than that of the unoxidized regions (III) and greater than that of unaged resin. Based on nanoindentation data, a profile of the modulus change across the oxidized (I), active (II), and unoxidized zones (III) of PMR-15 neat resin specimens is shown in Figs. 9.39 and 9.40 for specimens aged at 288 and 343°C, respectively. Based on data from Figs. 9.39 and 9.40, as well as other PMR-15 specimens, it is observed that the oxidized regions of the specimens can be represented as having a nearly constant modulus. The modulus change as a function of the state variable ϕ can then be expressed as

$$E(\phi,t) = E_{\text{non}} \exp\left(k_{\text{ox}} \frac{1-\phi}{1-\phi_{\text{ox}}} + k_{\text{un}} \frac{\phi-\phi_{\text{ox}}}{1-\phi_{\text{ox}}} \right), \tag{9.18}$$

where E_{non} is the nonaged (as opposed to unoxidized) modulus of the resin, parameter k_{ox} models oxidation-dependent modulus changes, and parameter k_{un} models the unoxidized time-dependent modulus changes due to physical/chemical aging [125]. Solid plots of the modulus profiles using (9.18), as they compare to the nanoindentation data, are shown in Figs. 9.39 and 9.40.

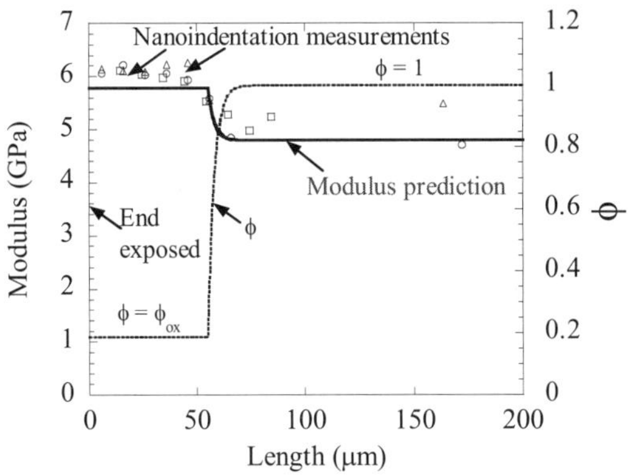

Fig. 9.39. Profile of oxidation layer modulus for PMR-15 aged at 288°C for 50 h

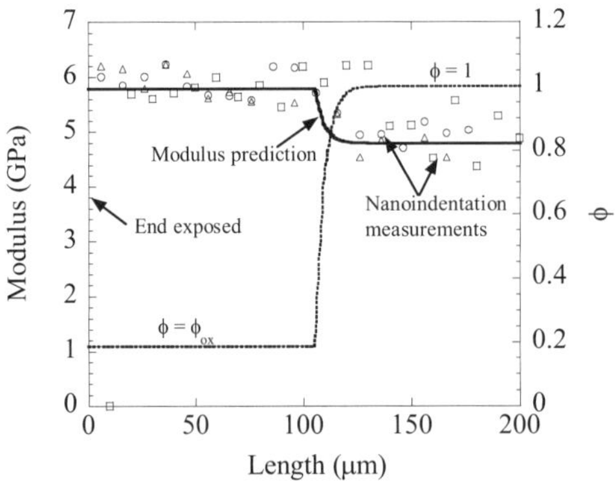

Fig. 9.40. Profile of oxidation layer modulus for PMR-15 aged at 343°C for 541 h

Oxidation simulations for neat polymer

The capability to obtain direct measurements of all of the properties of the PMR-15 for each of the three material regions that are needed to populate the parameters for the model is lacking. Therefore, many of the parameters are obtained through indirect measurement or correlation with model predictions. To determine the parameters of the outlined model for PMR-15 neat resin, predictions from the model are compared and correlated to the experimental observation of the oxidation layer growth obtained at 288°C. These comparisons provide quantitative values for various parameters of the model, especially the temperature-dependent saturation reaction rate $R_O(T)$, the behavior of the proportionality α between reaction rate and the weight loss, the polymer availability state variable ϕ, and the apparent diffusivity of the oxidized region D_{ij}^{ox} .

The diffusion–reaction system of equations is solved using numerical solutions to differential algebraic equations in the one-dimensional domain. These methods are computationally effective when performing parametric sensitivity analysis. This section describes the solution algorithms used to solve the diffusion–reaction system with the reaction rate described as follows

$$R(C,T) = \begin{cases} R_O(T)2\xi(1-\xi/2), & \phi > \phi_{ox}, \\ 0, & \phi = \phi_{ox}, \end{cases} \tag{9.19}$$

where

$$\xi = \frac{\beta C}{1+\beta C}, \tag{9.20}$$

as given by (9.11). The diffusion–reaction system is solved in MATLAB™ using a modified implementation of *ode15s* and `Pdepe` solvers. The *ode15s* is a variable order multistep solver based on the numerical differentiation formulas (NDFs) and uses the backward differentiation formulas (BDFs, also known as Gear's method). `Pdepe` solves initial–boundary value problems for systems of parabolic and elliptic partial differential equations (PDEs) in the one space variable x and time variable t. The ordinary differential equations (ODEs) resulting from discretization in space are integrated to obtain approximate solutions at times specified in t span. The `Pdepe` function returns values of the solution on a mesh provided in `xmesh`. `Pdepe` solves PDEs of the form

$$c(x,t,u,\partial u/\partial x)\frac{\partial u}{\partial t}=x^{m}\frac{\partial}{\partial x}(x^{m}f(x,t,u,\partial u/\partial x))+s(x,t,u,\partial u/\partial x). \tag{9.21}$$

The PDE in (9.21) is solved for a time interval of $t_i \le t \le t_f$ and over a spatial domain of $a \le x \le b$. The initial time is represented by t_i, the final time of the simulation as t_f, and the boundary points of the computational domain are $x = a$ and $x = b$. The interval $[a, b]$ must be finite, while m can be 0, 1, or 2, corresponding to slab, cylindrical, or spherical symmetry, respectively. If $m > 0$, then a must be ≥ 0. In the above equation, $f(x,t,u,\partial u/\partial x)$ is a flux term and $s(x,t,u,\partial u/\partial x)$ is a source term. The coupling of the partial derivatives with respect to time is restricted to multiplication by a diagonal matrix $c(x,t,u,\partial u/\partial x)$. The diagonal elements of this matrix are either identically zero or positive. The diffusion–reaction system given in (9.1) can be put into the form shown in (9.21) for the one-dimensional case by using $m = 0$ and the following functions:

$$\begin{aligned}
&u = C,\\
&c(x,t,u,\partial u/\partial x)=1,\\
&f(x,t,u,\partial u/\partial x)=D^{*}\mathrm{d}u/\mathrm{d}x,\\
&s(x,t,u,\partial u/\partial x)=-R^{*}(u).
\end{aligned} \tag{9.22}$$

For $(t - t_o)$ and all x, the solution components satisfy the initial conditions of the form

$$C(x,t_o)=C'(x), \tag{9.23}$$

where $C'(x)$ is the known initial state of oxygen concentration. For all t and either $x = a$ or $x = b$, the solution components satisfy a boundary condition of the form

$$p(x,t,u)+q(x,t)f(x,t,u,\partial u/\partial x)=0. \tag{9.24}$$

The boundary conditions for the present problem can be written as follows:

$$\begin{aligned}
&@\,x=a(=0)\begin{cases}p(0,t,u)=u-C^{S},\\ q(0,t)=0,\end{cases}\\
&@\,x=b(=L)\begin{cases}p(L,t,u)=0,\\ q(L,t)=1.\end{cases}
\end{aligned} \tag{9.25}$$

The oxidation-dependent diffusivity D_{ij} is determined at each computational point using (9.8). The oxidation state variable ϕ is determined at every time step during the computation using the weight loss relationship shown in (9.6)

$$\phi = \min\left\{\phi_{\text{ox}}, \left(1 - \int_0^t \alpha(\varsigma)R(\varsigma)\mathrm{d}\varsigma\right)\right\}. \tag{9.26}$$

For initial results, the proportionality constant α relating the weight loss and reaction rate is assumed constant with aging time. In later simulations, we consider the variation of α with time, as discussed below.

Influence of saturation reaction rate, R_o

Figure 9.41 shows the experimental observations and the predictions of oxidation layer (Zones I + II) growth at 288°C. The simulations shown are with a constant diffusivity ($D_{ij}^{\text{ox}} = D_{ij}^{\text{un}} = 53.6 \times 10^{-6}\ \text{mm}^2\ \text{min}^{-1}$) and a constant value of the proportionality constant ($\alpha = 0.01$). Several values of reaction rates (5.5, 3.5, and 0.5 mol (m^3 min)$^{-1}$) are considered. The oxidation layer growth predictions show little dependence on the reaction rate $R_O(T)$. Moreover, the simulations overpredict the oxidation layer growth at longer aging times. As mentioned earlier, the choice of saturation reaction rate mainly influences the size of the active reaction zone (II). A larger reaction rate results in a decrease in the size of the active zone (II), whereas a smaller value results in an increase in the size of the active reaction zone (II). The experimental measurements of Ripberger et al. [78] support greater reaction rates during the initial aging period which then decrease at longer aging times.

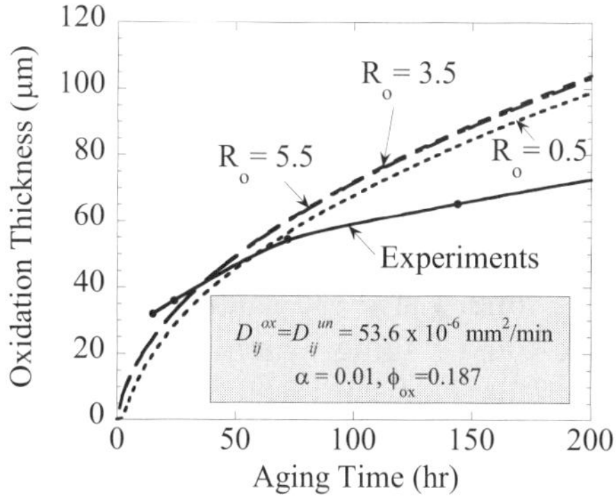

Fig. 9.41. Influence of R_O on oxidation layer growth predictions

Influence of polymer availability state, ϕ_{ox}

Figure 9.42 shows the influence of ϕ_{ox} on oxidation layer growth predictions using constant values of $D_{ij}^{ox} = D_{ij}^{un} = 53.6 \times 10^{-6}$ mm^2 min^{-1}, $R_O(T) = 3.5$ mol (m^3 min)$^{-1}$, and $\alpha = 0.01$. An average ϕ_{ox} value of 0.187 is determined using the weight loss data during the initial 100 h of aging. For this figure, two other constant ϕ_{ox} values, 0.1 and 0.3, were considered. Clearly, the oxidation layer thickness predictions are seen to increase with increasing values of ϕ_{ox}. A larger value of ϕ_{ox} implies earlier exhaustion of polymer availability for reaction leading to a more rapid advancement of the reaction zone (II) and ultimately a larger oxidation layer thickness.

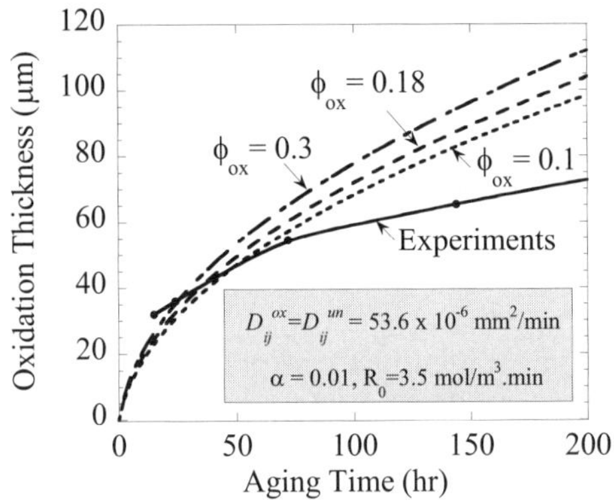

Fig. 9.42. Influence of ϕ_{ox} on oxidation layer growth predictions

Influence of proportionality factor, α

The proportionality constant α relates the molar reaction rates with observed weight loss with respect to aging time. Although we do not present a direct correlation with known physical or chemical mechanisms, the experimental results by Ripberger et al. [78] indicate that the initial weight loss rate is much more rapid than rates at longer aging times and approaches a constant rate for longer aging times. It is, therefore, reasonable to assume that the proportionality constant α is larger at the beginning of aging and decreases with aging time. For these simulations, a linear relation for α is assumed for the first 40 h of aging which leads to numerical predictions of oxidation layer growth that correlates well with experimental measurements.

Figure 9.43 shows the influence of α on oxidation layer growth predictions using a constant diffusivity ($D_{ij}^{ox} = D_{ij}^{un} = 53.6\times10^{-6}$ mm^2 min^{-1}) and a constant value of the saturation reaction rate ($R_O = 3.5$ mol (m^3 min)$^{-1}$). Initially, it is assumed that the proportionality factor α is constant with aging time. As seen in Fig. 9.43, the larger value of $\alpha = 0.01$ predicts the data well for early aging times but clearly overpredicts it at longer aging times. On the other hand, the smaller value of $\alpha = 0.0033$ vastly underpredicts at smaller aging times but provides a better correlation to the trends of the experimental data for longer aging time periods. This, therefore, indicates that the weight loss proportionality factor α is not a constant but is a function of the oxidative aging time.

For the third simulation shown in Fig. 9.43, α is assumed to vary linearly from a value of 0.01 to 0.0033 within the first 40 h and then remain constant after 40 h. With this assumed variation (which is in agreement with larger weight loss rates observed for early aging times), the numerical simulations of oxidation layer growth are still underpredicted but, qualitatively the predicted curve matches the experimental data. For this parametric study, simple functional relationships are used to represent α. However, more complex functional relationships can be used to qualitatively match the experimental trends for extended aging times.

The proportionality between the oxygen consumption rate and weight loss rate is established by Abdeljaoued [1] using simplifying assumptions of which some of them, e.g., existence of a relationship between termination rate constants, are rejected by Gillen et al. [45]. Thus, a value of α which does not remain constant with time, as seen from our simulations, bodes well with the findings of Gillen et al. [45].

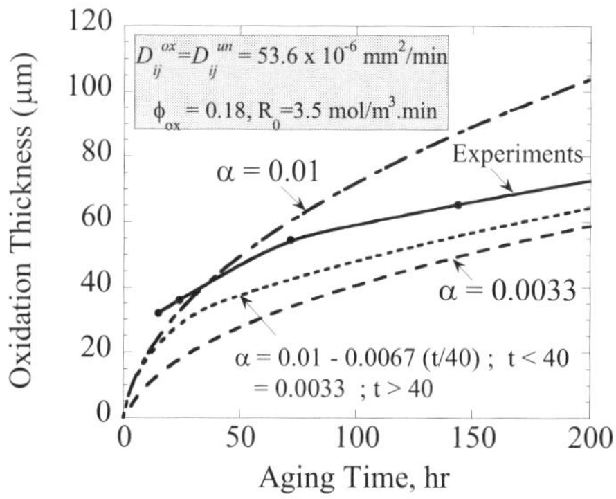

Fig. 9.43. Influence of α on oxidation layer growth predictions

Influence of diffusivities, D_{ij}^{ox} ***and*** D_{ij}^{un}

Thermooxidative aging of the polymeric material will change the chemical composition of the polymer and, hence, the physical properties of the material. In general, the diffusivities for each of three zones of the material (Regions I–III) will be different. Parametric analyses show that the diffusivity of the oxidized phase region (I) D_{ij}^{ox} controls the growth of the oxidation layer. Since a direct measurement of D_{ij}^{ox} is difficult to obtain and the value is not available in the literature, estimation is obtained from simulations.

Figure 9.44 shows the experimental oxidation layer growth data at 288°C compared with several simulations based on assuming different diffusivities. It is initially assumed that the diffusivity of the oxidized material D_{ij}^{ox} is the same as the diffusivity of the unoxidized material $D_{ij}^{un} = 53.6 \times 10^{-6}$ mm^2 min^{-1}. For the next simulation, the diffusivity for both the oxidized and unoxidized regions is both increased to 78 × 10^{-6} mm^2 min^{-1}, i.e., setting $D_{ij}^{ox} = D_{ij}^{un} = 78 \times 10^{-6}$ mm^2 min^{-1}, for which the predictions overestimate the oxidation layer even further at longer aging times. The predictions deviate even further from the experimental results using $D_{ij}^{ox} = D_{ij}^{un} = 100 \times 10^{-6}$ mm^2 min^{-1}. Increasing the diffusivity value in the oxidized region from 53.6 × 10^{-6} to 78 × 10^{-6} mm^2 min^{-1} and keeping $D_{ij}^{un} = 53.6 \times 10^{-6}$ mm^2 min^{-1} shows very little change from the earlier simulation with $D_{ij}^{ox} = D_{ij}^{un} = 78 \times 10^{-6}$ mm^2 min^{-1}. This illustrates that the oxidation region growth is predominantly controlled by the diffusivity of the oxidized region and is far less sensitive to the diffusivity of the unoxidized material. In the case of low conversion ratios, it is generally assumed that the diffusivity does not change during oxidation. This assumption has been checked experimentally for some linear polymers (PP, PE) and thermosets (epoxy), but the issue remains unresolved in the case of high oxidation conversion ratios. However, it can be reasonably considered, in a first approximation, that the diffusivity increases when the chemical structure significantly changes (presence of many oxygenated species) and the original chemical network is destroyed during oxidation (presence of many dangling chains). As a result, it can be assumed that diffusivity increases in the superficial layers where high conversion ratios are reached.

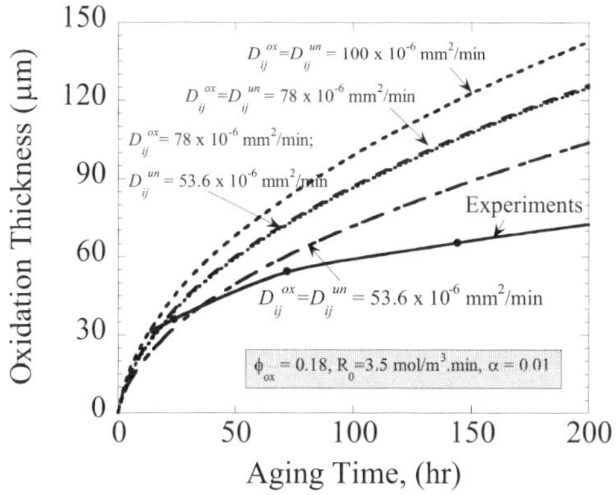

Fig. 9.44. Influence of diffusivity on oxidation layer growth predictions

Combined influence of controlling parameters

Figure 9.45a shows that, if the diffusivity of the resin is left unchanged with $D_{ij}^{un} = D_{ij}^{ox} = 53.6 \times 10^{-6}$ mm^2 min^{-1} but α is assumed to vary linearly from a value of 0.01 to 0.0033 within the first 40 h and kept constant beyond that time, the numerical simulations of oxidation layer growth are underpredicted but the shape of the predicted curve is approximately parallel to the experimental measurements. Next, if the diffusivity value in the oxidized region is increased from 53.6×10^{-6} to 78×10^{-6} mm^2 min^{-1} and kept $D_{ij}^{un} = 53.6 \times 10^{-6}$ mm^2 min^{-1} in conjunction with a linear variation of α for the first 40 h, a very good agreement between the predictions and the measurements of the oxidation layer (see Fig. 9.45b) can be obtained. This good agreement, therefore, gives credence to the assumptions that the diffusivity values within the oxidized region are larger compared to the values in the unoxidized region and that the value of α changes with aging with a larger value obtained initially, which is also in agreement with larger weight loss rates observed during the beginning of the experiments. With a diffusivity value of 78×10^{-6} mm^2 min^{-1} for the oxidized region at 288°C and assuming that the activation energy E^a remains unchanged in the oxidized region, the constant $D^o = 8.9 \times 10^{-11}$ m^2 s^{-1} can be evaluated for the oxidized region, as listed in Table 9.1.

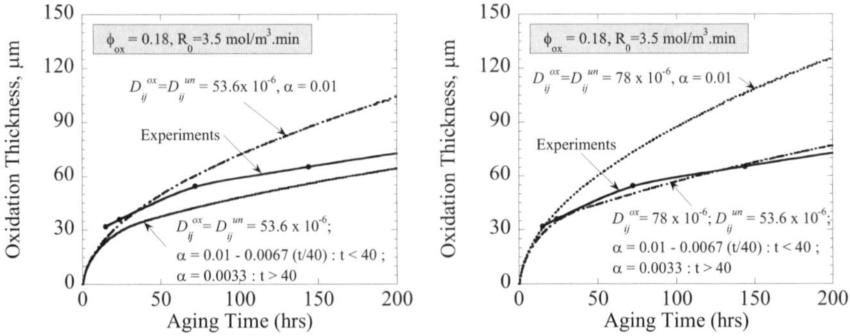

(a) **(b)**

Fig. 9.45. Combined influence of controlling parameters on oxidation layer growth predictions

Extrapolations to longer aging times

To project the oxidation layer growth to thousands of hours of aging, extrapolations from the simulation of the first 200 h of the oxidation layer growth are used. The oxidation layer size is fitted using a power law dependence with time, and the long-term predictions are obtained by extrapolation. The oxidation layer size, extrapolated to 1,500 h of aging, is compared with experimental measurements in Fig. 9.46. The values of the parameters used in the simulations are also listed in the figure. The results

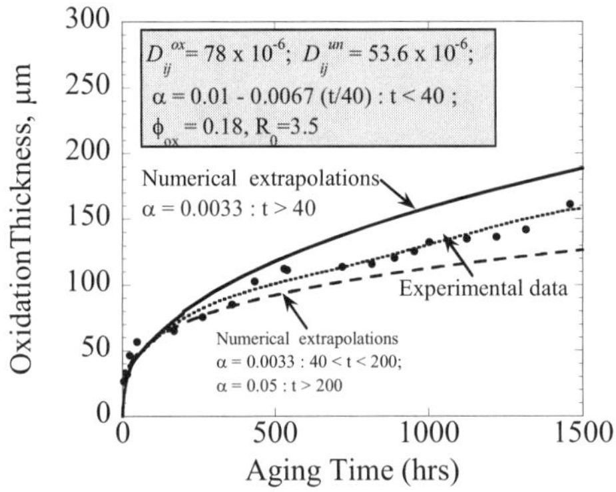

Fig. 9.46. Comparison of extrapolated oxidation layer growth predictions with measurements

indicate that the comparisons, using a constant $\alpha = 0.0033$ beyond 40 h of aging, are reasonable for 500 h of aging but are overestimated at longer time periods. However, when the layer growth by reducing is simulated the value of α from 0.0033 to 0.000165 for aging from 200 to 400 h, the predictions underestimate the oxidized layer thickness at longer aging times. These extrapolations and comparisons for longer aging times illustrate, therefore, that the proportionality factor (α) is not a constant but depends upon the aging time. Further work is needed to help establish this correspondence from the weight loss data.

Simulation of Arrhenius temperature dependence

The temperature dependence of the diffusivity D_{ij} (9.7) and the reaction rate $R(C, T)$ through the saturation reaction rate $R_O(T)$ are expressed by Arrhenius-type relationships. Along with experimental oxidation layer thickness measurements for neat resin PMR-15 specimens aged at 288°C (550°F), oxidation thickness measurements were also obtained for specimens aged at 316°C (600°F) and 343°C (650°F). Figure 9.47 shows predictions of the oxidation layer thicknesses for 288°C (550°F), 316°C (600°F), and 343°C (650°F) neat resin PMR-15 specimens based on 288°C (550°F) properties. The figure shows that the model accurately predicts the 316°C (600°F) and 343°C (650°F) oxidation layer thicknesses, and the temperature dependence of the reaction rate and diffusivity are accurately represented.

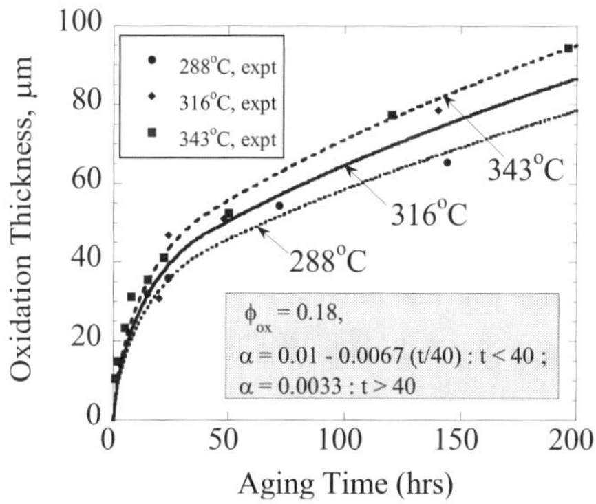

Fig. 9.47. Prediction of oxidation thickness at 316 and 343°C based on 288°C model parameters

Influence of pressure on oxidation growth

Thermal aging in a pressurized environment has been shown, in Figs. 9.27 and 9.28, to accelerate the oxidation layer growth rate and increase the weight loss. Since the partial pressure of a gas is directly proportional to the total pressure and, in general, the solubility of gases increases with increasing pressure, the sorption, as expressed by (9.2), is expected to increase with increasing pressure. Figure 9.48 shows the results of a parametric evaluation of how changes in the sorption affect the predicted oxidation layer thickness. Oxidation layer thickness for neat resin PMR-15 specimens aged at ambient and elevated pressures at 288°C is also shown in the figure. The predicted oxidation layer thickness increases with increasing oxygen concentration on the boundary and for the sorption value $C^S = 1.25$, there is a good correlation with the pressurized aging data for aging times up to 200 h. Figure 9.49 shows the predicted oxidation layer thickness for predictions of up to 800 h of aging time. The results show good agreement with the experimental data.

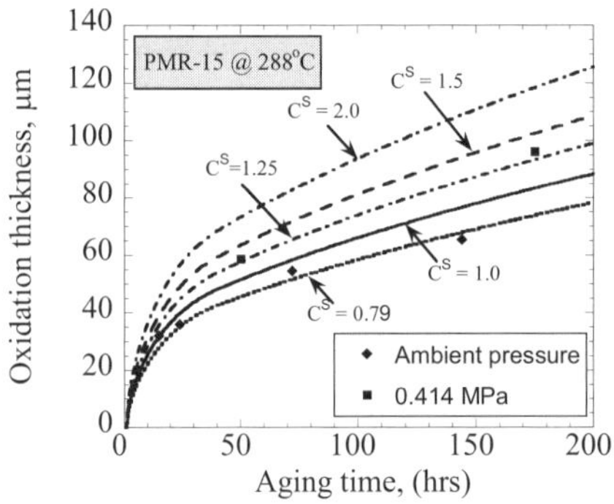

Fig. 9.48. Influence of sorption on the oxidation layer thickness

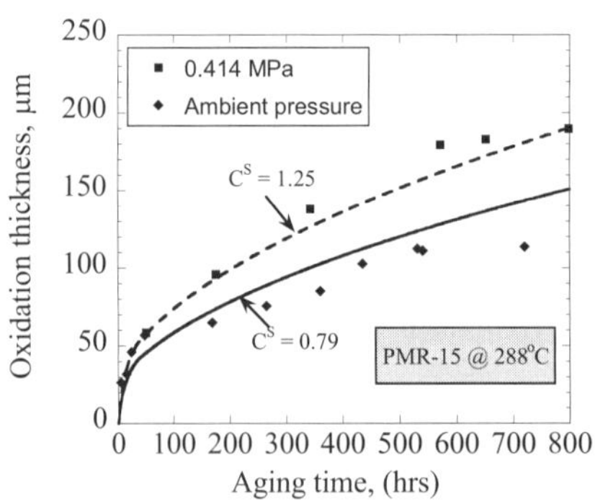

Fig. 9.49. Prediction of oxidation thickness for pressurized aging of PMR-15

9.4.2 Unit Cell Modeling

Both two- [101] and three-dimensional unit cell models capable of discrete crack analysis coupled with thermooxidative aged analysis can be used to model and identify the origins of degradation and mechanisms of failure initiation representative of that in unidirectional composites. Traditionally, the fibers are assumed to be chemically static for short-aging times (this assumption may not be valid for extended aging times), and residual stresses must be taken into account due to the strong coupling between stress, diffusion, and aging. Of primary importance is the characterization of the interface/interphase behavior that may have different reaction/diffusion kinetics than the polymer matrix phase of the composite. The fiber–matrix interphase region developed during the composite cure cycle is likely dependent on the fiber surface (unsized or sized) and the composition of the fiber sizing, if it is present [34]

Fiber-reinforced composites degrade several times more rapidly than their matrix and fiber constituents, as illustrated by Skontorp [95] in Fig. 9.50. The figure shows the weight loss (normalized with respect to the surface area) of a sample of IM6 fibers and an Avimid-N neat resin specimen for up to 200 h of aging at 343°C. A rule of mixtures is used to predict weight loss of a unidirectional Avimid-N/IM6 with 50% fiber volume based on the weight loss of the fiber and resin constituents and surface area fractions. The measured weight loss from the unidirectional specimen's axial surfaces S_3 is five times greater than the predicted rule-of-mixtures weight loss. The difference between the measured and predicted weight loss is attributed by Skontorp [95] to the fiber–matrix interface that is significantly more susceptible to oxidation than the fiber and resin constituents. As shown in Fig. 9.51, the intraply cracking typically occurs

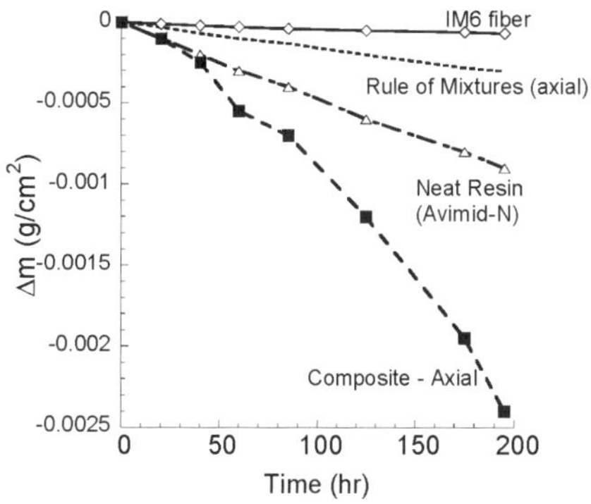

Fig. 9.50. Normalized weight loss for Avimid-N/IM6 at 650°F/343°C [95]

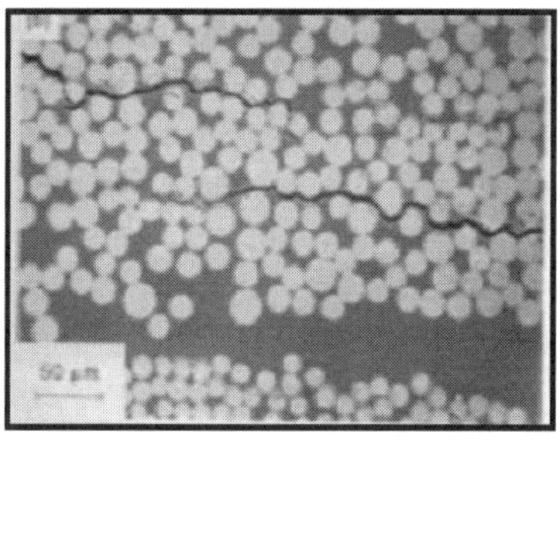

Fig. 9.51. Intraply cracks along fiber–matrix interface

along fiber–matrix interfaces providing pathways for oxidizing agents to penetrate deeper into the material, accelerating the degradation process. Incorporation of accurate polymer constitutive models into a unit cell micromechanics analysis, along with proper characterization and representation of the fiber–matrix interface/interphase region (including debonding), can provide a link between the degradation kinetics and chemical state of the polymer matrix and the performance of individual lamina or unidirectional composites.

The diffusion–reaction oxidation model is implemented into a three-dimensional Galerkin finite element analysis (GFEA). The GFEA method is more appropriate for studying structural scale problems than the one- and two-dimensional implementation, however the computational requirement of the three-dimensional analysis precludes its use in parametric analyses required for correlation with experimental results and parameter determination. The GFEA requires mesh sizes in the 1-μm scale and time increments in 1-s steps. A 200-h oxidation simulation with 100-μm oxidation zone size typically requires problem sizes in the order of 100,000 degrees of freedom (DOF) and 720,000 time steps. The domain Ω is discretized by a union of finite elements as $\Omega = \bigcup_{e=1}^{N_e} \Omega^{(e)}$, where $\Omega^i \cap \Omega^j = 0$, if $i \neq j$, and N_e is the total number of elements in Ω. The approximate element solution for the concentration $C(x,y,z,t)$ in the diffusion model is defined as

$$C(x,y,z,t) = [\Phi(x,y,z)]^{\mathrm{T}}[C(t)], \tag{9.27}$$

where $\Phi(x,y,z)$ is the spatial interpolation function and $C(t)$ is the time varying nodal concentration values. Using the Galerkin method, i.e., using a weight function the same as the interpolation Φ,

$$w = \Phi(x,y,z). \tag{9.28}$$

The residual for the diffusion–reaction equation (9.1) must vanish as given in (9.29)

$$\oint \left(\Phi\Phi^{\mathrm{T}} \frac{\partial C(t)}{\partial t} \right) \mathrm{d}V + \oint (\Phi R^{*}(C))\mathrm{d}V$$
$$+ \oint_{v} \left(D^{*}_{11} \left(\frac{\partial \Phi}{\partial x} \frac{\partial \Phi^{\mathrm{T}}}{\partial x} C(t) \right) + D^{*}_{22} \left(\frac{\partial \Phi}{\partial y} \frac{\partial \Phi^{\mathrm{T}}}{\partial y} C(t) \right) + D^{*}_{33} \left(\frac{\partial \Phi}{\partial z} \frac{\partial \Phi^{\mathrm{T}}}{\partial z} C(t) \right) \right) \mathrm{d}V = 0. \tag{9.29}$$

As before, the superscript * indicates the additional dependence of the diffusivity and reaction rate on the temperature T and oxidation state variable ϕ. After some mathematical manipulations, (9.29) can be written in matrix form as

$$[\mathbf{C}]\frac{\partial C(t)}{\partial t}+[\mathbf{B}]C(t)+\{\mathbf{R}\}=0, \tag{9.30}$$

where, assuming orthotropic diffusivity, the matrices, [**B**] and [**C**], and vector $\{\mathbf{R}\}$ are given by

$$[\mathbf{B}]=\oint_{v}\left(D_{11}^{*}\frac{\partial\Phi}{\partial x}\frac{\partial\Phi^{\mathrm{T}}}{\partial x}+D_{22}^{*}\frac{\partial\Phi}{\partial y}\frac{\partial\Phi^{\mathrm{T}}}{\partial y}+D_{33}^{*}\frac{\partial\Phi}{\partial z}\frac{\partial\Phi^{\mathrm{T}}}{\partial z}\right)\mathrm{d}V, \tag{9.31}$$

$$[\mathbf{C}]=\sum_{e}\int_{\Omega(e)}\Phi\Phi^{\mathrm{T}}\mathrm{d}\Omega, \tag{9.32}$$

$$\{\mathbf{R}\}=\sum_{e}\int_{\Omega(e)}\Phi R^{*}(C)\mathrm{d}\Omega. \tag{9.33}$$

Equation (9.30) is discretized by the backward Euler method in time domain, where $\Delta t=t^{n+1}-t^{n}$. Therefore, (9.30) can be denoted as a system of linear algebraic equations

$$[\mathbf{C}]\left\{\frac{C^{n+1}-C^{n}}{\Delta t}\right\}+[\mathbf{B}]\{C^{n+1}\}+\{\mathbf{R}\}=0. \tag{9.34}$$

In summary, the diffusion–reaction equations in their discretized form can be written as

$$[\mathbf{D}]\{C^{n+1}\}=[\mathbf{E}], \tag{9.35}$$

where

$$[\mathbf{D}]=[\mathbf{C}]+\Delta t[\mathbf{B}], \tag{9.36}$$

$$[\mathbf{E}]=[\mathbf{C}]\{C^{n}\}-\{\mathbf{R}\}\Delta t. \tag{9.37}$$

Equation (9.35) is solved using a direct solution with the Frontal technique and using an iterative solution with Jacobi conjugate gradient (JCG) iterative method [106] for problems with a large number of DOF.

Effective orthotropic diffusivity estimations for the lamina

The representative volume element (RVE) is used to determine the effective diffusivity of the lamina using homogenization. At the lamina scale, the homogenized diffusivity tensor is transversely isotropic or orthotropic depending upon the constituent diffusivities. Using a three-dimensional RVE, a cube of periodicity $(0,\varepsilon)\times(0,\varepsilon)\times(0,\varepsilon)$ is scaled into the unit cube $Q=(0,1)\times(0,1)\times(0,1)$, as shown in Fig. 9.52a, under the change of variables $\xi_1 = x_1/\varepsilon$, $\xi_2 = x_2/\varepsilon$, $\xi_3 = x_3/\varepsilon$, to find the effective coefficients $\hat{D}_{ij}$. Let **F** be the domain in the unit cell which corresponds to the fiber cross section and **M** is the domain occupied by the matrix. Set:

$$\hat{D}_{ij}(\xi_1,\xi_2,\xi_3) = \begin{cases} D_{ij\mathrm{f}}, & \text{for } (\xi_1,\xi_2,\xi_3)\in \mathbf{F}, \\ D_{ij\mathrm{m}}, & \text{for } (\xi_1,\xi_2,\xi_3)\in \mathbf{M}. \end{cases} \tag{9.38}$$

Let $C(\xi_1,\xi_2,\xi_3)$ be the concentration field variable that satisfies the Laplace equation in the fiber and matrix domains, and the field is computed explicitly using numerical methods under applied concentration gradients (9.39), periodic boundary conditions (9.40), and continuity conditions (9.41) imposed at the fiber–matrix interface

$$C|_{\xi_\mathrm{q}=0} = 0, \quad C|_{\xi_\mathrm{q}=1/2} = 1/2, \tag{9.39}$$

$$\left.\frac{\partial C}{\partial \xi_\mathrm{q}}\right|_{\xi_\mathrm{q}=0} = \left.\frac{\partial C}{\partial \xi_\mathrm{q}}\right|_{\xi_\mathrm{q}=1/2} = 0, \tag{9.40}$$

$$\sum_{i,j=1}^{3} \frac{\partial}{\partial \xi_i}\left(D_{ij\,\mathrm{f,m}}(\xi_1,\xi_2,\xi_3)\frac{\partial C}{\partial \xi_j}\right) = 0. \tag{9.41}$$

The three diagonal components of the orthotropic effective diffusivity tensor D_i are estimated from the concentration fields using the averaging method as given in (9.42)

$$D_i = 8\int_0^{1/2}\int_0^{1/2}\int_0^{1/2} \sum_{q=1}^{3} \hat{D}_{iq}(\xi_1,\xi_2,\xi_3)\frac{\partial C}{\partial \xi_\mathrm{q}}\,\mathrm{d}\xi_1\mathrm{d}\xi_2\mathrm{d}\xi_3, \tag{9.42}$$

where the factor "8" derives from the fact that symmetry on each orthogonal axis plane allows for the modeling of 1/8 of the fiber–matrix unit cell.

At the lamina scale, axial and transverse diffusivities are determined by applying unit concentration gradients across the unit cell in two separate simulations along the axial and transverse directions, respectively. Isotropic diffusivities denoted by D_f and D_m are assumed for the fibers and matrix, respectively, unless otherwise noted. The simulations are performed with $D_f = D_m/1{,}000$. Table 9.2 shows the effective diffusivity in the axial and the transverse directions for a unidirectional lamina with two volume fractions. While the effective diffusivity in the axial direction is approximately proportional to the volume fraction, the transverse diffusivity is inherently lower due to the microstructure geometry effects. The ratio D_2/D_1 of transverse to axial diffusivity is a measure of the effective anisotropic composite response.

Table 9.2. Fiber volume-dependent effective diffusivity for unidirectional lamina

	D_1/D_m	$D_2/D_m = D_3/D_m$	D_2/D_1
Unidirectional $V_f = 50.0\%$	0.50	0.32	0.64
Unidirectional $V_f = 59.4\%$	0.40	0.23	0.58

Interface/interphase issues

The fiber–matrix interface/interphase region of HTPMCs has long been suspected of being a major factor in the thermooxidative stability of the material. The fiber–matrix interphase region may have a different chemical structure and morphology due to the presence of fiber sizing or fiber coupling agent, the development of localized residual stresses, the coefficient of thermal expansion mismatch between the fiber and matrix, and other chemical interactions between the fiber and matrix during the cure processing cycle. Furthermore, the interphase region may have different reaction/diffusion kinetics than the polymer matrix phase of the composite. Typically, sizings on carbon fibers are on the order of a few molecules thick, which presumably results in an interphase region that is very thin compared to the fiber diameter. It follows that the interphase region represents a very small-volume percentage of the composite. The ability to directly characterize this region through testing is extremely limited due to its small size. Therefore, investigators typically rely on observations of the composite behavior to postulate what role the fiber–matrix interphase region has on the observed behavior. In most cases (including the simulations here), this is done without full knowledge of the role of the fiber and matrix phases of the composite. Although studies on neat bulk resins are numerous, characterization of the oxidative properties of the in situ resin and the role of the fibers in the diffusion process are not well understood.

To investigate the role of the interphase, parametric variation of the ratio of the interphase diffusivity to matrix diffusivity D_i / D_m is considered from a value of 1 to 100, using the RVE model shown in Fig. 9.52b. In addition, the interphase thickness is varied from $R_f/50$ (one-fiftieth of the radius of the fiber) to $R_f/10$. Figure 9.53 shows the effective transverse diffusivity (S_1 and S_2 surfaces) of the composite as a function of the interphase diffusivity and the thickness of the interphase. For these simulations, the diffusivity of the fiber is assumed to be zero. The figure shows that having a nondiffusive interphase results in a composite that has

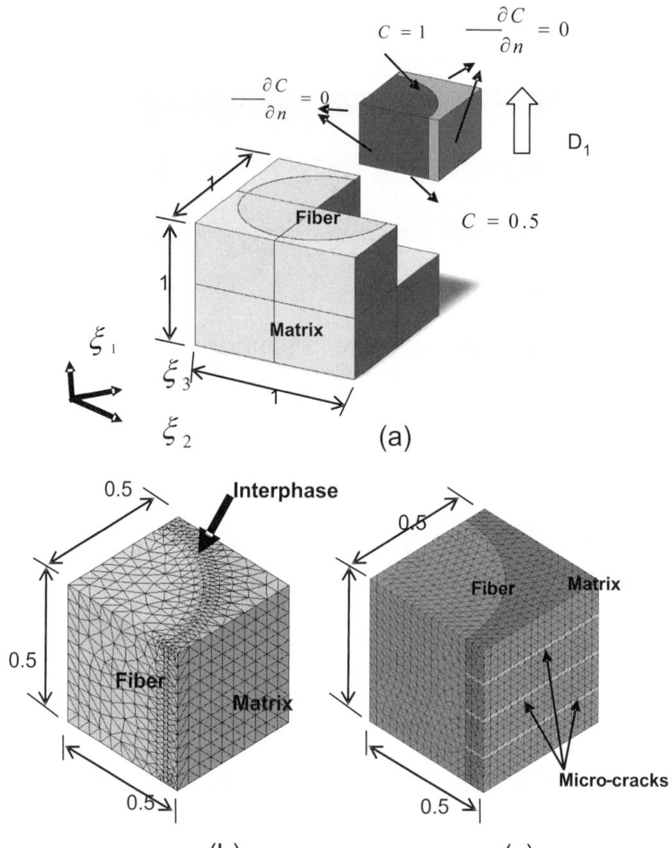

Fig. 9.52. RVE models (**a**) without interphase region, (**b**) with interphase region, and (**c**) with matrix microcracks

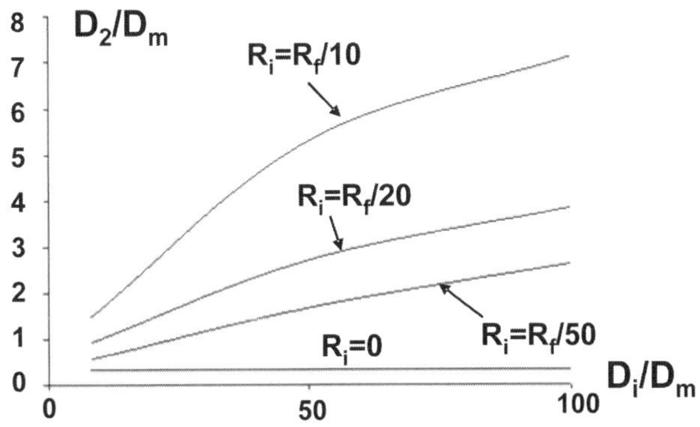

Fig. 9.53. Effect of interphase diffusivity on composite material diffusivity

an effective transverse diffusivity D_2 less than that of the diffusivity of the matrix D_m. As seen in Fig. 9.53, a highly diffusive interface can substantially increase the effective diffusivity of the composite. An interphase thickness of 5% that of the fiber radius and 100 times more diffusive than the matrix results in a composite (with 50% fiber volume fraction) with an effective transverse diffusivity of three times that of the matrix.

Effective diffusivity with discrete cracks in the RVE

A homogenization method is used to obtain the effective diffusivity of unit cells with discrete microcracks. Figure 9.52c shows the unit cell models of lamina with transverse cracks. The microcracks are modeled both as additional surfaces (with boundary sorption conditions) and highly permeable pseudomaterials with an artificial thickness assigned to them. The homogenization method described earlier is used to obtain the effective diffusivity of the unit cell in the presence of microcracking.

The fiber, matrix, and crack domains are individually modeled for a nominal fiber volume fraction of 50%, while the number of cracks modeled is varied. The matrix diffusivity $D_m = 3.45\ \mu m^2\ s^{-1}$, the fiber diffusivity $D_f = 0.00345\ \mu m^2\ s^{-1}$ ($D_f = D_m/1{,}000$), and the diffusivity across the crack faces is taken to be $D_c = 345\ \mu m^2\ s^{-1}$ ($D_c = 1{,}000 D_m$). The effective axial D_1 and transverse D_2 diffusivities with transverse matrix cracking are shown in Table 9.3. The transverse diffusivity is substantially affected by the presence of the microcracks, while the axial diffusivity is affected to a much lesser extent [35].

Table 9.3. Effective axial and transverse diffusivity with matrix cracking

	D_1/D_m (axial)	D_2/D_m (transverse)
No crack	0.498	0.323
Two cracks	0.553	3.509
Four cracks	0.621	6.709
Six cracks	0.709	9.902

Modeling oxidation propagation in the unit cell

Three-dimensional oxidation modeling is performed using GFEA on RVEs of the composite. Inputs to the model include material properties, simulation parameters, and the unit cell geometry. Unit cells, as shown in Fig. 9.54, describe the geometry of the constituents, volume fraction, and orientation of the constituents in the composite, while the model allows multifiber and

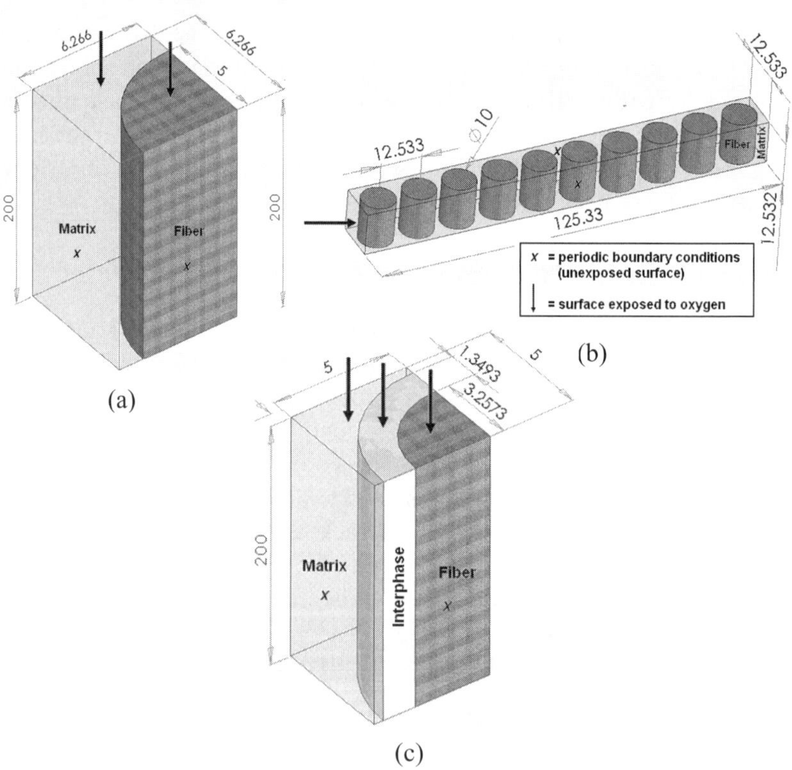

Fig. 9.54. Unit cells for modeling (**a**) axial oxidation, (**b**) transverse oxidation, and (**c**) the influence of interphase on oxidation growth in a unidirectional composite

multicell simulations to be performed. The unit cell geometry and element meshes are generated using commercially available computer-aided design (CAD) and FEM programs. For pure oxygen, the boundary concentration C^S is taken to be 3.74 mol m^{-3}. For air (as used for all simulations in this work), it is 0.79 mol m^{-3}. The model is meshed such that all constituents are perfectly bonded with continuity of the oxygen concentration imposed at the interface. The mesh size is varied between simulations. Two output files describing the concentration profiles (C) and the oxidation state (ϕ) in the model are produced at every time step.

For the majority of the PMC simulations presented, a 50% fiber volume fraction is assumed with no explicit representation of the interphase region, unless otherwise noted. It is assumed that the fibers are in a square packing arrangement, with each having a diameter of 10 μm and a spacing of 12.532 μm (center to center) between adjacent fibers. To reduce simulation time, the RVE model consists of only one quarter of a unit cell single fiber and the surrounding matrix as shown in Fig. 9.54a. The mesh has an element size of 1.28 μm, with a total of 13,226 nodes and 6,845 elements. The length of this model is 200 μm. Resin properties corresponding to PMR-15, as given in Table 9.1, are used.

The results of the oxidation simulation of the RVE with the surface exposed to the air being parallel to the fiber cross section are shown in Fig. 9.55. The figure shows the region where the ϕ values are within 5% of

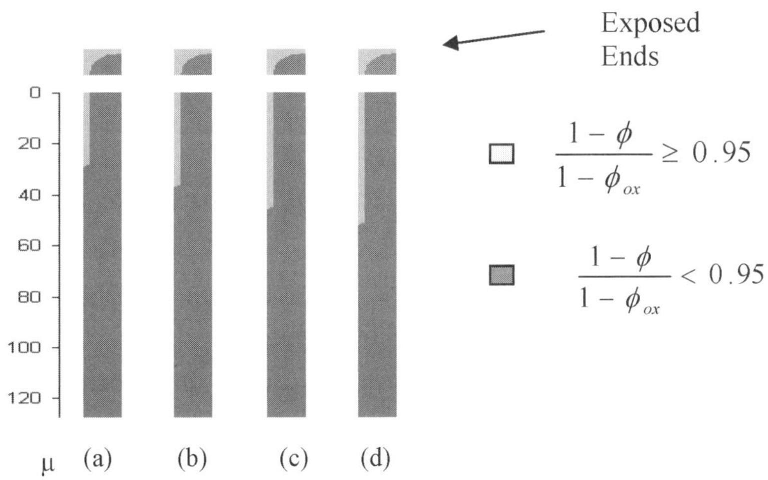

Fig. 9.55. Simulations of oxidation growth in axial direction at (**a**) 50 h, (**b**) 100 h, (**c**) 150 h, and (**d**) 200 h of exposure

completely oxidized (light gray, $\phi = \phi_{ox}$) and unoxidized (dark, $\phi = 1$) at 50, 100, 150, and 200 h of exposure. The fiber diffusivity (D_f) is taken to be 1/1,000 that of the unoxidized diffusivity of the matrix ($D_f = D_m^{un}/1{,}000$). The prediction of oxidation propagation along the fiber direction is less than the experimentally observed extent of oxidation, indicating that the diffusion of oxygen in the fiber direction is not fully accounted for in this model. Next, several simulations are performed to determine the impact of the fiber diffusivity on the oxidation of the matrix. The fiber diffusivity is varied such that $D_f = (1/1{,}000, 1/100, 1/10, 1, 10) \times D_m^{un}$. The results are shown in Fig. 9.56. The differences between the effective unit cell diffusivity – assuming that the fiber diffusivity D_f is 1/10, 1, and 10 – and matrix diffusivity D_m^{un} are significant, indicating that fiber diffusivity has a considerable effect on composite oxidation behavior. For fiber diffusivities below $1/10 \times D_m^{un}$, oxygen diffusion through the fiber becomes insignificant. There is little change in overall oxidation, and oxidation is driven by matrix diffusivity alone.

Simulations corresponding to the exposure of transverse surfaces of the lamina were also performed. Since the repeated unit is not a single fiber, a model consisting of ten fibers in a single row was created, as shown in Fig. 9.54b. As before, the fibers have a diameter of 10 μm and a spacing of 12.532 μm (center to center). The mesh has an element size of 1.86 μm, with a total of 34,300 nodes and 22,493 elements. As before, the fiber

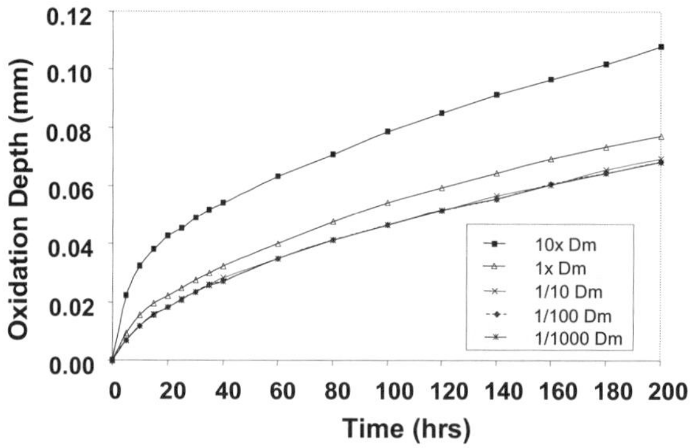

Fig. 9.56. Effect of fiber diffusivity on matrix oxidation

diffusivity is taken to be 1/1,000 that of the unoxidized diffusivity of the matrix ($D_f = D_m^{un}/1{,}000$). The resin parameters corresponding to that of PMR-15 in Table 9.1 are used. Figure 9.57 illustrates the simulation of oxidation growth transverse to the fibers, while Fig. 9.58 shows a detailed view of the active reaction zone as oxidation occurs around the fiber. The observed active reaction front (Zone II) is approximately one fiber diameter in size (10 μm), which is smaller than that observed in the neat resin. The smaller active reaction zone (II) is a result of lower effective diffusivity in the transverse direction due to the presence of the fiber.

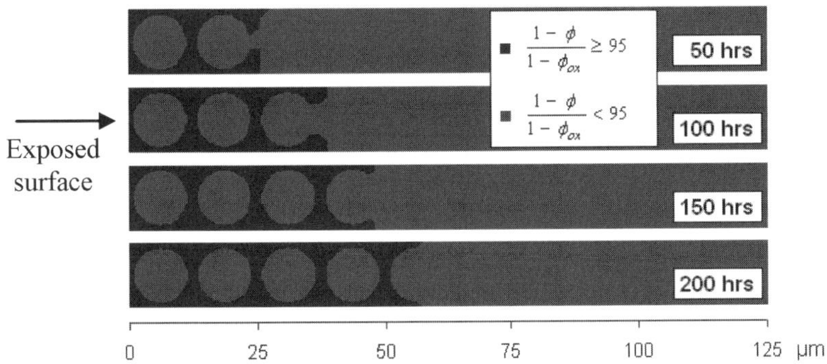

Fig. 9.57. Transverse oxidation growth in a composite lamina

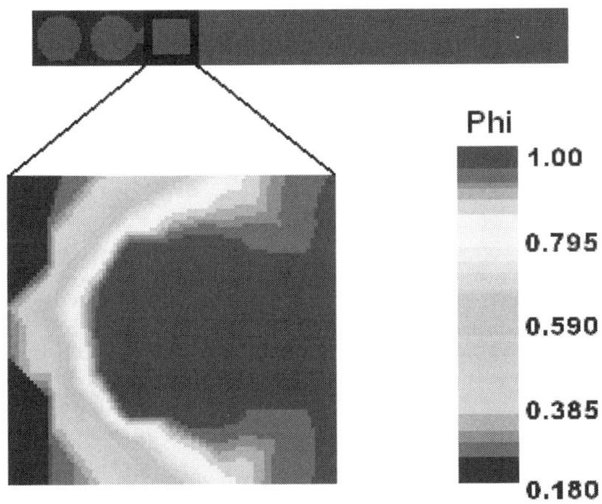

Fig. 9.58. Detail of active reaction zone from transverse simulation

A comparison of the anisotropic nature of the oxidation growth in the transverse and axial directions for 200 h of aging is shown in Fig. 9.59. The differences in the transverse and axial oxidation growth are observed to be small for the simulations performed. This is to be expected because the diffusion process for these simulations is dominated by the diffusivity of the matrix alone, since the fiber diffusivity is assumed to be negligible ($D_f = D_m^{un} / 1{,}000$). Moreover, the RVE unit cells shown in Fig. 9.54a,b do not include the fiber–matrix interphase region that can greatly contribute to the diffusion along the fiber axis direction. Neither have we accounted for the increase in effective diffusivity due to presence of damage such as fiber–matrix debonds and/or matrix microcracking.

To investigate the role of the interphase region, particularly if the interphase thickness and its diffusivity are large, a parametric study is conducted to understand the influence of the interphase parameters on the oxygen diffusivity and transport in the composite. This is accomplished by adding a concentric region around the fiber that has distinct material properties to represent the interphase region. A high interphase volume fraction (33%) and a low fiber volume fraction (33%) are chosen to exaggerate the effects of the presence of the interphase. For geometric and meshing convenience, the model assumes that the fiber diameter is 6.51 μm, the interphase thickness is 1.35 μm, and the fiber spacing is 10 μm (center to center), as shown in Fig. 9.54c. This produces a quarter model unit cell measuring 5 × 5 × 200 μm in size. PMR-15 properties (Table 9.1) are used for the matrix in the unit cell model. In addition, the

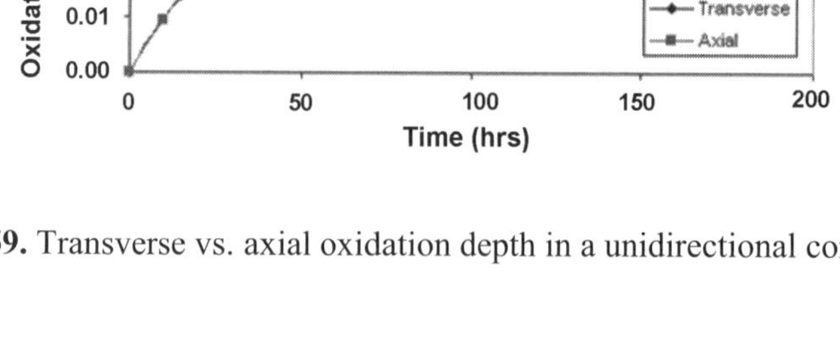

Fig. 9.59. Transverse vs. axial oxidation depth in a unidirectional composite

fiber is assigned an anisotropic diffusivity. For the initial simulations, the axial diffusivity is assumed to be equal to that of the resin ($D_{\mathrm{m}}^{\mathrm{un}}$) and the transverse diffusivity is taken to be $D_{\mathrm{m}}^{\mathrm{un}}/10$. The interphase layer is assigned diffusivity values of 100×, 1×, and 1/100× that of the unoxidized resin ($D_{\mathrm{m}}^{\mathrm{un}}$). Both the fiber and the interphase region are assumed to be nonoxidized, and only the matrix phase reacts with oxygen.

Figure 9.60 shows the parametric analysis of axial oxidation region growth for the three interphase diffusivity values. The simulation with the high interphase diffusivity value shows considerable oxidation layer thicknesses and accelerated growth even during the early hours of aging. The diffusion-controlled behavior is clearly evident as the oxidation process is accelerated due to the presence of additional diffusion paths through the interphase.

Where the path of the oxygen to the matrix/interphase region is due to the fiber–matrix debonds, the effective diffusivity and, hence, the oxidation process can be effectively modeled without a discrete representation of the debonds. This can be accomplished by assigning an artificially high diffusivity value to the fiber. In such a case, it is assumed that the path the oxygen takes to penetrate to the interior is inconsequential and that it is the consequence of having the oxygen available for the oxidation process in the interior of the composite that is important. Using this approach may eliminate the need to do costly modeling of the discrete fiber–matrix debonds in the analysis. Further analysis is in progress to parametrically estimate the increase in the oxidation layer depth in the matrix due to the presence of a diffusive fiber and interphase region.

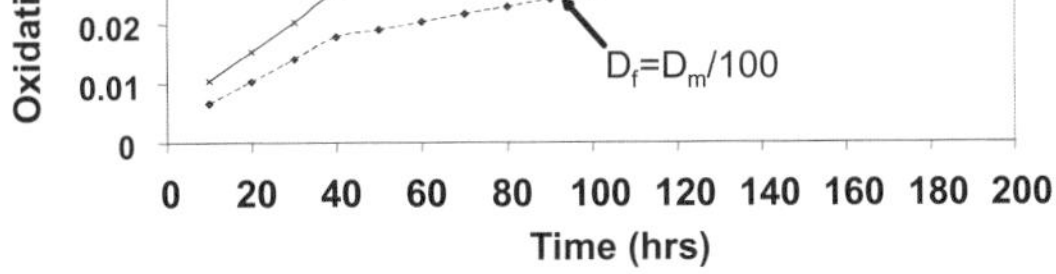

Fig. 9.60. Oxidation layer size variation with interface diffusivity

9.4.3 Lamina/Laminate Oxidation

Utilization of multiscale modeling to accurately predict the oxidative behavior of polymer composite and multidirectional laminates requires knowledge of the individual aging mechanisms, their synergistic effects, and the spatial variability of the thermooxidative degradation [78, 79, 90]. Integration of the unit cell models as input to the lamina and laminate scale models is a formidable task based on the fact that the effective tensorial properties at each material point evolve with aging time. Since the diffusion-limited oxidation process creates a large spatial variability of the oxidation state, particularly near the free surfaces, a given unit cell RVE is applicable only over material regions with common oxidation states, as well as common mechanical and thermal stresses. To use the unit cell model to predict the spatial variability of the ply properties, a finite element analysis method needs the capability to represent unique material properties for each integration point. In addition, it is well documented that multidirectional composite laminates degrade faster than unidirectional laminates [56], presumably due to the presence of ply-level residual stresses. To properly account for free-edge residual stresses and stresses due to thermomechanical loading, as well as to meet the requirements of identifying unique material properties for each integration point, a three-dimensional model in which each ply is represented discretely is warranted.

In addition, the development of damage in HTPMCs exacerbates the chemical and physical aging by introduction of stress concentrations that accelerate physical aging effects and exacerbates the chemical aging by introducing pathways for oxidants and other agents to advance deeper into the material. Damage typically takes the form of matrix cracks and fiber–matrix interface debonds with the micromechanical cracks and transverse ply cracks coalescing to form larger ply-level cracks. The mechanistic approach of modeling discrete damage, such as transverse ply cracks [58, 89], provides accurate assessments of the effective response for a given damage state but has limitations on the number of discrete cracks or amount of discrete damage that can be represented. Therefore, mechanistic damage modeling is often reserved for analyzing local details and small components or it can be used as a local model in a global analysis. Ultimately, for modeling large structures, phenomenological approaches of continuum damage mechanics [4, 100] can be used to represent the evolution of effective local damaged properties through the constitutive relationships, whereby widespread dispersed damage can be more easily represented.

Viscoelastic modeling of physical aging and thermomechanical loading

Although current aerospace applications for HTPMCs are typically limited to secondary or lightly loaded structures, the insertion of these materials into critical load-bearing structural applications will require the ability to model their creep/relaxation and long-term fatigue performance. Linear viscoelastic models are applicable in the cases where the stress levels are low such that the creep strain is proportional to the stress level. However, if the applied stress levels or the temperatures are high, nonlinear viscoelastic–plastic models may be required to adequately represent the long-term behavior [84]. To maintain acceptable factors of safety, particularly for human-rated vehicles, the use of HTPMCs in the nonlinear constitutive range for flight-critical structures will likely be limited. Nevertheless, to predict the behavior of HTPMCs for such cases as accelerated aging at higher than use temperatures and determination of the material response for load and temperature excursions, a nonlinear viscoelastic–plastic model may be needed.

Studies of the creep behavior of neat resin high-temperature polymers provide the basis for micromechanical unit cell modeling for HTPMCs. Marais and Villoutreix [57] show the effect of physical aging on the relaxation constants and relaxation times of thermostable PMR-15 resin. In that work [57], the viscoelastic behavior is described by the Kohlrausch function, and, combined with the Maxwell relation, enables the creep compliance to be modeled for short-loading times. However, it is not possible to establish an equivalence between time and temperature using Williams, Landel, and Ferry (WLF)-type shift factors [124]. The master curve, constructed graphically on the basis of linear viscoelastic strains, short times, and a reference temperature of 250°C, deviates significantly from the experimental curves at longer times. Thus, the long-loading time response could not be determined from the short-loading time viscoelastic response as the material undergoes a slow plastic deformation. Although such approaches may be appropriate and adequate for treating homogeneous materials, their applicability to the material designer is limited. The shift functions and master curves are needed for each composite system and can only be determined after extensive and expensive testing. Methodologies that construct the material behavior from the constituent's age-influenced response are scant and nascent.

The nonlinear viscoelastic theory by Schapery [84] has been substantially exploited for evaluating PMCs in high-temperature applications. Nicholson et al. [69] investigated the effect of molecular weight and temperature on the elastic and viscoelastic properties of thermoplastic polyimides. They found that the lower molecular weight materials have higher creep compliance and creep compliance rate and are more sensitive to temperature than the higher molecular weight materials.

As discussed previously, for a micromechanical unit cell analysis, the fiber, matrix, and interphase region can all be modeled explicitly. The polymer matrix is typically assumed to be viscoelastic while the fiber is assumed to be elastic [21, 72, 127]. Modeling of the time and temperature variation of the reinforced composite material's compliance typically utilizes effective properties derived from the unit cell analysis, or treats the problem at the lamina scale in the context of classical lamination theory. A general method for incorporating the influence of physical aging on the material stiffness has been to study the relaxation behavior of the composite material. Most of the approaches are empirical and treat composite as a homogeneous continuum. They characterize the relaxation moduli and relaxation times based on dynamic mechanical analyzer (DMA) testing, and add an additional physical, age-based shift to the time–temperature-shifted reduced time. Hashin [46] observed that, for unidirectional composites, the viscoelastic effects can be significant for axial shear, transverse shear, and transverse normal stresses (matrix-dominated properties), with the effects becoming more pronounced with increase in temperature and loading.

It is beyond the scope of this chapter to review all of the relevant research on this important subject. Reviews such as that by Scott et al. [93] can be referred to far more comprehensive treatment of the topic. In lieu of a comprehensive review of the topic, a review primarily focused on the viscoelastic models as they have been applied to HTPMCs will be given. Such a limited review would not be complete without inclusion of the significant body of work performed in the NASA HSR program [21, 42, 49, 68, 115, 116]. Much of this work focuses on the use of linear viscoelasticity theory using the time–temperature-based superposition principles to model long-term physical aging behavior based on short-term observations. A large volume of work on describing the short- and long-term physical aging of amorphous (glassy) polymers and PMCs is based on the free-volume concept by Struik [98]. Struik showed that Boltzmann's superposition principle could be used to describe physical aging in many polymers. This concept is widely accepted in the linear viscoelastic or small deformation range [21, 23, 36, 99]. In time aging, time superposition [98], based on the premise that the shape of the creep curves for a given material does not change with aging time, aging primarily affects the creep properties by way of changes in the relaxation times. Modeling one aspect of the HTPMC's aging response, namely physical aging, Brinson and Gates [21] described the momentary creep compliance of the material by a three-parameter fit model as

$$S(t) = S_o \exp(t / \tau(t_e))^\beta \tag{9.43}$$

with

$$\tau(t_e) = \tau(t_{\text{eref}}) / a_{t_e}, \tag{9.44}$$

where t_{eref} is the reference aging time, and the aging time shift factor a_{t_e} can be written as

$$a_{t_e} = (t_{\text{eref}} / t_e)^\mu \tag{9.45}$$

with the reference aging time shift factor set equal to unity. Numerous expressions for the shift factor have been presented in the literature including the temperature-dependent WLF empirically derived relationship that has been found to work well above the material's T_g [124]. Therefore, the parameters needed to describe the momentary creep compliance at any aging time are the initial compliance S_0, the shape parameter β, the shift rate μ, and the relaxation time τ [21]. Struik characterized aging by the double-logarithmic shift rate μ, defined as

$$\mu = -\frac{\mathrm{d}\left(\log a_{t_e}\right)}{\mathrm{d}(\log t_e)}, \tag{9.46}$$

where μ is constant over a wide range of aging times and is approximately unity when the material is far removed from thermodynamic equilibrium. Through horizontal shifting on the log scale, the momentary creep curves can superpose through changing the relaxation time. In the case where the testing time t is greater than the initial aging time t_e^0, such that the total aging time is given by $t + t_e^0$, the long-term response is expected to deviate from the momentary response. If the reference aging time is taken to be the initial aging time $t_{\text{eref}} = t_e^0$, the shift factor $a_{t_e^0}$ for any time t is given as

$$a_{t_e^0}(t) = (t_e^0 / (t_e^0 + t))^\mu. \tag{9.47}$$

Furthermore, the effective time is defined as the accumulation of the shift factors given by

$$\lambda = \int_0^t a_{t_e^0}(\xi) \mathrm{d}\xi. \tag{9.48}$$

Using the effective time in place of the testing time in (9.43) given by

$$S(t) = S_0 \exp(\lambda / \tau(t_e^0))^\beta \tag{9.49}$$

allows for the prediction of long-term response based on short-term observations [21]. Based on experimental observations [99] that the S_{22} and S_{66} terms of the compliance tensor are dominated by matrix properties, the viscoelastic response of a unidirectional laminate is typically modeled by representing S_{22} and S_{66} as viscoelastic and the other compliance terms as elastic. Brinson and Gates [21] determined that, for tension tests, the aging shift rate is the most critical parameter in predicting long-term performance. Later work focused on utilization of this approach for compression-loaded composites [115, 116].

During the thermal aging process, in which both physical and chemical aging are occurring, chemical oxidative aging results in changes to the thermomechanical properties of the material and damage evolution. To account for the aging time-dependent chemical/oxidative aging, characterization of the oxidation-dependent viscoelastic constitutive parameters may be required. That is, the viscoelastic constitutive properties/parameters as a function of the polymer availability state variable $\phi(t,T)$ for all times t and temperatures T of interest will be required. Additionally, ϕ will be path dependent if the range of temperatures of interest entails changes in the oxidation mechanisms. Such an experimental characterization program would be overwhelming. In lieu of such an experimental program, a capability to predict mechanical properties based on the oxidative state of the neat resin is needed. However, this capability does not currently exist. Various simplifying assumptions can be used to overcome these deficiencies. One alternative is to represent the oxidative degradation as smeared or discrete damage in the viscoelastic framework. Both linear [3, 85] and nonlinear [127] viscoelastic models with damage evolution have been used to account for both physical aging and the damage associated with oxidative aging.

These viscoelastic material representations at the micromechanical level and the lamina level can predict the effective response of the material subjected to physical aging. Generalizations of these models to include damage and in the case of Skontorp and Wang [96] to represent chemical aging have been presented. However, for PMCs subjected to environmental conditions in which the predominant aging effect is chemical oxidative aging, a satisfactory model to predict long-term performance has yet to be completed. It is likely that the viscoelastic models discussed here will be an integral part of a global–local model where the ply-level homogenized continuum representation of the effective anisotropic behavior and spatial property variability of degraded plies is determined through micromechanical models. However as mentioned earlier, for lack of simulation-based tools, composite material designers routinely perform weight loss studies for specimens aged in simulated operating and accelerated aging conditions to assess the thermooxidative stability of materials.

The relative oxidative stability of the composite is determined by comparing the percent weight lost during the exposure for various materials. While this is an effective way of comparing materials, the test yields little information that can be used to determine the quantitative response of the materials for in-service operating conditions. In general, the relationship between weight loss and property changes can be highly nonlinear and does not provide a measure of the spatial variability of degradation within the specimen. The reliability of predicting performance using methods based entirely on weight loss, in which small changes in weight can result in significant declines in mechanical properties, is questionable. However, in the absence of methods to predict end-of-life properties, weight loss has been an accepted method to compare the thermooxidative stability of different materials [94] and is discussed in subsequent sections.

9.4.4 Lamina Weight Loss Models for Oxidative Aging

One of the most widely used methods of characterizing the level of thermooxidative degradation is presented by Nam and Seferis [66] where specimen weight loss is correlated to mechanical performance. In this method, specimens are aged under isothermal conditions and weight loss measurements are recorded as a function of aging time. It is assumed that weight loss is primarily due to oxidation (neglecting moisture weight loss) and that oxidation is the dominant cause of changes in the properties of the polymer composite. The method provides valuable screening information on the thermooxidative stability of material systems and can be used to compare the stability of different material systems. Although the method has been used by numerous researchers [24, 30, 55, 97, 108], it only provides an effective measure of the degradation of the specimen as a whole and does not provide any measure of the spatial variability of degradation within the specimen. In addition, the correlation of weight loss to mechanical and dielectric properties combines all of the effects of aging (physical and chemical) into a single scalar quantity.

Weight loss observations

Since the weight loss as a function of isothermal aging time has traditionally been used to compare the thermooxidative stability of HTPMCs, it is necessary to understand the specimen parameters affecting the weight loss. As mentioned earlier, the anisotropy of oxidation (and in turn, weight loss behavior) of PMCs was first documented by Nelson [67] when he observed that the oxidation process is sensitive to the surface area for the different test specimens that he aged. He found that the materials degraded preferentially at the specimen surface perpendicular to the fiber (0° edges) and that the rate of oxidation is hastened by microcracks opening on the 0° edges. The enhanced weight loss from surfaces cut perpendicular to the fiber direction has also been noted and reported by numerous investigators, e.g., Bowles and Novak [12], Bowles [7], Bowles et al. [16], Skontorp et al. [97], Chung et al. [24], Bellenger et al. [6], and Kung [54]. Further, interlaminar residual curing stresses play a significant role in the thermooxidation process at the free-edge of composites, resulting in much greater weight loss in crossply laminates compared to that of unidirectional laminates [15].

In a recent work [91], four different unidirectional G30-500/PMR-15 composite specimen geometries representing different surface area ratios were selected, as reported in Table 9.4. The total surface areas for the R1, R2, and R4 specimens are all approximately equal (~800 mm^2), while the total surface area of the R3 specimens is nearly double that of the other specimens (~1,500 mm^2). In Table 9.5, the surface areas and the surface area ratios of the four specimen geometries are given. Figures 9.61–9.64 show the weight loss history for the four specimen geometries isothermally aged at 288°C in both air and argon gaseous environments. Comparing the four specimen types aged in an inert argon environment shows that the specimen geometry does not significantly influence the weight loss when normalized to the total surface area of the specimens. Volume changes and weight loss resulting from nonoxidative degradation are likely a result of degassing of low molecular weight components created by chain cleavage in the polymer [24]. Nonoxidative aging is expected to be independent of surface area and surface type and is a volumetric response. Therefore, the percentage weight change for the argon-aged specimens is expected to be independent of the specimen geometry. For all four geometries, the argon weight loss appears to approach an asymptote between 2,000 and 2,500 h. However, the weight loss data for the air-aged specimens for each of the specimen geometry show that there is a strong dependence of the weight loss on the surface area ratio $(S_1 + S_2)/S_3$, indicating that the weight loss is dependent on surface type. The lower the magnitude of the surface area ratio $(S_1 + S_2)/S_3$, the greater the normalized weight loss, as seen in Fig. 9.65. In addition, the weight loss in air appears to be nonlinear in time with an increase in oxidation rate for increasing aging times. This is consistent with weight loss increases due to surface cracking that increases the actual total surface area through which oxygen can diffuse.

Figure 9.66 shows plots of the nonoxidative weight loss as a percentage of the total weight loss in air for the four specimen geometries as a function of the aging time. Although such results are dependent on specimen volume-to-surface area ratios, some general trends can be inferred. For early aging times, the nonoxidative weight loss is a substantial percentage of the total weight loss of the composites. For the specimen geometries that were tested, the nonoxidative weight loss is in the range of 25–50% up to the first 500 h of aging. Since the nonoxidative weight loss tended toward an asymptote between 2,000 and 2,500 h and the overall weight loss tends to accelerate for long-aging times, the contribution of the nonoxidative aging decreases for long-aging times. It is also noted that the ratio of the weight loss increases with an increasing $(S_1 + S_2)/S_3$ ratio.

Table 9.4. Dimensions of four specimen geometries

Specimens (three each)	Width (mm) axial	Length (mm) transverse	Thickness (mm)
R1	37.77	7.57	2.47
R2	17.52	17.58	2.44
R3	25.09	25.08	2.39
R4	7.58	37.76	2.70

Table 9.5. Surface areas and area ratios for four specimen geometries

Specimens	S_1 (mm^2)	S_2 (mm^2)	S_3 (mm^2)	$S_1 + S_2 + S_3$ (mm^2)	$(S_1 + S_2)/S_3$
R1	571.0	37.3	186.3	794.6	3.27
R2	616.0	85.8	85.5	787.3	8.21
R3	1,258.2	119.7	119.7	1,497.6	11.51
R4	572.5	203.9	40.9	817.3	18.98

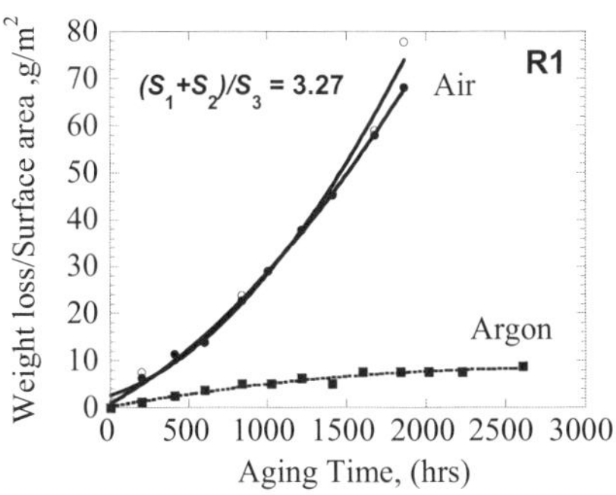

Fig. 9.61. Normalized weight loss for R1 specimens aged in air and argon

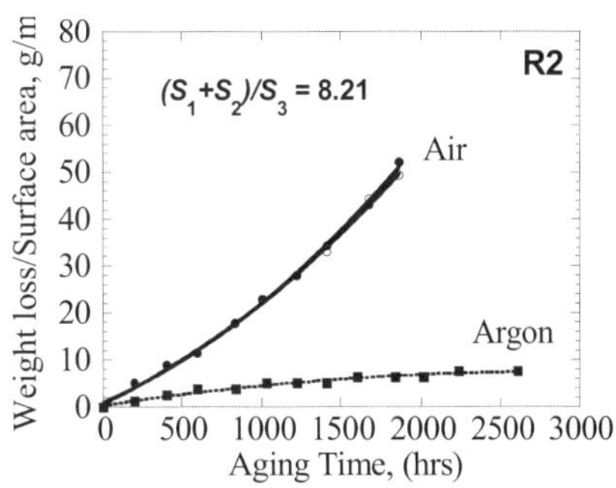

Fig. 9.62. Normalized weight loss for R2 specimens aged in air and argon

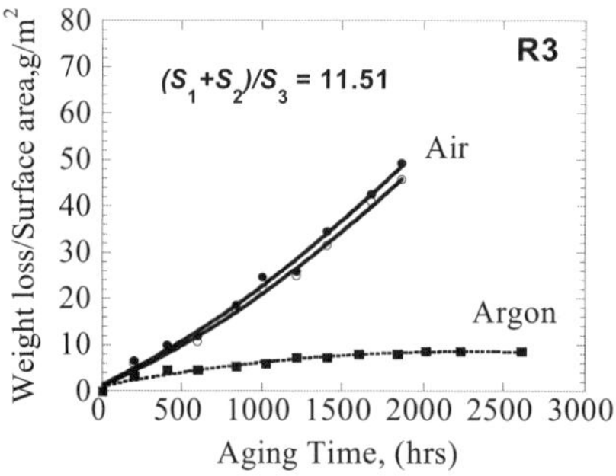

Fig. 9.63. Normalized weight loss for R3 specimens aged in air and argon

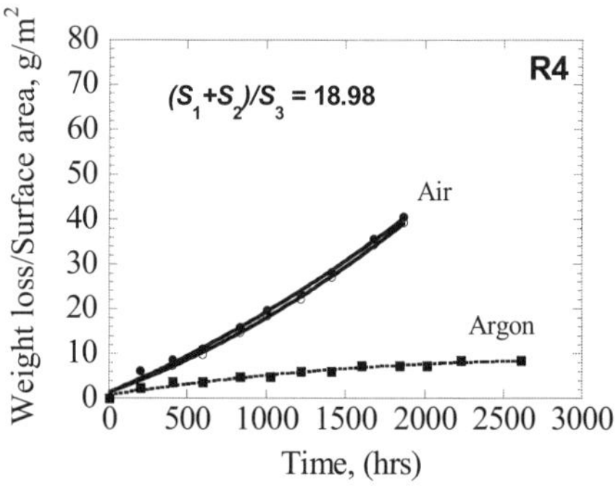

Fig. 9.64. Normalized weight loss for R4 specimens aged in air and argon

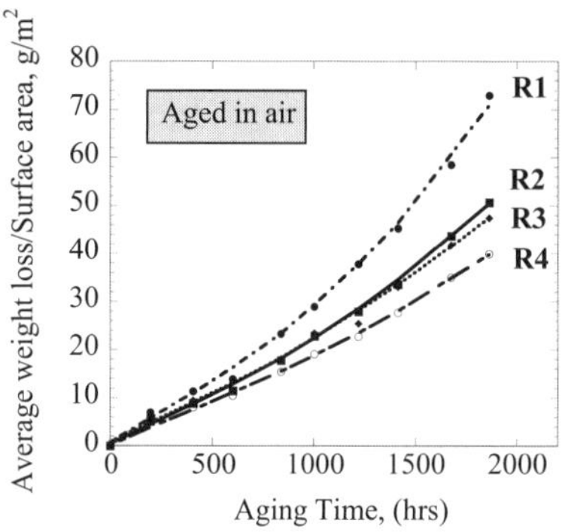

Fig. 9.65. Comparison of normalized weight loss in air for four specimen geometries

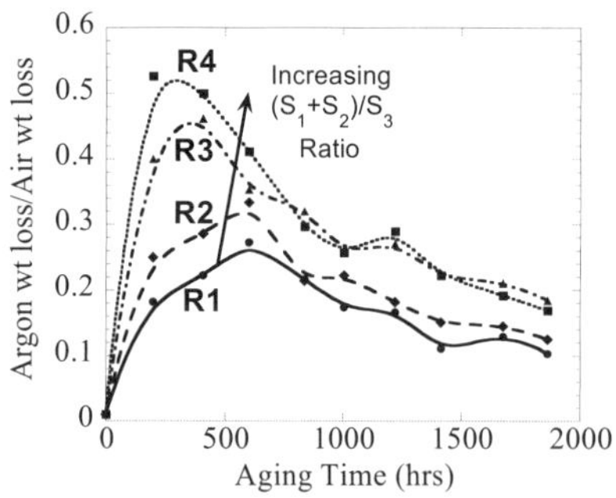

Fig. 9.66. Percentage of total weight loss attributed to nonoxidation weight loss

Empirically derived weight loss models

Led by the work of Bowles and Meyers [11], several weight loss models for predicting the long-term aging of HTPMCs based on weight loss surface fluxes have been developed [12, 24, 94, 95]. Most weight loss models found in the literature assume that weight loss is due to surface-controlled oxidation mechanisms and, to a lesser extent, to thermally induced reactions in the bulk of the material. However, it is shown in Fig. 9.66 that the surface-independent weight loss in the inert argon environment can account for a substantial part of the overall weight loss, particularly for short-aging times. Although the existing empirical weight loss models do not separately account for nonoxidative weight loss, its effect on the overall weight loss is inherently accounted for when determining the empirical weight loss parameters.

Bowles and Nowak [12] considered three different rates of weight loss from three different types of unidirectional composite surfaces shown in Fig. 9.15. The weight losses were expressed as weight loss fluxes or weight loss per unit surface area per unit of time. The composite weight loss is expressed as

$$Q(t) = At^{B} + C, \tag{9.50}$$

where A, B, and C are determined from an empirical curve fit. Further A is given as $A = S_1k_1 + S_2k_2 + S_3k_3$ and t is the aging time. The material parameters k_1, k_2, and k_3 are the respective weight loss fluxes or weight loss per unit surface area per unit of time. Knowing A from empirical curve fitting to the total weight loss for a given specimen k_1, k_2, and k_3 can be determined from three different specimen geometries. Bowles and Nowak [12] found that k_1 and k_2 were nearly equal, indicating that the weight loss rates from the transverse machined surface S_2 and the transverse tool surface S_1 are approximately the same. Further, the nonlinear time dependence of the flux rates is accounted for in the B exponent which is found to be time and temperature dependent. Based on testing composite specimens with different thickness, Bowles et al. [18] commented that only specimens of like dimensions can be directly compared on the basis of mass loss. That is, the model parameters, in particular the weight loss fluxes k_1, k_2, and k_3, and the parameter B, are dependent on the geometry of the specimen and B is not necessarily constant for a given aging temperature. Therefore, caution must be used when utilizing (9.50) to predict weight loss for specimen geometries and aging times that have not been tested.

As an alternative to using the surface weight loss fluxes (k_i, $i = 1, 2, 3$), Skontorp [95] and Wong et al. [126] approximated the total isothermal aging weight loss $Q(t)$ of a unidirectional composite by using weight loss per unit area (q_i, $i = 1, 2, 3$) as

$$Q(t) = (S_1 + S_2)q_1 + S_3 q_3, \qquad (9.51)$$

where the weight loss per unit area, q_1 and q_2, was assumed to be equivalent based on experimental data. Knowing S_1, S_2, and S_3, the weight loss per unit area parameters q_1 and q_3 can be determined empirically by aging two specimens with different ratios of $(S_1 + S_2)/S_3$ in identical environments and solving the resulting set of linear equations. For discrete times throughout the oxidative aging, these weight loss factors can be calculated to provide a history of the weight loss rates for each of the three surface types. Additionally, the parameters can be determined for a range of isothermal aging temperatures to obtain both their time and temperature dependence for a given material system.

Using a similar expression, Chung et al. [24] also expressed the total weight loss in terms of the weight loss rates as

$$Q(t) = S_1 q_1 + S_2 q_2 + S_3 q_3, \qquad (9.52)$$

where q_i is the weight loss per unit area defined using the shrinking core weight loss model [66], namely $q_i = D_{\mathrm{E}i} t^{n_i}$, where n_i are time exponent factors and $D_{\mathrm{E}i}$ are effective diffusion coefficients that may be described by Arrhenius expressions. The weight loss expression is similar to that of Bowles and Nowak [12], differing by the use of independent time exponential factors for each of the surface-specific weight loss parameters. Chung et al. [24] found it necessary to include an additional extrinsic term q_3' giving the final weight loss relationship as $Q(t) = S_1 q_1 + S_2 q_2 + S_3 (q_3 + \kappa q_3')$, where κ is a modeling parameter representing a dimensionless characteristic diffusion length and $q_3' = D_{\mathrm{E}3}' t / (D_{\mathrm{E}3}' t + 1)$ with an effective diffusion coefficient $D_{\mathrm{E}3}'$. This additional term accounted for the nonoxidation thermal degradation such as that highlighted in Fig. 9.66. The weight loss model (9.52) requires six independent parameters, including two extrinsic parameters.

The use of weight loss expressions given in (9.50)–(9.52) is based on weight loss rates from the three types of specimen surface areas. The surface areas S_1, S_2, and S_3 are the preaging surface areas of the specimens. Due to shrinking, material loss from exposed surfaces, and surface microcracking, the surface areas of the specimens change throughout the aging process. Therefore, surface areas S_1, S_2, and S_3 should more accurately be described as the initial areas based on the original specimen dimensions. Since the surface areas S_1, S_2, and S_3 are used for the duration of the prediction, the change in the surface areas due to aging damage is effectively accounted for in the weight loss rate parameters q_1, q_2, and q_3. The weight loss rate parameters q_3 and θ_3 (defined in a later section) are also strongly influenced by the fiber–matrix interface or interphase properties [95]. Theoretically, it is possible to measure the effect of different fiber sizings using weight loss rate parameters, however the practicality of this depends on the magnitude of the influence of the sizing on weight loss.

Weight loss predictions in unidirectional composites

In [91], the weight loss model used for the predictions is based on that of Skontorp [95]. The parameters q_1, q_2, and q_3 were calculated for aging times up to 1,864 h using the 288°C (550°F) air-aged weight loss data (Figs. 9.61–9.64) for the four specimen geometries. To determine the range of the parameters as a function of which three of the four specimen geometries were used for the calculations, four unique combinations of three of the four specimens were used. That is, four different sets of q_1, q_2, and q_3 parameters were calculated for the data set over the time range shown. Figure 9.67 shows the data for q_1, q_2, and q_3 based on the weight loss data for the four specimen geometries. This is referred to as the three-parameter model. It is shown that there is little variation of the parameters

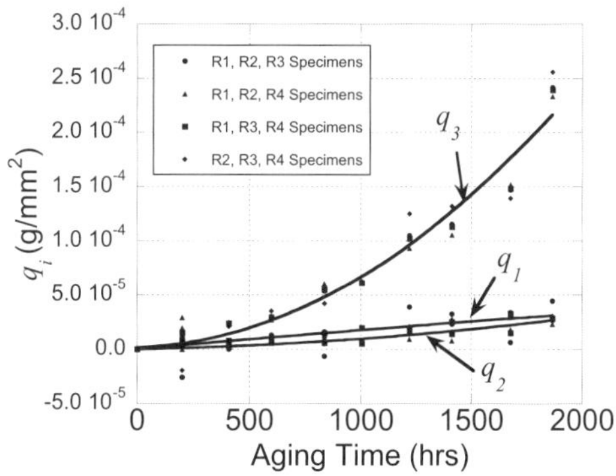

Fig. 9.67. Parameter determination using data from all four specimen geometries

as a function of which three of the four data sets were used for the calculations. Quadratic curves were fitted through the combined four data sets for each of the q_1, q_2, and q_3 parameters. It can be seen that the weight loss per unit surface area for the S_1 and S_2 surfaces is nearly identical for aging times up to 1,864 h. In addition, it can be seen that q_3 is increasing nonlinearly for increasing aging times (indicative of surface cracking), while q_2 and q_3 are increasing nearly linearly.

Next, q_1 and q_2 are assumed to be equal in (9.52) which results in the weight loss relationship expressed by (9.51). Equation (9.51) is then used to independently evaluate the q_1 and q_3 parameters. An advantage of using (9.51) rather than (9.52) is that it requires two rather than three specimen geometries to obtain the needed weight loss parameters. The data based on (9.51) are referred to as the two-parameter model. While there is some variation of the weight loss factors depending on the particular selection of two specimens used in their determination, the agreement for all specimen combinations is considered acceptable [91]. Further, the curve fits for the two-parameter model were found to be consistent with the curve fits for the three-parameter model. Finally, using the evaluated two-parameter curve fit functions, (9.51) is used to calculate the weight loss of the four specimen types. The experimental data (discrete points) and the weight loss determined by (9.51) (solid lines) are shown in Fig. 9.68. The figure demonstrates that specimens with different geometry and different overall surface areas can be characterized with single time-dependent weight loss per unit area expressions for q_1 and q_3.

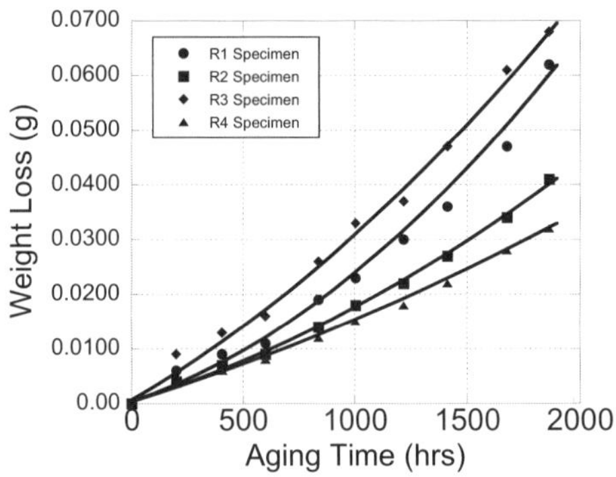

Fig. 9.68. Weight loss predictions for four specimens with four surface area ratios compared to experimental weight loss data

Weight loss predictions in woven composites

To demonstrate the predictive capability of (9.51) for multidirectional laminates, literature data of T650-35/PMR-15, 8-harness, satin-weave composites from Bowles et al. [20] are used. In that work [20], four specimen geometries with different $\Sigma_1/(\Sigma_2+\Sigma_3)$ surface area ratios aged at five different temperatures ranging from 204 to 343°C were examined. The specimen designation used by Bowles et al. [20], along with the area fraction ratios and total surface areas ($\Sigma_{total}=\Sigma_1+\Sigma_2+\Sigma_3$) of the specimens, is given in Table 9.6.

Table 9.6. Specimen geometries from Bowles et al. [20]

Specimen	$\Sigma_1/(\Sigma_2+\Sigma_3)$	Σ_{total} (mm^2)
T-3	68.66	19,898.5
T-5	34.16	20,385.5
T-12	13.68	21,962.8
T-50	6.01	3,233.6

The equivalent form of the weight loss relationship of (9.51) for the woven composites is given by $Q(t)=\Sigma_1\theta_1+(\Sigma_2+\Sigma_3)\theta_3$, where, in a manner similar to unidirectional composites, the weight loss per unit surface area for the warp and fill surface areas Σ_2 and Σ_3 is assumed to be equal. For each of the five test temperatures, two of the four specimen types were used to calculate θ_1 and θ_3. Figures 9.69 and 9.70 show the variation of the

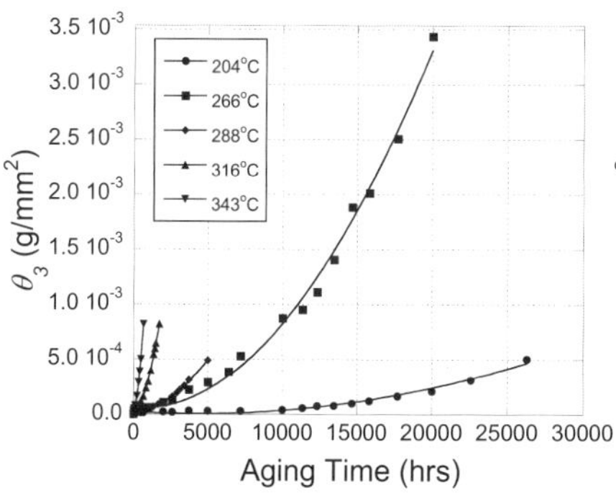

Fig. 9.69. Weight loss rate function θ_3 as a function of aging temperature (based on experimental data from Bowles et al. [20])

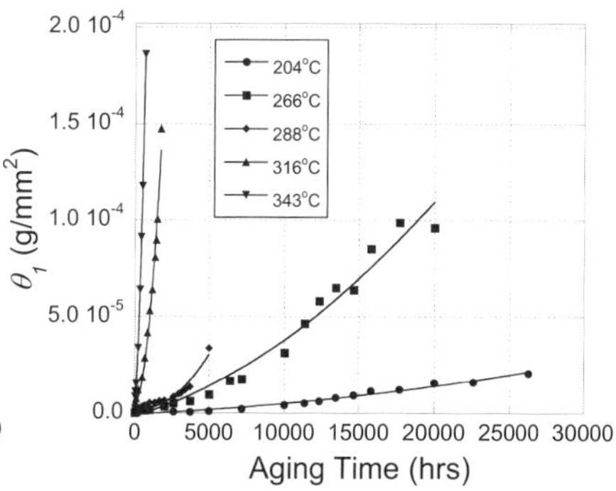

Fig. 9.70. Weight loss rate function θ_1 as a function of aging temperature (based on experimental data from Bowles et al. [20])

weight loss factors θ_1 and θ_3 as a function of aging temperature and aging time. As with the unidirectional PMR-15 composites, the weight loss rate on the edge surfaces of the specimen θ_3 is an order of magnitude greater than the weight loss rate on the top and bottom surfaces θ_1 for all aging temperatures. In addition, Fig. 9.69 illustrates the increase in the rate of θ_3 with an increase in aging time. This “apparent” rate increase is likely due to increases in the actual surface area due to aging cracks. Since the oxidation process is typically diffusion rate limited rather than reaction rate limited, the crack development provides pathways for oxygen to diffuse deeper into the composite increasing the overall oxidation rate. Figures 9.69 and 9.70 show the extraordinary temperature dependence of the weight loss rates. The maximum sustained use temperature of PMR-15 composites is typically considered to be 288°C and at this temperature, the weight loss rates θ_1 and θ_3 are relatively small compared to the rates at 316 and 343°C. In part, this can be explained by the early initiation of cracks in specimens aged at these higher temperatures.

Figure 9.71 shows the prediction of T-5 weight loss history at 204°C over 25,000 h of aging time. The T-5 specimen predictions are based on the functions θ_1 and θ_3 determined from the T-3 and T-12 specimen data for 204°C aging. There is an excellent correlation of the predicted weight loss of the T-5 specimens and the measured weight loss over the entire 25,000-h history. Thus, using an equivalent form of the weight loss relationship for woven composites, the empirically based model is also successfully used to predict the weight loss of 8-harness satin composites using data from the literature. Thus, the long-term isothermal weight loss response of both unidirectional and woven composites is successfully predicted.

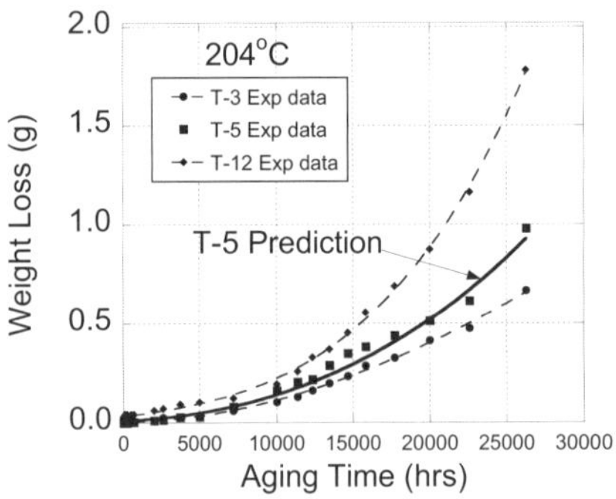

Fig. 9.71. Weight loss rate functions θ_3 and θ_1 (based on T-3 and T-12 specimen data) used to predict T-5 weight loss history at 204°C for aging times over 25,000 h (based on experimental data from Bowles et al. [20])

Design recommendation

The practice of using weight loss measurements on small specimens (4–5 cm^2) used by the aerospace industry to screen candidate PMCs for high-temperature applications could provide significantly more information than is currently being obtained. Modifying this practice by including one additional specimen with a different $(S_1 + S_2)/S_3$ surface area ratio would provide weight loss rates per unit area for the transverse q_1 and axial q_3 surfaces rather than an average weight loss per unit area for the entire specimen. This would allow designers to predict the performance of the components made from the candidate material for in-service environments. Whereas the weight loss from the axial surface may have a dominate contribution to the overall weight loss for small specimens, it may have a negligible contribution to the weight loss of aircraft components. This is due to the fact that only a very small percentage (or more likely none) of the exposed surface of components is the axial surface. Therefore, the weight loss rate per unit area for the transverse surface q_1 is what controls the thermooxidative weight loss of aircraft components and its value cannot be determined using a single specimen.

Accelerated aging behavior

The concept of equivalent property time (EPT) was established by Seferis [94] to understand degradation of polymers and composites, both for isothermal and dynamic elevated temperature exposures. According to Seferis [94], conversion may be defined as the ratio of actual weight loss to total weight loss corresponding to a given stage of the degradation process, i.e.,

$$\alpha = \frac{(M_0 - M)}{(M_0 - M_f)}, \tag{9.53}$$

where M, M_0, and M_f are the current, initial, and final weights of the specimen, respectively. When activation energy (E) and time (t_{ref}) to a certain conversion at reference temperature T_{ref} under isothermal conditions are known, the time to the same conversion at other temperatures can be obtained from

$$t = \text{EPT} = t_{ref} \exp\left[\frac{E}{R}\left(\frac{1}{T} - \frac{1}{T_{ref}}\right)\right], \tag{9.54}$$

where R is the universal gas constant, and the activation energy is assumed to be independent of temperature. In cases where the activation energy is not a constant over the entire temperature range of interest (implying that the weight loss mechanisms vary or differ over the temperature range), multiple subtemperature ranges have to be considered and constant activation energy E values evaluated over the temperature subdivisions for describing the weight loss behavior. The EPT concept is used to describe thermal gravimetric analysis (TGA) weight loss of bismaleimide and epoxy composites [94] and was demonstrated as a useful tool in designing aging experiments and assessing the lifetime of composite systems. An extension of this concept was developed to include equivalent cycle time (ECT), where ECT is defined as the characteristic time that equated properties of an isothermally aged material to that subjected to temperature cycles. Collectively, the EPT and ECT methodologies were shown to have the potential for correlating the data relating to the time and temperature equivalence of properties that continuously change.

9.5 Conclusions

Thermooxidative degradation of PMCs is clearly a subject of major importance as designers push the high-temperature endurance limits of polymer composites to improve the performance of aerospace systems. In recognition of recent developments of more thermally stable and durable high-temperature polyimides for replacing the legacy high-temperature polyimides such as PMR-15, the importance of a service life prediction capability for HTPMCs cannot be overstated.

The extraordinary cost of qualifying and developing empirically based material design allowables for new materials is prohibitive for today's low-rate production aircraft, particularly when affordability is a major concern. Robust analysis and design tools to predict service performance and service life enable the insertion of new materials that offer superior performance. This is accomplished by replacing the empirically based allowables generation methodology with one that relies primarily on predictive tools and strategic testing. Of course the challenge of implementing this methodology is to develop predictive tools with a level of robustness such that the analytical derived design allowables can be used with the same level of confidence and fidelity as the empirically based design allowables. Added to all of the tremendous challenges of modeling the performance and strength of composites in benign thermal environments are the difficult challenges of modeling the age-dependent properties of composites at high temperature. These added challenges are manifested even at the molecular scale at which an understanding of the basic chemical degradation mechanisms of the polymer of interest is lacking. Until robust predictive capabilities are developed, designers will continue to rely on expensive long-term testing to establish design allowables for thermooxidative stability of PMCs.

Recognition that the molecular level oxidative and hydrolytic degradation reactions of the matrix polymer of composites can ultimately influence the structural response prescribes that multiple scales need to be considered for analyzing their behavior. However, difficulties in relating the chemical state of a material to its physical and mechanical properties are a pervasive problem in the materials science community. This is particularly true when the initial state of the material may not be adequately known, and the material's chemical state continuously evolves by reacting with its environment. Although large-scale molecular dynamics simulations offer future potential to resolve these difficulties, application of molecular dynamics to highly crosslinked polyimides is beyond current modeling capabilities. Empirically derived relationships that relate the chemical state of polymers to physical and mechanical properties are a known alternative to the mechanistic modeling of molecular dynamics. However, experimental evaluation of the time-dependent evolution of the chemical state of polymers is particularly challenging for long-term aging.

The strong interaction of the physical, chemical, and mechanical aging behavior of HTPMCs is due to the significant contrast in the properties of the fiber and matrix constituents of the composite. These interactions are particularly apparent in the critical fiber–matrix interface/interphase where the interface/interphase has been shown to play a dominant role in the thermooxidation process. Difficulties associated with modeling the interface/interphase region include characterization of its properties (interface fracture toughness and strength [40, 102] and interphase thermal–physical–mechanical properties), determination of the size of the interphase region, and proper treatment of the coupling between the mechanical (damage) and chemical processes. Since direct measurements of the properties of the interphase are deficient, interphase properties and model parameters may be *estimated* based on indirect observations of the composite behavior. HTPMCs have high glass transition temperatures (high by polymer standards), and the temperatures at which they are cured may be on the order of 200°C greater than the cure temperature for epoxy composites. High curing temperatures exacerbate the detrimental fiber–matrix residual cure stresses (as well as the interlaminar residual cure stresses). The contribution and role of these micromechanical residual curing stresses at the interface/interphase to the thermooxidative stability require additional consideration.

As with the interface/interphase, experimental characterization of anisotropic physical and mechanical properties of carbon fibers is elusive. Although the degradation of the in situ fiber in the composite has long been considered to be negligible, the role of the fiber and its contribution to the oxidation process has not been fully explored. Specifically, little is known about the diffusivity of oxygen in the fiber and its contribution to the transport of oxygen to the interior of the composite. Alternative pathways for transport of oxygen into the interior of the composite are fiber–matrix debonds that propagate with the oxidation front. This possible mechanism for accelerated oxidation along the composite fiber direction is an excellent example of the intrinsic coupling of chemical oxidative aging and damage. Whereas cracks can accelerate oxidation by providing pathways for oxidants, oxidation can lead to cracking in the resin matrix and at fiber–matrix interfaces. The critical nature of the fiber–matrix interphase on degradation and failure processes in composites signifies the importance of proper representation of its behavior in predictive models. Since failure initiation is typically associated with the fiber–matrix interface/interphase region, the importance of modeling the coupling effects of damage and oxidation cannot be overemphasized.

Integration of constitutive models for the composite constituents into a unit cell micromechanical representation provides a foundation for understanding the mechanisms of degradation for long-term environmental exposure and loading. Although the accuracy of the unit cell analysis is certainly dependent on the accuracy of the constitutive models for the constituents, even rudimentary constitutive models can provide constructive insight into the primitive mechanisms of degradation. This is particularly true for short-aging times when the coupling effects of the physical and chemical aging and the damage may be less prominent. Indeed, in the absence of a robust predictive model for composite oxidation, researchers routinely apply models that represent only limited facets of the mechanistic behavior. Such is the case for the application of linear viscoelastic analyses in the NASA HSR program to model the long-term (60,000 h at 177°C) behavior of composites. Clearly, the linear viscoelastic models are expected to perform most favorable for cases in which oxidation is a secondary effect, e.g., lower temperatures and shorter aging times but inherent in determination of the viscoelastic model parameters are phenomenological contributions from the oxidative degradation if it is present. Although these models were used with success in the HSR program, these limited representations of long-term thermal aging can lead to large prediction errors particularly for extended aging times if used without discretion.

Of greatest importance to the use of predictive tools for development of material design allowables is the capability to accurately predict long-term performance based on short-term observations for the same loading environment. An associated issue is the capability to predict the long-term performance based on accelerated aging observations. Moreover, the latter capability would be particularly beneficial if the predictive tool could properly account for changes in the degradation mechanism for accelerated test methods. This would provide opportunities to expand the testing environment envelope allowing, for example, shorter duration accelerated aging test that would ultimately reduce the cost of generating material design allowables.

An overview of the experimentally observed thermooxidative behavior of PMR-15 neat polymer and carbon fiber/PMR-15 composites has been presented along with a theoretical treatment of oxidation development. Numerous shortcomings of existing analytical/numerical models and the challenges of experimentally characterizing or measuring the parameters required by the models have been pointed out. The listing of modeling and characterization challenges for HTPMCs provided herein is only a small subset of the issues that must be addressed to reach the goal of developing a robust modeling tool. Some of the other overarching areas for which only a cursory treatment is given include hygrothermal degradation, thermo-mechanical cyclic loading at high temperatures, and coupling of the mechanical response (including damage) with the chemical evolution of properties, to name just a few. To identify problem areas needing further study, one only needs to start with a review of the section titles of the topics covered here.

Acknowledgments

This work is supported by the Air Force Office of Scientific Research under the Materials Engineering for Affordable New Systems (MEANS) program sponsored by Dr. Charles Lee. Particular thanks are due to Mr. Erik Ripberger (AFRL/MLSA) and Mr. Josh Briggs (UDRI) for their help with aging experiments, Mr. Ken Goecke (UDRI) for mechanical testing, Dr. Sirina Putthanarat (UDRI) for nanoindentation measurements, and Stevens Institute of Technology students (Eva Yu, Harold Cook, and Tan Huan) for numerical analyses.

References

1. Abdeljaoued K (1999) Thermal oxidation of PMR-15 polymer used as a matrix in composite materials reinforced with carbon fibers. MS Thesis, Ecole Nationale Superieure des Arts et Metiers, Paris
2. Accelerated Test Methods for Durability of Composites, High Speed Research Materials Durability (Task 23), High Speed Research Program, March 1998
3. Ahci E, Talreja R (2006) Characterization of viscoelasticity and damage in high temperature polymer matrix composites. Composites Science and Technology 66:2506–2519
4. Allen DH, Harris CE, Groves SE, Norvell RG (1988) Characterization of stiffness loss in cross-ply laminates with curved matrix cracks. Journal of Composite Materials 22:71–80
5. Audouin L, Langlois V, Verdu J, De Bruijn JCM (1994) Review: role of oxygen diffusion in polymer aging: kinetic and mechanical aspects. Journal of Materials Science 29:569–583
6. Bellenger V, Decelle J, Huet N (2005) Aging of a carbon epoxy composite for aeronautic applications. Composites Part B: Engineering 36:189–194
7. Bowles K (1991) Effect of fiber reinforcements on thermo-oxidative stability and mechanical properties of polymer matrix composites. NASA TM 103648
8. Bowles KJ (1992) Effect of fiber reinforcements on thermo-oxidative stability and mechanical properties of polymer matrix composites. SAMPE Quarterly, April 2–12
9. Bowles KJ (1998) Thermal and mechanical durability of graphite-fiber-reinforced PMR-15 composites. NASA TM 113116/REV1
10. Bowles KJ (1999) Durability of graphite-fiber reinforced PMR-15 composites aged at elevated temperatures. Journal of Composites Technology & Research 21:127–132
11. Bowles KJ, Meyers A (1986) Specimen geometry effects on graphite/PMR-15 composites during thermo-oxidative aging. In: Proceedings 31st International SAMPE Symposium
12. Bowles KJ, Nowak G (1988) Thermo-oxidative stability studies of Celion 6000/PMR-15 unidirectional composites, PMR-15, and Celion 6000 fiber. Journal of Composite Materials 22:966–985
13. Bowles KJ, Jayne D, Leonhardt TA (1992) Isothermal aging effects on PMR-15 resin. NASA TM 105648
14. Bowles KJ, Jayne D, Leonhardt TA (1993) Isothermal aging effects on PMR-15 resin. SAMPE Quarterly 24:2–9
15. Bowles KJ, Jayne D, Leonhardt TA, Bors D (1993) Thermal stability relationships between PMR-15 resin and its composites. NASA Technical Memorandum 106285
16. Bowles KJ, Jayne D, Leonhardt TA, Bors D (1993) Thermal stability relationships between PMR-15 resin and its composites. NASA TM 106285

17. Bowles KJ, Madhukar M, Papadopolous DS, Inghram L, McCorkle L (1995) The effects of fiber surface modification and thermal aging on composite toughness and its measurement. NASA TM-106765
18. Bowles KJ, McCorkle L, Ingraham L (1998) Comparison of graphite fabric reinforced PMR-15 and Avimid N composites after long term isothermal aging at various temperatures. NASA/TM-1998-107529
19. Bowles KJ, Papadopoulos DS, Inghram, LL, McCorkle LS, Klan OV (2001) Longtime durability of PMR-15 matrix polymer at 204, 260, 288 and 316°C. NASA/TM-2001-210602
20. Bowles KJ, Tsuji L, Kamvouris J, Roberts GD (2003) Long-term isothermal aging effects on weight loss, compression properties, and dimensions of T650-35 fabric-reinforced PMR-15 composites – data. NASA TM-2003-211870
21. Brinson LC, Gates TS (1995) Effects of physical aging on long term creep of polymers and polymer matrix composites. International Journal of Solids and Structures 32:827
22. Celina M, Wise J, Ottesen DK, Gillen KT, Clough RL (2000) Correlation of chemical and mechanical property changes during oxidative degradation of neoprene. Polymer Degradation and Stability 68:171–184
23. Chung K, Seferis JC (2001) Evaluation of thermal degradation on carbon fiber/cyanate ester composites. Polymer Degradation and Stability 71:425
24. Chung K, Seferis JC, Nam J-D (2000) Investigation of thermal degradation behavior of polymeric composites: prediction of thermal cycling effect from isothermal data. Composites Part A: Applied Science and Manufacturing 31:945–957
25. Ciutacu S, Budrugeac P, Niculae I (1991) Accelerated thermal aging of glass-reinforced epoxy resin under oxygen pressure. Polymer Degradation and Stability 31:365–372
26. Colin X (2000) Kinetic modeling of thermal oxidation of high performance polymer and composite materials. MS Thesis, Ecole Nationale Superieure des Arts et Metiers, Paris
27. Colin X, Verdu J (2003) Thermal aging and lifetime prediction for organic matrix composites. Plastics Rubber and Composites 32:349–356
28. Colin X, Verdu J (2005) Strategy for studying thermal oxidation of organic matrix composites. Composites Science and Technology 65:411–419
29. Colin X, Marais C, Favre JP (1999) Damage/weight loss relationship of polymer matrix composites under thermal aging. In: Proceedings of ICCM-12, ICCM Europe Publications, Massard T, Vautrin A, Eds
30. Colin X, Marais C, Cochon JL, Verdu J (1999) Damage/weight loss relationship for bismaleimide matrix under isothermal oxidation aging. In: Proceedings of the International Conference on Recent Developments in Durability Analysis of Composite Systems, Brussels University, Belgium
31. Colin X, Marais C, Verdu J (2001) Thermal oxidation kinetics for a poly(bismaleimide). Journal of Applied Polymer Science 82:3418–3430

32. Colin X, Marais C, Verdu J (2001) A new method for predicting the thermal oxidation of thermoset matrices: application to an amine cross-linked epoxy. Polymer Testing 20:795–803
33. Colin X, Marais C, Verdu J (2002) Kinetic modeling and simulation of gravimetric curves: application to the oxidation of bismaleimide and epoxy resins. Polymer Degradation and Stability 78:545–553
34. Colin X, Marais C, Verdu J (2005) Kinetic modeling of the stabilizing effects of carbon fibers on thermal aging of thermoset matrix composites. Composites Science and Technology 65:117–127
35. Colin X, Mavel A, Marais C, Verdu J (2005) Interaction between cracking and oxidation in organic matrix composites. Journal of Composite Materials 39:1371–1389
36. Crasto AS, Kim RY (1995) Creep behavior of high-temperature composites. In: Proceedings of ICCM-10, Whistler, BC, Canada IV:423
37. Cunningham RA (1996) High temperature degradation mechanisms in polymer matrix composites. MS Thesis, Massachusetts Institute of Technology, Cambridge, MA
38. Decelle J, Huet N, Bellenger V (2003) Oxidation induced shrinkage for thermally aged epoxy networks. Polymer Degradation and Stability 81:239–248
39. Eckstein BH (1981) The oxidation of carbon fibers in air between 230°C and 375°C. Fiber Science and Technology 14:139–156
40. Foster DC, Tandon GP, Zoghi M (2006) Evaluation of failure behavior of transversely loaded unidirectional model composites. Experimental Mechanics 46:217–243
41. Gates TS (2003) On the use of accelerated test methods for characterization of advanced composite materials. NASA/TP-2003-212407
42. Gates TS, Feldman M (1994) The effects of physical aging at elevated temperatures on the viscoelastic creep of IM7/K3B. NASA-TM-109114
43. Gibbs HH, Wendt RC, Wilson FC (1979) Carbon fiber structure and stability studies. Polymer Engineering Science 19:342–349
44. Gibson T, Price W, Arnold F, Uttermohlen D, Storage T, Jacques A (2006) Physical and mechanical properties of AFR-PE-4 and BIM-15 with S-2 and Astroquartz-III composites. Presented at 26th High Temple Workshop, Austin, TX
45. Gillen KT, Wise J, Clough RL (1995) General solution for the basic autoxidation scheme. Polymer Degradation and Stability 47:149–161
46. Hashin Z (1966) Viscoelastic fiber reinforced materials. AIAA Journal 4:1141–1147
47. Hergenrother PM, Smith GJ (1997) Chemistry and properties of imide oligomers end-capped with phenylethynylphthalic anhydrides. Polymer 35:4857–4864
48. Iarve EV (1996) Spline variational three-dimensional stress analysis of laminated composite plates with open holes. International Journal of Solids and Structures 44:2095–2118

49. Johnson TF, Gates TS (2001) High temperature polyimide materials in extreme temperature environments. In: AIAA Paper 2001-1214 42nd AIAA/ASME/ASCE/AHS/ASC Structures, Structural Dynamics and Materials Conference and Exhibition, Seattle, WA
50. Johnson LL, Eby RK, Meador MAB (2003) Investigation of oxidation profile in PMR-15 polyimide using atomic force microscope (AFM). Polymer 44:187–197
51. Klett JW, Ervin VJ, Edie DD (1999) Finite element modeling of heat transfer in carbon/carbon. Composites Science and Technology 59:593–607
52. Kulkarni R, Ochoa O (2006) Transverse and longitudinal CTE measurements of carbon fibers and their impact on interfacial residual stresses in composites. Journal of Composite Materials 40:733–754
53. Kung HK (1996) High-temperature oxidation, aging and creep in carbon-fiber reinforced polyimide composites during thermal fatigue. Ph.D. Thesis, University of Houston, TX
54. Kung HK (2006) Effects of surface roughness on high-temperature oxidation of carbon-fiber-reinforced polyimide composites. Journal of Composite Materials 39:1677–1687
55. Leung CL, Ghaffarian R, Leung, KC (1997) Thermo-oxidative stability of polyimides. I. Effect of prepolymer molecular weight. Polymer Degradation and Stability 58:1–14
56. Lincoln JE (2001) Structure–property–processing relationships and the effects of physical structure on the hygrothermal durability and mechanical response of polyimides. Ph.D. Thesis, Michigan State University
57. Marais C, Villoutreix G (1998) Analysis and modeling of the creep behavior of the thermostable PMR-15 polyimide. Journal of Applied Polymer Science 69:1983–1991
58. McCartney LN (1995) A recursive method of calculating stress transfer in multiple-ply cross-ply laminates subjected to biaxial loading. NPL Report DMMA(A)150
59. McKenna GB (1992) Aging of polymeric resins: implications for composite performance. In: Proceedings of the Conference of the American Chemical Society
60. McMahon PE (1978) Oxidative resistance of carbon fibers and their composites. In: Advanced Composite Materials – Environmental Effects, ASTM STP 658, Vinson JR, Ed., American Society for Testing and Materials, pp 254–266
61. McManus HL, Cunningham RA (1995) Coupled materials and mechanics analyses of durability tests for high temperature polymer matrix composites. NASA/CR-95-112619
62. McManus HL, Foch BJ, Cunningham RA (2000) Mechanism-based modeling of long-term degradation. Journal of Composites Technology & Research 22:146–152
63. Meador MA, Lowell CE, Cavano PJ, Herrera-Fierro P (1996) On the oxidative degradation of nadic endcapped polyimides. I. Effect of thermocycling on weight loss and crack formation. High Performance Polymers 8:363–379

64. Meador MA, Johnston JC, Cavano PJ, Frimer AA (1997) On the oxidative degradation of nadic endcapped polyimides. 2. Evidence for reactions occurring at high temperature. Macromolecules 30:3215–3223
65. Meador MA, Johnston JC, Frimer AA, Gilinsky-Sharon P (1999) On the oxidative degradation of nadic endcapped polyimides. 3. Synthesis and characterization of model compounds for end-cap degradation products. Macromolecules 32:5532–5538
66. Nam JD, Seferis JC (1992) Anisotropic thermo-oxidative stability of carbon fiber reinforced polymeric composites. SAMPE Quarterly 24:10–18
67. Nelson JB (1983) Thermal aging of graphite polyimide composites. In: Long-Term Behavior of Composites, ASTM STP 813, O'Brien TK, Ed., American Society for Testing and Materials, Philadelphia, pp 206–221
68. Nicholson LM, Whitley KS, Gates TS (2000) The combined influence of molecular weight and temperature on the aging and viscoelastic response of a glassy thermoplastic polyimide. NASA-2000-TM210312
69. Nicholson LM, Whitley KS, Gates TS (2002) The role of molecular weight and temperature on the elastic and viscoelastic properties of glassy thermoplastic polyimide. International Journal of Fatigue 24:185–195
70. Paipetis A, Galiotis C (1996) Effect of fiber sizing on stress transfer efficiency in carbon/epoxy model composites. Composites Part A: Applied Science and Manufacturing 27A:755–767
71. Paris O, Loidl D, Peterlik H, M ller M, Lichtenegger H, Fratzl P (2000) The internal structure of single carbon fibers determined by simultaneous small and wide-angle x-ray scattering. Journal of Applied Crystallography 33:695–699
72. Pasricha A, Tuttle ME, Emery AF (1996) Time-dependent response of IM7/5260 composites subjected to cyclic thermo-mechanical loading. Composites Science and Technology 56:55–62
73. Pochiraju KV, Tandon GP (2004) Time dependent composite material behavior under thermo-oxidative environmental conditions. In: Proceedings of IMECE, Anaheim, CA
74. Pochiraju KV, Tandon GP (2006) Modeling thermo-oxidation layer growth in high temperature resins. Journal of Engineering Materials and Technology 128:107–116
75. Popov AA, Krysyuk BE, Zaikov GY (1980) Translation of vysokomol soyed. Polymer Science, USSR, 22
76. Reifsnider KL, Lesko J (1998) Mechanics of long-term behavior of high temperature polymer composites. Report AFRL-SR-BL-TR-98-0835
77. Reynolds WN (1973) Structure and physical properties of carbon fiber. Chemistry and Physics of Carbon 11:1–67
78. Ripberger E, Tandon GP, Schoeppner GA (2004) Characterizing oxidative degradation of PMR-15 resin. In: Proceedings of the Spring SAMPE 2004 Symposium/Exhibition, Long Beach, CA
79. Ripberger E, Tandon GP, Schoeppner GA (2004) Characterizing oxidative layer development in AFR-PE-4 resin. In: Proceedings of the Fall SAMPE 2004 Symposium/Exhibition, Long Beach, CA

80. Ripberger ER, Tandon GP, Schoeppner GA (2005) Experimental techniques for characterizing thermo-oxidative behavior in high temperature polyimide composites. In: Proceedings of the 2005 SEM Annual Conference & Exposition, Portland, OR
81. Roy S, Vengadassalam K, Wang Y, Park S, Liechti KM (2006) Characterization and modeling of strain assisted diffusion in an epoxy adhesive layer. International Journal of Solids and Structures 43:27–52
82. Saff C, Hahn G, Griffith J, Ingle R, Nelson K (2005) Accelerated Insertion of Materials – Composites AIAA-2005-2165 46th AIAA/ASME/ASCE/AHS/ASC Structures, Structural Dynamics and Materials Conference, Austin, TX
83. Salin IM, Seferis JC (1993) Anisotropic effects in thermogravimetry of polymeric composites. Journal of Polymer Science 31:1019–1027
84. Schapery RA (1966) A theory of non-linear thermoviscoelasticity based on irreversible thermodynamics. In: Proceedings of the 5th US National Congress of Applied Mechanics, ASME, pp 511–530
85. Schapery RA (2002) Homogenized constitutive equations for linear viscoelastic unidirectional composites with growing transverse cracks. Mechanics of Time-Dependent Materials 6:101–131
86. Schapery RA, Sicking DL (1995) On nonlinear constitutive equations for elastic and viscoelastic composites with growing damage. Mechanical Behavior of Materials 47:45–76
87. Schieffer A, Maire J-F, L v que D (2002) A coupled analysis of mechanical behavior and aging for polymer-matrix composites. Composites Science and Technology 62:543–549
88. Schoeppner GA, Curliss DB (2002) Model-based design for composite materials life management. In: Proceedings of the 9th AIAA/ISSMO Symposium on Multidisciplinary Analysis and Optimization Conference, 4–6 September, Atlanta, GA, AIAA-2002-5516
89. Schoeppner GA, Pagano NJ (1999) 3D thermoelastic moduli and saturation crack density for cross-ply laminates with transverse cracks. International Journal of Damage Mechanics 8:273–309
90. Schoeppner GA, Tandon GP (2003) Aging and durability of PMR-15 high temperature polyimide. In: Proceedings of 35th International SAMPE Technical Conference, Dayton, OH
91. Schoeppner GA, Tandon GP, Ripberger ER (2007) Anisotropic oxidation and weight loss in PMR-15 composites. Composites Part A: Applied Science and Manufacturing 38:890–904
92. Scola DA, Vontell JH (1988) High temperature polyimides, chemistry and properties. Polymer Composites 9:443–452
93. Scott DW, Lai JS, Zureick A-H (1995) Creep behavior of fiber reinforced polymeric composite: a review of the technical literature. Journal of Reinforced Plastics and Composites 14:588–617
94. Seferis JC (1999) Aging analyses of polymer composites through time–temperature equivalence. Journal of Composites Technology & Research 21:173–179

95. Skontorp A (1995) Isothermal high-temperature oxidation, aging and creep of carbon-fiber/polyimide composites. Ph.D. Thesis, University of Houston, TX
96. Skontorp A, Wang SS (1995) High temperature creep with physical and chemical aging, and associated viscoelastic constitutive equations of polyimide-matrix composites. In: Recent Advances in Composite Materials, White SR, Hahn HT, Jones WF, Eds., ASME MD, Vol. 56, American Society of Mechanical Engineers, New York, NY, pp 57–70
97. Skontorp A, Wong M-S, Wang S-S (1995) High-temperature anisotropic thermal oxidation of carbon-fiber reinforced polyimide composites. Theory and experiment. In: Proceedings of ICCM-10, Whistler, BC, Canada, IV:375
98. Struik LCE (1978) Physical Aging in Amorphous Polymers and Other Materials. Elsevier, Amsterdam
99. Sullivan JL (1990) Creep and physical aging of composites. Composites Science and Technology 39:207
100. Talreja R (1985) Transverse cracking and stiffness reduction in composite laminates. Journal of Composite Materials 19:355–375
101. Tandon GP (1995) Use of composite cylinder model as representative volume element for unidirectional fiber composites. Journal of Composite Materials 29:388–409
102. Tandon GP, Kim RY, Bechel VT (2000) Evaluation of interfacial normal strength in SCS-0/epoxy composite using cruciform specimen. Composites Science and Technology 60:2281–2295
103. Tandon GP, Ripberger ER, Schoeppner GA (2005) Accelerated aging of PMR-15 resin at elevated pressure and/or temperature. In: Proceedings of the SAMPE 2005 Symposium and Exhibition, Seattle, WA
104. Tandon GP, Pochiraju KV, Schoeppner GA (2006) Modeling of oxidative development in PMR-15 resin. Polymer Degradation and Stability 91:1861–1869
105. Tandon GP, Briggs J, Schoeppner GA (2006) Thermo-oxidative degradation of carbon fibers for high-temperature polymer matrix composites. In: Proceedings of American Society for Composites, 21st Technical Conference, University of Michigan-Dearborn, Dearborn, MI
106. Tang LQ, Pochiraju K, Chassapis C, Manoochehri S (1998) A computer-aided optimization approach for the design of injection mold cooling systems. Journal of Mechanical Design: Transactions of the ASME 120:165–174
107. Thorp KE (2000) Hydrolytic degradation of norbornene-terminated addition polyimides: a mechanistic and kinetic study. Ph.D. Thesis, University of Dayton
108. Tsotsis TK (1995) Thermo-oxidative aging of composite materials. Journal of Composite Materials 29:410–422
109. Tsotsis TK (1998) Long-term thermo-oxidative aging in composite materials: experimental methods. Journal of Composite Materials 32:1115–1135

110. Tsotsis TK, Lee SM (1998) Long-term thermo-oxidative aging in composite materials: failure mechanisms. Composites Science and Technology 58:355–368
111. Tsotsis TK, Keller S, Bardis J, Bish J (1999) Preliminary evaluation of the use of elevated pressure to accelerate thermo-oxidative aging in composites. Polymer Degradation and Stability 64:207–212
112. Tsotsis TK, Keller S, Lee K, Bardis J, Bish J (2001) Aging of polymeric composite specimens for 5000 hours at elevated pressure and temperature. Composites Science and Technology 61:75–86
113. Tsuji LC, McManus HL, Bowles KJ (1998) Mechanical properties of degraded PMR-15 resin. NASA Technical Report, 1998-208487, pp 1–18
114. Upadhyaya D, Tsakiropoulos P (1995) Evaluation of the effect of sizing levels on transverse flexural and shear strengths of carbon/epoxy composites. Journal of Materials Processing Technology 54:17–20
115. Veazie DR, Gates TS (1995) Physical aging effects of the compressive linear viscoelastic creep of IM7/K3B composite. NASA-95-TM110224
116. Veazie DR, Gates TS (1998) Effects of aging-time reference on the long term behavior of the IM7/K3B composite. In: Proceedings of 39th AIAA/ASME/ASCE/AHS/ASC Structures, Structural Dynamics, and Materials Conference Paper AIAA 98-1962
117. Verdu S, Verdu J (1997) A new kinetic model for polypropylene thermal oxidation at moderate temperatures. Macromolecules 30:2262–2267
118. Waldron WK Jr. and McKenna GB (1995) The nonlinear viscoelastic response and apparent rejuvenation of an epoxy glass. Journal of Rheology 39:471–497
119. Wang SS, Chen X (2006) Computational micromechanics for high-temperature constitutive equations of polymer-matrix composites with oxidation reaction, damage, and degradation. Journal of Engineering Materials and Technology 128:81–89
120. Wang SS, Chen X, Skontorp A (2003) High temperature mechanics modeling and experiments of thermal oxidation, degradation and damage evolution in carbon fiber/polyimide composites. In: Proceedings of American Society for Composites 18th Technical Conference, University of Florida, Gainesville, FL
121. Wang SS, Chen X, Skontorp A (2003) High-temperature mechanics modeling and experiments of thermal oxidation, degradation and damage evolution in carbon/fiber/polyimide composites. Technical Report CEAC-TR-03-0106, University of Houston
122. Weitsman Y (1987) Stress assisted diffusion in elastic and viscoelastic materials. Journal of the Mechanics and Physics of Solids 35:73–97
123. Whitley KS, Collins TJ (2006) Mechanical properties of T650/AFR-PE-4 at elevated temperatures for lightweight aeroshell designs. In: AIAA-2006-2202 47th AIAA/ASME/ASCE/AHS/ASC Structures, Structural Dynamics, and Materials Conference, Newport, Rhode Island, AIAA-2006-2202

124. Williams ML, Landel RF, Ferry JD (1955) The temperature dependence of relaxation mechanisms in amorphous polymers and other glass-forming liquids. Journal of the American Chemical Society 77:3701–3707
125. Wise J, Gillen KT, and Clough RL (1997) Quantitative model for the time development of diffusion-limited oxidation profiles. Polymer 38:1929–1944
126. Wong MS, Skontorp A, Wang SS (1994) Thermal oxidation of carbon fibers and carbon-fiber reinforced high-temperature polyimide composite at elevated temperature. In: Proceedings of the American Society for Composites 9th Technical Conference, Technomic Publishing Co., Lancaster, PA, pp 458–467
127. Zhang Y, Xia Z, Ellyin F (2005) Nonlinear viscoelastic micromechanical analysis of fibre-reinforced polymer laminates with damage evolution. International Journal of Solids and Structures 42:591–604

Chapter 10: Modeling of Stiffness, Strength, and Structure–Property Relationship in Crosslinked Silica Aerogel

Samit Roy and Awlad Hossain

Department of Aerospace Engineering and Mechanics, The University of Alabama, Tuscaloosa, AL 35487, USA

10.1 Introduction

Mechanically stable forms of lightweight materials with porosities up to 98% were first introduced in the form of silica aerogels in the 1930s. Recently, interest in aerogels and other lightweight materials in engineering applications have increased tremendously. Native silica aerogels are chemically inert, low-density, nanostructured porous materials with poor mechanical properties. They are the product of the sol–gel process whose final step involves extracting the pore-filled solvent with liquid carbon dioxide through supercritical drying. Practical applications of native aerogels are somewhat limited as they are brittle and hygroscopic, absorbing moisture from the environment which eventually leads to aerogel collapse due to capillary forces in the pores. Nevertheless, it has been recently discovered that crosslinking the nanoparticle building blocks of silica aerogels with polymeric tethers increases both modulus and strength significantly [4]. Along these lines, a novel, multifunctional, crosslinked silica aerogel, to be

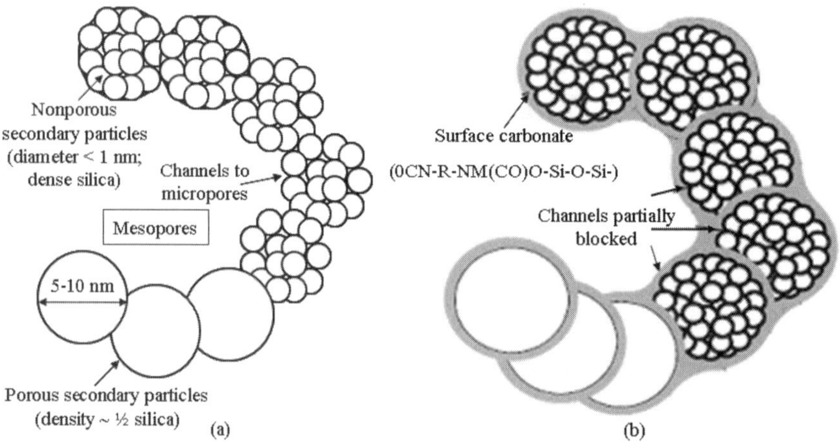

Fig. 10.1. (**a**) Silica aerogel structure before crosslinking and (**b**) x-aerogel structure after crosslinking [6]

referred to as *x-aerogel*, is derived by coating and encapsulating the skeletal framework of amine-modified silica aerogels with polyurea as depicted in Fig. 10.1.

Aerogels are reported to be one of the best thermal insulators. When sandwiched between two glass layers, aerogels reduce heat loss coefficient by more than a factor of 10, while preserving the capability of moderately high light transmission [5]. Aerogels are being considered for different aerospace applications, such as a thermal protection system (TPS), catalyst support, or as hosts for a variety of functional materials for chemical, optical, and electronic devices.

Cylindrical samples of x-aerogel manufactured in the author's laboratory are shown in Fig. 10.2. It was observed from the mechanical characterization tests that x-aerogel has very good compressive, tensile, and shear properties, in addition to its low thermal conductivity [4, 6, 8]. Recently, manufacturing of a lightweight cryogenic propellant tank with low thermal conductivity has been proposed using novel x-aerogel material. While the use of composite sandwich panels for cryotanks is not novel, it is feasible that the delamination of facesheet from the core due to cryopumping, analogous to the failure that occurred in the X-33 prototype, could perhaps be prevented through the use of x-aerogel core instead of a standard honeycomb core.

Fig. 10.2. Cylindrical samples of x-aerogel

It is envisioned that aerogel material can be used as the central core bonded between two facesheets of a sandwich plate. As aerogels are highly porous, facesheets will be necessary to make the sandwich composite panels impermeable for storing cryogenic fuels. Facesheets can be made of carbon fiber-reinforced polymer (CFRP) having high tensile load-bearing capability. Schematics of a typical sandwich plate with a standard honeycomb core and an x-aerogel core at the center are shown in Fig. 10.3. As another example of a practical application of x-aerogel, a conceptual design of a prototype cryogenic propellant tank using x-aerogel material is shown in Fig. 10.4.

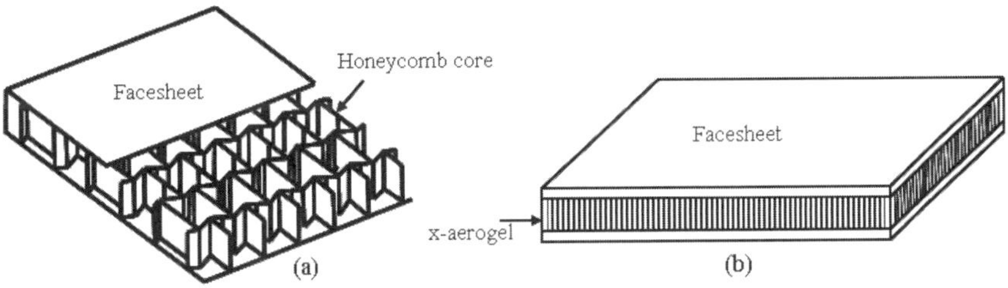

Fig. 10.3. (**a**) Traditional sandwich panel with honeycomb core and (**b**) novel sandwich panel with x-aerogel as central core material

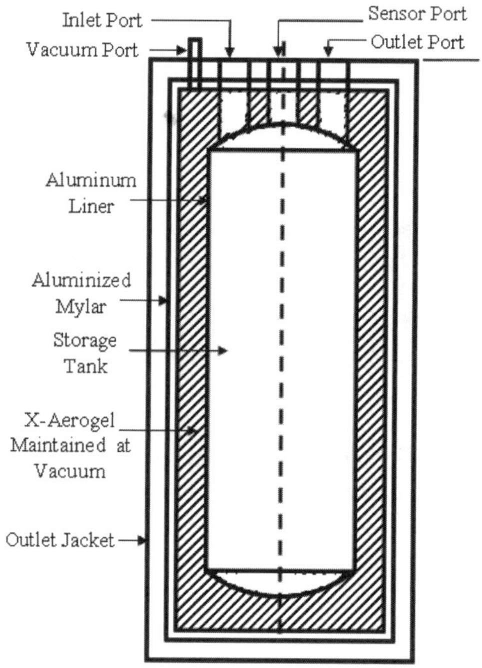

Fig. 10.4. Prototype cryotank design concept with x-aerogel material

10.2 Nanostructural Features of Silica Aerogel

Aerogel is a class of monolithic material that possesses porous structure. Scanning electron microscopy (SEM) [1] and transmission electron microscopy (TEM) [9] are widely used to produce direct images of mesoporous structures. For the crosslinked silica aerogel, SEM imaging was conducted for different loading stages, as shown in Fig. 10.5. In this figure, the clusters of secondary nanoparticles are clearly visible along with the mesopores.

The mechanical, thermal, electrical, and optical properties exhibited by x-aerogels are related to their mesoporous cluster assemblies, as depicted earlier in Fig. 10.1. The sol–gel manufacturing process can control the geometry, porosity, and physical properties of mesoporous silica aerogels by manipulating its chemistry and processing parameters. The stiffness and strength of x-aerogels strongly depend on their microstructural features, such as particle connectivity. However, there is no experimental technique currently available to measure this connectivity directly. As an alternative metric, the self-similar characteristics of aerogel structures can be investigated by evaluating their fractal dimension from geometric correlations.

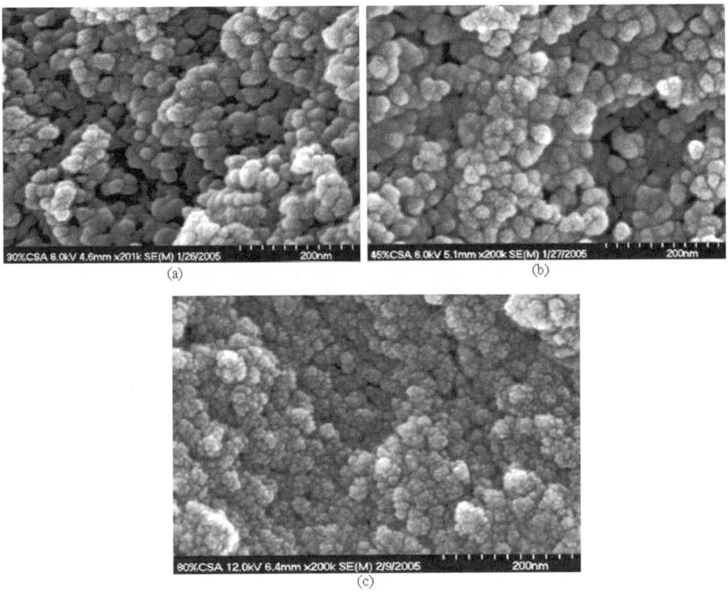

Fig. 10.5. SEM images of crosslinked silica aerogels. The clusters of secondary nanoparticles (*round particles*) and the mesopores (*dark spots*) are clearly visible: (**a**) 30% strain – no appreciable change in mesoporous structure, (**b**) 45% strain – gradual decrease in the mesoporosity, few dark spots, and (**c**) 77% strain – appreciable loss of porosity, particles are squeezed closer to one another

The shape of clusters or cluster configuration, the existence of voids of all sizes, and the gradual loss of connectivity among mesoporous particles suggest that a fractal dimension can be attributed to the x-aerogel structures as a useful descriptive parameter [5]. The fractal dimension of a mesoporous structure can be determined from its particle orientation within a sphere of a given radius or from the slope of a radial distribution function. It was reported in the literature [5] that not only the mass of aerogel but also other properties, such as vibrational dynamics, scale according to aerogel's fractal dimension. In general, due to its inherent porosity, the aerogel morphology represents a fractal dimension of less than 3; and its fractal dimension decreases with decreasing cluster densities, as presented later in this chapter.

In this study, a three-dimensional distinct element analysis (DEA) simulation was performed to determine the structure–property relationship of nanostructured x-aerogel material. The model attempted to incorporate microscale effects – such as particle bond stiffness, bond strength, particle frictional coefficient, initial cluster porosity (or density), and density of secondary silica particles – into a macroscale structure–property relationship for the prediction of Young's modulus and strength. In addition, numerical analyses were carried out to determine the fractal dimension of the aerogel structure while varying its initial cluster porosity or density. Modeling methodology will provide insights for both stiffening and strengthening mechanisms and how these mechanisms can be optimized with minimum weight penalty. Therefore, it is envisioned that numerical modeling will greatly reduce the number of "trial-and-error" experiments necessary to further enhance the properties of this novel material.

10.3 Particle Mechanics for Numerical Modeling of Aerogels

Various researchers have developed different cluster aggregation algorithms for simulating structural characterizations of mesoporous materials. Diffusion-limited cluster aggregation (DLCA) and reaction-limited cluster aggregation (RLCA) algorithms are some examples [7]. A DLCA technique was first developed by a research group at Harvard University to interpret scattering experiments and subsequently used for understanding different phenomenon related to porous media, such as gelation, fractal studies, and scattering spectroscopy. The DLCA algorithm proceeds with random filling of a three-dimensional cubic volume with nonintersecting spheres. The diameters of these spheres are chosen from a Gaussian distribution function. These spheres are then set to diffuse inside the cubic boundary. When in motion, a particle is tested against overlapping with neighboring particles. If an overlap is detected, then that particle is merged to neighboring particles to form a new cluster. The diffusive motion is completed once all the particles have merged to form a single cluster. The process of aggregation is depicted schematically in Fig. 10.6. A cluster of three particles is moved and tested for an overlap with neighboring particle as shown in Fig. 10.6a. Once an overlap is detected, this cluster is aligned to the neighboring particle as shown in Fig. 10.6b. Then, this cluster and particle are merged to form another cluster of four particles as shown in Fig. 10.6c. This process continues until all particles have merged to form a final single cluster. After the network connectivity has been determined among the particles using DLCA, this information is transferred to the DEA software. Finally, the DEA can be used to determine the structure–property relationship through numerical simulation.

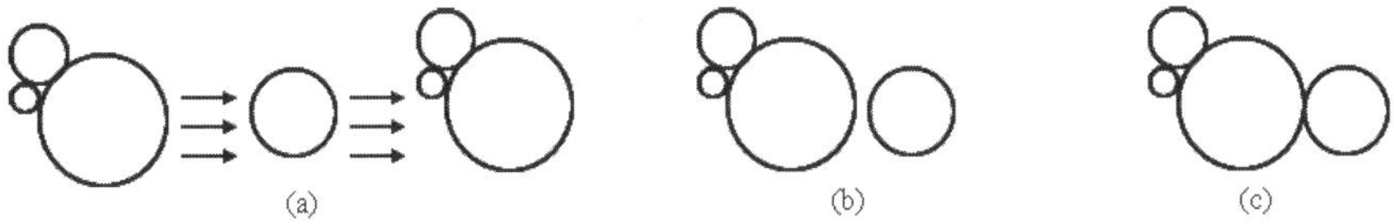

Fig. 10.6. Aggregation process for particles in DLCA

A particle flow code in three dimensions (PFC^{3D}) [3] simulates mechanical behavior of mesoporous structures when a bonded assembly of spherical particles is available from DLCA, as described earlier. PFC^{3D} is classified as a discrete or distinct element analysis code as it allows finite displacements and rotations of discrete bodies and recognizes new contacts automatically. PFC models are categorized as direct damage-type numerical models in which deformation is not a function of prescribed relationships between stress and strain but of changing microstructure. The numerical model is composed of distinct particles that displace independently from one another and interact only at contacts. Newton's laws of motion provide the fundamental relations between particle motion and forces. The complex nature of mesoporous structures can be modeled by bonding particles together at their contact points and allowing the bond breakage for excessive loading to exceed the bond strength. The PFC^{3D} conducts a particle flow model with the following assumptions:

1. The particles are treated as rigid bodies and are spherical in shape. The deformation of a packed-particle assembly results primarily from the sliding and rotation of rigid particles and not from the individual particle deformation.
2. Particles are in contact with each other. Contacts among the particles occur over a vanishingly small area, i.e., at a point, where bonds can exist.
3. Behavior at contacts uses a soft-contact approach wherein the particles are allowed to overlap one another at contact points.
4. The magnitude of the overlap is related with contact force via the force–displacement law, and all overlaps are small enough compared to particle sizes.

In addition to spherical particles (referred to as "balls"), the PFC^{3D} includes "walls" to apply velocity boundary conditions for compaction and confinement of particle assemblies. The balls and walls interact with one another via forces that arise at contacts. The equations of motion are satisfied for each ball; however, they are not satisfied for each wall, i.e., forces acting on a wall do not influence its motion. Instead, its motion is specified by the user and remains constant regardless of the contact forces acting upon it.

The calculation cycle in PFC^{3D} is a time-stepping algorithm that requires repeated applications of the laws of motion to each particle, a force–displacement law to each contact, and an updating of wall positions. Contact among particles forms and breaks automatically during the course of a simulation. The calculation cycle is shown in Fig. 10.7. At the start of each time step, a set of contacts is updated from known particle and wall positions. The force–displacement law is then applied to each contact to update contact forces. Next, the law of motion is applied to each particle to update its velocity and position. The constitutive behavior used in PFC^{3D} is mainly represented by contact models which, in essence, describe physical behavior at each contact by stiffness, slips, and bonding models.

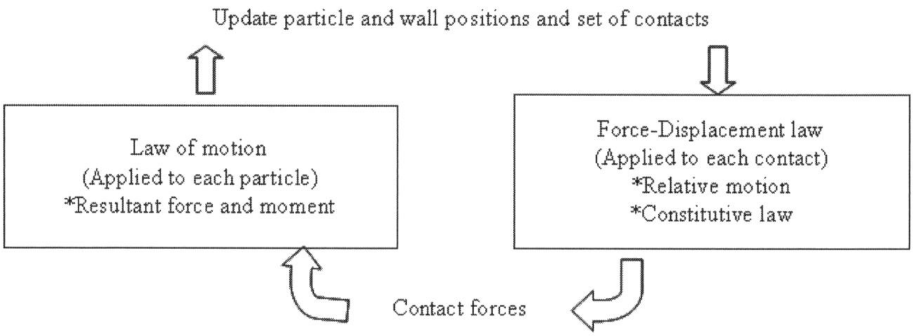

Fig. 10.7. Calculation cycle used in PFC^{3D} [3]

The contact stiffness relates contact forces and relative displacements in normal and shear directions. The normal stiffness is a secant stiffness since it relates total normal force to total normal displacement. The shear stiffness represents a tangent stiffness as it relates the shear force and displacement in incremental form. The normal and shear stiffness are expressed in (10.1) and (10.2), respectively

$$P_{\mathrm{n}} = k_{\mathrm{n}} v_{\mathrm{n}}, \tag{10.1}$$

$$\Delta P_{\mathrm{s}} = k_{\mathrm{s}} \Delta v_{\mathrm{s}}. \tag{10.2}$$

Here, P, k, and v indicate force, stiffness, and particle velocity, respectively. Subscripts n and s represent normal and shear components, respectively.

If the contact normal stiffness is altered during the course of simulation, there will be an immediate effect upon the entire assembly. Whereas, if the shear stiffness is altered, it will only affect the new increment of shear force. Two stiffness models, linear and Hertz–Mindlin, are available in PFC^{3D} for representing linear and nonlinear relations between force and displacement, respectively. The slip model in PFC^{3D} allows two entities in contact to slide relative to one another. A separation occurs if they are not bonded and a tensile force develops between them. The slip condition exists when the shear component of force reaches its maximum limit.

PFC^{3D} allows particles to be bonded together at contacts and supports two types of bonding models: a contact-bond model and a parallel-bond model. Both bonds can be envisioned as a kind of glue joining two particles. The contact-bond glue is of a vanishingly small radius that acts only at the contact point, while the parallel-bond glue is of a finite radius that acts over a circular cross-section lying between the particles. The contact bond can only transmit a force, while the parallel bond can transmit both a force and a moment. Both types of bonds may be active at the same time; however, the presence of a contact bond inactivates the slip model. The bonding logic is illustrated in Fig. 10.8.

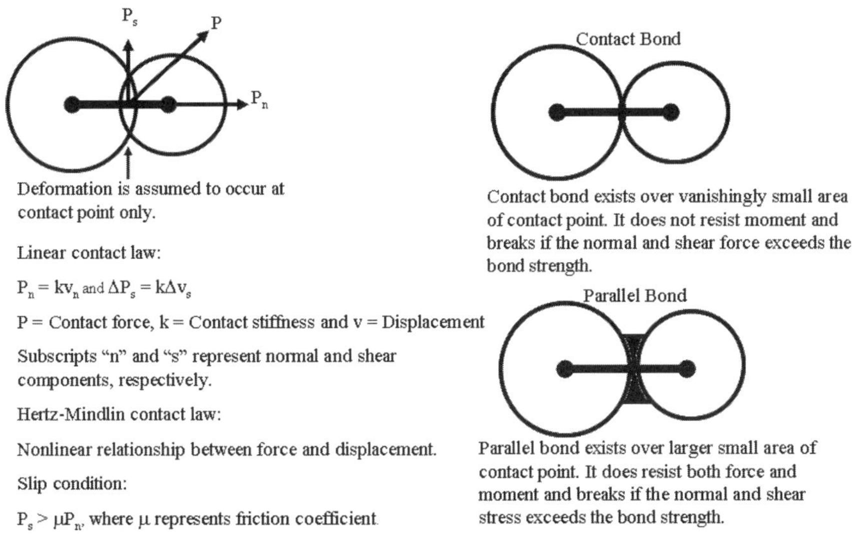

Fig. 10.8. Contact and parallel bonding logics used in PFC^{3D} [3]

Properties related to particles and their corresponding contacts are required to perform a simulation in PFC^{3D}. The response of a mesoporous material is mainly affected by particle size and packing arrangement. Therefore, the model parameters cannot be related directly to a set of relevant material properties. The relation between PFC model parameters and commonly measured material properties is only known a priori for certain simple packing arrangements. In case of arbitrary packing or particle assemblage, the relation is found by means of a calibration process where repeated simulations are required to mimic the true material responses. The user needs to specify parameters related with particle contact stiffness, particle friction coefficients, bond strengths, and others to simulate a corresponding set of macroresponses, such as elastic constants and peak strength envelope, etc. To get a rough estimate of particle contact stiffness and bond strength, PFC^{3D} provides the following two equations

$$E = \frac{k_n}{4R}, \tag{10.3}$$

$$\sigma_t = \frac{s_n}{4R^2}, \tag{10.4}$$

where E is the Young's modulus of the particle assembly as obtained from laboratory tests, k_n is the normal contact stiffness of the particles, R is the particle radius, σ_t is the measured tensile strength of the particle assembly, and s_n is the normal bond strength in particle contacts. The shear components of stiffness and bond strength k_s and s_s are taken to be some fraction or equal to their respective normal components. These two relations are derived for a cubic array of particles, which may not be a correct representation of the actual particle arrangement in laboratory samples. Nevertheless, these equations provide useful information that could be used to obtain first estimates of the micromechanical parameters. A flowchart is provided in Fig. 10.9 to help the reader better understand the simulation process.

10.4 Materials Characterization of X-Aerogel Through Compression Experiments

Compression experiments were performed to obtain material responses of x-aerogel specimens, which were subsequently used to verify the numerical results obtained from PFC^{3D}. During the experiments, aerogel samples were

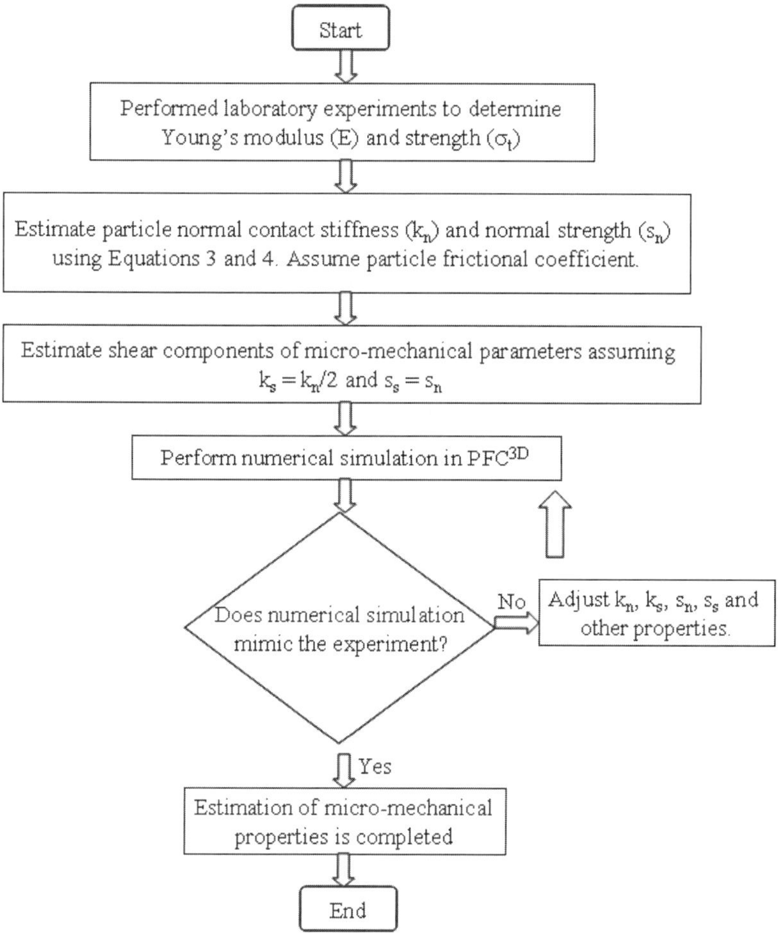

Fig. 10.9. Flowchart of the simulation process in PFC^{3D}

prepared following the procedures described in [8]. Compression experiments were conducted under different temperature conditions to study their corresponding effects on the measured properties. Cylindrical specimens consistent with ASTM D695-02a standard were used in an MTS machine equipped with a 55,000 lb load cell. Five replicate samples were tested. A typical experimental setup is shown in Fig. 10.10. Separate sets of samples were loaded at strain rates of 0.0035, 0.035, and 0.35 s^{-1} to investigate strain rate effects in x-aerogel.

Fig. 10.10. Uniaxial compression testing of crosslinked silica aerogel

Figure 10.11 represents the average stress–strain response for crosslinked silica aerogel specimens under compressive loading at room temperature at a strain rate of 0.0035 s^{-1}. Data for compressive yield strength, compressive stress at ultimate failure, and Young's modulus for individual sample calculated at room temperature, as well as the average values with their standard deviations, are summarized in Table 10.1. The ultimate compressive strength with other properties of native (uncrosslinked) silica aerogel is given in Table 10.2 for comparison.

During the compression test, crosslinked aerogels were found to behave as linearly elastic under small strains (<4%) and then exhibited yield (until ~40% compressive strain), followed by densification and inelastic hardening. Aerogel samples ultimately failed at approximately 77% compressive strain, yielding an ultimate compressive strength of 186 ± 22 MPa. The average yield stress defined at the beginning of densification was 4.26 ± 0.25 MPa and occurred at approximately 4% strain.

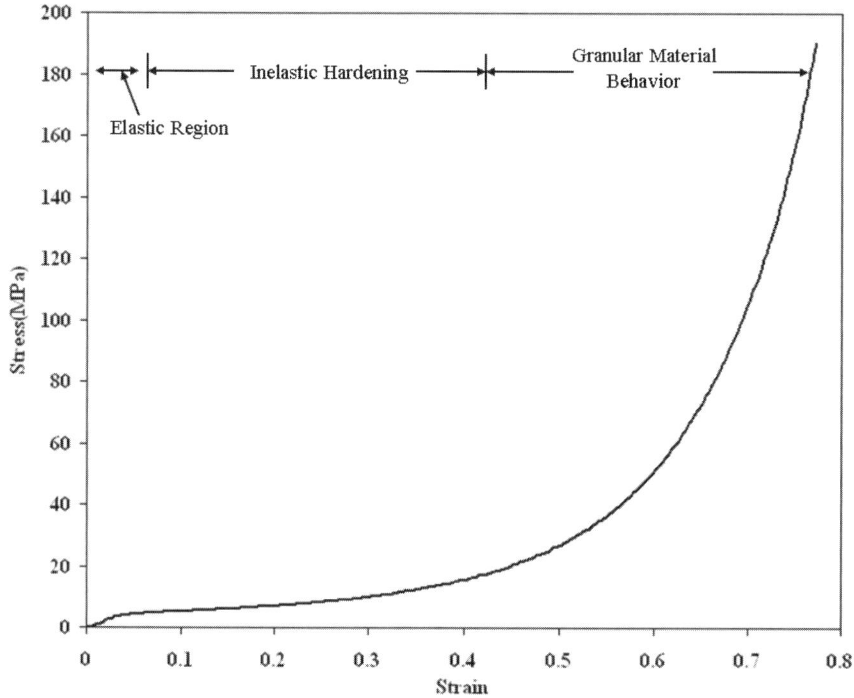

Fig. 10.11. Stress–strain curve in uniaxial compression on crosslinked silica aerogel

Table 10.1. Summary of compressive strength data of isocyanate crosslinked silica aerogels at a strain rate of 0.0035 s^{-1}

Sample number	Density (g cc^{-1})	Compressive yield strength (MPa)	Compressive strength (MPa)	Failure strain (%)	Young's modulus (MPa)
1	0.48	4.19	190.33	77.22	122.85
2	0.47	4.25	222.08	77.04	119.58
3	0.48	3.88	168.55	77.20	126.27
4	0.48	4.47	173.72	76.10	135.27
5	0.48	4.50	173.20	78.35	138.94
Average	0.478 ± 0.004	4.26 ± 0.25	186 ± 22	77.2 ± 0.8	129 ± 8

Table 10.2. Summary of compressive strength data of native silica aerogels at a strain rate of 0.0035 s^{-1}

Sample number	Density (g cc^{-1})	Ultimate compressive strength (MPa)	Failure strain (%)	Young's modulus (MPa)
1	0.18	4.0	5.31	96.7
2	0.20	4.1	6.00	86.9
Average	0.19 ± 0.01	4.1 ± 0.07	5.66 ± 0.49	92 ± 7

Table 10.3 compares the absolute and specific compressive strengths of crosslinked aerogel with other materials. The specific compressive strength of crosslinked aerogel is higher than that of steel, aluminum, and fiberglass, and is comparable to that of aerospace grade graphite composite. The relatively high specific compressive strength at ultimate failure combined with low thermal conductivity would render crosslinked silica aerogel attractive as multifunctional material for various space applications, such as cryogenic fuel tanks.

Table 10.3. Comparison of compressive strength of crosslinked silica aerogel with other composite materials for engineering applications

Material	Density (g cc^{-1})	Compressive strength (MPa)	Specific compressive strength (N m kg^{-1})
E-glass epoxy[a]	1.94	550	283,000
Kevlar-49 epoxy[a]	1.30	280	215,000
T 300 epoxy[a]	1.47	830	564,000
VSB-32 epoxy[a]	1.63	690	423,000
GY-70 epoxy[a]	1.61	620	385,000
2024 T3 Al	2.87	345	120,000
7075T6	2.80	475	169,000
4130 steel	7.84	1,100	140,000
x-aerogel	0.48	186	389,000

[a]Fiber volume fraction, $V_f = 0.6$.

Young's modulus in compression was evaluated from the slope of initial linear portion (at <4% strain) of the stress vs. strain curve. The Young's modulus of the crosslinked aerogel was 129 ± 8 MPa compared to 92 ± 7 MPa of the native silica framework. The recovery (or spring back) behavior of crosslinked silica aerogels was characterized by conducting loading–unloading cycles along the stress vs. strain curve. A typical cylindrical sample was loaded and unloaded six times, first in the elastic region where deformation was completely recovered as shown in Fig. 10.12. Then it was unloaded and reloaded in the inelastic region where a hysteresis loop was always observed, presumably due to energy loss for bond breaking and slippage of polymer chains [2]. It was further noted that the percent strain recovery became progressively lower at higher strains, while in the inelastic hardened range (approximately 60% strain) there was virtually no strain recovery.

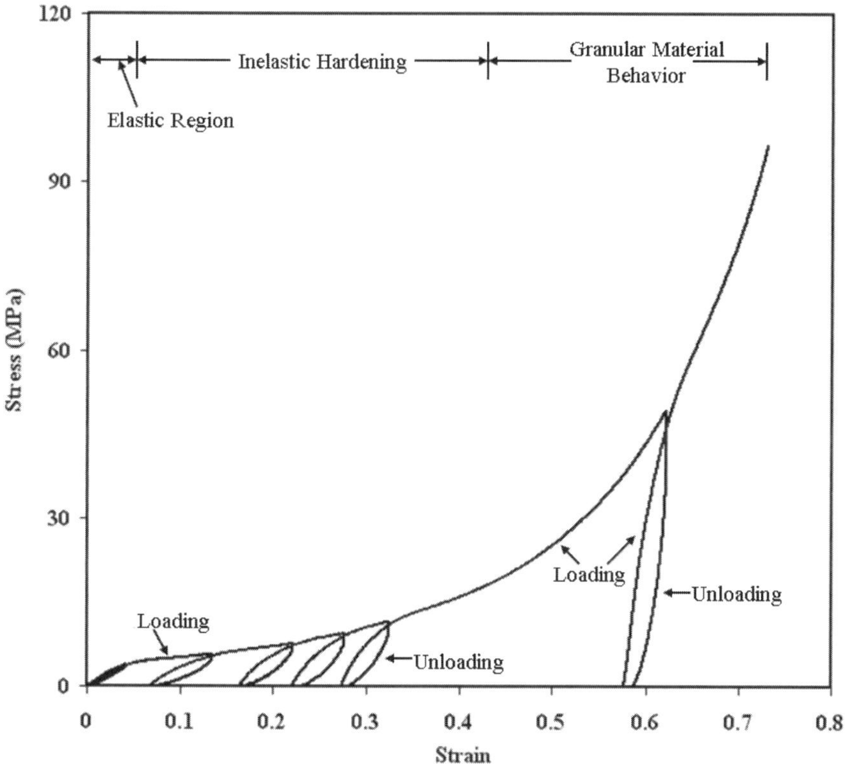

Fig. 10.12. Loading–unloading response of crosslinked silica aerogel

The changes in morphology of x-aerogel material under inelastic compression were evaluated by SEM, as shown earlier in Fig. 10.5. In this case, samples were loaded up to a predetermined strain. Then the load was removed and the samples were analyzed. Curiously, under SEM, the material did not show any noticeable difference in microstructure from its original state to 45% strain. Ultimately, SEM micrograph showed clear signs of collapse of the aerogel mesopores due to compaction, i.e., a nearly total loss of porosity, at failure (77% strain).

Finally, the effects of strain rate and temperature on ultimate compressive strength were evaluated. Figure 10.13 shows stress vs. strain curves at three different compressive strain rates of 0.0035, 0.035, and 0.35 s^{-1}. Although Young's modulus increased significantly with increasing strain rates (128, 160, and 205 MPa at strain rates of 0.0035, 0.035, and 0.35 s^{-1}, respectively), the shape of the overall stress vs. strain curves did not

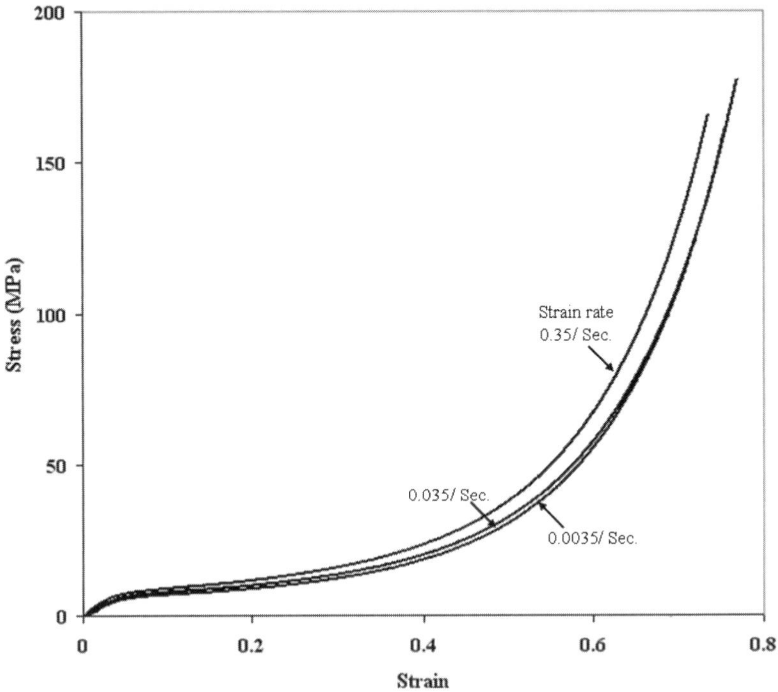

Fig. 10.13. Effect of strain rate on the behavior of crosslinked silica aerogel under uniaxial compression

change significantly. Therefore, within the strain rate range investigated, it can be stated that increasing strain rate does not have a deleterious effect on the energy-absorption behavior of crosslinked aerogels.

Figure 10.14 presents the compressive stress vs. strain curves for crosslinked aerogel samples that were tested at various temperatures. Mechanical response is essentially invariant of the temperature in the range between 21 and –55°C. However, the material stiffens significantly (the elastic modulus increases to ~450 MPa) and suffers premature compressive failure at cryogenic temperatures, e.g., –196°C. It should be noted that, due to a shortage of x-aerogel specimens, only one sample was available for this test. Additional cryogenic temperature tests are planned in the near future.

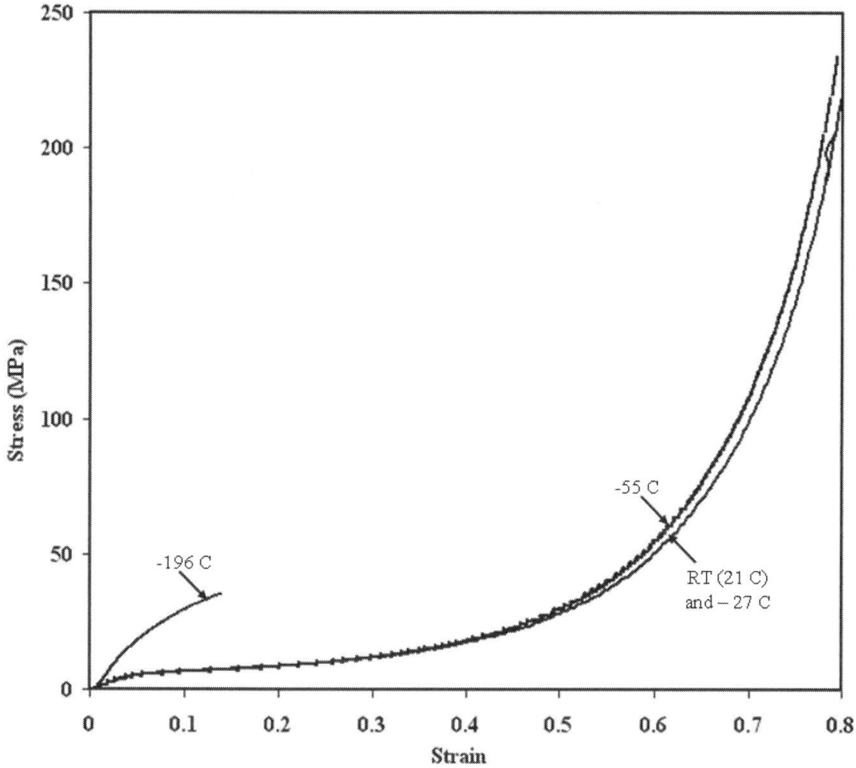

Fig. 10.14. Effect of temperature on the behavior of crosslinked silica aerogel under uniaxial compression

10.5 Numerical Simulation of Compression Experiment Using PFC^{3D}

Numerical simulation of compression experiment of crosslinked silica aerogel has been performed by using a PFC^{3D}, as described earlier. First, a cluster assemblage of spherical particles was generated, as shown in Fig. 10.15, using the DLCA algorithm. This algorithm randomly generates nonintersecting spherical particles according to a Gaussian distribution function. The DLCA provides the radius and location of each particle in a three-dimensional cubic space with a specified porosity. These values are then exported to PFC^{3D} (DEA code) as input data files.

The following section summarizes the necessary procedures or steps involved in a typical compression simulation using PFC^{3D}. Two separate data files can be used and executed in the command-driven mode during the simulation process. The first data file performs the following operations:

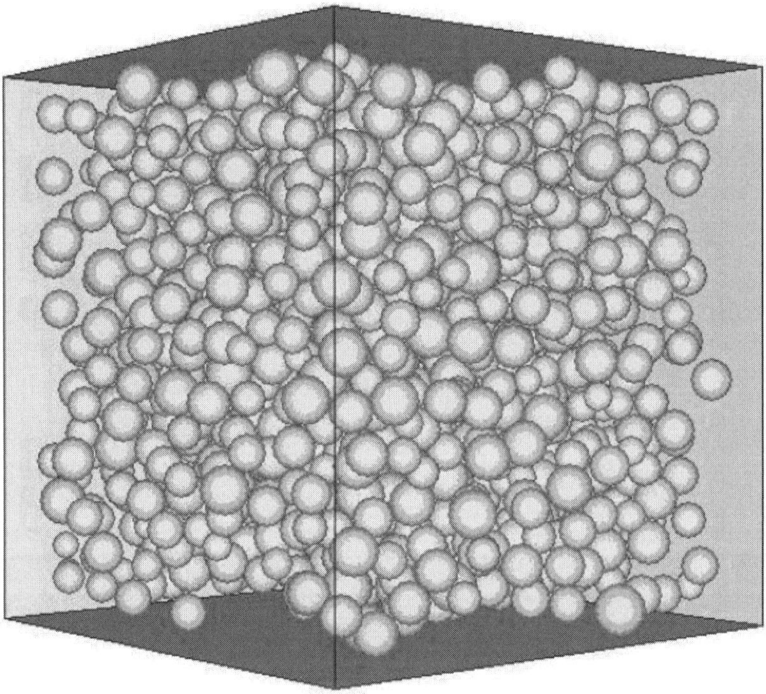

Fig. 10.15. Cluster assemblage of spherical particles generated by DLCA algorithm

- Define the boundary of the model, a three-dimensional cubic space, by defining the geometric locations of six walls or sides.
- Insert the cluster of nonintersecting spherical particles (balls) created by DLCA within the model boundary.
- Assign properties of spherical balls and boundary walls, which include ball density, friction coefficient, contact stiffness in normal and shear directions, contact-bond strength in normal and shear directions (in force unit), parallel-bond stiffness in normal and shear directions, parallel-bond strength in normal and shear directions, and wall stiffness in normal and shear directions.
- Apply gravitational force, if required.
- Assign the computational cycle number.
- Perform a number of calculation steps which monitor the ball movements and the change in unbalanced force as the assemblage is compacted. An equilibrium state has been reached when the unbalanced force converges to a value lower than the specified tolerance limit.
- Save the equilibrium state of assemblage that will be used by the second data file.

When the particle assemblage has reached an equilibrium state, the following operations are performed by the second data file:

- Restore the equilibrium state of particle assemblage confined by six fixed walls.
- Delete the fixed walls and define new top and bottom walls (moving) to compress the particle assemblage.
- Define the formulation for calculating average stress and strain based on force exerted on walls and wall movement, respectively.
- Assign wall properties, which include contact stiffness in normal and shear directions, friction coefficients, and wall velocities, to apply load in preferred direction.
- Save the history of average stress and strain. Plot the result and provide output file in desired format.

PFC^{3D} can be used in a fully predictive mode where enough data of high quality are available, or it can be used as a numerical laboratory to test design ideas in a data-limited system. Most of the aerogel modeling in this study is based on limited available data. In this study, various particle clusters were generated in DLCA varying aerogel initial cluster porosity from 70 to 80%. When the particle assemblage had been formed, it was compressed by two walls moving at opposite directions. All properties related with PFC^{3D} simulation are presented in Table 10.4 with their nominal values.

Table 10.4. Parameters used in PFC^{3D} simulation with nominal values

Properties	Description	Value
	Mass density	480 kg m^{-3}
	Contact stiffness in normal direction	150 N m^{-1}
	Contact stiffness in shear direction	150 N m^{-1}
	Radius multiplier for parallel bond	30×10^{-9}
	Parallel-bond normal stiffness per unit area	30×10^{15} N m^{-3}
Ball and bond	Parallel-bond shear stiffness per unit area	30×10^{15} N m^{-3}
	Parallel-bond normal strength	190×10^{6} Pa
	Parallel-bond shear strength	190×10^{6} Pa
	Contact-bond normal strength	3×10^{4} N
	Contact-bond shear strength	3×10^{4} N
	Friction coefficient	0.5
Fixed wall	Normal stiffness	1×10^{6} N m^{-1}
	Shear stiffness	1×10^{6} N m^{-1}
Moving wall	Normal stiffness	1×10^{15} N m^{-1}
	Shear stiffness	1×10^{15} N m^{-1}
	Friction coefficient	0.001
	Velocity in vertical (z) direction	0.1×10^{-10} m per step
Porosity	Initial cluster porosity	70–80%
Other Parameters	Number of particles (balls)	2,000–2,600
	Type of particles distribution	Gaussian
	Cell (3D box) size	100 × 100 × 200 nm

Snapshots of PFC^{3D} simulation at different compression stages are shown in Fig. 10.16. The bonds between particles break as the compressive load increases. Subsequently, secondary particles located at dead ends separate from the core assembly, as shown in Fig. 10.16. Numerical results for stress vs. strain were calculated and compared with experimental values. An excellent agreement was observed between the numerical and experimental results as shown in Fig. 10.17.

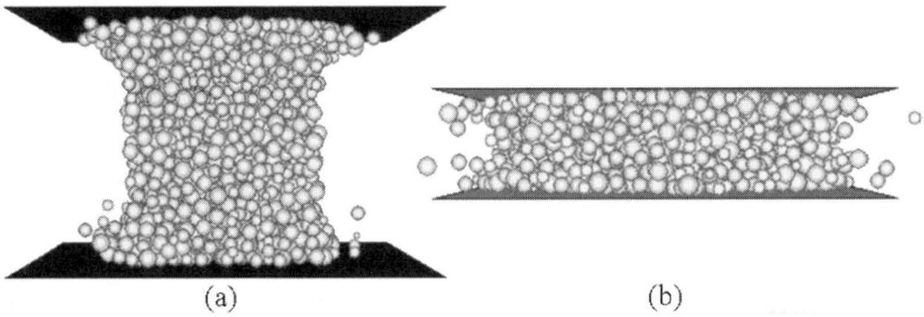

Fig. 10.16. Uniaxial compression simulation of aerogel (**a**) at 45% strain and (**b**) at 77% strain

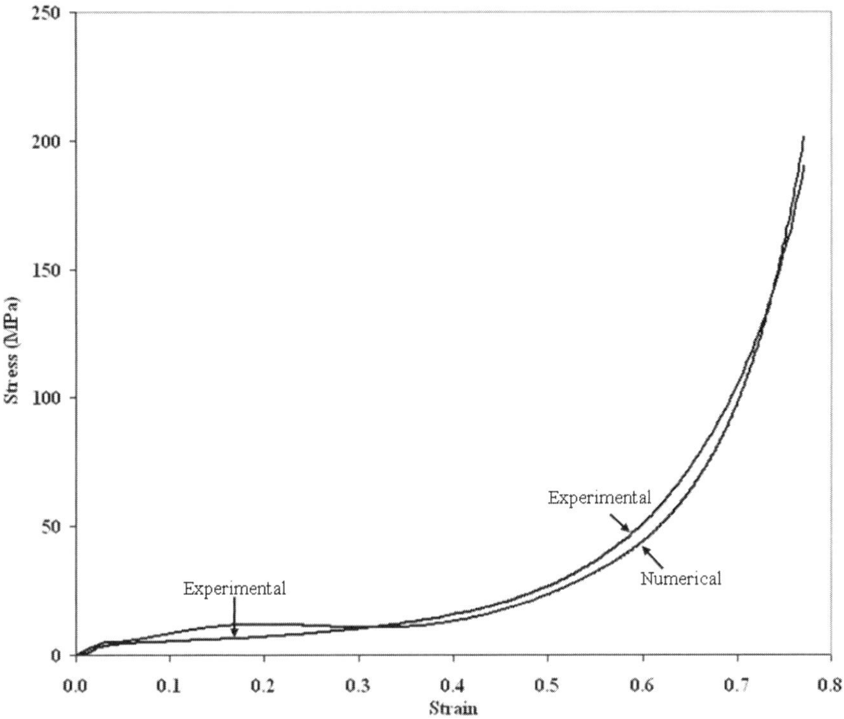

Fig. 10.17. Comparison between the experimental and numerical results

10.6 Parametric Sensitivity Analyses

Parametric sensitivity analyses were performed on the numerical model by varying one parameter at a time, while keeping the other parameters constant. The key parameters affecting the behavior of the numerical model were identified by performing a few trial cases. The particle contact stiffness was found to be the most important parameter affecting the numerical results. Figure 10.18 shows the effect of increasing the contact stiffness, assuming both normal and shear stiffness are equal, on the stress vs. strain behavior. Not surprisingly, the modulus predicted by the numerical model was found to increase with increasing particle contact stiffness.

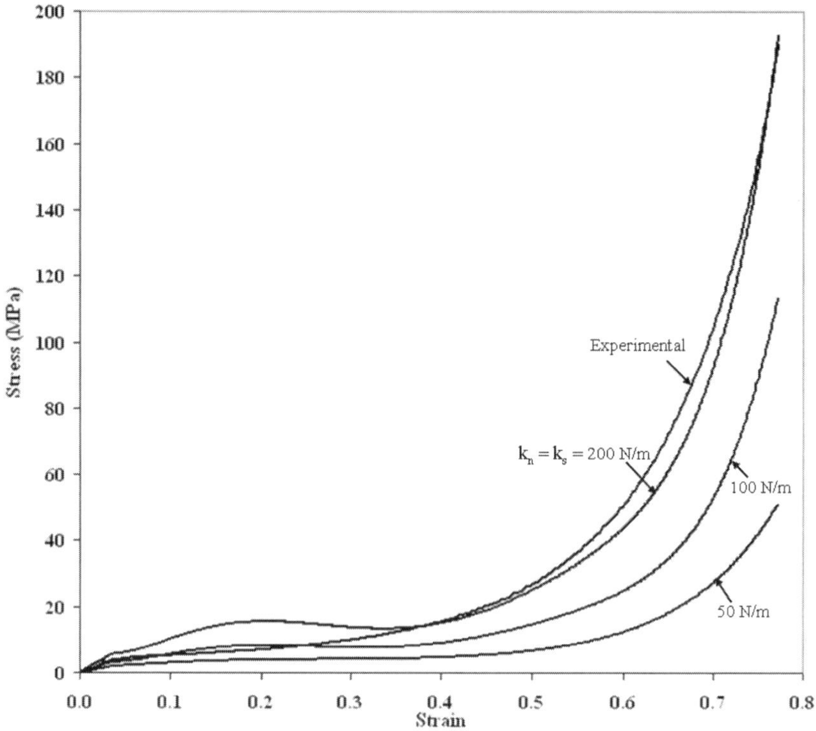

Fig. 10.18. Parametric sensitivity analyses with changing particle contact stiffness (N m^{-1}) equally in normal and shear directions

Parametric sensitivity analyses were then performed by changing the particle contact stiffness independently along the normal and shear directions. First, contact stiffness in the normal direction (k_n) was held constant at 150 N m^{-1} while the stiffness in shear direction (k_s) was varied. As shown in Fig. 10.19, numerical results did not change significantly with changes in the normal to shear stiffness ratio. However, a noticeable

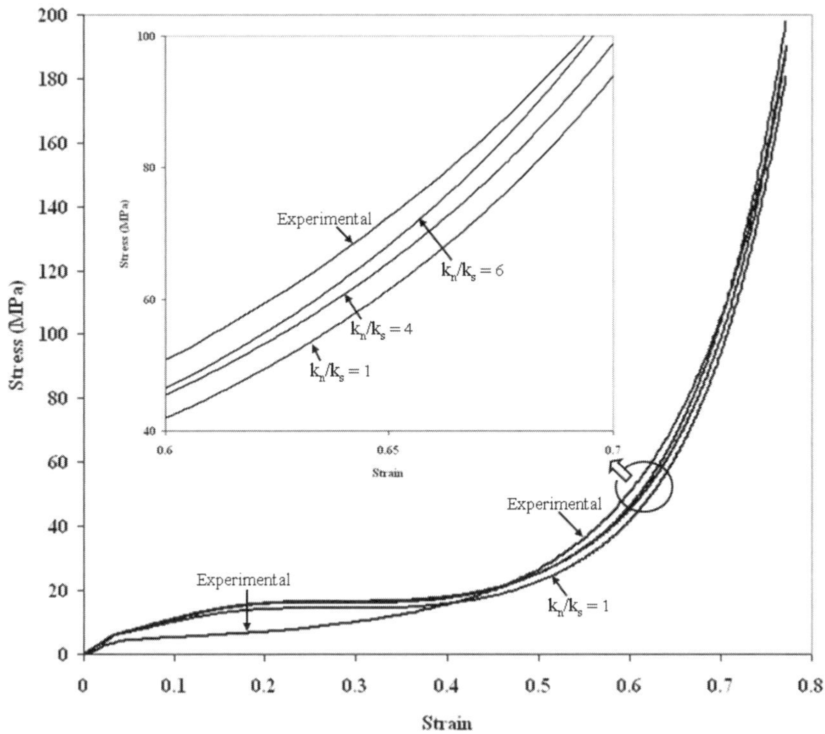

Fig. 10.19. Parametric sensitivity analyses with changing particle contact stiffness in shear direction while stiffness in normal direction was kept constant at 150 N m^{-1}

change was observed in stress vs. strain behavior when the particle contact stiffness in the shear direction (k_s) was held constant at 150 N m^{-1} and stiffness in the normal direction (k_n) was varied. As shear to normal stiffness ratio was increased, the numerical model was found to predict a significantly softer response, as shown in Fig. 10.20.

Particles used in PFC are not bonded among themselves by default. Particles in contact within a given range become bonded by assigning a bond strength. A bond breaks when force equals or exceeds the bond strength. In this parametric sensitivity study, the numerical simulation was repeated with varying bond strength. Bond breakage was found to be significantly reduced with increasing parallel-bond strength, as shown in Fig. 10.21, where each black patch represents a broken bond.

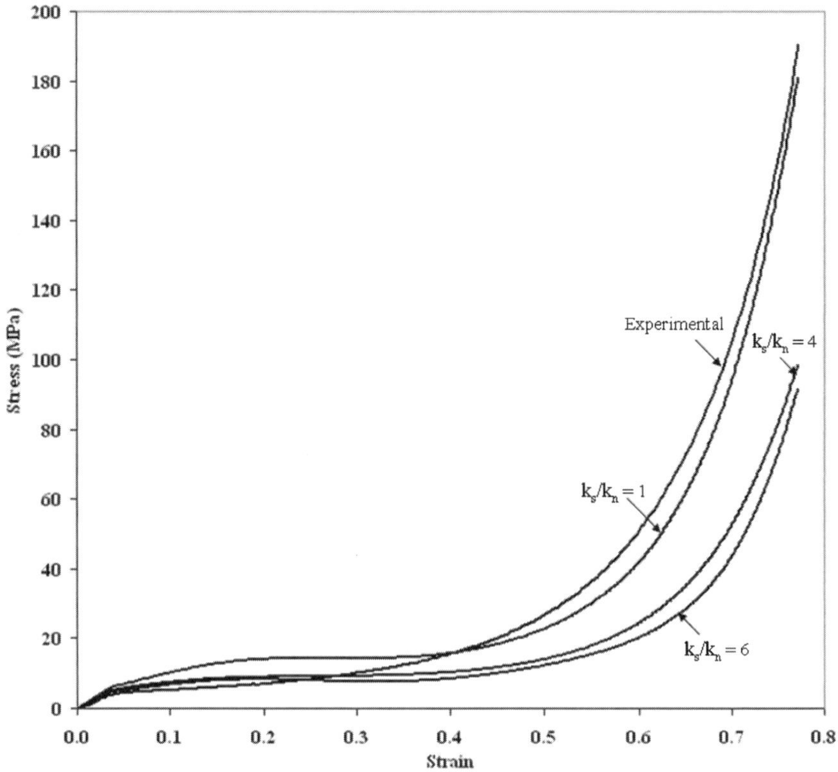

Fig. 10.20. Parametric sensitivity analyses with changing particle contact stiffness in normal direction while stiffness in shear direction was kept constant at 150 N m^{-1}

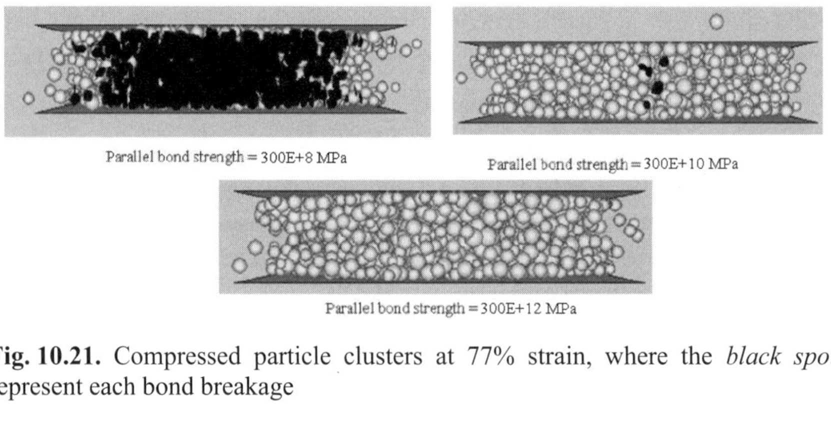

Fig. 10.21. Compressed particle clusters at 77% strain, where the *black spots* represent each bond breakage

Parametric sensitivity analyses were also performed to investigate the stress–strain response with changes in the particle frictional coefficient ranging from 0.01 to 1.0. The slope of the stress vs. strain curve in the granular region was found to increase as the frictional coefficient increased, as shown in Fig. 10.22. As the particle frictional coefficient increases, more force is required to compress the cluster, thereby stiffening the numerical model. Not surprisingly, there was very little effect of increasing particle mass density on the response of the stress vs. strain curve as shown in Fig. 10.23.

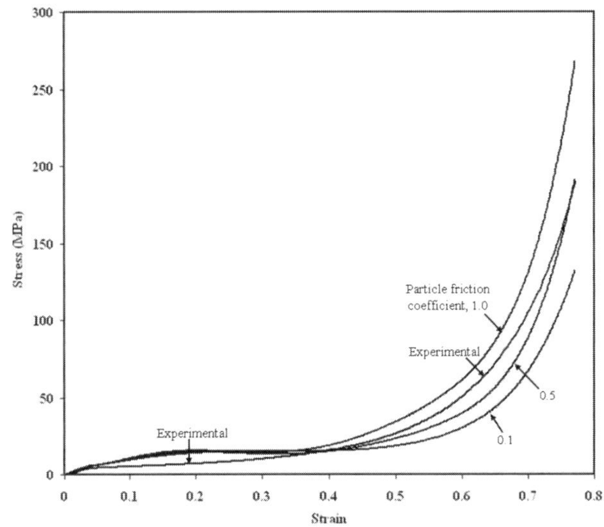

Fig. 10.22. Parametric sensitivity analyses with changing particle frictional coefficient

As mentioned earlier, the network connectivity of mesoporous aerogel structures consisting of nanosize particles can be investigated by evaluating their fractal dimension from initial cluster geometry. In this study, effort was directed to determining the fractal dimension of an aerogel structure by calculating the total number of particles within a specified spherical region [5], as expressed by (10.5) below:

$$d_{\mathrm{f}} = \frac{\mathrm{d} \ln N(r)}{\mathrm{d} \ln r}. \tag{10.5}$$

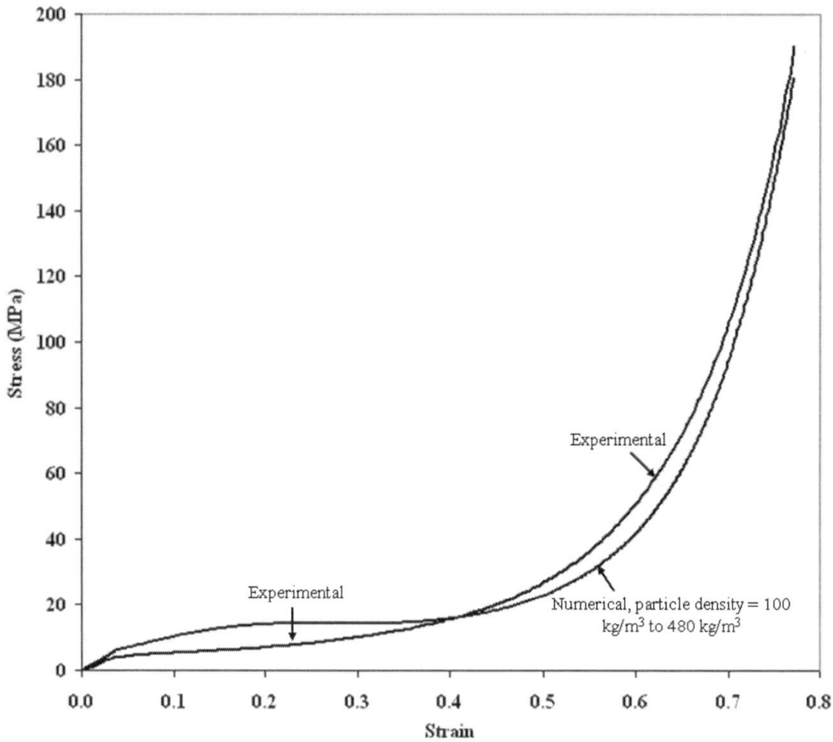

Fig. 10.23. Parametric sensitivity analyses with changing particle mass density (kg m^{-3})

Here, d_f represents the fractal dimension and $N(r)$ represents the total number of particles within a sphere of radius r, as shown in Fig. 10.24.

In Fig. 10.25, the total number of particles ($N(r)$) surrounding a reference point as a function of radial distance (r) in a double-logarithmic scale have been plotted. The different curves correspond to different initial cluster porosities, ranging from 70 to 80%, with clusters generated using DLCA. The fractal dimension (d_f) is evaluated as the slope of each line, as described in (10.5), and is indicated at the top of Fig. 10.25. This figure shows that aerogel structures with different initial cluster porosities have different fractal dimension. Therefore, each aerogel structure signifies a distinct network connectivity among the secondary nanoparticles.

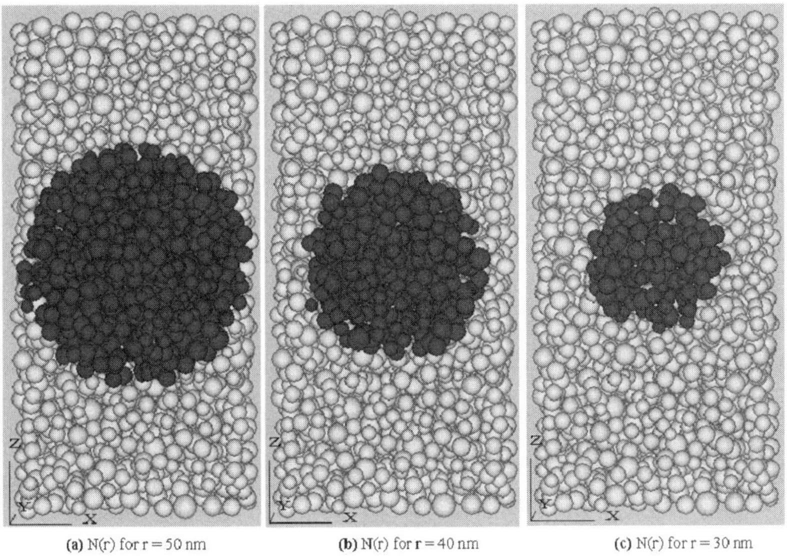

Fig. 10.24. Plan view for identifying secondary nanoparticles (*dark in color*) in aerogel cluster within varying radii of spherical regions

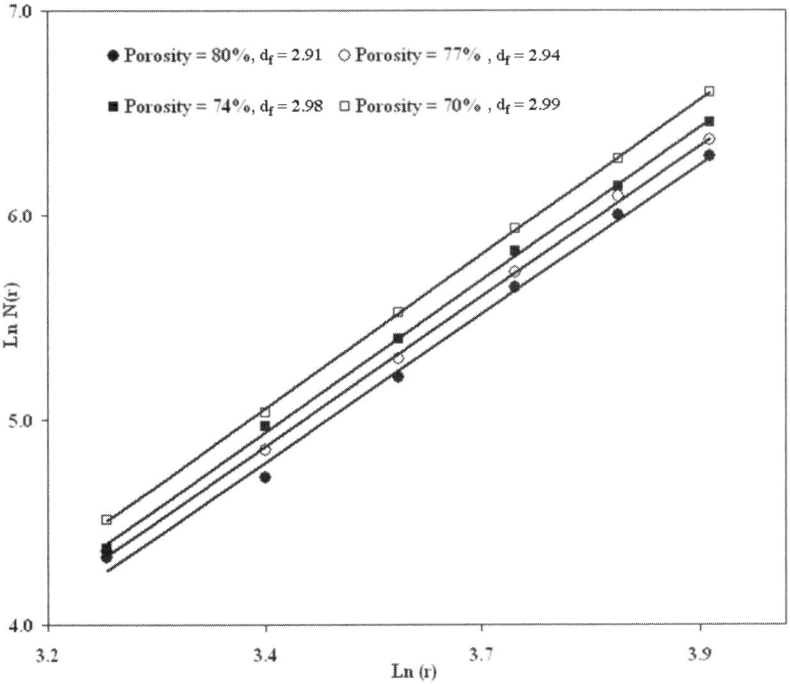

Fig. 10.25. Aerogel secondary particles distribution as a function of radial distance for clusters generated from DLCA with different initial cluster porosity

The fractal dimension (d_f) of the aerogel structure was found to decrease as the initial cluster porosity increases, as shown in Fig. 10.26. The figure confirms that aerogel structures become more fractal with increasing cluster porosity, as might be expected. This figure also reveals that the fractal dimension of an aerogel structure will asymptotically approach 3 with decreasing initial cluster porosity. As porosity represents the extent of void space, the aerogel cluster density decreases with increasing cluster porosity, which is intuitive. The trends observed in this study of fractal dimension of mesoporous aerogel structure are corroborated by similar observations for silica glass, reported in [5].

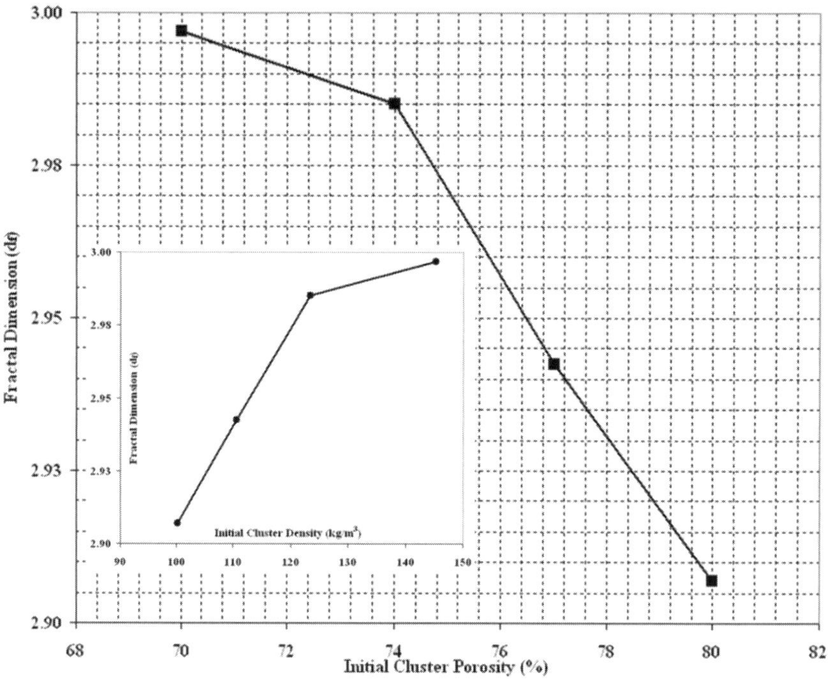

Fig. 10.26. Variation of fractal dimension of aerogel structure with initial cluster porosity (the *inset* shows the variation of fractal dimension with initial cluster density)

Finally, parametric sensitivity analyses were performed with changing x-aerogel cluster porosity. As shown in Fig. 10.27, numerical simulation of x-aerogel with 80% initial cluster porosity agrees well with the experimental response up to a strain level of 35%. At higher strain level, numerical results with 70% initial cluster porosity closely match with the experimental curve. During the compression experiment, the cluster porosity should decrease gradually, as depicted earlier in Fig. 10.5. The numerical model was unable to exactly replicate this porosity variation. Therefore, a slight mismatch between the experimental and numerical model, as shown in Fig. 10.27, is not surprising. The numerical model was found to offer a softer response with increasing cluster porosity, as might be expected.

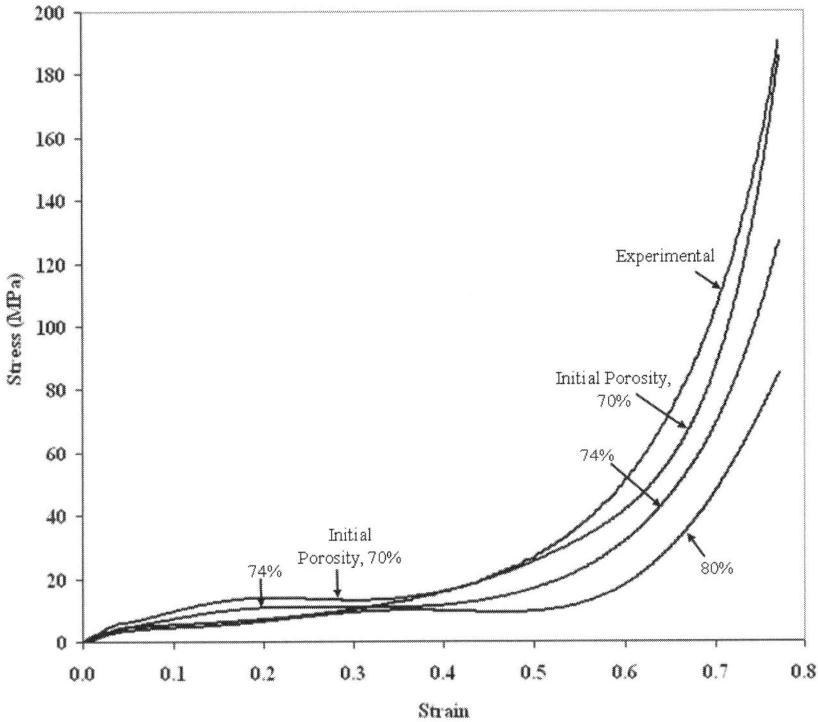

Fig. 10.27. Sensitivity analyses with changing initial cluster porosity

A summary of the parametric sensitivity analyses, discussed above, is presented in concise form in Table 10.5 below.

Table 10.5. Parametric sensitivity analyses through numerical simulation of aerogel

Variable parameter	Variation range (% change)	Variation of max. stress at 77% strain level (% change)
Particle contact stiffness, equal increment both in normal (k_n) and shear (k_s) directions	50–200 N m^{-1} (300%)	55–195 MPa (255%)
Ratio of normal stiffness to shear stiffness (k_n/k_s), where normal stiffness (k_n) was kept constant at 150 N m^{-1}	1–6 (500%)	No significant change
Ratio of shear stiffness to normal stiffness (k_s/k_n), where shear stiffness (k_s) was kept constant at 150 N m^{-1}	1–6 (500%)	190–90 MPa (–53%)
Bond strength	300×10^8 to 300×10^{12} MPa	Significantly reduced bond breakage
Friction coefficient	0.1–1.0 (900%)	130–270 MPa (108%)
Particle mass density	100–480 kg m^{-3} (380%)	No significant change
Cluster porosity	70–80% (14.3%)	185–85 MPa (–55%)

Only one parameter was changed at a time, while other parameters were held constant (see Table 10.4).

10.7 Conclusions

Aerogel is reported as one of the lightweight materials which can be used for different engineering applications. This chapter briefly describes the improvement of mechanical properties of silica aerogel by crosslinking its nanosize secondary particles with polymeric tethers. However, the stiffness and strength of x-aerogel material strongly depend on its microstructural features, such as mesoporous cluster assemblies and particle connectivity, which can be evaluated by determining the fractal dimension. Each particle cluster generated from DLCA signified distinct network connectivity in terms of fractal dimension. A mesoporous aerogel structure was found to become more fractal with increasing initial cluster porosity.

Uniaxial compression experiment was conducted on x-aerogels to collect the basic mechanical properties. During the compression test, x-aerogels were found to behave as linearly elastic under very small strains (<4%) and then exhibited yielding followed by densification and inelastic hardening. The compressive strength of x-aerogels was found to be greatly improved to 186 MPa compared with native silica aerogels of 4 MPa, with only a 2.5 times density increment. The specific compressive strength of x-aerogels was also found to be higher than most other conventional materials. Effects of strain rate and temperature on the compression experimental response were also studied. The mechanical behavior of x-aerogels was unaffected within the strain rate range investigated.

In this study, numerical simulations were performed using particle mechanics to develop the structure–property relationship of nanostructured x-aerogel. First, a cluster of nanoparticles was generated in DLCA and then imported into PFC^{3D} for simulating the compression experiment. Good agreement was observed between the experimental and numerical results in terms of stress and strain data. The numerical model was then used to conduct a series of parametric sensitivity analyses by varying different parameters, including particle contact stiffness, bond strength, friction coefficient, particle mass density, cluster density, and cluster porosity. The particle contact stiffness was found to be an important factor affecting the failure mechanism. The numerical model offered stiffer response as the particle contact stiffness increased, predominantly in the normal direction. Reduced bond breakage was observed during the compression as the bond strength increased. The numerical model offered stiffer response with increasing particle frictional coefficient. The secondary particle density itself was found ineffective in the compression response. As porosity represents the amount of void space, the numerical model of the aerogel particle clusters predicted a softer response with increasing initial cluster porosity, and, consequently, with decreasing aerogel fractal dimension.

Acknowledgments

This study was funded by a research contract from NASA-Glenn Research Center. The contributions of Mr. Nilesh Shimpi, Mr. Atul Katti, Dr. Nicholas Leventis, and Dr. Hongbing Lu are hereby acknowledged.

References

1. Bouaziz J, Bout-ret D, Sivade A, Grill C (1992) Doping of partially densified aerogels: impregnation by solutions or by xerogel. Journal of Non-Crystalline Solids 145, 71–74
2. Ferry JD (1980) Viscoelastic Properties of Polymers, 3rd edition, Wiley: New York
3. Itasca (2003) Particle flow code in 3 dimensions. Version 3.0, User's Manual, Itasca Consulting Group
4. Katti A, Shimpi N, Roy S, Lu H, Fabrizio EF, Dass A, Capadona LA, Leventis N (2006) Chemical, physical, and mechanical characterization of isocyanate crosslinked amine-modified silica aerogels. Chemistry of Materials 18, 285–296
5. Kieffer J, Angell CA (1998) Generation of fractal structures by negative pressure rupturing of SiO_2 glass. Journal of Non-Crystalline Solids 106, 336–342
6. Leventis N, Sotiriou-Leventis C, Zhang C, Rawashdeh A (2002) Nanoengineering strong silica aerogels. Nano Letters 2, 957–960
7. Meakin P, Family F (1987) Structure and dynamics of reaction-limited aggregation. Physical Review A 36, 5498–5501
8. Roy S, Shimpi N, Katti A, Lu H, Rahman M (2006) Mechanical characterization and modeling of isocyanate-crosslinked nanostructured silica aerogels. In: 47th AIAA/ASME/ASCE/AHS/ASC Structures, Structural Dynamics, and Materials Conference, 14th AIAA/ASME/AHS Adaptive Structures Conference, 7th AIAA Gossamer Spacecraft Forum, 2nd AIAA Multidisciplinary Design Optimization Specialist Conference, 8th AIAA Non-Deterministic Approaches Conference, Newport, RI
9. Ruben GC, Hrubesh LW, Tillotson TM (1995) High resolution transmission electron microscopy nanostructure of condensed-silica aerogels. Journal of Non-Crystalline Solids 186, 209–218

Chapter 11: Multiscale Modeling of the Evolution of Damage in Heterogeneous Viscoelastic Solids

David H. Allen and Roberto F. Soares

University of Nebraska-Lincoln, Lincoln, NE 68520, USA

11.1 Introduction

It has long been recognized that one of the primary failure modes in solids is due to crack growth, whether it be a single or multiple cracks. It is known, for instance, that Da Vinci [14] proposed experiments of this type in the late fifteenth century. Indeed, modern history is replete with accounts of events wherein fracture-induced failure of structural components has caused the loss of significant life. Such events are common in buildings subjected to acts of nature, such as earthquakes, aircraft subjected to inclement weather, and even human organs subjected to aging. Therefore, it would seem self-evident that cogent models capable of predicting such cata- strophic events could be utilized to avoid much loss of life. However, despite the fact that such events occur regularly, the ability to predict the evolution of cracks, especially in inelastic media, continues to elude scientists and engineers. This appears to be at least due, in part, to two as yet unresolved issues (1) there is still no agreed upon model for predicting crack extension in inelastic media and (2) the prediction of the extension of multiple cracks simultaneously in the same object is as yet untenable.

While it would be presumptuous to say that the authors have resolved these two outstanding issues, there is at least a glimmer of hope that these two issues may be resolved by using an approach not unlike that proposed herein. This chapter outlines an approach for predicting the evolution of multiple cracks in heterogeneous viscoelastic media that ultimately leads to failure of the component to perform its intended task. Examples of such components would include geologic formations, cementitious roadways, human organs, and advanced structures, including composite aircraft components and defensive armor such as that used on tanks. Implicit in the need for deploying such a model as that proposed herein are the following two requirements (1) at least some subdomain of the medium must be inelastic and (2) cracks must grow on at least two significantly different length scales prior to failure of the component.

The model that is proposed herein for addressing this problem is posed entirely within the confines of the fundamental assumption embodied in continuum mechanics, i.e., that the mass density of a body is continuously differentiable in spatial coordinates on all length scales of interest so that cracks that initiate on the scale of single atoms or molecules cannot be modeled by this approach, implying that the smallest scale that can be considered is of the order of tens to hundreds of nanometers.

This chapter opens with a short historical review of developments that have led up to the current state of knowledge on this subject, followed by a detailed description of the methodology proposed by the authors for addressing this problem. This will be followed by a few example problems that are meant to illustrate how the approach described herein can be utilized to make predictions of practical significance.

11.2 Historical Review

The discipline of mechanics, the study of the motion of bodies, dates to the ancients. Chief among these is Archimedes [32], who enunciated the principle of the lever among other achievements. However, the first systematic study of the mechanics of bodies is attributed to Galileo [19] in the early seventeenth century. These accomplishments were not withstanding, it was not until the early nineteenth century that concerted efforts were made to study the motions of *deformable* bodies within the context of continuum mechanics. These efforts appear to have been initiated with the study of plates by Germain [20] and were followed shortly thereafter by the seminal papers by Navier [26] and Cauchy [10] on the prediction of deformations in elastic bodies. These formulations utilized Newton's laws of motion [27], together with definitions of strain and the necessary idea of the constitution of an elastic body, first enunciated by Hooke [8], a contemporary of Newton. These initial formulations did not encompass the notion of dissipation of energy, so the prediction of failure was not a component of these models. However, over the course of the succeeding century, the formulation of fundamental concepts of thermodynamics led to the first cogent theory of fracture by Griffith [21] in 1920.

Griffith proposed that a crack would extend in an elastic body whenever

$$G \geq G_C, \tag{11.1}$$

where G is the energy released per unit area of crack produced and G_C is assumed to be a material constant called the *critical energy release rate*.

Though succeeding progress has been slow to develop, this monumental proposition seems to have been the key step that was necessary to begin to make somewhat accurate predictions of crack growth. Two obstacles lay in the way before the usefulness of Griffith's proposition could be ascertained. The first obstacle was centered around the right-hand side of inequality (11.1): how to measure the material property required to make cogent predictions. The answer to this question was suggested in a paper by Rice [29] and proven mathematically a decade later by Gurtin [22]. Subsequently, techniques have been developed for quite accurately measuring the critical energy release rate for a broad range of materials. The other obstacle arose due to the left-hand side of inequality (11.1): how to accurately *calculate* the available energy in a body necessary to produce new crack surface area. This issue is complicated by the fact that, in an imaginary elastic body, it is necessary for the stresses at a crack tip to be singular in order for there to be a nonzero energy available for crack extension. This problem has been studied in significant detail over the past half-century with some success. However, it would be presumptive to say that the subject is resolved; because in reality, it is not possible for the stresses at a crack tip to be singular.

Initial experimental results for brittle materials indicated that Griffith's proposition was accurate. However, when experimental results were obtained for ductile materials, such as crystalline metals, experimental results compared less favorably to predictions. For some time, efforts were made to improve upon the calculations of the available energy for crack growth in ductile materials; and to make these calculations, researchers turned to the more advanced constitutive theory, such as that embodied in plasticity theory [24]. However, it is now widely understood that Griffith's proposition is not accurate for some ductile materials due to the fact that energy dissipation occurs in a variety of ways other than crack extension, and in ways that depend on the history of loading of the body. In these circumstances, it may be more appropriate to envision the critical energy release rate G_c as a *history-dependent material property* rather than a material constant. In the meantime, other approaches have been developed, such as *cohesive zone models* [7, 16], that do not require the concept of a critical energy release rate to predict crack extension (although energy release rates *can* be calculated by this approach); and these have met some success in modeling crack growth in ductile media.

Simultaneously, over the past half-century, two more or less contiguous developments have led to significant improvements in calculating the available energy for crack extension in both elastic and a variety of inelastic (including elastoplastic, viscoplastic, and both linear and nonlinear viscoelastic) media. One of these developments was the rise of the high-speed computer, whose power has made it possible to make billions of calculations of the type needed to estimate the energy required for crack extension, even in bodies of quite complicated geometry and material makeup. The other development is the finite element method, which grew out of the so-called *flexibility method* used in the aerospace and civil engineering communities in the first half of the twentieth century. This methodology came under scrutiny by the applied math community after World War II and was subsequently identified as a member of the *method of weighted residuals* for solving sets of coupled partial differential equations. Today, quite a few finite element codes are available for calculating stresses in both elastic and inelastic bodies.

11.3 The Current State of the Art

While significant progress has been made in the ability to predict when a crack will grow and where it will go, the subject has not yet been completely closed. As mentioned above, there is still no completely agreed upon way of predicting when a crack will grow in a ductile medium. Furthermore, when there are multiple cracks, the computational requirements needed to utilize the finite element method go up significantly. Even with today's high-speed computers, it is not yet possible to predict, with sufficient accuracy, the available energy for crack extension for the physical circumstance wherein a few cracks are simultaneously imbedded in a body. And yet, *it is known from experimental observation that many, many cracks can occur simultaneously in all manner of structural components and that these cracks can coalesce into a single crack that leads to structural failure*. It can be said here without reservation that the state of the art of fracture mechanics is not to the point where the evolution of large numbers of cracks of evenly distributed sizes in a single inelastic body can be predicted. However, there is one case involving multiple cracks that may be a tenable problem at this time. That is the case wherein the cracks in the body are distributed by size into widely separated length scales, with a small number of cracks observed at the largest scale, termed the global or macroscale, upon which failure ultimately occurs. This, then, is the subject of this chapter: *to develop a modeling approach for predicting the evolution of multiple cracks on widely separated length scales in heterogeneous viscoelastic bodies*. To affect a solution technique, the problem will be solved by using the concept of *multiscaling*, as described below.

The concept of multiscaling in continuous media is an old one that is based on classical elasticity theory. In this approach, constitutive properties of the elastic object are required to predict deformations, stresses, and strains in a structural part. To obtain these properties, a *constitutive test* is performed on a specimen made of the material of interest. For the test to be valid, not only should the state of stress and strain in the body be measurable by observing boundary displacements of the object when it is loaded, but also it is necessary that the object be "statistically homogeneous." This is a sometimes ill-defined term; but what is meant by the term is that any asperities in the test specimen are several orders of magnitude smaller than the specimen itself, so that the spatial variations in the magnitudes of the observed stresses and strains in the test specimen are small compared to the mean stresses and strains observed during the test to obtain the constitutive properties. This type of experiment essentially embodies the concept of multiscaling. By *assuming* that the response of the test specimen is statistically homogeneous, the smaller length scale on which asperities might be observed is separated from the larger scale of the structural component.

This separation of length scales has long been understood, having been considered in some detail by nineteenth-century scientists such as Maxwell and Boltzmann, as well as in the early twentieth century by Einstein, to explain macroscale observations (visible to the naked eye) of molecular phenomena in liquids and gases. Capitalizing on this approach, a number of researchers developed rigorous mathematical techniques in the 1960s for bounding the elastic properties of multiphase elastic continua [17, 23, 25]. Such methods earned the descriptor "micromechanics," although this designator is perhaps not the best terminology, since the observed heterogeneity is often not microscopic. Nevertheless, this approach has gained acceptance as a means of estimating the elastic properties of objects composed of multiple elastic phases which are small compared to the size of the body of interest. The advantage of such models (over the experimental approach described above) for measuring elastic properties is that the volume fractions (as well as shape, orientations, etc.) of the constituents can be changed without the necessity of redoing sometimes costly constitutive experiments. Thus, this approach, that inherently involves multiscaling, has become quite popular in the engineering field. Furthermore, because the resulting body is elastic, the analyses on the smaller and larger scales can be performed independently of one another, so that no coupling between the two-length scales is necessary.

In the case of inelastic media, this, unfortunately, is not the case. When materials undergo load-induced energy dissipation, such as that occurs in elastoplastic or viscoelastic media, the micromechanical description does not decouple from the analysis to be performed on the larger scale. In other words, the material properties become spatially variable and dependent on the load history, so that coupling between the macro- and microscale is unavoidable. Therefore, it becomes essential to develop modeling approaches that account for this fundamental increase in the level of complexity of the problem if there is to be any hope of achieving accuracy of prediction.

For the better part of the last half of the twentieth century, efforts to account for this complexity in inelastic media centered on development of ever more complicated constitutive theories for the microscale, similar to that used successfully to model heterogeneous elastic media, as described above. This had the pragmatic basis that one could perform a finite element analysis on a single length scale, which was just about the limit that computers of that time could handle. However, as it became apparent that microscale cracking would have to be included in constitutive models of heterogeneous media at the macroscale, efforts began to bog down and become very complicated indeed. To account for observed behavior in test specimens with time-dependent microcracking, more and more (often unexplained) phenomenological parameters had to be introduced into models. This approach developed the name "continuum damage mechanics." It also inherited the unpalatable complication that sometimes many experimentally measured material parameters were required, especially when it became necessary to model *evolving* microcracks.

Enter the twenty-first century and more and more powerful computers. What required a supercomputer 10 years ago now requires only a desktop computer. Therefore, it is now possible to conceive of algorithms that obviate the necessity to perform many complicated experiments at the microscale. Furthermore, these new algorithms have the added advantage that, by performing simultaneous computations on both the micro- and global scales, they possess the flexibility to include heretofore unmanageable design variables at the microscale in the global design process, and without recourse to expensive constitutive testing.

11.4 Multiscale Modeling in Inelastic Media with Damage

In this section, a multiscale model is proposed for predicting the evolution of damage on multiple scales in inelastic media. The formulation is taken from [5].

11.4.1 Microscale Model

Consider an approach proposed herein that can be used on any number of length scales l_μ observed in a solid object. The number of scales n utilized is determined by the physics of the problem on the one hand and the amount of computational speed and size available on the other hand. To that end, consider a solid object with a region wherein microcracks are evolving on the smallest length scale considered l_1, as shown in Fig. 11.1.

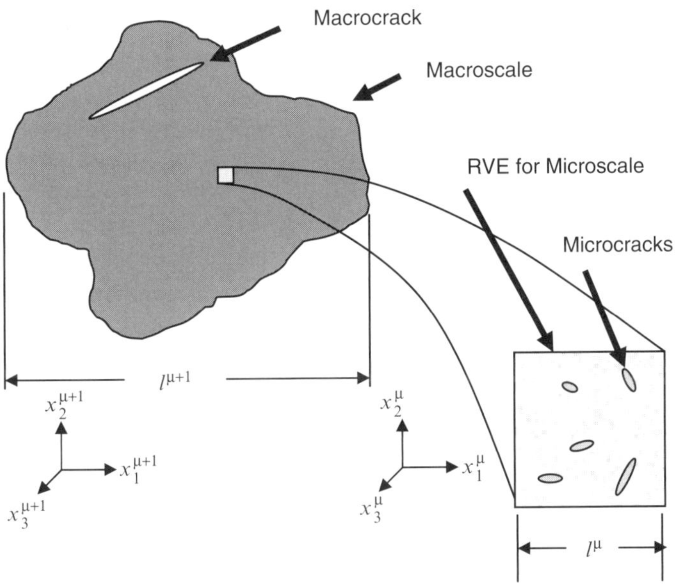

Fig. 11.1. Scale problem with cracks on both length scales

While it is not necessary (or even always correct) that a representative volume of the object on this length scale be accurately modeled by continuum mechanics, it is assumed that this is the case in this chapter to simplify the discussion. Suppose that the object can be treated as linear viscoelastic, again for simplicity, so that the following initial–boundary value problem (IBVP) may be posed.

Conservation of linear momentum

$$\vec{\nabla} \cdot \tilde{\sigma}_{\mu} + \rho \vec{f} = 0, \quad \forall \vec{x}_{\mu} \in V_{\mu}, \tag{11.2}$$

where $\tilde{\sigma}_{\mu}$ is the Cauchy stress tensor defined on length scale μ, ρ is the mass density, and $\vec{f}$ is the body force vector per unit mass. Note that inertial effects have been neglected, implying that the length scale of interest is small compared to the next larger length scale, thus neglecting the effects of waves at this scale on the next scale up. Ultimately, it will be convenient within this context to model waves only on the largest, or global, scale.

Strain–displacement equations

$$\tilde{\varepsilon}_{\mu} \equiv \frac{1}{2}[\vec{\nabla}\vec{u}_{\mu} + (\vec{\nabla}\vec{u}_{\mu})^{\mathrm{T}}], \tag{11.3}$$

where $\tilde{\varepsilon}_{\mu}$ is the strain tensor on the length scale μ and $\vec{u}_{\mu}$ is the displacement vector on the length scale μ. Note that the linearized form of the strain tensor has been taken for simplicity, although a nonlinear form may be employed without loss of generality.

Constitutive equations

$$\tilde{\sigma}_{\mu}(\vec{x}_{\mu}, t) = \Omega_{\tau=-\infty}^{\tau=t}\{\tilde{\varepsilon}_{\mu}(\vec{x}_{\mu}, \tau)\}, \tag{11.4}$$

where $\vec{x}_{\mu}$ is the coordinate location in the object on the length scale μ, which has interior V_{μ} and boundary ∂V_{μ}.

The above description implies that the entire history of strain at any point in the body is mapped into the current stress, which is termed a *viscoelastic material model*. Because only the value of strain (the symmetric part of the deformation gradient is used in this model) is required at the point of interest, it is sometimes called a simple (or local) model [15]. Note that a local elastic material model, such as Hooke's law [32], is a special case of (11.4).

Equations (11.2)–(11.4) must apply in the body, together with appropriate initial and boundary conditions. These are then adjoined with a fracture criterion that is capable of predicting the growth of new or existing cracks anywhere in the object. There are multiple possibilities but, for example, the Griffith criterion given by inequality (11.1) can be taken. The above then constitutes a well-posed boundary value problem, albeit nonlinear due to the crack growth criterion (perhaps as well as the constitutive model (11.4)).

Obtaining solutions for this problem, even for simple geometries, is in itself a difficult challenge, as anyone who has every attempted to do so will attest. Nevertheless, assume that by some means (most likely computational) a solution can be obtained for the boundary conditions, geometry, and precise form of the constitutive (11.4) at hand. Assume, furthermore, that the cracks that are predicted within the model dissipate so much energy locally that they may have further deleterious effects on the response at the next larger length scale. As an example, the so-called *microcracks* may in some way influence the development or extension of one or more macrocracks on the next larger length scale l_2. It will be assumed that the cracks on the next larger length scale are much larger than those on the current scale and that this restriction applies to all length scales for cracks in the object of interest

$$l_{\mu+1} \gg l_{\mu}, \quad \mu = 1,\ldots,n, \tag{11.5}$$

where n is the number of different length scales observed in the solid.

Note that the above restriction is a *necessary condition* (but not sufficient) for the multiscale methodology proposed herein to produce reasonably accurate predictions on the larger length scale(s). If this condition is not satisfied, as in the case of a so-called *localization problem*, then there may indeed be no alternative to performing an exhaustive analysis at a single scale that takes into account all of the asperities simultaneously.

11.4.2 Homogenization Principle Connecting the Microscale to the Macroscale

To perform an analysis of the solid on the next length scale up from the local scale (termed the macroscale herein for simplicity), it is necessary to find a means of linking the state variables predicted on the microscale to those on the macroscale. Of course, the state variables at the microscale are predicted at an infinite collection of material points in the local domain $V_\mu + \partial V_\mu$, so that there is plenty of information available to supply to the next larger length scale. However, the objective herein is to find an efficient means of constructing this link without sacrificing too much accuracy. In other words, it is propitious to utilize the minimum data obtained at the local scale necessary to make a sufficiently accurate prediction at the macroscale. One way is to link the microscale to the macroscale via the use of mean fields. To see how this might work, consider the following mathematical expansion for the macroscale stress in terms of the microscale stress

$$\tilde{\sigma}_{\mu+1} = \bar{\tilde{\sigma}}_\mu + \sum_{j=1}^{\infty} \frac{1}{V_\mu \left|\vec{\bar{x}}\right|^j} \int_{V_\mu} \left(\tilde{\sigma}_\mu - \bar{\tilde{\sigma}}\right) \left|\vec{x}\right|^j \,\mathrm{d}V, \tag{11.6}$$

where

$$\bar{\tilde{\sigma}}_\mu \equiv \frac{1}{V_\mu} \int_{V_\mu} \tilde{\sigma}_\mu \mathrm{d}V \tag{11.7}$$

is the volume averaged (or mean) stress at the microscale, and it is assumed that the local coordinate system is set at the geometric centroid of the microscale volume.

Note that, since the microscale domain $V_\mu + \partial V_\mu$ can be placed arbitrarily within the domain on the next larger length scale $V_{\mu+1} + \partial V_{\mu+1}$, the mean stress $\bar{\tilde{\sigma}}_\mu$ is a continuously varying function of coordinates $\vec{x}_{\mu+1}$ on the next larger length scale $\mu + 1$, as shown in Fig. 11.1. Note also that the terms within the summation in (11.6) represent higher area moments of the stress tensor.

Now, it may be said without loss of generality that microscale conservation of momentum (11.2) also applies to the macroscale (assuming that quasistatic conditions still hold at this length scale)

$$\vec{\nabla} \cdot \tilde{\sigma}_{\mu+1} + \rho \vec{f} = 0, \quad \forall \vec{x}_{\mu+1} \in V_{\mu+1}. \tag{11.8}$$

By using (11.6), it can be shown that

$$\lim_{l_\mu / l_{\mu+1} \to 0} (\tilde{\sigma}_{\mu+1}) = \bar{\tilde{\sigma}} \tag{11.9}$$

and (11.8) reduces to the following:

$$\vec{\nabla} \cdot \bar{\tilde{\sigma}}_\mu + \rho \vec{f} = 0, \quad \forall \vec{x}_{\mu+1} \in V_{\mu+1}. \tag{11.10}$$

The similarity between (11.2) and (11.10) is sufficiently striking that one is immediately tempted to use the same modeling algorithm on both length scales (note that if the momentum terms on the right-hand side of (11.8) are not negligible at the macroscale, then a different algorithm must be used at this length scale, as will be discussed below). This indeed is the approach that will be taken herein; but it must necessarily be said that (11.10) is only exact in the limit, i.e., (11.9) is a *sufficient condition* for (11.10) to be exact. However, in all real circumstances, (11.9) cannot be satisfied, so that some error must necessarily be introduced by utilizing approximate (11.10) in lieu of exact (11.10).

The use of (11.10) is termed herein a "mean field theory" because the higher-order terms that are dropped from (11.6) are essentially higher area moments of the microscale stress. Thus, the macroscale analysis is performed only in terms of the mean stress $\bar{\tilde{\sigma}}$. Note that, in cases wherein there is localization induced by damage or large strain gradients, one or more of the higher-order terms will not be negligible. In this case, a mean field theory is no longer accurate; and a nonlocal approximation (including one or more of the higher-order terms in (11.6)) or even a full field analysis performed simultaneously on all length scales may be necessary to obtain reasonable accuracy. However, the necessity for converting to this procedure may be monitored by calculating the higher-order terms in (11.6) after each time step during the local scale analysis.

Now consider the standard deviation of the microscale stress, given by

$$\tilde{\sigma}_{\mu}^{\mathrm{SD}} \equiv \frac{1}{V_{\mu}} \int_{V_{\mu}} \left(\tilde{\sigma}_{\mu} - \bar{\tilde{\sigma}}_{\mu} \right)^{2} \mathrm{d}V. \tag{11.11}$$

An object in which the standard deviation of all of the state variables is small compared to their respective means is termed, in this chapter, "statistically homogeneous" (this, of course, implies that the effects of any singularities are bounded when integrated over the volume). It can also be shown that, when (11.9) is satisfied, the standard deviation of the microscale stress, given by (11.11), goes to zero. Therefore, in many cases it is sufficient for the object to be statistically homogeneous at the microscale in order for (11.10) to be an accurate representation at the macroscale. One implication of this result is that the microcracks contained within the microscale volume must be statistically homogeneous in location and orientation. If this is not the case, then higher-order moments will necessarily have to be included at the macroscale [1].

Now note that, as long as any tractions on the crack faces are self-equilibrating, (11.2) may be used to show that [6, 9, 13]

$$\bar{\tilde{\sigma}}_\mu = \frac{1}{V_\mu} \int_{\partial V_\mu} (\tilde{\sigma}_\mu \cdot \vec{n}_\mu) \vec{x}_\mu \mathrm{d}S, \tag{11.12}$$

where $\vec{n}_\mu$ is the unit outer normal vector on the local boundary ∂V_μ.

Note that the boundary averaged stress given in (11.12) actually is physically more palatable than the volume averaged stress given in (11.7), as it is commensurate with the original definition of stress, as defined by Cauchy [10], to act on a surface.

The fact that the volume averaged stress is equivalent to the boundary averaged stress is of little importance when there are no cracks. However, when cracks grow and evolve with time, it becomes a very important aspect of the homogenization process, as will now be shown by considering the homogenization process for the strain tensor. It can be shown by careful employment of the divergence theorem that

$$\bar{\tilde{\varepsilon}}_\mu = \tilde{\varepsilon}_{\mu+1} + \tilde{\alpha}_{\mu+1}, \tag{11.13}$$

where

$$\bar{\tilde{\varepsilon}}_\mu = \frac{1}{V_\mu} \int_{V_\mu} \tilde{\varepsilon} \, \mathrm{d}V \tag{11.14}$$

is the mean strain at the local scale,

$$\tilde{\varepsilon}_{\mu+1} = \frac{1}{V_\mu} \int_{\partial V_\mu^{\mathrm{E}}} \frac{1}{2} [\vec{u}_\mu \vec{n}_\mu + (\vec{u}_\mu \vec{n}_\mu)^{\mathrm{T}}] \mathrm{d}S \tag{11.15}$$

is the boundary averaged strain on the initial (external) boundary of the local volume $\partial V_\mu^{\mathrm{E}}$, and

$$\tilde{\alpha}_{\mu+1} = \frac{1}{V_\mu} \int_{\partial V_\mu^{\mathrm{I}}} \frac{1}{2} [\vec{u}_\mu \vec{n}_\mu + (\vec{u}_\mu \vec{n}_\mu)^{\mathrm{T}}] \mathrm{d}S \tag{11.16}$$

is the boundary averaged strain on the newly created (internal) boundary due to cracking $\partial V_\mu^{\mathrm{I}}$ and is called a *damage parameter* [17, 33].

Since kinematic equation (11.15) is consistent with kinetic equation (11.12), it is reasonable to construct constitutive equations at the macro-scale in terms of these two variables, rather than in terms of volume averages. This is in striking contrast to the approach taken when there are no microcracks. In this case, there is no difference between boundary averages and volume averages, as can be seen from the above equations. Nevertheless, using (11.15) and the divergence theorem, it can be shown that

$$\tilde{\varepsilon}_{\mu+1} = \frac{1}{2}[\vec{\nabla}\vec{u}_{\mu+1} + (\vec{\nabla}\vec{u}_{\mu+1})^{\mathrm{T}}], \tag{11.17}$$

which can be seen to be similar in form to local equation (11.3).

The construction of a homogenized macroscale IBVP, similar to that posed in (11.2)–(11.4), is now nearly complete, as (11.10) replaces (11.2), and (11.17) replaces (11.3) at the macroscale. It remains to construct constitutive equations at the macroscale. Where one to utilize the continuum damage mechanics approach, it would be sufficient to simply postulate constitutive equations of the form:

$$\tilde{\sigma}_{\mu+1}(\vec{x}_{\mu+1},t) = \Omega_{\tau=-\infty}^{\tau=t}\{\tilde{\varepsilon}_{\mu+1}(\vec{x}_{\mu+1},\tau), \alpha(\vec{x}_{\mu+1},\tau)\}. \tag{11.18}$$

The precise nature of this equation would then be determined by some curve-fitting scheme either to experimental data provided from macroscale experiments or the predictions made at the local scale. While this approach may be taken, as mentioned above, it removes the input parameters at the local scale from the design process. Therefore, it is preferable to take a multiscaling approach.

Instead, (11.18) is obtained by direct substitution of the microscale constitutive (11.4) into the volume averaged stress (11.7). The precise nature of the resulting equation will depend on the choice of a constitutive model. As an example, consider the case wherein the microscale constitutive behavior is linear nonaging viscoelastic

$$\tilde{\sigma}_{\mu}(\vec{x}_{\mu},t) = \int_{-\infty}^{t} \tilde{\tilde{E}}_{\mu}(\vec{x}_{\mu},t-\tau)\frac{\partial\tilde{\varepsilon}_{\mu}(\vec{x}_{\mu},\tau)}{\partial\tau}\mathrm{d}\tau, \tag{11.19}$$

where $\tilde{\tilde{E}}(\vec{x}_{\mu},t)$ is the relaxation modulus at the microscale.

Direct substitution of (11.19) into (11.7), and subsequent careful utilization of (11.6), (11.9), and (11.12)–(11.16), will result in a constitutive description at the macroscale that is of the following form [31]

$$\tilde{\sigma}_{\mu+1}(\vec{x}_{\mu+1},t) = \int_{-\infty}^{t} \tilde{\tilde{E}}_{\mu+1}(\vec{x}_{\mu+1},t-\tau)\frac{\partial\tilde{\varepsilon}_{\mu+1}}{\partial\tau}\mathrm{d}\tau, \tag{11.20}$$

where

$$\tilde{\tilde{E}}_{\mu+1}(\vec{x}_{\mu+1},t) \equiv \int_{V_{\mu}} \tilde{\tilde{E}}(\vec{x}_{\mu},t-\tau)\mathrm{d}V \tag{11.21}$$

is the volume average of the relaxation modulus at the microscale and is dependent on the damage incurred in the representative volume element at this scale, thereby implying that the material model described in (11.20) is nonlinear.

It is now apparent that macroscale equations (11.10), (11.13)–(11.16), and (11.19) correspond to microscale equations (11.2)–(11.4), so that a similar algorithm may be utilized for the analysis on both scales. The significant difference is that the introduction of cracks at the local scale results in a more complex and inherently nonlinear formulation of the constitutive equations at the macroscale. This then completes the description of the homogenization process and the resulting macroscale IBVP.

11.4.3 Cohesive Zone Model for Predicting Crack Growth on Each Length Scale

As mentioned in the historical review, there are several shortcomings of the Griffith criterion. First, it is often found to be inaccurate for viscoelastic media. Second, it is not convenient to utilize in a computational algorithm, which may be a necessary byproduct of modeling multiple cracks simultaneously. For these reasons, a different approach is taken herein for predicting crack growth in viscoelastic media. In this chapter, a cohesive zone model is utilized instead of the Griffith criterion. Models of this type are not new, having been introduced many years ago by Dugdale [16] and Barenblatt [7]. Initially, at least, a primary motivation of these models was to account for ductility that occurs in many materials, a phenomenon that is not generally captured well by the Griffith criterion. Unfortunately, cohesive zone models suffer from several shortcomings that have inhibited their deployment until recently. These are essentially related to the inability to measure directly the material parameters necessary to characterize a particular cohesive zone model. Furthermore, a cohesive zone model is normally deployed in such a way that it is necessary to know where the crack will propagate a priori. For these reasons, cohesive zone models are only now finding widespread usage.

On the other hand, cohesive zone models are endowed with several significant strengths. Firstly, they are quite conveniently deployable into a finite element code by simply joining two or more subdomains with self-equilibrating tractions, so that the domain may be treated as simply connected and then allowing the tractions to relax to zero as a function of one or more observed state variables during problem solution, thereby resulting in the production of new surface area. Secondly, cohesive zone models can be formulated in such a way that they can more accurately capture fracture phenomena in some media than can the Griffith criterion. For example, it is often observed in viscoelastic media that the critical energy release rate required for crack extension is both rate and history dependent.

Recently, Allen and Searcy [2, 3, 4, 30] have produced a cohesive zone model for some viscoelastic media that is formulated in such a way that the material parameters required to characterize the cohesive zone model can be obtained directly from microscale experiments. Furthermore, this model is inherently two scale in nature, in that it utilizes the solution to a microscale continuum mechanics problem, together with a homogenization theorem, to produce a cohesive zone model on the next larger length scale. The model has also been shown to be consistent with advanced fracture mechanics, in that the cohesive zone requires a nonstationary critical energy release rate in order for a crack to propagate [11, 12, 33].

This model will not be reviewed in detail herein since it has already been reported in the literature; however, a brief review is given here. As shown in Fig. 11.2, the cohesive zone is postulated to be represented by a fibrillated or crazed zone that is small compared to the total cohesive zone area.

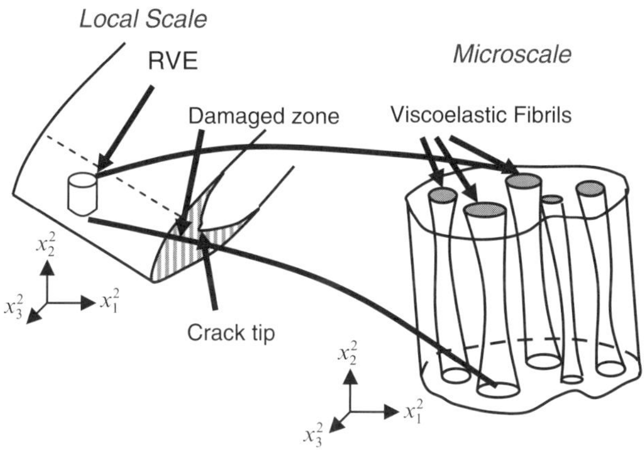

Fig. 11.2. Two-scale problem showing a cohesive zone at the microscale

The length scale of this IBVP is one-length scale below that of the smallest local scale required in the multiscale problem. In this chapter, the value $\mu = 1$ has been arbitrarily assigned to this length scale. The solution to this IBVP (with geometry as shown in Fig. 11.2 and governing equations identical to (11.2)–(11.4)) has been obtained and homogenized, thus leading to the following traction–displacement relation in the cohesive zone [3]

$$\vec{T} = \frac{\vec{\delta}}{\delta_0}(1-\alpha)\int_{\tau=0}^{\tau=1} E(t-\tau)\frac{\partial \lambda}{\partial \tau}\mathrm{d}\tau, \quad \forall \vec{x} \in \partial V_1, \tag{11.22}$$

where $E(t)$ is the uniaxial viscoelastic relaxation modulus of the undamaged cohesive zone material, ∂V_1 is the part of the boundary on which cohesive zones are active, $\vec{\delta}$ is the crack opening displacement vector in the coordinate system, aligned with the crack faces, λ is the Euclidean norm of the crack opening displacement vector, and α is the damage parameter, which in this case degenerates to a scalar, defined by

$$\alpha \equiv \frac{A_0 - \sum_{k=1}^{\mathrm{nf}} A_k}{A_0}, \tag{11.23}$$

where A_0 is the undamaged planform cross-sectional area of a representative area of the cohesive zone and nf is the number of fibrils contained in the representative area.

It can be seen that, when all of the fibrils in a representative area fracture, the damage parameter α goes to unity; and the traction vector in (11.20) becomes zero, thereby inducing crack propagation. Note that the damage parameter α does not exist on the smallest length scale. It appears as a natural byproduct of the homogenization process linking this scale to the next larger scale. This concept is not unlike the concept of temperature, which does not exist at the molecular scale but arises as an outcome of kinetic motions averaged up to the continuum scale. Thus, both are representations of the kinematics associated with entropy generation.

Note that herein the damage parameter for this scale is a scalar, unlike that produced at the other length scales, as defined in (11.16). This is due to the fact that, for the case of a cohesive zone, the homogenization process must be slightly altered to perform an area average rather than a volume average, as described in Sect. 11.4.2. In this case, the limit is taken as the dimension normal to the plane of the cohesive zone, which goes to zero, thereby reducing the *homogenized* cohesive zone to a traction–displacement relation rather than a stress–strain relation.

11.4.4 Formulation of Multiscale Algorithms

The approach detailed above may be used to develop multiscale algorithms for obtaining approximate solutions to problems containing multiple cracks growing simultaneously on widely differing length scales. This is accomplished by constructing a time-stepping algorithm in which the global solution is first obtained for a small time step, assuming some initially damaged (or undamaged) state, as shown in Fig. 11.3. The global solution for this time step is then utilized to obtain solutions for each integration point at the local scale, using the state variables obtained as output from the global analysis to obtain the solution at the local scale. The results for each integration point are then homogenized to produce the global constitutive equations to be used on the next time step at the global scale. This procedure is essentially an operator splitting technique, assuming that there is one-way coupling between the two-length scales. Sufficient accuracy can usually be obtained by this method if successively smaller time steps are employed until convergence is obtained. Details of this approach may be found in [18, 34].

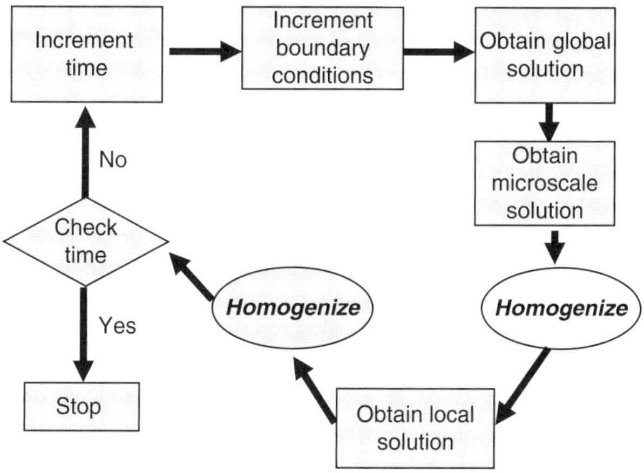

Fig. 11.3. Flowchart showing multiscale computational algorithm

In principle, the approach described herein can be utilized on as many (continuum) length scales as necessary to solve complex problems. However, the limits of continuum scales in nature ($10^{-10}\,\text{m} < l < 10^{3}\,\text{m}$), and the requirement that the length scales be broadly separated, as given by inequality (11.5), lead to the conclusion that only about five, or perhaps six, length scales are physically possible. On the other hand, depending on the complexity of the given problem, only about three *computational* scales are practical with current computer capacities. Fortunately, there are few problems of current technological significance that require more than about three computational scales (there is generally no limitation on the number of analytic scales, as these require little computation; but analytic solutions, unlike the cohesive zone model described in Sect. 11.4.3, are not often attainable). Allen and coworkers have been able to obtain solutions on a desktop computer by this technique using as many as four scales simultaneously (although it must be admitted that two of the scales were analytical) [28]. For simplicity, a three-scale problem is illustrated in Fig. 11.4.

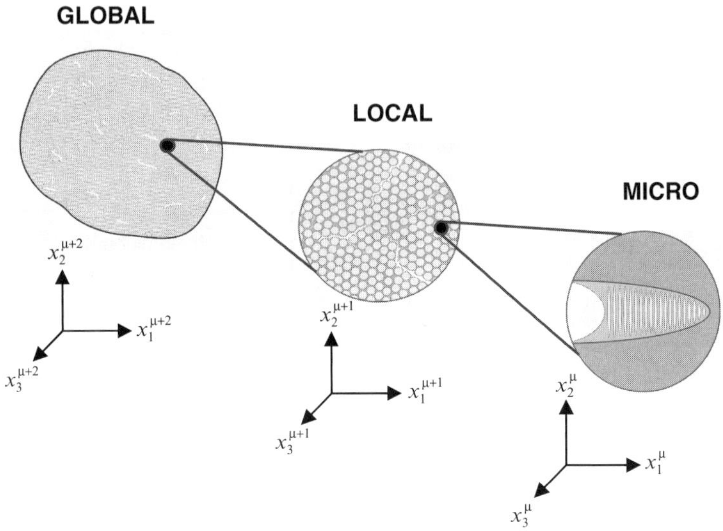

Fig. 11.4. Example of three-scale problem

11.5 Example Problems

In this section, two example problems are presented to demonstrate the technique of multiscaling with damage.

11.5.1 Tapered Bar Problem

The first problem to be considered is a uniaxial bar 10 m long and 2 m in depth with a linear varying cross-sectional area. Figure 11.5a shows the geometry of this tapered bar. The right end of the bar is subjected to a monotonically increasing load of 200 N in the x-direction, and displacements on the left end are restricted. In the local scale, the structure is represented by a repeating unit cell, a square element with 0.025 m of side. This represents a quarter of one aggregate surrounded by asphalt material, shown in Fig. 11.5b. Cohesive zone elements are introduced in the interface between the aggregate and binder. As the load increases, the cohesive zones weaken and accumulate damage, leading to eventual crack growth on the local scale.

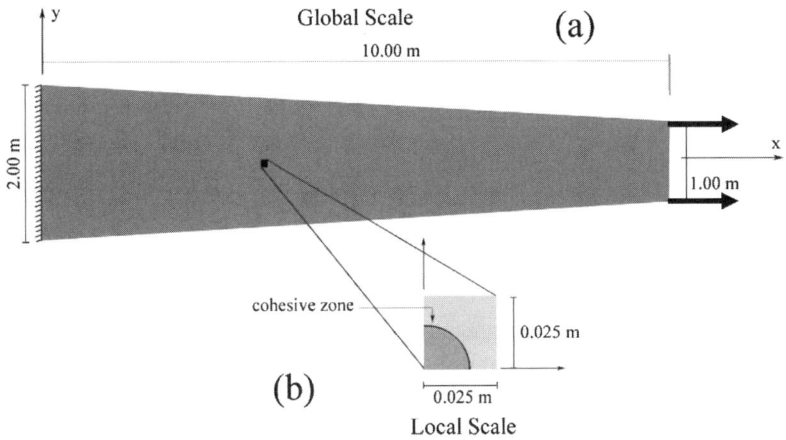

Fig. 11.5. (**a**) Global scale geometry and (**b**) local scale geometry

Symmetry along the x-direction in the global scale allows modeling of only half of the problem. The global scale finite element mesh is shown in Fig. 11.6a. The bar is discretized into 20 elements on the global scale; and each one of them is designated as a multiscale element, thus requiring a separate local analysis for each element in the global scale. The domain of the local scale is then partitioned into 12 triangular elements. A simple local mesh is then created with cohesive zone elements introduced in the interface between the aggregate and binder. The 20 undeformed local scale meshes at $t = 0$ are shown in Fig. 11.6b. Note that these are identical in the initial state but become different from one another as the damage accumulates.

As the bar is loaded, local elements experience differing damage accumulation according to their location, because the macroscale stresses increase toward the loaded end of the bar. Due to the fact that the bar is tapered, local elements close to the load undergo more damage than the ones away from the load.

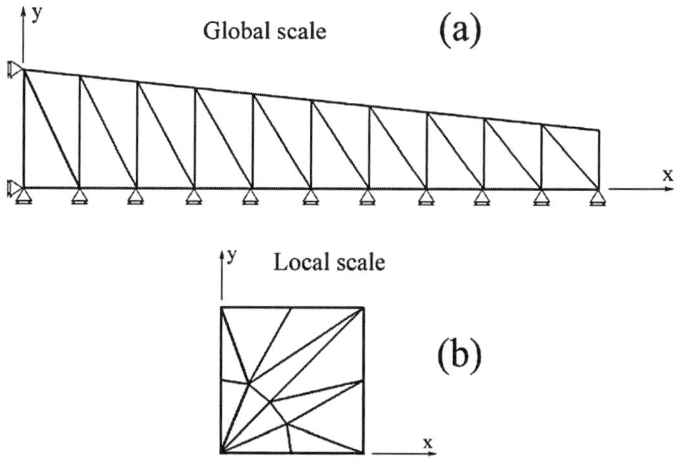

Fig. 11.6. Finite element global (**a**) and local (**b**) meshes of tapered bar

Table 11.1 shows the material properties for local and global meshes, as well as cohesive zone parameters. The bar is made of a hypothetical material model. All materials are assumed to be isotropic linear elastic, where δ_t and δ_n are material length parameter, A and n are damage parameters, and σ_n and σ_t are assumed to be zero.

Table 11.1. Material properties

Global		Local			
Bulk properties		Bulk properties		Cohesive zone properties	
E (Pa)	3.00×10^8	E (Pa)	3.00×10^8	E_∞ (Pa)	5.00×10^7
ν	0.35	ν	0.35	E_1 (Pa)	5.00×10^7
				η_1 (Pa s)	1.00×10^2
				ν	0.40
				δ_t (m)	1.00×10^{-3}
				δ_n (m)	1.00×10^{-3}
				A	0.1
				n	15

The applied force is illustrated in Fig. 11.7. A time increment of 1.0 s is used in this problem.

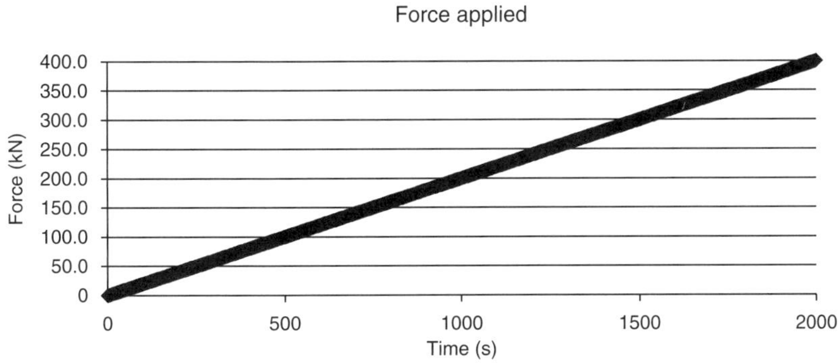

Fig. 11.7. Load applied

To illustrate the local behavior of the elastic tapered bar after the load is applied, the deformation of four different local meshes positioned at different locations along the bar are shown in different times. Four unit cells, chosen strategically within the global bar, are shown in Fig. 11.8. Those unit cells are shown in increments of 250 s, up to 1,000 s, in Fig. 11.9.

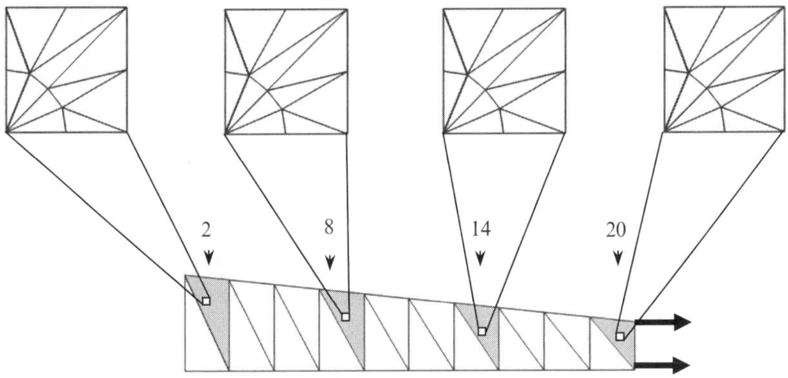

Fig. 11.8. Illustrated multiscale elements

The stress legend is positioned on the left side. It can be seen that all local elements start with zero stresses and no displacements. After 250 s, a cohesive zone opening can be seen in all but Element 2. At 500 s, a crack

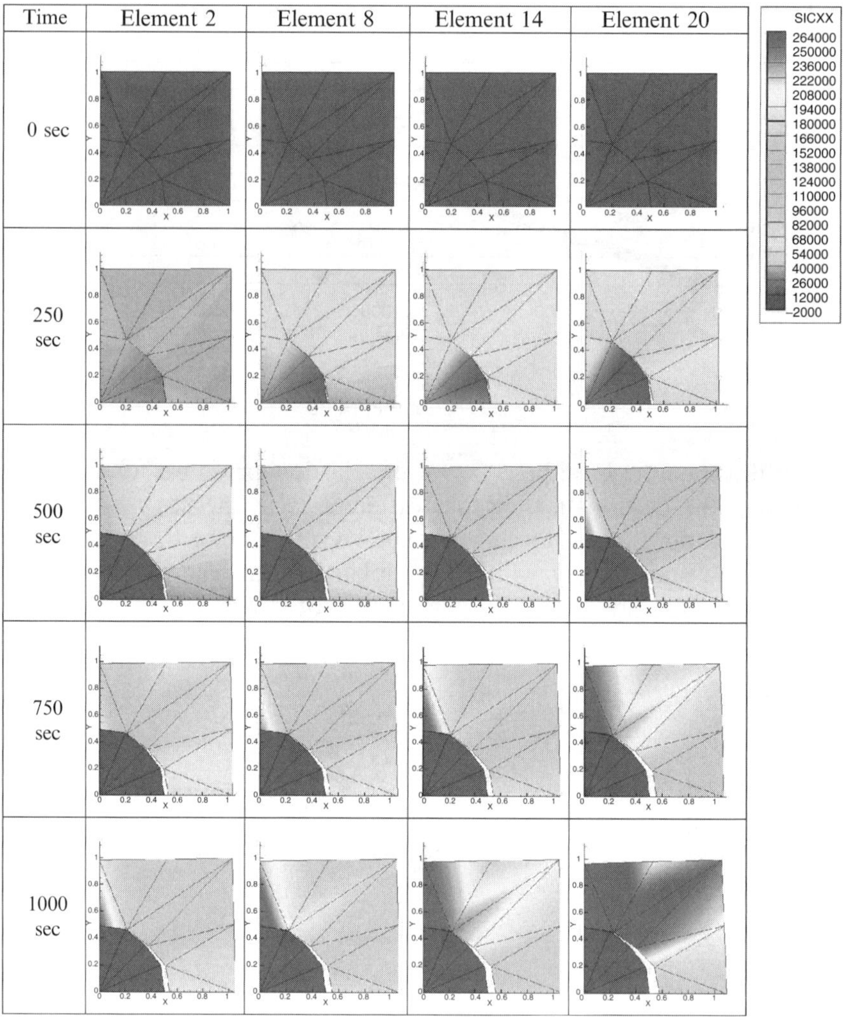

Fig. 11.9. Deformation of elastic tapered bar

has developed in all elements; and stresses are higher in Element 20. As the global force continues to increase and the time approaches 1,000 s, the stresses are still higher in Element 20; and the crack opening is also larger in this element. This is expected to happen; since the cross-sectional area of the tapered bar is smaller on the right end, the global axial stresses are higher at this end. Figure 11.10 shows how the global scale evolves with time.

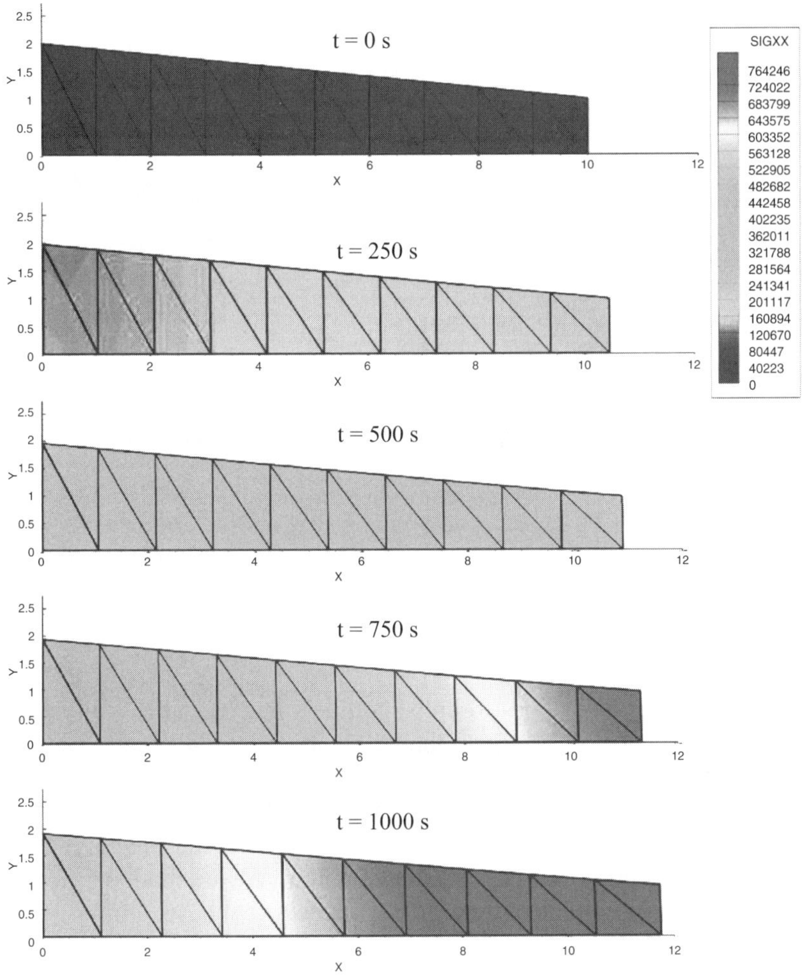

Fig. 11.10. Global deformation of elastic tapered bar

11.5.2 Roadway Problem

The next problem is an asphaltic roadway problem. This is an interesting multiscale problem which considers two different scales. The first step is to consider the geometry of the problem. Consider a typical two-lane asphalt roadway that contains a symmetry line down the middle of it, as shown in Fig. 11.11. For simplicity, only the right half of the pavement will be modeled. The road is 12 m wide with two 3.50 m traffic lanes and two shoulders of 2.50 m.

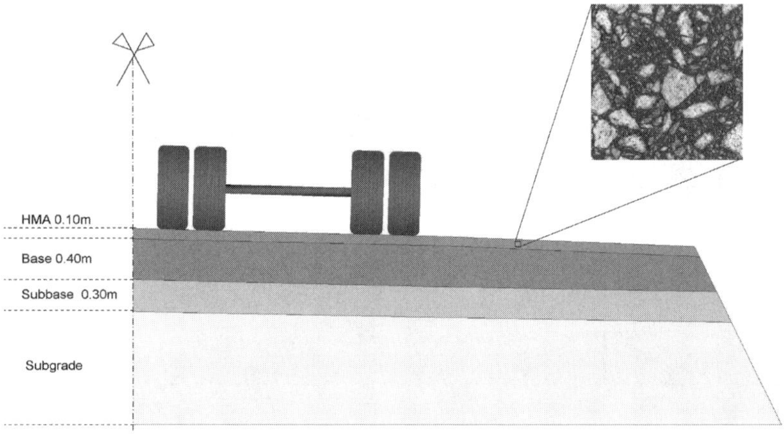

Fig. 11.11. Pavement geometry

The travel lane has slopes to both shoulders at 2% grade, while paved shoulders have slopes of 4%. Even though the pavement geometry can vary from case to case and require additional layers if necessary, the selected pavement for this analysis contains four layers. The top layer is a 10 cm hot mix asphalt, or HMA. The asphalt concrete layer is the final layer to be built on top of the other layers. The subsequent layers are made of granular material: a 40 cm granular base, a 30 cm granular subbase, and an in situ subgrade, which is the graded natural terrain and has a depth of 1.10 m in the model. All layers are modeled as isotropic elastic media. Table 11.2 shows material properties for all the layers.

Table 11.2. Material properties for each layer

	HMA	Base	Subbase	Subgrade
E (MPa)	14,500	4,000	800	200
ν	0.40	0.35	0.35	0.35

The local scale geometry is defined by scanning actual asphalt samples designed to perform laboratory tests. The global domain is discretized into 1,423 elements; the finite element mesh is shown in Fig. 11.12. Four global scale elements have been chosen for multiscale analysis. These elements are located in the surrounding area of the tire load.

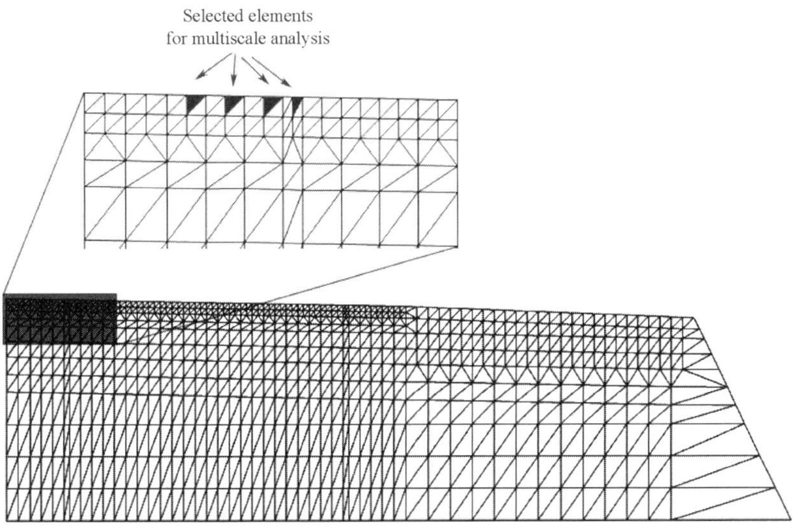

Fig. 11.12. Selected elements for multiscale analysis

The local scale problem was divided into 404 elements and is depicted in Fig. 11.13 in its undeformed configuration. The asphalt material and aggregate are both modeled as linear elastic material. Fracture, in the form of discrete cracks, is introduced at the local scale by 930 cohesive zone elements. All cohesive zone elements are located within the asphalt material; therefore, the interior of each aggregate does not possess any cohesive elements. Cohesive zone elements located in the interface of aggregate and asphalt imply that fracture will occur around the aggregate boundaries or by defibrillation of the asphalt.

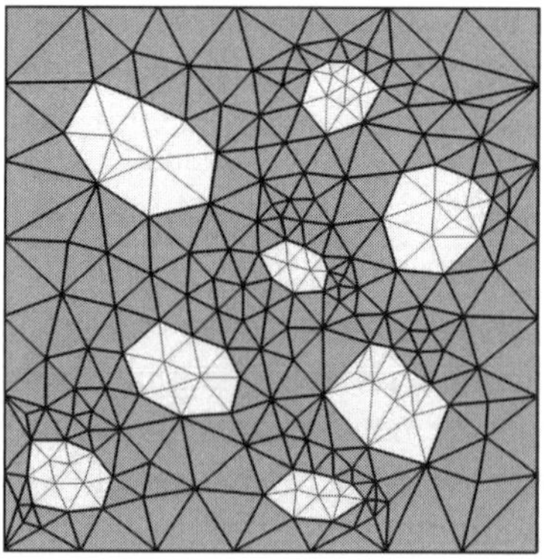

Fig. 11.13. Local scale mesh

Table 11.3 shows the material properties for all constituents of the local scale.

Table 11.3. Material properties for all constituents of the local scales

Local			
Bulk properties		Cohesive zone properties	
Asphalt		E_∞ (Pa)	5.00×10^7
E (Pa)	3.00×10^7	E_1 (Pa)	5.00×10^7
ν	0.35	η_1 (Pa s)	1.00×10^2
Rock		ν	0.40
E (Pa)	8.00×10^8	δ_t (m)	1.00×10^{-3}
ν	0.20	δ_n (m)	1.00×10^{-3}
		A	0.1
		n	15

For this analysis, the response of a cyclic load imposed by a truck on the pavement is simulated. The truck applies a static load of 151 kN with a tandem axle. A 60-ft. long truck traveling at 70 mph takes less than 1 s to pass through a fixed point on the pavement. However, for simplicity, 1 s of simulation was considered to simulate the passage of all five axles; ramp functions for axle loading and unloading were implemented in the code, with a period of 0.2 s for each passing axle or 0.1 s for loading and 0.1 s for unloading, into a peak load of 18,900 kPa. The truck takes a total of five cycles of 0.2 or 1 s for all five axles, followed by a 19-s interval of rest, which is the time until the next truck passes, totaling a period of 20 s. Three trucks per minute or 180 trucks per hour are considered. Figure 11.14 shows the five axle loads and a rest period of up to 5 s, although the rest period still goes up to 20 s. For scale reasons, it would be hard to see the five cycles with a larger scale.

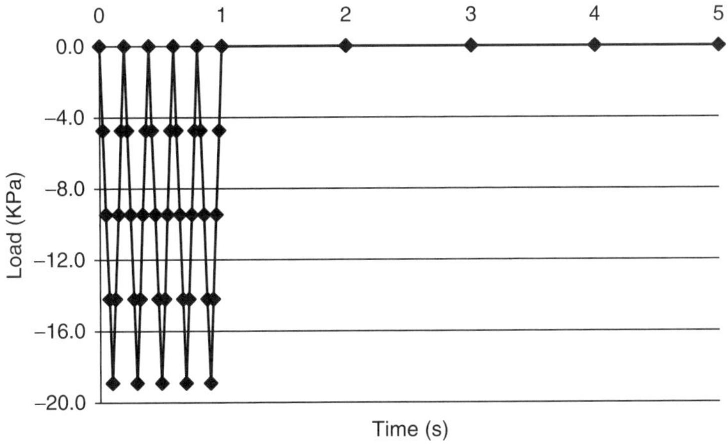

Fig. 11.14. Load history

Let us turn attention to the results of this multiscale problem. As shown in Fig. 11.12, four elements are selected for multiscale analysis; and the deformation of each of those elements as an outcome of the load applied is featured in Fig. 11.15.

It can be seen that, as time increases, elements on the edge of the load (Elements 11 and 23) accumulate less damage than the others with more direct action from the load (Elements 15 and 19). At $t = 0$, all unit cells have zero stresses; and as time progresses, Elements 15 and 19 suffer more damage because they are subjected to higher stresses. Figure 11.16 presents the global mesh with stress contours. It is possible to see that higher compressive stresses occur where the load is applied by the truck.

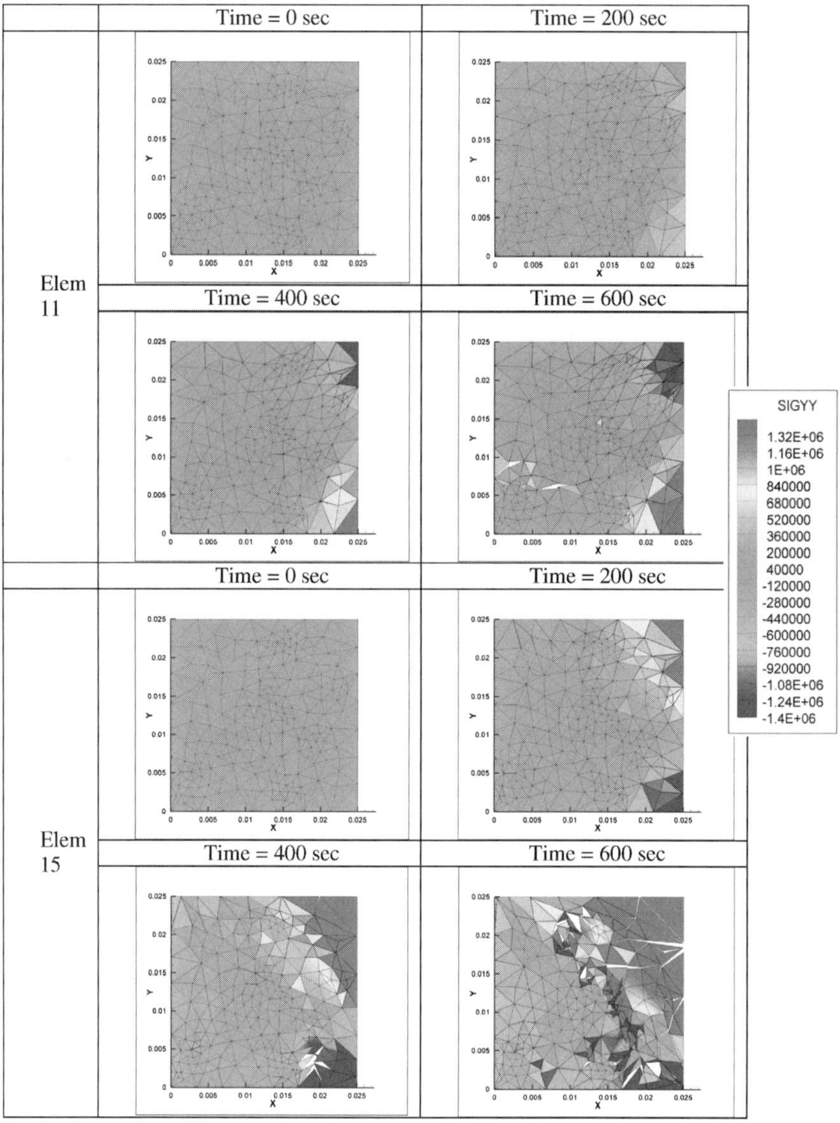

Fig. 11.15. Local deformation

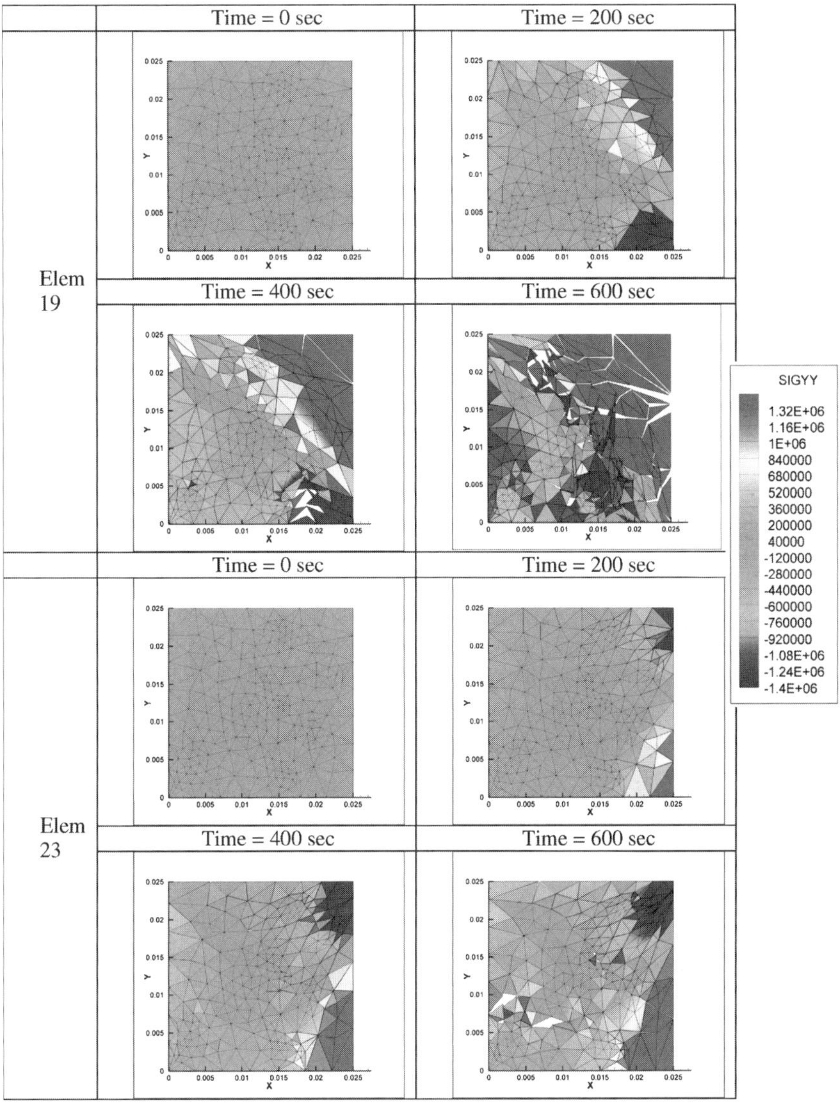

Fig. 11.15. *(cont'd)*

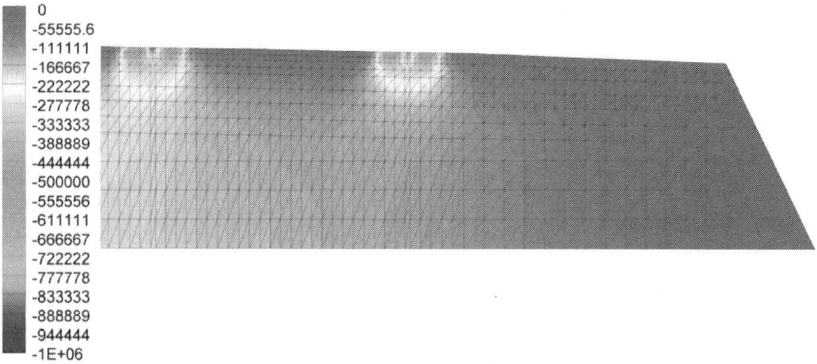

Fig. 11.16. Stress in global scale (y component)

11.6 Conclusion

A multiscale method has been developed for analysis of structural components that exhibit two or more length scales due to heterogeneity and/or evolving damage. The model is implemented into a finite element formulation, and the code employs a micromechanic, alloy-based, viscoelastic cohesive zone model to predict rate-dependent damage evolution.

Two simple example problems have been presented to facilitate the understanding of how the multiscale method works. Although further research is surely needed before this approach can be demonstrated to be accurate, it possesses the potential advantages that (1) material properties need be supplied only on the constituent scale, thereby simplifying the evaluation of material properties and (2) because material properties are specified at the constituent scale, variables, such as volume fraction of aggregate, can be readily incorporated into the design process.

Acknowledgment

The authors are grateful for funding received for this research from the U.S. Army Research Laboratory under contract No. W911NF-04-2-0011.

References

1. Allen, DH, Damage Evolution in Laminates, *Damage Mechanics of Composite Materials*, Talreja, R., Ed., Amsterdam: Elsevier, pp 79–114, 1994
2. Allen, DH, Homogenization Principles and Their Application to Continuum Damage Mechanics, *Composites Science and Technology* 61, 2223–2230, 2002
3. Allen, DH and CR Searcy, A Micromechanical Model for a Viscoelastic Cohesive Zone, *International Journal of Fracture* 107, 159–176, 2001
4. Allen, DH and CR Searcy, A Micromechanically-Based Model for Predicting Damage Evolution in Ductile Polymers, *Mechanics of Materials* 33, 177–184, 2001
5. Allen, DH and CR Searcy, A Model for Predicting the Evolution of Multiple Cracks on Multiple Length Scales in Viscoelastic Composites, *Journal of Materials Science* 41(20), 6510–6519, 2006
6. Allen, DH and C Yoon, Homogenization Techniques for Thermoviscoelastic Solids Containing Cracks, *International Journal of Solids and Structures* 35, 4035–4054, 1998
7. Barenblatt, GI, The Mathematical Theory of Equilibrium Cracks in Brittle Fracture, *Advances in Applied Mechanics* 7, 55–129, 1962
8. Bennett, J, Cooper, M, Hunter, M, and L Jardine, *London's Leonardo: The Life and Work of Robert Hooke*, Oxford: Oxford University Press, 2003
9. Boyd, JG, Costanzo, F, and DH Allen, A Micromechanics Approach for Constructing Locally Averaged Damage Dependent Constitutive Equations in Inelastic Composites, *International Journal of Damage Mechanics* 2, 209–228, 1993
10. Cauchy, AL, Recherches sur l'équilibre et le mouvement intérieur des corps solides ou fluides, élastiques ou non élastiques. *Bulletin de la Soci t philomatique* 9–13, 1823
11. Costanzo, F and DH Allen, A Continuum Mechanics Approach to Some Problems in Subcritical Crack Propagation, *International Journal of Fracture* 63(1), 27–57, 1993
12. Costanzo, F and DH Allen, A Continuum Thermodynamic Analysis of Cohesive Zone Models, *International Journal of Engineering Science* 33(15), 2197–2219, 1996

13. Costanzo, F, Boyd, JG, and DH Allen, Micromechanics and Homogenization of Inelastic Composite Materials with Growing Cracks, *Journal of the Mechanics and Physics of Solids* 44(3), 333–370, 1996
14. Da Vinci, Leonardo, *The Notebooks of Leonardo Da Vinci*, Richter, J.P., Ed., London: Jonathan Cape, 1970
15. Day, WA, *The Thermodynamics of Simple Materials with Fading Memory*, Springer Tracts in Natural Philosophy, Berlin Heidelberg New York: Springer, 1972
16. Dugdale, DS, Yielding of Steel Sheets Containing Slits, *Journal of the Mechanics and Physics of Solids* 8, 100–104, 1960
17. Eshelby, JD, The Determination of the Elastic Field of an Ellipsoidal Inclusion and Related Problems, *Proceedings of the Royal Society A* 421, 376–396, 1957
18. Foulk, JW, Allen, DH, and KLE Helms, Formulation of a Three Dimensional Cohesive Zone Model for Application to a Finite Element Algorithm, *Computer Methods in Applied Mechanics and Engineering* 183, 51–66, 2000
19. Galileo, G, *Dialogues Concerning Two New Sciences*, New York: Dover, 1636
20. Germain, S, *Recherches sur la th orie des surfaces lastiques*, Paris: Hurard-Courcier, 1821
21. Griffith, AA, The Phenomena of Rupture and Flow in Solids, *Philosophical Transactions of the Royal Society of London A* 221, 163–197, 1920
22. Gurtin, ME, On the Energy Release Rate in Quasi-static Elastic Crack Propagation, *Journal of Elasticity* 9, 187–195, 1979
23. Hashin, Z, Theory of Mechanical Behaviour of Heterogeneous Media, *Applied Mechanics Reviews* 17, 1–9, 1964
24. Hill, R, *The Mathematical Theory of Plasticity*, Oxford: Clarendon, 1950
25. Hill, R, A Self-Consistent Mechanics of Composite Materials, *Journal of the Mechanics and Physics of Solids* 13, 213–222, 1965
26. Navier, CLMH, Memoire sur les lois de l'equilibre et du mouvement des corps solides elastiques, *Mémoires de l'Académie Royale des Sciences* 7, 375–394, 1827
27. Isaac Newton, I, Bernard Cohen and Anne Whitman, *Principia*, Mathematical principles of Natural Philosophy, University of California Press, pp. 994, 1999

28. Phillips, ML, Yoon, C, and DH Allen, A Computational Model for Predicting Damage Evolution in Laminated Composite Plates, *Journal of Engineering Materials and Technology* 21, 436–444, 1999
29. Rice, JR, A Path Independent Integral and Approximate Analysis of Strain Concentration by Notches and Cracks, *Journal of Applied Mechanics* 35(2), 379–386, 1968
30. Searcy, CR, A Multiscale Model for Predicting Damage Evolution in Heterogeneous Media, Ph.D. Thesis, Texas A&M University, 2004
31. Timoshenko, SP, *History of Strength of Materials*, New York: McGraw-Hill, 1972
32. Vakulenko, AA, and ML Kachanov, Continuum Model of Media with Cracks, *Mekhanika Tverdogo Tela* 4, 159–166, 1971
33. Yoon, C and DH Allen, Damage Dependent Constitutive Behavior and Energy Release Rate for a Cohesive Zone in a Thermoviscoelastic Solid, *International Journal of Fracture* 96, 56–74, 1999
34. Zocher, MA, Allen, DH, and SE Groves, A Three Dimensional Finite Element Formulation for Thermoviscoelastic Orthotropic Media, *International Journal for Numerical Methods in Engineering* 40, 2267–2288, 1997

Chapter 12: Multiscale Modeling for Damage Analysis

Ramesh Talreja and Chandra Veer Singh

Department of Aerospace Engineering, Texas A&M University, College Station, TX 77843-3141, USA

12.1 Introduction

The increased computational power and programming capabilities in recent years have given impetus to the so-called *multiscale modeling*, which implements the largely intuitive notion that physical phenomena occurring at a lower length or size scale determine the observed response at a higher scale. A logical outcome of this thought is an organization of differentiated scales – from the lowest, such as nanometer scale, to the highest scale typical of the part or structure in mind – giving a hierarchy of scales. Working up the scales produces a hierarchical multiscale modeling, in which the essential challenge consists of "bridging" the scales. The simulation techniques, such as molecular dynamics simulation (MDS), succeed mostly in revealing phenomena from one scale to the next; but proceeding to three or more scales often necessitates unrealistic computing power even with the most versatile facilities available. In addition, the limitation of independent physical validation of the simulated results questions the wisdom of total reliance on the multiscale hierarchical modeling strategy.

When it comes to subcritical (prefailure) damage in composites, the multiscale modeling concept needs closer examination, firstly, because the length scales of constituents and heterogeneities are fixed while those of damage evolve progressively, and secondly, because the mechanisms of damage tend to segregate in modes with individual characteristic scales. All this is the subject of this chapter, which will first describe and clarify the damage mechanisms in common types of composites followed by the induced response observed at the macroscale. The hierarchical modeling approach will be discussed against this knowledge; and a different approach, named *synergistic multiscale modeling*, will be advocated. Assessment will be offered of the current state of this modeling, and future activities aimed at accomplishing its objectives will be outlined.

The following treatment of multiscale modeling will draw upon a recent paper by Talreja [61] as well as other previous works.

12.2 Phenomenon of Damage in Composite Materials

Engineered structures must be capable of performing their functions throughout a specified lifetime while being exposed to a series of events that include loading, environment, and damage threats. These events, either individually or in combination, can cause structural degradation, which, in turn, can affect the ability of the structure to perform its function. The performance degradation in structures made of composites is quite different when compared to metallic components because the failure is not uniquely defined in composite materials. To understand how composites may lose the ability to perform satisfactorily, some basic definitions related to damage of composite materials must be reviewed. Section 12.2.1 contains a brief overview of significant mechanisms that can degrade a composite material.

In a conventional sense, *fracture* is understood to be "breakage" of material, or at a more fundamental level, breakage of atomic bonds, which manifests itself in formation of internal surfaces. Examples of fractures in composites are fiber fragmentation, cracks in matrix, fiber/matrix debonding, and separation of bonded plies (delamination). The field of fracture mechanics concerns itself with conditions for enlargement of the surfaces of material separation.

Damage refers to a collection of all the irreversible changes brought about by energy dissipating mechanisms, of which atomic bond breakage is an example. Unless specified differently, damage is understood to refer to distributed changes. Examples of damage are multiple fiber-bridged matrix cracking in a unidirectional composite, multiple intralaminar cracking in a laminate, local delamination distributed in an interlaminar plane, and fiber/matrix interfacial slip associated with multiple matrix cracking. These damage mechanisms are explained in some detail in Sect. 12.2.1. The field of damage mechanics deals with conditions for initiation and progression of distributed changes as well as consequences of those changes on the response of a material (and by implication, a structure) to external loading.

Failure is defined as the inability of a given material system (and consequently, a structure made from it) to perform its design function. Fracture is one example of a possible failure; but, generally, a material could fracture (locally) and still perform its design function. Upon suffering damage, e.g., in the form of multiple cracking, a composite material may still continue to carry loads and, thereby, meet its load-bearing requirement but fail to deform in a manner needed for its other design requirements, such as vibration characteristics and deflection limits.

Structural integrity is defined as the ability of a load-bearing structure to remain intact or functional upon the application of loads. In contrast to metals, remaining intact (not breaking up in pieces) for composites is not necessarily the same as remaining functional. Composites can lose their functionality by suffering degradation in their stiffness properties while still carrying significant loads.

12.2.1 Mechanisms of Damage

Due to extreme levels of anisotropy and inhomogeneity of composites, a variety of damage mechanisms cause degradation in the material behavior. These can occur separately or in combination. A short description of each damage mechanism follows.

Multiple matrix cracking

Matrix cracks are usually the first observed form of damage in composite laminates [45]. These are intralaminar or ply cracks, transverse to loading direction, traversing the thickness of the ply and running parallel to the fibers in that ply. The terms *matrix microcracks*, *transverse cracks*, *intra-laminar cracks*, and *ply cracks* are invariably used to refer to this very same phenomenon. Matrix cracks are observed during tensile loading, fatigue loading, changes in temperature, and thermocycling. Figure 12.1

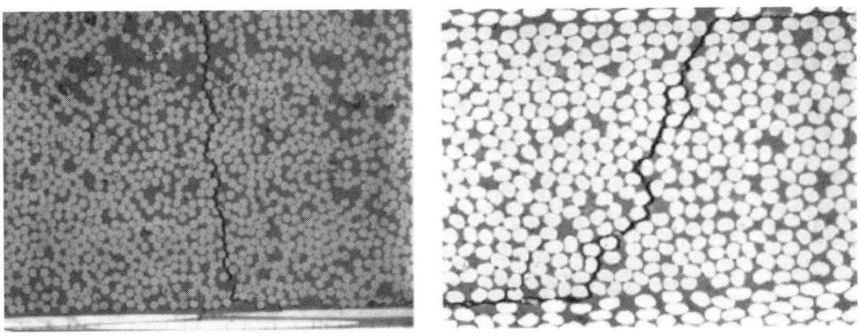

a {…0°/90°$_2$/0°…} b {…+45°/90°/-45°…}

Fig. 12.1. Examples of matrix cracks observed on the free edges induced due to fatigue loading in composite laminates [37] illustrates matrix cracks observed on the free edges induced due to fatigue loading in composite laminates [37]. Although matrix cracking does not cause structural failure by itself, it can result in significant degradation in material stiffness and also can induce more severe forms of damage, such as delamination and fiber breakage [44]. Numerous studies of microcracking initiation were performed in the 1970s and early 1980s [4, 13–15, 29, 48, 49]. It was observed that the strain to initiate microcracking increases as the thickness of 90° plies decreases. Also, these microcracks form almost instantaneously across the width of the specimen.

The first attempt to predict the strain to first microcrack used the *first ply failure* theory [18] where it is assumed that the first crack develops when the strain in the plies reaches the strain to failure in the plies. The predictions were not in agreement with the experimental observations since the first ply failure theory predicts that the strain to initiate microcracking will be independent of the ply thickness. The experimental observations on laminates with a 90° layer on the surface $[90_n/0_m]_s$ show that the strain to initiate microcracking is lower for laminates with cracks in surface plies than for laminates with cracks in central plies [52, 54].

The simplest way to model transverse matrix cracks in composite laminates is to completely neglect the transverse stiffness of cracked plies, called the *ply discount method.* This method underestimates the stiffness of cracked laminates, since cracked plies, in reality, can take some loading. Another simple way is *shear lag analysis*, wherein the load transfer between plies is assumed to take place in shear layers between neighboring plies. The normal stress in the external load direction is assumed to be constant over the ply thickness. The thicknesses and stiffness of these shear layers are generally unknown, and the variations in the thickness direction of local ply stresses and strains are also neglected in the shear lag theory. The shear lag theory has limited success for crossply laminates [19, 25, 39, 62]. For crossply laminates, the most successful approach is the *variational method.* By application of the principle of minimum complementary potential energy, Hashin [21, 22] derived estimates for thermomechanical properties and local ply stresses, which were in good agreement with experimental data. Varna and Berglund [65] later made improvements to the Hashin model by use of more accurate trial stress functions. A disadvantage of the variational method is that it is extremely difficult to use for laminate layups other than crossplies. McCartney [43] used Reissner's energy function to derive governing equations similar to Hashin's model. He applied this approach to doubly cracked crossply laminates assuming that the in-plane normal stress dependence on the two in-plane coordinates is given by two independent functions. Gudmundson and coworkers [16, 17] considered laminates with general layup and used the homogenization technique to derive expressions for stiffness and thermal expansion coefficient of laminates with cracks in layers of three-dimensional (3D) laminates. These expressions correlate damaged laminate thermoelastic properties with parameters characterizing crack behavior: the average crack opening displacement (COD) and the average crack face sliding. These parameters follow from the solution of the local boundary value problem, and their determination is a very complex task. Also, the effect of neighboring layers on crack face displacements was neglected; and the displacements were determined assuming a periodic system of cracks in an infinite homogeneous, transversely isotropic medium (90° layer). The application of their methodology by other researchers has been rather limited due to the fairly complex form of the presented solutions.

An alternative way to describe the mechanical behavior of matrix-cracked laminates is to apply concepts of damage mechanics. Generally speaking, the continuum damage mechanics (CDM) approaches [1, 2, 56, 57] may be used to describe the stiffness of laminates with intralaminar cracks in off-axis plies of any orientation. The damage is represented by internal state variables (ISVs), and the laminate constitutive equations are expressed in general forms containing ISV and a certain number of material constants. These constants must be determined for each laminate configuration considered either experimentally, measuring stiffness for a laminate with a certain crack density, or using finite element (FE) analysis. This limitation is partially removed in synergistic damage mechanics (SDM) suggested by Talreja [60], which incorporates micromechanics information in determining the material constants. The SDM approach has proved to be quite efficient for a variety of laminate layups and material systems. The present chapter builds on this methodology, and relevant details will be discussed later.

Interfacial debonding

The performance of a composite is markedly influenced by the properties of the interface between the fiber and matrix resin. The adhesion bond at the interfacial surface affects the macroscopic mechanical properties of the composite. The interface plays a significant role in stress transfer between fiber and matrix. Controlling interfacial properties thus leads to the control of composite performance. In unidirectional composites, debonding occurs at the interface between fiber and matrix when the interface is weak. The longitudinal interfacial debonding behavior of single-fiber composites has been studied in detail by the use of the pullout [26, 38, 73] and fragmentation [10, 12, 24, 72] tests. The mechanics of interfacial debonding in a unidirectional fiber-reinforced composite are depicted in Fig. 12.2. When fracture strain of the fiber is greater than that of the matrix, i.e., $\varepsilon_f > \varepsilon_m$, a crack originating at a point of stress concentration, e.g., voids, air bubbles, or inclusions, in the matrix is either halted by the fiber, if the stress is not high enough, or it may pass around the fiber without destroying the interfacial bond (Fig. 12.2a). As the applied load increases, the fiber and matrix deform differentially, resulting in a buildup of large local stresses in the fiber. This causes local Poisson contraction; and eventually shear force developed at the interface exceeds the interfacial shear strength, resulting in interfacial debonding at the crack plane that extends some distance along the fiber at the interface (Fig. 12.2c).

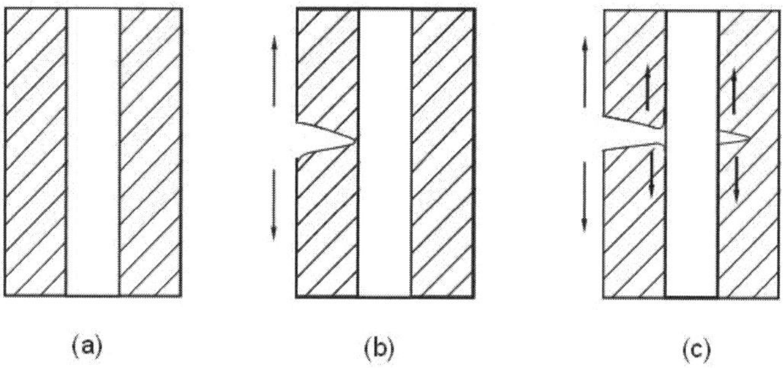

Fig. 12.2. Mechanics of interfacial debonding in a simple composite [20]

Interfacial sliding

Interfacial sliding between constituents in a composite can take place by differential displacement of the constituents. One example of this is when fibers and matrix in a composite are not bonded together adhesively but by a "shrink-fit" mechanism, due to difference in thermal expansion properties of the constituents. On thermomechanical loading, the shrink-fit (residual) stresses can be removed, leading to a relative displacement (sliding) at the interface. The relief of interfacial normal stress can also occur when a matrix crack tip approaches or hits the interface.

When the two constituents are bonded together adhesively, interfacial sliding can occur subsequent to debonding if a compressive normal stress on the interface is present. The debonding can be induced by a matrix crack, or it can result from growth of interfacial defects. Thus, interfacial sliding that follows debonding can be a separate damage mode or it can be a damage mode coupled with matrix damage.

When the interface between the matrix and the fiber debonds, this relieves the tensile residual stresses in the matrix. Due to different stresses in the matrix and the fiber at the interface, the fibers slide on the interfacial surface. Subsequently, the sliding surfaces cause degradation of material due to frictional wear at the interface. Pullout and pushback tests are useful in determining the stress required to cause interfacial sliding. This mostly depends upon the strength of the adhesive bond between the matrix and the fiber at the interface.

Fiber microbuckling

When a unidirectional composite is loaded in compression, the failure is governed by the matrix and occurs through a mechanism known as *microbuckling of fibers*. There are two basic modes of microbuckling deformation: "extensional" and "shear" modes [51], as shown in Fig. 12.3, depending upon whether the fibers deform "out of phase" or "in phase." The compressive strength corresponds to the onset of instability and is given as

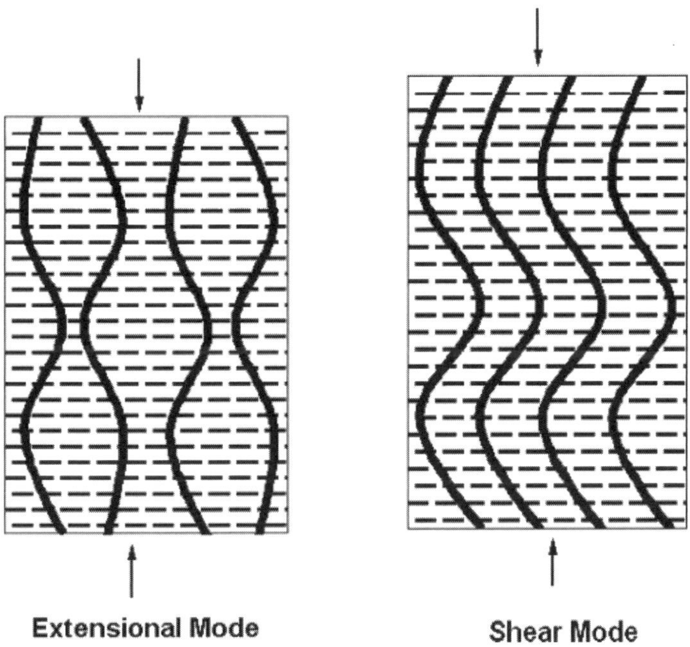

Fig. 12.3. Extensional and shear modes of microbuckling [51]

$$\sigma_{\mathrm{c}} = 2V_{\mathrm{f}}\sqrt{\frac{V_{\mathrm{f}}E_{\mathrm{f}}E_{\mathrm{m}}}{3(1-V_{\mathrm{f}})}} \tag{12.1}$$

for the extension mode and

$$\sigma_{\mathrm{c}} = \frac{G_{\mathrm{m}}}{1-V_{\mathrm{f}}} \tag{12.2}$$

for the shear mode, where E and G denote Young's modulus and shear modulus, respectively, and subscripts "f" and "m" designate fiber and matrix, respectively.

Although these expressions are based on energy balance, they do not agree with experimental observations. As an alternative, it has been argued that manufacturing of composites tends to cause misalignment of fibers, which can induce localized kinking of fiber bundles. The kinking process is driven by local shear, which depends on the initial misalignment angle ϕ_0 [3]. The critical compressive stress corresponding to instability is given by

$$\sigma_{\mathrm{c}} = \frac{\tau_{\mathrm{y}}}{\phi_0}, \tag{12.3}$$

where τ_{y} represents the interlaminar shear strength. Budiansky [6] considered the kink band geometry and derived the following estimate for the kink band angle β in terms of the transverse modulus E_{T} and shear modulus G of a two-dimensional (2D) composite layer:

$$(\sqrt{2}-1)^2\frac{G-\sigma_{\mathrm{c}}}{E_{\mathrm{T}}} < \tan^2\beta < \frac{G-\sigma_{\mathrm{c}}}{E_{\mathrm{T}}}. \tag{12.4}$$

To account for shear deformation effects, Niu and Talreja [46] modeled the fiber as a generalized Timoshenko beam with the matrix as an elastic foundation. It was observed that not only an initial fiber misalignment but also any misalignment in the loading system can affect the critical stress for kinking.

Delamination

Delamination as a result of low-velocity impact loading is a major cause of failure in fiber-reinforced composites [7, 9, 40]. Delamination can occur below the surface of a composite structure with a relatively light impact, such as that from a dropped tool, while the surface remains undamaged to visual inspection [9, 28, 50]. The growth of delamination cracks under the subsequent application of external loads leads to a rapid deterioration of the mechanical properties and may cause catastrophic failure of the composite structure [55]. Delamination is a substantial problem because the composite laminates, although having strength in the fiber direction, lack strength in the through-thickness direction. This essentially limits the strength of a traditional 2D composite to the properties of the brittle matrix alone [71]. The development of interlaminar stresses is the primary cause of delamination in laminated fibrous composites. Delamination occurs when the interlaminar stress level exceeds the interlaminar strength. The interlaminar stress level is associated with the specimen geometry and loading parameters, while the interlaminar strength is related to the material properties [40, 71]. From an energy point of view, delamination cracks will grow when the energy required to overcome the cohesive force of the atoms is equal to the dissipation of the strain energy that is released by the crack [11]. The delamination can be reduced by either improving the fracture toughness of the material or modifying the fiber architecture [8, 42].

Typically, a low-speed impact overstresses the matrix material, producing local subcritical cracking (microcracking). This does not necessarily produce fracture; however, it will result in stress redistribution and the concentration of energy and stress at the interply regions where large differences in material stiffness exist. The onset and rapid propagation of a crack results in sudden variations in both section properties and load paths within the composite local to the impactor. This requires an adaptive method to track the progression of damage and fracture growth.

Fiber fracture

As the applied load is increased, progressive matrix cracks lead to fiber/matrix interfacial debonding and delamination; and the stress state inside laminate material becomes quite complex. Ultimately, when the laminate strain reaches fiber failure strain, the fibers start to fail; and multiple cracks develop in the fibers. The multiple fiber cracks also develop due to stress transfer in the regions where the matrix is not able to take any more load.

Since at this load level other damage modes are also present, the real reason for ultimate failure is often not clear. At ultimate failure load, the matrix is shattered; and, evidently, the fibers carry the full failure load. The composites usually support large load and deformation at failure, although the measured ultimate strength clearly may not be reliable in actual applications [47]. All fibers are not of the same strength, and a statistical variation of strength between fibers and along fiber lengths is used. In addition to strength and modulus, another important property of a fiber-reinforced composite is its resistance to fracture. The fracture toughness of a composite depends not only on the properties of the constituents but also significantly on the efficiency of bonding across the interface [33].

The damage mechanisms described above have different characteristics depending on a variety of geometric and material parameters. Each mechanism has different governing length scales and evolves differently when the applied load is increased. Interactions between individual mechanisms further complicate the damage picture. As the loading increases, stress transfer takes place from a region of high damage to that of low damage, and the composite failure results from the criticality of the last load-bearing element or region. For clarity of treatment, the full range of damage can be separated into damage modes, treating them individually followed by examining their interactions. This approach will be discussed in detail in later sections with respect to ceramic matrix composites (CMCs) and polymer matrix composites (PMCs).

12.2.2 Damage-Induced Response of Composites

The presence of damage in a composite induces *permanent* changes in the response with respect to the virgin state. One objective of multiscale modeling is to relate these changes to the damage, specifically taking into account the scale(s) at which damage mechanisms operate. In this section, a simple case of unidirectional continuous fiber composites, which respond linear elastically in the virgin state, will be examined to illustrate how the response can be varied when multiple matrix cracking damage exists. Two cases will be considered (1) a constrained PMC loaded in tension transverse to fibers and (2) an unconstrained CMC loaded in tension along fibers.

Constrained PMC loaded in tension transverse to fibers

When a unidirectional PMC is loaded in uniform tension normal to fibers, it responds linear elastically until failure initiates from matrix or interfacial cracking. However, if this composite is bonded to stiff elastic elements and then loaded, still transverse to fibers, its failure changes from single fracture to multiple matrix cracking as described above. The response of the combined composite and the stiff elements changes as the multiple cracking progresses, i.e., its intensity, measured by, e.g., crack number density, increases. The changes in response induced by cracking depend on the constraining effect of the stiff elements. This phenomenon is conveniently illustrated in Fig. 12.4 by an axially loaded crossply composite $[0_m/90_n]_s$ in which the degree of constraint to transverse ply cracking can be varied by selecting the m/n ratio. Considering the strain ε_{FPF} at which first cracking occurs in the constrained transverse ply, Talreja [56] classified the constraint in four categories (Fig. 12.5) (A) no constraint, (B)

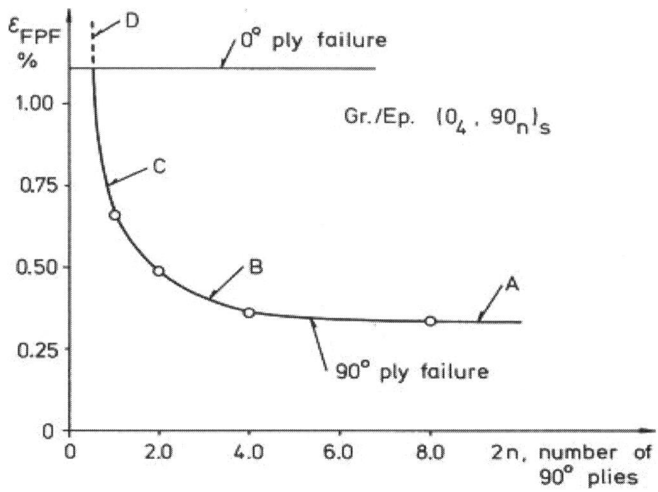

Fig. 12.4. The strain at first ply failure as a function of the number of transverse plies in $[0_4/90_n]_s$ laminate

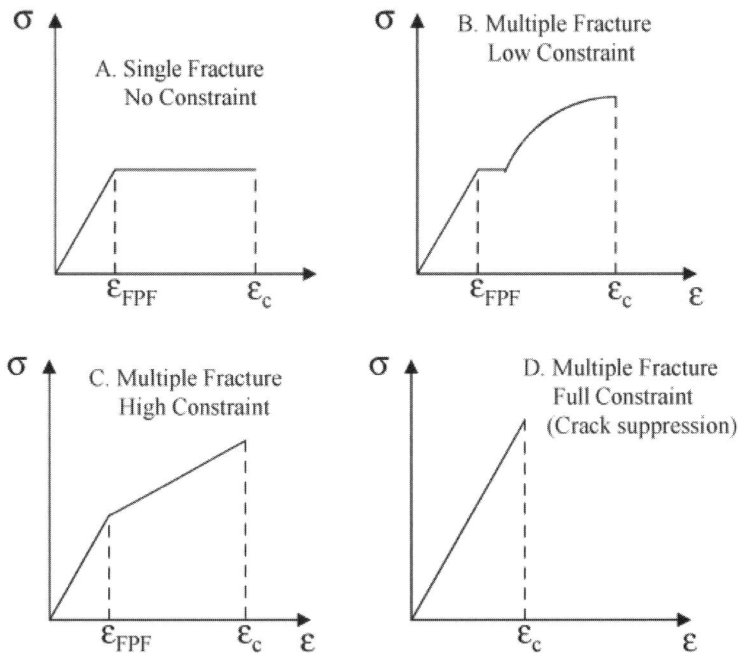

Fig. 12.5. Stress–strain response at different constraints to transverse cracking in crossply laminates

low constraint, (C) high constraint, and (D) full constraint. In the case of $[0_4/90_n]_s$ laminate, ε_{FPF} varies with the number $2n$ of 90° plies. As this number increases, the constraint of the 0° plies becomes increasingly insignificant and ε_{FPF} approaches the failure strain of the unconstrained 90° plies, i.e., the failure strain normal to fibers. On the other extreme, as the constraint of the 0° plies becomes effective, this strain increases; and multiple cracking occurs. This process continues to higher constraint; and at some point, the ε_{FPF} exceeds the fiber failure strain, at which point the constraining plies fail.

Unconstrained CMC loaded in tension along fibers

The stress–strain response of a unidirectional CMC loaded in axial tension is described in Fig. 12.6. This response develops in stages as the matrix cracking progresses, as evidenced by the set of micrographs obtained by S rensen and Talreja [53] shown in Fig. 12.7. The micrograph taken at 0.15% axial strain shows the matrix cracks normal to the (horizontal) fiber axis that do not span the complete specimen cross section. As strain increases, more cracks form and quickly span the whole specimen width. Finally, the cracking saturates, i.e., no more cracks form on increasing the load. This stage of progressive matrix cracking represents Stage II, extending from 0.13 to 0.5% axial strain in Fig. 12.6. The preceding stage (Stage I) consists of linear elastic behavior before the onset of cracking. Beyond 0.5%

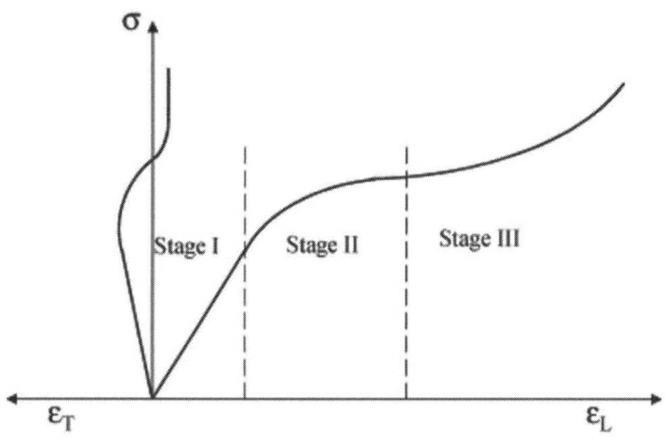

Fig. 12.6. The three stages of stress–strain response in a SiC fiber-reinforced glass-ceramic composite [53]

strain, the frictional sliding at the fiber/matrix interface becomes significant. Finally, beyond 0.7% strain, a progressive fiber breakage takes place leading to localization of damage and catastrophic failure. The Stage II progressive cracking was treated by S rensen and Talreja [53]. In Sect. 12.3, two cases of damage will be used to illustrate the multiscale nature of damage and discuss how the scales can be incorporated into a damage mechanics framework.

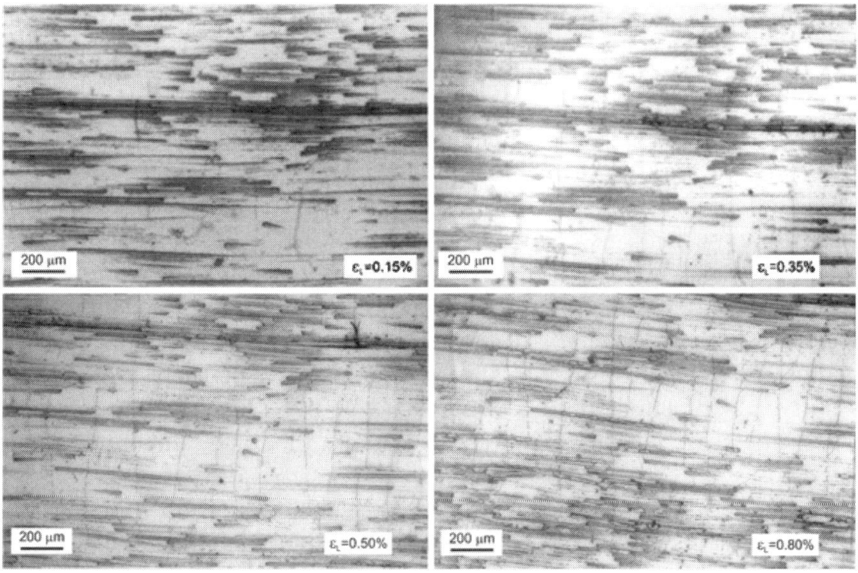

Fig. 12.7. Surface micrographs of a SiC fiber-reinforced glass-ceramic composite at different axial strains. Tensile loading was in the (*horizontal*) fiber direction [53]

12.3 Multiscale Nature of Damage

As described in Sect. 12.2, the damage in composites occurs due to a variety of dissipative mechanisms which cause permanent changes in the internal microstructure of the material and decrease the energy storing capacity of the material. The most basic scale at which these mechanisms occur depends upon the size of inhomogeneities in the microstructure of the material. As an example, nanocomposites may show dissipative mechanisms at the nanometer scale. In reality, however, identifying this scale is limited by the ability to observe as well as to model and analyze the mechanisms at the observed scale. The so-called microscale is a reference to the scale at which entities or features within a material are observable by a certain type of microscope. Thus, for example, the microscale can be a few micrometers, if an electron microscope is used to observe entities, such as cracks or crystalline slip within grains or at grain boundaries. The scale reduces by an order of magnitude if one focuses on dislocations observed by a transmission electron microscope. Today, the use of nano-scale elements (particles, fibers, tubes, etc.) has moved the basic scale further down to the atomic scale. At this scale, the basic notions of continuum mechanics fail; and it is necessary to develop modeling tools that can bridge the discrete-level descriptions (quantum mechanics) to continuum-type (smeared-out) descriptions.

In an engineering approach, the purpose at hand should guide the choice of the basic scale. Thus, if the overall (effective) characteristics of inelastic response are of interest, it would suffice to incorporate the energy dissipating mechanisms in a model, directly or indirectly, in an appropriate average sense; while if, for instance, the aim is a particular material failure characteristic, the analysis may need to be conducted at the local physical scale of the relevant details of the mechanisms. On the other hand, if the purpose is to design a material, i.e., to engineer its response or to provide it with certain functionalities, then it would be necessary to address scales where the material (micro) structure can be modified, manipulated, or intruded.

In composite materials, the scales of inhomogeneities (reinforcements, additives, second phases, etc.) embedded in the baseline material (matrix) determine the characteristic scales of operation of the mechanisms of energy dissipation. Although energy dissipation may also be occurring at other (smaller) scales, e.g., the scale of the matrix material's microstructure, the dissipative mechanisms associated with the inhomogeneities have usually an overriding influence on the composite behavior. For instance, in short-fiber PMCs, the size of fiber diameter manifests the scale at which matrix cracks form, although energy dissipation may also occur at the matrix polymer's molecular scale.

The complexity introduced by inhomogeneities in composite damage is in the form of multiple scales of dissipative mechanisms depending on the geometrical features of the inhomogeneities. In the case of short fibers, for instance, the matrix cracking from the fiber ends and the fiber/matrix debonding occurs at two length scales, determined by the fiber diameter and fiber length, respectively. For composite laminates, the thickness of identically oriented plies sets the scale for development of intralaminar cracking, while for formation of these cracks the appropriate scale is given by the fiber diameter. Thus, in modeling of a composite material's behavior, one faces a complex situation concerning the length scales; and taking a hierarchical approach may not be efficient, as will be discussed later. In the following, the multiscale nature of damage in composite materials will be illustrated by examining the two cases pertaining to damage in unidirectional CMCs and ply cracking in laminates.

12.3.1 Unidirectional CMCs

In the case of CMCs, the fiber/matrix interfacial region has a strong influence on the thermomechanical response. The properties of the interfacial region determine whether a matrix crack front approaching a fiber advances into the fiber or bypasses it by causing interfacial slip and/or debonding. The damage configurations at the microscopic level thus generated govern the macroscopic (overall) response of a composite. There are three basic mechanisms of damage in CMCs: matrix cracking, interfacial sliding, and interfacial debonding. They can occur independently or interactively. The experimental evidence indicates that the interfacial damage (slip and debonding) occurs primarily in conjunction with matrix cracking [41]. Talreja [59] used these damage configurations to characterize damage, as shown in Fig. 12.8a–d. Figure 12.8b,d shows interfacial slip and debonding, both in conjunction with matrix cracking. This situation (b) will result if the fibers are held in the matrix by frictional forces at the interfaces, whilc (d) is likely to result from a nonuniform interfacial bond strength [59].

The characterization of damage is done by regarding damage entities as internal structure of the homogeneous body. The internal structure changes with loading and causes changes of the overall response of the composite. The internal structure of a continuum is described using the so-called *internal variables*. These variables are some appropriately defined quantities representing the geometry, i.e., size, shape, orientation, etc., of the internal structure as well as the influence of the internal structure on the response considered. The quantities chosen depend on the geometrical characteristics of the entities involved in the internal structure constitution and the nature of the influence of these entities on the response of the composite. The elementary damage entities present in the damage configurations treated here are cracks, debonds, and slipped surfaces. The characterization used for cracks and debonds is different from that used for slipped surfaces. These two types are, therefore, treated separately in the following.

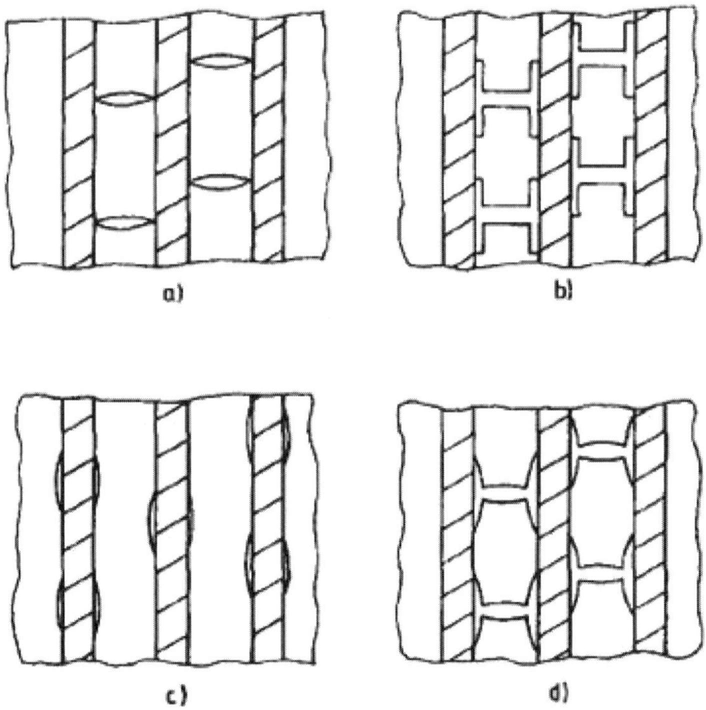

Fig. 12.8. Distributed damage configurations in CMCs: (**a**) matrix cracking, (**b**) interfacial slip in conjunction with matrix cracking, (**c**) debonding, and (**d**) debonding in conjunction with matrix cracking [59]

Matrix cracking

A matrix crack can be viewed as a pair of internal surfaces in a composite that are able to perturb the stress state in a region around the surfaces by conducting displacement, i.e., separation of surfaces, from the undeformed configuration. The surface separation per unit of applied external load depends on the size and shape of the surfaces as well as on the constraint, if any, imposed by the surroundings. For a matrix crack in a unidirectional CMC, the constraint comes from the bridging fibers as well as from the stiffening effect of fibers in the matrix surrounding the crack.

The description for matrix cracks follows a second-order tensor characterization as suggested first by Vakulenko and Kachanov [64] and described in further detail by Kachanov [31]. Talreja [58] used a diad *an* to characterize a damage entity of a finite volume bounded by a surface *S*. In this characterization, *n* is the unit outward normal to the surface at a point on the surface and a represents the extent and direction of some "influence," e.g., the disturbance of the strain fields, due to presence of the damage entity, referred to the same point on the surface. A "damage entity tensor" is then defined as

$$d_{ij} = \int_S a_i n_j \mathrm{d}S, \tag{12.5}$$

with reference to a Cartesian coordinate system X_i. The influence vector can be resolved along the normal and tangential directions with respect to the crack surface. For the type of crack considered here, it is reasonable to assume that only the normal (crack opening) displacement matters, allowing a_i to be expressed as

$$a_i = an_i, \tag{12.6}$$

where the quantity a now represents a measure of the crack influence. From dimensional analysis, with d_{ij} taken to be dimensionless, a has dimensions of length. Drawing upon fracture mechanics, this length is in proportion to the crack length. For a fiber-bridged matrix crack, the crack length l can be expressed in multiples of the average interfiber spacing. Thus,

$$l = kd\frac{1-\sqrt{v_{\mathrm{f}}}}{\sqrt{v_{\mathrm{f}}}}, \tag{12.7}$$

where k is a constant, d is the fiber diameter, and v_{f} is the fiber volume fraction. The expression in (12.7) is based on a hexagonal fiber arrangement. Similar expression will result from another assumption of fiber distribution in the cross section. It can now be inferred that the microstructural *length scale* for matrix microcracking is the fiber diameter. Note that, for an irregularly shaped crack surface, the interfiber spacing and, therefore, the fiber diameter will still be the length scale.

The consequence of the presence of a matrix crack is generally in changing the composite's deformational response, which is defined and measured at a larger length scale, e.g., the characteristic length of a volume containing a representative sample of the cracks. This volume is called a *representative volume element* (RVE). For the Stage II stress–strain response [53], the model proposed in [59] was used. Accordingly, assuming the influence vector magnitude a to be proportional to the crack length,

$$a = \kappa l, \tag{12.8}$$

where l is a constant representing the constraint to the crack surface displacement. This constant equals zero when the constraint allows no crack separation, while it increases as the constraint reduces. From (12.5), (12.6), and (12.8), the damage entity tensor for matrix cracking is

$$d_{ij}^{\mathrm{mc}} = \kappa l^2 t n_i n_j, \tag{12.9}$$

where t is the specimen thickness (or the through-thickness characteristic dimension of the crack).

The macrolevel deformational response is derived from a strain energy density function that depends on the strain and damage states. The matrix-cracking damage state is characterized by Talreja [59]

$$D_{ij}^{\mathrm{mc}} = \frac{1}{V}\sum d_{ij}^{\mathrm{mc}}, \tag{12.10}$$

where V is the RVE volume of an RVE over which the summation is conducted. Substituting (12.5) in (12.6), one obtains

$$D^{\mathrm{mc}} = \kappa\eta A\langle fl\rangle, \tag{12.11}$$

where $D^{\mathrm{mc}} = D_{11}^{\mathrm{mc}}$ is the only surviving component of the damage mode tensor, f is the fraction of RVE width spanned by a crack, η is the crack number density, i.e., the number of cracks per unit volume, and A is the cross-sectional area. The quantity within the brackets $\langle\rangle$ is averaged over the RVE volume.

The matrix crack length (12.7) appears in the damage descriptor (12.11). As shown in [59], the crack length also governs the elastic constants at a given crack density η. For instance, the axial Young's modulus can be written as

$$E_{11} = E_{11}^0(1 - c\eta l), \tag{12.12}$$

where c is a constant and the superscript 0 is for the initial value.

In characterizing matrix cracks as a damage mode, no specific account is made of the associated fiber/matrix debonding and sliding mechanisms. These can be considered separately and then accounted for by their interactions with the matrix cracks [59]. Discussions of these mechanisms follow.

Interfacial debonding

The fiber/matrix interface can debond due to several causes. Essentially, a stress normal to fibers or a shear stress along fibers, or a combination of the two, must exist for the bond to fail. These stresses can be generated by a fiber break or brought into play by an approaching matrix crack. Alternatively, a preexisting flaw at the fiber surface or an imperfect fiber, or its misalignment, can produce those stresses. If debonds are produced without interaction with matrix cracks, then they can be characterized in a manner similar to matrix cracks. A characterization of such distributed debonding is given in [59] based on certain simplifying assumptions. The only surviving damage mode tensor component for this case is D_{22}, and its form is the same as that of D_{mc} in (12.11). Thus, the debond length and the debond number density enter into the damage mode description. The debond length will depend on the characteristic flaw length, which in turn depends on the manufacturing process. Unless the ability of the manufacturing process to produce interfacial flaws somehow depends on the composite microstructure, no microstructural length scale can be identified for the debonding mechanism.

In the case of a matrix crack initiating debonding and then merging with the debond crack, more force on the advancing debond crack comes from the opening displacement of the matrix crack. The damage configuration of interest, then, is not the debond crack itself but a combined matrix-debond crack. The latter can be viewed as a fiber-bridged matrix crack, discussed above, with the constraint to its surface displacement now modified by the presence of debonding. Then, the constant α in (12.8), (12.9), and (12.11) may be changed to another value, resulting in a change of the constant c in (12.12).

Thus, for debonding that occurs in conjunction with matrix cracking, the determining length associated with the damage mode is still the matrix crack length l, although with a modified influence. This length can still be expressed by (12.7), giving the fiber diameter as the microstructural length scale.

Specific treatments of debonding by itself and debonding in conjunction with matrix cracking are given in [59]. Based on that work, the axial modulus for the latter case can be modified from (12.12) to be

$$E_{11} = E_{11}^{0}(1 - c^{'}\eta l), \tag{12.13}$$

Where

$$c^{'} = c + k_1 d_1, \tag{12.14}$$

where d_1 is the ratio of the debond length to the crack length and k_1 is a constant. Here, a fixed ratio of the number of debonds per unit of matrix crack length has been assumed.

From (12.13) and (12.14), it can be seen that the debond length does not enter into the RVE response directly but via its ratio to the crack length, suggesting that the governing length for this response is the crack length.

Interfacial sliding

Interfacial sliding occurs when fibers and matrix remain in contact after debonding of the interface and undergo unequal displacements. Talreja [59] defined a measure of the slip at the interface as the area swept off by the relative displacement of one constituent over the other and expressed this measure in terms of a slippage vector. A slippage tensor was then constructed as a dyadic product of the slippage vector with itself to account for the insensitivity of the material response to the direction of slip. As in the case of debonding, discussed previously, when sliding occurs in conjunction with matrix cracking, the slip damage tensor, which represents this damage mode averaged over the RVE, turns out to depend on the average matrix crack length. In fact, it depends explicitly on the average COD, which in turn depends on the average crack length. The only surviving component of the slip damage tensor can be written as [59]

$$D^{\mathrm{sl}} = \frac{\pi^2 d^4 \eta^2}{64 v_{\mathrm{f}}^2} \langle c_{\mathrm{d}}^2 \rangle, \tag{12.15}$$

where d is the fiber diameter, v_{f} is the fiber volume fraction, and c_{d} is the COD; and the quantity within the brackets $\langle\,\rangle$ is averaged over the RVE volume.

Assuming the COD to be proportional to the crack length, (12.15) may be rewritten as

$$D^{\mathrm{sl}} = \xi \eta^2 \langle d^4 l^2 \rangle, \tag{12.16}$$

where ξ is a constant depending on the fiber volume fraction and fiber stiffness. The fiber diameter is placed within the brackets to allow for its variation. Equation (12.16) indicates that this damage mode depends directly and strongly on the fiber diameter in addition to depending on the matrix crack length, which in turn is expressible in terms of the fiber diameter, as in (12.7). Thus, the microstructural length scale in this case is the fiber diameter. Note that the fiber length over which sliding occurs is not a characteristic dimension of the mechanism when it occurs in conjunction with matrix cracking.

12.3.2 Ply Cracking in Laminates

Figure 12.9 shows an X-ray radiograph of a carbon–epoxy crossply laminate after being subjected to tension–tension cycling. In this 2D view, the horizontal lines are images of cracks in the 90° plies, while the vertical lines indicate cracks (also called *axial splits*) that lie in the 0° plies [27].

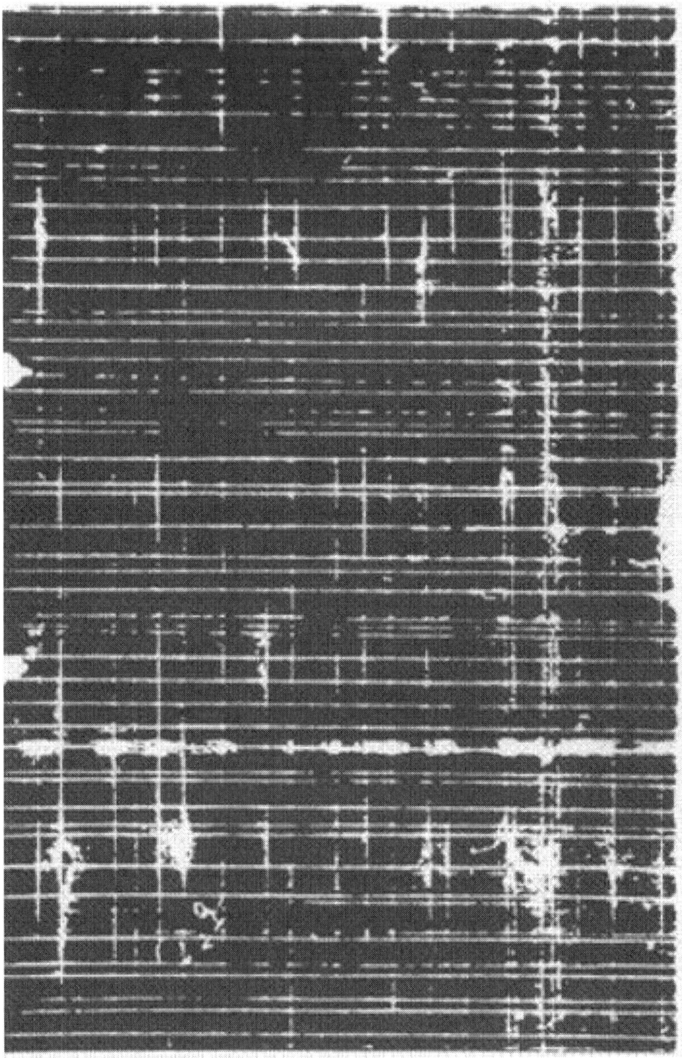

Fig. 12.9. An X-ray radiograph showing transverse cracks, axial cracks, and delaminations in a crossply laminate after fatigue [27]

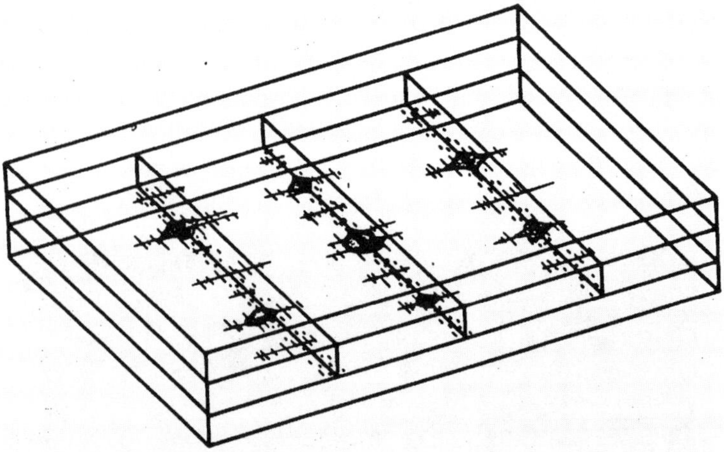

Fig. 12.10. The cracks and delaminations seen in the X-ray radiograph (see Fig. 12.9)

The shaded areas are sites of interlaminar cracks (delaminations), which are depicted in Fig. 12.10. For the sake of this discussion on length scales of damage, the focus will primarily be on ply cracking.

Figure 12.11 illustrates multiple matrix cracking in a ply of an arbitrary orientation θ with respect to the 90° direction. The cracks are shown at a mutual spacing s which represents the average crack spacing in an RVE. Using a second-order tensor characterization for this mode of damage [58] gives

$$D_{ij}^{\mathrm{pc}} = \frac{\kappa t_{\mathrm{c}}^2}{st\cos\theta} n_i n_j, \tag{12.17}$$

where the superscript pc stands for ply cracking, κ is a ply constraint parameter, t_{c} is the thickness of the cracked ply, and t is the laminate thickness. The components n_i of the unit vector normal on a crack surface are given by

$$n_i = (\cos\theta, \sin\theta, 0), \tag{12.18}$$

where θ, as shown in Fig. 12.11, is the crack inclination.

The laminate stiffness matrix in the presence of a fixed state of ply cracking is given by Talreja [58]

$$C_{pq} = C_{pq}^{0} - C_{pq}^{\mathrm{D}}, \tag{12.19}$$

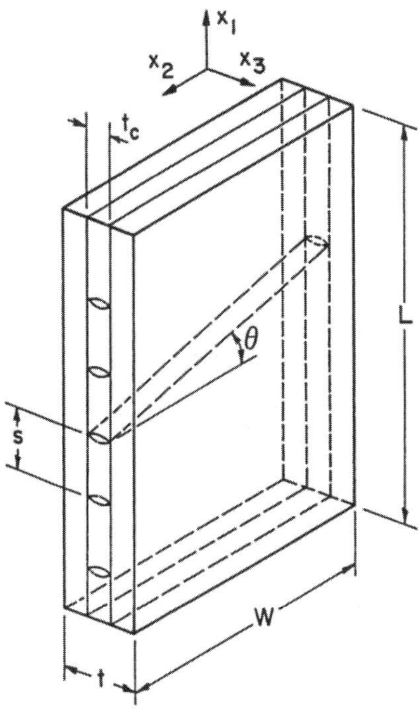

Fig. 12.11. Multiple cracking in a general off-axis ply of a laminate

where the indices p and q take values from 1 to 6 in accordance with the Voigt notation. The superscript 0 on the stiffness matrix indicates initial value while D indicates the contribution due to damage. Following classical laminate theory, the stiffness matrix for the virgin laminate can be written in terms of elastic moduli as below:

$$C_{pq}^{0} = \begin{bmatrix} \dfrac{E_1^0}{1-\nu_{12}^0\nu_{21}^0} & \dfrac{\nu_{12}^0 E_2^0}{1-\nu_{12}^0\nu_{21}^0} & 0 \\ & \dfrac{E_2^0}{1-\nu_{12}^0\nu_{21}^0} & 0 \\ \text{Symm} & & G_{12}^0 \end{bmatrix}. \tag{12.20}$$

The stiffness change due to damage depends upon the laminate ply layup. For instance, consider damage in $[0/\pm\theta_m/0_n]_s$ laminate. The ply cracks develop in $\pm\theta$ layers and the change in stiffness matrix is given as

$$C_{pq}^{\mathrm{D}}(\pm\theta) = C_{pq}^{\mathrm{D}}(+\theta) + C_{pq}^{\mathrm{D}}(-\theta) = \frac{\kappa_\theta t_{\mathrm{c}}^2}{st} \sin\theta \begin{bmatrix} 2a_1 & a_4 & 0 \\ & 2a_2 & 0 \\ \text{Symm} & & 2a_3 \end{bmatrix}, \quad (12.21)$$

where κ_θ is the constraint parameter for ply orientation equal to θ and a_i are phenomenological constants. For the particular case of damage in $[0/\pm\theta_m/0]_s$ laminates, the crack spacing in both $+\theta$ and $-\theta$ plies is assumed to be the same, as the damage response for both orientations would be nearly the same on external loading. Thus, these two damage modes, by virtue of their response behavior, effectively act like a single damage mode. Clearly, the above equation involves four unknown constants. These constants can be evaluated either experimentally or though numerical FE simulations. The procedure is outlined as follows. Experimentally observe degradation in stiffness properties for a reference laminate configuration such as a crossply laminate. Fit a straight line to the experimental data normalized with regard to stiffness of the virgin laminate. Evaluate $(C_{pq})_{\exp}$, i.e., compute E_1, E_2, υ_{12}, and $\upsilon_{21} = (E_2 / E_1)\upsilon_{12}$ at a certain crack spacing s_0. Using the equations above with $\theta = 90°$, the following is obtained

$$C_{pq} = (C_{pq})_{\exp} \Rightarrow$$

$$\begin{bmatrix} \frac{E_1^0}{1-\nu_{12}^0\nu_{21}^0} & \frac{\nu_{12}^0 E_2^0}{1-\nu_{12}^0\nu_{21}^0} & 0 \\ & \frac{E_2^0}{1-\nu_{12}^0\nu_{21}^0} & 0 \\ \text{Symm} & & G_{12}^0 \end{bmatrix} - \frac{\kappa_{90} t_{\mathrm{c}}^2}{s_0 t} \begin{bmatrix} 2a_1 & a_4 & 0 \\ & 2a_2 & 0 \\ \text{Symm} & & 2a_3 \end{bmatrix} = \begin{bmatrix} \frac{E_1}{1-\nu_{12}\nu_{21}} & \frac{\nu_{12} E_2}{1-\nu_{12}\nu_{21}} & 0 \\ & \frac{E_2}{1-\nu_{12}\nu_{21}} & 0 \\ \text{Symm} & & G_{12} \end{bmatrix}. \quad (12.22)$$

One important aspect pertaining to these constants is that though these constants are determined for the reference laminate, for other laminate configurations they are assumed to remain unaffected by angle θ for a given ply material. From experimental observations on carbon/epoxy [60] and glass/epoxy [68] laminates, this assumption is found to hold true. This is because the damage constants are primarily determined by the constituent ply properties and are not very dependent on ply orientation. Of course the influence of the ply orientation on the constraint posed by undamaged plies over damaged plies is important and is suitably carried by the "constraint parameter" κ through changes in COD.

The length scale variable entering the damage descriptor (12.17) and, consequently, the elastic response change (12.21) is the crack dimension, which for fully developed ply cracks, as assumed here, is the ply thickness t_c. The other crack dimension along the fiber axis in the ply extends as far as the imposed stress acts and is, therefore, not the characteristic length scale of the cracking mechanism. Expressed differently, the crack surface displacement, which is the cause of stress perturbations and, thereby, the elastic response changes, depends on the crack dimension through the ply thickness.

Although the ply cracks are assumed for simplicity to be sharp tipped, as illustrated in Fig. 12.11, in reality they must get blunted by merging with the local delamination, i.e., the separation of plies at the interface, caused by the intense stress field carried by the approaching ply crack fronts. The extent of the delamination cracks along the ply interfaces must depend on the interfacial bond strength as well as on the ply crack length t_c. In fact this situation is analogous to the fiber/matrix debonding in conjunction with matrix cracking in unidirectional CMCs, discussed previously. Drawing upon that analogy, it can be deduced that the delamination length enters the analysis not directly but via its ratio to the ply crack length t_c. Thus, once again, the relevant length scale variable is the total cracking ply thickness t_c.

When cracking occurs in more than one-ply orientation, multiple length scales result with each length scale variable equal to the combined thickness of the set of consecutive cracking plies of the corresponding orientation. Also, the delamination associated with each ply cracking contributes to the effect on the elastic response via the ratio of the associated delamination length to the ply crack length. Figure 12.12 illustrates the ply (matrix) cracking and delamination in an angle ply laminate.

Finally, return to the delamination mode observed in the fatigue of cross-ply laminates depicted in Figs. 12.9 and 12.10. As illustrated in Fig. 12.10, this delamination occurs locally at the intersection of cracks in the two orthogonal orientations in adjacent plies. The cause of this delamination and the effect of its presence have not been adequately analyzed. Therefore, any inference regarding its characteristic length scale is speculative at present. It appears, however, that the growth of the delamination is mainly along the two orthogonal ply crack directions, suggesting, therefore, two length scales. These length scales may be described as the two principal directions of an ellipse, which may be taken to approximate the delamination geometry.

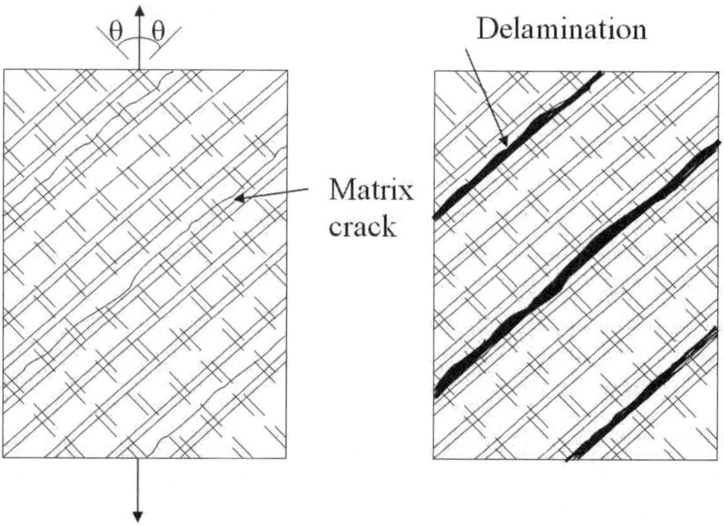

Fig. 12.12. Matrix cracks (*left*) and delamination (*right*) in an angle ply laminate

12.4 Multiscale Modeling of Damage for Elastic Response

12.4.1 Local vs. Nonlocal Description of Damage

The traditional way of describing the effect of damage entities on material behavior is quite similar to the definition of stress and strain, i.e., it is defined at a point by taking an infinitesimal volume around the point into consideration and taking a limit as volume goes to zero. However, this localized definition of damage is not consistent with the real behavior of the damage process. The damage entities affect the stress and strains in the neighborhood. Moreover, at the microscale level, the size of these damage entities, such as matrix cracks, debonds, etc., or flaws, such as voids, inclusions, etc., is finite and cannot be neglected. The third aspect to be considered is the variety of scales involved with different damage entities. The length scale aspects involved with different damage entities will be dealt with in Sect. 12.4.2.

The key aspect to be discussed here is the evolution of damage entities. A solid that is highly heterogeneous at the mesoscale is considered an effective homogeneous continuum at the macroscale. Macroscopic damage variables are judiciously selected to reflect the effects of mesostructural-level irreversible processes on macroscale material behavior. Such damage descriptors are usually obtained through a "low-order" homogenization (spatial average) of individual damage entities, neglecting details of the distribution of damage throughout the RVE. Whereas effective moduli are somewhat insensitive to the distribution of microcracks, damage evolution is highly dependent on the local fluctuations in crack arrangement within the RVE used for stiffness calculations [32]. Bazant and Chen [5] discussed the scale dependence of energy release in fracture of heterogeneous, quasibrittle solids.

In the homogenization process, critical information regarding the largest flaw size, minimum distance between flaws, and distribution of damage within an RVE is irrevocably lost. Such information is crucial to the development of viable damage evolution equations. Current CDM approaches have been generally limited to the case of dilute (noninteracting) damage. This limitation suggests the need for a higher-order continuum description of damage that retains key aspects of the damage distribution within an RVE.

The choice of damage variable is either macroscopic or micromechanic based. The damage descriptors could be scalar or tensor, scalar descriptions being too simplistic in nature. In general, both macroscopically measurable and micromechanically inspired damage variables neglect the varying effects of nonlocal or "nearest neighbor" influences, e.g., shielding and enhancement associated with adjacent flaws, that are essential to formulate damage evolution laws. Inclusion of such effects represents, perhaps, one of the greatest challenges in the development of a robust CDM formulation. For these reasons, more careful consideration of appropriate ISV measures of damage is warranted.

The RVE is commonly defined as a cube of material with dimension L_{RVE} subject to the following conditions [34, 35]

$$\begin{aligned} &\frac{d}{L_{\mathrm{RVE}}} \ll 1, \quad L_{\mathrm{c}} \le L_{\mathrm{RVE}} \le L, \\ &\left|\frac{\partial \sigma_{ij}}{\partial x_k}\right| L_{\mathrm{RVE}} \ll \sigma_{ij}^{0}. \end{aligned} \tag{12.23}$$

where d is the characteristic size of microconstituents, L_{c} is the heterogeneity correlation length, L is a characteristic macroscopic structural dimension, σ_0 is the mean field (volume averaged) stress, and x_k (x_1, x_2, x_3)

are the components of a Cartesian basis. Statistical homogeneity, for general purposes, requires that all response functions of interest at the scale of the observation volume window (Helmholtz free energy, Cauchy stress tensor, stiffness tensor, etc.) are essentially invariant with respect to window position [36]. It is very important to properly select suitable observation windows for averaging response functions and to analyze the influence of observation window boundary conditions, e.g., uniform vs. "random-periodic" traction and displacement boundary conditions, on the statistical homogeneity of response functions for a given window size. Due to multiple scales involved with evolving damage mechanisms,

$$\mathrm{RVE}_{\mathrm{evolution}} \neq \mathrm{RVE}_{\mathrm{stiffness}}. \tag{12.24}$$

For further discussion on RVE-level damage characterization, the reader is referred to [36].

12.4.2 Microstructural Length Scales

In Sect. 12.3, the governing dimensions of the damage entities are examined from the viewpoint of elastic response in the presence of damage. Two specific cases of damage in unidirectional CMCs and PMC laminates have been examined. The framework within which the issue of characteristic lengths has been analyzed is CDM using characterization of damage with second-order tensors. This particular representation of damage, in the form used here, provides a consistent characterization of the basic damage entity involved in each case, e.g., a matrix crack in a CMC and a ply crack in a PMC laminate.

The crux of the characterization is the "influence" vector, which provides a relevant measure of the action induced by the presence of the damage entity. By expressing the magnitude of this vector in terms of the characteristic and governing dimension of the damage entity, the "essence" of the damage entity is carried into the damage entity tensor. This dimension, when properly identified and related to the microstructural entities, provides the length scale associated with the damage mechanism considered.

The relevance of the length scale to determining the elastic response affected by that damage mode becomes clear when the response measured over the RVE is examined. For this, a damage mode tensor, which acts as an ISV in a continuum damage framework, is considered. The damage mode tensor has been examined in a simple form, such as that in (12.10), which is the volume average of the damage entity tensor over the RVE.

This does not account for the damage entity distribution and can, therefore, be used only for the RVE-averaged, i.e., the mesoscale, response and not for describing damage evolution. Thus, the length scales of damage within the context of the elastic response have been examined.

The next question to address is: What is the significance of the length scales of damage? The basic concept behind length scales appears to be the intuitive idea that effects seen at a given observation "window," e.g., RVE size, must be determined by events occurring at dimensions smaller than the window size. Implicit is the assumption that those events are associated with certain discrete entities, such as grains in a polycrystalline material, and that the action of those entities and interactions between them, when averaged over the window size, provides the "response" variables applicable at that scale.

Carrying this logic one step "behind" would suggest that the response at the scale of the discrete entities in an RVE could be given by the subentities lying within those entities. Thus, if a grain is viewed as an entity, then the dislocations within the grain could be the subentities. This move to smaller and smaller size scales could, in principle, have no end other than the limit set by the tools of observation and analysis available at a given time.

In engineering science, in contrast to "pure" sciences, one takes a pragmatic approach driven by the application or need at hand. From this point of view, one must consider the purpose first and go as far down in scales as needed. Thus, if the purpose is to determine the elastic response changes induced by damage in a composite material, then one must go as much down in length scales as necessary to determine the reversible deformation (or stress) related effects but no further.

The next question is whether a hierarchy of length scales can be identified. What has been illustrated by the discussion of the two cases of composites with damage is that a simple hierarchy of length scales does not exist. Instead, a complex damage mode may involve more than one governing length, e.g., the matrix crack length and the interfacial sliding length for a fiber-bridged crack. Also, multiple damage modes may operate simultaneously and interactively, leading to multiple length scales, e.g., in the case of multiple off-axis plies in a laminate. These considerations suggest that an alternative is needed to the strategy of starting at the smallest length scale and working up the scale hierarchy. Talreja [60] proposed one such strategy, the so-called "synergistic" damage mechanics. The following discussion will address the two strategies.

12.4.3 Hierarchical Multiscale Modeling

There are three levels of modeling of composites with a hierarchical approach, as depicted in Fig. 12.13:

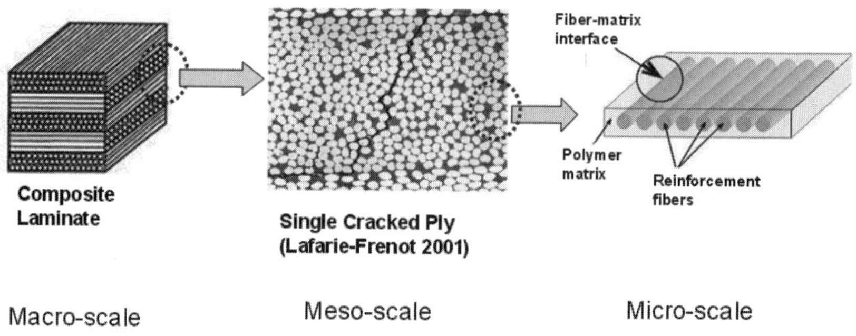

Fig. 12.13. Three scales involved in hierarchical modeling of composite laminates

- *Microscopic level.* This is the lowest level of observation, wherein fiber and matrix phases are modeled separately and the average properties of a single reinforced layer are determined from individual constituent properties by a suitable homogenization technique. With regard to damage in composites, micromechanics includes matrix cracks inside a layer, called *microdamage mechanics* (MDM). Hashin type of analyses belongs to the set of MDM approaches.
- *Mesolevel.* At this level, the ply is considered homogeneous and the virgin (undamaged) material is regarded as either orthrotropic or transversely isotropic. This scale is very useful in describing and predicting the damage/failure of composites.
- *Macroscopic level.* This refers to the structural level wherein the whole structure is considered as a homogeneous continuum and material behavior is described by an anisotropic constitutive law. The traditional concepts of continuum mechanics work quite remarkably here; and, thus, the overall structural behavior to external loading can be studied using suitable FE modeling or by solving a boundary value problem with effective material properties.

It would be fair to say at the outset that the hierarchical multiscale approach is intuitively logical. For a complex composite architecture, which is quite often the case in practical applications, one can think of starting with the smallest basic unit – a discrete fiber embedded in a matrix – and proceed to the level of a representative unit of a collective fiber arrangement. The basic unit can be analyzed as a piecewise homogeneous continuum, with two regions, if fiber and matrix are considered, or three, if an interfacial layer is added. The result (stress, strain, temperature, etc.) can then be averaged in some sense over a representative unit to get a description for the homogenized medium. Several models for doing this exist, e.g., the Mori–Tanaka method. These models aim at bridging the two scales: the scale of the basic unit and the RVE scale.

Generally, the issue of uniqueness remains unresolved in the sense of representation of the collective fiber effect. There is as yet no precise and rigorous definition of a representative unit for a general case, which is the source of the lack of uniqueness. A logical extension of the discrete-to-collective bridging of the fiber–matrix case to higher scales produces the hierarchical approach in multiscale modeling. One can argue whether or not this approach is efficient, in spite of its logic. Historically, the hierarchical approach has not preceded other approaches. A structural analyst has worked with macrolevel descriptions of material behavior, e.g., the classical laminate plate theory, and has looked for microlevel information as needed. A materials developer, on the other hand, has focused on effects of constituents and their microstructural arrangements on properties. In recent years, the seemingly abundant computer power has motivated the hierarchical approach with the hope of integrating materials design and structural analysis.

The objective here is to examine approaches for multiscale analysis of damage in composites. A first thought would be to conduct damage initiation and progression analysis as a part of the hierarchical multiscale approach. It turns out not to be that straightforward. The issues confronting this approach are discussed below, along with the merits of an alternative approach.

An overall view of the multiscale approach is reviewed first. Figure 12.14 illustrates, from left, an object of structural integrity assessment within which a region of potential criticality (failure) exists. This region (a substructure) is analyzed to determine the loading on its boundary. The next step is to examine how this loading induces damage. This step requires analyzing heterogeneities (microstructure, generally), which govern initiation of damage. Simple examples are debonding of fibers from the matrix and matrix cracking from broken fiber ends. The analysis of local stress/strain fields to determine such microfailures is commonly referred to as *micromechanics*. Micromechanics could be conducted at multiple scales. An example is fiber/matrix debonding at the fiber diameter scale and coalescence of the debond cracks at the scale of a representative number of fibers.

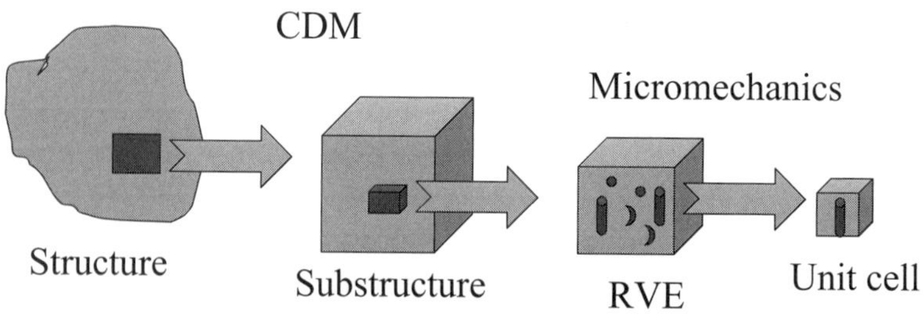

Fig. 12.14. A multiscale approach starting from the structural scale, moving down to lower scales

In Fig. 12.14, the direction of the arrows indicates moving from the structural (macro) scale downward to decreasing length scales. Until the microstructural entities are explicitly included in an analysis, the regime of analysis is characterized as "continuum," beyond which it is known as micromechanics. In the context of damage, the continuum regime is called *continuum damage mechanics*, while the micromechanics is typically not given an additional characterization (except, perhaps, occasionally as micro-damage mechanics) [23].

Historically, the fields of CDM and micromechanics have developed independently, CDM going back to [31], while micromechanics originated in various works; but its characterization as a coherent field may be credited to Budiansky [6], who defined it as "the mechanics of very small things." In recent years, the upsurge of computational mechanics has also boosted micromechanics, adding the aspects of numerical simulation and length scale-based characterizations such as "nanomechanics."

The increasing confidence in the power of computation has led to the notion of the hierarchical approach, with the implicit assumption that "basic" laws, when placed into a simulation scheme, will lead to physically correct results. Thus, once the microstructure, at any chosen level of length scale, has been codified in a simulation scheme, the results of the computation will describe the behavior at the next higher level, the assumption goes. In the context of damage mechanics, this may raise a few issues, as discussed below.

The first issue in a hierarchical approach is the choice of length scales. As discussed in Sect. 12.3, the microstructural length scales are relatively straightforward; and, consequently, setting up a hierarchy of scales and procedures for bridging between them can be accomplished relatively easily. However, the *microstructural configuration and driving forces for damage initiation and progression determine the length scales of damage.*

Thus, length scales of damage and their hierarchy are not fixed but are subject to evolution. To illustrate this, consider ply cracking in a laminate. In the early stage, individual ply cracks initiate from debonding of fibers, giving the damage length scale in terms of the fiber diameter. When the ply cracks are fully grown through the ply thickness, the mechanism of interest is the multiplication of cracks. At this stage, the damage length scale is crack spacing, which in turn depends on the ply thickness as well as the constraint to surface displacement of the ply cracks. The two-stage behavior and the evolving nature of damage complicate any hierarchical scheme for predicting response.

Another issue in a hierarchical approach is the multiplicity of damage modes. If more than one damage mode operates at a time and there is interaction between the modes, then a hierarchy of length scales becomes questionable. Consider ply cracking in a commonly used quasi-isotropic $[0/\pm45/90]_s$ laminate in axial tension. Multiple matrix cracking occurs in 90° plies, followed by the same in plies of –45° and 45° orientations. The three-ply cracking modes progress interactively and, at some stage, concurrently. The length scales associated with the three damage modes do not show hierarchy. Consequently, bridging the scales by some averaging scheme becomes irrelevant. Further complicating the hierarchical scale arrangement is the interlaminar cracking that results from the cracking in individual off-axis plies.

12.4.4 Synergistic Multiscale Modeling

An alternative to the hierarchical approach is the SDM approach proposed by Talreja [60]. Conceptually, the approach combines the strengths of CDM and MDM. In CDM, the material microstructure, e.g., distributed fibers, and the distributed damage, which may be called the *microdamage structure*, are treated as smeared-out fields. This homogenization is illustrated in Fig. 12.15 as a two-step process, where the material microstructure is viewed as consisting of "stationary" entities, e.g., fibers and plies, and the microdamage structure is considered as a family of evolving entities, e.g., cracks and voids. A set of response functions are expressed in terms of the field variables (stress, strain, and temperature) and internal variables, which represent the smeared-out field of evolving damage entities. The internal variables, although being field quantities, actually have an RVE associated with them at each material point. Strictly speaking, there is another RVE associated with the stationary microstructure; but it is customary in continuum treatments to bypass it by requiring that the quantities, such as the elastic moduli of virgin material, be measured at a scale much larger than the scale of the individual stationary microstructure entities. There is a tendency to do that for the RVE associated with damage as well, as evidenced in finite element-based analyses where reduced (damage induced) properties are assigned at nodal points. The reduced properties are meaningful only at the RVE scale, which depends on the length scales of damage discussed above.

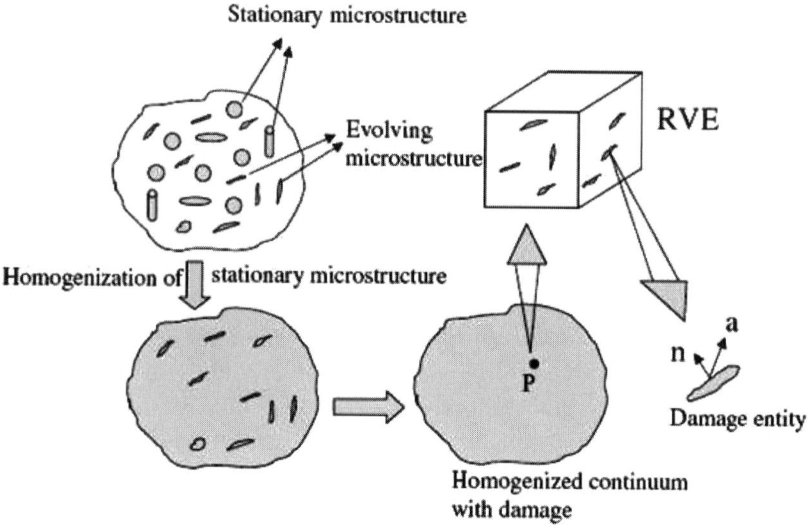

Fig. 12.15. The two-step homogenization process for composites with damage

Although the hierarchical approach seems to be the most common choice, it disregards many of underlying issues. Another suitable approach is to combine the micromechanics with continuum damage mechanics, called *synergistic damage mechanics* [60]. To illustrate the structure of SDM, consider the Helmholtz free energy function for isothermal mechanical response as

$$\phi = f(\varepsilon, D), \tag{12.25}$$

where the strain tensor ε and damage variable D, generally also a tensor, are independent variables representing the material state. The variable D is viewed as an internal variable, representing some measure of the collective presence of damage entities in an RVE at the considered point where the material response is sought. In Fig. 12.15, the RVE at a point is shown as a finite-sized cube of material containing a representative sample of damage entities.

The internal state (damage) in a general case may contain multiple modes, such as the ply cracking modes in a $[0/\pm45/90]_s$ laminate. In the conventional CDM approach, the response function R, and any function derived from it, is expressed in terms of D, which is formulated to represent a measure of the intensity of damage. Examples of such measures are void volume fraction and crack number density. Such "passive" measures end the CDM at the RVE level, i.e., the role of CDM gets limited to generating constitutive relationships at the RVE (meso) level that are then used for analyzing structural (macro) response. In the SDM approach, one proceeds down from the RVE level to one or more microlevels as warranted by the situation at hand. This is accomplished by developing a characterization of damage entities that is "active" in the sense that the presence of damage entities is accounted for by including the "influence" of damage entities.

In contrast, the passive characterization is limited to only accounting for the "presence" of damage entities by measures such as crack number density, as noted above. The characterization of influence is accomplished by assigning a two-vector representation to a damage entity, as illustrated in Fig. 12.15. The vector a carries the influence through its magnitude and direction. The magnitude of the vector represents a measure of how much the damage entity is able to affect its surroundings, while the direction of the vector indicates the orientation in which this effect acts. For instance, if there is concern about the deformational response of a composite, then clearly the surface of a given damage entity must conduct a displacement to affect this response. Imagine, for instance, a transverse crack in 90° plies of a $[0/\pm 45/90]_s$ laminate. The degree to which this crack opens under an imposed axial load increment will determine how much the axial elastic modulus of the composite will reduce. If the axial stiffness of the sublaminate [0/±45] is high, then the COD will be low; and, consequently, the modulus reduction will be small. In a passive damage characterization, where only the crack number density enters, no distinction can be made between the presence of cracks in different constraining environments.

In the SDM approach, the constraint to the damage entity influence is represented in a constraint parameter, such as α in (12.8) for a fiber-bridged matrix crack. The determination of the constraint parameter, and generally any influence function, is accomplished by a micromechanics analysis at levels warranted by the length scales of damage.

The SDM approach has been illustrated in [69] for the elastic response of $[\pm\theta/90_4]_s$ laminates and in [70] for the linear viscoelastic response of crossply laminates of different 0/90° ply mix. In each case, the transverse cracking in 90° plies was considered as the damage mode subjected to different constraints. Thus, in the $[\pm\theta/90_4]_s$ laminates, θ is varied to vary the constraint, while no cracking is considered in the $\pm\theta$ plies. The objective in both of the works just cited was to demonstrate SDM for the case of one damage mode with varying constraint and varying mesoscale (RVE size). This addresses the first of the two issues in the hierarchical multiscale modeling discussed previously. This author and his associates in ongoing work are treating the other issue of multiple damage modes. The main ideas in dealing with the first issue are discussed next.

Let us consider the elastic response of $[\pm\theta/90_4]_s$ laminates. At a damage state where multiple transverse cracks of average spacing s exist, an elasticity response function (modulus) derived from the free energy function (12.25) can be expressed as [56]

$$\frac{R}{R_0} = 1 - \kappa \frac{t_c}{s} f_1 f_2, \tag{12.26}$$

where R_0 is the initial (undamaged) value of the response, t_c is the thickness of the cracked plies, κ is the constraint parameter, and f_1 and f_2 are normalized functions of the laminate geometry (ratio of cracked to uncracked plies) and ply properties, respectively. The expression in (12.26) results from a linearized theory; more terms of higher order in t_c / s will appear in a higher-order theory. In [69], it was shown, based on MDM analysis, that the constraint parameter κ could be approximated as a function of θ by a polynomial function of ply properties and ratio of thicknesses of cracked and uncracked (constraining) plies. Thus, with input from MDM, the CDM framework could be applied to the class of $[\pm\theta/90_4]_s$ laminates. Note that, in the conventional CDM framework, the response function R must be calculated separately for each θ value.

In the case of linear viscoelastic response, R, κ, f_1, and f_2 are all functions of time. In [70], it was shown that the functions f_1 and f_2 are normalized functions of laminate geometry and relaxation moduli of undamaged plies, respectively, while the time variation of κ was found by parametric studies of $[0/90_n]_s$ to be given by a polynomial function of the ratio of axial relaxation moduli of the cracked plies to that of the uncracked plies. Once again, an MDM analysis allowed predicting visco-elastic response (for a fixed crack density) for a class of composites with a CDM framework without experimentally determining material constants for each laminate configuration.

12.4.5 Multiple Damage Modes

The authors are presently working toward developing SDM methodology for multiple damage modes. The multiple damage modes may appear due to separate damage mechanisms, such as matrix cracks, debonding, delaminations, etc. Alternatively, if the matrix cracks appear in more than one orientation, these can also be treated through multiple damage modes by assigning one damage tensor to cracks of a particular orientation. SDM methodology has been successfully used to predict stiffness degradation due to matrix cracking in laminate configuration. Work is ongoing to predict damage behavior for more general laminate layups. Based on the results available for the case of multimode damage using SDM, it is evident that a purely MDM approach or a conventional CDM framework will not provide results without excessive computation (for MDM) or tedious experiments (for CDM). Section 12.4.6 presents some of the recent advances pertaining to the SDM approach.

12.4.6 SDM Characterization of Damage in Off-Axis Plies

In the case of ply cracking, the most direct measure of damage is the separation of crack surfaces, known as *crack opening displacement*. These surface displacements do not occur freely due to the constraints from the adjoining undamaged plies. These "constraint effects" are suitably incorporated in the CDM formulation through the constraint parameter κ. Using COD, there are four common ways to characterize damage in the laminates:

1. *MDM/micromechanics*. For a given laminate system, the ply cracks inside the RVE can be characterized by defining an equivalent boundary value problem; and its solution would yield expression for COD in terms of laminate geometry and material properties. This type of analysis has been performed [66] for crossply laminates. This methodology yields exact bounds on stiffness degradation due to ply cracking. Unfortunately, it is limited to only crossply laminates. Solving a boundary value problem for off-axis plies is impossible due to too many unknowns in the formulation.
2. *Gudmundson model*. The second approach is to use the Gudmundson model [16] in which he characterized damage using average COD. Although the stiffness degradation relations are quite accurate in this model, the COD is calculated assuming a system of infinite cracks in a homogeneous medium. Hence, it does not include the "constraint effects" of the adjoining undamaged plies.
3. *Hierarchical approach*. This approach solves the boundary value problem inside RVE using FE simulation and then integrates it to the higher scale to obtain the overall macrolevel response of the structure. This methodology can be used to solve any laminate

layup and looks quite straightforward but has important limitations, which have already been discussed in detail.

4. *Synergistic damage mechanics.* SDM follows the regular CDM formulation except that the constraint effects area is evaluated using numerical simulations at the micromechanics level. This approach can be applied to a variety of laminate layups. It captures the physics of dissipating mechanisms in an accurate manner and is more computationally efficient than MDM or the hierarchical approach. It can also include the multiscale nature of damage efficiently and can solve problems involving multiple damage modes. In the following paragraphs, its applicability to characterize damage in off-axis plies is described.

The procedure for analyzing damage behavior in off-axis plies using SDM methodology is explained in Fig. 12.16. The example illustrated here is of damage in $[0/\pm\theta_4/0_{1/2}]_s$ laminates. The approach needs both micromechanics and CDM for a complete evaluation of structural response.

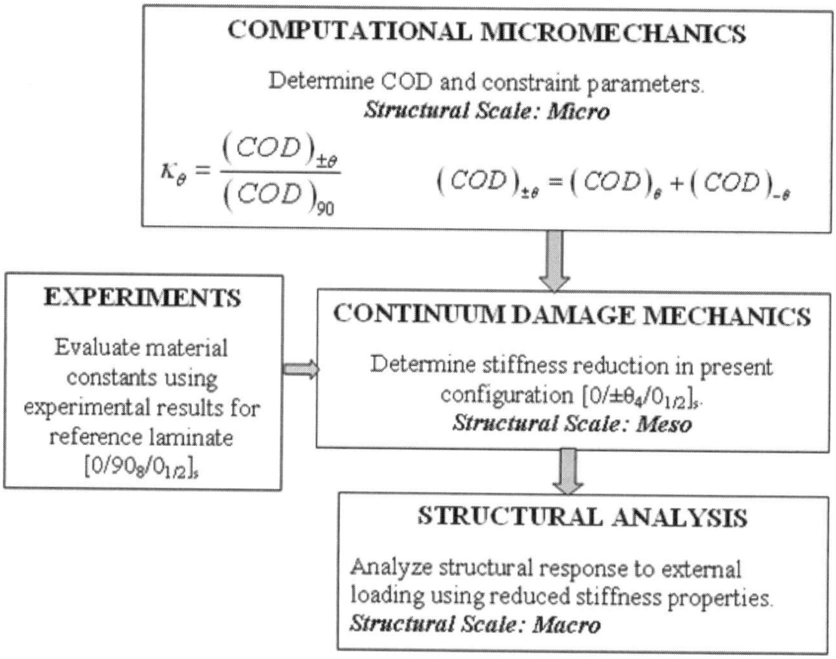

Fig. 12.16. Flowchart showing the synergistic multiscale methodology for analyzing damage behavior in a general symmetric laminate $[0/\theta_1/\theta_2]_s$ with matrix cracks in multiple orientations. The example taken here is $[0/\pm\theta_4/0_{1/2}]_s$ laminate with transverse matrix cracks in and $+\theta$ and $-\theta$ plies

Micromechanics involves analysis over a unit cell (or RVE) to determine COD values and the constraint effect due to adjoining uncracked plies. These constraint effects are carried over in CDM formulation through the "constraint parameter" κ. In a separate step, the material constants a_i appearing in expressions for damaged laminate stiffness relationships are determined from data for a reference laminate, such as $[0/90_8/0_{1/2}]_s$ for the present study. CDM expressions, given in (12.19), are then employed to predict stiffness degradation with crack density. The subsequent structural behavior in response to external loading is analyzed through a suitable FE model with material input as degraded stiffness properties at a given crack density (or corresponding applied strain).

To evaluate the constraint effect of undamaged plies over damaged plies, detailed 3D FE analyses over RVE, as shown in Fig. 12.17, were performed for angles, θ = 25°, 40°, 55°, 70°, and 90°. Specifically, CODs were determined as a function of ply orientation, ply thickness, and stiffness ratio of constrained to cracked plies. In literature, many analyses can be found which use 2D generalized FE models even for analyzing damage in off-axis laminates, e.g., $[0/90/\pm45]_s$ laminates [63]. However, a 2D FE modeling may not be accurate for analyzing structural behavior when damage is present in multiple modes as different sections in the width direction no longer behave in a similar manner. Only the key results are presented here.

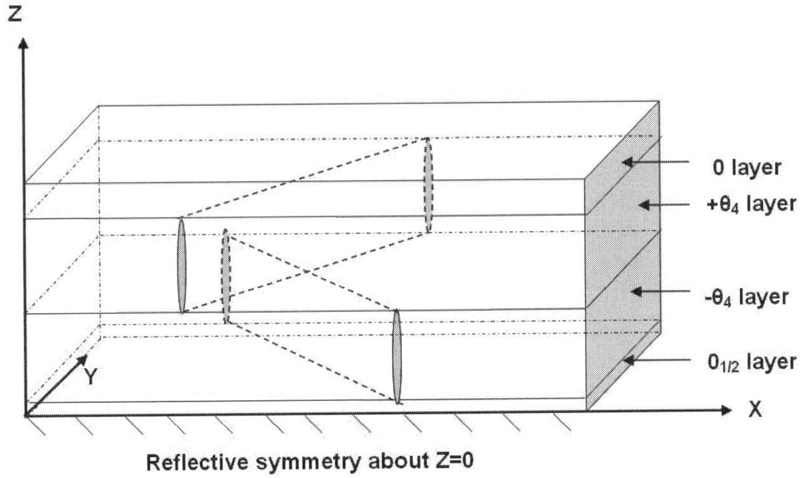

Fig. 12.17. A representative unit cell for $[0/\pm\theta_4/0_{1/2}]_s$ laminate configuration used in FE modeling. This shows the symmetry about the laminate midplane as the laminate is symmetric

Figure 12.18 shows the comparison between the experimental results and the 3D FE analysis for COD values evaluated at 0.5% axial strain. The FE results match very well with the experimental values. The experimental results are already published in [65, 66]. The experimental values are shown for ply orientations greater than 40° as there were no surface cracks observed below this angle. Average COD increases with ply orientation, but the rate of increase decreases and almost flattens out for crossply laminates ($\theta = 90°$). Thus, the effect of damage due to transverse matrix cracking is substantially lower in off-axis plies than in crossply laminates. This illustrates one advantage of using off-axis laminates, especially in complex loading. The profile of the crack opening was observed to be similar to that of a single crack in an infinite isotropic elastic medium subjected to a uniform far-field stress. The constraint due to adjoining plies makes it somewhat flat.

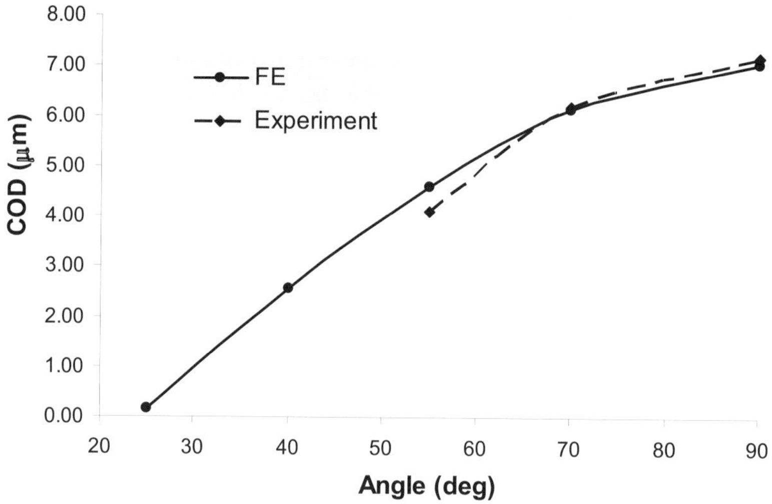

Fig. 12.18. Comparison of average crack opening displacements with experimental results for $[0/\pm\theta_4/0_{1/2}]_s$ laminate for $\varepsilon_{axial} = 0.5\%$

To understand the constraint effects further, a parametric study was performed by varying the stiffness of the constraining plies; and its effect on COD was analyzed. Figure 12.19 shows the effect of stiffness ratio on the COD of crack $\pm\theta$ plies. The COD values are normalized with the thickness of the cracked layer. For all cracked-ply orientations, normalized COD decreases as the constraining plies become stiffer than the cracked plies. The normalized COD values can be fitted to the following power law

$$\Delta\overline{u}_y = A + B\left(\frac{E_x^{\pm\theta}}{E_x^{90}}\right)^{-n}, \qquad (12.27)$$

where A, B, and n are the constants determined from the FE simulations, $E_x^{\pm\theta}$ represents the elasticity modulus of $\pm\theta$ plies in the laminate longitudinal direction, and the corresponding elasticity modulus of 90° ply in the longitudinal direction. The effects of ply thickness over CODs were also studied. To evaluate the stiffness degradation, the damage constants appearing in (12.21) were evaluated based on experimental plots for the reference laminate $[0/90_8/0_{1/2}]_s$. Figure 12.20 shows the plot of predicted longitudinal Young's modulus and Poisson's ratio normalized by their virgin state (undamaged) values with respect to the crack density for $\theta = 70°$. The stiffness reduction in laminates with damage in off-axis plies is found to be less significant than in damage in crossply laminates.

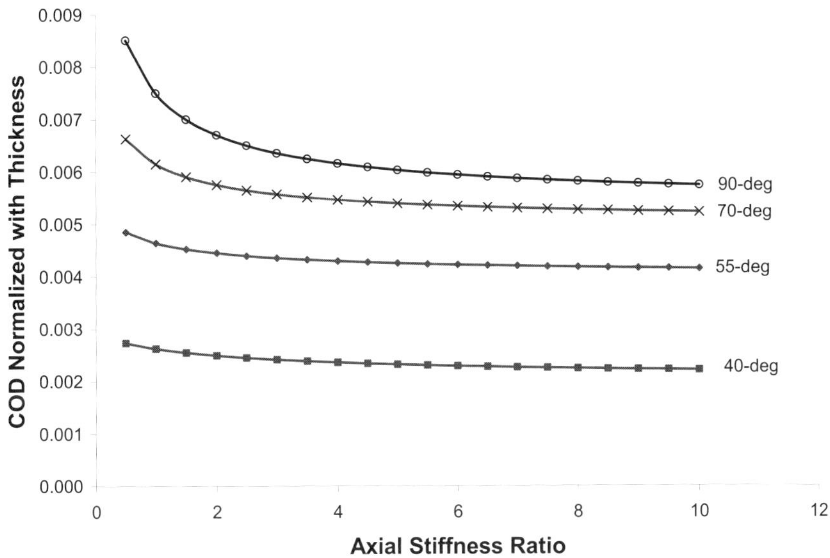

Fig. 12.19. Variation of COD for $[0/\pm\theta_n/0_{1/2}]_s$ laminate with axial stiffness ratio

To expand the scope of the SDM application, the authors are currently studying the damage behavior in $[0_m/\pm\theta_n/90_p]_s$ laminates, which involves matrix cracks in three different orientations and is characterized using CDM formulation with a separate damage mode tensor for each orientation. This will be taken up in a future publication.

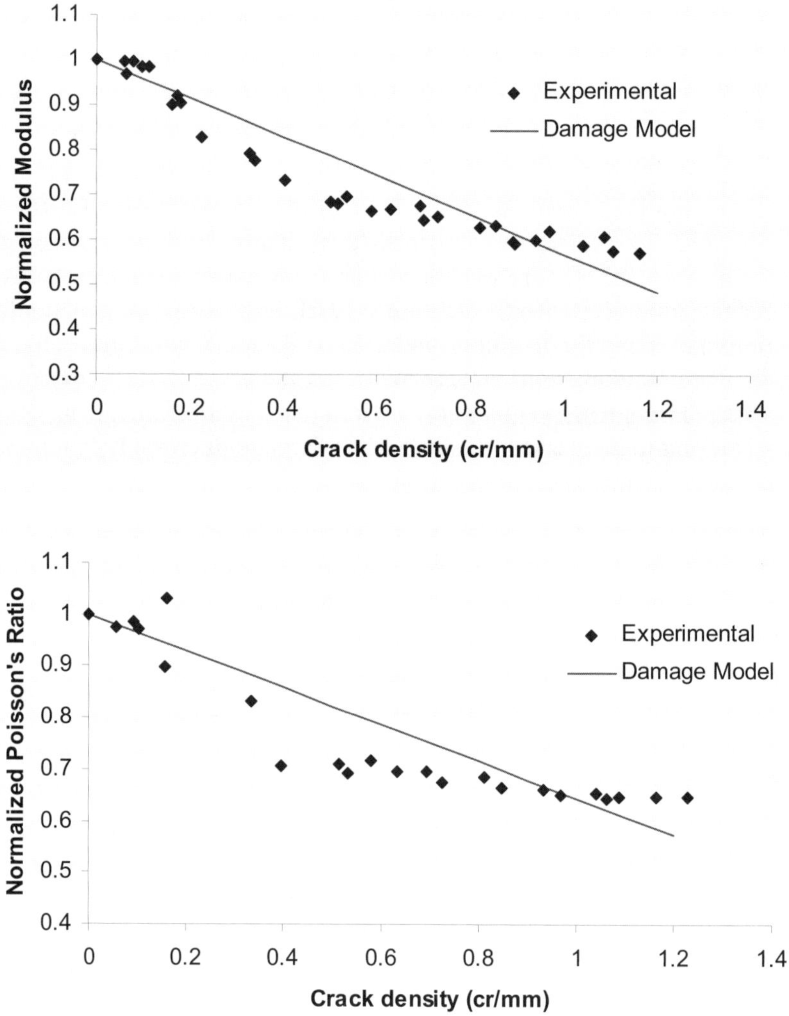

Fig. 12.20. Predicted stiffness reduction for $[0/\pm70_4/0_{1/2}]_s$ laminate compared with experimental results

12.5 Structural Integrity and Durability Assessment

In composite damage modeling and engineering research generally, it is important to have the ultimate goal in mind to develop the right strategy and to be clear about the context. For the authors, the goal is to assess integrity and durability of composite structures. This goal is not new; it has been in sight for most individuals involved in materials modeling. Figure 12.21 shows the "big picture" in which structural integrity and durability assessment are embedded. The starting place in the iterative process illustrated in the figure is manufacturing. One selects a process, e.g., liquid compression molding, and quantifies its process parameters, which along with other manufacturing details involved, such as machining and assembly, determine the material state in the component manufactured. The material state is characterized by a set of properties, e.g., elastic moduli, strength, and fracture toughness. These properties undergo evolution in the service environment due to phenomena such as fatigue, creep/viscoelasticity, and aging. The performance evaluation for the expected component life involves assessment of structural integrity and durability. Finally, a tradeoff study of cost against performance is conducted to assess the cost effectiveness. Most cost drivers lie in the manufacturing process, whose parameter variation allows moving toward the optimal design.

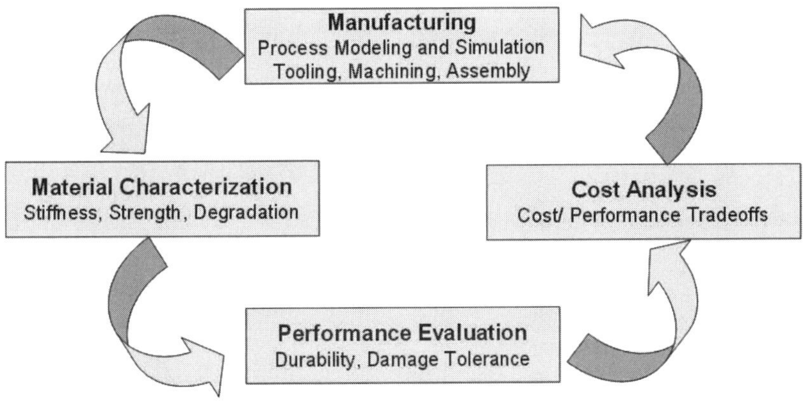

Fig. 12.21. The "big picture" considerations for cost-effective design of composite structures

Returning to the structural integrity and durability assessment, the role of materials modeling for composites is described in Fig. 12.22. Here the starting place is structural stress analysis of a given component, often by a finite element code. The input to this is the loading environment along with a deformational model, which is taken to be that of the initial material state (as produced by the manufacturing process). In most cases, prior experience guides in identifying critical sites in the component that are prone to failure. The local stress states in those sites determine the initiation and evolution of damage, also called *subcritical failure*. The mechanisms of damage depend additionally on the "microstructure," i.e., the fiber architecture, ply configuration, fiber/matrix interface, etc. In the discussion above, two cases of damage have been covered in two widely different microstructure scenarios. Along with this, the damage mechanics approaches, CDM and MDM, have been discussed as well as the hierarchical vs. SDM strategies. The outputs of these modeling efforts are either deformational changes, expressed as stiffness–damage relationships, or strength (failure criticality) or both. The stiffness change result also provides incremental input to the stress analysis, updating the deformational model. The final goal of life prediction (durability) can be reached by either a stiffness criterion or a strength criterion, depending on the performance requirement.

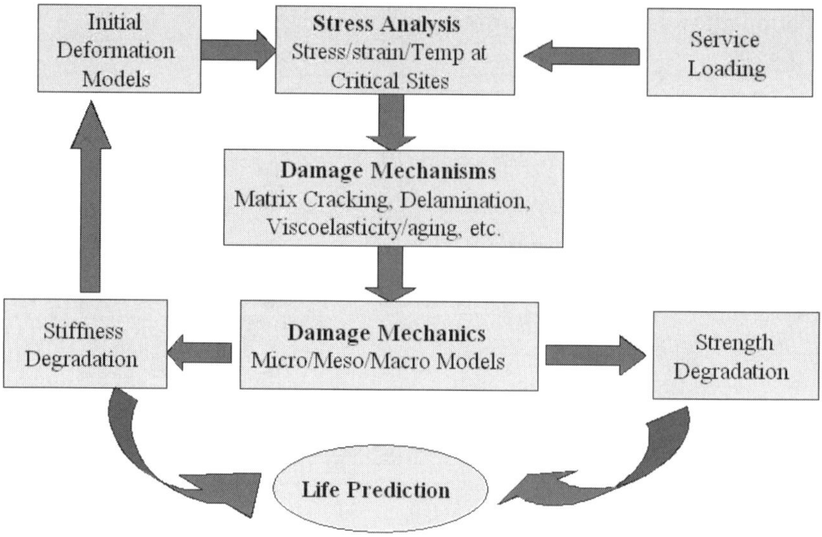

Fig. 12.22. Integrity and durability assessment procedure for composite structures

The multiscale modeling approach discussed here has been focused on the deformational response. Other considerations are needed for treating the local-to-global failure. The length scale issues are substantially different for failure than for deformational response. Discussion of these calls for a separate, focused treatment is reserved for a future work.

12.6 Conclusions

Multiscale modeling of composite materials generally, and of damage in these materials particularly, is still in the early stages of development. The overall field of multiscale modeling is undergoing intense development and is strongly driven by efforts to uncover phenomena that do not seem to be amenable to analytical treatments or are not yet feasible to study by direct or indirect observation. The need to enrich structural-level modeling with details of lower scales at which physical mechanisms operate forms a part of the multiscale modeling field. The discussion offered here has been targeted at this part.

The objectives of structural-level analysis of composites undergoing damage are to determine deformational response and assess structural integrity and durability. The treatment above has attempted to make the case that starting at a small scale, such as the fiber diameter, and working up the scales to the structural level is not an effective way to meet these objectives. This so-called *hierarchical multiscale strategy* will not, on its own, produce the observed features of damage, such as multiple cracking and multiple modes of multiple cracking. As discussed above, the characteristic scales change from a single damage entity to a damage mode and from a single damage mode to multiple damage modes. Furthermore, the length scales in each damage mode evolve as damage progresses, rendering the hierarchical approach a frustrating exercise. It should be realized that efforts in hierarchical multiscale modeling usually address phenomena at one scale and seek their outcome in terms of parameters valid at the next higher scale. In principle, of course, such efforts can be expanded to address multiple evolving scales, but it does not appear to lead to an effective strategy. Once again, this assessment is in the context of the set ultimate objective of analyzing behavior at the structural level.

An alternative strategy is to take the approach advocated here as SDM. In this approach, one begins at the structural level, i.e., the macrolevel, and formulates the material response in terms suited for structural analysis, e.g., by a finite element method. The scale at which the material response is addressed, i.e., the scale over which averaging is performed to determine the response functions, is the scale of an RVE, also called the *mesoscale*. The choice of the mesoscale is determined by the operating damage modes, which are known from experimental observations. In new situations, where experiments may not be feasible, a separate numerical simulation may be performed to uncover damage modes. In any case, the idea is not to indulge in an indiscriminate microlevel simulation but specifically to seek information needed at the mesolevel, as dictated by the macrolevel formulation.

In the examples discussed above, the relevant information is the crack surface displacement as affected by constraints of the material surrounding a crack in the mesolevel volume.

If the field of multiscale modeling is to advance beyond academic interests, it must address practical structures. Practical composite structures are manufactured by techniques that inevitably introduce defects such as voids in matrix, misaligned fibers, and interfacial disbonds. A multiscale modeling strategy must incorporate such defects in a judicial manner without increasing complexity to an impractical level. Work along these lines is ongoing in the authors' research group and will be reported as it progresses.

References

1. Allen DH, Harris CE, Groves SE (1987) A thermomechanical constitutive theory for elastic composites with distributed damage. I. Theoretical development. Int J Solids Struct 23:1301–1318
2. Allen DH, Harris CE, Groves SE (1987) A thermomechanical constitutive theory for elastic composites with distributed damage. II. Application to matrix cracking in laminated composites. Int J Solids Struct 23:1319–1338
3. Argon A (1972) Fracture of composites. In: Herman H (ed) Treatise on Materials Science and Technology, Vol. 1. Academic, New York, pp 79–114
4. Bader MG, Bailey JE, Curtis PT, Parvizi A (1979) Proceedings of International Conference on Mechanical Behavior of Materials, Cambridge, UK, pp 227–239
5. Bazant ZP, Chen EB (1997) Scaling of structural failure. ASME Appl Mech Rev 50:593
6. Budiansky B (1983) Micromechanics. Comput Struct 16:3–12
7. Cantwell WJ, Morton J (1991) The impact resistance of composite materials – a review. Composites 22:347–362
8. Chan WS (1991) Design approaches for edge delamination resistance in laminated composites. J Compos Tech Res 14:91–96
9. Chung WC, Jang BZ, Chang TC, Hwang LR, Wilcox RC (1989) Fracture behavior in stitched multidirectional composites. Mater Sci Eng A112:157–173
10. Drzal LT (1990) The effect of polymer matrix mechanical properties on the fiber–matrix interfacial strength. Mater Sci Eng A126:2890–2893
11. Elder DJ, Thomson RS, Nguyen MQ, Scott ML (2004) Review of delamination predictive methods for low speed impact of composite laminates. Compos Struct 66:677–683
12. Fabre JP, Sigety P, Jacques D (1991) Stress transfer by shear in carbon fiber model composites. J Mater Sci 26:189–195

13. Flaggs DL, Kural MH (1982) Experimental determination of the in situ transverse lamina strength in graphite/epoxy laminates. J Compos Mater 16:103–116
14. Garrett KW, Bailey JE (1977) Effect of resin failure strain on the tensile properties of glass fiber-reinforced polyester cross-ply laminates. J Mater Sci 12(11):2189–2194
15. Garrett KW, Bailey JE (1977) Multiple transverse fracture in 90 degree cross-ply laminate of a glass-reinforced polyester. J Mater Sci 12(1):157–168
16. Gudmundson P, stlund S (1992) First order analysis of stiffness reduction due to matrix cracking. J Compos Mater 26:1009–1030
17. Gudmundson P, Zang W (1993) A universal model for thermoelastic properties of macro cracked composite laminates. Int J Solids Struct 30:3211–3231
18. Hahn HT, Tsai SW (1974) On the behavior of composite laminates after initial failures. J Compos Mater 8:288–305
19. Han YM, Hahn HT (1989) Ply cracking and property degradations of symmetric balanced laminates under general in-plane loading. Compos Sci Technol 35:377–397
20. Harris B (1980) Metal Sci 14:351
21. Hashin Z (1985) Analysis of cracked laminates: a variational approach. Mech Mater 4:121–136
22. Hashin Z (1988) Thermal expansion coefficients of cracked laminates. Compos Sci Technol 31:247–260
23. Hashin Z (1990) Analysis of damage in composite materials In: Boehler JP (ed) Yielding, Damage, and Failure of Anisotropic Solids. Mechanical Engineering Publications, London, pp 3–32
24. Henstenburg RB, Phoenix SL (1989) Interfacial shear strength studies using single filament composite test. Polym Compos 10:389–408
25. Highsmith AL, Reifsnider KL (1982) Stiffness-reduction mechanisms in composite laminates, Damage in Composite Materials. ASTM STP 115:103–117
26. Hsueh CH (1992) Interfacial debonding and fiber pull-out stresses of fiber-reinforced composites. Mater Sci Eng A154:125–132
27. Jamison RD, Schulte K, Reifsnider KL, Stinchcomb WW (1984) Characterization and analysis of damage mechanisms in tension–tension fatigue of graphite/epoxy laminates. In: Effects of Defects in Composite Materials, ASTM STP 836, American Society for Testing and Materials, Philadelphia, pp 21–55
28. Jang BZ, Cholakara M, Jang BP, Shih WK (1991) Mechanical properties in multidimensional composites. Polym Eng Sci 31:40–46
29. Jones FR, Wheatley AR, Bailey JE (1981) In: Marshall IH (ed) Composite Structures. Applied Science Publishers, Barking, UK, pp 415–429
30. Kachanov LM (1958) On the creep fracture time (in Russian). Izv AN SSSR, Ofd Tekhn Nauk 8:26–31
31. Kachanov M (1980) Continuum model of medium with cracks. J Eng Mech Div ASCE 106(EM5):1039–1051
32. Kachanov M (1994) Elastic solids with many cracks and related problems. Adv Appl Mech 30:259

33. Kim JK, Mai Y (1991) High strength, high fracture toughness fiber composites with interface control – a review. Compos Sci Technol 41(4):333–378
34. Krajcinovic D (1996) Damage Mechanics. Elsevier, Amsterdam
35. Krajcinovic D (1996) Essential structure of the damage mechanics theories. In: Tatsumi T, Watanabe E, Kambe T (eds) Theoretical and Applied Mechanics. Elsevier, Amsterdam, pp 411–426
36. Lacy TE, McDowell DL, Talreja R (1999) Gradient concepts for evolution of damage. Mech Mater 31:831–861
37. Lafarie-Frenot MC, H naff-Gardin C, Gamby D (2001) Matrix cracking induced by cyclic ply stresses in composite laminates. Compos Sci Technol 61(15):2327–2336
38. Lauke BW, Beckert W, Singletary J (1996) Energy release rate and stress field calculation for debonding crack extension at the fiber–matrix interface during single-fiber pull-out. Compos Interface 3:263–273
39. Lim SG, Hong CS (1989) Prediction of transverse cracking and stiffness reduction in cross-ply laminate composites. J Compos Mater 23:695–713
40. Liu D (1990) Delamination resistance in stitched and unstitched composite planes subjected to composite loading. J Reinf Plast Compos 9:59–69
41. Marshall DB, Evans AG (1985) Failure mechanisms in ceramic-fiber/ceramic-matrix composites. J Am Ceram Soc 68:225
42. Mayadas A, Pastore C, Ko FK (1985) Tensile and shear properties of composites by various reinforcement concepts. In: Proceedings of 30th International SAMPE Syrup, pp 1284–1293
43. McCartney LN (1992) Theory of stress transfer in 0–90–0 crossply laminate containing a parallel array of transverse cracks. J Mech Phys Solids 40:27–68
44. Nairn JA (2000) Matrix microcracking in composites. In: Talreja R, M nson JAE (eds) Polymer Matrix Composites, Comprehensive Composite Materials, Vol. 2. Elsevier Science, Amsterdam, pp 403–432
45. Nairn JA, Hu S (1994) Micromechanics of damage: a case study of matrix microcracking. In: Talreja R (ed) Damage Mechanics of Composite Materials, Chapter 6. Elsevier, The Netherlands, pp 117–138
46. Niu K, Talreja R (1999) Modeling of wrinkling in sandwich panels under compression. J Eng Mech 125:875–883
47. Pagano NJ (1998) On the micromechanical failure modes in a class of ideal brittle matrix composites. Part 1. Coated-fiber composites. Compos Part B: Eng 29(2):93–119
48. Parvizi A, Bailey JE (1978) On multiple transverse cracking in glass fiber epoxy cross-ply laminates. J Mater Sci 13(10):2131–2136
49. Parvizi A, Garrett KW, Bailey JE (1978) Constrained cracking in glass fibre-reinforced epoxy cross-ply laminates. J Mater Sci 13:195–201
50. Prichard JC, Hogg PJ (1990) The role of impact damage in post-impacted compression testing. Composites 21:503–511
51. Rosen BW (1964) Tensile failure of fibrous composites. AIAA J 2:1985–1991
52. Smith PA, Boniface L, Glass NFC (1998) Comparison of transverse cracking phenomena in (0/90)s and (90/0)s CFRP laminates. Appl Compos Mater 5:11–23

53. S rensen BF, Talreja R (1993) Analysis of damage in a ceramic matrix composite. Int J Damage Mech 2:246–271
54. Stinchcomb WW, Reifsnider KL, Yeung P, Masters J (1981) Effect of ply constraint on fatigue damage development in composite material laminates. ASTM STP 723:64–84
55. Su KB (1989) Delamination resistance of stitched thermoplastic matrix composite laminates. ASTM STP 1044:279–300
56. Talreja R (1985) Transverse cracking and stiffness reduction in composite laminates. J Compos Mater 19:355–374
57. Talreja R (1986) Stiffness properties of composite laminates with matrix cracking and internal delaminations. Eng Fract Mech 25:751–762
58. Talreja R (1990) Internal variable damage mechanics of composite materials, Invited Paper. In: Boehler JP (ed) Yielding, Damage and Failure of Anisotropic Solids. Mechanical Engineering Publications, London, pp 509–533
59. Talreja R (1991) Continuum modeling of damage in ceramic matrix composites. Mech Mater 12:165–180
60. Talreja R (1996) A synergistic damage mechanics approach to durability of composite material systems. In: Cardon A, Fukuda H, Reifsnider K (eds) Progress in Durability Analysis of Composite Systems. A.A. Balkema, Rotterdam, pp 117–129
61. Talreja R (2006) Multiscale modeling in damage mechanics of composite materials. J Mater Sci 41(20):6800–6812
62. Tan SC, Nuismer RJ (1989) A theory for progressive matrix cracking in composite laminates. J Compos Mater 23:1029–1047
63. Tong J, Guild FJ, Ogin SL, Smith PA (1997) On matrix crack growth in quasi-isotropic laminates. II. Finite element analysis. Compos Sci Technol 57(11):1537–1545
64. Vakulenko AA, Kachanov ML (1971) Continuum theory of a medium with cracks (in Russian). Izv AN SSSR, Mekh Tverdogo Tela 6:159
65. Varna J, Berglund LA (1991) Multiple transverse cracking and stiffness reduction in cross-ply laminates. J Compos Technol Res 13:97–106
66. Varna J, Krasnikovs A (1998) Transverse cracks in cross-ply laminates. 2. Stiffness degradation. Mech Compos Mater 34(2):153–170
67. Varna J, Akshantala NV, Talreja R (1999) Crack opening displacement and the associated response of laminates with varying constraints. Int J Damage Mech 8:174–193
68. Varna J, Joffe R, Akshantala NV, Talreja R (1999) Damage in composite laminates with off-axis plies. Compos Sci Technol 59:2139–2147
69. Varna J, Joffe R, Talreja R (2001) A synergistic damage mechanics analysis of transverse cracking in $[\pm\theta, 90_4]_s$ laminates. Compos Sci Technol 61:657–665
70. Varna J, Krasnikovs A, Kumar R, Talreja R (2004) A synergistic damage mechanics approach to viscoelastic response of cracked cross-ply laminates. Int J Damage Mech 13(4):301–334

71. Verpoest I, Wevers M, DeMeester P, Declereq P (1989) 2D and 3D fabrics for delamination resistant composite laminates and sandwich structure. SAMPE J 25:51–56
72. Whitney JM, Drzal LT (1987) Axisymmetric stress distribution around an isolated fiber fragment. In: Johnson NJ (ed) Toughened Composites, ASTM STP 937. American Society for Testing and Materials, Philadelphia, pp 176–196
73. Zhou LM, Kim JK, Mai YW (1992) Interfacial debonding and fiber pull-out stresses. J Mater Sci 27:3155–3166

Chapter 13: Hierarchical Modeling of Deformation of Materials from the Atomic to the Continuum Scale

Namas Chandra

University of Nebraska-Lincoln, Lincoln, NE 68520, USA

13.1 Introduction

During the last few years, novel structures, phenomena, and processes have been observed at the nanoscale. In addition, the very nature of thermal and mechanical responses of materials is governed by the phenomena occurring at the atomic nanoscale. With the development in nanoscale systems, there is an urgent need for theory, modeling, and computational tools to understand and accelerate development and applications. Modeling efforts in nanoscale systems have predominantly used atomistic simulations based on molecular dynamics or other refined techniques, such as density functional and tight binding theories, which help clarify the issues involved at the macroscale. Though atomistic simulations provide insightful details into many problems of interest, the exceedingly high computational requirements of atomistic simulations place a stringent limitation on length and time scales for the problem of interest.

Recent advances in massive parallel computing have increased the simulation scales by a few orders of magnitude; but even with these advances, simple pair potential systems of up to 10^3–10^4 nm^3 can only be simulated for a few hundred nanoseconds. On the other hand, continuum-based methodologies solve for fewer degrees of freedom and, hence, are computationally more efficient.

These factors call for the development of a method that combines continuum and atomistic models so that the understanding of phenomena at the atomic scale can be attained using lesser computational power. As an example, grain boundaries and interfaces play important roles in the strengthening and deformation of metallic materials. There is a need not only to understand the atomic scale behavior of the grain boundaries but also to use this information in a continuum framework to enable the development of materials with engineered grain boundaries with greatly enhanced properties. This is achieved in multiscale modeling using lesser computing resources.

The fundamental idea behind multiscale modeling is that intense atomic calculations are required only at the core of the lattice defects, whereas continuum-based solutions can be used away from the core. Similar ideas have been used in the atomic simulation of cracks and dislocations with the flexible border technique [30, 66]. In this method, outer atoms surrounding cracks or dislocations are updated based on continuum elastic solutions. The finite element atomistic simulation method by Mullins and Dokainsh [47] generalizes the flexible border methods by employing the finite element method (FEM) in the outer regions of atomic simulation of crack systems. Kohlhoff et al. [34] use a similar method; but in their case, an interface between molecular dynamics (MD) and finite element regions is made up of a one-to-one correspondence between nodes and atoms. A more recent hand shaking or coupling of length scales (CLS) method adds increased sophistication to such calculations by using tight binding along with MD and finite element regions [1, 38]. Here again, the interface coupling is achieved by kinematic restraints.

Dynamic crack propagation has been studied by Rafii-Tabor and coworkers [52, 53] using a three-scale model. In this model, a macroscale FEM is connected to microscale molecular dynamics using a mesoscale finite element mesh with a one-to-one correspondence between nodes and atoms. Boundary conditions for MD are generated by the mesoscale, and the stochastic diffusion constant obtained at the atomic scale is fed into a macroscale calculation.

Another interesting multiscale method using MD near the crack core and micromechanics to define the outer region has been developed with the idea of permitting the motion of dislocations across the scales [23, 24]. Coarse-grained molecular dynamics (CGMD) [56] is another finite element-based multiscale method which reduces to MD when finite element mesh is refined to the atomistic limit. In the case of a coarser mesh, the nodal variables are determined as a weighted average of the atomic variables.

A finite element-based atomistic method, known as the quasicontinuum method, has been developed by Tadmor and coworkers [67, 68]. In this method, the total energy of the system is computed using atomic potentials minimized in a finite element framework. Energy is explicitly calculated for atoms of interest near the defect, known as nonlocal atoms. Homogeneous deformation is assumed in atoms away from the core, and energy is computed from a deformation gradient by invoking the Cauchy–Born rule. The quasicontinuum method has been successfully used to model crack problems [62], nanoindentation [67], grain boundaries, and dislocations [41]. Shenoy et al. [62] have modified this method to exactly mimic molecular statics in the nonlocal region. The quasicontinuum method has been applied to deformation of nanotubes by modifying the Cauchy–Born rule for planar curved systems [3].

There have been a number of multiscale methods developed to address two or more scales; but all of them involve a single modeling concept, e.g., continuum-to-continuum modeling using multiscale FEMs. This chapter, however, deals with the problem wherein one scale involves discrete atomic scale modeling that links with the continuous macroscopic scale.

13.2 Methodology

The issue of multiscale modeling is addressed using two special methods: one involving asymptotic expansion homogenization (AEH) and the other involving hierarchical modeling.

13.2.1 Asymptotic Expansion Homogenization

In certain situations, nano- and microsystems can be modeled as containing a periodic heterogeneity manifested at the atomic scale (grain boundaries or dislocations), yet influencing the overall macroscopic behavior. In such cases, it becomes necessary to consider the heterogeneity effects of the atomic scale within the framework of the macroscale problem. AEH is a mathematically rigorous approach to homogenization of periodic structures. It has been used in the study of heterogeneous materials consisting of two natural scales, e.g., composite materials and porous media.

AEH decouples a heterogeneous multiscale problem into a macroscale homogeneous problem and a microscale problem. In this approach, a methodology for multiscale simulation by adapting AEH to the atomic scale is used. Molecular dynamics and statics are used to study the energetics and deformation characteristics of symmetric tilt grain boundaries (STGB) of aluminum. This section demonstrates the great usefulness of molecular dynamics/molecular statics (MD/MS) in understanding the effect of atomistic scale variations on thermomechanical properties. However, the scale of heterogeneity and the scale at which properties are manifested are still within a few thousand atoms; but the real world problems span a few hundredths of a meter to a few meters, for which atomistic simulation methods are impractical. A theoretical formulation to apply AEH to the atomic scale is presented next.

Atomistic simulation of grain boundaries

Both experimental and computational studies of the structure and deformation characteristics of grain boundaries have assumed greater importance with the advent of powerful characterization tools such as the high-resolution transmission electron microscope, and computational methods such as molecular dynamics. A large volume of research exists on the equilibrium grain boundary structures, but relatively less exists on modeled grain boundary sliding. Deymier and Kalonji [17] investigated the effect of critical temperature and grain boundary melting on sliding. Molteni et al. [42] conducted ab initio simulations of grain boundary sliding on Ge in a quasistatic way by applying constant strain increments. Chandra and Dang [9] investigated the effect of applied stress and displacement on STGB in aluminum. Kurtz and coworkers [36] studied the effect of dislocations on the grain boundary sliding and observed that sliding occurs more easily in the presence of dislocations. MD can be used to study the effect of crystal misorientation on energy and sliding characteristics of the grain boundaries under mechanical loading conditions.

Molecular dynamics simulations are used to compute grain boundary energy for various [110] STGB of aluminum. Grain boundary sliding is studied by applying an external state of shear stress on certain boundaries, and a relationship between grain boundary energy and extent of sliding is observed. Embedded atom method [16] potentials implicitly include many body interactions and are more reliable than conventional pair potentials in representing the atomic interactions in metals. The EAM potential for a single atom is given by

$$E_i = F(\rho) + \frac{1}{2}\phi(r_{ij}). \tag{13.1}$$

The first term in (13.1) represents the energy required to embed an atom in a cloud of electron density, and the second term represents pair interactions. In molecular statics, the total energy is minimized to give the equilibrium position of all the atoms in the given crystal at 0 K. Molecular dynamics is used to study the time-related phenomena for crystals subjected to external loads.

Ercolessi–Adams' [19] force matching potentials for aluminum are used in this work since they are known to predict more accurate stacking fault energies and elastic constants than conventional embedded atom potentials.

For each of the boundaries studied, the computational crystal was generated based on the orientation of the given grain and on the symmetry between that and the adjacent grain across the boundary plane. It is necessary to remove or add atoms at the grain boundary because multiple energy minima may exist with a similar structure [72]. For studying the grain boundary energy, crystals of about 6,000 atoms are used, which are periodic in X and Z directions and free in the Y direction. Figure 13.1 illustrates the designations of STGB used for statics simulations.

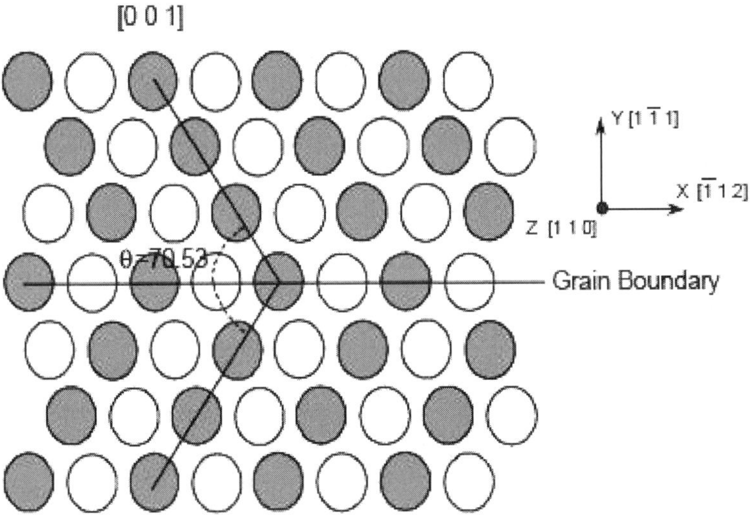

Fig. 13.1. STGB. $70.5° [110]\Sigma3(111)$

Most of the experiments on bicrystal sliding [22, 59, 74] are performed with the grain boundary oriented at 45° to the tensile (or compressive) axis. To simulate these conditions, grain boundaries for dynamics simulations are also formed with similar boundary conditions.

Figure 13.2 describes the crystal generation procedure. These boundaries are periodic in the Z direction and free in the X and Y directions. They contain about 14,000–15,000 atoms. Similar boundary conditions have been used in the study of intergranular crack propagation [20]. Both shape and volume of the simulation cell are allowed to change. A state of shear stress is applied on the grain boundary by applying an external normal stress in the XYZ reference frame. The stresses are applied by introducing an additional force term corresponding to the applied stress. In the simulation, the boundaries are considered dynamic variables driven by the imbalance between externally applied stress and the internal stress tensor. The amount of sliding is calculated as difference between the average *X* displacements in the bottom and top crystals in the *XYZ* reference frame. Stresses of about 10% of Voigt average shear modulus for this potential are employed in these studies. All the simulations are carried out at a temperature of 448 K for about 5 ps. Sliding in six tilt boundaries $\Sigma3(1\bar{1}1)$, $\Sigma9(2\bar{2}1)$, $\Sigma11(1\bar{1}3)$, $\Sigma11(3\bar{3}4)$, $\Sigma43(5\bar{5}6)$, and $\Sigma51(5\bar{5}1)$ is studied in this work.

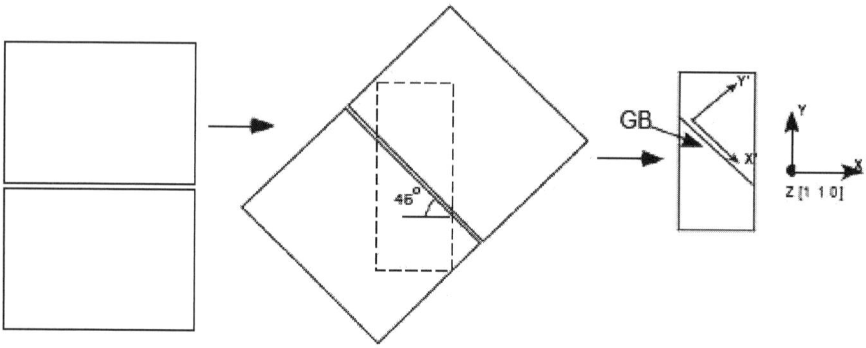

Fig. 13.2. Generation of crystal for simulation of sliding. Free boundary conditions in *X* and *Y* directions, periodic boundary conditions in *Z* direction

Grain boundary energy

Seventeen tilt [110] CSL grain boundaries $\Sigma3(1\bar{1}1)$, $\Sigma(1\bar{1}2)$, $\Sigma9(2\bar{2}1)$, $\Sigma9(1\bar{1}4)$, $\Sigma11(1\bar{1}3)$, $\Sigma11(3\bar{3}2)$, $\Sigma17(3\bar{3}4)$, $\Sigma19(3\bar{3}1)$, $\Sigma27(1\bar{1}5)$, $\Sigma27(5\bar{5}2)$, $\Sigma33(2\bar{2}5)$, $\Sigma33(4\bar{4}1)$, $\Sigma33(1\bar{1}8)$, $\Sigma41(4\bar{4}3)$, $\Sigma43(5\bar{5}6)$, $\Sigma43(3\bar{3}5)$, and $\Sigma51(5\bar{5}1)$ have been examined for grain boundary energy. Grain boundary energy is calculated as the difference between energy of all the atoms in the grain boundary region and a perfect crystal containing the same number of atoms, divided by the area of the grain boundary.

Figure13.3 describes the grain boundary energies obtained as a function of misorientation angle for a series of [110] bicrystals. The shape of the curve and the energy cusps match well with those obtained in earlier studies, but the absolute values of energy in Fig. 13.3 are slightly higher than those obtained in earlier studies using pair potentials [75] and conventional EAM potential [9]. This tendency could possibly be ascribed to the fact that the E–A force matching potential predicts higher (more accurate) stacking fault energy compared with the earlier potentials.

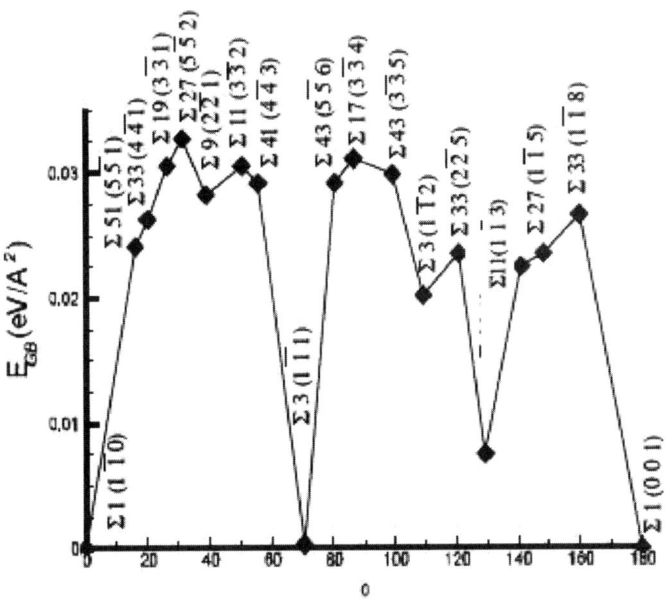

Fig. 13.3. Grain boundary energy E_{GB} as a function of the misorientation angle

Grain boundary sliding

A variation of grain boundary sliding with grain boundary energy is shown in Fig. 13.4. Maximum sliding is observed in $\Sigma 17(3\bar{3}4)$ and minimum in $\Sigma 3(1\bar{1}1)$. There exists a correlation between grain boundary energy and sliding.

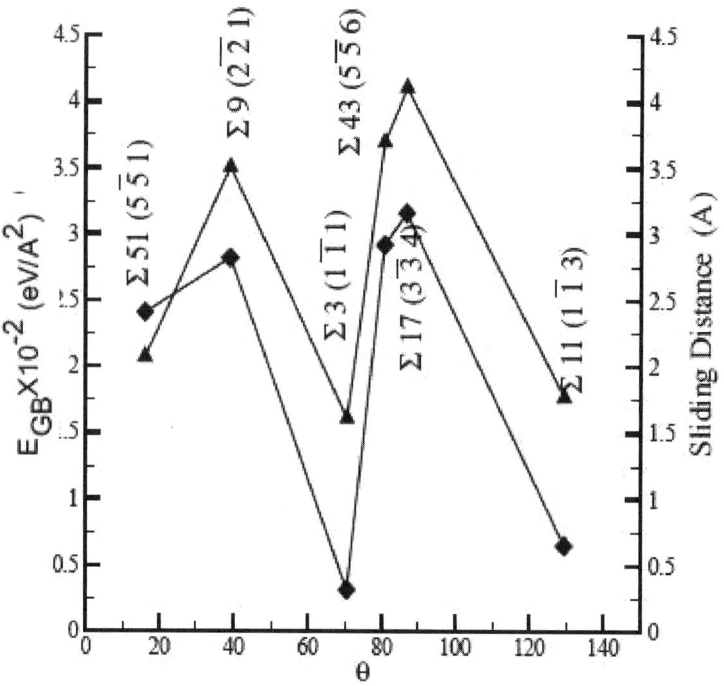

Fig. 13.4. Extent of sliding and grain boundary energy vs. misorientation angle

Grain boundary sliding is more in the boundary, which has higher grain boundary energy. Monzen and Takada [45] have observed a similar variation of grain boundary energy and tendency to slide by measuring nanometer scale sliding in copper. They report that there is no significant dislocation activity.

To study the directional dependence of the sliding state, shear stress was applied in the opposite direction for $\Sigma 3(1\bar{1}1)$, $\Sigma 9(2\bar{2}1)$, and $\Sigma 17(3\bar{3}4)$ grain boundaries. There was a decrease in the extent of sliding in the case of $\Sigma 9(2\bar{2}1)$ and $\Sigma 17(3\bar{3}4)$ boundaries but no noticeable difference in $\Sigma 3(1\bar{1}1)$ grain boundary. $\Sigma 9(2\bar{2}1)$ and $\Sigma 17(3\bar{3}4)$ boundaries slid by 3.46 and 4.11 in contrast to 3.50 and 4.23 in the other direction. It is interesting to note that directional dependence of crack propagation was observed experimentally in $\Sigma 9(2\bar{2}1)$ bicrystals of copper [71].

There are two mechanisms which explain grain boundary sliding at the atomic scale. One of them attributes sliding in STGB to the glide of interfacial dislocations and migration to the climb [59]. This mechanism is supported by experiments on Zn bicrystals, where ratio of migration to sliding matches with that predicted by theory [60]. Kurtz et al. [36] have noted, through molecular dynamics simulations on Al crystals, that sliding is much easier in the presence of interfacial dislocations. The second mechanism is based on geometrical analysis given by Ashby [4] in which atomic scale sliding is controlled by boundary diffusion.

In this simulation, grain boundary sliding is manifested as relative displacement of atoms across the grain boundaries; the top crystal displaces more compared to the bottom while the same state of stress is applied equally to the top and bottom crystals. During the process of sliding, no dislocations are observed; and this is similar to Cu bicrystal experiments by Monzen [43, 44]. Observation indicates that the natural tendency of the grain boundary to slide depends on the disorder of the grain boundary as represented by the grain boundary energy. A higher energy grain boundary like $\Sigma 17(3\bar{3}4)$ has a more disordered structure; hence, there is a higher stress concentration and, consequently, a higher tendency to slide than with low energy boundaries such as $\Sigma 3(1\bar{1}1)$.

Application of AEH to atomic simulation

The previous section clearly demonstrates that grain boundary energy and sliding characteristics are affected by the misorientation angle. Earlier work [48] also shows a minor impurity content as well. A natural desire is to use this information in predicting the behavior of real material systems with hundreds of grains, grain orientations, grain boundary characteristics, and triple points. Such a system cannot be simulated by MD/MS alone.

Here is where the AEH method is invoked. Many of the existing multiscale methods use the idea that intense atomic calculations are required only at the core of the defect while atoms away from the core can be modeled using the continuum approach. The finite element atomistic simulation method by Mullins and Dokainsh [47] employs an FEM in the outer regions of atomic simulation of a crack system. Atomistic forces are mapped onto the inner nodes of a finite element mesh through shape functions, and the nodal displacements are extrapolated to outer atoms. Kohlhoff and coworkers [34] use a similar method; but in their case, an interface between MD and the finite element regions is made up of one-to-one correspondence between nodes and atoms. Kinematic restraints are applied on the interface atoms and nodes so that the outermost atoms move in correspondence with the nodes and the inner atoms move along with the nodes. The energy of the entire system is minimized by zeroing the forces.

A more recent hand shaking or coupling of length scales method [38] adds increased sophistication to such calculations by using tight binding along with MD and finite element regions. Here again the interface coupling is achieved by kinematic restraints. Dynamic crack propagation has been studied by Rafii-Tabor and coworkers [52], using a three-scale model. In this model, macroscale FEM is connected to microscale molecular dynamics using a mesoscale finite element mesh with a one-to-one correspondence between nodes and atoms.

Boundary conditions for MD are generated by the mesoscale, and the stochastic diffusion constant obtained at the atomic scale is fed into the macroscale calculation. Crack velocity is computed in an independent simulation. This method has been applied for two-dimensional crack propagation. Another interesting multiscale method with MD near the crack core and micromechanics to define the outer region has been developed with the idea of permitting the motion of dislocations across the scales. Various multiscale methods used in the study of fracture are reviewed by Cleri et al. [14].

A finite element-based atomistic method known as the quasicontinuum method has also been developed [67]. In that method, the total energy of the system computed using atomic potentials is minimized in a finite element framework. Energy is explicitly calculated for atoms of interest near the defect, known as nonlocal atoms. Homogeneous deformation is assumed in atoms away from the core, and energy is computed from the deformation gradient by invoking the Cauchy–Born rule. The quasicontinuum method has been successfully used to model crack problems [41], nano indentations [64], grain boundaries, and dislocations [63]. Shenoy [61] has modified this method to exactly mimic molecular statics in the nonlocal region. With this method as well, the local and nonlocal regions are interfaced by using artificial kinematic restraints. This and related methods are reviewed by Ortiz and Phillips [50]. CGMD [27] is another finite element-based multiscale method which reduces to MD when finite element mesh is refined to the atomistic limit. In the case of a coarser mesh, the nodal variables are determined as the weighted average of the atomic variables.

From the preceding discussion on existing multiscale techniques, the following features clearly emerge. Firstly, most of the methods concentrate on solving problems in which there is a localized area of interest requiring atomic resolution, e.g., a crack problem. Micro- and nanosystems would require an atomic level treatment all through the system or structure, but this is very difficult to model entirely using atomic simulations. Secondly, most of the methods use unphysical kinematic restraints to link the microscopic and macroscopic areas of interest. This is necessary to ensure continuity of the displacement field, but it may cause discontinuity in the force field resulting in unphysical forces, such as ghost forces. There is a need for a multiscale simulation method that would take into consideration atomic effects throughout the system which could be used as a design and modeling tool for nano- and microscale systems and structures.

There is another set of multiscale problems relevant to heterogeneous materials, such as composites and porous materials. These materials naturally consist of multiple spatial scales, a microscale containing local heterogeneities, and a macroscale comprising the overall structure. These problems are typically solved by homogenization of a representative volume element (RVE) over the whole structure [70]. AEH is a mathematically rigorous method which has been successfully employed to solve linear elastic as well as plastic and viscous problems in composite materials with periodic microstructures [13] (Fig. 13.5).

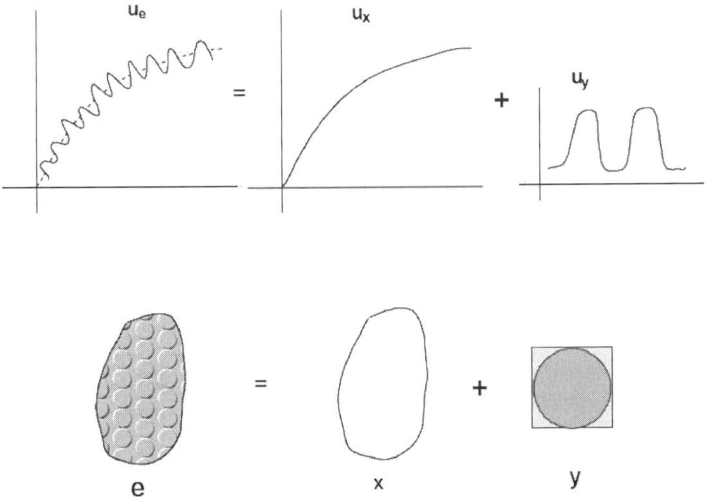

Fig. 13.5. The idea of AEH

AEH decouples the problem into microscale and macroscale problems by employing asymptotic expansion of the field variables. So far, AEH has been used only for the situations in which both the micro- and macroscales are described by continuum mechanics. The functional analysis aspects of AEH are well explained in the classical texts by Bensoussan et al. [6] and Sanchez-Palencia [58].

In spite of early mathematical developments, application of AEH to engineering problems is of relatively recent origin. Homogenization for linear problems is well established [21]. Further, there have been many improvisations in the application of AEH to linear problems, such as adaptive remeshing [7] and applications in biomechanics [27] and beam theory [31], etc. AEH has also been applied to nonlinear problems in composite materials. Fish and coworkers [21] employ this method in the study of failure and other inelastic effects, such as plasticity. Application to viscous problems has been studied by Chung et al. [12] and Koishi et al. [35]. Ghosh and coworkers [26] use AEH in conjunction with the Voronoi cell FEM to obtain the effects of realistic microstructures of composites. Apart from composite materials, AEH has been applied to problems of transport phenomena in porous materials [32].

Despite the fact that there has been extensive research in homogenization techniques, no attempt has been made to apply AEH to the atomic scale. Most engineering applications of AEH have focused on linking disparate scales, both of which are based on continuum theory. A theoretical framework for application of homogenization to the atomic scale problem is presented here. Hierarchical sets of equations describing the micro- and macroscales are derived to facilitate the computational solution procedure. The material under question consists of two natural scales: a micro-Y scale, which is described by atoms interacting through an interatomic potential, and a macro-X scale, described by continuum constitutive relations. The two scales are linked through a scaling parameter ε. The atomic scale may consist of atoms in arbitrary configurations containing inhomogeneities, such as dislocations and grain boundaries; but this scale is considered to be periodic. The equations are used based on well-developed AEH formulations for the elastic problem and incorporate atomic scale parameters to solve these equations. This multiscale approach enables the solution of a homogenized macroscale problem using atomic scale information.

The field equations, namely equilibrium, definition of strain, and constitutive relation for the overall material in space Ω and the relevant boundary conditions are shown below.

$$\frac{\partial \sigma^{\varepsilon}}{\partial x^{\varepsilon}} + f = 0 \quad \text{on } \Omega, \tag{13.2}$$

$$\sigma = C e^{\varepsilon} \quad \text{on } \Omega, \tag{13.3}$$

$$e^{\varepsilon} = \frac{\partial u^{\varepsilon}}{\partial x} \quad \text{on } \Omega. \tag{13.4}$$

The boundary conditions are

$$u^{\varepsilon} = \overline{u} \quad \text{on } \Gamma_u \quad \text{and} \quad \sigma^{\varepsilon} n = \overline{t} \quad \text{on } \Gamma_t. \tag{13.5}$$

The main assumption in AEH is that the primary variable displacement can be expanded as an asymptotic expansion in ε as shown in (13.6)

$$u^{\varepsilon} = u^{(0)}(x,y) + \varepsilon u^{(1)}(x,y) + \varepsilon^2 u^{(2)}(x,y) + \cdots \tag{13.6}$$

The functions $u^{(i)}(x,y)$ are Y periodic in variable y, assuming the existence of functions $u^{(i)}$ independent of the scaling parameter ε. Using the following identity for any function $g(x,y)$

$$\frac{\partial g(x,y)}{\partial x^{\varepsilon}} = \frac{\partial g(x,y)}{\partial x} + \frac{1}{\varepsilon}\frac{\partial g(x,y)}{\partial y}, \tag{13.7}$$

$$e(u^{\varepsilon}) = \frac{1}{\varepsilon}\left(\frac{\partial u^{(0)}}{\partial y} + \left(\frac{\partial u^{(0)}}{\partial x} + \frac{\partial u^{(1)}}{\partial x} + \varepsilon\left(\frac{\partial u^{(1)}}{\partial x} + \frac{\partial u^{(2)}}{\partial y}\right) + \cdots\right)\right). \tag{13.8}$$

In a similar manner, stress can be expanded as

$$\sigma^{\varepsilon}(x,y) = \varepsilon^{-1}\sigma^{(-1)}(x,y) + \sigma^{(0)}(x,y) + \varepsilon^{1}\sigma^{(1)}(x,y) + \cdots \tag{13.9}$$

Using these expansions in the equilibrium (13.2) and constitutive equation (13.3) and separating the coefficients of the powers of ε in three hierarchical equations (13.10)–(13.12) as shown below:

$$\frac{\partial}{\partial y}\left(C\frac{\partial u^{(0)}}{\partial y}\right) = 0, \tag{13.10}$$

$$\frac{\partial}{\partial y}\left(C\left(\frac{\partial u^{(0)}}{\partial x} + \frac{\partial u^{(1)}}{\partial y}\right) + \frac{\partial}{\partial x}\left(C\frac{\partial u^{(0)}}{\partial y}\right)\right) = 0, \tag{13.11}$$

$$\frac{\partial}{\partial y}\left(C\left(\frac{\partial u^{(1)}}{\partial x} + \frac{\partial u^{(2)}}{\partial y}\right)\right) + \frac{\partial}{\partial x}\left(C\left(\frac{\partial u^{(0)}}{\partial x} + \frac{\partial u^{(1)}}{\partial y}\right)\right) + f = 0. \tag{13.12}$$

Equations (13.10) and (13.11) represent the microscale or the Y scale, and (13.12) represents the macroscale or X scale equilibrium. Because of the Y periodicity of $u^{(0)}$ and (13.10), it can be observed that $u^{(0)}$ is a constant in y. In the case of linear elasticity, superposition and introduction of an elastic corrector function χ can be used as shown below:

$$u^{(1)} = \chi \frac{\partial u^{(0)}}{\partial x}. \tag{13.13}$$

With the help of this substitution, (13.12) now reduces to

$$\frac{\partial}{\partial y}\left(C\left(\delta \cdot \delta + \frac{\partial \chi}{\partial y}\frac{\partial u^{(0)}}{\partial x} \right)\right) = 0. \tag{13.14}$$

This can be solved in variational form as

$${}_Y C \frac{\partial \chi}{\partial y}\frac{\partial v}{\partial y} \mathrm{d}Y = {}_Y v \frac{\partial C}{\partial y} \mathrm{d}Y. \tag{13.15}$$

It may be observed that χ here is not a vector but a series of vectors. It denotes the corrector term in the macroscale due to microscale perturbations.

The Y scale here is composed of atoms interacting through an interatomic potential. If a finite element mesh is refined to atomic scale in the Y region, then $\left[\frac{\partial^2 W}{\partial q \partial q}\right.$ would denote the atomic level stiffness matrix, where W is the total strain energy density of the Y scale and q denotes the displacements of individual atoms. It is proposed that (13.15) can be solved using atomic scale parameters as

$$\left[\frac{\partial^2 W}{\partial q \partial q}\right.[\chi] = [B^{\mathrm{T}} C^{\mathrm{loc}}. \tag{13.16}$$

The right-hand side of (13.16) is the spatial variation of the elasticity matrix. B^{T} is the gradient of the shape function matrix in the FEM. C^{loc} are the local elastic constants in the atomic region, i.e., C^{loc} is the elasticity matrix in various regions of atomic scale which could be of interest, for example in grain boundaries. The procedure for evaluation of local elastic constants will be discussed in the next section.

The expression for determination of atomic level stiffness matrix

$$\left[\frac{\partial^2 W}{\partial q \partial q}\right]$$

for EAM potential has been developed by Tadmor and coworkers [69]. This procedure can, however, be adapted to other interatomic potentials apart from EAM.

Now consider the macroscale equation (13.12). If the mean operator is applied on this equation, then by virtue of Y periodicity of $u^{(2)}$, (13.12) reduces to

$$\frac{\partial}{\partial x}\left(C^{\mathrm{H}}\left(\frac{\partial u^0}{\partial x}\right)+f\right)=0. \tag{13.17}$$

C^{H} is the homogenized elasticity matrix for the overall region given by

$$C^{\mathrm{H}}=\frac{1}{|Y|}\int_Y\left(\delta\cdot\delta+\frac{\partial\chi}{\partial y}\right)\mathrm{d}y, \tag{13.18}$$

where $\delta\cdot\delta$ is the fourth-order identity tensor. It must be noted that though the elasticity matrix is homogenized here, stress or strain can also be homogenized [11].

Equation (13.17) can be solved by the FEM using appropriate boundary conditions and would give solution $u^{(0)}$ corrected for the atomic Y scale.

Local elastic constants

The formula for determining local elastic constants using molecular dynamics has been developed by Kluge and coworkers [33]. It is based on defining the local stress tensor at the atomic scale. Consider a system of N interacting atoms in a parallelepiped whose edges are described by vectors a, b, and c. The size and shape of the system can be described in a matrix, H, with column vectors a, b, and c. By using Parinello–Rehman variable cell molecular dynamics [51], stresses or strains can be applied on the system. Application of homogeneous strain to the system consists of changing the value of H from an initial value of H_0 to $H_0+\delta H$. The global stress is then defined by

$$\sigma_{\alpha\beta} = \frac{1}{\Omega}\frac{\partial U}{\partial r_{ij\alpha}} r_{ij\beta}, \tag{13.19}$$

where α and β are the components of the stress tensor, $\Omega = \det(H)$ is the volume of the system, r_{ij} is the distance between ith and jth atoms, and U is the interatomic potential function. Lutsko [40] has devised a local stress tensor for molecular dynamics given by

$$\sigma_{\alpha\beta} = \frac{1}{\Omega}\frac{\partial U}{\partial r_{ij}} r_{ij\alpha} r_{ij\beta} \frac{g(r_i, r_j, s)}{r_{ij}}, \tag{13.20}$$

where

$$g(r_i, r_j, s) = \delta(s - R_{ij})_{\perp} \begin{bmatrix} \theta(s - r_i)_{\|} \\ -\theta(-(s - r_i))_{\|} - \theta(s - r_j)_{\|} \\ +\theta(-(s - r_i))_{\|} \end{bmatrix}. \tag{13.21}$$

Here, θ is a unit step function with a value of 0 for arguments less than 0 and a value of 1 for arguments greater than 0, δ is the Dirac delta function, R_{ij} denotes the center of mass of particles i and j. $\perp$ and $\|$ denote components of vector perpendicular and parallel to vector r_{ij}.

Using this definition of local stress, local elastic constants over a volume V can be evaluated as

$$C^{\mathrm{loc}}_{\alpha\beta\gamma\phi} = -\left.\frac{\partial(\sigma_{\alpha\beta}\mathrm{d}V)}{\partial e_{\gamma\theta}}\right|_{F_i=0}, \tag{13.22}$$

where all the derivatives are taken at a constant force of zero on each atom. This method for determining local elastic constants has been applied to grain boundaries using EAM and pair potentials [73].

Local elastic constants at grain boundaries and other defect structures obtained in this manner can be used in the AEH formulation.

13.2.2 Hierarchical Modeling of Atoms to Continuum

While AEH attempts to model continuum to atomistic in a single step, the hierarchical model presented below links the two scales through information exchange in terms of traction–displacement relations in the form of cohesive zone models (CZMs).

Cohesive zone modeling

The CZM has been proposed by Barrenblatt [5] and extended to ductile fracture by Dugdale [18] and quasibrittle materials by Hillerborg et al. [29] (called fictitious crack model used for fracture mechanics of concrete). The basic concept of the cohesive zone method is illustrated in Fig. 13.6.

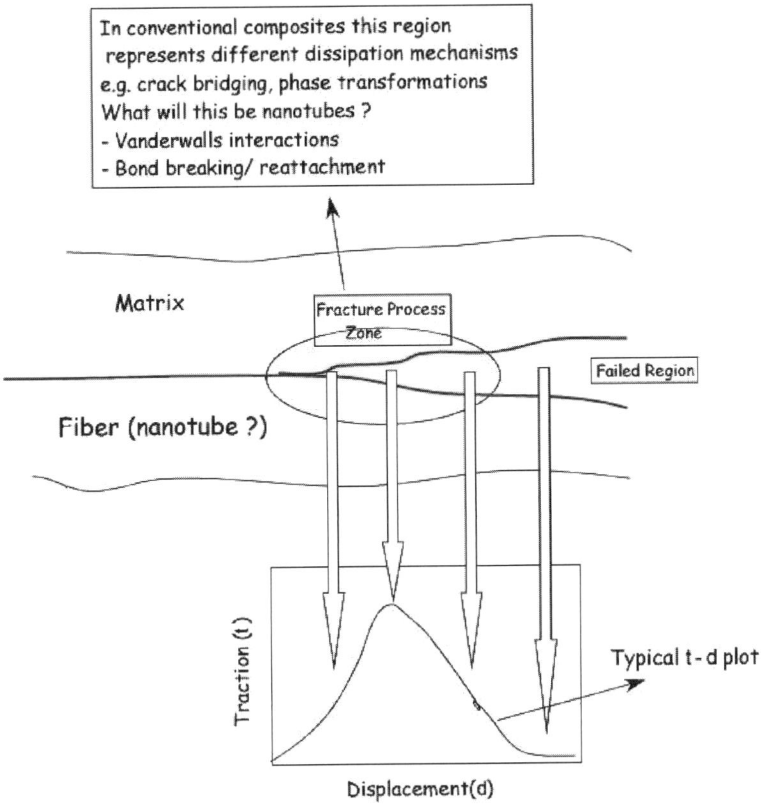

Fig. 13.6. Conceptual framework of CZM

There are several interesting concepts in the CZM. Firstly, this method is micromechanistic in nature, i.e., it considers all the possible micromechanisms that cause damage and fracture. The fracture process zone is particularly modeled through this approach; consequently, there are regions describing complete failure, perfect material, and material undergoing fracture. This region exhibits stress according to the traction–displacement curve. The advantage of the cohesive zone method is clear by the fact that the length of the fracture process zone (displacement to failure in τ–δ curve), the stress to failure (peak stress of τ–δ curve), and the nature of stress distribution in the process zone (through the shape and features of τ–δ curve) can be specified. The area under the traction–displacement curve represents energy for the failure process.

Interfaces, specifically in composites, have been modeled in various ways, e.g., using springs and contact finite elements. In contrast, the CZM-based formulation has the interesting capability of not only modeling bonded interfaces but also separating interfaces without the use of any ad hoc stress/strain/energy-based fracture criterion. Since CZMs have the ability to retain the continuity conditions mathematically, despite the physical separation, the methodology is very appealing. However, before CZMs can be used to model real structures under real loading conditions, the parameters that represent the CZMs should be properly identified and evaluated. In all the CZMs, the traction–separation relations for the interfaces are such that with increasing interfacial separation, the traction across the interface reaches a maximum, then decreases, and eventually vanishes, permitting a complete decohesion.

The main difference in various models lies in the shape and the constants that describe that shape. CZM consolidates the effects of a number of intrinsic and extrinsic toughening (or softening) mechanisms occurring within and in the immediate neighborhood of separating surfaces (process zone). The operative mechanisms depend on the type of material (ductile, brittle, semibrittle), microstructure (monolithic, composites), and temperature and rate of loadings (static, dynamic, cyclic). Some of the commonly used CZMs are shown in Fig. 13.7.

The cohesive zone method readily lends itself to numerical implementation in the FEM. In spite of early theoretical developments, use of this method has increased significantly only in recent times, owing to the developments in computing. A plethora of work has been recently reported covering a number of material systems and a variety of loading conditions and CZMs [8, 10, 25, 49, 54, 55, 65, 69, 73].

The cohesive zone method is used to study interface mechanics of nanotube composites in a two-scale model. Molecular dynamics simulations are first used to simulate fiber pullout tests in nanotube composites. The results of this model are then used to generate a traction–displacement curve used for the CZMs. The FEM with atomically informed cohesive zone parameters is then used for modeling the composite material.

In the following section, atomic scale effects, e.g., interfaces, are analyzed not only to study the thermomechanical response of an interface but also to evaluate composite properties where these effects become important.

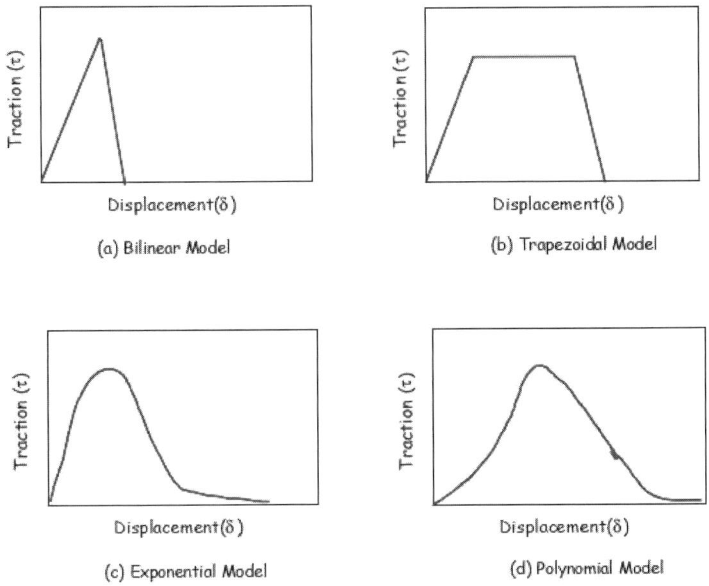

Fig. 13.7. Four commonly used cohesive zone models: (**a**) bilinear model, (**b**) trapezoidal model, (**c**) exponential model, and (**d**) polynomial model

Development of CZM for nanotube interface problem

In general, continuum models of materials are stated as boundary value problems in which quantities, such as displacements, velocities, stresses, or temperatures, are sought at each point in a given domain. The governing equations of the continuum are identified with the balance laws of classical physics, such as conservation of energy, mass, and momentum. The geometries of deformation are prescribed in terms of a kinematic relationship between strain and displacements or deformation gradients. To complete the description of the boundary value problem, a constitutive relationship between the kinematic quantity (strain, deformation gradient) and the kinetic quantity (stress, stress rate) is postulated.

Atomic simulations provide a means to study the material systems in their most pristine form based on forces (kinetic measure) and displacements (kinematic measure). Constitutive equations are phenomenological in nature and are typically determined using experiments. For nanoscale systems, such as carbon nanotubes (CNTs), atomic simulations can provide an alternative to experiments for determining mechanical response of materials which can be used in an average sense in continuum models.

In modeling of composite materials, the boundary value problem is distinct in the sense that there are two continuous media separated by an interface. This interface forms a discontinuity in at least some of the material parameters, e.g., strains, hence it needs to be independently characterized for studying aspects involving interfaces, such as load transfer. In the case of CNT-based composites, there is an additional complication of representing CNTs by the continuum model. In this work, it is assumed that CNTs can be represented by the linear elastic model concentrating on the interface behavior. This assumption is justified because a number of studies based on atomic simulations and experiments report linear elastic behavior for CNTs. In addition, CNTs are shown to exhibit elastic behavior for large strains (up to 30% in some reports). This extends the range of applicability of this model. The constitutive model for interface for a generic interface problem can now be developed.

Consider two solid bodies, Ω_1 and Ω_2, separated by a common boundary S as shown in Fig. 13.8, where S can be considered as the same surface, $S_1 \in \Omega_1$ and $S_2 \in \Omega_2$, in the initial configuration, i.e., it is preferable to define S as an infinitesimally thin 3D domain with surfaces S_1 and S_2 being the part of Ω_1 and Ω_2, respectively, before the separation occurs. For all practical purposes, the surface S_1 or S_2 can be identified as a single surface as a part of either of the domains. A material particle

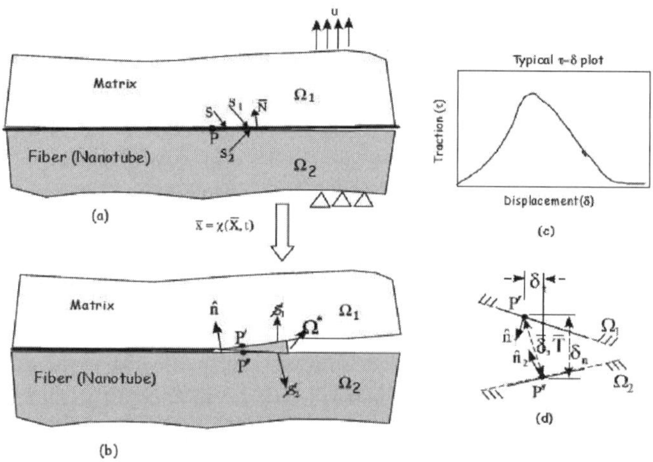

Fig. 13.8. Conceptual framework for cohesive zone model

initially located (within either of the domains, Ω_1 or Ω_2), at some position $\bar{X}$, moves to a new location $\bar{x}$, with a one-to-one correspondence between $\bar{x}$ and $\bar{X}$ given by the equation of motion $\bar{x} = \chi(\bar{X})$.

For either of the domains, Ω_1 or Ω_2, the continuum description is fairly clear. The deformed configuration can be related to the undeformed configuration as $\bar{x} = \bar{X} + \mathrm{d}\bar{X}$, alternately the deformation gradient $(\tilde{F})$ can be defined as

$$\tilde{F} = \frac{\mathrm{d}\bar{x}}{\mathrm{d}\bar{X}}. \tag{13.23}$$

The deformation gradient can be related to its kinetic conjugate, first Piola–Kirchoff stress $(\tilde{P})$, through the appropriate constitutive relation. In the case of elastic deformation, $\tilde{P}$ and $\tilde{F}$ can be related as

$$\tilde{P} = \tilde{C}\tilde{F}, \tag{13.24}$$

where $\tilde{C}$ is the elasticity tensor. This relation can be employed in the equilibrium equation

$$\frac{\partial}{\partial \bar{X}}(\tilde{C}\tilde{F}) + \bar{f} = 0, \tag{13.25}$$

which can be solved variationally with appropriate boundary conditions in a finite element formulation. This elastic formulation can be extended to an inelastic formulation by employing the corresponding constitutive relation.

Now consider the domain *S*. In a generic sense, *S* defines the interface between any two domains. In the problem of nanotube composites, Ω_1 may be considered to represent the nanotube and Ω_2 the matrix (see Fig. 13.8). Then, *S* represents the interface between the CNT and the matrix. When a fracture occurs, interface *S* separates two new surfaces, S_1 and S_2; and, consequently, the one-to-one relation between $\bar{x}$ and $\bar{X}$ is violated in this region. Because of the creation of these new surfaces, the rules of continuum are violated. The surface *S* represented by the unit normal $\bar{N}$ ($\bar{N}_1 \in S_1$ and $\bar{N}_2 \in S_2$) acts along the boundary separating the domain in an undeformed configuration, as shown in Fig. 13.8. After deformation, it can be represented by two unit normals, $\hat{n}_1$ and $\hat{n}_2$, acting along surfaces S_1 and S_2, respectively. This problem can be overcome by considering the interface to be a separate entity and prescribing an independent constitutive equation to it. The interface can be considered to consist of an extremely soft glue which can be shrunk to an infinitesimally thin surface and can be extended to a 3D domain after deformation. Now consider this domain to be Ω^*, then $\bar{N}_1 \in \Omega^*$ and $\bar{N}_2 \in \Omega^*$, based on the kinematic relation for the domain Ω^*. The constitutive equation for the domain Ω^* can be written in terms of shear and normal tractions and their corresponding displacements. A typical constitutive relation for Ω^* is given by $\bar{T}-\bar{\delta}$ (traction–displacement relation), as shown in Fig. 13.8c such that

$$\text{if } |\bar{\delta}| < |\bar{\delta}_{\text{sep}}|, \quad \bar{\sigma}\hat{n} = \bar{T}. \tag{13.26}$$

Beyond a separation distance of $|\bar{\delta}| < |\bar{\delta}_{\text{sep}}|$, the traction is identically zero within Ω^*,

$$\text{if } |\bar{\delta}| > |\bar{\delta}_{\text{sep}}|, \quad \bar{\sigma}\hat{n} = \bar{T} = 0. \tag{13.27}$$

It can also be construed that when $|\bar{\delta}| > |\bar{\delta}_{\text{sep}}|$ in the domain Ω^*, the stiffness $C_{ijkl} = 0$. To implement the vectorial inequalities given in (13.26) and (13.27), typically two separate identities are postulated for the normal and tangential components with limits set for each of them.

The above discussion forms the basis of CZMs. Though theoretical developments in CZM have been dated to the early 1970s, it was first implemented within the framework of the FEM by Needleman in 1987. Numerous investigations for various applications involving fracture and homogeneous and heterogeneous interfaces have been reported since. In the next section, obtaining a cohesive traction–displacement constitutive equation for nanotube-based composite interfaces from molecular dynamics simulations will be discussed.

Atomistic simulation for a pullout test for carbon nanotubes

Evaluation of typical interfacial properties in composite materials is done using single fiber pullout and pushout tests. The objective of these tests is to extract the qualitative and quantitative information regarding the nature of the interface between the fiber and matrix.

Figure 13.9 shows the schematic for pullout and pushout tests. The typical output of these tests is the force required for debonding the fiber from the matrix as well as shear stress distribution along the interface. Pushout and pullout tests have been used to understand composite behavior in terms of various issues, such as thermal stresses, characteristic length of fiber–matrix interfaces, etc. In addition to experimental investigations, there have been a number of finite element-based numerical investigations of pullout and pushout tests to understand the thermomechanical behavior of interfaces [2, 10].

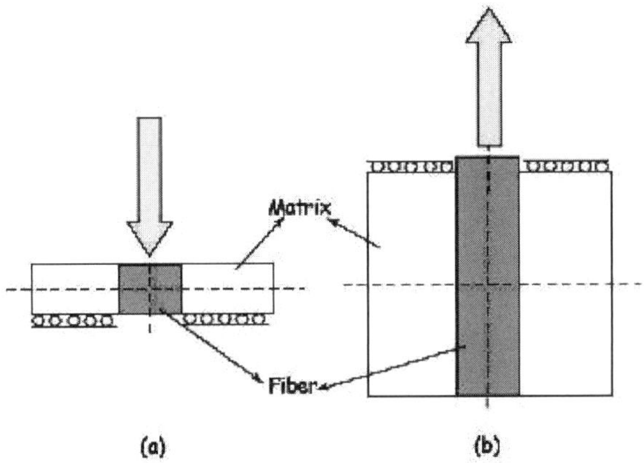

Fig. 13.9. Pushout and pullout tests used for determining interface properties in composites

In the case of nanotube-reinforced composites, the nanometer size of fibers makes applying mechanical loads experimentally on single fibers extremely difficult. There has been only one experimental determination of interface strength by a pullout test using a scanning probe microscope [15]. Conducting these tests on regular basis in different materials for different boundary conditions is beyond current capabilities. With this perspective, atomic simulations based on molecular dynamics and statics become a natural choice for studying the interface behavior by simulating pullout tests.

In general, interfacial strength can be caused both due to the non-bonded Van der Waal's interaction between matrix and fiber and due to bonded chemical interactions. Lordi and Yao [39] and Liao and Li [37] have simulated pullout tests with nonbonded interactions using molecular statics, but there has not been any detailed investigation into the case of bonded interactions. Consider the situation where the matrix and nanotube are chemically bounded by hydrocarbons chains. The interaction of hydrocarbon and hydrogen on the surface of CNT, in terms of changes in hybridization, bond length, and energetics, is studied first.

To study the effect of chemical bonding on CNTs, linear hydrocarbon chains are attached to various (*n*, 0) and (*n*, *n*) types of CNTs. The Tersoff–Brenner potential is used to calculate C–C and C–H chemical interactions based on molecular dynamics and statics simulations. Linear hydrocarbons can form an SP^3 bond with a pi orbital on the CNT surface. For calculating bond lengths and energies, the attachment of the methyl group ($-CH_3$) and hydrogen on the surface of various nanotubes is examined. Bond energies are calculated as differences in energies between chemically bonded structures and nonbonded structures. All structures are first equilibrated by energy minimization using molecular statics.

When a hydrogen atom is bonded on the surface of a graphene sheet, the bond length of the C–H bond is found to be 1.092 . The bond energy of this bond is –2.29 eV. The negative free energy shows the thermodynamic stability of this bond. When the hydrogen atom is bonded to a (10,0) CNT, a bond length of 1.087 is obtained, with a bond energy of –3.0 eV. Lowering of the bond length and energy are indicative of higher stability of a C–H bond in a nanotube compared to a graphene sheet. This can be explained based on geometry. Because of the curvature of nanotubes, the changes in bond angles with chemical attachments are lesser in CNT than in planar graphene sheets; consequently, the bonds are more energetically stable. It has been reported that bond energy of the C–H bond is still lower in fullerene C_{60} structures due to the same reason.

Figure 13.10 shows the chemical attachment of a $-CH_3$ group to a (10,10) CNT. The bond length of a C–C bond at the site of chemical attachment is found to be 1.54 in the CNT and 1.58 in planar graphene sheets. The bond energy for C–C bond attachments to (9,0) nanotubes is –2 eV for CH_3 chemical attachment. It is interesting to note that this bond energy does not significantly change when the length of the chemical attachment is

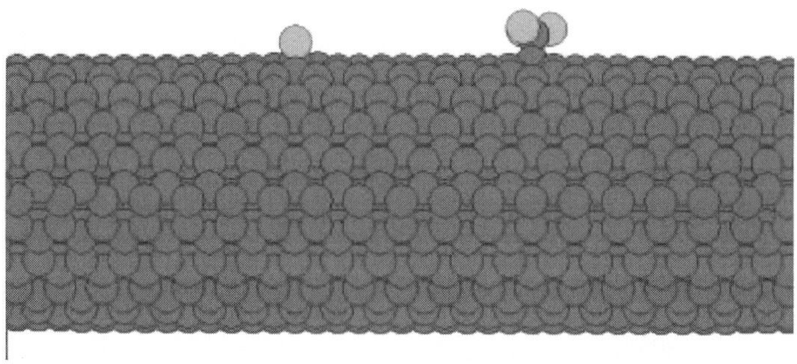

Fig. 13.10. H and $-CH_3$ attachments to (10,10) CNT

increased. However, there is a decrease in bond energy to 1.91 eV when the methyl group is bonded to a graphene sheet. There is also a local increase in the radius of the CNT by about 0.7 due to the change in bonding structure. It can be concluded from the interaction of chemical groups that a CNT is more stable than a graphene sheet, possibly due to the curvature of CNTs.

The interface behavior when simulating the pullout tests is studied next. For this purpose, a (10,10) CNT is considered a typical CNT. This is motivated by the fact that larger proportions of CNTs of this diameter were obtained in laser synthesis of CNTs [57]. Further, various hydrocarbon attachments of varying lengths (with two, four, and five carbon atoms in hydrocarbon functional groups) are attached in varying numbers at random locations on the CNT. The length of the CNT considered here is about 122 , and it consists of 100 repeat units. The choice of hydrocarbon functional attachments is based on the capability of the Tersoff–Brenner potential to model hydrocarbon chains effectively. In addition, there are reports of functionalization of CNTs with hydrocarbon functional groups indicating that this is experimentally feasible [75]. In case of longer hydrocarbon chains, stereoisomers are used, which correspond to a stretched configuration. The boundary conditions applied to the system are shown schematically in Fig. 13.11.

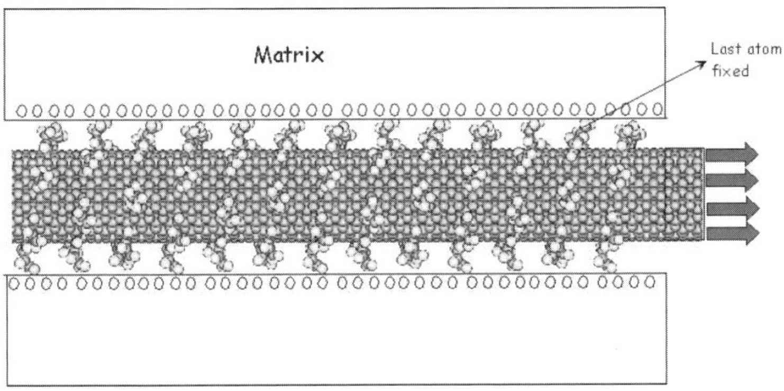

Fig. 13.11. Boundary conditions applied for atomic simulation of pullout test

The corner atoms of the hydrocarbon attachments were fixed indicating that they are connected to the matrix at those locations. Displacements are applied to one end of the nanotube about 15 in length to simulate the effect of pullout. The atoms in this region are fixed and displaced by 0.02 every 1,500 time steps and then equilibrated for the duration of 1,500 time steps. Each time step is of 0.2 fs duration. The simulations were carried out till some of the hydrocarbon chains fail. Typically, simulations lasted for 500,000–800,000 time steps.

To study the effect of chemical bonding, the number of hydrocarbon chains were varied from 3 to 85 chemical attachments. For studying the effect of the length of hydrocarbon chains, varying lengths of 2, 4, and 5 carbon atom hydrocarbon attachments were used. The reaction force on the fixed atoms was monitored throughout the simulation and averaged over 100 time steps before the next set of displacements was applied.

Figure 13.12 shows typical force vs. displacement for any hydrocarbon attachment. The force in the figure is the average (averaged over 100 time steps) reaction force experienced by the fixed atom in the *Y* direction, i.e., along the length of the nanotube, and corresponds to shear. The displacement is the change in position experienced by the atom, which is attached to the CNT with respect to its initial position. Though there are statistical variations for different chemical attachments, the general shape of the force–displacement plot is as shown.

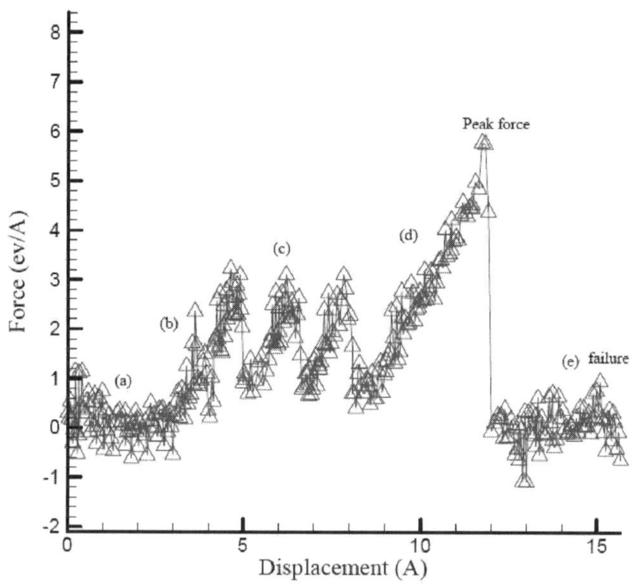

Fig. 13.12. Plot showing typical reaction force vs. displacement in pullout test

The initial region of the curve is flat (parallel to *X*-axis), marked as region (a). This region corresponds to stretching of the hydrocarbon attachment. The flat region shows that there is minimal load transfer in this portion of the curve, and it is similar to the mechanical analogue of loose strings becoming taut. The length of this flat portion is directly dependant on the length of the hydrocarbon chain, i.e., the flat region is longer for hydrocarbon chains with four and five carbon atoms than with two carbon atoms.

After this flat region, there is a gradual increase in the reaction force corresponding to region (b) of the curve. In this part, the carbon chain contributes to the load transfer. Though there are statistical variations from plot to plot, the typical force experienced in this portion of the curve is about 3 eV $^{-1}$ (4.8×10^{-3} μN). This value of force is very small but it must be noted that the area on which this force acts is of the order of angstroms; consequently, the resulting shear stress is very high.

Region (c) of the plot consists of a region of fluctuations in the reaction force. This is due to an interesting behavior of bond detachment and rejoining with adjacent atoms. When separation occurs, there is a sudden drop in force; but this is followed by a rejoining of the hydrocarbon chain with adjacent atoms of the CNT. Because of this, there is an increase in the reaction force corresponding to the increased load transfer between the matrix and fiber.

After a series of jagged regions in the force–displacement curve, there is a sudden increase in reaction force, as shown in region (d); and then total failure occurs. The force at which the failure occurs is about 6 eV $^{-1}$ ($\sim 10^{-2}$ μN). This is the force required to break one chemical attachment; the overall force required to break all of the chemical attachments is much higher and is the sum of all individual reaction forces.

The area under the force–displacement curve denotes the energy required for detaching of the hydrocarbon group from the CNT. Calculations based on molecular statics indicate that the energy required for separation is of the order of 3 eV; however, the area under the force–displacement curve is much higher and is of the order of 20 ± 4 eV for various attachments. This shows that the dynamic process of nanotube pullout requires much higher energy than that predicted based on statics. The main reason for the increase in the energy is due to the bonding and rebonding process associated with the nanotube pullout. Energy of 3 eV is associated with each time the bond is broken. As this process is repeated a number of times for each chemical attachment, the total energy consumed in the separation process is much higher.

In most of the cases, the bond failure occurs at the site of attachment of the hydrocarbon group to the nanotube. To understand more clearly the process of pullout and the energetics involved, simulations are conducted with a single hydrocarbon attachment. The end atom of the single chain was subject to displacement independently in the normal and shear directions (as shown in the Fig. 13.13), and the position energetics of the CNT are closely monitored. The temperature in these simulations is also kept constant at 300 K.

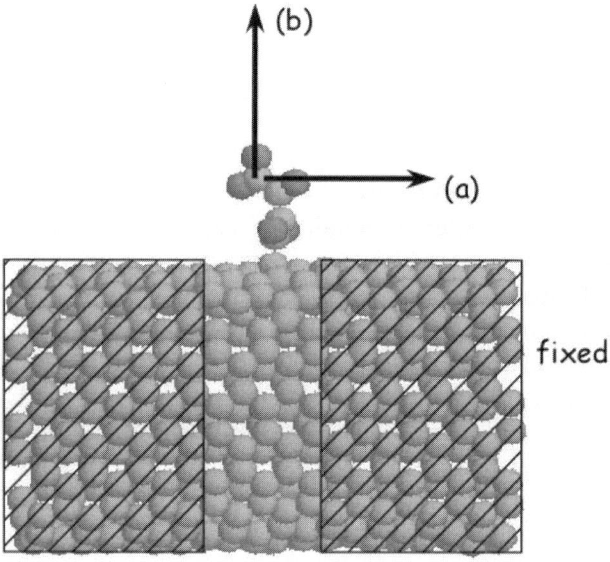

Fig. 13.13. Boundary conditions for simulation: (**a**) displacement applied in shear and (**b**) normal directions

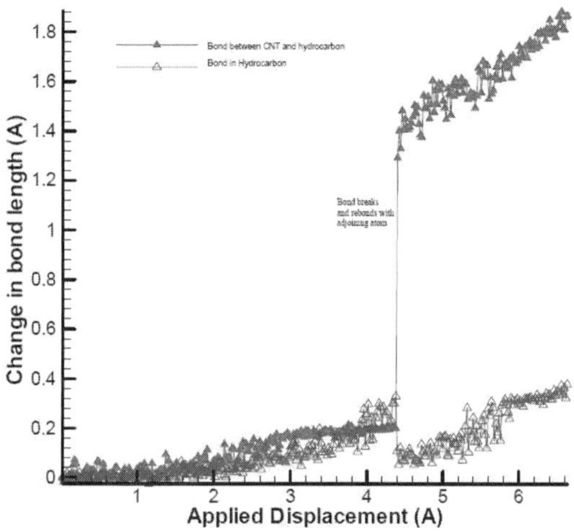

Fig. 13.14. Change in bond length vs. applied displacement for shear loading

Figure 13.14 shows the variation of the bond lengths between the hydrocarbon and the CNT, and another bond in between hydrocarbon atoms in the functional attachment in the case of shear loading. It can be observed that the bond lengths increase steadily up to a certain value and then suddenly jump. This jump corresponds to the debonding of the hydrocarbon attachment and rejoining of that adjacent atom.

Figure 13.15 shows similar bond length variation when normal load is applied. In this case there is no possibility of rebonding, because once detached, the hydrocarbon moves further away from CNT with applied displacement. The hydrocarbon chain fails at the central region leaving a portion of the hydrocarbon attached to the CNT. If the two plots are compared, the increase in bond length is higher in the case of normal loading. This is expected because at the same rate of applied displacement, the component of displacement resulting in bond length increase is higher in the case of normal displacement. In addition, the pattern of the plots is similar for both of the cases until bond separation or rejoining occurs. However, it can be observed that in the case of shear loading, bond rejoining occurs before the bond in the hydrocarbon can increase in length by more than 0.2 . This indicates that if rejoining did not occur, shear loading would be similar to normal loading.

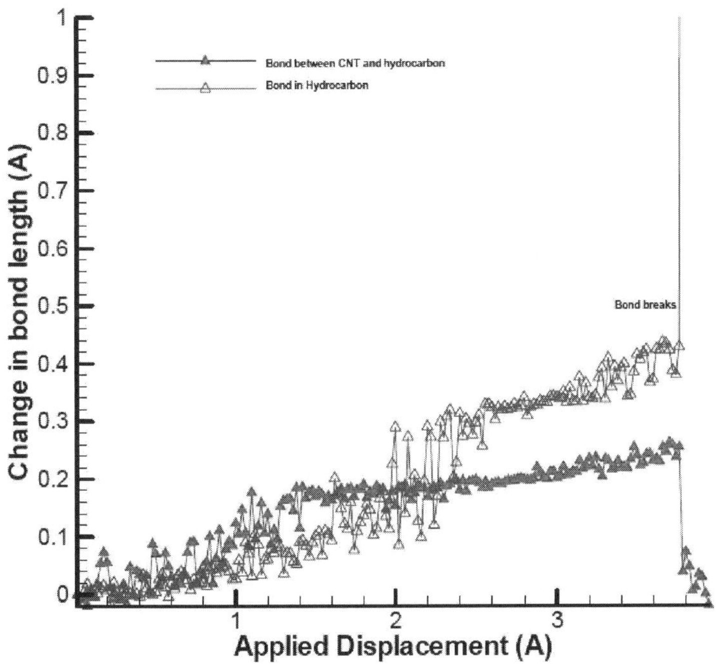

Fig. 13.15. Change in bond length vs. applied displacement for normal loading

The difference in energies between initial configuration and loaded configuration can be considered as strain energy. This strain energy is indicative of the extent of load transfer between the CNT and the matrix. Figures 13.16 and 13.17 show the strain energy per atom as a function of applied displacement. In both the figures, the jump in strain energy can be observed. In the case of normal separation, this corresponds to breaking of the bond; while in the case of shear separation, this corresponds to debonding and rejoining with the adjacent atom. It can be clearly noted that the strain energy increase is much higher in normal loading where bond failure occurs compared to shear loading. This shows that the energy required for debonding and rebonding is much lower than that required for bond separation.

Similar observation in the pullout test in Fig. 13.12 indicates that the reaction force when bond separation occurs is much higher than during the debonding/rebonding phase. To observe the increase in energy when more than one hydrocarbon group is attached, simulations were performed with boundary conditions shown in Fig. 13.16 but with three hydrocarbon attachments. The strain energy per atom variation with applied displacement is shown in Fig. 13.16b. It can be observed that the initial region of the two curves with one hydrocarbon attachment and that with three attachments

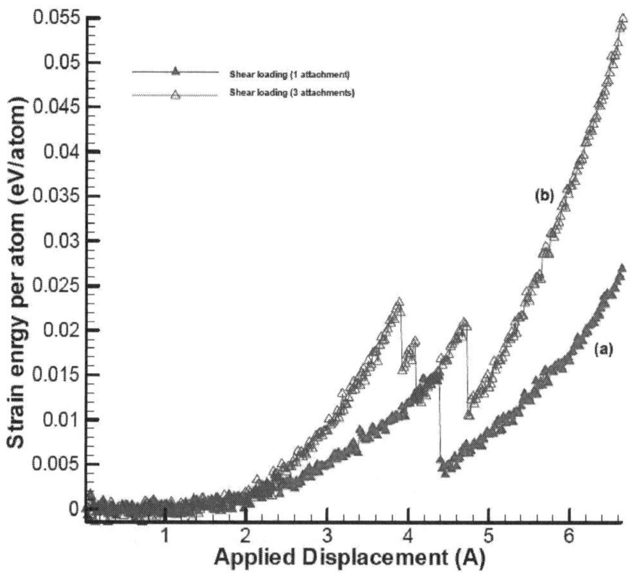

Fig. 13.16. Strain energy per atom variation with applied displacement in shear loading conditions: (**a**) one chemical attachment and (**b**) three chemical attachments

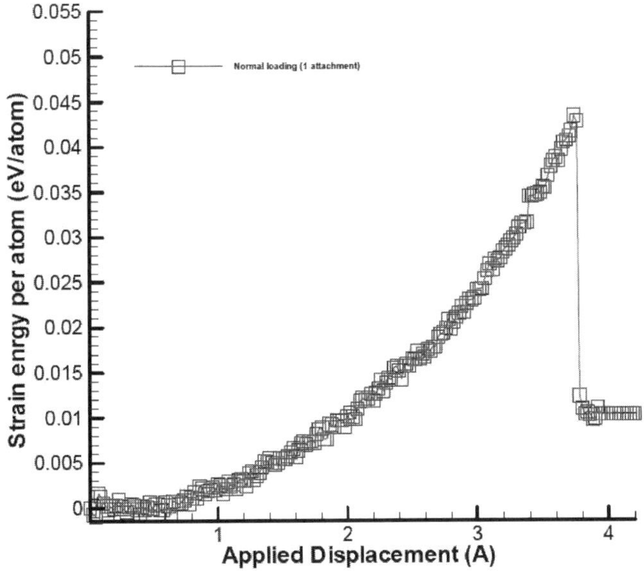

Fig. 13.17. Strain energy per atom variation with applied displacement in normal loading conditions

coincides. This corresponds to rearrangement of the hydrocarbon attachment without significant extension in bond lengths (Fig. 13.16). In this region, there is no significant increase in strain energy indicating minimal load transfer. In plot (b), there are three drops in energy corresponding to debonding and rebonding in three hydrocarbon attachments. In addition, the increase in strain energy is higher than that with a single attachment. This shows that the increased extent of chemical bonding contributes to the greater extent of load transfer; however, this phenomenon is not linearly additive, i.e., attaching three chains does not increase the energy (and thereby load transfer) by three times. One more interesting aspect is that the increase in energy with three hydrocarbon attachments in shear loading is less than that with one hydrocarbon attachment in normal loading. This shows that the amount of energy consumed (thereby load transfer) is much higher when bond breaking occurs. This signifies that loading conditions affect the load transfer more significantly than the extent of chemical bonding.

The results of pullout tests for nanotubes with varying degrees of chemical bonding can now be addressed. This effect is studied by varying the number of hydrocarbon chemical attachments. For a (10,10) CNT of 122 length, there are 85 chemical attachments if one chemical attachment is attached per repeat unit of the CNT, excluding the end of the CNT which is subject to displacement. Different numbers of chemical attachments 85, 45, 20, 10, and 5 are attached at circumferentially random and cylindrically equidistant locations, and pullout tests are simulated with all the other variables such as temperature and displacement rate being constant. The general shape of the force–displacement curve remains similar to that shown in Fig. 13.16. The values of points of inflexion also remain similar within statistical variations; however, there is a tendency for extended debonding and rebonding regions when the density of chemical attachments is lowered.

The debonding/rebonding behavior observed in the pullout test is interesting since it is an example of phenomena which occur only at the nanoscale and is generally not observed in conventional composites at the macroscopic scale. Macroscopic fracture is generally considered to be irreversible – that the energy required in creation of new surface during fracture is not equal to the energy required to join two surfaces and heal the damage.

At the atomic scale, fracture is defined purely by breaking of chemical bonds. In this case, the energy required to break two bonds is exactly equal to the energy released to join the same two atoms to form a chemical bond. Whether two atoms will join to form a chemical bond is determined based on the distance between the atoms and the kinetics of the process. This phenomenon is not entirely new, for example, similar observations have been made during atomic simulations of stick slip friction and that of metal cutting processes. Zhang and Tanaka [76] have performed simulations of friction on copper surfaces; they observed a variation in frictional force in an oscillatory manner similar to that of the reaction force observed in this project's simulations. Further, frictional force during rolling of a nanotube on a graphene sheet has been shown to produce similar periodic variations [76]. However, significance of this phenomenon is increased here as it affects the load transfer considerably.

The next section discusses the extension of these observations at the atomic scale to the continuum scale using CZMs.

Traction–displacement equation for interfaces in nanotube-based composites

One of the major bottlenecks in the application of CZMs is the ad hoc nature of traction–displacement relations used by most investigators. The points of inflection in traction–displacement curves have been often selected in an ad hoc manner. Chandra and coworkers [10] have emphasized the importance of the shape of the traction–displacement plot in addition to maximum stress and separation distance. Obtaining traction–displacement plots from experiments would be the correct approach to the problem. In the case of nanotube-based composites, the nanoscale size of fibers limits controlled experimentation; hence the next best approach for atomistic simulations, a nanotube pullout test, was used.

In the previous section, the atomistic aspects of the pullout test were discussed. To apply the results of those simulations to the CZM, a traction–displacement plot needs to be constructed to describe the nanoscale interface. Before considering the construction of the traction–displacement law, it is necessary to understand the fundamental idea of the CZM. The CZM is primarily a micromechanistic approach to study interfaces. All of the micromechanical details occurring during the load transfer and failure are assumed to be lumped together and quantitatively described in a traction–displacement relation. For example, in the study of interfaces in conventional composites, various micromechanical processes of energy dissipation leading to fracture (or affecting load transfer), such as crack deflection at precipitates, contact wedging, frication, inelasticity-induced energy dissipation, and compressive deformation-induced crack closure, e.g., cyclic loading, are considered to account for processes such as microcracking, void coalescence, and phase transformation-induced energy dissipation, crack deflection along grain boundaries, or second phase, etc. In the context of nanotube-based composites, various factors, such as the effect of rebonding/debonding, effect nanotube waviness and curvature. Van der Waal's interaction should be incorporated into the traction–displacement relation. A few of these factors were considered while formulating the traction–displacement relation.

There is an additional complexity involved in obtaining cohesive zone information from atomic simulations. As a result of discrete numerical implementation of the FEM, any particular element is considered to experience constitutive response as a single unit (albeit based on an interpolation) according to the traction–displacement curve. The element dimensions can be refined to the order of few nanometers; however, it would still physically correspond to a large number of functional group attachments. One way of circumventing this problem would be to use a finite element mesh refined to atomic proportions, but that would defeat the initial objective of the multiscale mode – to achieve high computational efficiency. A way is needed to homogenize the force–displacement relations shown in the preceding section to obtain the traction–displacement plots which are representative of the entire system. Simple averaging over all of the chemical chains will not serve the purpose since the effect of length and density would be ignored. In other words, if only a few chemical attachments near the loading end take up most of the load, then the traction–displacement plot obtained by averaging over the entire length of the CNT would give an erroneous underestimation.

Figure 13.17 shows a variation of reaction forces for the specific atoms in the hydrocarbon attachments (across the length of the nanotube) at various simulation times. Figure 13.18 shows the plot in the case of 85 attachments that is about one chemical attachment for every repeat unit of a (10,10) CNT. It can be observed from this figure that the variation of

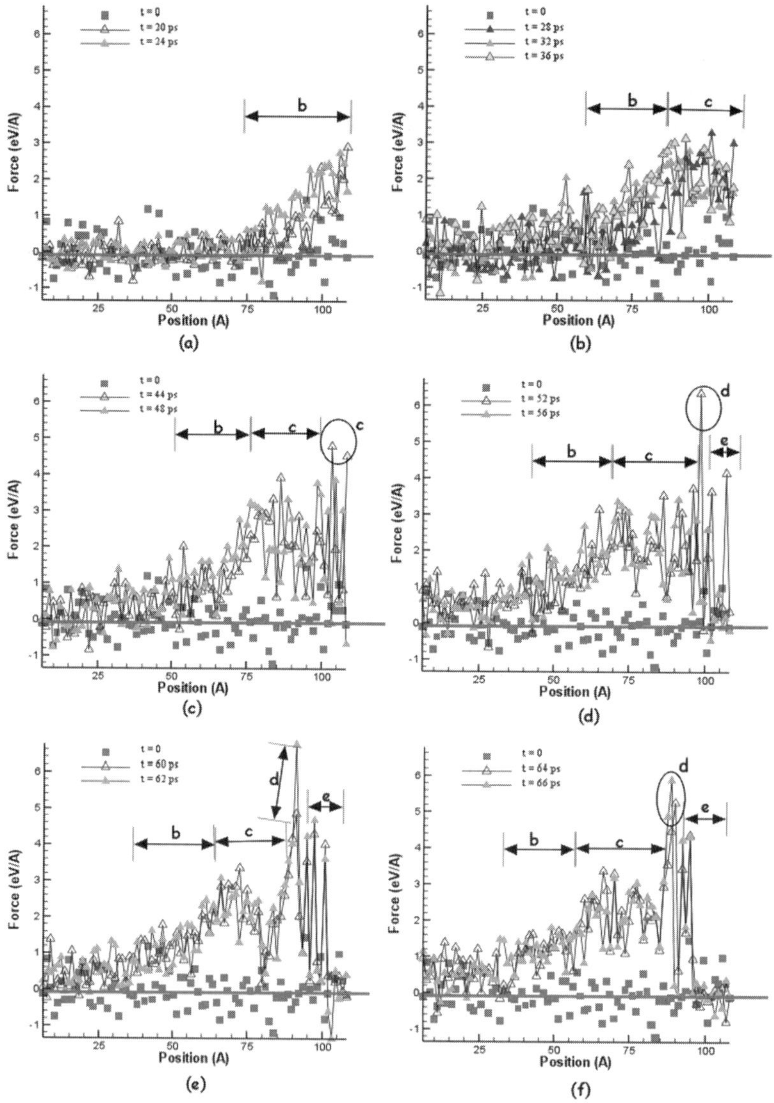

Fig. 13.18. Variation of reaction force along the length of a (10,10) CNT with 85 chemical attachments at different simulation times

force occurs in a sequential manner. If the typical force–displacement is divided into various parts (a) initial loading, (b) debonding/rebonding region, (c) maximum loading region, (d) and failure denoted by (e) (Fig. 13.18), the force–displacement plots of each chemical attachment go through these regions one after the other until they fail. As expected, the chemical attachments near the loading region reach the peak stress and fail sooner than chemical attachments away from the loading end. Once the chemical attachments detach completely from the CNT, it fails to take any more load; thus, the peak stress region moves progressively from the loading end to the other end. This process can be observed in Fig. 13.18. Figure 13.18a at 20 and 24 ps shows atoms near the loading end taking up load; in Fig. 13.18b the chemical attachments near the loading end reach maximum load and failure (region d).

The maximum load region moves away from the loading end as shown in Fig. 13.18c,d. At any point in time, only about 40 chemical attachments take load; and others have either failed or yet to take any load. It has been observed earlier that the force–displacement plots for any chemical attachment are similar within statistical variations. Based on the above observations, the average of these plots was calculated with applied displacement for 40 chemical attachments and divided by the corresponding area to obtain the traction–displacement plot shown in Fig. 13.19.

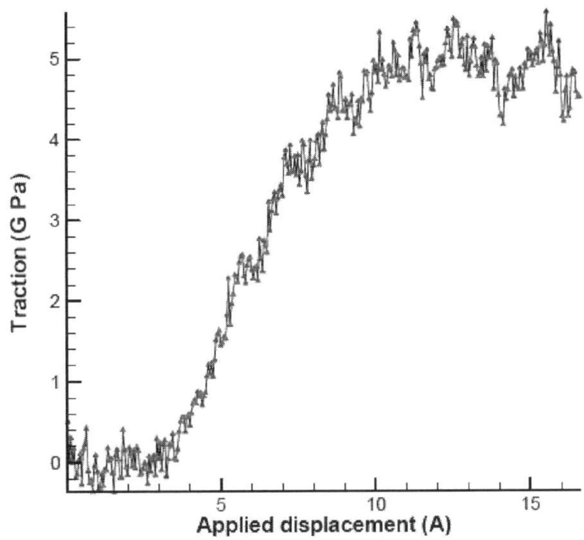

Fig. 13.19. Calculated traction–displacement plot for the case of 85 chemical attachments

The earlier discussion was for the CNT with a high density of chemical attachments with a matrix. Now consider the case of the nanotube with only five chemical attachments in the same length. The reaction force variation along the length of the CNT at various simulation times is shown in Fig. 13.20. It can be observed that all of the chemical attachments take similar loads at any time. In this case, the traction–displacement plot was

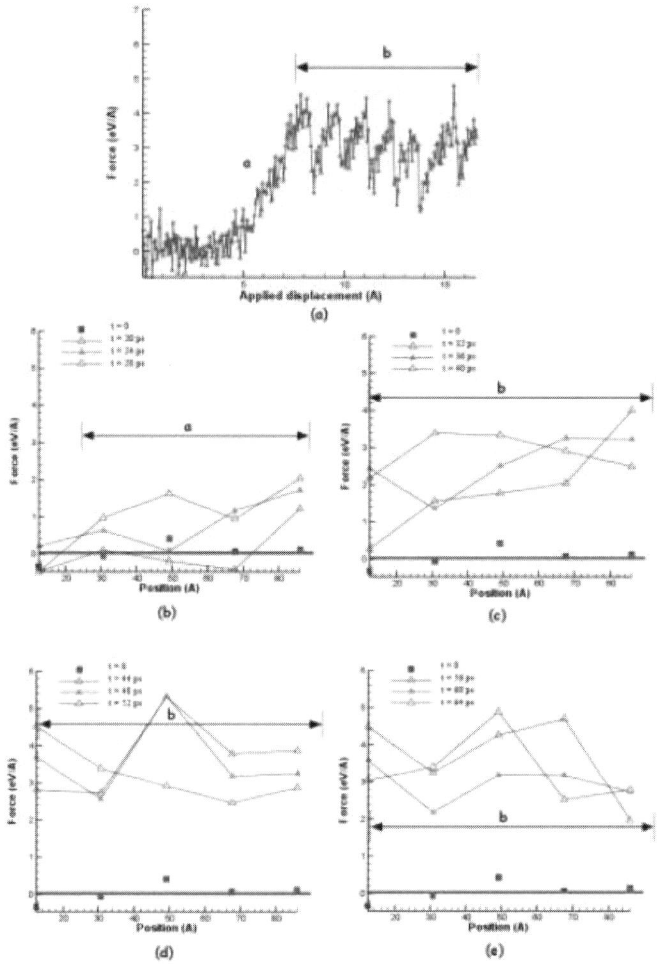

Fig. 13.20. Variation of reaction force along the length of a (10,10) CNT with five chemical attachments at different simulation times

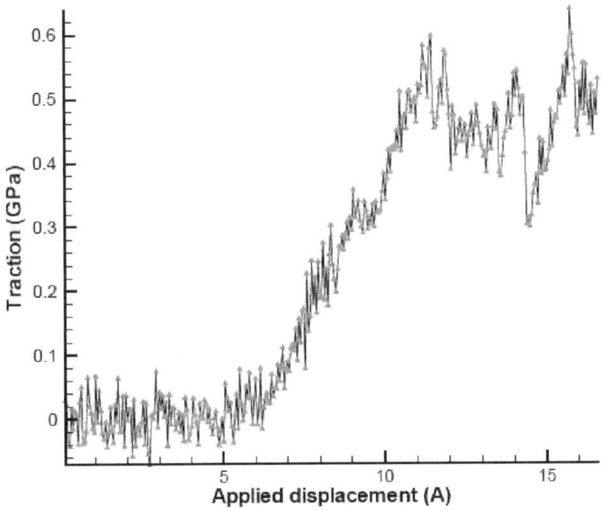

Fig. 13.21. Calculated traction–displacement plot in the case of five chemical attachments

obtained by averaging the forces in all of the chemical attachments and dividing by the entire surface area of the CNT. The corresponding traction–displacement plot is shown in Fig. 13.21.

The traction–displacement plots for use in cohesive zones are obtained by extrapolating figures to complete failure as shown schematically in Fig. 13.22. The most interesting aspect of this investigation is the very high interfacial stresses obtained when there is chemical bonding between the matrix and the CNT. For a high degree of chemical bonding (85 chemical attachments to a CNT of 100), interfacial stress as high as 5 GPa was obtained, which is about two orders of magnitude higher than that typically observed in conventional polymeric composites. In the case of five hydrocarbon attachments distributed over 100 , an interfacial peak stress of 500 MPa was obtained. It is interesting to note that experimental measurements of interfacial traction by Cooper et al. [15] report interfacial stress of same order. Further, Wagner and coworkers indicate that this high value might be due to a chemical interaction between the CNTs and the urethane matrix. An interfacial peak stress of about 50 MPa was obtained with three chemical attachments in 200 length of a CNT.

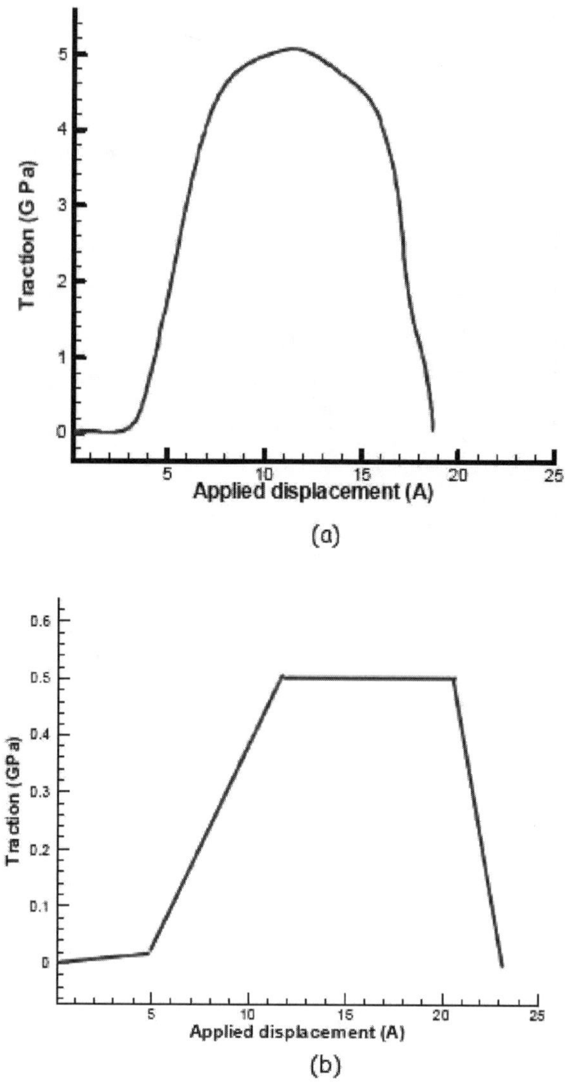

Fig. 13.22. Traction–displacement plots used in the finite element model

Application of multiscale model

The atomically informed traction–displacement plots are now applied in finite element simulations of CNT-reinforced composites. To demonstrate the effectiveness of the multiscale model, the elastic behavior of the composite and the effect of interfacial strength on the stiffness is studied. It is common practice to study the mechanical behavior of conventional composites using an RVE. Because of their ability to fill square or hexagonal spaces, RVEs are popular in formulations where the RVE approach is extended to generalized problems using global–local or similar approaches. Cylindrical RVEs are generally used to reduce the computational load by employing an axisymmetric formulation.

The finite element mesh and the boundary conditions are shown in Fig. 13.23.

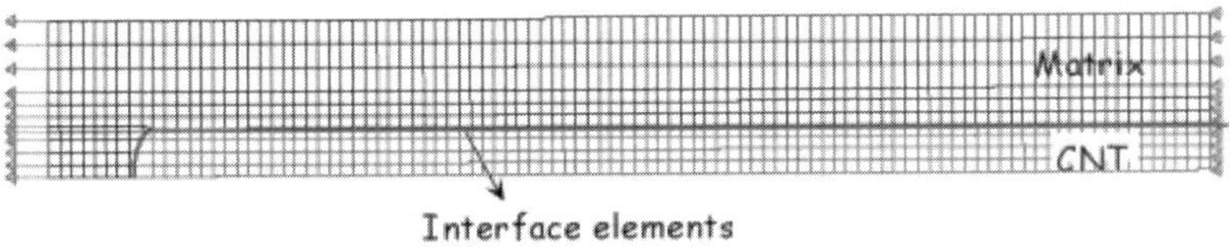

Fig. 13.23. Typical mesh used in finite element simulations

Axisymmetric cohesive zone elements, each having four nodes and zero thickness in the direction normal to the interface, are used to model the interface behavior. The traction–displacement plots discussed in the previous section are used to generate the stiffness matrix for these elements. Four axisymmetric node elements are employed for both the matrix and the CNT. Duplicate nodes are used at the interface. The models typically consist of 1,000 elements (varied for different fiber content), and the interface consists of 93 cohesive zone elements. The CNT length is 450 with a diameter of 20 . In these computations, a CNT is considered as a solid cylinder. This can be justified by assuming a multiwalled nanotube with the innermost tube having a very small radius, e.g., a (5,0) nanotube. Both the matrix and the CNT are assumed to be continuous media exhibiting homogeneous linear elastic isotropic behavior with a given Young's modulus and Poisson's ratio. A Young's modulus of 1,000 GPa and Poisson's ratio of 0.3 are employed for the nanotube.

The general purpose commercial code ABAQUS, Version 6.3, is employed to carry out the analysis because of its flexibility in allowing user-defined subroutines to be linked to the main program. The cohesive element model is input as a user-defined element subroutine, UEL, into ABAQUS. Parametric studies for different interfacial strengths, matrix, and fiber properties were performed to understand the effect of interfacial strength on the stiffness and load transfer of the composite.

Two extreme cases of load transfer occur when the interfacial stress transfer is perfect and when there is zero load transfer between the matrix and fiber. In terms of CZM, these two cases would correspond to infinite and zero interfacial traction, respectively. In the case of zero interfacial traction, the composite elastic properties become equal to that of the matrix material, as no load transfer occurs, and progressively increase as the peak traction of CZM increases. For very high value of peak stress, the load transfer should be near maximum possible load transfer and, correspondingly, the elastic modulus of the composite should be near that of the ideal interface. This concept is illustrated in Fig. 13.24.

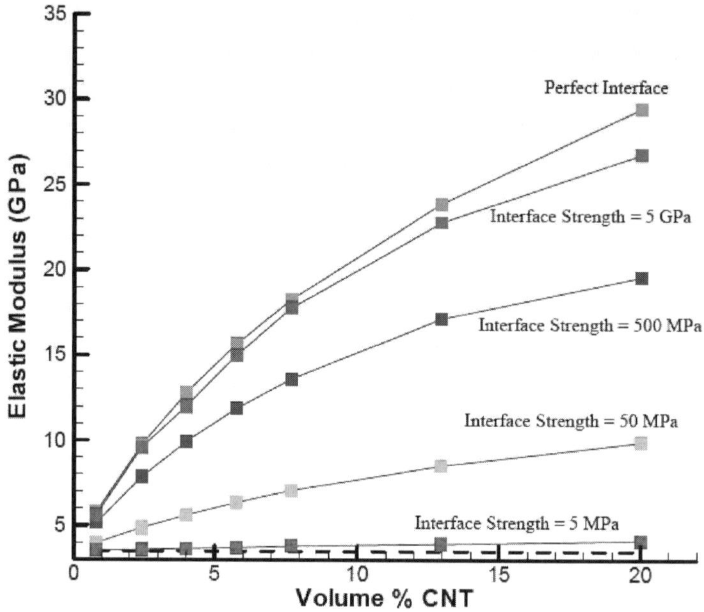

Fig. 13.24. Variation of composite stiffness with volume percentage CNT for different interfacial strengths. Matrix elastic moduli of 3.5 GPa are used in the simulations

Figure 13.24 shows the variation of composite elastic modulus with volume percentage for the CNT for various CZMs described in the earlier section. When the matrix and CNT are connected by a large number of chemical attachments (one per repeat unit of CNT), the interfacial strength is very high; and the peak traction is of the order of 5 GPa. To provide a perspective, typical values of interfacial strength in polymeric composites are about 50 MPa. This high interfacial strength leads to stiffness values near that of composite with the perfectly bonded interface as seen in Fig. 13.24. On the other extreme, when the peak traction is 5 MPa (this is a fictitious traction–displacement plot and does not correspond to any atomic simulation result), the composite stiffness is close to that of matrix material (denoted by dotted line in Fig. 13.24). The more likely values of interfacial peak traction are between 50 and 500 MPa, which correspond to three bonds for a 200 CNT and five bonds for a 100 CNT. In these cases, composite stiffness is about 80 and 60% of the maximum possible value (perfect interface).

Figure 13.25 shows the variation of composite stiffness with matrix stiffness for various CZMs. Typical values of elastic moduli for polymeric systems range from 1 to 10 GPa. Four matrix elastic moduli of 1, 3.5, 5, and 10 GPa are considered in this study.

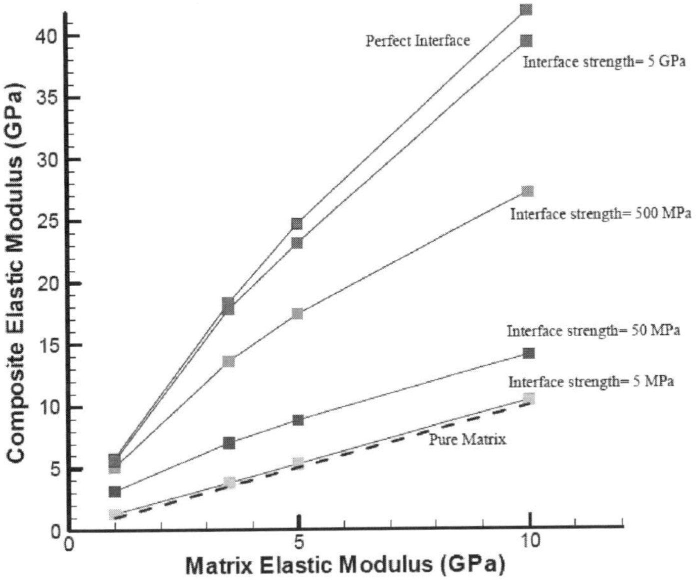

Fig. 13.25. Variation of composite stiffness with changes in matrix stiffness for different interfacial strengths

The composite stiffness increases with an increase in matrix stiffness and decreases with lowering of the interfacial strength, as expected. Figure 13.25 shows the variation of composite elastic moduli with different fibers corresponding to glass fibers ($E = 85$ GPa), carbon fibers ($E = 300$ GPa), SiC ($E = 400$ GPa), and CNT ($E = 1{,}000$ GPa). These computations are again performed for different interface models. One interesting observation here is that there is a sort of flattening of stiffness values as the stiffness of fiber increases. Though the stiffness values for CNT-based composites are higher than that of other fibers, they are not as high as one would expect, specifically when the interfacial strength is low. This emphasizes the importance of interface load transfer in determining the stiffness of nanotube-reinforced composites.

References

1. Abraham, F, Broughton, J, Bernstien, N, and E Kaxarias, Spanning the length scales in dynamic simulation, *Computers in Physics*, Vol. 12, p. 538, 2001
2. Ananth, CR and N Chandra, Numerical modeling of fiber pushout testing metallic ceramic and internetallic matrix composites – mechanics of the failure processes, *Journal of Composite Materials*, Vol. 29, p. 1488, 1995
3. Arroyo, M and T Belytschko, An atomistic-based finite deformation membrane for single layer crystalline films, *Journal of the Mechanics and Physics of Solids*, Vol. 50, p. 1941, 2002
4. Ashby, MF, Boundary defects, and atomistic aspects of boundary sliding and diffusional creep, *Surface Science*, Vol. 31, p. 498, 1972
5. Barrenblatt, GI, The formation of equilibrium cracks in brittle fracture. General ideas and hypothesis. Axially-symmetric cracks, *Prikladnaya Matematika i Mekhanika*, Vol. 23, p. 434, 1959
6. Bensoussan, A, Lions, JL, and G Papanicolaou, *Asymptotic analysis for periodic structures*, New York: North-Holland, 1978
7. Buannic, N and P Cartraud, *International Journal of Solids and Structures*, Vol. 38, p. 7139, 2001
8. Camacho, GT and M Ortiz, Computational modeling of impact damage in brittle materials, *International Journal of Solids and Structures*, Vol. 33, p. 2899, 1990
9. Chandra, N and P Dang, Atomistic simulation of grain boundary sliding and migration, *Journal of Materials Science*, Vol. 34, p. 655, 1999
10. Chandra, N, Li, H, Shet, C, and H Ghonem, Some issues in the application of cohesive zone models for metal ceramic interfaces, *International Journal of Solids and Structures*, Vol. 39, p. 2827, 2002
11. Chung, PW, Tamma, KK, and RR Namburu, *Structural Dynamics and Materials Conference*, Vol. 1, St Louis, MO, 1999, pp. 411–421
12. Chung, PW, Tamma, KK, and RR Namburu, *Composites Science and Technology*, Vol. 60, p. 2233, 2000

13. Chung, PW, Tamma, KK, and RR Namburu, *Composites: Part A*, Vol. 32, p. 1291, 2001
14. Cleri, F, Philpot, SR, Wolf, D, and S Yip, Atomistic simulations of materials fracture and the link between atomic and continuum length scales, *Journal of the American Ceramic Society*, Vol. 81, p. 500, 1998
15. Cooper, CA, Cohen, SR, Barber, AH, and HD Wagner, Detachment of nanotubes from a polymer matrix, *Applied Physics Letters*, Vol. 81, p. 3873, 2002
16. Daw, MS and I Baskes, *Physical Review Letters*, Vol. 50, p. 1285, 1983
17. Deymier, P and G Kalonji, Effect of grain-boundary melting on sliding in bicrystals – a molecular-dynamics study, *Scripta Materialia*, Vol. 20, p. 13, 1984
18. Dugdale, DS, Yielding of steel sheets containing silts, *Journal of the Mechanics and Physics of Solids*, Vol. 8, p. 100, 1959
19. Ercolessi, F and JB Adams, *Europhysics Letters*, Vol. 26, p. 583, 1994
20. Farakas, D, *Philosophical Magazine Letters*, Vol. 80, p. 229, 2000
21. Fish, J, Shek, K, Pandheeradi, M, and MS Shepard, *Computer Methods in Applied Mechanics and Engineering*, Vol. 148, p. 53, 1997
22. Fukutomi, H, Yamamot, T, Nonomura, V, and K Takada, *Interface Science*, Vol. 7, p. 141, 1999
23. Furuya, Y and H Noguchi, A method of seamlessly combining a crack tip molecular dynamics enclave with a linear elastic outer domain in simulating elastic–plastic crack advance, *International Journal of Fracture*, Vol. 87, p. 309, 1997
24. Furuya, Y and H Noguchi, A combined method of molecular dynamics with micromechanics improved by moving the molecular dynamics region successively in the simulation of elastic–plastic crack propagation, *International Journal of Fracture*, Vol. 94, p. 17, 1998
25. Geubelle, PH and J Baylor, Impact-induced delamination of laminated composites: a 2D simulation, *Composites*, Vol. 29B, p. 589, 1998
26. Ghosh, S, Lee, K, and S Moorthy, *International Journal of Solids and Structures*, Vol. 38, p. 2335, 2001
27. Guedes, JM and N Kikuchi, *Computer Methods in Applied Mechanics and Engineering*, Vol. 83, p. 143, 1990
28. Gumbsch, P, An atomistic study of brittle-fracture – toward explicit failure criteria from atomistic modeling, *Journal of Materials Research*, Vol. 10, p. 2897, 1995
29. Hillerborg, A, Modeer, M, and PE Petersson, Analysis of crack formation and crack growth in concrete by means of fracture mechanics and finite elements, *Cement and Concrete Research*, Vol. 6, p. 773, 1976
30. Hoagland, RG, Hirth, JP, and PC Gehlen, Atomic simulation of dislocation core structure and Peirels stress in alkali-halide, *Philosophical Magazine*, Vol. 34, p. 413, 1976
31. Hollister, SJ, Fyhrie, DP, Jepsen, KJ, and SA Goldstein, *Journal of Biomechanics*, Vol. 24, p. 825, 1991

32. Hornung, U, *Homogenization and porous media*, Berlin Heidelberg New York: Springer, 1997
33. Kluge, MD, Wolf, D, Lutsko, JF, and SR Philpot, *Journal of Applied Physics*, Vol. 67, p. 2370, 1990
34. Kohlhoff, P, Gumbsch, P, and HF Fischmeister, Crack-propagation in bcc crystals studied with a combined finite-element and atomistic model, *Philosophical Magazine A*, Vol. 64, p. 851, 1991
35. Koishi, M, Shiratori, M, Miyoshi, T, and K Kabe, *JSME International Journal: Series A*, Vol. 40, p. 306, 1997
36. Kurtz, RJ, Hoagland, RG, and JP Hirth, Effect of extrinsic grain-boundary defects on grain-boundary sliding resistance, *Philosophical Magazine A*, Vol. 79, p. 665, 1999
37. Liao, K and S Li, Interfacial characteristics of a carbon nanotube-polystyrene composite system, *Applied Physics Letters*, Vol. 79, p. 4225, 2001
38. Lidorikis, E, Bachlechner, ME, Kalia, RK, Nakano, A, Vashishta, P, and GJ Voyiadjis, Coupling length scales for multiscale atomistics-continuum simulations: atomistically induced stress distributions in Si/Si3N4 nanopixels, *Physical Review Letters*, Vol. 87, p. 086104, 2001
39. Lordi, V and N Yao, Molecular mechanics of binding in carbon-nanotube-polymer composites, *Journal of Materials Research*, Vol. 15, p. 2770, 2000
40. Lutsko, JF, *Journal of Applied Physics*, Vol. 64, p. 1152, 1988
41. Miller, R, Ortiz, M, Philips, R, Shenoy, VB, and EB Tadmor, Quasicontinuum models of fracture and plasticity, *Engineering Fracture Mechanics*, Vol. 61, p. 427, 1998
42. Molteni, C, Francis, GP, Payne, MC, and V Heine, *Material Science and Engineering B*, Vol. 37, p. 121, 1996
43. Monzen, R, Futakuchi, M, and T Suzuki, *Scripta Metallurgica et Materialia*, Vol. 32(8), p. 1277, 1995
44. Monzen, R and Y Sumi, *Philosophical Magazine A*, Vol. 70(5), p. 805, 1994
45. Monzen, R and N Takada, Nanometer sliding of [001] symmetric tilt boundaries in copper with boundary precipitates, *Materials Science and Engineering A*, Vol. 234, p. 202, 1997
46. Mullins, M, Computer-simulation of fracture using long-range pair potentials, *Acta Materialia*, Vol. 32, p. 381, 1984
47. Mullins, M and MA Dokainsh, Simulation of the (001) plane crack in alpha-iron employing a new boundary scheme, *Philosophical Magazine A*, Vol. 46, p. 771, 1982
48. Namilae, S, Chandra, N, and TG Neih, Atomisic simulation of grain boundary sliding in pure and magnesium doped aluminum bicrystals, *Scripta Materialia*, Vol. 46, p. 49, 2002
49. Needleman, A, An analysis of tensile decohesion along an interface, *Journal of the Mechanics and Physics of Solids*, Vol. 38, p. 289, 1990
50. Ortiz, M and R Phillips, *Advances in Applied Mechanics*, Vol. 36, p. 1, 2000

51. Parrinello, M and A Rahman, *Journal of Applied Physics*, Vol. 52, p. 7182, 1981
52. Rafii-Tabor, H, Hua, L, and M Cross, A multiscale atomistic-continuum modelling of crack propagation in a two-dimensional macroscopic plate, *Journal of Physics: Condensed Matter*, Vol. 10, p. 2375, 1998
53. Rafii-Tabor, H, Hua, L, and M Cross, *Materials Science and Technology*, Vol. 14, p. 554, 1998
54. Rahulkumar, P, Jagota, A, Bennison, SJ, and S Saigal, Cohesive element modeling of viscoelastic fracture: application to peel testing of polymers, *International Journal of Solids and Structures*, Vol. 37, p. 1873, 2000
55. Rice, JR and J Wang, Embrittlement of interfaces by solute segregation, *Materials Science and Engineering A*, Vol. 107, p. 23, 1989
56. Rudd, RE and JQ Broughton, Concurrent coupling of length scales in solid state systems, *Physica Status Solidi B*, Vol. 73, p. 1529, 1996
57. Saito, R, Dresselhaus, G, and MS Dresselhaus, *Physical properties of carbon nanotubes*. London: Imperial College Press, 1998
58. Sanchez-Palencia, E, *Non-homogeneous media vibration theory: lecture notes in physics*, Vol. 127, Berlin Heidelberg New York: Springer, 1980
59. Sheikh-Ali, AD, *Acta Materialia*, Vol. 45(8), p. 3109, 1997
60. Sheikh-Ali, AD and J Szpunar, *Materials Science and Engineering A*, Vol. 245(1), pp. 49–54, 1998
61. Shenoy, VB, Ph.D. Thesis, Brown University, Providence, RI, 1999
62. Shenoy, VB, Miller, R, Tadmor, EB, and M Ortiz, An adaptive finite element approach to atomic-scale mechanics – the quasicontinuum method, *Journal of the Mechanics and Physics of Solids*, Vol. 47, p. 611, 1999
63. Shenoy, VB, Miller, R, Tadmor, EB, and M Ortiz, Quasicontinuum models of interfacial structure and deformation, *Physical Review Letters*, Vol. 80, p. 742, 1999
64. Shenoy, VB, Miller, R, Tadmor, EB, and M Ortiz, Nucleation of dislocations beneath a plane strain indenter, *Journal of the Mechanics and Physics of Solids*, Vol. 48, p. 649, 2000
65. Shet, C and N Chandra, Analysis of energy balance when using cohesive zone models to simulate fracture processes, *Journal of Engineering Materials and Technology*, Vol. 124, p. 440, 2002
66. Sinclair, J, Gehlen, PC, Hogland, RG, and JP Hirth, Flexible boundary conditions and nonlinear geometric effects in atomic dislocation modeling, *Journal of Applied Physics*, Vol. 49, p. 3890, 1977
67. Tadmor, EB, Ortiz, M, and R Philips, Hierarchical modeling in the mechanics of materials, *International Journal of Solids and Structures*, Vol. 37, p. 379, 2000
68. Tadmor, EB, Philips, R, and M Ortiz, Quasicontinuum analysis of defects in solids, *Philosophical Magazine A*, Vol. 44(3), pp. 333–370, 1996

69. Tvergaard, V, Effect of fibre debonding in a whisker-reinforced metal, *Materials Science and Engineering A*, Vol. 125, p. 203, 1990
70. Voleti, RS, Ph.D. Thesis, Florida State University, Tallahassee, FL, 1997
71. Wang, JS and PM Anderson, *Acta Metallurgica et Materialia*, Vol. 39(5), p. 779, 1991
72. Wang, GJ, Sutton, AP, and V Vitek, *Acta Metallurgica*, Vol. 32, p. 1093, 1994
73. Wappling, D, Gunnars, J, and P Stahle, *International Journal of Fracture*, Vol. 89, p. 223, 1998
74. Watanabe, T and P Davies, *Philosophical Magazine A*, Vol. 37(5), p. 649, 1978
75. Wolf, D, *Journal of Applied Physics*, Vol. 68, p. 185, 1990
76. Zhang, L and H Tanaka, Towards a deeper understanding of wear and friction on the atomic scale – a molecular dynamics analysis, *Wear*, Vol. 211, p. 44, 1997

Index